EUCLIDEAN QUANTUM GRAVITY

EUCLIDEAN QUANTUM GRAVITY

EDITORS

G W GIBBONS
S W HAWKING

Department of Applied Mathematics
and Theoretical Physics
University of Cambridge
England

World Scientific
Singapore • New Jersey • London • Hong Kong

Published by

World Scientific Publishing Co. Pte. Ltd.
5 Toh Tuck Link, Singapore 596224
USA office: 27 Warren Street, Suite 401-402, Hackensack, NJ 07601
UK office: 57 Shelton Street, Covent Garden, London WC2H 9HE

British Library Cataloguing-in-Publication Data
A catalogue record for this book is available from the British Library.

The editors and publisher are grateful to the authors and the following publishers for their assistance and their permission to reproduce the articles found in this volume:

The American Physical Society (*Phys. Rev. D* and *Phys. Rev. Lett.*)
Elsevier Science Publishers (*Nucl. Phys. B* and *Phys. Lett.*)
Springer-Verlag (*Commun. Math. Phys.*)
The American Mathematical Society
Cambridge University Press
Gauther-Villars
Lehigh University
Plenum Press
The Royal Society of London

EUCLIDEAN QUANTUM GRAVITY

ISBN-13 978-981-02-0515-7
ISBN-10 981-02-0515-5

ISBN-13 978-981-02-0516-4 (pbk)
ISBN-10 981-02-0516-3 (pbk)

Printed in Singapore

CONTENTS

III. QUANTUM COSMOLOGY

IV. WORMHOLES

V. GRAVITATIONAL INSTANTONS

INTRODUCTION

Historical Origins

The 1960's and early 1970's was a period of rapid development for the classical theory of general relativity. Stimulated by the experimental discoveries of quasars, pulsars and the cosmic microwave background, attention turned to the problem of the gravitational collapse of stars and more massive bodies and to the problem of the big bang. This period culminated in the proofs of the singularity theorems (see e.g. [1]) showing that under rather general conditions spacetime singularities are, according to the classical theory, inevitable both in the collapse of stars to form what became known as black holes and at some time in the past during the very earliest phases of the universe. Shortly afterwards a number of important results were obtained which provided a broad and satisfactory picture of the formation of black holes and their basic properties. These early results and the resulting picture were described in a coherent form for the first time in the 1972 Les Houches Summer School Proceedings [2]. Among the most intriguing results, and as it turned out, most significant for future research, were the clear analogies between the properties of black holes according to classical general relativity and thermodynamics [3].

It was natural, therefore, to attempt to extend these purely classical results to the realm of the quantum theory. Since at that time, and indeed to this day, there was no satisfactory fully quantized theory of gravity, the first steps in this direction were taken using quantum field theory on a purely classical background. It had been realized for some time previously [4] that one of the most important physical effects that the quantum theory introduces is the production of particle pairs by strong gravitational fields.

When applied to black holes for the first time in [5, 6] it was discovered that the previously noted analogy between the properties of black holes and thermodynamics could be extended to a complete correspondence since a black hole in free space was shown to radiate thermally with a temperature T given by

$$T_H = \frac{\kappa}{2\pi}$$

where κ is the surface gravity. As a consequence one should be able to assign an entropy S to a black hole given by

$$S_H = \frac{1}{4} A_H ,$$

where A is the surface area of the black hole.

Shortly afterwards the same result was rederived, still within the context of field theory on a fixed background, using a path integral formulation [7] which showed for the first time how the thermal character of the emission could be understood in terms of the complex geometry of the Schwarzschild metric. In particular, in order that the path integral be better defined, it was found convenient to pass to a pure imaginary time coordinate $\tau = it$. The regularity of the resulting positive definite, or Euclidean as opposed to Lorentzian metric, demanded that τ be identified modulo

$8\pi M = 1/T_H$. It was soon realized that this periodicity in imaginary time was related to the thermal character of the resulting Green's functions [8, 9] and was not restricted to the Schwarzschild case but could be extended quite generally to cover all time-independent horizons including what has turned out to be of considerable importance in connection with inflationary cosmology — the case of cosmological event horizons [9].

These early advances provided the needed clue on how to incorporate the quantized gravitational field. A formalism was required which allowed one both to exploit to the full the topological and geometrical character of Einstein's general relativity and to perform the calculations efficiently. It had, by that time, become clear that the path integral method, despite its considerable mathematical problems, provided the most direct way to quantize the Yang–Mills theory. It was found that important nonperturbative information could be extracted, as well as the structure of the divergences of the theory, by considering Yang–Mills connections defined on flat Euclidean four-space. Nonsingular and nontrivial solutions of the classical equations of motion, called "instantons" allowed one to calculate nonperturbative amplitudes.

The extension of these Euclidean methods to quantum gravity was initiated in [11] where it was shown that, about the topologically trivial case of R^4, the coefficients of the one-loop divergences for pure gravity vanish. The first application of these ideas to the topologically nontrivial case of black-hole backgrounds was contained in [12] where it was shown how the evaluation of the classical Einstein action with the appropriate and essential boundary term [13] allowed a derivation of the relation between entropy and event horizon surface area.

There followed a period of rapid development of the general formalism of Euclidean quantum gravity and since then it has been applied to a number of other important physical problems. It has also spawned a number of purely mathematical developments, for example the study of gravitational instantons which have, in turn been applied to problems in other areas of physics.

The present volume is intended to provide a selection of some of the most influential and important papers in this field. We have arranged the material in five sections, each dealing with a particular aspect of the theory. We shall describe them in more detail below. We do not pretend that Euclidean quantum gravity is a complete or entirely mathematically consistent quantum theory of gravity. The well-known divergences which not even supergravity theories have succeeded in eliminating suggest that a field theory of gravity like Einstein's general relativity is a purely low-energy or large-distance approximation to some more fundamental underlying theory. It is too early at present to say what such a theory might be. Despite its considerable successes in apparently eliminating divergences it is clear, for example, that superstring theory is not at present sufficiently well developed to tackle the physical questions which the Euclidean formulation addresses. Moreover it seems very likely that, whatever the true fundamental quantum theory of gravity turns out to be, it will have Einstein's theory of general relativity as its classical limit and very probably some form of path integral formulation as its semi-classical limit. Certainly we know of no other formalism capable at present of dealing with the problems raised by gravitational collapse either in black holes or in cosmology. Indeed it is interesting to note how many attempts to apply quantum mechanical ideas to gravity invoke not quite explicitly the basic ideas behind Euclidean quantum gravity.

Outline of this Volume

We turn now to a more detailed description of the contents of this reprint volume. The first section is concerned with the general formalism. It begins with the important paper by 't Hooft and Veltmann which initiated a whole series of investigations into the divergences and one-loop corrections to quantum gravity by showing how these could be calculated in a path integral formalism analogous to that used for the Yang–Mills gauge theory. The next two papers contain reviews of the first stage of the subject and the initial applications to black-hole theory and gravitational instantons, the third paper containing applications to supergravity theories.

The fourth paper of Section I deals with a central problem in this approach and one which has yet to be completely resolved: that of the indefiniteness of the gravitational action. By means of a conformal rescaling the gravitational action can be made arbitrarily negative. Thus one cannot merely integrate over all metrics with positive definite signature subject to suitable boundary conditions; rather what appears to be needed is a complex contour in the space of all metrics. In the fourth paper the proposal was presented that, at least for asymptotically Euclidean (or ALE) metrics, this contour should run over all positive definite AE metrics with vanishing Ricci-scalar and over all conformal rescalings of such metrics, the rescalings being in the purely imaginary direction. The reasonableness of this proposal in the purely perturbative case would seem to be confirmed by the results of the seventh paper of Section I. Its correctness in the nonperturbative regime depends, at the very least, on the correctness of a conjecture concerning the Einstein action of asymptotically Euclidean metrics with vanishing Ricci scalar which generalizes the positivity of the ADM mass in classical general relativity. A proof of this important and mathematically difficult conjecture was announced in the fifth paper of Section I. The perturbative method for calculating functional integrals involves the evaluation of certain "determinants" of differential operators. An appropriate method is by means of their "zeta functions" and this technique is described in the sixth paper of Section I. For certain tunneling amplitudes, the eigenvalues of these differential operators may be negative. The correct method for evaluating them, at least in flat spacetime, is described in the eighth paper of Section I. The Positive Action Theorem is a purely mathematical result about the Euclidean action of asymptotically and Euclidean four-metrics. For application to quantum cosmology and the theory of wormholes, it is more appropriate to consider compact four-manifolds with no boundary. A highly relevant mathematical result is the theorem of Bishop described in the penultimate paper of Section I which provides us with a strict lower boundary for the action of a compact metric with constant Ricci scalar analogous to the result of Schoen and Yau. The last paper of Section I illustrates the relevance of this mathematical result to one of the most challenging problems confronting any quantum theory of gravity — the cosmological constant problem.

The second section of this reprint volume is concerned with what is perhaps the most successful area for the application of the basic formalism, and which, as described above, stimulated the entire development — the subject of black holes. In addition to the papers whose significance has been described above, we have included as paper five of this section an attempt to relate the Euclidean Schwarzschild solution in a tunneling calculation. The Euclidean formulation of black-hole thermodynamics allows the calculation of expectation values of the energy–momentum tensor of quantum fields on the Schwarzschild background $\langle \hat{T}_{\mu\nu} \rangle$. We have included as the penultimate paper of Section II, one describing a particularly useful approximation for $\langle \hat{T}_{\mu\nu} \rangle$ for a scalar field whose somewhat unexpected accuracy was confirmed by detailed numerical studies described in the last paper of Section II.

Perhaps the most ambitious area to which quantum gravity has been applied is the entire Universe. This is the subject of Section III. The first paper describes the basic quantum properties of De Sitter. Apart from its intrinsic interest in helping to elucidate various conceptual and technical problems, this has turned out to be of great importance for the theory of inflation. However the cosmological term that is believed to be present during inflation arises as a (classical) expectation value, and may change during a phase transition. If this change is forbidden classically it may happen by quantum tunneling. This requires an appropriate formalism and in the second and third papers of Section III it is shown how Euclidean quantum gravity may provide such a formalism. We know of no other. A different viewpoint was inaugurated in the fourth paper of Section III and further developed in paper five, where it is shown how the Euclidean path integral can provide solutions of the Wheeler–De Witt equation, wave functions for the universe. A proposal is made for the wave function of the universe which is suggested by its simplicity and naturalness in the Euclidean framework. The final paper of Section III shows how one may extract detailed cosmological predictions, such as the spectrum of fluctuations from the basic formalism together with the non-boundary boundary condition.

The fourth section of this reprint collection deals with one of the most recent developments in Euclidean quantum gravity — the possible relevance of topologically nontrivial Euclidean manifolds in the path integral. It has always been a major goal of Euclidean quantum gravity to incorporate the nonperturbative effects of quantum fluctuations of such metrics. Of particular interest are metrics in which apparently distant flat regions of the manifold are connected by short "wormholes." Some examples and an account of the effects they might give rise to was first given in the first paper of Section IV. If one considers asymptotically Euclidean vacuum metrics, the results of Schoen and Yau and Witten show that the only stationary point of the classical action is flat space. The inclusion of particular types of matter can alter this fact. In the second paper of Section IV, it is shown how axion fields such as those occurring in string theory coupled to general relativity allow the existence of simple wormhole solutions. In the third paper of Section IV, it is shown how such wormholes may have preferred consequences for our understanding of the origin of the values of the observed coupling constants of nature. The most important example is the cosmological constant but the last three papers of Section IV show that even elementary particle coupling constants such as θ_{QCD} may be affected as well. This offers the prospect of a much more intimate connection between Planck scale physics and observable physics than had hitherto been envisaged.

The last section of this reprint collection is more mathematical in nature, and is concerned with complete nonsingular Ricci-flat four-metrics or "gravitational instantons." As mentioned above there are no asymptotically Euclidean (AE) metrics of this sort but Eguchi and Hanson (in the first paper of Section V) provided the first explicit nonsingular example of what turned out to be an asymptotically locally Euclidean (or ALE) metric. In the second paper of Section V, a generalization to a multi-instanton metric which is also asymptotically locally Euclidean is made.

More information about asymptotically Euclidean metrics is contained in the third paper of Section V. The multi-instanton metrics of paper two of this section were rederived using Twistor techniques by Hitchin in the fourth paper of Section V and the existence of other examples related to each of the discrete sub-groups of SU(2) was conjectured. The confirmation of this conjecture by means of an explicit construction is provided by Kronheimer's work in the fifth paper of Section V. The programme of using techniques from classical general relativity to find complete non-singular Einstein metrics (with cosmological constant) gave rise to a construction by Page of the first, and to date only, explicitly or implicitly known inhomogeneous Einstein metric with positive Ricci curvature. This is given in the sixth paper of Section V. In the seventh paper, a description of the geometry and possible physical significance of most of the known gravitational instantons is provided. Finally we have included what might at first sight appear to be a completely unrelated paper on the low-energy scattering of non-Abelian monopoles. This may be represented by geodesic motion on a self-dual gravitational instanton (The existence of this solution had, regrettably, been overlooked in the third paper of this section). This represents one of a number of "spin-offs" arising from the study of gravitational instantons — another notable example being their application to the Kaluza-Klein theory and Kaluza-Klein monopoles.

References

Items marked with an asterisk are reprinted in this volume.

[1] S. W. Hawking & G. F. R. Ellis, *The large scale structure of spacetime* (Cambridge University Press, Cambridge, 1973).

[2] C. Dewitt & B. S. DeWitt eds., *Black Holes* (Gordon & Breach, New York, 1973).

[3] J. M. Bardeen, B. Carter & S. W. Hawking, *Commun. Math. Phys.* **31**, 181 (1973).

[4] E. Schrödinger, *Physica* **6**, 899 (1939).

[5] S. W. Hawking, *Nature* **248**, 30 (1974).

[6*] S. W. Hawking, *Commun. Math. Phys.* **43**, 199 (1975).

[7*] J. B. Hartle & S. W. Hawking, *Phys. Rev.* **D13**, 2188 (1976).

[8] G. W. Gibbons & M. J. Perry, *Phys. Rev. Letts.* **36**, 985 (1976).

[9*] G. W. Gibbons & M. J. Perry, *Proc. Roy. Soc. London.* **A358**, 467 (1978).

[10*] G. W. Gibbons & S. W. Hawking, *Phys. Rev.* **D15**, 2738 (1977).

[11*] G. 't Hooft & M. Veltman, *Ann. Inst. Henri Poincaré* **20**, 69 (1974).

[12*] G. W. Gibbons & S. W. Hawking, *Phys. Rev.* **D15**, 2752 (1977).

[13] J. York, *Phys. Rev. Letts.* **6**, 1656 (1972).

I. GENERAL FORMALISM

Ann. Inst. Henri Poincaré. Section A :
Vol. XX, n° 1, 1974, p. 69-94. *Physique théorique.*

One-loop divergencies in the theory of gravitation

par

G. 't HOOFT (*) and M. VELTMAN (*)

C. E. R. N., Geneva.

ABSTRACT. — All one-loop divergencies of pure gravity and all those of gravitation interacting with a scalar particle are calculated. In the case of pure gravity, no physically relevant divergencies remain; they can all be absorbed in a field renormalization. In case of gravitation interacting with scalar particles, divergencies in physical quantities remain, even when employing the socalled improved energy-momentum tensor.

1. INTRODUCTION

The recent advances in the understanding of gauge theories make a fresh approach to the quantum theory of gravitation possible. First, we now know precisely how to obtain Feynman rules for a gauge theory [*1*]; secondly, the dimensional regularization scheme provides a powerful tool to handle divergencies [*2*]. In fact, several authors have already published work using these methods [*3*], [*4*].

One may ask why one would be interested in quantum gravity. The foremost reason is that gravitation undeniably exists; but in addition we may hope that study of this gauge theory, apparantly realized in nature, gives insight that can be useful in other areas of field theory. Of course, one may entertain all kinds of speculative ideas about the role of gravitation in elementary particle physics, and several authors have amused themselves imagining elementary particles as little black holes etc. It may well be true that gravitation functions as a cut-off for other interactions; in view of the fact that it seems possible to formulate all known

(*) On leave from the University of Utrecht, Netherlands.

interactions in terms of field-theoretical models that show only logarithmic divergencies, the smallness of the gravitational coupling constant need not be an obstacle. For the time being no reasonable or convincing analysis of this type of possibilities has been presented, and in this paper we have no ambitions in that direction. Mainly, we consider the present work as a kind of finger exercise without really any further underlying motive.

Our starting point is the linearized theory of gravitation. Of course, much work has been reported already in the literature [*5*], in particular we mention the work of B. S. Dewitt [*6*]. For the sake of clarity and completeness we will rederive several equations that can be found in his work. It may be noted that he also arrives at the conclusion that for pure gravitation the counterterms for one closed loop are of the form R^2 or $R_{\mu\nu}R^{\mu\nu}$; this really follows from invariance considerations and an identity derived by him. This latter identity is demonstrated in a somewhat easier way in appendix B of this paper.

Within the formalism of gauge theory developed in ref. 7, we must first establish a gauge that shows clearly the unitarity of the theory. This is done in section 2. The work of ref. 7, that on purpose has been formulated such as to encompass quantum gravity, assures us that the S-matrix remains invariant under a change of gauge.

In section 3 we consider the one loop divergencies when the gravitational field is treated as an external field. This calculation necessitates a slight generalization of the algorithms recently reported by one of us [*8*]. From the result one may read off the known fact that there are fewer divergencies if one employs the so-called improved energy-momentum tensor [*9*]. Symanzik's criticism [*10*] applies to higher order results, see ref. 11. In the one loop approximation we indeed find the results of Callan *et al.* [*9*].

Next we consider the quantum theory of gravity using the method of the background field [*6*], [*12*]. In ref. 8 it has already been shown how this method can be used fruitfully within this context. In sections 4, 5 and 6 we apply this to the case of gravitation interacting with a scalar field in the conv ntional way. The counter Lagrangian for pure gravity can be deduced immediately.

In section 7 finally we consider the use of the « improved » energy-momentum tensor. Some results are quoted, the full answer being unprintable.

Appendix A quotes notations and conventions. Appendix B gives the derivation of a well-known [*6*] but for us very important result. An (also well-known) side results is the fact that the Einstein-Hilbert Lagrangian is meaningless in two dim nsions. This shows up in the form of factors $1/(n-2)$ in the graviton propagator, as noted by Neveu [*13*] and Capper *et al.* [*4*].

2. UNITARITY

One of the remarkable aspects of gravitation is the freedom of choice in the fundamental fields. In conventional renormalizable field theory the choice of the fields is usually such as to produce the smoothest possible Green's functions. In the case of quantum gravity there seems to be no clear choice based on such a criterium. For instance, one may use as basic field the metric tensor $g_{\mu\nu}$, or its inverse $g^{\mu\nu}$, or any other function of the $g_{\mu\nu}$ involving, say, the Riemann tensor. From the point of view of the S-matrix many choices give the same result, the Jacobian of the transformation being one (provided dimensional regularization is applied).

We chose as basic fields the $h_{\mu\nu}$, related to the $g_{\mu\nu}$ by $g_{\mu\nu} = \delta_{\mu\nu} + h_{\mu\nu}$. This is of course the conventional choice. The Lagrangian that we start from is the Einstein-Hilbert Lagrangian, viz:

$$\mathscr{L} = -\sqrt{g}\mathrm{R} \tag{2.1}$$

(see appendix A for symbols and notations). This Lagrangian is invariant under the infinitesimal gauge transformation (or rather it changes by a total derivative)

$$g_{\mu\nu} \rightarrow g_{\mu\nu} + g_{\alpha\nu}\partial_\mu\eta^\alpha + g_{\mu\alpha}\partial_\nu\eta^\alpha + \eta^\alpha\partial_\alpha g_{\mu\nu}, \tag{2.2}$$

or

$$h_{\mu\nu} \rightarrow h_{\mu\nu} + \mathrm{D}_\mu\eta_\nu + \mathrm{D}_\nu\eta_\mu. \tag{2.3}$$

In here the η_μ are four independent infinitesimal functions of space-time. The D_μ are the usual covariant derivatives.

In order to define Feynman rules we must supplement the Lagrangian (2.1) with a gauge breaking term $-\frac{1}{2}\mathrm{C}_\mu^2$ and a Faddeev-Popov ghost Lagrangian (historically, the name Feynman-DeWitt ghost Lagrangian would be more correct). In order to check unitarity and positivity of the theory we first consider the (non-covariant) Prentki gauge which is much like the Coulomb gauge in quantum-electrodynamics:

$$\sum_{i=1}^{3} \partial_i h_{i\mu} = 0, \qquad \mu = 1, \ldots, 4. \tag{2.4}$$

In the language of ref. 7 we take correspondingly:

$$\mathrm{C}_\mu = b \sum_{i=1}^{3} \partial_i h_{i\mu}, \qquad b \rightarrow \infty. \tag{2.5}$$

With this choice for C the part quadratic in the $h_{\mu\nu}$ is (comma denotes differentiation):

$$-\frac{1}{4}h_{\alpha\beta,\mu}h_{\alpha\beta,\mu}+\frac{1}{4}h_{\alpha\alpha,\mu}h_{\beta\beta,\mu}-\frac{1}{2}h_{\alpha\alpha,\beta}h_{\beta\mu,\mu}$$
$$+\frac{1}{2}h_{\beta\mu,\alpha}h_{\beta\alpha,\mu}-\frac{1}{2}b^2h_{i\mu,i}h_{j\mu,j}. \qquad (2.6)$$

This can be written as $\frac{1}{2}h_{\alpha\beta}V_{\alpha\beta\mu\nu}h_{\mu\nu}$. The Fourier transform of V is:

$$V_{\alpha\beta\mu\nu}=\frac{1}{2}k^2(\delta_{\alpha\beta}\delta_{\mu\nu}-\delta_{\alpha\mu}\delta_{\beta\nu})-k_\mu k_\nu\delta_{\alpha\beta}+k_\beta k_\nu\delta_{\alpha\mu}-b^2\vec{k}_\beta\vec{k}_\nu\delta_{\alpha\mu}. \qquad (2.7)$$

Calculations in the theory of gravitation are as a rule cumbersome, and the calculation of the graviton propagator gives a foretaste of what is to come. In principle things may be done as follows. First, symmetrize V with respect to $\alpha \leftrightarrow \beta$, $\mu \leftrightarrow \nu$ and $\alpha\beta \leftrightarrow \mu\nu$ interchange. Then find the propagator P from the equation $V.P = -\mathbb{1}$, where

$$\mathbb{1}_{\alpha\beta,\mu\nu}=\frac{1}{2}(\delta_{\alpha\mu}\delta_{\beta\nu}+\delta_{\alpha\nu}\delta_{\beta\mu}). \qquad (2.8)$$

Subsequently the limit $b^2 \to \infty$ must be taken. It is of advantage to go in the coordinate system where $k_1 = k_2 = 0$. Alternatively, write $\pi_1 = h_{11}$, $\pi_2 = h_{22}$, $\pi_3 = h_{33}$, $\pi_4 = h_{44}$, $\pi_5 = h_{34}$, $\pi_6 = h_{12}$, $\pi_7 = h_{13}$, $\pi_8 = h_{14}$, $\pi_9 = h_{23}$, $\pi_{10} = h_{24}$. Then V can be rewritten as a rather simple 10×10 matrix, that subsequently must be symmetrized. In the limit $b^2 \to \infty$ one may in a row or column containing a b^2 neglect all elements, except of course the b^2 term itself. Inversion of that matrix is trivial, and providing a minus sign the result is the propagator in the Prentki gauge:

$$P_{\mu\nu,\alpha\beta}(k)=\frac{1}{k^2}\left(\bar{\delta}_{\mu\alpha}\bar{\delta}_{\nu\beta}+\bar{\delta}_{\mu\beta}\bar{\delta}_{\nu\alpha}-\frac{2}{n-2}\bar{\delta}_{\mu\nu}\bar{\delta}_{\alpha\beta}\right)$$
$$+\frac{1}{\vec{k}^2}\left(\bar{\delta}_{\mu\alpha}\delta_{\nu\beta4}+\bar{\delta}_{\mu\beta}\delta_{\nu\alpha4}+\bar{\delta}_{\nu\alpha}\delta_{\mu\beta4}+\bar{\delta}_{\nu\beta}\delta_{\mu\alpha4}\right.$$
$$\left.-\frac{2}{n-2}\bar{\delta}_{\mu\nu}\delta_{\alpha\beta4}-\frac{2}{n-2}\bar{\delta}_{\alpha\beta}\delta_{\mu\nu4}\right)+\frac{2n-6}{n-2}\frac{k^2}{\vec{k}^4}\delta_{\mu\nu\alpha\beta4}. \qquad (2.9)$$

In here n is the dimensionality of space-time. Further

$$\vec{k}^2=k^2-k_4^2,\qquad \bar{\delta}_{\mu\nu}=(1-\delta_{\mu4})(1-\delta_{\nu4})\left(\delta_{\mu\nu}-\frac{k_\mu k_\nu}{\vec{k}^2}\right),$$
$$\delta_{\mu\nu4}=\delta_{\mu4}\delta_{\nu4},\qquad \delta_{\mu\nu\alpha\beta4}=\delta_{\mu4}\delta_{\nu4}\delta_{\alpha4}\delta_{\beta4}. \qquad (2.10)$$

The first part of eq. (2.9) corresponds in 4 dimensions (i. e. $n = 4$) to the propagation of two polarization states of a mass zero spin 2 particle (see ref. 14, in particular section 3). The second and third part have no pole, they are non-local in space but simultaneous (or « local ») in time. They describe the $1/r$ behaviour of the potential in this gauge. In any case, they do not contribute to the absorptive part of the S-matrix.

In addition to the above we must also consider the Faddeev-Popov ghost. Subjecting C_μ of eq. (2.5) to the gauge transformation (2.3) and working in zero'th order of the field $h_{\mu\nu}$ we obtain the quadratic part of the F-P ghost Lagrangian:

$$\mathcal{L}_{FP} = \varphi_\mu^*(\delta_{\mu\nu}\vec{\partial}^2 + \vec{\partial}_\nu\partial_\mu)\varphi_\nu \tag{2.11}$$

where the arrow indicates the 3-dimensional derivative. The propagator resulting from this has no pole, therefore does not contribute to the absorptive part of the S-matrix.

The above may be formulated in a somewhat neater way by means of the introduction of a fixed vector with zero space components. For instance:

$$\begin{aligned} \partial_i h_{i\mu} &= \partial_\nu h_{\nu\mu} + \vartheta_\alpha\vartheta_\beta\partial_\alpha h_{\beta\mu}, \\ \vartheta_\alpha &= (0, 0, 0, i). \end{aligned} \tag{2.12}$$

The continuous dimension regularization method can now be applied without further difficulty.

3. EXTERNAL GRAVITATIONAL FIELD

In this section we assume that the reader is acquainted with the work of ref. 8. The principal result is the following. Let there be given a Lagrangian

$$\mathcal{L} = -\frac{1}{2}\partial_\mu\varphi_i\partial_\mu\varphi_i + \varphi_i N^\mu_{ij}\partial_\mu\varphi_j + \frac{1}{2}\varphi_i M_{ij}\varphi_j \tag{3.1}$$

where N and M are functions of external fields etc., but do not depend on the quantum fields φ_i. The counter-Lagrangian $\Delta\mathcal{L}$ that eliminates all one loop divergencies is

$$\Delta\mathcal{L} = \frac{1}{\varepsilon}\,\mathrm{Tr}\left(\frac{1}{4}X^2 + \frac{1}{24}Y_{\mu\nu}Y_{\mu\nu}\right). \tag{3.2}$$

The trace is with respect to the indices $i, j, \ldots$ Further

$$\begin{aligned} X &= M - N^\mu N^\mu, \\ Y_{\mu\nu} &= \partial_\mu N^\nu - \partial_\nu N^\mu + N^\mu N^\nu - N^\nu N^\mu, \\ \frac{1}{\varepsilon} &= \frac{1}{8\pi^2(n-4)}. \end{aligned} \tag{3.3}$$

It is assumed that N^μ is antisymmetric and M symmetric in the indices i, j.

For our purposes it is somewhat easier to work with complex fields. Given the Lagrangian

$$\mathcal{L} = -\partial_\mu \varphi_i^* \partial_\mu \varphi_i + 2\varphi_i^* \mathcal{N}_{ij}^\mu \partial_\mu \varphi_j + \varphi_i^* \mathcal{M}_{ij} \varphi_j \tag{3.4}$$

one easily derives

$$\begin{aligned} \Delta\mathcal{L} &= \frac{1}{\varepsilon} \operatorname{Tr} \left(\frac{1}{2} \mathcal{X}^2 + \frac{1}{12} \mathcal{Y}_{\mu\nu} \mathcal{Y}_{\mu\nu} \right), \\ \mathcal{X} &= \mathcal{M} - \mathcal{N}^\mu \mathcal{N}^\mu - \partial_\mu \mathcal{N}^\mu \\ \mathcal{Y}_{\mu\nu} &= \partial_\mu \mathcal{N}^\nu - \partial_\nu \mathcal{N}^\mu + \mathcal{N}^\mu \mathcal{N}^\nu - \mathcal{N}^\nu \mathcal{N}^\mu. \end{aligned} \tag{3.5}$$

Writing $\varphi = (A + iB)/\sqrt{2}$ with real fields A and B, it is seen that (3.2) is contained twice in (3.5) provided $\mathcal{M} - \partial_\mu \mathcal{N}_\mu$ and $\mathcal{N}_\mu$ are symmetric and antisymmetric respectively. Eq. (3.5) is valid independently of these symmetry properties.

There is now a little theorem that says that the counter-Lagrangian remains unchanged if in the original Lagrangian, eq. (3.4), everywhere φ_i^* is replaced by $\varphi_k^* Z_{ki}$ where Z is a (possibly space-time dependent) matrix. To see that we consider the following simplified case:

$$\mathcal{L} = \varphi_i^* \partial^2 \varphi_i + \varphi_i^* Z_{ij} \partial^2 \varphi_j + \varphi_i^* \mathcal{M}_{ij} \varphi_j. \tag{3.6}$$

The Feynman rules are

$\dfrac{\delta_{ij}}{k^2 - i\varepsilon}$ φ-propagator,

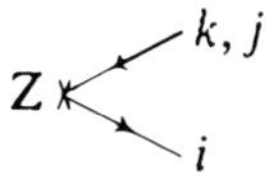

$-k^2 Z_{ij}$ Z-vertex,

$\mathcal{M}_{ij}$ $\mathcal{M}$-vertex.

Any $\mathcal{M}$ vertex may be followed by 0, 1, 2, ... vertices of the Z-type. Now it is seen that the Z-vertex contains a factor that precisely cancels the propagator attached at one side. So one obtains a geometric series of the form

$$\mathcal{M} - Z\mathcal{M} + Z^2\mathcal{M} - Z^3\mathcal{M} \ldots = \frac{1}{1+Z}\mathcal{M}.$$

Clearly, the results are identical in case we had started with the Lagrangian

$$\mathcal{L} = \varphi_i^* \partial^2 \varphi_i + \varphi^* \left(\frac{1}{1+Z} \right) \mathcal{M} \varphi, \tag{3.7}$$

which is related to the above Lagrangian by the replacement

$$\varphi^* \to \varphi^* \frac{1}{1+Z}.$$

In the following we need the generalization of the above to the case that the scalar product contains the metric tensor $g_{\mu\nu}$. This tensor, in the applications to come, is a function of space-time, but not of the fields φ. Thus we are interested in the Lagrangian:

$$\mathscr{L} = \sqrt{g}(-\partial_\mu\varphi^* g^{\mu\nu}\partial_\nu\varphi + 2\varphi^*\mathscr{N}^\mu\partial_\mu\varphi + \varphi^*\mathscr{M}\varphi). \qquad (3.8)$$

This Lagrangian is invariant under general coordinate transformations, and so will be our counter-Lagrangian. This counter-Lagrangian is given in eq. (3.35), and we will sketch the derivation.

Taking into account that $\Delta\mathscr{L}$ will contain terms of a certain dimensionality only, we find as most general form:

$$\begin{aligned}\Delta\mathscr{L} = \sqrt{g}(&a_1R^2 + a_2R_{\mu\nu}R^{\mu\nu} + b_1\mathscr{M}R + b_2D_\mu D^\mu\mathscr{M} \\ &+ b_3\mathscr{M}\mathscr{N}^\mu\mathscr{N}_\mu + b_4\mathscr{M}D_\mu\mathscr{N}^\mu + b_5\mathscr{M}^2 \\ &+ c_1D_\mu\mathscr{N}^\mu R + c_2\mathscr{N}^\mu\mathscr{N}_\mu R + c_3\mathscr{N}^\mu\mathscr{N}^\nu R_{\mu\nu} \\ &+ c_4D_\mu\mathscr{N}^\nu D^\mu\mathscr{N}_\nu + c_5D_\mu\mathscr{N}^\nu D_\nu\mathscr{N}^\mu + c_6\mathscr{N}^\mu\mathscr{N}_\mu D_\nu\mathscr{N}^\nu \\ &+ c_7\mathscr{N}^\nu\mathscr{N}_\mu D_\nu\mathscr{N}^\mu + c_8\mathscr{N}^\mu\mathscr{N}_\mu\mathscr{N}^\nu\mathscr{N}_\nu + c_9\mathscr{N}^\mu\mathscr{N}^\nu\mathscr{N}_\mu\mathscr{N}_\nu). \qquad (3.9)\end{aligned}$$

The term $R_{\mu\nu\alpha\beta}R^{\mu\nu\alpha\beta}$ need not be considered; see Appendix B. As usual, $\mathscr{N}_\mu = g_{\mu\nu}\mathscr{N}^\nu$ etc. Several coefficients can readily be determined by comparison with the special case $g_{\mu\nu} = \delta_{\mu\nu}$:

$$b_2 = 0,\ b_3 = -1,\ b_4 = -1,\ b_5 = \frac{1}{2}.$$

$$c_4 = \frac{1}{6},\ c_5 = \frac{1}{3},\ c_6 = \frac{2}{3},\ c_7 = -\frac{2}{3},\ c_8 = \frac{1}{3},\ c_9 = \frac{1}{6}. \qquad (3.10)$$

The remaining coefficients are determined in two steps. First, we take the special case

$$g^{\mu\nu} = \frac{\delta^{\mu\nu}}{F},\ g_{\mu\nu} = \delta_{\mu\nu}F,\ \sqrt{g} = F^2,\ F = 1 - f. \qquad (3.11)$$

In here f is an arbitrary function of space-time. The Lagrangian (3.8) becomes:

$$\mathscr{L} = \varphi^*F\partial^2\varphi + 2\varphi^*\left(F^2\mathscr{N}^\mu + \frac{1}{2}\partial_\mu F\right)\partial_\mu\varphi + \varphi^*F^2\mathscr{M}\varphi. \qquad (3.12)$$

The replacement $\varphi^* \to \varphi^*F^{-1}$ leaves the counter-Lagrangian unchanged; we have thus the equivalent Lagrangian

$$\mathscr{L} = \varphi^*\partial^2\varphi + 2\varphi^*\left(F\mathscr{N}^\mu + \frac{1}{2}F^{-1}\partial_\mu F\right)\partial_\mu\varphi + \varphi^*F\mathscr{M}\varphi. \qquad (3.13)$$

This is precisely of the form studied before; the answer is as given in eq. (3.5), but now with the replacements

$$\mathcal{M} \to F\mathcal{M},$$
$$\mathcal{N}^\mu \to F\mathcal{N}^\mu + \frac{1}{2}F^{-1}\partial_\mu F = \mathcal{N}_\mu + \frac{1}{2}F^{-1}\partial_\mu F. \tag{3.14}$$

To proceed, it is necessary to compute a number of quantities for this choice of $g_{\mu\nu}$. Using the notations

$$f_\alpha = \partial_\alpha f, \qquad f_{\alpha\beta} = \partial_\alpha\partial_\beta f,$$

we have (on the right hand side there is no difference between upper and lower indices):

$$\Gamma^\alpha_{\mu\nu} = -\frac{1}{2}F^{-1}(\delta^\alpha_\mu f_\nu + \delta^\alpha_\nu f_\mu - \delta_{\mu\nu}f^\alpha). \tag{3.15}$$

For any contravariant vector Z^α:

$$\begin{aligned} D_\mu Z^\alpha &= \partial_\mu Z^\alpha + \Gamma^\alpha_{\mu\nu}Z^\nu \\ &= \partial_\mu Z^\alpha - \frac{1}{2}F^{-1}(\delta^\alpha_\mu f_\nu Z^\nu + f_\mu Z^\alpha - f^\alpha Z_\mu). \end{aligned}$$

$$D_\mu Z^\mu = \partial_\mu Z^\mu - 2F^{-1}f_\mu Z^\mu.$$

$$\begin{aligned} R^\mu_{\nu\alpha\beta} = &-\frac{1}{2}F^{-1}(\delta^\mu_\beta f_{\alpha\nu} - \delta^\mu_\alpha f_{\beta\nu} - \delta_{\nu\beta}f^\mu_\alpha + \delta_{\nu\alpha}f^\mu_\beta) \\ &-\frac{1}{4}F^{-2}(3\delta^\mu_\beta f_\alpha f_\nu - 3\delta^\mu_\alpha f_\beta f_\nu - 3\delta_{\nu\beta}f_\alpha f^\mu \\ &+ 3\delta_{\nu\alpha}f_\beta f^\mu + \delta_{\beta\nu}\delta^\mu_\alpha f^\gamma f_\gamma - \delta_{\alpha\nu}\delta^\mu_\beta f^\gamma f_\gamma). \end{aligned} \tag{3.16}$$

$$R_{\nu\alpha} = -\frac{1}{2}F^{-1}(2f_{\alpha\nu} + \delta_{\alpha\nu}f_{\gamma\gamma}) - \frac{3}{2}F^{-2}f_\alpha f_\nu. \tag{3.17}$$

$$R = -3F^{-2}f_{\gamma\gamma} - \frac{3}{2}F^{-3}f_\gamma f_\gamma. \tag{3.18}$$

$$\begin{aligned} R_{\nu\alpha}R^{\nu\alpha} = &F^{-4}(f_{\alpha\nu}f_{\alpha\nu} + 2f_{\alpha\alpha}f_{\nu\nu}) \\ &+ F^{-5}\left(3f_{\alpha\nu}f_\alpha f_\nu + \frac{3}{2}f_{\alpha\alpha}f_\nu f_\nu\right) + \frac{9}{4}F^{-6}f_\alpha f_\alpha f_\nu f_\nu. \end{aligned} \tag{3.19}$$

$$R^2 = 9F^{-4}f_{\alpha\alpha}f_{\nu\nu} + 9F^{-5}f_{\nu\nu}f_\alpha f_\alpha + \frac{9}{4}F^{-6}f_\alpha f_\alpha f_\nu f_\nu. \tag{3.20}$$

We leave it to the reader to verify that

$$\sqrt{g}(R_{\mu\nu\alpha\beta}R^{\mu\nu\alpha\beta} - 4R_{\nu\alpha}R^{\nu\alpha} + R^2) = \text{total derivative.} \tag{3.21}$$

In this particular case, unfortunately, also another identity holds:

$$\sqrt{g}\left(R_{\mu\nu}R^{\mu\nu} - \frac{1}{3}R^2\right) = \text{total derivative.} \tag{3.22}$$

We must now try to write the counter-Lagrangian in terms of covariant objects. First:

$$\left(F\mathcal{N}^\mu + \frac{1}{2}F^{-1}\partial_\mu F\right)^2 + \partial_\mu\left(F\mathcal{N}^\mu + \frac{1}{2}F^{-1}\partial_\mu F\right)$$

$$= F\mathcal{N}^\mu\mathcal{N}_\mu - \mathcal{N}^\mu f_\mu + \frac{1}{4}F^{-2}f_\mu f_\mu + F\partial_\mu\mathcal{N}^\mu$$

$$- f_\mu\mathcal{N}^\mu - \frac{1}{2}F^{-2}f_\mu f_\mu - \frac{1}{2}F^{-1}f_{\mu\mu}$$

$$= F\mathcal{N}^\mu\mathcal{N}_\mu + FD_\mu\mathcal{N}^\mu + \frac{1}{6}FR. \tag{3.23}$$

Similarly:

$$\partial_\mu\left(\mathcal{N}_\nu + \frac{1}{2}F^{-1}\partial_\nu F\right) - \partial_\nu\left(\mathcal{N}_\mu + \frac{1}{2}F^{-1}\partial_\mu F\right)$$

$$+ \left(\mathcal{N}_\mu + \frac{1}{2}F^{-1}\partial_\mu F\right)\left(\mathcal{N}_\nu + \frac{1}{2}F^{-1}\partial_\nu F\right)$$

$$- \left(\mathcal{N}_\nu + \frac{1}{2}F^{-1}\partial_\nu F\right)\left(\mathcal{N}_\mu + \frac{1}{2}F^{-1}\partial_\mu F\right)$$

$$= D_\mu\mathcal{N}_\nu - D_\nu\mathcal{N}_\mu + \mathcal{N}_\mu\mathcal{N}_\nu - \mathcal{N}_\nu\mathcal{N}_\mu. \tag{3.24}$$

This equation looks more complicated then it is; one has

$$D_\mu\mathcal{N}_\nu = \partial_\mu\mathcal{N}_\nu - \Gamma^\alpha_{\mu\nu}\mathcal{N}_\alpha.$$

Now Γ is symmetrical in the two lower indices, and therefore

$$D_\mu\mathcal{N}_\nu - D_\nu\mathcal{N}_\mu = \partial_\mu\mathcal{N}_\nu - \partial_\nu\mathcal{N}_\mu.$$

The various F dependent terms all cancel out.

The result for the special case $g_{\mu\nu} = \delta_{\mu\nu}(1 - f)$ is:

$$\Delta\mathcal{L} = \frac{\sqrt{g}}{\varepsilon}\,\mathrm{Tr}\left\{\frac{1}{12}\mathcal{Y}^{\mu\nu}\mathcal{Y}_{\mu\nu} + \frac{1}{2}\left(\mathcal{M} - \mathcal{N}^\mu\mathcal{N}_\mu - D_\mu\mathcal{N}^\mu - \frac{1}{6}R\right)^2\right\}. \tag{3.25}$$

Inspecting the general form eq. (3.9) and remembering the identity (3.22) we see that we have determined $\Delta\mathcal{L}$ up to a term

$$\sqrt{g}a_0\left(R_{\mu\nu}R^{\mu\nu} - \frac{1}{3}R^2\right). \tag{3.26}$$

To determine the coefficient a_0 we consider the special case

$$\mathcal{L} = \sqrt{g}(-\partial_\mu\varphi^* g^{\mu\nu}\partial_\nu\varphi). \tag{3.27}$$

With $g_{\mu\nu} = \delta_{\mu\nu} + h_{\mu\nu}$ we need to expand up to first order in h

$$\mathcal{L} = -\partial_\mu\varphi^*\partial_\mu\varphi + \partial_\mu\varphi^*\left(h^{\mu\nu} - \frac{1}{2}\delta^{\mu\nu}h^{\alpha\alpha}\right)\partial_\nu\varphi. \tag{3.28}$$

With $s_{\mu\nu} = h^{\mu\nu} - \frac{1}{2}\delta^{\mu\nu}h^{\alpha\alpha}$ we need only compute the selfenergy type of graph with two s vertices. Note that the terms of second order in h in the Lagrangian do not contribute because they give rise to a vertex with two h, and they can only contribute to $\Delta\mathscr{L}$ in order h^2 by closing that vertex into itself:

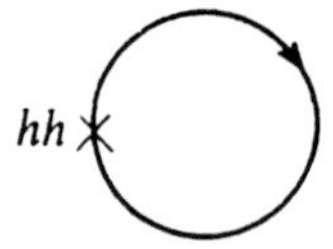

These tadpole type diagrams give integrals of the form

$$\int d_n p \, \frac{p_\mu p_\nu}{p^2},$$

and these are zero in the continuous dimension method.

The computation of the pole part of the graph

is not particularly difficult. The result is:

$$\Delta\mathscr{L} = \frac{1}{\varepsilon}\left[\frac{1}{480}\partial^2 s_{\mu\mu}\partial^2 s_{\alpha\alpha} + \frac{1}{240}\partial^2 s_{\mu\alpha}\partial^2 s_{\mu\alpha} + \frac{1}{60}\partial^2 s_{\mu\mu}\partial_\alpha\partial_\beta s_{\alpha\beta} - \frac{1}{120}\partial_\alpha\partial_\nu s_{\mu\nu}\partial_\alpha\partial_\beta s_{\mu\beta} + \frac{1}{60}\partial_\mu\partial_\nu s_{\mu\nu}\partial_\alpha\partial_\beta s_{\alpha\beta}\right], \quad (3.29)$$

$$s_{\mu\nu} = h_{\mu\nu} - \frac{1}{2}\delta_{\mu\nu}h_{\alpha\alpha}.$$

Working to second order in h one has

$$R_{\nu\alpha} = \frac{1}{2}(\partial_\alpha\partial_\nu h_{\mu\mu} - \partial_\nu\partial_\mu h_{\alpha\mu} - \partial_\alpha\partial_\mu h_{\nu\mu} + \partial^2 h_{\nu\alpha}). \quad (3.30)$$

$$R_{\alpha\beta}R^{\alpha\beta} = \frac{1}{4}(\partial^2 h_{\mu\mu}\partial^2 h_{\alpha\alpha} + \partial^2 h_{\mu\alpha}\partial^2 h_{\mu\alpha} - 2\partial^2 h_{\mu\mu}\partial_\alpha\partial_\beta h_{\alpha\beta} - 2\partial_\alpha\partial_\nu h_{\mu\nu}\partial_\alpha\partial_\beta h_{\mu\beta} + 2\partial_\mu\partial_\nu h_{\mu\nu}\partial_\alpha\partial_\beta h_{\alpha\beta}). \quad (3.31)$$

$$R = \partial^2 h_{\mu\mu} - \partial_\alpha\partial_\mu h_{\alpha\mu}. \quad (3.32)$$

$$R^2 = \partial^2 h_{\mu\mu}\partial^2 h_{\alpha\alpha} - 2\partial^2 h_{\mu\mu}\partial_\alpha\partial_\beta h_{\alpha\beta} + \partial_\alpha\partial_\beta h_{\alpha\beta}\partial_\mu\partial_\nu h_{\mu\nu}. \quad (3.33)$$

The result is:

$$\Delta\mathscr{L} = \frac{\sqrt{g}}{\varepsilon}\left[\frac{1}{72}R^2 + \frac{1}{60}\left(R_{\mu\nu}R^{\mu\nu} - \frac{1}{3}R^2\right)\right]. \quad (3.34)$$

The first term has been found before (see eq. 3.25).

All coefficients have now been determined and we can write the final result. To the Lagrangian (3.8) corresponds the counter-Lagrangian:

$$\Delta\mathcal{L} = \frac{\sqrt{g}}{\varepsilon}\,\mathrm{Tr}\left\{ \frac{1}{12}\mathcal{Y}^{\mu\nu}\mathcal{Y}_{\mu\nu} + \frac{1}{2}\left(\mathcal{M} - \mathcal{N}^{\mu}\mathcal{N}_{\mu} - \mathrm{D}_{\mu}\mathcal{N}^{\mu} - \frac{1}{6}\mathrm{R}\right)^2 + \frac{1}{60}\left(\mathrm{R}_{\mu\nu}\mathrm{R}^{\mu\nu} - \frac{1}{3}\mathrm{R}^2\right)\right\}. \quad (3.35)$$

Note that a trace is to be taken; the last term has as factor the unit matrix. As before

$$\mathcal{Y}_{\mu\nu} = \mathrm{D}_{\mu}\mathcal{N}_{\nu} - \mathrm{D}_{\nu}\mathcal{N}_{\mu} + \mathcal{N}_{\mu}\mathcal{N}_{\nu} - \mathcal{N}_{\nu}\mathcal{N}_{\mu}. \quad (3.36)$$

To obtain the result for real fields, write

$$2\varphi^*\mathcal{N}^{\mu}\partial_{\mu}\varphi = \varphi^*\mathcal{N}^{\mu}\partial_{\mu}\varphi - \partial_{\mu}\varphi^*\mathcal{N}^{\mu}\varphi - \varphi^*\partial_{\mu}\mathcal{N}^{\mu}\varphi, \quad (3.37)$$

and substitute

$$\varphi = \frac{1}{\sqrt{2}}(\mathrm{A} + i\mathrm{B}), \qquad \varphi^* = \frac{1}{\sqrt{2}}(\mathrm{A} - i\mathrm{B}). \quad (3.38)$$

The result is then as follows. To the Lagrangian

$$\mathcal{L} = \sqrt{g}\left(-\frac{1}{2}\partial_{\mu}\varphi g^{\mu\nu}\partial_{\nu}\varphi + \varphi\mathrm{N}^{\mu}\partial_{\mu}\varphi + \frac{1}{2}\varphi\mathrm{M}\varphi\right), \quad (3.39)$$

corresponds the counter-Lagrangian

$$\Delta\mathcal{L} = \frac{\sqrt{g}}{\varepsilon}\,\mathrm{Tr}\left\{ \frac{1}{24}\mathrm{Y}^{\mu\nu}\mathrm{Y}_{\mu\nu} + \frac{1}{4}\left(\mathrm{M} - \mathrm{N}^{\mu}\mathrm{N}_{\mu} - \frac{1}{6}\mathrm{R}\right)^2 + \frac{1}{120}\left(\mathrm{R}_{\mu\nu}\mathrm{R}^{\mu\nu} - \frac{1}{3}\mathrm{R}^2\right)\right\}, \quad (3.40)$$

$$\mathrm{Y}_{\mu\nu} = \mathrm{D}_{\mu}\mathrm{N}_{\nu} - \mathrm{D}_{\nu}\mathrm{N}_{\mu} + \mathrm{N}_{\mu}\mathrm{N}_{\nu} - \mathrm{N}_{\nu}\mathrm{N}_{\mu}.$$

Eq. (3.40) contains a well known result. The gravitational field enters through $g_{\mu\nu}$ and the Lagrangian (3.39) describes the interaction of bosons with gravitation, whereby gravity is treated in the tree approximation. If one adds now to the Lagrangian (3.39) the term

$$\frac{1}{12}\mathrm{R}\varphi\varphi,$$

then the « unrenormalizable » counterterms of the form MR and $\mathrm{N}^{\mu}\mathrm{N}_{\mu}\mathrm{R}$ disappear. Indeed, the energy-momentum tensor of the expression

$$\sqrt{g}\left(-\frac{1}{2}\partial_{\mu}\varphi g^{\mu\nu}\partial_{\nu}\varphi + \frac{1}{12}\mathrm{R}\varphi^2\right),$$

is precisely the « improved » energy-momentum tensor of ref. 9. We see also that closed loops of bosons introduce nasty divergencies quadratic in the Riemann tensor. This unpleasant fact remains if we allow also for closed loops of gravitons. This is the subject of the following sections.

4. CLOSED LOOPS INCLUDING GRAVITONS

We now undertake the rather formidable task of computing the divergencies of one loop graphs including gravitons. The starting point is the Lagrangian

$$\mathscr{L} = \sqrt{\bar{g}}\left(- \bar{\mathrm{R}} - \frac{1}{2}\partial_\mu\bar{\varphi}\bar{g}^{\mu\nu}\partial_\nu\bar{\varphi}\right) \tag{4.1}$$

$\bar{\mathrm{R}}$ is the Riemann scalar constructed from $\bar{g}_{\mu\nu}$. In section 7 we will include other terms, such as $\bar{\mathrm{R}}\bar{\varphi}^2$.

Again using the background field method [6], [12] we write

$$\begin{aligned} \bar{\varphi} &= \tilde{\varphi} + \varphi, \\ \bar{g}_{\mu\nu} &= g_{\mu\nu} + h_{\mu\nu}. \end{aligned} \tag{4.2}$$

If we take the c-number quantities $\tilde{\varphi}$ and $g_{\mu\nu}$ such that they obey the classical equations of motion then the part of $\mathscr{L}$ linear in the quantum fields φ and $h_{\mu\nu}$ is zero. The part quadratic in these quantities determines the one loop diagrams. We have:

$$\mathscr{L} = \mathscr{L}^{\mathrm{cl}} + \underline{\mathscr{L}} + \underline{\underline{\mathscr{L}}} + \mathscr{L}^{\mathrm{rest}}. \tag{4.3}$$

$\underline{\mathscr{L}}$ and $\underline{\underline{\mathscr{L}}}$ are linear and quatratic in the quantum fields φ and h respectively. The higher order terms contained in $\mathscr{L}^{\mathrm{rest}}$ play a role only in multiloop diagrams.

At this point we may perhaps clarify our notations. In the following we will meet quantities like the Riemann tensor $\mathrm{R}_{\mu\nu}$. This is then the tensor made up from the classical field $g_{\mu\nu}$. In the end we will use the classical equations of motion for this tensor. All divergencies that are physically irrelevant will then disappear. In fact, using these classical equations of motion is like putting the external lines of the one loop diagrams on mass-shell, with physical polarizations. Note that we still allow for trees connected to the loop. Only the very last branches of the trees must be physical.

To obtain $\underline{\mathscr{L}}$ and $\underline{\underline{\mathscr{L}}}$ we must expand the various quantities in eq. (4.1) up to second order in the quantum fields. We list here a number of subresults. Note that

$$h^\alpha_\beta = g^{\alpha\gamma}h_{\gamma\beta}. \tag{4.4}$$

Thus indices are raised and lowered by means of the classical field $g_{\mu\nu}$.

The following equations hold up to terms of third and higher order in h and φ:

$$\bar{g}_{\mu\nu} = g_{\mu\nu} + h_{\mu\nu} = g_{\mu\alpha}(\delta^{\alpha}_{\nu} + h^{\alpha}_{\nu}). \tag{4.3}$$

$$\bar{g}^{\mu\nu} = g^{\mu\nu} - h^{\mu\nu} + h^{\mu}_{\alpha}h^{\alpha\nu}. \tag{4.4}$$

Using

$$\begin{aligned}\sqrt{\bar{g}} = \sqrt{\det(\bar{g})} &= \exp\left(\frac{1}{2}\,\mathrm{Tr}\,\ln \bar{g}\right) \\ &= \sqrt{g}\,\exp\left(\frac{1}{2}\,\mathrm{Tr}\,\ln(\delta^{\alpha}_{\nu} + h^{\alpha}_{\nu})\right) \\ &= \sqrt{g}\,\exp\left(\frac{1}{2}\,\mathrm{Tr}\left(h^{\alpha}_{\nu} - \frac{1}{2}h^{\alpha}_{\beta}h^{\beta}_{\nu}\right)\right) \\ &= \sqrt{g}\,\exp\left(\frac{1}{2}h^{\alpha}_{\alpha} - \frac{1}{4}h^{\alpha}_{\beta}h^{\beta}_{\alpha}\right),\end{aligned}$$

we find

$$\sqrt{\bar{g}} = \sqrt{g}\left(1 + \frac{1}{2}h^{\alpha}_{\alpha} - \frac{1}{4}h^{\alpha}_{\beta}h^{\beta}_{\alpha} + \frac{1}{8}h^{\alpha 2}_{\alpha}\right). \tag{4.5}$$

Further

$$\bar{\Gamma}^{\alpha}_{\mu\nu} = \Gamma^{\alpha}_{\mu\nu} + \underline{\Gamma}^{\alpha}_{\mu\nu} + \underline{\underline{\Gamma}}^{\alpha}_{\mu\nu}. \tag{4.6}$$

$$\underline{\Gamma}^{\alpha}_{\mu\nu} = \frac{1}{2}(h^{\alpha}_{\nu,\mu} + h^{\alpha}_{\mu,\nu} - h_{\mu\nu}^{\;\;\;,\alpha}). \tag{4.7}$$

$$\underline{\underline{\Gamma}}^{\alpha}_{\mu\nu} = -\frac{1}{2}h^{\alpha\gamma}(h_{\gamma\nu,\mu} + h_{\mu\gamma,\nu} - h_{\mu\nu,\gamma}). \tag{4.8}$$

$$\underline{\Gamma}^{\alpha}_{\mu\alpha} = \frac{1}{2}h^{\alpha}_{\alpha,\mu}, \qquad \underline{\underline{\Gamma}}^{\alpha}_{\mu\alpha} = -\frac{1}{2}h^{\alpha}_{\beta}h^{\beta}_{\alpha,\mu}. \tag{4.9}$$

We used here the fact that the co- or contra-variant derivative of $g_{\mu\nu}$ is zero; therefore

$$h^{\alpha}_{\mu,\alpha} = \mathrm{D}_{\alpha}h^{\alpha}_{\mu} = g^{\alpha\beta}\mathrm{D}_{\alpha}h_{\mu\beta} = \mathrm{D}^{\alpha}h_{\mu\alpha} = h_{\mu\alpha}^{\;\;\;,\alpha}. \tag{4.10}$$

Note that we employ the standard notation to denote the co- and contra-variant derivatives:

$$h^{\alpha}_{\nu,\mu} = \mathrm{D}_{\mu}h^{\alpha}_{\nu}, \qquad h_{\nu\beta,\;\;\alpha}^{\;\;\;\;,\mu} = \mathrm{D}_{\alpha}\mathrm{D}^{\mu}h_{\nu\beta}, \text{ etc.} \tag{4.11}$$

Observe that the order of differentiation is relevant. The D symbol involves the Christoffel symbol Γ made up from the classical field $g_{\mu\nu}$.

$$\bar{\mathrm{R}}^{\mu}_{\nu\alpha\beta} = \mathrm{R}^{\mu}_{\nu\alpha\beta} + \underline{\mathrm{R}}^{\mu}_{\nu\alpha\beta} + \underline{\underline{\mathrm{R}}}^{\mu}_{\nu\alpha\beta}. \tag{4.12}$$

$$\begin{aligned}\underline{\mathrm{R}}^{\mu}_{\nu\alpha\beta} = \mathrm{D}_{\alpha}\underline{\Gamma}^{\mu}_{\nu\beta} - \mathrm{D}_{\beta}\underline{\Gamma}^{\mu}_{\nu\alpha} &= \frac{1}{2}(h^{\mu}_{\beta,\nu\alpha} - h_{\nu\beta,\;\;\alpha}^{\;\;\;\;,\mu} - h^{\mu}_{\alpha,\nu\beta} + h_{\nu\alpha,\;\;\beta}^{\;\;\;\;,\mu}) \\ &\quad + \frac{1}{2}\mathrm{R}^{\mu}_{\gamma\alpha\beta}h^{\gamma}_{\nu} + \frac{1}{2}\mathrm{R}^{\gamma}_{\nu\beta\alpha}h^{\mu}_{\gamma}.\end{aligned} \tag{4.13}$$

In the derivation we used the identity

$$\mathrm{D}_\alpha \mathrm{D}_\beta h^\mu_\nu - \mathrm{D}_\beta \mathrm{D}_\alpha h^\mu_\nu = \mathrm{R}^\mu_{\gamma\alpha\beta} h^\gamma_\nu + \mathrm{R}^\gamma_{\nu\beta\alpha} h^\mu_\gamma . \tag{4.14}$$

Further:

$$\underline{\underline{\mathrm{R}}}^\mu_{\nu\alpha\beta} = \mathrm{D}_\alpha \underline{\underline{\Gamma}}^\mu_{\nu\beta} - \mathrm{D}_\beta \underline{\underline{\Gamma}}^\mu_{\nu\alpha} + \underline{\Gamma}^\gamma_{\beta\nu} \underline{\Gamma}^\mu_{\gamma\alpha} - \underline{\Gamma}^\gamma_{\alpha\nu} \underline{\Gamma}^\mu_{\gamma\beta} . \tag{4.15}$$

$$\bar{\mathrm{R}}_{\nu\alpha} = \mathrm{R}_{\nu\alpha} + \underline{\mathrm{R}}_{\nu\alpha} + \underline{\underline{\mathrm{R}}}_{\nu\alpha} . \tag{4.16}$$

$$\begin{aligned} \underline{\mathrm{R}}_{\nu\alpha} &= \frac{1}{2}\left(h^\beta_{\beta,\nu\alpha} - h^\beta_{\nu,\beta\alpha} - h^\beta_{\alpha,\nu\beta} + h_{\nu\alpha,}{}^{\beta}{}_{\beta}\right) + \frac{1}{2} h^\beta_{\nu,\beta\alpha} - \frac{1}{2} h^\beta_{\nu,\alpha\beta} \\ &= \frac{1}{2}\left(h^\beta_{\beta,\nu\alpha} - h^\beta_{\nu,\alpha\beta} - h^\beta_{\alpha,\nu\beta} + h_{\nu\alpha,}{}^{\beta}{}_{\beta}\right). \end{aligned} \tag{4.17}$$

$$\begin{aligned} \underline{\underline{\mathrm{R}}}_{\nu\alpha} = & -\frac{1}{2}\mathrm{D}_\alpha (h^\beta_\mu h^\mu_{\beta,\nu}) + \frac{1}{2}\mathrm{D}_\beta \left\{ h^\beta_\gamma (h^\gamma_{\nu,\alpha} + h^\gamma_{\alpha,\nu} - h_{\alpha\nu}{}^{,\gamma}) \right\} \\ & + \frac{1}{4}(h^\gamma_{\beta,\nu} + h^\gamma_{\nu,\beta} - h_{\beta\nu}{}^{,\gamma})(h^\beta_{\gamma,\alpha} + h^\beta_{\alpha,\gamma} - h_{\gamma\alpha}{}^{,\beta}) \\ & - \frac{1}{4}(h^\gamma_{\alpha,\nu} + h^\gamma_{\nu,\alpha} - h_{\nu\alpha}{}^{,\gamma}) h^\beta_{\beta,\gamma} . \end{aligned} \tag{4.18}$$

$$\bar{\mathrm{R}} = \mathrm{R} + \underline{\mathrm{R}} + \underline{\underline{\mathrm{R}}} \tag{4.19}$$

$$\underline{\mathrm{R}} = h^{\beta,\alpha}_{\beta,\alpha} - h^{\beta,\alpha}_{\alpha,\ \beta} - \mathrm{R}^\alpha_\nu h^\nu_\alpha . \tag{4.20}$$

$$\begin{aligned} \underline{\underline{\mathrm{R}}} = & -\frac{1}{2}\mathrm{D}_\alpha (h^\beta_\mu h^{\mu,\alpha}_\beta) + \frac{1}{2}\mathrm{D}_\beta \left\{ h^\beta_\nu (2h^{\nu\alpha}_{,\alpha} - h^{\alpha,\nu}_\alpha) \right\} \\ & + \frac{1}{4}(h^\nu_{\beta,\alpha} + h^\nu_{\alpha,\beta} - h_{\beta\alpha}{}^{,\nu})(h^{\beta,\alpha}_\nu + h^{\beta\alpha}_{,\nu} - h^{\alpha,\beta}_\nu) \\ & - \frac{1}{4}(2h^{\nu\alpha}_{,\alpha} - h^{\alpha,\nu}_\alpha) h^\beta_{\beta,\nu} - \frac{1}{2} h^{\nu\alpha} h^\beta_{\beta,\nu\alpha} \\ & + \frac{1}{2} h^\nu_\alpha \mathrm{D}_\beta (h^{\beta,\alpha}_\nu + h^{\beta\alpha}_{,\nu} - h^{\alpha,\beta}_\nu) + h^\nu_\beta h^\beta_\alpha \mathrm{R}^\alpha_\nu . \end{aligned} \tag{4.21}$$

Inserting the various quantities in eq. (4.1), we find

$$\begin{aligned} \underline{\mathscr{L}} = \sqrt{g}\Big(& -\frac{1}{2} h^\alpha_\alpha \mathrm{R} - \frac{1}{4}\partial_\mu \tilde{\varphi} g^{\mu\nu} \partial_\nu \tilde{\varphi} h^\alpha_\alpha - h^{\beta,\alpha}_{\beta,\alpha} + h^{\beta\alpha}_{,\alpha\beta} \\ & + \mathrm{R}^\alpha_\beta h^\beta_\alpha - \partial_\mu \varphi g^{\mu\nu} \partial_\nu \tilde{\varphi} + \frac{1}{2}\partial_\mu \tilde{\varphi} \partial_\nu \tilde{\varphi} h^{\mu\nu} \Big). \end{aligned} \tag{4.22}$$

This expression will eventually supply us with the equations of motion for the classical fields $\tilde{\varphi}$ and $g_{\mu\nu}$. Allowing partial (co- and contra-variant) integration and omitting total derivatives:

$$\begin{aligned} \underline{\mathscr{L}} = \sqrt{g}\Big(& -\frac{1}{2} h^\alpha_\alpha \mathrm{R} - \frac{1}{4} h^\alpha_\alpha \partial_\mu \tilde{\varphi} g^{\mu\nu} \partial_\nu \tilde{\varphi} + h^\beta_\alpha \mathrm{R}^\alpha_\beta \\ & + \frac{1}{2} h^{\mu\nu} \partial_\mu \tilde{\varphi} \partial_\nu \tilde{\varphi} - \partial_\mu \varphi g^{\mu\nu} \partial_\nu \tilde{\varphi} \Big). \end{aligned} \tag{4.23}$$

Further:

$$\mathscr{L} = \sqrt{g}\Big[-\frac{1}{2}\partial_\mu\varphi\partial_\nu\tilde{\varphi}(g^{\mu\nu}h^\alpha_\alpha - 2h^{\mu\nu}) - \frac{1}{2}\partial_\mu\tilde{\varphi}\partial_\nu\tilde{\varphi}\Big(h^\mu_\alpha h^{\alpha\nu} - \frac{1}{2}h^\alpha_\alpha h^{\mu\nu}\Big)$$
$$- \frac{1}{2}\partial_\mu\varphi g^{\mu\nu}\partial_\nu\varphi - \Big(\frac{1}{8}(h^\alpha_\alpha)^2 - \frac{1}{4}h^\alpha_\beta h^\beta_\alpha\Big)\Big(R + \frac{1}{2}\partial_\mu\tilde{\varphi}g^{\mu\nu}\partial_\nu\tilde{\varphi}\Big)$$
$$- \frac{1}{2}h^\alpha_\alpha(h^{\beta,\mu}_{\beta,\mu} - h^{\beta,\mu}_{\mu,\ \beta} - R^\beta_\nu h^\nu_\beta) + \frac{1}{2}D_\alpha(h^\nu_\mu h^{\mu,\alpha}_\nu)$$
$$- \frac{1}{2}D_\beta\Big\{ h^\beta_\nu(2h^{\nu\alpha}_{,\alpha} - h^{\alpha,\nu}_\alpha) \Big\}$$
$$- \frac{1}{4}(h^\nu_{\beta,\alpha} + h^\nu_{\alpha,\beta} - h^{;\nu}_{\beta\alpha})(h^{\beta,\alpha}_\nu + h^{\beta\alpha}_{,\nu} - h^{\alpha,\beta}_\nu)$$
$$+ \frac{1}{4}(2h^{\nu\alpha}_{,\alpha} - h^{\alpha,\nu}_\alpha)h^\beta_{\beta,\nu} + \frac{1}{2}h^{\nu\alpha}h^\beta_{\beta,\nu\alpha} - \frac{1}{2}h^\nu_\alpha D_\beta(h^{\beta,\alpha}_\nu + h^{\beta\alpha}_{,\nu} - h^{\alpha,\beta}_\nu)$$
$$- h^\nu_\beta h^\beta_\alpha R^\alpha_\nu\Big]. \tag{4.24}$$

Performing partial integration and omitting total derivatives:

$$\mathscr{L} = \sqrt{g}\Big[-\frac{1}{2}\partial_\mu\varphi\partial_\nu\tilde{\varphi}(g^{\mu\nu}h^\alpha_\alpha - 2h^{\mu\nu}) - \frac{1}{2}\partial_\mu\tilde{\varphi}\partial_\nu\tilde{\varphi}\Big(h^\mu_\alpha h^{\alpha\nu} - \frac{1}{2}h^\alpha_\alpha h^{\mu\nu}\Big)$$
$$- \frac{1}{2}\partial_\mu\varphi g^{\mu\nu}\partial_\nu\varphi - \Big(\frac{1}{8}(h^\alpha_\alpha)^2 - \frac{1}{4}h^\alpha_\beta h^\beta_\alpha\Big)\Big(R + \frac{1}{2}\partial_\mu\tilde{\varphi}g^{\mu\nu}\partial_\nu\tilde{\varphi}\Big)$$
$$- h^\nu_\beta h^\beta_\alpha R^\alpha_\nu + \frac{1}{2}h^\alpha_\alpha h^\nu_\beta R^\beta_\nu - \frac{1}{4}h^\beta_{\alpha,\nu}h^{\alpha,\nu}_\beta + \frac{1}{4}h^\alpha_{\alpha,\mu}h^{\beta,\mu}_\beta$$
$$- \frac{1}{2}h^\alpha_{\alpha,\beta}h^{\beta,\mu}_\mu + \frac{1}{2}h^{\nu,\alpha}_\beta h^\beta_{\alpha,\nu}\Big] \tag{4.25}$$

The Lagrangian $\mathscr{L}$ is invariant to the gauge transformation of φ and

$$h_{\mu\nu} \rightarrow h_{\mu\nu} + (g_{\alpha\nu} + h_{\alpha\nu})D_\mu\eta^\alpha + (g_{\mu\alpha} + h_{\mu\alpha})D_\nu\eta^\alpha + \eta^\alpha D_\alpha h_{\mu\nu}. \tag{4.26}$$

To have Feynman rules we must supplement the Lagrangian eq. (4.25) with a gauge fixing part $-\frac{1}{2}(C_\mu)^2$ and a Faddeev-Popov ghost Lagrangian. In section 2 we have shown that there exists a gauge that allows easy verification of unitarity; the work of ref. 7 tells us that other choices of C are physically equivalent, and describe therefore also a unitary theory.

We will employ the following C:

$$C_\alpha = \sqrt[4]{g}\Big(h^\nu_{\mu,\nu} - \frac{1}{2}h^\nu_{\nu,\mu} - \varphi\partial_\mu\tilde{\varphi}\Big)t^{\mu\alpha}. \tag{4.27}$$

The quantity $t^{\mu\alpha}$ is the root of the tensor $g^{\mu\nu}$:

$$t^{\mu\alpha}t^{\alpha\nu} = g^{\mu\nu}. \tag{4.28}$$

It has what one could call « mid-indices », but it will not play any substantial role. Using (4.27) we find:

$$-\frac{1}{2}C_\mu^2 = -\frac{1}{2}\sqrt{g}\left(h^\nu_{\mu,\nu} - \frac{1}{2}h^\nu_{\nu,\mu}\right)\left(h^{\alpha\mu}_{,\alpha} - \frac{1}{2}h^{\alpha,\mu}_\alpha\right)$$
$$+ \sqrt{g}\left(h^{\nu\mu}_{,\nu} - \frac{1}{2}h^{\nu,\mu}_\nu\right)\varphi\partial_\mu\tilde{\varphi} - \frac{1}{2}\sqrt{g}\varphi^2\partial_\mu\tilde{\varphi}\partial_\nu\tilde{\varphi}g^{\mu\nu}. \quad (4.29)$$

With this choice for C we obtain:

$$\underline{\underline{\mathscr{L}}} - \frac{1}{2}C_\mu^2 = \sqrt{g}\left(-\frac{1}{4}h^\beta_{\alpha,\nu}h^{\alpha,\nu}_\beta + \frac{1}{8}h^\alpha_{\alpha,\nu}h^{\beta,\nu}_\beta - \frac{1}{2}\partial_\mu\varphi g^{\mu\nu}\partial_\nu\varphi\right.$$
$$\left. + \frac{1}{2}h^\alpha_\beta X^{\beta\mu}_{\alpha\nu}h^\nu_\mu + \varphi Y^\alpha_\beta h^\beta_\alpha + \frac{1}{2}\varphi Z\varphi\right), \quad (4.30)$$

with

$$X^{\beta\mu}_{\alpha\nu} = 2\left(-\frac{1}{2}\delta^\beta_\nu D^\mu\tilde{\varphi}D_\alpha\tilde{\varphi} + \frac{1}{4}\delta^\beta_\alpha D^\mu\tilde{\varphi}D_\nu\tilde{\varphi} - \frac{1}{16}\delta^\beta_\alpha\delta^\mu_\nu D_\gamma\tilde{\varphi}D^\gamma\tilde{\varphi}\right.$$
$$+ \frac{1}{8}\delta^\beta_\nu\delta^\mu_\alpha D_\gamma\tilde{\varphi}D^\gamma\tilde{\varphi} - \frac{1}{8}\delta^\beta_\alpha\delta^\mu_\nu R + \frac{1}{4}\delta^\beta_\nu\delta^\mu_\alpha R$$
$$\left. - \frac{1}{2}\delta^\beta_\nu R^\mu_\alpha + \frac{1}{2}\delta^\beta_\alpha R^\mu_\nu + \frac{1}{2}R^{\beta\mu}{}_{\alpha\nu}\right), \quad (4.31)$$

$$Y^\alpha_\beta = \frac{1}{2}\delta^\alpha_\beta D_\nu D^\nu\tilde{\varphi} - D_\beta D^\alpha\tilde{\varphi}, \quad (4.32)$$

$$Z = -D_\mu\tilde{\varphi}D^\mu\tilde{\varphi}. \quad (4.33)$$

The Lagrangian eq. (4.30) is formally of the same forms as considered in the previous section, with fields φ_i written for the h^β_α. Even if the result of the previous section was very simple it still takes a considerable amount of work to evaluate the counter Lagrangian. This will be done in the next section. There also the ghost Lagrangian will be written down.

5. EVALUATION OF THE COUNTER LAGRANGIAN

To evaluate the counter Lagrangian we employ first the doubling trick. In addition to the fields h and φ we introduce fields h' and φ' that interact with one another in the identical way as the h and φ. That is, to the expression (4.30) we add the identical expression but with h' and φ' instead of h and φ. Obviously our counter Lagrangian will double, because in addition to any closed loop with h and φ particles we will have the same closed loop but with h' and φ' particles.

After doubling of eq. (4.30) and some trivial manipulations we obtain:

$$\underline{\underline{\mathscr{L}}}_d = \sqrt{g}\left\{ h^*_{\alpha\beta}P^{\alpha\beta\mu\nu}D_\gamma D^\gamma h_{\mu\nu} + \frac{1}{2}h^*_{\alpha\beta}(X^{\alpha\beta\mu\nu} + X^{\mu\nu\alpha\beta})h_{\mu\nu} \right.$$
$$\left. + \varphi^* D_\mu D^\mu \varphi + \varphi^* Y^{\alpha\beta} h_{\alpha\beta} + h^*_{\alpha\beta} Y^{\alpha\beta}\varphi + \varphi^* Z\varphi \right\}, \quad (5.1)$$

$$P^{\alpha\beta\mu\nu} = \frac{1}{2}g^{\alpha\mu}g^{\beta\nu} - \frac{1}{4}g^{\alpha\beta}g^{\mu\nu}. \quad (5.2)$$

The counter Lagrangian is invariant to the replacement

$$h^*_{\alpha\beta} \to h^*_{\mu\nu}P^{-1}_{\mu\nu\alpha\beta}, \quad (5.3)$$

with (compare eq. 2.8):

$$P^{-1}_{\mu\nu\alpha\beta} = g_{\mu\alpha}g_{\nu\beta} + g_{\mu\beta}g_{\nu\alpha} - g_{\mu\nu}g_{\alpha\beta}. \quad (5.4)$$

It is to be noted that the replacement (5.3) is not a covariant replacement. But at this point the transformation properties of the $h_{\mu\nu}$ are no more relevant, they are treated simply as certain fields φ_i as occurring in the equations of section 3.

We so arrive at a Lagrangian of which the part containing two derivatives (with respect to the fields h and φ) is of the form

$$\sqrt{g}\varphi^*_i D_\mu D^\mu \varphi_i = -\sqrt{g}\partial_\mu \varphi^*_i g^{\mu\nu}\partial_\nu \varphi_i. \quad (5.5)$$

Note that $D_\mu D^\mu h_{\alpha\beta}$ is *not* the same thing as $D_\mu D^\mu \varphi_i$, treating the φ_i as scalars. We must rewrite $D_\mu D^\mu h_{\alpha\beta}$ in terms of derivatives $\bar{D}$ that do not work on the indices α, β. We have:

$$D_\mu D_\nu h_{\alpha\beta} = \frac{1}{2}\bar{D}_\mu \bar{D}_\nu h_{\alpha\beta} - 2\Gamma^\gamma_{\mu\alpha}\partial_\nu h_{\gamma\beta} - (\partial_\mu \Gamma^\gamma_{\nu\alpha})\delta^\pi_\beta h_{\gamma\pi}$$
$$+ \Gamma^\pi_{\mu\nu}\Gamma^\gamma_{\pi\alpha}\delta^\vartheta_\beta h_{\gamma\vartheta} + \Gamma^\pi_{\mu\alpha}\Gamma^\gamma_{\nu\pi}\delta^\vartheta_\beta h_{\gamma\vartheta} + \Gamma^\pi_{\mu\beta}\Gamma^\gamma_{\nu\alpha}h_{\gamma\pi}$$
$$+ (\text{same, but with } \alpha \leftrightarrow \beta). \quad (5.6)$$

In this way one obtains

$$\sqrt{g}h^*_{\alpha\beta}D_\mu D^\mu h_{\alpha\beta} = \sqrt{g}(h^*_{\alpha\beta}\bar{D}_\mu \bar{D}^\mu h_{\alpha\beta} + 2h^*_{\alpha\beta}\mathscr{N}^{\mu\gamma\nu}_{\alpha\beta}\bar{D}_\mu h_{\gamma\nu} + h^*_{\alpha\beta}\mathscr{T}^{\mu\nu}_{\alpha\beta}h_{\mu\nu}), \quad (5.7)$$

with

$$\mathscr{N}^{\mu\nu\gamma}_{\alpha\beta} = (-2g^{\mu\pi}\Gamma^\nu_{\pi\alpha}\delta^\gamma_\beta)_{\text{symm } \alpha,\beta;\,\nu,\gamma}. \quad (5.8)$$

We have written for symplicity only one term. the subscript « symm » denotes that only the part symmetrical with respect to α, β exchange, as well as ν, γ exchange is to be taken. Further:

$$\mathscr{T}^{\mu\nu}_{\alpha\beta} = D_\gamma \mathscr{N}^{\gamma\mu\nu}_{\alpha\beta} + \Gamma^\gamma_{\gamma\pi}\mathscr{N}^{\pi\mu\nu}_{\alpha\beta} + \mathscr{N}^{\pi\delta\epsilon}_{\alpha\beta}\mathscr{N}^{\gamma\mu\nu}_{\delta\epsilon}g_{\pi\gamma}, \quad (5.9)$$

or symbolically

$$\mathscr{T} = D_\mu \mathscr{N}^\mu + \mathscr{N}_\mu \mathscr{N}^\mu, \quad (5.10)$$

where the covariant derivative « sees » only the index explicity written.

We can now apply the equations of section 3. The fields h_{11}, h_{22}, etc. may be renamed $\varphi_1, \ldots, \varphi_{10}$, the remaining fields φ as φ_{11} etc. The matrix $\mathcal{N}$ of section 3 can be identified with eq. (5.8), that is non-zero only in the first 10×10 submatrix. The matrix $\mathcal{M}$ is of the form

$$\mathcal{M} = \begin{pmatrix} \mathcal{T} + P^{-1}\bar{X} & P^{-1}\bar{Y} \\ \bar{Y} & Z \end{pmatrix}. \tag{5.11}$$

Here

$$\bar{X}^{\alpha\beta\mu\nu} = \frac{1}{2} g^{\alpha\gamma} g^{\nu\pi} X^{\beta\mu}_{\gamma\pi} + \frac{1}{2} g^{\beta\gamma} g^{\mu\pi} X^{\alpha\nu}_{\gamma\pi}. \tag{5.12}$$

$$\bar{Y}^{\alpha\beta} = g^{\alpha\nu} Y^{\beta}_{\nu}, \tag{5.13}$$

with X, Y, Z given in eqs. (4.31-33) and P^{-1} in eq. (5.4). The counter Lagrangian due to all this is given in eq. (3.35). Note that the $\mathcal{T}$ in eq. (5.11) cancels out (see eq. 5.10). A calculation of a few lines gives:

$$\begin{aligned} \mathrm{Tr}\,(\mathcal{Y}^{\mu\nu}\mathcal{Y}_{\mu\nu}) &= 6 g^{\mu\gamma} g^{\nu\beta} R^{\pi}_{\alpha\mu\nu} R^{\alpha}_{\pi\gamma\beta} \\ &= -\,6 R_{\alpha\beta\mu\nu} R^{\alpha\beta\mu\nu} = 6R^2 - 24 R_{\alpha\beta} R^{\alpha\beta}. \end{aligned} \tag{5.14}$$

See appendix B for the last equality (apart from total derivatives). From eq. (3.35) we see that we must evaluate

$$\left(\mathcal{M} - \mathcal{N}^{\mu}\mathcal{N}_{\mu} - D_{\mu}\mathcal{N}^{\mu} - \frac{1}{6}R\right)^2.$$

The various pieces contributing to this are

$$\mathrm{Tr}\,(P^{-1}\bar{X}P^{-1}\bar{X}) = 2R^2 + 6R_{\alpha\beta}R^{\alpha\beta} + 3(\partial_\mu\tilde{\varphi} g^{\mu\nu}\partial_\nu\tilde{\varphi})^2. \tag{5.15}$$

$$\mathrm{Tr}\left(-\frac{1}{3}P^{-1}\bar{X}R + \frac{1}{36}R^2\right) = -\frac{31}{18}R^2. \tag{5.16}$$

In evaluating terms like $\mathrm{Tr}(R^2)$ remember that one takes the trace of a 10×10 matrix.

$$2\mathrm{Tr}(\bar{Y}P^{-1}\bar{Y}) = 2(D_\mu D^\mu \tilde{\varphi})^2 + 4R^{\alpha\beta}(\partial_\alpha\tilde{\varphi}\partial_\beta\tilde{\varphi}). \tag{5.17}$$

$$\mathrm{Tr}\left(Z - \frac{1}{6}R^2\right) = \frac{1}{3}R(\partial_\nu\tilde{\varphi} g^{\mu\nu}\partial_\mu\tilde{\varphi}) + \frac{1}{36}R^2 + (\partial_\mu\tilde{\varphi} g^{\mu\nu}\partial_\nu\tilde{\varphi})^2. \tag{5.18}$$

As a final step we must compute the contribution due to the Faddeev-Popov ghost. The Faddeev-Popov ghost Lagrangian is obtained by subjecting C_μ to a gauge transformation. Without any difficulty we find (note $\varphi \to \varphi + \eta^\alpha D^\alpha(\tilde{\varphi} + \varphi)$):

$$\mathcal{L}_{\text{ghost}} = \sqrt{g}\,\eta^*_\mu \{\, \eta_{\mu;\alpha}^{;\alpha} - R_{\alpha\mu}\eta^\alpha - (\partial_\alpha\tilde{\varphi}\partial_\mu\tilde{\varphi})\eta^\alpha \,\}. \tag{5.19}$$

Terms containing h or φ can be dropped, because we are not splitting up η in classical and quantum part. The Faddeev-Popov ghost is never external. In deriving (5.19) we used an equation like (4.14), and transformed a factor $\sqrt[4]{gt}$ away by means of an η^* substitution. Again, it is necessary to work out the covariant derivatives; the contribution due to that is:

$$\mathrm{Tr}\,(\mathscr{Y}^{\mu\nu}\mathscr{Y}_{\mu\nu})_{\text{ghost}} = -\,R_{\alpha\beta\mu\nu}R^{\alpha\beta\mu\nu} = R^2 - 4R_{\alpha\beta}R^{\alpha\beta}. \qquad (5.20)$$

The evaluation of the rest is not particularly difficult; the total result is:

$$(\Delta\mathscr{L})_{\text{ghost}} = -\frac{\sqrt{g}}{\varepsilon}\left\{\frac{1}{6}R(\partial_\mu\tilde{\varphi}g^{\mu\nu}\partial_\nu\tilde{\varphi}) + \frac{17}{60}R^2 + \frac{7}{30}R_{\alpha\beta}R^{\alpha\beta} + R^{\alpha\beta}(\partial_\alpha\tilde{\varphi}\partial_\beta\tilde{\varphi}) + \frac{1}{2}(\partial_\mu\tilde{\varphi}g^{\mu\nu}\partial_\nu\tilde{\varphi})^2\right\}. \qquad (5.21)$$

Notice the minus sign that is to be associated with F-P ghost loops.

Adding all pieces together, not forgetting the factor 2 to undo the doubling of the non-ghost part, gives the total result (remember also the last part of eq. 3.35 for the non-ghost part; one must add 11 times that part):

$$\Delta\mathscr{L} = \frac{\sqrt{g}}{\varepsilon}\left\{\frac{9}{720}R^2 + \frac{43}{120}R_{\alpha\beta}R^{\alpha\beta} + \frac{1}{2}(\partial_\mu\tilde{\varphi}g^{\mu\nu}\partial_\nu\tilde{\varphi})^2 - \frac{1}{12}R(\partial_\mu\tilde{\varphi}g^{\mu\nu}\partial_\nu\tilde{\varphi}) + 2(D_\mu D^\mu\tilde{\varphi})^2\right\}. \qquad (5.22)$$

The obtain the result for pure gravitation we note that contained in eq. (5.22) are the contributions due to closed loops of φ-particles. But this part is already known, from our calculations concerning a scalar particle in an external gravitational field. It is obtained from eq. (3.40) with $M = N = 0$:

$$\frac{\sqrt{g}}{\varepsilon}\left(\frac{1}{144}R^2 + \frac{1}{120}R_{\alpha\beta}R^{\alpha\beta} - \frac{1}{360}R^2\right). \qquad (5.23)$$

Subtracting this from eq. (5.22) and setting $\tilde{\varphi}$ equal to zero gives the counter Lagrangian for the case of pure gravity:

$$\Delta\mathscr{L}_{\text{grav}} = \frac{\sqrt{g}}{\varepsilon}\left(\frac{1}{120}R^2 + \frac{7}{20}R_{\alpha\beta}R^{\alpha\beta}\right). \qquad (5.24)$$

6. EQUATIONS OF MOTION

From eq. (4.23) we can trivially read off the equations of motion that the classical fields must obey in order that the first order part $\underline{\mathscr{L}}$ disappears:

$$D_\mu D^\mu\tilde{\varphi} = 0, \qquad (6.1)$$

$$\left(-\frac{1}{2}R - \frac{1}{4}D_\alpha\tilde{\varphi}D^\alpha\tilde{\varphi}\right)\delta^\mu_\nu + R^\mu_\nu + \frac{1}{2}D_\nu\tilde{\varphi}D^\mu\tilde{\varphi} = 0. \qquad (6.2)$$

Taking the trace of eq. (6.2) we find

$$R = -\frac{1}{2}(D_\alpha\tilde{\varphi}D^\alpha\tilde{\varphi}).$$

Substituting this back into eq. (6.2) gives us the set:

$$\begin{aligned} D_\mu D^\mu\tilde{\varphi} &= 0, \\ R_{\mu\nu} &= -\frac{1}{2}(D_\mu\tilde{\varphi})(D_\nu\tilde{\varphi}), \\ R &= -\frac{1}{2}(D_\mu\tilde{\varphi})(D^\mu\tilde{\varphi}). \end{aligned} \tag{6.3}$$

For pure gravity we have simply

$$R^\mu_\nu = 0, \qquad R = 0. \tag{6.4}$$

Inserting eq. (6.3) into eq. (5.22) and (5.24) gives:

$$\Delta\mathscr{L}_{\text{grav,scal}} = \frac{\sqrt{g}}{\varepsilon}\frac{203}{80}R^2, \tag{6.5}$$

$$\Delta\mathscr{L}_{\text{grav}} = 0. \tag{6.6}$$

If one were to approach the theory of gravitation just as any other field theory, one recognizes that the counterterm eq. (6.5) is not of a type present in the original Lagrangian eq. (4.1), and is therefore of the non-renormalizable type.

The question arises if the counterterm can be made to disappear by modification of the original Lagrangian. This will be investigated in the next section.

7. THE « IMPROVED » ENERGY-MOMENTUM TENSOR

The Lagrangian eq. (4.1) can be modified by inclusion of two extra terms:

$$\mathscr{L} = \sqrt{g}\left(-R - \frac{1}{2}\partial_\mu\varphi g^{\mu\nu}\partial_\nu\varphi + aR\varphi^2 + bR^{\mu\nu}\partial_\mu\varphi\partial_\nu\varphi\right). \tag{7.1}$$

The last term cannot improve the situation, because it has not the required dimension. So, we have not considered the case $b \neq 0$. Concerning the coefficient a we know already that the choice $a = \frac{1}{12}$ reduces divergencies of diagrams without internal gravitons. These are not present in eq. (7.1) because the φ-field has no non-gravitational interactions of the type $\alpha\varphi^4$, say. Actually this same choice for a seems of some help in the more general case, but it still leaves us with divergencies.

The essential tool in the study of more complicated theories is the Weyl transformation (see for example ref. 15). This concerns the behaviour under the transformation

$$g_{\mu\nu} \to g_{\mu\nu} f, \tag{7.2}$$

where f is any function of space-time. By straightforward calculation one establishes that under this transformation

$$\begin{aligned} R^{\mu}_{\nu\alpha\beta} \to R^{\mu}_{\nu\alpha\beta} &- \frac{1}{2}\delta^{\mu}_{\alpha} D_{\beta} s_{\nu} + \frac{1}{2}\delta^{\mu}_{\beta} D_{\alpha} s_{\nu} + \frac{1}{2} D_{\beta}(g_{\nu\alpha} s^{\mu}) \\ &- \frac{1}{2} D_{\alpha}(g_{\nu\beta} s^{\mu}) + \frac{1}{4}\delta^{\mu}_{\alpha} s_{\beta} s_{\nu} - \frac{1}{4}\delta^{\mu}_{\beta} s_{\alpha} s_{\nu} \\ &+ \frac{1}{4}(g_{\beta\nu} s_{\alpha} s^{\mu} - g_{\beta\nu}\delta^{\mu}_{\alpha} s^2 - g_{\alpha\nu} s_{\beta} s^{\mu} + g_{\alpha\nu}\delta^{\mu}_{\beta} s^2), \end{aligned} \tag{7.3}$$

with

$$s_{\alpha} = \frac{\partial_{\alpha} f}{f} = \partial_{\alpha}(\ln f). \tag{7.4}$$

From this:

$$R_{\nu\alpha} \to R_{\nu\alpha} + D_{\alpha} s_{\nu} + \frac{1}{2} g_{\nu\alpha} D_{\beta} s^{\beta} - \frac{1}{2} s_{\alpha} s_{\nu} + \frac{1}{2} g_{\alpha\nu} s^2, \tag{7.5}$$

and

$$\begin{aligned} R = g^{\nu\alpha} R_{\nu\alpha} \to \frac{1}{f} R + \frac{1}{f}\left(3 D_{\alpha} s^{\alpha} + \frac{3}{2} s_{\alpha} s^{\alpha}\right) \\ = \frac{1}{f} R + \frac{3}{f} D_{\alpha} \frac{f^{\alpha}}{f} + \frac{3}{2 f^3} f_{\alpha} f^{\alpha}. \end{aligned} \tag{7.6}$$

Conversely

$$\begin{aligned} g_{\mu\nu} &\to g_{\mu\nu} \frac{1}{f}, \\ R &\to f R - 3 f D_{\alpha} \frac{f^{\alpha}}{f} + \frac{3}{2 f} f_{\alpha} f^{\alpha}. \end{aligned} \tag{7.7}$$

Consider now the Lagrangian

$$\mathscr{L} = \sqrt{g}\left(- f R - \frac{1}{2}\partial_{\mu}\varphi g^{\mu\nu}\partial_{\nu}\varphi - \frac{1}{2} m^2 \varphi^2\right), \tag{7.8}$$

with $f = 1 - a\varphi^2$. Now perform the transformation (7.7). We obtain

$$\begin{aligned} \mathscr{L} = \sqrt{g}\Big(- R + 3 D_{\alpha}\frac{f^{\alpha}}{f} - \frac{3}{2 f^2} f_{\alpha} f^{\alpha} - \frac{2}{2 f}\partial_{\mu}\varphi g^{\mu\nu}\partial_{\nu}\varphi \\ - \frac{1}{2 f^2} m^2 \varphi^2\Big). \end{aligned} \tag{7.9}$$

This Lagrangian belongs to the general class

$$\mathscr{L} = \sqrt{g}\left\{ - R + \frac{1}{2}\partial_{\mu}\varphi g^{\mu\nu}\partial_{\nu}\varphi f_1(\varphi) + f_2(\varphi) \right\}. \tag{7.10}$$

It is not too difficult to see the changes with respect to the treatment of the previous sections. However, it becomes quite cumbersome to work out the quantity $(\mathcal{M})^2$, and we have taken recourse to the computer and the Schoonschip program [16]. Roughly speaking the following obtains. The required Lagrangian (7.1) has $f = 1 - a\varphi^2$. As is clear from eq. (7.9) the Lagrangian written in the form eq. (7.10) becomes a power series in the field φ, with coefficients depending on a. The counter Lagrangian becomes also a power series in $\tilde{\varphi}$ with non-trivial coefficients. Putting the coefficient a to $\frac{1}{12}$ is of little help, the final result seems not to be of any simple form.

8. CONCLUSIONS

The one loop divergencies of pure gravitation have been shown to be such that they can be transformed away by a field renormalization. This depends crucially on the well-known identity (see appendix B)

$$R_{\alpha\beta\mu\nu}R^{\alpha\beta\mu\nu} - 4R_{\alpha\beta}R^{\alpha\beta} + R^2 = \text{total derivative},$$

which is true in four-dimensional space only.

In case of gravity interacting with scalar particles divergencies of physically meaningful quantities remain. They cannot be absorbed in the parameters of the theory.

Modification of the gravitational interaction, such as would correspond to the use of the improved energy-momentum tensor is of help only with respect to a certain (important) class of divergencies, but unrenormalizable divergencies of second order in the gravitational coupling constant remain. We do not feel that this is the last word on this subject, because the situation as described in section 7 is so complicated that we feel less than sure that there is no way out. A certain exhaustion however prevents us from further investigation, for the time being.

APPENDIX A

NOTATIONS AND CONVENTIONS

Our metric is that corresponding to a purely imaginary time coordinate. In flat space $g_{\mu\nu} = \delta_{\mu\nu}$. Units are such that the gravitational coupling constant is one.

The point of view we take is that gravitation is a gauge theory. Under a gauge transformation scalars, vectors, tensors are assigned the following behaviour under infinitesimal transformations

$$\varphi' = \varphi + \eta^\alpha \partial_\alpha \varphi \qquad \text{scalar}$$
$$A'_\mu = A_\mu + A_\alpha \partial_\mu \eta^\alpha + \eta^\alpha \partial_\alpha A_\mu \qquad \text{covariant vector}$$
$$A'^\mu = A^\mu - A^\alpha \partial_\alpha \eta^\mu + \eta^\alpha \partial_\alpha A^\mu \qquad \text{contravariant vector}$$
$$B'_{\mu\nu} = B_{\mu\nu} + B_{\alpha\nu}\partial_\mu \eta^\alpha + B_{\mu\alpha}\partial_\nu \eta^\alpha + \eta^\alpha \partial_\alpha B_{\mu\nu}. \qquad \text{covariant two-tensor}$$

Note that dot-products such as $A^\mu B_\mu$ are not invariant but behave as a scalar.

Let now $B_{\mu\nu}$ be an arbitrary two-tensor. It can be established that under a gauge transformation

$$\sqrt{\det B'} = \sqrt{\det B} + \partial_\alpha(\eta^\alpha \sqrt{\det B}) \tag{A.1}$$

A Lagrangian of the form

$$\int d_4x \sqrt{\det B}\,.\,\varphi$$

where φ is a scalar (in the sense defined above) is invariant under gauge transformations. One finds:

$$\int d_4x \sqrt{\det B'}\varphi' = \int d_4x \left\{ \sqrt{\det B}\varphi + \partial_\alpha(\eta^\alpha \sqrt{\det B}\varphi) \right\} \tag{A.2}$$

The second term is a total derivative and the integral of that term vanishes (under proper boundary conditions).

Further invariants may be constructed in the usual way:

$$D_\nu \varphi = \partial_\nu \varphi \tag{A.3}$$
$$D_\nu A_\mu = \partial_\nu A_\mu - \Gamma^\alpha_{\nu\mu} A_\alpha, \text{ etc.} \tag{A.4}$$

with

$$\Gamma^\alpha_{\mu\nu} = \frac{1}{2} g^{\alpha\beta}(\partial_\nu g_{\mu\beta} + \partial_\mu g_{\nu\beta} - \partial_\beta g_{\mu\nu}). \tag{A.5}$$

In here $g_{\mu\nu}$ may be any symmetric two tensor possessing an inverse $g^{\mu\nu}$, but in practice one encounters here only the metric tensor. The quantities Γ do not transform under gauge transformations as its indices indicate; in fact

$$\Gamma'^\mu_{\nu\alpha} = (\Gamma'^\mu_{\nu\alpha})_{\text{tensor}} + \partial_\nu \partial_\alpha \eta^\mu. \tag{A.6}$$

The quantities $D_\nu A_\mu$ behave under gauge transformations as a covariant two tensor. Similarly

$$D_\mu B^\nu = \partial_\mu B^\nu + \Gamma^\nu_{\mu\alpha} B^\alpha \tag{A.7}$$

behaves as a mixed two tensor.

Let now $g_{\mu\nu}$ be the tensor used in the definition of the covariant derivatives. Then it is easy to show that

$$D_\alpha g_{\mu\nu} = 0. \tag{A.8}$$

Another useful equation relates the Christoffel symbol Γ and the determinant of the tensor used in its construction:

$$\Gamma^{\alpha}_{\mu\alpha} = \frac{1}{\sqrt{\det g}} \partial_\mu \sqrt{\det g}. \tag{A.9}$$

This leads to the important equation

$$\sqrt{\det g} D_\mu A^\mu = \partial_\mu(\sqrt{\det g} A^\mu). \tag{A.10}$$

As a consequence one can perform partial differentiation much like in the usual theories:

$$\int d_4x \sqrt{g}(\varphi D_\mu A^\mu) = - \int d_4x \sqrt{g}(D_\mu \varphi A^\mu). \tag{A.11}$$

Given any symmetric two tensor having an inverse one can construct the associated Riemann tensor:

$$R^{\mu}_{\nu\alpha\beta} = - \partial_\beta \Gamma^{\mu}_{\nu\alpha} + \partial_\alpha \Gamma^{\mu}_{\nu\beta} + \Gamma^{\mu}_{\gamma\alpha}\Gamma^{\gamma}_{\nu\beta} - \Gamma^{\mu}_{\gamma\beta}\Gamma^{\gamma}_{\nu\alpha}. \tag{A.12}$$

We use the convention $R^{\mu}_{\nu\alpha\beta} = R^{\mu}{}_{\nu\alpha\beta}$, which is of importance in connection with the raising and lowering of indices.

The Riemann tensor has a number of symmetry properties. With, as usual

$$R_{\mu\nu\alpha\beta} = g_{\mu\gamma} R^{\gamma}_{\nu\alpha\beta}, \qquad R_{\nu\beta} = R^{\mu}_{\nu\beta\mu}, \qquad R = R_{\nu\beta} g^{\nu\beta} \tag{A.13}$$

one has

$$R_{\mu\nu\alpha\beta} = - R_{\nu\mu\alpha\beta}, \quad R_{\mu\nu\alpha\beta} = - R_{\mu\nu\beta\alpha}, \quad R_{\mu\nu\alpha\beta} = R_{\alpha\beta\mu\nu},$$
$$R_{\mu\nu\alpha\beta} + R_{\mu\beta\nu\alpha} + R_{\mu\alpha\beta\nu} = 0, \qquad R_{\nu\beta} = R_{\beta\nu}, \tag{A.14}$$

The Bianchi identities are

$$D_\gamma R_{\alpha\beta\mu\nu} + D_\nu R_{\alpha\beta\gamma\mu} + D_\mu R_{\alpha\beta\nu\gamma} = 0. \tag{A.15}$$

The generalisation of the completely antisymmetric four-tensor is

$$\eta^{\alpha\beta\mu\nu} = \frac{1}{\sqrt{g}} \varepsilon^{\alpha\beta\mu\nu}, \qquad \eta_{\alpha\beta\mu\nu} = \sqrt{g} \varepsilon_{\alpha\beta\mu\nu}$$
$$\varepsilon^{\alpha\beta\mu\nu} = \varepsilon_{\alpha\beta\mu\nu} = \begin{cases} 1 \text{ if } \alpha, \beta, \mu, \nu = 1, 2, 3, 4 \\ \text{antisymmetric under exchange of any two indices.} \end{cases} \tag{A.16}$$

Note that

$$\eta^{\alpha\beta\mu\nu} \eta^{\kappa\gamma\delta\vartheta} g_{\alpha\kappa} g_{\beta\gamma} g_{\mu\delta} g_{\nu\vartheta} = 1. \tag{A.17}$$

It is easily shown that

$$D_\gamma \eta^{\alpha\beta\mu\nu} = 0 \tag{A.18}$$

In the derivation one uses the fact that in four dimensions a totally antisymmetric tensor with five indices is necessarily zero. Thus:

$$\Gamma^{\gamma}_{\kappa\delta} \eta^{\alpha\beta\mu\nu} = \Gamma^{\alpha}_{\kappa\delta} \eta^{\gamma\beta\mu\nu} + \Gamma^{\beta}_{\kappa\delta} \eta^{\alpha\gamma\mu\nu} + \Gamma^{\mu}_{\kappa\delta} \eta^{\alpha\beta\gamma\nu} + \Gamma^{\nu}_{\kappa\delta} \eta^{\alpha\beta\mu\gamma} \tag{A.19}$$

Finally, the Bianchi identities lead directly to the following equation

$$D_\gamma R_{\alpha\beta\mu\nu} \eta^{\gamma\mu\nu\kappa} = 0. \tag{A.20}$$

APPENDIX B

PRODUCTS OF RIEMANN TENSORS [6]

Let us consider the following Lagrangian

$$\int d_4 x \sqrt{g} R_{\pi\vartheta\alpha\beta} R_{\mu\nu\gamma\delta} \eta^{\pi\vartheta\mu\nu} \eta^{\alpha\beta\gamma\delta}. \tag{B.1}$$

Subjecting $g_{\mu\nu}$ to a small variation

$$g_{\mu\nu} \rightarrow g_{\mu\nu} + h_{\mu\nu},$$

we will show that the integral remains unchanged (the integrand changes by a total derivative).

In section 4 the equations showing the variation of the various objects under a change of the $g_{\mu\nu}$ have been given. One finds:

$$\begin{aligned} \delta(\sqrt{g} RR\eta\eta) = & -\frac{1}{2} g^{\alpha\beta} h_{\alpha\beta} (\sqrt{g} RR\eta\eta) \\ & - 4\sqrt{g} g_{\gamma\mu} (D_\beta \underline{\Gamma}^\mu_{\nu\alpha}) R_{\rho\vartheta\pi\delta} \eta^{\gamma\nu\pi\delta} \eta^{\rho\vartheta\alpha\beta} \\ & + 2\sqrt{g} h_{\gamma\mu} g^{\mu\pi} R_{\pi\vartheta\alpha\beta} R_{\rho\tau\nu\delta} \eta^{\gamma\vartheta\nu\delta} \eta^{\rho\tau\alpha\beta}. \end{aligned}$$

The last term can be treated by means of the same identity as we used at the end of appendix A, eq. (A.19); i. e. this term is equal to the sum of the four terms obtained by interchanging π with γ, ϑ, ν and δ respectively. The last three terms are equal to minus the original term, and one obtains, after some fiddling with indices:

$$g^{\mu\pi} R_{\pi\vartheta\alpha\beta} R_{\rho\tau\nu\delta} \eta^{\gamma\vartheta\nu\delta} \eta^{\rho\tau\alpha\beta} = \frac{1}{4} g^{\mu\gamma} R_{\pi\vartheta\alpha\beta} R_{\rho\tau\nu\delta} \eta^{\pi\vartheta\nu\delta} \eta^{\rho\tau\alpha\beta}. \tag{B.2}$$

The final result is

$$\delta(\sqrt{g} RR\eta\eta) = -4\sqrt{g} g_{\mu\nu} (D_\beta \underline{\Gamma}^\mu_{\vartheta\alpha}) R_{\pi\tau\gamma\delta} \eta^{\nu\vartheta\gamma\delta} \eta^{\pi\tau\alpha\beta}. \tag{B.3}$$

Using the Bianchi identities in the form (A.20), and exploiting the fact that the covariant derivatives of $g_{\mu\nu}$ and $\eta^{\mu\nu\alpha\beta}$ are zero we see that

$$\delta(\sqrt{g} RR\eta\eta) = -4\sqrt{g} D_\beta (g_{\nu\mu} \underline{\Gamma}^\mu_{\vartheta\alpha} R_{\pi\tau\gamma\delta} \eta^{\nu\vartheta\gamma\delta} \eta^{\pi\tau\alpha\beta}) \tag{B.4}$$

which is the desired result.

The consequence of this work is that in a Lagrangian the expression (B.1) can be omitted. Now (B.1) can be worked out by using the well known identity

$$\begin{aligned} \varepsilon^{\mu\nu\alpha\beta} \varepsilon_{\pi\vartheta\gamma\delta} = & \delta^\mu_\pi \delta^\nu_\vartheta \delta^\alpha_\gamma \delta^\beta_\delta - \delta^\mu_\pi \delta^\nu_\vartheta \delta^\beta_\gamma \delta^\alpha_\delta \\ & + \text{(all permutations of the upper indices with the approprate sign).} \end{aligned} \tag{B.5}$$

One obtains

$$R_{\pi\vartheta\alpha\beta} R_{\mu\nu\gamma\delta} \eta^{\pi\vartheta\mu\nu} \eta^{\alpha\beta\gamma\delta} = 4(R_{\mu\nu\alpha\beta} R^{\mu\nu\alpha\beta} - 4R_{\mu\nu} R^{\mu\nu} + R^2). \tag{B.6}$$

The above derivation can be generalized to an arbitrary number of dimensions. The recipe is simple: take two totally antisymmetric objects and saturate them with the Riemann four tensor. The resulting expression is such that its variation is a total derivative. For instance in two dimensions:

$$R_{\alpha\beta\mu\nu} \eta^{\alpha\beta} \eta^{\mu\nu} = R_{\alpha\beta}{}^{\mu\nu} (\delta^\alpha_\mu \delta^\beta_\nu - \delta^\beta_\mu \delta^\alpha_\nu) = -2R. \tag{B.7}$$

Clearly, the Einstein-Hilbert Lagrangian is meaningless in two dimensions. This fact shows up as an n-dependence in the graviton propagator; in the Prentki gauge as shown in eq. (2.9). Also in other gauges factors $1/(n-2)$ appear, as found by Neveu [*13*], and Capper *et al.* [*4*].

ACKNOWLEDGMENTS

The authors are indebted to Prof. S. Deser, for very useful comments.

REFERENCES

[*1*] G. 'T HOOFT, *Nuclear Phys.*, **35B**, 1971, p. 167.
[*2*] G. 'T HOOFT and M. VELTMAN, *Nuclear Phys.*, **44B**, 1972, p. 189; C. G. BOLLINI and J. J. GIAMBIAGI, *Il Nuovo Cim.*, **12B**, 1972, p. 20.
[*3*] D. H. CAPPER and G. LEIBBRANDT, *Lett. Nuovo Cim.*, t. **6**, 1973, p. 117; M. BROWN, *Nuclear Phys.*, **56B**, 1973, p. 194.
[*4*] D. H. CAPPER, G. LEIBBRANDT and M. RAMON MEDRANO, ICTP Trieste preprint 73/76, April 1973.
[*5*] R. ARNOWITT, S. DESER and C. MISNER, *Phys. Rev.*, t. **113**, 1959, p. 745; *Phys. Rev.*, t. **116**, 1959, p. 1322; *Phys. Rev.*, t. **117**, 1960, p. 1595;
R. P. FEYNMAN, *Acta Phys. Polon.*, t. **24**, 1963, p. 697;
S. MANDELSTAM, *Phys. Rev.*, t. **175**, 1968, p. 1580, 1604;
L. D. FADDEEV and V. N. POPOV, *Phys. Lett.*, **B25**, 1967, p. 29;
E. S. FRADKIN and I. V. TYUTIN, *Phys. Rev.*, **2D**, 1970, p. 2841;
C. J. ISHAM, A. SALAM and J. STRATHDEE, *Phys. Rev.*, **3D**, 1971, p. 867; *Lett. Nuovo Cim.*, t. **5**, 1972, p. 969.
[*6*] Relativity, Groups and Topology, Les Houches, 1963; B. S. DEWITT, *Phys. Rev.*, t. **162**, 1967, p. 1195, 1239.
[*7*] G. 'T HOOFT and M. VELTMAN, *Nuclear Phys.*, **50B**, 1972, p. 318.
[*8*] G. 'T HOOFT, *Nuclear Phys.*, **62B**, 1973, p. 444.
[*9*] C. G. CALLAN, S. COLEMAN and R. JACKIW, *Annals of Physics*, t. **59**, 1970, p. 42.
[*10*] K. SYMANZIK, private communication cited in ref. 9; *Comm. Nath. Phys.*, t. **18**, 1970, p. 227.
[*11*] J. H. LOWENSTEIN, *Comm. Math. Phys.*, t. **24**, 1971, p. 1; B. SCHROER, *Lett. Nuovo Cim.*, t. **2**, 1971, p. 867.
[*12*] J. HONERKAMP, *Nuclear Phys.*, **48B**, 1972, p. 269; Proc. Marseille Conf., 19-23 June 1972, C. P. Korthals-Altes Ed.
[*13*] A. NEVEU, *Inst. For Adv. Study*, Princeton, preprint, dec. 1972.
[*14*] H. VAN DAM and M. VELTMAN, *Nuclear Phys.*, **22B**, 1970, p. 397.
[*15*] B. ZUMINO, Brandeis Lectures July 1970.
[*16*] SCHOONSCHIP, A CDC 6000 program for symbolic evaluation of algebraic expressions. M. Veltman, CERN preprint 1967. For further information: H. Strubbe, Div. DD, CERN.

(Manuscrit reçu le 4 septembre 1973).

15. The path-integral approach to quantum gravity

S. W. HAWKING

15.1 Introduction

Classical general relativity is a very complete theory. It prescribes not only the equations which govern the gravitational field but also the motion of bodies under the influence of this field. However it fails in two respects to give a fully satisfactory description of the observed universe. Firstly, it treats the gravitational field in a purely classical manner whereas all other observed fields seem to be quantized. Second, a number of theorems (see Hawking and Ellis, 1973) have shown that it leads inevitably to singularities of spacetime. The singularities are predicted to occur at the beginning of the present expansion of the universe (the big bang) and in the collapse of stars to form black holes. At these singularities, classical general relativity would break down completely, or rather it would be incomplete because it would not prescribe what came out of a singularity (in other words, it would not provide boundary conditions for the field equations at the singular points). For both the above reasons one would like to develop a quantum theory of gravity. There is no well defined prescription for deriving such a theory from classical general relativity. One has to use intuition and general considerations to try to construct a theory which is complete, consistent and which agrees with classical general relativity for macroscopic bodies and low curvatures of spacetime. It has to be admitted that we do not yet have a theory which satisfies the above three criteria, especially the first and second. However, some partial results have been obtained which are so compelling that it is difficult to believe that they will not be part of the final complete picture. These results relate to the conection between black holes and thermodynamics which has already been described in chapters 6 and 13 by Carter and Gibbons. In the present article it will be shown how this relationship between gravitation and thermodynamics appears also when one quantizes the gravitational field itself.

There are three main approaches to quantizing gravity:

1 The operator approach

In this one replaces the metric in the classical Einstein equations by a distribution-valued operator on some Hilbert space. However this would not seem to be a very suitable procedure to follow with a theory like gravity, for which the field equations are non-polynomial. It is difficult enough to make sense of the product of the field operators at the same spacetime point let alone a non-polynomial function such as the inverse metric or the square root of the determinant.

2 The canonical approach

In this one introduces a family of spacelike surfaces and uses them to construct a Hamiltonian and canonical equal-time commutation relations. This approach is favoured by a number of authors because it seems to be applicable to strong gravitational fields and it is supposed to ensure unitarity. However the split into three spatial dimensions and one time dimension seems to be contrary to the whole spirit of relativity. Moreover, it restricts the topology of spacetime to be the product of the real line with some three-dimensional manifold, whereas one would expect that quantum gravity would allow all possible topologies of spacetime including those which are not products. It is precisely these other topologies that seem to give the most interesting effects. There is also the problem of the meaning of equal-time commutation relations. These are well defined for matter fields on a fixed spacetime geometry but what sense does it make to say that two points are spacelike-separated if the geometry is quantized and obeying the Uncertainty Principle?

For these reasons I prefer:

3 The path-integral approach

This too has a number of difficulties and unsolved problems but it seems to offer the best hope. The starting point for this approach is Feynman's idea that one can represent the amplitude

$$\langle g_2, \phi_2, S_2 | g_1, \phi_1, S_1 \rangle,$$

to go from a state with a metric g_1 and matter fields ϕ_1 on a surface S_1 to a state with a metric g_2 and matter fields ϕ_2 on a surface S_2, as a sum over all field configurations g and ϕ which take the given values on the surfaces S_1

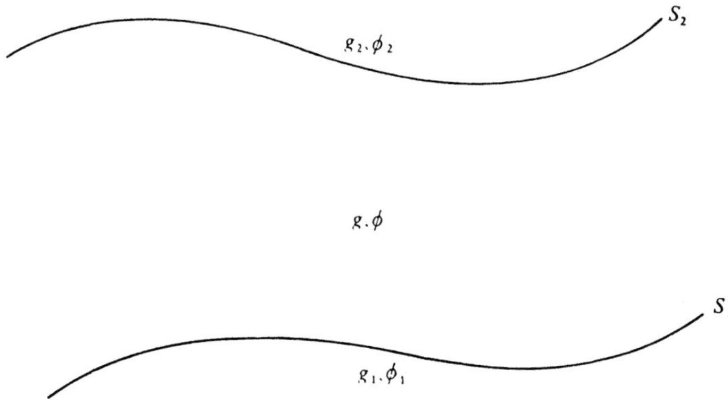

Figure 15.1. The amplitude $\langle g_2, \phi_2, S_2 | g_1, \phi_1, S_1 \rangle$ to go from a metric g_1 and matter fields ϕ_1, on a surface S_1 to a metric g_2 and matter fields ϕ_2 on a surface S_2 is given by a path integral over all fields g, ϕ which have the given values on S_1 and S_2.

and S_2 (figure 15.1). More precisely

$$\langle g_2, \phi_2, S_2 | g_1, \phi_1, S_1 \rangle = \int D[g, \phi] \exp(iI[g, \phi]),$$

where $D[g, \phi]$ is a measure on the space of all field configurations g and ϕ, $I[g, \phi]$ is the actfon of the fields, and the integral is taken over all fields which have the given values on S_1 and S_2.

In the above it has been implicitly assumed either that the surfaces S_1 and S_2 and the region between them are compact (a 'closed' universe) or that the gravitational and matter fields die off in some suitable way at spatial infinity (the asymptotically flat space). To make the latter more precise one should join the surfaces S_1 and S_2 by a timelike tube at large radius so that the boundary and the region contained within it are compact, as in the case of a closed universe. It will be seen in the next section that the surface at infinity plays an essential role because of the presence of a surface term in the gravitational action.

Not all the components of the metrics g_1 and g_2 on the boundary are physically significant, because one can give the components $g^{ab}n_b$ arbitrary values by diffeomorphisms or gauge transformations which move points in the interior, M, but which leave the boundary, ∂M, fixed. Thus one need specify only the three-dimensional induced metric h on ∂M and that only up to diffeomorphisms which map the boundary into itself.

In the following sections it will be shown how the path integral approach can be applied to the quantization of gravity and how it leads to the concepts of black hole temperature and intrinsic quantum mechanical entropy.

15.2 The action

The action in general relativity is usually taken to be

$$I=\frac{1}{16\pi G}\int (R-2\Lambda)(-g)^{1/2}\,\mathrm{d}^4x+\int L_m(-g)^{1/2}\,\mathrm{d}^4x, \qquad (15.1)$$

where R is the curvature scalar, Λ is the cosmological constant, g is the determinant of the metric and L_m is the Lagrangian of the matter fields. Units are such that $c=\hbar=k=1$. G is Newton's constant and I shall sometimes use units in which this also has a value of one. Under variations of the metric which vanish and whose normal derivatives also vanish on ∂M, the boundary of a compact region M, this action is stationary if and only if the metric satisfies the Einstein equations:

$$R_{ab}-\tfrac{1}{2}g_{ab}R+\Lambda g_{ab}=8\pi G T_{ab}, \qquad (15.2)$$

where $T^{ab}=\frac{1}{2}(-g)^{-1/2}(\delta L_m/\delta g_{ab})$ is the energy-momentum tensor of the matter fields. However this action is not an extremum if one allows variations of the metric which vanish on the boundary but whose normal derivatives do not vanish there. The reason is that the curvature scalar R contains terms which are linear in the second derivatives of the metric. By integration by parts, the variation in these terms can be converted into an integral over the boundary which involves the normal derivatives of the variation on the boundary. In order to cancel out this surface integral, and so obtain an action which is stationary for solutions of the Einstein equations under all variations of the metric that vanish on the boundary, one has to add to the action a term of the form (Gibbons and Hawking, 1977*a*):

$$\frac{1}{8\pi G}\int K(\pm h)^{1/2}\,\mathrm{d}^3x+C, \qquad (15.3)$$

where K is the trace of the second fundamental form of the boundary, h is the induced metric on the boundary, the plus or minus signs are chosen according to whether the boundary is spacelike or timelike, and C is a term which depends only on the boundary metric h and not on the values of g at the interior points. The necessity for adding the surface term (15.3) to the action in the path-integral approach can be seen by considering the

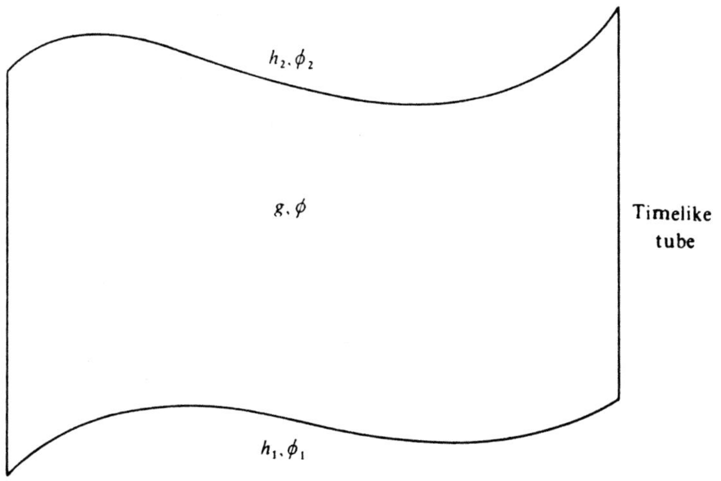

Figure 15.2. Only the induced metric h need be given on the boundary surface. In the asymptotically flat case the initial and final surfaces should be joined by a timelike tube at large radius to obtain a compact region over which to perform the path integral.

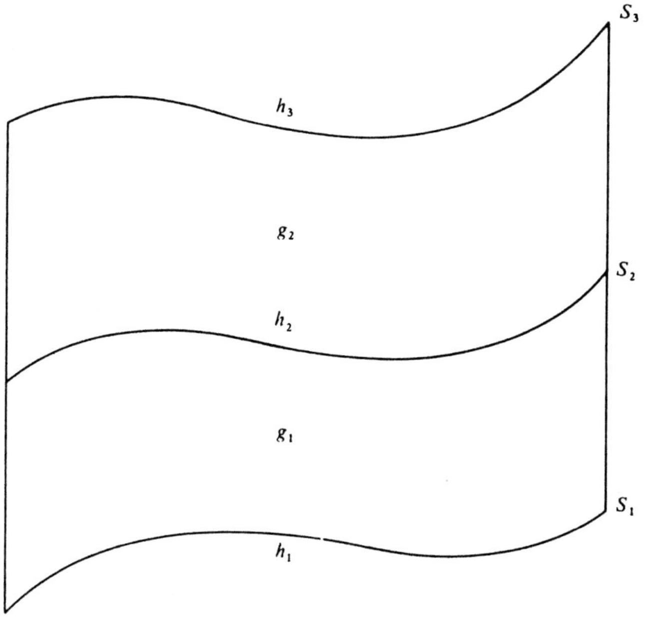

Figure 15.3. The amplitude to go from the metric h_1 on the surface S_1 to the metric h_3 on the surface S_3 should be the sum of the amplitude to go by all metrics h_2 on the intermediate surface S_2. This will be true only if the action contains a surface term.

situation depicted in figure 15.3, where one considers the transition from a metric h_1, on a surface S_1, to a metric h_2 on a surface S_2 and then to a metric h_3 on a later surface S_3. One would want the amplitude to go from the initial to the final state to be obtained by summing over all states on the intermediate surface S_2, i.e.

$$\langle h_3, S_3 | h_1, S_1 \rangle = \sum_{h_2} \langle h_2, S_2 | h_1, S_1 \rangle \langle h_3, S_3 | h_2, S_2 \rangle. \tag{15.4}$$

This will be true if and only if

$$I[g_1 + g_2] = I[g_1] + I[g_2], \tag{15.5}$$

where g_1 is the metric between S_1 and S_2, g_2 is the metric between S_2 and S_3, and $[g_1 + g_2]$ is the metric on the regions between S_1 and S_3 obtained by joining together the two regions. Because the normal derivative of g_1 at S_2 will not in general be equal to that of g_2 at S_2, the metric $[g_1 + g_2]$ will have a δ-function in the Ricci tensor of strength $2(K^1_{ab} - K^2_{ab})$, where K^1_{ab} and K^2_{ab} are the second fundamental forms of the surface S_2 in the metrics g_1 and g_2 respectively, defined with respect to the future-directed normal. This means that the relation (15.5) will hold if and only if the action is the sum of (15.1) and (15.3), i.e.

$$I = \frac{1}{16\pi G} \int (R - 2\Lambda)(-g)^{1/2}\, d^4x + \int L_m (-g)^{1/2}\, d^4x$$

$$+ \frac{1}{8\pi G} \int K (\pm h)^{1/2}\, d^3x + C. \tag{15.6}$$

The appearance of the term C in the action is somewhat awkward. One could simply absorb it into the renormalization of the measure $D[g, \phi]$. However, in the case of asymptotically flat metrics it is natural to treat it so that the contribution from the timelike tube at large radius is zero when g is the flat-space metric, η. Then

$$C = -\frac{1}{8\pi G} \int K^0 (\pm h)^{1/2}\, d^3x, \tag{15.7}$$

where K^0 is the second fundamental form of the boundary imbedded in flat space. This is not a completely satisfactory prescription because a general boundary metric h cannot be imbedded in flat space. However in an asymptotically flat situation one can suppose that the boundary will become asymptotically imbeddable as one goes to larger and larger radii. Ultimately I suspect that one should do away with all boundary surfaces and should deal only with closed spacetime manifolds. However, at the present state of development it is very convenient to use non-compact, asymptotically flat metrics and to evaluate the action using a boundary at large radius.

A metric which is asymptotically flat in the three spatial directions but not in time can be written in the form

$$ds^2 = -(1-2M_t r^{-1})\,dt^2 + (1+2M_s r^{-1})\,dr^2 + r^2(d\theta^2 + \sin^2\theta\, d\phi^2) + O(r^{-2}). \tag{15.8}$$

If the metric satisfies the vacuum Einstein equations ($\Lambda = 0$) near infinity then $M_t = M_s$, but in the path integral one considers all asymptotically flat metrics, whether or not they satisfy the Einstein equation. In such a metric it is convenient to choose the boundary ∂M to be the t-axis times a sphere of radius r_0. The area of ∂M is

$$\int (-h)^{1/2}\, d^3x = 4\pi r_0^2 \int (1 - M_t r_0^{-1} + O(r_0^{-2}))\, dt. \tag{15.9}$$

The integral of the trace of the second fundamental form of ∂M is given by

$$\int K(-h)^{1/2}\, d^3x = \frac{\partial}{\partial n} \int (-h)^{1/2}\, d^3x, \tag{15.10}$$

where $\partial/\partial n$ indicates the derivative when each point of ∂M is moved out along the unit normal. Thus

$$\int K(-h)^{1/2}\, d^3x = \int (8\pi r_0 - 4\pi M_t - 8\pi M_s + O(r_0^{-2}))\, dt. \tag{15.11}$$

For the flat space metric, η, $K^0 = 2r_0^{-1}$. Thus

$$\frac{1}{8\pi G}\int (K - K^0)(-h)^{1/2}\, d^3x = \frac{1}{2G}\int (M_t - 2M_s)\, dt. \tag{15.12}$$

In particular for a solution of the Einstein equation with mass M as measured from infinity, $M_s = M_t = M$ and the surface term is

$$-\frac{M}{2G}\int dt + O(r_0^{-1}). \tag{15.13}$$

15.3 Complex spacetime

For real Lorentzian metrics g (i.e. metrics with signature $-+++$) and real matter fields ϕ, the action $I[g, \phi]$ will be real and so the path integral will oscillate and will not converge. A related difficulty is that to find a field configuration which extremizes the action between given initial and final surfaces, one has to solve a hyperbolic equation with initial and final boundary values. This is not a well-posed problem: there may not be any solution or there may be an infinite number, and if there is a solution it will not depend smoothly on the boundary values.

In ordinary quantum field theory in flat spacetime one deals with this difficulty by rotating the time axis 90° clockwise in the complex plane, i.e. one replaces t by $-i\tau$. This introduces a factor of $-i$ into the volume integral for the action I. For example, a scalar field of mass m has a Lagrangian

$$L = -\tfrac{1}{2}\phi_{,a}\phi_{,b}g^{ab} - \tfrac{1}{2}m^2\phi^2. \tag{15.14}$$

Thus the path integral

$$Z = \int D[\phi]\exp(iI[\phi]) \tag{15.15}$$

becomes

$$Z = \int D[\phi]\exp(-\hat{I}[\phi]), \tag{15.16}$$

where $\hat{I} = -iI$ is called the 'Euclidean' action and is greater than or equal to zero for fields ϕ which are real on the Euclidean space defined by real τ, x, y, z. Thus the integral over all such configurations of the field ϕ will be exponentially damped and should therefore converge. Moreover the replacement of t by an imaginary coordinate τ has changed the metric η^{ab} from Lorentzian (signature $-+++$) to Euclidean (signature $++++$). Thus the problem of finding an extremum of the action becomes the well-posed problem of solving an elliptic equation with given boundary values.

The idea, then, is to perform all path integrals on the Euclidean section (τ, x, y, z real) and then analytically continue the results anticlockwise in the complex t-plane back to Lorentzian or Minkowski section (t, x, y, z real). As an example consider the quantity

$$Z[J] = \int D[\phi]\exp-(\tfrac{1}{2}\phi A\phi + J\phi)\,dx\,dy\,dz\,d\tau, \tag{15.17}$$

where A is the second-order differential operator $-\Box + m^2$, $\Box$ is the four-dimensional Laplacian and $J(x)$ is a prescribed source field which dies away at large Euclidean distances. The path integral is taken over all fields ϕ that die away at large Euclidean distances. One can write $Z[J]$

symbolically as

$$Z[J] = \exp\left(\tfrac{1}{2}JA^{-1}J\right)\int D[\phi]\exp\left(-\tfrac{1}{2}(\phi - A^{-1}J)A(\phi - A^{-1}J)\right), \quad (15.18)$$

where $A^{-1}(x_1, x_2)$ is the unique inverse or Green's function for A that dies away at large Euclidean distances,

$$A^{-1}J(x) = \int A^{-1}(x, x')J(x')\,\mathrm{d}^4x' \quad (15.19)$$

$$JA^{-1}J = \iint J(x)A^{-1}(x, x')J(x')\,\mathrm{d}^4x\,\mathrm{d}^4x'. \quad (15.20)$$

The measure $D[\phi]$ is invariant under the translation $\phi \to \phi - A^{-1}J$. Thus

$$Z[J] = \exp\left(\tfrac{1}{2}JA^{-1}J\right)Z[0]. \quad (15.21)$$

Then one can define the Euclidean propagator or two-point correlation function

$$\langle 0|\phi(x_2)\phi(x_1)|0\rangle = \left.\frac{\delta^2 \log Z}{\delta J(x_1)\,\delta J(x_2)}\right|_{J=0} = A^{-1}(x_2, x_1). \quad (15.22)$$

One obtains the Feynman propagator by analytically continuing $A^{-1}(x_2, x_1)$ anticlockwise in the complex $t_2 - t_1$-plane.

It should be pointed out that this use of the Euclidean section has enabled one to define the vacuum state by the property that the fields ϕ die off at large positive and negative imaginary times τ. The time-ordering operation usually used in the definition of the Feynman propagator has been automatically achieved by the direction of the analytic continuation from Euclidean space, because if $\mathrm{Re}\,(t_2 - t_1) > 0$, $\langle 0|\phi(x_2), \phi(x_1)|0\rangle$ is holomorphic in the lower half $t_2 - t_1$-plane, i.e. it is positive-frequency (a positive-frequency function is one which is holomorphic in the lower half t-plane and which dies off at large negative imaginary t).

Another use of the Euclidean section that will be important in what follows is to construct the canonical ensemble for a field ϕ. The amplitude to propagate from a configuration ϕ_1 on a surface at time t_1 to a configuration ϕ_2 on a surface at time t_2 is given by the path integral

$$\langle \phi_2, t_2|\phi_1, t_1\rangle = \int D[\phi]\exp(\mathrm{i}I[\phi]). \quad (15.23)$$

Using the Schrödinger picture, one can also write this amplitude as

$$\langle \phi_2|\exp(-\mathrm{i}H(t_2 - t_1))|\phi_1\rangle.$$

Put $t_2 - t_1 = -\mathrm{i}\beta$, $\phi_2 = \phi_1$ and sum over a complete orthonormal basis of configurations ϕ_n. One obtains the partition function

$$Z = \sum \exp(-\beta E_n) \tag{15.24}$$

of the field ϕ at a temperature $T = \beta^{-1}$, where E_n is the energy of the state ϕ_n. However from (15.23) one can also represent Z as a Euclidean path integral

$$Z = \int D[\phi] \exp(-\mathrm{i}\hat{I}[\phi]), \tag{15.25}$$

where the integral is taken over all fields ϕ that are real on the Euclidean section and are periodic in the imaginary time coordinate τ with period β. As before one can introduce a source J and obtain a Green's function by functionally differentiating $Z[J]$ with respect to J at two different points. This will represent the two-point correlation function or propagator for the field ϕ, not this time in the vacuum state but in the canonical ensemble at temperature $T = \beta^{-1}$. In the limit that the period β tends to infinity, this thermal propagator tends to the normal vacuum Feynman propagator.

It seems reasonable to apply similar complexification ideas to the gravitational field, i.e. the metric. For example, supposing one was considering the amplitude to go from a metric h_1 on a surface S_1 to a metric h_2 on a surface S_2, where the surfaces S_1 and S_2 are asymptotically flat, and are separated by a time interval t at infinity. As explained in section 15.1, one would join S_1 and S_2 by a timelike tube of length t at large radius. One could then rotate this time interval into the complex plane by introducing an imaginary time coordinate $\tau = \mathrm{i}t$. The induced metric on the timelike tube would now be positive-definite so that one would be dealing with a path integral over a region M on whose boundary the induced metric h was positive-definite everywhere. One could therefore take the path integral to be over all positive-definite metrics g which induced the given positive-definite metric h on ∂M. With the same choice of the direction of rotation into the complex plane as in flat-space Euclidean theory, the factor $(-g)^{1/2}$ which appears in the volume element becomes $-\mathrm{i}(g)^{1/2}$, so that the Euclidean action, $\hat{I} = -\mathrm{i}I$, becomes

$$\hat{I} = -\frac{1}{16\pi G}\int (R - 2\Lambda)(g)^{1/2}\,\mathrm{d}^4x - \frac{1}{8\pi G}\int (K - K^0)(h)^{1/2}\,\mathrm{d}^3x$$
$$-\int L_{\mathrm{m}}(g)^{1/2}\,\mathrm{d}^4x. \tag{15.26}$$

The problem arising from the fact that the gravitational part of this Euclidean action is not positive-definite will be discussed in section 15.4.

The state of the system is determined by the choice of boundary conditions of the metrics that one integrates over. For example, it would seem reasonable to expect that the vacuum state would correspond to integrating over all metrics which were asymptotically Euclidean, i.e. outside some compact set as they approached the flat Euclidean metric on R^4. Inside the compact set the curvature might be large and the topology might be different from that of R^4.

As an example, one can consider the canonical ensemble for the gravitational fields contained in a spherical box of radius r_0 at a temperature T, by performing a path integral over all metrics which would fit inside a boundary consisting of a timelike tube of radius r_0 which was periodically identified in the imaginary time direction with period $\beta = T^{-1}$.

In complexifying the spacetime manifold one has to treat quantities which are complex on the real Lorentzian section as independent of their complex conjugates. For example, a charged scalar field in real Lorentzian spacetime may be represented by a complex field ϕ and its complex conjugate $\bar{\phi}$. When going to complex spacetime one has to analytically continue $\bar{\phi}$ as a new field $\tilde{\phi}$ which is independent of ϕ. The same applies to spinors. In real Lorentzian spacetime one has unprimed spinors λ_A which transform under SL(2, C) and primed spinors $\mu_{A'}$ which transform under the complex conjugate group $\overline{\text{SL}(2, \text{C})}$. The complex conjugate of an unprimed spinor is a primed spinor and vice versa. When one goes to complex spacetime, the primed and unprimed spinors become independent of each other and transform under independent groups SL(2, C) and $\widetilde{\text{SL}}(2, \text{C})$ respectively. If one analytically continues to a section on which the metric is positive-definite and restricts the spinors to lie in that section, the primed and unprimed spinors are still independent but these groups become SU(2) and $\widetilde{\text{SU}}(2)$ respectively. For example, in a Lorentzian metric the Weyl tensor can be represented as

$$C_{AA'BB'CC'DD'} = \psi_{ABCD}\varepsilon_{A'B'}\varepsilon_{C'D'} + \bar{\psi}_{A'B'C'D'}\varepsilon_{AB}\varepsilon_{CD}. \quad (15.27)$$

When one complexifies, $\bar{\psi}_{A'B'C'D'}$ is replaced by an independent field $\tilde{\psi}_{A'B'C'D'}$. In particular one can have a metric in which $\psi_{ABCD} \neq 0$, but $\tilde{\psi}_{A'B'C'D'} = 0$. Such a metric is said to be conformally self-dual and satisfies

$$C_{abcd} = {}^*C_{abcd} = \tfrac{1}{2}\varepsilon_{abef}C^{ef}{}_{cd}. \quad (15.28)$$

The metric is said to be self-dual if

$$R_{abcd} = {}^*R_{abcd}$$

which implies

$$R_{ab} = 0, \quad C_{abcd} = {}^*C_{abcd}. \tag{15.29}$$

A complexified spacetime manifold M with a complex self-dual or conformally self-dual metric g_{ab} may admit a section on which the metric is real and positive definite (a 'Euclidean' section) but it will not admit a Lorentzian section, i.e. a section on which the metric is real and has a signature $-+++$.

15.4 The indefiniteness of the gravitational action

The Euclidean action for scalar or Yang–Mills fields is positive-definite. This means that the path integral over all configurations of such fields that are real on the Euclidean section converges, and that only those configurations contribute that die away at large Euclidean distances, since otherwise the action would be infinite. The action for fermion fields is not positive-definite. However, one treats them as anticommuting quantities (Berezin, 1966) so that the path integral over them converges. On the other hand, the Euclidean gravitational action is not positive-definite even for real positive-definite metrics. The reason is that although gravitational waves carry positive energy, gravitational potential energy is negative because gravity is attractive. Despite this, in classical general relativity it seems that the total energy or mass, as measured from infinity, of any asymptotically flat gravitational field is always non-negative. This is known as the *positive energy conjecture* (Brill and Deser, 1968; Geroch, 1973). What seems to happen is that whenever the gravitational potential energy becomes too large, an event horizon is formed and the region of high gravitational binding undergoes gravitational collapse, leaving behind a black hole of positive mass. Thus one might expect that the black holes would play a role in controlling the indefiniteness of the gravitational action in quantum theory and there are indications that this is indeed the case.

To see that the action can be made arbitrarily negative, consider a conformal transformation $\tilde{g}_{ab} = \Omega^2 g_{ab}$, where Ω is a positive function which is equal to one on the boundary ∂M.

$$\tilde{R} = \Omega^{-2} R - 6\Omega^{-3} \Box \Omega \tag{15.30}$$

$$\tilde{K} = \Omega^{-1} K + 3\Omega^{-2} \Omega_{;a} n^a, \tag{15.31}$$

where n^a is the unit outward normal to the boundary ∂M. Thus

$$\hat{I}[\tilde{g}] = -\frac{1}{16\pi G}\int_M (\Omega^2 R + 6\Omega_{;a}\Omega_{;b}g^{ab} - 2\Lambda\Omega^4)(g)^{1/2}\,\mathrm{d}^4x - \frac{1}{8\pi G}\int_{\partial M}\Omega^2(K - K^0)(h)^{1/2}\,\mathrm{d}^3x. \tag{15.32}$$

One sees that $\hat{I}$ may be made arbitrarily negative by choosing a rapidly varying conformal factor Ω.

To deal with this problem it seems desirable to split the integration over all metrics into an integration over conformal factors, followed by an integration over conformal equivalence classes of metrics. I shall deal separately with the case in which the cosmological constant Λ is zero but the spacetime region has a boundary ∂M, and the case in which Λ is nonzero but the region is compact without boundary.

In the former case, the path integral over the conformal factor Ω is governed by the conformally invariant scalar wave operator, $A = -\Box + \frac{1}{6}R$. Let $\{\lambda_n, \phi_n\}$ be the eigenvalues and eigenfunctions of A with Dirichlet boundary conditions, i.e.

$$A\phi_n = \lambda_n\phi_n,\ \phi_n = 0 \text{ on } \partial M.$$

If $\lambda_1 = 0$, then $\Omega^{-1}\phi_1$ is an eigenfunction with zero eigenvalue for the metric $\tilde{g}_{ab} = \Omega^2 g_{ab}$. The nonzero eigenvalues and corresponding eigenfunctions do not have any simple behaviour under conformal transformation. However they will change continuously under a smooth variation of the conformal factor which remains positive everywhere. Because the zero eigenvalues are conformally invariant, this shows that the number of negative eigenvalues (which will be finite) remains unchanged under a conformal transformation Ω which is positive everywhere.

Let $\Omega = 1 + y$, where $y = 0$ on ∂M. Then

$$\begin{aligned}\hat{I}[\tilde{g}] &= -\frac{6}{16\pi g}\int (yAy + 2Ry)(g)^{1/2}\,\mathrm{d}^4x + \hat{I}[g]\\ &= -\frac{6}{16\pi G}\int \{(y - A^{-1}R)A(y - A^{-1}R)\}(g)^{1/2}\,\mathrm{d}^4x\\ &\quad + \frac{6}{16\pi G}RA^{-1}R + \hat{I}[g]\\ &= \frac{6}{16\pi G}RA^{-1}R + \hat{I}[g] - \frac{6}{16\pi G}\int \gamma A\gamma(g)^{1/2}\,\mathrm{d}^4x,\end{aligned} \tag{15.33}$$

where $\gamma = (y - A^{-1}R)$.

Thus one can write

$$\hat{I}[\tilde{g}] = I^1 + I^2,$$

where I^1 is the first and second term on the right of (15.33) and I^2 is the third term.

I^1 depends only on the conformal equivalence class of the metric g, while I^2 depends on the conformal factor. One can thus define a quantity X to be the path integral of $\exp(-I^2)$ over all conformal factors in one conformal equivalence class of metrics.

If the operator A has no negative or zero eigenvalues, in particular if g is a solution of the Einstein equations, the inverse, A^{-1}, will be well defined and the metric $g'_{ab} = (1 + A^{-1}R)^2 g_{ab}$ will be a regular metric with $R' = 0$ everywhere. In this case I^1 will equal $\hat{I}[g']$, which in turn will be given by a surface integral of K' on the boundary. It seems plausible to make the *positive action conjecture*: any asymptotically Euclidean, positive-definite metric with $R = 0$ has positive or zero action (Gibbons, Hawking and Perry, 1978). There is a close connection between this and the positive energy conjecture in classical Lorentzian general relativity. This claims that the mass or energy as measured from infinity of any Lorentzian, asymptotically flat solution of the Einstein equations is positive or zero if the solution develops from a non-singular initial surface, the mass being zero if and only if the metric is identically flat. Although no complete proof exists, the positive energy conjecture has been proved in a number of restricted cases or under certain assumptions (Brill, 1959; Brill and Deser, 1968; Geroch, 1973; Jang and Wald, 1977) and is generally believed. If it held also for classical general relativity in five dimensions (signature $-++++$), it would imply the positive action conjecture, because a four-dimensional asymptotically Euclidean metric with $R = 0$ could be taken as time-symmetric initial data for a five-dimensional solution and the mass of such a solution would be equal to the action of the four-dimensional metric. Page (1978) has obtained some results which support the positive action conjecture. However he has also shown that it does not hold for metrics like the Schwarzschild solution which are asymptotically flat in the spatial directions, but are not in the Euclidean time direction. The significance of this will be seen later.

Let g_0 be a solution of the field equations. If I^1 increases under all perturbations away from g_0 that are not purely conformal transformations, the integral over conformal classes will tend to converge. If there is some non-conformal perturbation, δg, of g_0 which reduces I^1, then in order to make the path integral converge one will have to integrate over the metrics of the form $g_0 + i\delta g$. This will introduce a factor i into Z for each mode of non-conformal perturbations which reduces I^1. This will be discussed in the next section. For metrics which are far from a solution of the field equation, the operator A may develop zero or negative eigenvalues. When an eigenvalue passes through zero, the inverse, A^{-1}, will become undefined and I^1 will become infinite. When there are negative eigenvalues but not zero eigenvalues, A^{-1} and I^1 will be well defined, but the conformal factor $\Omega = 1 + A^{-1}R$, which transforms g to the metric g' with $R' = 0$, will pass through zero and so g' will be singular. This is very similar to what happens with three-dimensional metrics on time-symmetric initial surfaces (Brill, 1959). If h is a three-dimensional positive-definite metric on the initial surface, one can make a conformal transformation $\tilde{h} = \Omega^4 h$ to obtain a metric with $\tilde{R} = 0$ which will satisfy the constraint equations. If the three-dimensional conformally invariant operator $B = -\Delta + R/8$ has no zero or negative eigenvalues (which will be the case for metrics h sufficiently near flat space) the conformal factor Ω needed will be finite and positive everywhere. If, however, one considers a sequence of metrics h for which one of the eigenvalues of B passes through zero and becomes negative, the corresponding Ω will first diverge and then will become finite again but will pass through zero so that the metric $\tilde{h}$ will be singular. The interpretation of this is that the metric h contained a region with so much negative gravitational binding energy that it cut itself off from the rest of the universe by forming an event horizon. To describe such a situation one has to use initial surfaces with different topologies.

It seems that something analogous may be happening in the four-dimensional case. In some sense one could think that metrics g for which the operator A had negative eigenvalues contained regions which cut themselves off from the rest of the spacetime because they contained too much curvature. One could then represent their effect by going to manifolds with different topologies. Anyway, metrics for which A has negative eigenvalues are in some sense far from solutions of the field equations, and we shall see in the next section that one can in fact evaluate path integrals only over metrics near solutions of the field equations.

The operator A appears in I^2 with a minus sign. This means that in order to make the path integral over the conformal factors converge at a solution of the field equations, and in particular at flat space, one has to take γ to be purely imaginary. The prescription, therefore, for making the path integral converge is to divide the space of all metrics into conformal equivalence classes. In each equivalence class pick the metric g' for which $R'=0$. Integrate over all metrics $\tilde{g}=\Omega^2 g'$, where Ω is of the form $1+i\xi$. Then integrate over conformal equivalence classes near solutions of the field equations, with the non-conformal perturbation being purely imaginary for modes which reduce I^1.

The situation is rather similar for compact manifolds with a Λ-term. In this case there is no surface term in the action and no requirement that $\Omega=1$ on the boundary. If $\tilde{g}=\Omega^2 g$,

$$\hat{I}[\tilde{g}]=-\frac{6}{16\pi G}\int(\Omega^2 R+6\Omega_{;a}\Omega_{;b}g^{ab}-2\Lambda\Omega^4)(g)^{1/2}\,d^4x. \quad (15.34)$$

Thus quantum gravity with a Λ-term on a compact manifold is a sort of average of $\lambda\phi^4$ theory over all background metrics. However unlike ordinary $\lambda\phi^4$ theory, the kinetic term $(\nabla\Omega)^2$, appears in the action with a minus sign. This means that the integration over the conformal factors has to be taken in a complex direction just as in the previous case.

One can again divide the space of all the positive-definite metrics g on the manifold M into conformal equivalence classes. In each equivalence class the action will have one extremum at the vanishing metric for which $\Omega=0$. In general there will be another extremum at a metric g' for which $R'=4\Lambda$, though in some cases the conformal transformation $g'=\Omega^2 g$, where g is a positive-definite metric, may require a complex Ω. Putting $\tilde{g}=(1+y)^2 g'$, one obtains

$$\hat{I}[\tilde{g}]=-\frac{\Lambda V}{8\pi G}-\frac{6}{16\pi G}\int(6y_{;a}y_{;b}g^{ab}-8y^2\Lambda-8y^3\Lambda-2y^4\Lambda)(g')^{1/2}\,d^4x, \quad (15.35)$$

where $V=\int(g')^{1/2}\,d^4x$.

If Λ is negative and one neglects the cubic and quartic terms in y, one obtains convergence in the path integral by integrating over purely imaginary y in a similar manner to what was done in the previous case. It therefore seems reasonable to adopt the prescription for evaluating path integrals with Λ-terms that one picks the metric g' in each conformal equivalence class for which $R'=4\Lambda$, and one then integrates over conformal factors of the form $\Omega=1+i\xi$ about g'.

If Λ is positive, the operator $-6\Box-8\Lambda$, which acts on the quadratic terms in ξ, has at least one negative eigenvalue, $\xi=$ constant. In fact it seems that this is the only negative eigenvalue. Its significance will be discussed in section 15.10.

15.5 The stationary-phase approximation

One expects that the dominant contribution to the path integral will come from metrics and fields which are near a metric g_0, and fields ϕ_0 which are an extremum of the action, i.e. a solution of the classical field equations. Indeed this must be the case if one is to recover classical general relativity in the limit of macroscopic systems. Neglecting for the moment, questions of convergence, one can expand the action in a Taylor series about the background fields g_0, ϕ_0,

$$\hat{I}[g, \phi] = \hat{I}[g_0, \phi_0] + I_2[\bar{g}, \bar{\phi}] + \text{higher-order terms}, \qquad (15.36)$$

where

$$g_{ab} = g_{0ab} + \bar{g}_{ab}, \quad \phi = \phi_0 + \bar{\phi},$$

and $I_2[\bar{g}, \bar{\phi}]$ is quadratic in the perturbations $\bar{g}$ and $\bar{\phi}$. If one ignores the higher-order terms, the path integral becomes

$$\log Z = -\hat{I}[g_0, \phi_0] + \log \int D[\bar{g}, \bar{\phi}] \exp(-I_2[\bar{g}, \bar{\phi}]). \qquad (15.37)$$

This is known variously as the stationary-phase, WKB or one-loop approximation. One can regard the first term on the right of (15.37) as the contribution of the background fields to $\log Z$. This will be discussed in sections 15.7 and 15.8. The second term on the right of (15.37) is called the one-loop term and represents the effect of quantum fluctuations around the background fields. The remainder of this section will be devoted to describing how one evaluates it. For simplicity I shall consider only the case in which the background matter fields, ϕ_0, are zero. The quadratic term $I_2[\bar{g}, \bar{\phi}]$ can then be expressed as $I_2[\bar{g}] + I_2[\phi]$ and

$$\log Z = -\hat{I}[g_0] + \log \int D[\phi] \exp(-I_2[\phi]) + \log \int D[\bar{g}] \exp(-I_2[\bar{g}]). \qquad (15.38)$$

I shall consider first the one-loop term for the matter fields, the second term on the right of (15.38). One can express $I_2[\phi]$ as

$$I_2[\phi] = \tfrac{1}{2} \int \phi A \phi (g_0)^{1/2} \, \mathrm{d}^4 x, \qquad (15.39)$$

where A is a differential operator depending on the background metric g_0. In the case of boson fields, which I shall consider first, A is a second-order differential operator. Let $\{\lambda_n, \phi_n\}$ be the eigenvalues and the corresponding eigenfunctions of A, with $\phi_n = 0$ on ∂M in the case where there is a boundary surface. The eigenfunctions, ϕ_n, can be normalized so that

$$\int \phi_n \phi_m \cdot (g_0)^{1/2} \, d^4x = \delta_{mn}. \tag{15.40}$$

One can express an arbitrary field ϕ which vanishes on ∂M as a linear combination of these eigenfunctions:

$$\phi = \sum_n y_n \phi_n. \tag{15.41}$$

Similarly one can express the measure on the space of all fields ϕ as

$$D[\phi] = \prod_n \mu \, dy_n. \tag{15.42}$$

Where μ is a normalization factor with dimensions of mass or $(\text{length})^{-1}$. One can then express the one-loop matter term as

$$\begin{aligned} Z_\phi &= \int D[\phi] \exp(-I_2[\phi]) \\ &= \prod_n \int \mu \, dy_n \exp(-\tfrac{1}{2}\lambda_n y_n^2) \\ &= \prod_n (2\pi\mu^2\lambda_n^{-1})^{1/2} \\ &= (\det(\tfrac{1}{2}\pi^{-1}\mu^{-2}A))^{-1/2}. \end{aligned} \tag{15.43}$$

In the case of a complex field ϕ like a charged scalar field, one has to treat ϕ and the analytic continuation $\tilde{\phi}$ of its complex conjugate as independent fields. The quadratic term then has the form

$$I_2[\phi, \tilde{\phi}] = \tfrac{1}{2} \int \tilde{\phi} A \phi (g_0)^{1/2} \, d^4x. \tag{15.44}$$

The operator A will not be self-adjoint if there is a background electromagnetic field. One can write $\tilde{\phi}$ in terms of eigenfunctions of the adjoint operator $A^\dagger$:

$$\tilde{\phi} = \sum_n \tilde{y}_n \tilde{\phi}_n. \tag{15.45}$$

The measure will then have the form

$$D[\phi, \tilde{\phi}] = \prod_n \mu^2 \, dy_n \, d\tilde{y}_n. \tag{15.46}$$

Because one integrates over y_n and $\tilde{y}_n$ independently, one obtains

$$Z_\phi = (\det(\tfrac{1}{2}\pi^{-1}\mu^{-2}A))^{-1}. \tag{15.47}$$

To treat fermions in the path integrals one has to regard the spinor ψ and its independent adjoint $\tilde{\psi}$ as anticommuting Grassman variables (Berezin, 1966). For a Grassman variable x one has the following (formal) rules of integration

$$\int \mathrm{d}x = 0, \quad \int x\,\mathrm{d}x = 1. \tag{15.48}$$

These suffice to determine all integrals, since x^2 and higher powers of x are zero by the anticommuting property. Notice that (15.48) implies that if $y = ax$, where a is a real constant, then $\mathrm{d}y = a^{-1}\,\mathrm{d}x$.

One can use these rules to evaluate path integrals over the fermion fields ψ and $\tilde{\psi}$. The operator A in this case is just the ordinary first-order Dirac operator. If one expands $\exp(-I_2)$ in a power series, only the term linear in A will survive because of the anticommuting property. Integration of this respect to $\mathrm{d}\psi$ and $\mathrm{d}\tilde{\psi}$ gives

$$Z_\psi = \det(\tfrac{1}{2}\mu^{-2}A). \tag{15.49}$$

Thus the one-loop terms for fermion fields are proportional to the determinant of their operator while those for bosons are inversely proportional to determinants.

One can obtain an asymptotic expansion for the number of eigenvalues $N(\lambda)$ of an operator A with values less than λ:

$$N(\lambda) \sim \tfrac{1}{2}B_0\lambda^2 + B_1\lambda + B_2 + O(\lambda^{-1}), \tag{15.50}$$

where B_0, B_1 and B_2 are the 'Hamidew' coefficients referred to by Gibbons in chapter 13. They can be expressed as $B_n = \int b_n(g_0)^{1/2}\,\mathrm{d}^4x$, where the b_n are scalar polynomials in the metric, the curvature and its covariant derivatives (Gilkey, 1975). In the case of the scalar wave operator, $A = -\Box + \xi R + m^2$, they are

$$b_0 = \frac{1}{16\pi^2} \tag{15.51}$$

$$b_1 = \frac{1}{16\pi^2}((1/6 - \xi)R - m^2) \tag{15.52}$$

$$b_2 = \frac{1}{2880\pi^2}(R^{abcd}R_{abcd} - R_{ab}R^{ab} + (6 - 30\xi)\Box R + \tfrac{5}{2}(6\xi - 1)^2R^2$$
$$+ 30m^2(1 - 6\xi)R + 90m^4). \tag{15.53}$$

When there is a boundary surface ∂M, this introduces extra contributions into (15.50) including a $\lambda^{1/2}$-term. This would seem an additional reason for trying to do away with boundary surfaces and working simply with closed manifolds.

From (15.50) one can see that the determinant of A, the product of its eigenvalues, is going to diverge badly. In order to obtain a finite answer one has to regularize the determinant by dividing out by the product of the eigenvalues corresponding to the first two terms on the right of (15.50) (and those corresponding to a $\lambda^{1/2}$-term if it is present). There are various ways of doing this – dimensional regularization (t'Hooft and Veltman, 1972), point splitting (DeWitt, 1975), Pauli–Villars (Zeldovich and Starobinsky, 1972) and the zeta function technique (Dowker and Critchley, 1976; Hawking, 1977). The last method seems the most suitable for regularizing determinants of operators on a curved space background. It will be discussed further in the next section.

For both fermion and baryon operators the term B_0 is $(nV/16\pi^2)$, where V is the volume of the manifold in the background metric, g_0, and n is the number of spin states of the field. If, therefore, there are an equal number of fermion and boson spin states, the leading divergences in Z produced by the B_0-terms will cancel between the fermion and boson determinants without having to regularize. If in addition the B_1-terms either cancel or are zero (which will be the case for zero-rest-mass, conformally invariant fields), the other main divergence in Z will cancel between fermions and bosons. Such a situation occurs in theories with supersymmetry, such as supergravity (Deser and Zumino, 1976; Freedman, van Nieuwenhuizen and Ferrara, 1976) or extended supergravity (Ferrara and van Nieuwenhuizen, 1976). This may be a good reason for taking these theories seriously, in particular for the coupling of matter fields to gravity.

Whether or not the divergences arising from B_0 and B_1 cancel or are removed by regularization, the net B_2 will in general be nonzero, even in supergravity, if the topology of the spacetime manifold is non-trivial (Perry, 1978). This means that the expression for Z will contain a finite number (not necessarily an integer) of uncancelled eigenvalues. Because the eigenvalues have dimensions $(\text{length})^{-2}$, in order to obtain a dimensionless result for Z each eigenvalue has to be divided by μ^2, where μ is the normalization constant or regulator mass. Thus Z will depend on μ. For renormalizable theories such as $\lambda\phi^4$, quantum electrodynamics or Yang–Mills in flat spacetime, B_2 is proportional to the action of the field. This means that one can absorb the μ-dependence into an effective coupling constant $g(\mu)$ which depends on the scale at which it is measured. If $g(\mu)\to 0$ as $\mu\to\infty$, i.e. for very short length scales or high energies, the theory is said to be asymptotically free.

In curved spacetime however, B_2 involves terms which are quadratic in the curvature tensor of the background space. Thus unless one supposes that the gravitational action contains terms quadratic in the curvature (and this seems to lead to a lot of problems including negative energy, fourth-order equations and no Newtonian limit (Stelle, 1977, 1978)) one cannot remove the μ-dependence. For this reason gravity is said to be unrenormalizable because new parameters occur when one regularizes the theory.

If one tried to regularize the higher-order terms in the Taylor series about a background metric, one would have to introduce an infinite sequence of regularization parameters whose values could not be fixed by the theory. However it will be argued in section 15.9 that the higher-order terms have no physical meaning and that one ought to consider only the one-loop quadratic terms. Unlike $\lambda\phi^4$ or Yang–Mills theory, gravity has a natural length scale, the Planck mass. It might therefore seem reasonable to take some multiple of this for the one-loop normalization factor μ.

15.6 Zeta function regularization

In order to regularize the determinant of an operator A with eigenvalues and eigenfunctions $\{\lambda_n, \phi_n\}$, one forms a generalized zeta function from the eigenvalues

$$\zeta_A(s)=\sum \lambda_n^{-s}. \tag{15.54}$$

From (15.50) it can be seen that ζ will converge for $\operatorname{Re} s>2$. It can be analytically extended to a meromorphic function of s with poles only at $s=2$ and $s=1$. In particular it is regular at $s=0$. Formally one has

$$\zeta'_A(0)=-\sum \log \lambda_n. \tag{15.55}$$

Thus one can *define* the regularized value of the determinant of A to be

$$\det A=\exp\left(-\zeta'_A(0)\right). \tag{15.56}$$

The zeta function can be related to the kernel $F(x, x', t)$ of the heat or diffusion equation

$$\frac{\partial F}{\partial t}+A_x F=0, \tag{15.57}$$

where A_x indicates that the operator acts on the first argument of F. With the initial condition

$$F(x, x', 0) = \delta(x, x'), \tag{15.58}$$

F represents the diffusion over the manifold M, in a fifth dimension of parameter time t, of a point source of heat placed at x' at $t = 0$. The heat equation has been much studied by a number of authors including DeWitt (1963), McKean and Singer (1967) and Gilkey (1975). A good exposition can be found in Gilkey (1974).

It can be shown that if A is an elliptic operator, the heat kernel $F(x, x', t)$ is a smooth function of x, x', and t, for $t > 0$. As $t \to 0$, there is an asymptotic expression for $F(x, x, t)$:

$$F(x, x, t) \sim \sum_{n=0}^{\infty} b_n t^{n-2}, \tag{15.59}$$

where again the b_n are the 'Hamidew' coefficients and are scalar polynomials in the metric, the curvature and its covariant derivatives of order $2n$ in derivatives of the metrics.

One can represent F in terms of the eigenfunctions and eigenvalues of A

$$F(x, x', t) = \sum \phi_n(x)\phi_n(x') \exp(-\lambda_n t). \tag{15.60}$$

Integrating this over the manifold, one obtains

$$Y(t) = \int F(x, x, t)(g_0)^{1/2}\, \mathrm{d}^4x = \sum \exp(-\lambda_n t). \tag{15.61}$$

The zeta function can be obtained from $Y(t)$ by an inverse Mellin transform

$$\zeta(s) = \frac{1}{\Gamma(s)} \int_0^{\infty} Y(t) t^{s-1}\, \mathrm{d}t. \tag{15.62}$$

Using the asymptotic expansion for F, one sees that $\zeta(s)$ has a pole at $s = 2$ with residue B_0 and a pole at $s = 1$ with residue B_1. There would be a pole at $s = 0$ but it is cancelled by the pole in the gamma function. Thus $\zeta(0) = B_2$. In a sense the poles at $s = 2$ and $s = 1$ correspond to removing the divergences caused by the first two terms in (15.50).

If one knows the eigenvalue explicitly, one can calculate the zeta function and evaluate its derivative at $s = 0$. In other cases one can obtain some information from the asymptotic expansion for the heat kernel. For example, suppose the background metric is changed by a constant scale factor $\tilde{g}_0 = k^2 g_0$, then the eigenvalues, λ_n, of a zero-rest-mass operator A will become $\tilde{\lambda}_n = k^{-2}\lambda_n$. Thus

$$\zeta_{\tilde{A}}(s) = k^{2s}\zeta_A(s)$$

and

$$\zeta'_{\tilde{A}}(0) = 2\log k\zeta_A(0) + \zeta'_A(0), \tag{15.63}$$

therefore

$$\log(\det \tilde{A}) = -2\zeta(0)\log k + \log(\det A). \tag{15.64}$$

Because B_2, and hence $\zeta(0)$, are not in general zero, one sees that the path integral is not invariant under conformal transformations of the background metric, even for conformally invariant operators A. This is known as a conformal anomaly and arises because in regularizing the determinant one has to introduce a normalization quantity, μ, with dimensions of mass or inverse length. Alternatively, one could say that the measure $D[\phi] = \prod \mu \, dy_n$ is not conformally invariant.

Further details of zeta function regularization of matter field determinants will be found in Hawking (1977), Gibbons (1977*c*), and Lapedes (1978).

The zeta function regularization of the one-loop gravitational term about a vacuum background has been considered by Gibbons, Hawking and Perry (1978). I shall briefly describe this work and generalize it to include a Λ-term.

The quadratic term in the fluctuations $\bar{g}$ about a background metric, g_0, is

$$I_2[\bar{g}] = \tfrac{1}{2}\int \bar{g}^{ab}A_{abcd}\bar{g}^{cd}(g_0)^{1/2}\,d^4x, \tag{15.65}$$

where

$$g^{ab} = g_0^{ab} + \bar{g}^{ab} \tag{15.66}$$

and

$$\begin{aligned}16\pi A_{abcd} = {}& \tfrac{1}{4}g_{cd}\nabla_a\nabla_b - \tfrac{1}{4}g_{ac}\nabla_d\nabla_b + \tfrac{1}{8}(g_{ac}g_{bd} + g_{ab}g_{cd})\nabla_e\nabla^e + \tfrac{1}{2}R_{ad}g_{bc} \\ & - \tfrac{1}{4}R_{ab}g_{cd} + \tfrac{1}{16}Rg_{ab}g_{cd} - \tfrac{1}{8}Rg_{ac}g_{bd} - \tfrac{1}{8}\Lambda g_{ab}g_{cd} + \tfrac{1}{4}\Lambda g_{ac}g_{bd} \\ & + (a \leftrightarrow b) + (c \leftrightarrow d) + (a \leftrightarrow b, c \leftrightarrow d).\end{aligned} \tag{15.67}$$

One cannot simply take the one-loop term to be $(\det(\tfrac{1}{2}\pi^{-1}\mu^{-1}A))^{1/2}$, because A has a large number of zero eigenvalues corresponding to the fact that the action is unchanged under an infinitesimal diffeomorphism

(gauge transformation)

$$x^a \to x^a + \varepsilon\xi^a$$
$$g_{ab} \to g_{ab} + 2\varepsilon\xi_{(a;b)}. \tag{15.68}$$

One would like to factor out the gauge freedom by integrating only over gauge-inequivalent perturbations $\bar{g}$. One would then obtain an answer which depended on the determinant of A on the quotient of all fields $\bar{g}$ modulo infinitesimal gauge transformations. The way to do this has been indicated by Feynman (1972), DeWitt (1967) and Fade'ev and Popov (1967). One adds a gauge-fixing term to the action

$$I_f = \tfrac{1}{2}\int \bar{g}^{ab}B_{abcd}\bar{g}^{cd}(g_0)^{1/2}\,\mathrm{d}^4x. \tag{15.69}$$

The operator B is chosen so that for any sufficiently small perturbation $\bar{g}$ which satisfies the appropriate boundary condition there is a unique transformation, ξ^a, which vanishes on the boundary such that

$$B_{abcd}(\bar{g}^{cd} + 2\xi^{(c;d)}) = 0. \tag{15.70}$$

I shall use the harmonic gauge in the background metric

$$16\pi B_{abcd} = \tfrac{1}{4}g_{bd}\nabla_a\nabla_c - \tfrac{1}{8}g_{cd}\nabla_a\nabla_b - \tfrac{1}{8}g_{ab}\nabla_c\nabla_d$$
$$+\tfrac{1}{16}g_{ab}g_{cd}\Box + (a\leftrightarrow b) + (c\leftrightarrow d) + (a\leftrightarrow b,\, c\leftrightarrow d). \tag{15.71}$$

The operator $(A+B)$ will in general have no zero eigenvalues. However, det $(A+B)$ contains the eigenvalues of the arbitrarily chosen operator B. To cancel them out one has to divide by the determinant of B on the subspace of all $\bar{g}$ which are pure gauge transformations, i.e. of the form $\bar{g}^{ab} = 2\xi^{(a;b)}$ for some ξ which vanishes on the boundary. The determinant of B on this subspace is equal to the square of the determinant of the operator C on the space of all vector fields which vanish on the boundary, where

$$16\pi C_{ab} = -g_{ab}\Box - R_{ab}. \tag{15.72}$$

Thus one obtains

$$\log Z = -\hat{I}[g_0] - \tfrac{1}{2}\log\det(\tfrac{1}{2}\pi^{-1}\mu^{-2}(A+B)) + \log\det(\tfrac{1}{2}\pi^{-1}\mu^{-2}C). \tag{15.73}$$

The last term is the so-called ghost contribution.

In order to use the zeta function technique it is necessary to express $A+B$ as $K-L$ where K and L each have only a finite number of negative eigenvalues. To do this, let

$$A+B=-F+G, \tag{15.74}$$

where

$$F=-\tfrac{1}{16}(\nabla_a\nabla^a+2\Lambda), \tag{15.75}$$

which operates on the trace, ϕ, of $\bar{g}$, $\phi=\bar{g}^{ab}g_{0ab}$

$$G_{abcd}=-\tfrac{1}{8}(g_{ac}g_{bd}+g_{ad}g_{bc})\nabla^e\nabla_e-\tfrac{1}{4}(C_{dcab}+C_{dbac})+\tfrac{1}{6}\Lambda g_{ab}g_{cd}, \tag{15.76}$$

which operates on the trace-free part, $\tilde{\phi}$, of $\bar{g}$, $\tilde{\phi}^{ab}=\bar{g}^{ab}-\frac{1}{4}g_0{}^{ab}\phi$.

If $\Lambda\leq 0$, the operator F will have only positive eigenvalues. Therefore in order to make the one-loop term converge, one has to integrate over purely imaginary ϕ. This corresponds to integrating over conformal factors of the form $\Omega=1+i\xi$. if $\Lambda>0$, F will have some finite number, p, of negative eigenvalues. Because a constant function will be an eigenfunction of F with negative eigenvalue (in the case where there is no boundary), p will be at least one. In order to make the one-loop term converge, one will have to rotate the contour of integration of the coefficient of each eigenfunction, with a negative eigenvalue to lie along the real axis. This will introduce a factor of i^p into Z.

If the background metric g_0 is flat, the operator G will be positive-definite. Thus one will integrate the trace-free perturbations $\tilde{\phi}$ along the real axis. This corresponds to integrating over real conformal equivalence classes. However for non-flat background metrics, G may have some finite number, q, of negative eigenvalues because of the Λ and Weyl tensor terms. Again one will have to rotate the contour of integration for these modes (this time from real to imaginary) and this will introduce a factor of i^{-q} into Z.

The ghost operator is

$$C_{ab}=-g_{ab}(\nabla^e\nabla_e+\Lambda). \tag{15.77}$$

If $\Lambda>0$, C will have some finite number, r, of negative eigenvalues. Because it is the determinant of C that appears in Z rather than its square root, the negative eigenvalues will contribute a factor $(-1)^r$.

One has

$$\log Z=-\hat{I}[g_0]+\tfrac{1}{2}\zeta'_F(0)+\tfrac{1}{2}\zeta'_G(0)-\zeta'_C(0)$$
$$+\tfrac{1}{2}\log(2\pi\mu^2)(\zeta_F(0)+\zeta_G(0)-2\zeta_C(0)). \tag{15.78}$$

From the asymptotic expansion for the heat kernel one has to evaluate the zeta functions at $s = 0$. From the results of Gibbons and Perry (1979) one has

$$\zeta_F(0)+\zeta_G(0)-2\zeta_C(0)=\int\left(\frac{53}{720\pi^2}C_{abcd}C^{abcd}+\frac{763}{540\pi^2}\Lambda^2\right)(g_0)^{1/2}\,\mathrm{d}^4x. \tag{15.79}$$

From this one can deduce the behaviour of the one-loop term under scale transformations of the background metric. Let $\tilde{g}_{0ab} = k^2 g_{0ab}$, then

$$\log\tilde{Z} = \log Z + (1-k^2)\hat{I}[g_0] + \tfrac{1}{2}\gamma\log k, \tag{15.80}$$

where γ is the right-hand side of (15.79). Providing $\hat{I}[g_0]$ is positive, $\tilde{Z}$ will be very small for large scales, k. The fact that γ is positive will mean that it is also small for very small scales. Thus quantum gravity may have a cut-off at short length scales. This will be discussed further in section 15.10.

15.7 The background fields

In this section I shall describe some positive-definite metrics which are solutions of the Einstein equations in vacuum or with a Λ-term. In some cases these are analytic continuations of well-known Lorentzian solutions, though their global structure may be different. In particular the section through the complexified manifold on which the metric is positive-definite may not contain the singularities present on the Lorentzian section. In other cases the positive-definite metrics may occur on manifolds which do not have any section on which the metric is real and Lorentzian. They may nevertheless be of interest as stationary-phase points in certain path integrals.

The simplest non-trivial example of a vacuum metric is the Schwarzschild solution (Hartle and Hawking, 1976; Gibbons and Hawking, 1977*a*). This is normally given in the form

$$\mathrm{d}s^2 = -\left(1-\frac{2M}{r}\right)\mathrm{d}t^2+\left(1-\frac{2M}{r}\right)^{-1}\mathrm{d}r^2+r^2\,\mathrm{d}\Omega^2. \tag{15.81}$$

Putting $t = -\mathrm{i}\tau$ converts this into a positive-definite metric for $r > 2M$. There is an apparent singularity at $r = 2M$ but this is like the apparent singularity at the origin of polar coordinates, as can be seen by defining a new radial coordinate $x = 4M(1-2Mr^{-1})^{1/2}$. Then the metric becomes

$$\mathrm{d}s^2 = \left(\frac{x}{4M}\right)^2\mathrm{d}\tau^2+\left(\frac{r^2}{4M^2}\right)^2\mathrm{d}x^2+r^2\,\mathrm{d}\Omega^2.$$

This will be regular at $x=0$, $r=2M$, if τ is regarded as an angular variable and is identified with period $8\pi M$ (I am using units in which the gravitational constant $G=1$). The manifold defined by $x \geqslant 0$, $0 \leqslant \tau \leqslant 8\pi M$ is called the Euclidean section of the Schwarzschild solution. On it the metric is positive-definite, asymptotically flat and non-singular (the curvature singularity at $r=0$ does not lie on the Euclidean section).

Because the Schwarzschild solution is periodic in imaginary time with period $\beta = 8\pi M$, the boundary surface ∂M at radius r_0 will have topology $S^1 \times S^2$ and the metric will be a stationary-phase point in the path integral for the partition function of a canonical ensemble at temperature $T = \beta^{-1} = (8\pi M)^{-1}$. As shown in section 15.2, the action will come entirely from the surface term, which gives

$$\hat{I} = \tfrac{1}{2}\beta M = 4\pi M^2. \tag{15.82}$$

One can find a similar Euclidean section for the Reissner–Nordström solution with $Q^2 + P^2 < M^2$, where Q is the electric charge and P is the magnetic monopole charge. In this case the radial coordinate has the range $r_+ \leqslant r < \infty$. Again the outer horizon, $r = r_+$, is an axis of symmetry in the $r-\tau$-plane and the imaginary time coordinate, τ, is identified with period $\beta = 2\pi\kappa^{-1}$, where κ is the surface gravity of the outer horizon. The electromagnetic field, F_{ab}, will be real on the Euclidean section if Q is imaginary and P is real. In particular if $Q = \mathrm{i}P$, the field will be self-dual or anti-self-dual,

$$F_{ab} = \pm {}^*F_{ab} = \tfrac{1}{2}\varepsilon_{abcd}F^{cd}, \tag{15.83}$$

where ε_{abcd} is the alternating tensor. If F_{ab} is real on the Euclidean section, the operators governing the behaviour of charged fields will be elliptic and so one can evaluate the one-loop terms by the zeta function method. One can then analytically continue the result back to real Q just as one analytically continues back from positive-definite metrics to Lorentzian ones.

Because $R = 0$, the gravitational part of the action is unchanged. However there is also a contribution from the electromagnetic Lagrangian, $-(1/8\pi)F_{ab}F^{ab}$. Thus

$$\hat{I} = \tfrac{1}{2}\beta(M - \Phi Q + \psi P), \tag{15.84}$$

where $\Phi = Q/r_+$ is the electrostatic potential of the horizon and $\psi = P/r_+$ is the magnetostatic potential.

In a similar manner one can find a Euclidean section for the Kerr metric provided that the mass M is real and the angular momentum J is imaginary. In this case the metric will be periodic in the frame that co-rotates with the horizon, i.e. the point (τ, r, θ, ϕ) is identified with $(\tau+\beta, r, \theta, \phi+i\beta\Omega)$ where Ω is the angular velocity of the horizon (Ω will be imaginary if J is imaginary). As in the electromagnetic case, it seems best to evaluate the one-loop terms with J imaginary and then analytically continue to real J. The presence of angular momentum does not affect the asymptotic metric to leading order to that the action is

$$\hat{I} = \tfrac{1}{2}\beta M \quad \text{with } \beta = 2\pi\kappa^{-1},$$

where κ is the surface gravity of the horizon.

Another interesting class of vacuum solutions are the Taub–NUT metrics (Newman, Unti and Tamburino, 1963; Hawking and Ellis, 1973). These can be regarded as gravitational dyons with an ordinary 'electric' type mass M and a gravitational 'magnetic' type mass N. The metric can be written in the form

$$ds^2 = -V\left(dt + 4N\sin^2\frac{\theta}{2}\,d\phi\right)^2 + V^{-1}\,dr^2 + (r^2+N^2)(d\theta^2 + \sin^2\theta\,d\phi^2), \tag{15.85}$$

where $V = 1-(2Mr+N^2)/(r^2+N^2)$. This metric is regular on half-axis $\theta = 0$ but it has a singularity at $\theta = \pi$ because the $\sin^2(\theta/2)$ term in the metric means that a small loop around the axis does not shrink to zero length as $\theta = \pi$. This singularity can be regarded as the analogue of a Dirac string in electrodynamics, caused by the presence of a magnetic monopole charge. One can remove this singularity by introducing a new coordinate

$$t' = t + 4N\phi. \tag{15.86}$$

The metric then becomes

$$ds^2 = -V\left(dt' - 4N\cos^2\frac{\theta}{2}\,d\phi\right)^2 + V^{-1}\,dr^2 + (r^2+N^2)(d\theta^2 + \sin^2\theta\,d\phi^2). \tag{15.87}$$

This is regular at $\theta = \pi$ but not at $\theta = 0$. One can therefore use the (t, r, θ, ϕ) coordinates to cover the north pole ($\theta = 0$) and the (t', r, θ, ϕ) co-ordinates to cover the south pole ($\theta = \pi$). Because ϕ is identified with period 2π, (15.86) implies that t and t' have to be identified with period $8\pi N$. Thus if ψ is a regular field with t-dependence of the form $\exp(-i\omega t)$, then ω must satisfy

$$4N\omega = \text{an integer}. \tag{15.88}$$

This is the analogue of the Dirac quantization condition and relates the 'magnetic' charge, N, of the Taub–NUT solution to the 'electric' charge or energy, ω, of the field ψ. The process of removing the Dirac string singularity by introducing coordinates t and t' and periodically identifying, changes the topology of the surfaces of constant r from $S^2 \times R^1$ to S^3 on which $(t/2N)$, θ and ϕ are Euler angle coordinates.

The metric (15.85) also has singularities where $V = 0$ or ∞. As in the Schwarzschild case $V = \infty$ corresponds to an irremovable curvature singularity but $V = 0$ corresponds to a horizon and can be removed by periodically identifying the imaginary time coordinate. This identification is compatible with the one to remove the Dirac string if the two periods are equal, which occurs if $N = \pm \mathrm{i}M$. If this is the case, and if M is real, the metric is real and is positive-definite in the region $r > M$ and the curvature is self-dual or anti-self-dual

$$R_{abcd} = \pm {}^*R_{abcd} = \pm \tfrac{1}{2}\varepsilon_{abef}R^{ef}{}_{cd}. \qquad (15.89)$$

The apparent singularity at $r = M$ becomes a single point, the origin of hyperspherical coordinates, as can be seen by introducing new radial and time variables

$$x = 2(2M(r - M))^{1/2}, \qquad \psi = -\frac{\mathrm{i}t}{2M}. \qquad (15.90)$$

The metric then becomes

$$\mathrm{d}s^2 = \frac{Mx^2}{2(r+M)}(\mathrm{d}\psi + \cos\theta\, \mathrm{d}\phi)^2 + \frac{r+M}{2M}\mathrm{d}x^2 + \frac{x^2(r+M)}{8M}(\mathrm{d}\theta^2 + \sin^2\theta\, \mathrm{d}\phi^2). \qquad (15.91)$$

Thus the manifold defined by $x \geqslant 0$, $0 \leqslant \psi \leqslant 4\pi$, $0 \leqslant \theta \leqslant \pi$, $0 \leqslant \phi \leqslant 2\pi$, with ψ, θ, ϕ interpreted as hyperspherical Euler angles, is topologically R^4 and has a non-singular, positive-definite metric. The metric is asymptotically flat in the sense that the Riemann tensor decreases as r^{-3} as $r \to \infty$ but it is not asymptotically Euclidean, which would require curvature proportional to r^{-4}. The surfaces of the constant r are topologically S^3 but their metric is that of a deformed sphere. The orbits of the $\partial/\partial\psi$ Killing vector define a Hopf fibration π; $S^3 \to S^2$, where the S^2 is parametrized by the coordinates θ and ϕ. The induced metric on the S^2 is that of a

2-sphere of radius $(r^2 - M^2)^{1/2}$, while the fibres are circles of circumference $8\pi M V^{1/2}$. Thus, in a sense the boundary at large radius is $S^1 \times S^2$ but is a twisted product.

It is also possible to combine self-dual Taub–NUT solutions (Hawking, 1977). The reason is that the attraction between the electric type masses M is balanced by the repulsion between the imaginary magnetic type masses N. The metric is

$$ds^2 = U^{-1}(d\tau + \boldsymbol{\omega} \cdot d\boldsymbol{x})^2 + U\, d\boldsymbol{x} \cdot d\boldsymbol{x}, \qquad (15.92)$$

where

$$U = 1 + \sum \frac{2M_i}{r_i}$$

and

$$\operatorname{curl} \boldsymbol{\omega} = \operatorname{grad} U. \qquad (15.93)$$

Here r_i denotes the distance from the ith 'NUT' in the flat, three-dimensional metric $d\boldsymbol{x} \cdot d\boldsymbol{x}$. The curl and grad operations refer to this 3-metric, as does the vector $\boldsymbol{v}$. Each NUT has $N_i = iM_i$.

The vector fields $\boldsymbol{\omega}$ will have Dirac string singularities running from each NUT. If the masses M_i are all equal, these string singularities and the horizon-type singularities at $r_i = 0$ can all be removed by identifying τ with period $8\pi M$. The boundary surface at large radius is then a lens space (Steenrod, 1951). This is topologically an S^3 with n points identified in the fibre S^1 of the Hopf fibration $S^3 \to S^2$, where n is the number of NUTs.

The boundary surface cannot be even locally imbedded in flat space so that one cannot work out the correction term K^0 in the action. If one tries to imbed it as nearly as one can, one obtains the value of $4\pi n M^2$ for the action, the same as Schwarzschild for $n = 1$ (Davies, 1978). In fact the presence of a gravitational magnetic mass alters the topology of the space and prevents it from being asymptotically flat in the usual way. One can, however, obtain an asymptotically flat space containing an equal number, n, of NUTs ($N = iM$) and anti-NUTs ($N = -iM$). Because the NUTs and the anti-NUTs attract each other, they have to be held apart by an electromagnetic field. This solution is in fact one of the Israel–Wilson metrics (Israel and Wilson, 1972; Hartle and Hawking, 1972). The gravitational part of the action is $8\pi n M^2$, so that each NUT and anti-NUT contributes $4\pi M^2$.

I now come on to positive-definite metrics which are solutions of the Einstein equations with a Λ-term on manifolds which are compact without boundary. The simplest example is an S^4 with the metric induced by imbedding it as a sphere of radius $(3\Lambda^{-1})^{1/2}$ in five-dimensional Euclidean space. This is the analytic continuation of de Sitter space (Gibbons and Hawking, 1977*b*). The metric can be written in terms of a Killing vector $\partial/\partial\tau$:

$$ds^2 = (1-\tfrac{1}{3}\Lambda r^2)\,d\tau^2 + (1-\tfrac{1}{3}\Lambda r^2)^{-1}\,dr^2 + r^2\,d\Omega^2. \qquad (15.94)$$

There is a horizon-type singularity at $r=(3\Lambda^{-1})^{1/2}$. This is in fact a 2-sphere of area $12\pi\Lambda^{-1}$ which is the locus of zeros of the Killing vector $\partial/\partial\tau$. The action is $-3\pi\Lambda^{-1}$.

One can also obtain black hole solutions which are asymptotically de Sitter instead of asymptotically flat. The simplest of these is the Schwarzschild–de Sitter (Gibbons and Hawking, 1977*b*). The metric is

$$ds^2 = V\,d\tau^2 + V^{-1}\,dr^2 + r^2\,d\Omega^2, \qquad (15.95)$$

where

$$V = 1 - 2Mr^{-1} - \tfrac{1}{3}\Lambda r^2.$$

If $\Lambda < (9M^2)^{-1}$, there are two positive values of r for which $V=0$. The smaller of these corresponds to the black hole horizon, while the larger is similar to the 'cosmological horizon' in de Sitter space. One can remove the apparent singularities at each horizon by identifying τ periodically. However, the periodicities required at the two horizons are different, except in the limiting case $\Lambda=(9M^2)^{-1}$. In this case, the manifold is $S^2\times S^2$ with the product metric and the action is $-2\pi\Lambda^{-1}$.

One can also obtain a Kerr–de Sitter solution (Gibbons and Hawking, 1977*b*). This will be a positive-definite metric for values of r lying between the cosmological horizon and the outer black hole horizon, if the angular momentum is imaginary. Again, one can remove the horizon singularities by periodic identifications and the periodicities will be compatible for a particular choice of the parameters (Page, 1978). In this case one obtains a singularity-free metric on an S^2 bundle over S^2. The action is $-0.9553\,(2\pi\Lambda^{-1})$.

One can also obtain Taub–de Sitter solutions. These will have a cosmological horizon in addition to the ordinary Taub–NUT ones. One can remove all the horizon and Dirac string singularities simultaneously in a limiting case which is CP^2, complex, projective 2-space, with the standard Kaehler metric (Gibbons and Pope, 1978). The action is $-\tfrac{9}{4}\pi\Lambda^{-1}$.

One can also obtain solutions which are the product of two two-dimensional spaces of constant curvature (Gibbons, 1977*b*). The case of

$S^2 \times S^2$ has already been mentioned, and there is also the trivial flat torus $T^2 \times T^2$. In the other examples the two spaces have genera g_1 and $g_2 > 1$ and the Λ-term has to be negative. The action is $-(2\pi/\Lambda)(g_1-1)(g_2-1)$.

Finally, to complete this catalogue of known positive-definite solutions on the Einstein equations, one should mention $K3$. This is a compact four-dimensional manifold which can be realized as a quartic surface in CP^3, complex projective 3-space. It can be given a positive-definite metric whose curvature is self-dual and which is therefore a solution of the Einstein equation with $\Lambda = 0$ (Yau, 1977). Moreover $K3$ is, up to identifications, the only compact manifold to admit a self-dual metric. The action is 0.

There are two topological invariants of compact four-dimensional manifolds that can be expressed as integrals of the curvature:

$$\chi = \frac{1}{128\pi^2} \int R_{abcd} R_{efgh} \varepsilon^{abef} \varepsilon^{cdgh} (g)^{1/2} \, d^4x, \tag{15.96}$$

$$\tau = \frac{1}{96\pi^2} \int R_{abcd} R^{ab}{}_{ef} \varepsilon^{cdef} (g)^{1/2} \, d^4x. \tag{15.97}$$

χ is the Euler number of the manifold and is equal to the alternating sum of the Betti numbers:

$$\chi = B_0 - B_1 + B_2 - B_3 + B_4. \tag{15.98}$$

The pth Betti number, B_p, is the number of independent closed p-surfaces that are not boundaries of some $p+1$-surface. They are also equal to the number of independent harmonic p-forms. For a closed manifold, $B_p = B_{4-p}$ and $B_1 = B_4 = 1$. If the manifold is simply connected, $B_1 = B_3 = 0$, so $\chi \geq 2$.

The Hirzebruch signature, τ, has the following interpretation. The B_2 harmonic 2-forms can be divided into B_2^+ self-dual and B_2^- anti-self-dual 2-forms. Then $\tau = B_2^+ - B_2^-$. It determines the gravitational contribution to the a [illegible]l-current anomaly (Eguchi and Freund, 1976; Hawking, 1977; Hawking and Pope, 1978).

S^4 has $\chi = 2$ and $\tau = 0$; CP^2 has $\chi = 3$, $\tau = 1$; the S^2 bundle over S^2 has $\chi = 4$, $\tau = 0$; $K3$ has $\chi = 24$, $\tau = 16$ and the product of two-dimensional spaces with genera g_1, g_2 has $\chi = 4(g_1-1)(g_2-1)$, $\tau = 0$.

In the non-compact case there are extra surface terms in the formulae for χ and τ. Euclidean space and the self-dual Taub–NUT solution has $\chi = 1$, $\tau = 0$ and the Schwarzschild solution has $\chi = 2$, $\tau = 0$.

15.8 Gravitational thermodynamics

As explained in section 15.3, the partition function

$$Z = \sum \exp(-\beta E_n)$$

for a system at temperature $T = \beta^{-1}$, contained in a spherical box of radius r_0, is given by a path integral over all metrics which fit inside the boundary, ∂M, with topology $S^2 \times S^1$, where the S^2 is a sphere of radius r_0 and the S^1 has circumference β. By the stationary-phase approximation described in section 15.5, the dominant contributions will come from metrics near classical solutions g_0 with the given boundary conditions. One such solution is just flat space with the Euclidean time coordinate identified with period β. This has topology $R^3 \times S^1$. The action of the background metric is zero, so it makes no contribution to the logarithm of the partition function. If one neglects small corrections arising from the finite size of the box, the one-loop term also can be evaluated exactly as Z_g

$$\log Z_g = \frac{4\pi^5 r_0^3 T^3}{135}. \tag{15.99}$$

This can be interpreted as the partition function of thermal gravitons on a flat-space background.

The Schwarzschild metric with $M = (8\pi T)^{-1}$ is another solution which fits the boundary conditions. It has topology $R^2 \times S^2$ and action $\hat{I} = \beta^2/16\pi = 4\pi M^2$. The one-loop term has not been computed, but by the scaling arguments given in section 15.6 it must have the form

$$\frac{106}{45} \log\left(\frac{\beta}{\beta_0}\right) + f(r_0\beta^{-1}) \tag{15.100}$$

where β_0 is related to the normalization constant μ. If $r_0\beta^{-1}$ is much greater than 1, the box will be much larger than the black hole and one would expect $f(r_0\beta^{-1})$ to approach the flat-space value (15.99). Thus f should have the form

$$f(r_0\beta^{-1}) = \frac{4\pi^5 r_0^3}{135\beta^3} + O(r_0^2\beta^{-2}). \tag{15.101}$$

From the partition function one can calculate the expectation value of the energy

$$\langle E \rangle = \frac{\sum E_n \exp(-\beta E_n)}{\exp(-\beta E_n)}$$

$$= -\frac{\partial}{\partial \beta} \log Z. \tag{15.102}$$

Applying this to the contribution $(-\beta^2/16\pi)$ to $\log Z$ from the action of the action of the Schwarzschild solution, one obtains $\langle E\rangle = M$, as one might expect. One can also obtain the entropy, which can be defined to be

$$S = -\sum p_n \log p_n, \tag{15.103}$$

where $p_n = Z^{-1} \exp(-\beta E_n)$ is the probability that the system is in the nth state. Then

$$S = \beta\langle E\rangle + \log Z. \tag{15.104}$$

Applying this to the contribution from the action of the Schwarzschild metric, one obtains

$$S = 4\pi M^2 = \tfrac{1}{4}A, \tag{15.105}$$

where A is the area of the event horizon.

This is a remarkable result because it shows that, in addition to the entropy arising from the one-loop term (which can be regarded as the entropy of thermal gravitons on a Schwarzschild background), black holes have an intrinsic entropy arising from the action of the stationary-phase metric. This intrinsic entropy agrees exactly with that assigned to black holes on the basis of particle-creation calculations on a fixed background and the use of the first law of black hole mechanics (see chapters 6 and 13 by Carter and Gibbons). It shows that the idea that gravity introduces a new level of unpredictability or randomness into physics is supported not only by semi-classical approximation but by a treatment in which the gravitational field is quantized.

One reason why classical solutions in gravity have intrinsic entropy while those in Yang–Mills or $\lambda\phi^4$ do not is that the actions of these theories are scale-invariant, unlike the gravitational action. If g_0 is an asymptotically flat solution with period β and action $\hat{I}[g_0]$, then $k^2 g_0$ is a solution with a period $k\beta$ and action $k^2\hat{I}$. This means that the action, $\hat{I}$, must be of the form $c\beta^2$, where c is a constant which will depend on the topology of the solution. Then $\langle E\rangle = 2c\beta$, $\beta\langle E\rangle = 2c\beta^2$, while $\log Z = -\hat{I} = -c\beta^2$. Thus $S = c\beta^2$. The reason that the action $\hat{I}$ is equal to $\tfrac{1}{2}\beta\langle E\rangle$ and not $\beta\langle E\rangle$, as one would expect for a single state with energy $\langle E\rangle$, is that the topology of the Schwarzschild solution is not the same as that of periodically identified flat space. The fact that the Euler number of the Schwarzschild solution is 2 implies that the time-translation Killing vector, $\partial/\partial\tau$, must be zero on some set (in fact a 2-sphere). Thus the surfaces of a constant τ have two boundaries: one at the spherical box of radius r_0 and the other at the horizon $r = 2M$. Consider now the region of

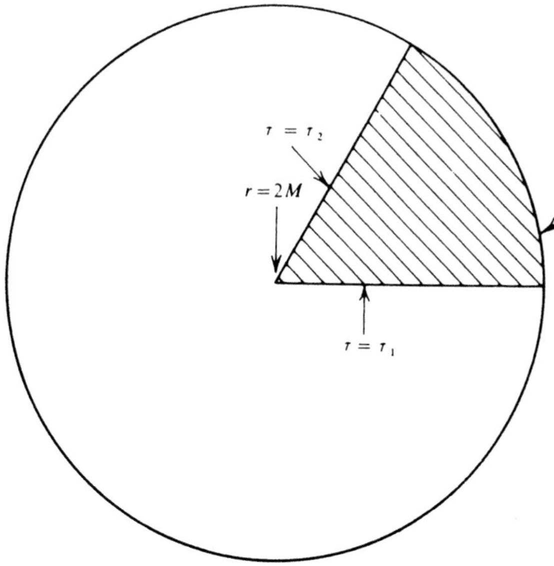

Figure 15.4. The τ–r plane of the Schwarzschild solution. The amplitude $\langle\tau_2|\tau_1\rangle$ to go from the surface τ_1 to the surface $\tau = \tau_2$ is dominated by the action of the shaded portion of the Schwarzschild solution.

the Schwarzschild solution bounded by the surfaces $\tau = \tau_1$, $\tau = \tau_2$ and $r = r_0$ (figure 15.4). The amplitude $\langle\tau_2|\tau_1\rangle$ to go from the surface τ_1 to the surface τ_2 will be given by a path integral over all metrics which fit inside this boundary, with the dominant contribution coming from the stationary-phase metric – which is just the portion of the Schwarzschild solution bounded by these surfaces. The action of this stationary-phase metric will be given by the surface terms because $R = 0$. The surface terms from the surfaces $\tau = \tau_1$ and $\tau = \tau_2$ will cancel out. There will be a contribution of $\frac{1}{2}M(\tau_2 - \tau_1)$ from the surface $r = r_0$. However there will also be a contribution from the 'corner' at $r = 2M$ where the two surfaces $\tau = \tau_1$ and $\tau = \tau_2$ meet, because the second fundamental form, K, of the boundary will have a δ-function behaviour there. By rounding off the corner slightly one can evaluate this contribution, and it turns out to be $\frac{1}{2}M(\tau_2 - \tau_1)$. Thus the total action is $\langle E\rangle(\tau_2 - \tau_1)$ and $\langle\tau_2|\tau_1\rangle = \exp(-\langle E\rangle(\tau_2 - \tau_1))$, as one would expect for a single state with energy $E = \langle E\rangle$. However, if one considers the partition function one simply has the boundary at $r = r_0$ and so the action equals $\frac{1}{2}\beta E$ rather than βE. This difference, which is equal to $\frac{1}{4}A$, gives the entropy of the black hole.

From this one sees that qualitatively new effects arise from the fact that the gravitational field can have different topologies. These effects would not have been found using the canonical approach, because such metrics as the Schwarzschild solution would not have been allowed.

The above derivation of the partition function and entropy of a black hole has been based on the use of the canonical ensemble, in which the system is in equilibrium with an infinite reservoir of energy at temperature T. However the canonical ensemble is unstable when black holes are present because if a hole were to absorb a bit more energy, it would cool down and would continue to absorb more energy than it emitted. This pathology is reflected in the fact that $\langle \Delta E^2\rangle = \langle E^2\rangle - \langle E\rangle^2 = (1/Z)(\partial^2 Z/\partial\beta^2) - (\partial \log Z/\partial B)^2 = -1/8\pi$, which is negative. To obtain sensible results with black holes one has to use the micro-canonical ensemble, in which a certain amount of energy E is placed in an insulated box and one considers all possible configurations within that box which have the given energy. Let $N(E)\,\mathrm{d}E$ be the number of states of the gravitational field with energies between E and $E+\mathrm{d}E$ in a spherical box of radius r_0. The partition function is given by the Laplace transform of $N(E)$,

$$Z(\beta) = \int_0^\infty N(E)\exp(-\beta E)\,\mathrm{d}E. \tag{15.106}$$

Thus, formally, the density of states is given by an inverse Laplace transform,

$$N(E) = \frac{1}{2\pi i}\int_{-\mathrm{i}\infty}^{\mathrm{i}\infty} Z(\beta)\exp(E\beta)\,\mathrm{d}\beta. \tag{15.107}$$

For large β, the dominant contribution to $Z(\beta)$ comes from the action of the Schwarzschild metric, and is of the form $\exp(-\beta^2/16\pi)$. Thus the right-hand side of (15.107) would diverge if the integral were taken up the imaginary β-axis as it is supposed to be. To obtain a finite value for (15.107) one has to adopt the prescription that the integral be taken along the real β-axis. This is very similar to the procedure used to evaluate the path integral in the stationary-phase approximation, where one rotated the contour of integration for each quadratic term, so that one would obtain a convergent Gaussian integral. With this prescription the factor $1/2\pi\mathrm{i}$ in (15.107) would give an imaginary value for the density of states $N(E)$ if the partition function $Z(\beta)$ were real. However, as mentioned in section 5.6, the operator G which governs non-conformal or trace-free perturbations has one negative eigenvalue in the Schwarzschild metric. This contributes a factor i to the one-loop term for Z. Thus the partition function is purely imaginary but the density of space is real. This is what one might expect: the partition function is pathological because the canonical ensemble is not well defined but the density of states is real and positive because the micro-canonical ensemble is well behaved.

It is not appropriate to go beyond the stationary-phase approximation in evaluating the integral in (15.107) because the partition function, Z, has been calculated in this approximation only. If one takes just the contribution $\exp(-\beta^2/16\pi)$ from the action of the background metric, one finds that a black hole of mass M has a density of states $N(M)= 2\pi^{-1/2}\exp(4\pi M^2)$. Thus the integral in (15.106) does not converge unless one rotates the contour integration to lie along the imaginary E-axis. If one includes the one-loop term Z_g, the stationary-phase point in the β integration in (15.107) occurs when

$$E=\frac{-\partial \log Z_g}{\partial\beta} \tag{15.108}$$

for the flat background metric, and

$$E=\frac{\beta}{8\pi}-\frac{\partial \log Z_g}{\partial\beta} \tag{15.109}$$

for the Schwarzschild background metric. One can interpret these equations as saying that E is equal to the energy of the thermal graviton and the black hole, if present. Using the approximate form of Z_g one finds that if the volume, V, of the box satisfies

$$E^5<\frac{\pi^2}{15}(8354.5)V, \tag{15.110}$$

the dominant contribution to N comes from the flat-space background metric. Thus in this case the most probable state of the system is just thermal gravitons and no black hole. If V is less than the inequality (15.110), there are two stationary-phase points for the Schwarzschild background metric. The one with a lower value of β gives a contribution to N which is larger than that of the flat-space background metric. Thus the most probable state of the system is a black hole in equilibrium with thermal gravitons. These results confirm earlier derivations based on the semi-classical approximations (Hawking, 1976; Gibbons and Perry, 1978).

15.9 Beyond one loop

In section 15.5 the action was expanded in a Taylor series around a background field which was a solution of the classical field equations. The path integral over the quadratic terms was evaluated but the higher-order terms were neglected. In renormalizable theories such as quantum electrodynamics, Yang–Mills or $\lambda\phi^4$ one can evaluate these higher or 'interaction' terms with the help of the differential operator A appearing in the quadratic or 'free' part of the action. One can express their effect by Feynman diagrams with two or more closed loops, where the lines in the diagram represent the propagator or Green's function A^{-1} and the vertices correspond to the interaction terms, three lines meeting at a cubic term and so on. In these renormalizable theories the undetermined quantities which arise from regularizing the higher loops turn out to be related to the undetermined normalization quantity, μ, of the single loop. They can thus all be absorbed into a redefinition of the coupling constant and any masses which appear in the theory.

The situation in quantum gravity is very different. The single-loop term about a flat or topologically trivial vacuum metric does not contain the normalization quantity, μ. However, about a topologically non-trivial background one has $\log Z_g$ proportional to $(106/45)\chi \log \mu$, where Z_g is the one-loop term and χ is the Euler number. One can express this as an addition to the action of an effective topological term $-k(\mu)\chi$, where $k(\mu)$ is a scale-dependent topological coupling constant. One cannot in general provide such a topological interpretation of the μ-dependence of the one-loop term about a background metric which is a solution of the field equations with nonzero matter fields. However one can do it in the special case where the matter fields are related to the gravitational field by local supersymmetry or spinor-dependent gauge transformations. These are the various supergravity and extended supergravity theories (Freedman, Van Nieuwenhuizen and Ferrara, 1976; Deser and Zumino, 1976).

Two loops in supergravity, and maybe also in pure gravity, do not seem to introduce any further undetermined quantities. However it seems likely that, both in supergravity and in pure gravity, further undetermined quantities will arise at three or more loops, though the calculations needed to verify this are so enormous that no-one has attempted them. Even if by some miracle no further undetermined quantities arose from the regularization of the higher loop, one would still not have a good procedure for evaluating the path integral, because the perturbation expansion around a given background field has only a very limited range of validity in gravity, unlike the case in renormalizable theories such as Yang–Mills or $\lambda\phi^4$. In the latter theory the quadratic or 'free' term in the action $\int (\nabla\phi)^2 \, d^4x$ bounds the interaction term $\lambda \int \phi^4 \, d^4x$. This means that one can evaluate the expectation value of the interaction term in the measure $D[\phi]\exp(-\int(\nabla\phi)^2\,d^4x)$ or, in other words, using Feynman diagrams where the lines correspond to the free propagator. Similarly in quantum electrodynamics or Yang–Mills theory, the interaction term is only cubic or quartic and is bounded by the free term. However, in the gravitational case the Taylor expansion about a background metric contains interaction terms of all orders in, and quadratic in derivatives of, the metric perturbations. These interaction terms are not bounded by the free, quadratic term so their expectation values in the measure given by the quadratic term are not defined. In other words, it does not make any sense to represent them by higher-order Feynman diagrams. This should come as no surprise to those who have worked in classical general relativity rather than in quantum field theory. We know that one cannot represent something like a black hole as a perturbation of flat space.

In classical general relativity one can deal with the problem of the limited range of validity of perturbation theory by using matched asymptotic expansions around different background metrics. It would therefore seem natural to try something similar in quantum gravity. In order to ensure gauge-invariance it would seem necessary that these background metrics should be solutions of the classical field equations. As far as we know, in a given topology and with given boundary conditions there is only one solution of the field equations or, at the most, a finite-dimensional family. Thus solutions of a given topology could not be dense in the space of metrics of that topology. However the Einstein action, unlike that of Yang–Mills theory, does not seem to provide any barrier to passing from fields of one topology to another.

One way of seeing this is to use Regge calculus (Regge, 1961). Using this method, one decomposes the spacetime manifold into a simplical complex. Each 4-simplex is taken to be flat and to be determined by its edge (i.e. 1-simplex) lengths. However the angles between the faces (i.e. 2-simplices) are in general such that the 4-simplices could not be joined together in flat four-dimensional space. There is thus a distortion which can be represented as a δ-function in the curvature concentrated on the faces. The total action is $(-1/8\pi)\sum A_i\delta_i$ taken over all 2-simplices, where A_i is the area of the ith 2-simplex and δ_i is the deficit angle at that 2-simplex, i.e. δ_i equals 2π minus the sum of the angles between those 3-simplices which are connected by the given 2-simplex.

A complex in which the action is stationary under small variations of the edge length can be regarded as a discrete approximation to a smooth solution of the Einstein equations. However, one can also regard the

Regge calculus as defining the action of a certain class of metrics without any approximations. This action will remain well defined and finite even if the edge lengths are chosen so that some of the simplices collapse to simplices of lower dimension. For example if a, b, c are the edge lengths of a triangle (a 2-simplex) then they must satisfy the inequalities $a < b + c$ etc. If $a = b + c$, the 2-simplex collapses to a 1-simplex. In general, the simplical complex will not remain a manifold if some of the simplices collapse to lower dimensions. However the action will still be well defined. One can then blow up some of the simplices to obtain a new manifold with a different topology. In this way one can pass continuously from one metric topology to another.

The idea is, therefore, that there can be quantum fluctuations of the metric not only within each topology but from one topology to another. This possibility was first pointed out by Wheeler (1963) who suggested that spacetime might have a 'foam-like' structure on the scale of the Planck length. In the next section I shall attempt to provide a mathematical framework to describe this foam-like structure. The hope is that by considering metrics of all possible topologies one will find that the classical solutions are dense in some sense in the space of all metrics. One could then hope to represent the path integral as a sum of background and one-loop terms from these solutions. One would hope to be able to pick out some finite number of solutions which gave the dominant contributions.

15.10 Spacetime foam

One would like to find which topologies of stationary-phase metrics give the dominant contribution to the path integral. In order to do this it is convenient to consider the path integral over all compact metrics which have a given spacetime volume V. This is not to say that spacetime actually is compact. One is merely using a convenient normalization device, like periodic boundary conditions in ordinary quantum theory: one works with a finite volume in order to have a finite number of states and then considers the values of various quantities per unit volume in the limit that the volume is taken to infinity.

In order to consider path integrals over metrics with a given 4-volume V one introduces into the action a term $\Lambda V/8\pi$, where Λ is to be regarded as a Lagrange multiplier (the factor $1/8\pi$ is chosen for convenience). This term has the same form as a cosmological term in the action but the motivation for it is very different as is its value: observational evidence shows that any cosmological Λ would have to be so small as to be practically negligible whereas the value of the Lagrange multiplier will turn out to be very large, being of the order of one in Planck units.

Let

$$Z[\Lambda]=\int D[g]\exp\left(-\hat{I}[g]-\frac{\Lambda}{8\pi}V[g]\right), \tag{15.111}$$

where the integral is taken over all metrics on some compact manifold. One can interpret $Z[\Lambda]$ as the 'partition function' for what I shall call the *volume canonical ensemble*, i.e.

$$Z[\Lambda]=\sum_n\left\langle\phi_n\middle|\exp-\left(\frac{\Lambda V}{8\pi}\right)\middle|\phi_n\right\rangle, \tag{15.112}$$

where the sum is taken over all states $|\phi_n\rangle$ of the gravitational field. From $Z[\Lambda]$ one can calculate $N(V)\,\mathrm{d}V$, the number of the gravitational fields with 4-volumes between V and $V+\mathrm{d}V$:

$$N(V)=\frac{1}{16\pi^2\mathrm{i}}\int_{-\mathrm{i}\infty}^{\mathrm{i}\infty}Z[\Lambda]\exp(\Lambda V)\,\mathrm{d}\Lambda \tag{15.113}$$

In (15.113), the contour of integration should be taken to the right of any singularities in $Z[\Lambda]$ on the imaginary axis.

One wants to compare the contributions to N from different topologies. A convenient measure of the complexity of the topology is the Euler number χ. For simply connected manifolds it seems that χ and the signature τ characterize the manifold up to homotopy and possibly up to homeomorphisms, though this is unproved. In the non-simply connected case there is no possible classification: there is no algorithm for deciding whether two non-simply connected 4-manifolds are homeomorphic or homotopic. This would seem a good reason to restrict attention to simply connected manifolds. Another would be that one could always unwrap a non-simply connected manifold. This might produce a non-compact manifold, but one would expect that one could then close it off at some large volume V with only a small change in the action per unit volume.

By the stationary-phase approximation one would expect the dominant contributions to the path integral Z to come from metrics near solutions of the Einstein equations with a Λ-term. From the scaling behaviour of the action it follows that for such a solution

$$\Lambda=-8\pi cV^{-1/2},\quad \hat{I}=-\frac{8\pi c^2}{\Lambda} \tag{15.114}$$

where c is a constant (either positive or negative) which depends on the solution and the topology, and where the action $\hat{I}$ now includes the Λ-term. The constant c has a lower bound of $-(\frac{3}{8})^{1/2}$ which corresponds to its value for S^4. An upper bound can be obtained from (15.96) and (15.97) for χ and τ. For solutions of the Einstein equations with a Λ-term these take the form

$$\chi = \frac{1}{32\pi^2} \int (C_{abcd}C^{abcd} + 2\tfrac{2}{3}\Lambda^2)(g)^{1/2}\, \mathrm{d}^4x, \tag{15.115}$$

$$\tau = \frac{1}{48\pi^2} \int C_{abcd}{}^*C^{abcd}(g)^{1/2}\, \mathrm{d}^4x. \tag{15.116}$$

From (15.115) one sees that there can be a solution only if χ is positive. However this will be the case for simply connected manifolds because then $\chi = 2 + B_2$, where B_2 is the second Betti number. Combining (15.115) and (15.116) one obtains the inequality

$$2\chi - 3|\tau| \geqslant \frac{32c^2}{3}. \tag{15.117}$$

From (15.115) one can see that, for large Euler number, at least one of the following must be true:

(*a*) c^2 is large

(*b*) $\int C_{abcd}C^{abcd}(g)^{1/2}\, \mathrm{d}^4x$ is large.

In the former case c must be positive (i.e. Λ must be negative) because there is a lower bound of $-(\frac{3}{8})^{1/2}$ on c. In the latter case the Weyl tensor must be large. As in ordinary general relativity, this will have a converging effect on geodesics similar to that of a positive Ricci tensor. However, between any two points in space there must be a geodesic of minimum length which does not contain conjugate points. Therefore, in order to prevent the Weyl curvature from converging the geodesics too rapidly, one has to put in a negative Ricci tensor or Λ-term of the order of $-C_{abcd}C^{abcd}L^2$, where L is some typical length scale which will be of the order of $V^{1/4}\chi^{-1/4}$, the length per unit of topology. One would then expect the two terms in (15.115) to be of comparable magnitude and c to be of the order of $d\chi^{1/2}$, where $d \leqslant 3^{1/2}/4$.

This is borne out by a number of examples for which I am grateful to N. Hitchin. For products of two-dimensional manifolds of constant curvature one has $d = \frac{1}{4}$. For algebraic hypersurfaces one has $2^{1/2}/8$.

Hitchin has obtained a whole family of solutions lying between these limits. In addition, if the solution admits a Kähler structure one has the equality

$$3\tau + 2\chi = 32c^2. \tag{15.118}$$

One can interpret these results as saying that one has a collection of the order of χ 'gravitational instantons' each of which has action of the order of L^2, where L is the typical size and is of the order of $V^{1/4}\chi^{-1/4}$. One also has to estimate the dependence of the one-loop curve Z_g on Λ and χ. The dependence on Λ comes from the scaling behaviour and is of the form

$$Z_g \propto \Lambda^{-\gamma},$$

where

$$\gamma = \int \left(\frac{53}{720\pi^2} C_{abcd}C^{abcd} + \frac{763}{540\pi^2}\Lambda^2\right)(g_0)^{1/2}\, \mathrm{d}^4x. \tag{15.119}$$

One can regard γ as the number of extra modes from perturbations about the background metric, over and above those for flat space. From (15.119) one can see that it is of the same order as χ. One can therefore associate a certain number of extra modes with each 'instanton'.

From the above it seems reasonable to make the estimate

$$Z[\Lambda] = \left(\frac{\Lambda}{\Lambda_0}\right)^{-\gamma} \exp(b\chi\Lambda^{-1}), \tag{15.120}$$

where $b = 8\pi d^2$ and Λ_0 is related to the normalization constant μ. Using (15.120) in (15.113), one can do the contour integral exactly and obtain

$$N(V) = \Lambda_0^{\gamma}\left(\frac{8\pi b\chi}{V}\right)^{1-\gamma/2} I_{\gamma-1}\left(\frac{Vb\chi}{2\pi}\right) \tag{15.121}$$

for $V \geqslant 0$.

However the qualitative dependence on the parameters is seen more clearly by evaluating (15.113) approximately by the stationary-phase method. In fact it is inappropriate to do it more precisely because $Z[A]$ has been evaluated only in the stationary-phase approximation. The stationary-phase point occurs for

$$\Lambda_s = 4\pi\frac{\gamma \pm (\gamma^2 + Vb\chi/2\pi)^{1/2}}{V}. \tag{15.122}$$

Because the contour should pass to the right of the singularity at $\Lambda = 0$, one should take the positive sign of the square root.

The stationary-phase value of Λ is always positive even though $Z[\Lambda]$ was calculated using background metrics which have negative Λ for large Euler number. This means that one has to analytically continue Z from negative to positive Λ. This analytic continuation is equivalent to multiplying the metric by a purely imaginary conformal factor, which was necessary anyway to make the path integral over conformal factors converge.

From the stationary-phase approximation one has

$$N(V) = Q(\Lambda_s) \equiv \left(\frac{\Lambda_s}{\Lambda_0}\right)^{-\gamma} \exp\left(b\chi\Lambda_s^{-1} + \frac{V\Lambda_s}{8\pi}\right). \qquad (15.123)$$

The dominant contribution to $N(V)$ will come from those topologies for which $dQ/d\chi = 0$. If one assumes $\gamma = a\chi$, where a is constant, one finds that this is satisfied if

$$-a \log\left(\frac{\Lambda_s}{\Lambda_0}\right) + b\Lambda_s^{-1} = 0. \qquad (15.124)$$

If $\Lambda_0 \geqslant 1$, this will be satisfied by $\Lambda_s \approx \Lambda_0$. If $\Lambda_0 < 1$, $\Lambda_s \approx \Lambda_0^{a/b}$. Equation (15.122) then implies that $\chi = hV$, where the constant of proportionality, h, depends on Λ_0. In other words the dominant contribution to $N(V)$ comes from metrics with one gravitational instanton per volume h^{-1}.

What observable effects this foam-like structure of spacetime would give rise to has yet to be determined, but it might include the gravitational decay of baryons or muons, caused by their falling into gravitational instantons or virtual black holes and coming out again as other species of particles. One would also expect to get non-conservation of the axial-vector current caused by topologies with non-vanishing signature τ.

EUCLIDEAN QUANTUM GRAVITY

Stephen W. Hawking

University of Cambridge, D.A.M.T.P.

Silver Street, Cambridge, England

1. INTRODUCTION

In these lectures I am going to describe an approach to Quantum Gravity using path integrals in the Euclidean regime i.e. over positive definite metrics. (Strictly speaking, Riemannian would be more appropriate but it has the wrong connotations). The motivation for this is the belief that the topological properties of the gravitational fields play an essential role in Quantum Theory. Attempts to quantize gravity ignoring the topological possibilities and simply drawing Feynman diagrams corresponding to perturbations around flat space have not been very successful: there seem to be an infinite sequence of undetermined renormalization parameters. The situation is slightly better with supergravity theories; the undetermined renormalization parameters seem to come in only at the third and higher loops around flat space but perturbations around metrics that are topologically non-trivial introduce undetermined parameters even at the one loop level [1] [27] **as I shall show later on.**

It seems to me that the fault lies not with the pure gravity or supergravity theories themselves but with the uncritical application of perturbation theory to them. In classical general relativity we have found that the perturbation theory has only a limited range of validity. One can ot describe a black hole as a perturbation around flat space. Yet this is what writing down a string of Feynman diagrams amounts to. On a technical level, the failure of perturbation theory can be traced to the fact that the "free" quadratic part of the action for general relativity does not bound the higher "interaction" terms unlike in Yang-Mills theory or Q.E.D. The free action would bound the interaction terms if one added quadratic terms in the curvature to the action. However such additions **alter** the nature of the theory and lead to fourth order equations, tachyons and ghosts, though I shall mention a way in which they can be used as conformal gauge fixing terms.

I must admit that I do not have an answer to the breakdown of the perturbation expansion but, like the man who looked for his key under the lamp-post because that was the only place in which he had any chance of finding it, I feel that if there is an answer, it must involve the topological structure of their gravitational field. I use the path integral approach because that seems to be the only way to handle topological questions. When one does path integrals for non-gravitational fields in flat spacetime one normally performs a Wick rotation of the time axis by replacing t by - it. This converts Minkowski space with Lorentzian metric (signature -+++) into Euclidean space (signature ++++). One reason for doing this is that it makes the path integral better defined. The path integral for (say) a scalar field ϕ in Minkowski space has the form

$$\int D[\phi] \exp (iI[\phi]) \tag{1.1}$$

where $D[\phi]$ is a measure on the space of all field configurations ϕ and $I[\phi]$ is the action of the field ϕ. For real fields on real Minkowski space this integral oscillates and does not converge. However, if one performs a Wick rotation to Euclidean space the path integral becomes

$$\int D[\phi] \exp (-\hat{I}[\phi]) \tag{1.2}$$

where $\hat{I}[\phi] = -iI[\phi]$ is the "Euclidean" action of the field ϕ and is real for fields ϕ that are real on the Euclidean section of complexified spacetime. Since I is usually positive definite, the path integral over all such fields will tend to converge. One can then analytically continue the resultant expression back to Minkowski space. This analytic continuation automatically incorporates the concepts of positive frequency and time ordering. For instance the Feynman propagator

$$\langle 0|T\phi(x)\ \phi(y)|0\rangle \tag{1.3}$$

is positive frequency in $t = x^o - y^o$, i.e. holomorphic in the lower half plane for Re $t > 0$ and negative frequency i.e. holomorphic in the upper half plane for Re $t < 0$. It is therefore holomorphic on the Euclidean space obtained by rotating the time axis 90° clockwise in the complex plane. In fact one could take the attitude that quantum theory and indeed the whole of physics is really defined in the Euclidean region and that it is simply a consequence of our perception that we interpret it in the Lorentzian regime.

I feel that one should adopt a similar Euclidean approach in quantum gravity and supergravity. Of course one cannot simply replace the time coordinates by imaginary quantities because there is no preferred set of time coordinates in general relativity. Instead I think one should perform the path integrals over all positive definite metrics, most of which will not admit a section on which the metric is real and Lorentzian, and then analytically continue the result of the path integral, if necessary. In order to restrict the path integral to positive definite metrics and to exclude integration over metrics with Lorentzian or ultra hyperbolic signatures, one should probably integrate not over the components of the metric g_{ab} but over the components $e^a{}_m$ of a tetrad. This can be regarded as the square root of the metric

$$g_{ab} = e^m{}_a e_{bm} \tag{1.4}$$

spacetime indices a, b ... are raised or lowered by the metric g^{ab} and g_{ab} and tetrad indices m, n... are raised or lowered by the Euclidean metric δ^{mn} and δ_{mn}. Thus the group of tetrad rotations is $SO(4) = \frac{SU(2) \times SU(2)}{Z_2}$ rather than Lorentzian group SO(1,3) that one normally uses.

The metric g_{ab} defined by equation 1.4 for arbitrary real $e_a{}^m$ will be positive semidefinite and will be degenerate where $\det(e_a{}^m) = 0$. I think that one must include such degenerate metrics in the path integral. They provide a way in which one can pass continuously from one spacetime topology to another. I shall return to this point later on.

Tetrads are also essential for dealing with fermion fields. I shall use two component spinors but in a positive definite metric rather than the usual Lorentzian one. In the Lorentzian case one has unprimed spinors λ_A which transform under SL(2,C) and primed spinors $\mu_{A'}$ which transform under the complex conjugate group $\overline{SL(2,C)}$. The complex conjugate operation takes unprimed spinors to primed spinors and vice versa i.e. $\bar{\lambda}_{A'}$ is a primed spinor. When one analytically continues to a positive definite metric, the primed and unprimed spinors transform under independent groups SU(2) and $\overline{SU(2)}$. The complex conjugate operation now takes unprimed to unprimed and primed to primed and either raises the index or lowers it with a minus sign. Quantities that were complex are conjugates of each other in the Lorentzian case, like the Weyl spinor Ψ_{ABCD} and $\bar{\Psi}_{A'B'C'D'}$, are analytically continued to the Euclidean region as independent fields ψ_{ABCD} and $\tilde{\psi}_{A'B'C'D'}$. Thus it is possible to have a metric in which $\psi_{ABCD} \neq 0$ but $\tilde{\psi}_{A'B'C'D'} = 0$. Such a metric is conformally self dual i.e. $C_{abcd} = {}^*C_{abcd} = \frac{1}{2}\,\varepsilon_{abef}\,C^{ef}{}_{cd}$. If the Ricci spinor $\Phi_{ABC'D'}$ and the curvature scalar Λ are also 0, the metric is self-dual, $R_{abcd} = {}^*R_{abcd}$.

A Dirac 4 spinor ψ can be represented by two 2 component spinors in a column vector

$$\psi = \begin{pmatrix} \lambda_A \\ \bar{\mu}^{A'} \end{pmatrix} \tag{1.5}$$

The adjoint field $\bar{\psi}$ is represented by a row vector

$$\bar{\psi} = (\mu_A, \bar{\lambda}^{A'}) \tag{1.6}$$

when one goes to Euclidean space the 4 spinor and its adjoint become independent fields ψ and $\tilde{\psi}$ which are represented by four independent 2 component spinors λ_A, $\tilde{\lambda}^A$, $\tilde{\mu}_{A'}$ and μ^A. A Majorana 4 spinor ψ in Lorentzian space is represented by the column vector

$\psi = \begin{pmatrix} \rho_A \\ \bar{\rho}^{A'} \end{pmatrix}$. When one goes to the Euclidean region one has two independent 2 components spinors ρ_A and $\tilde{\rho}^{A'}$. Thus, contrary to what is often stated, one can deal with Majorana spinors in the Euclidean regime.

The action for the gravitational field is usually taken to be

$$I = \frac{1}{16\pi G} \int R(-g)^{\frac{1}{2}} d^4x \tag{1.7}$$

I shall use units in which $G = c = \hbar = 1$. When one performs the Wick rotation to the Euclidean region the volume element $(-g)^{\frac{1}{2}}$ becomes $-i(g)^{\frac{1}{2}}$. Thus the Euclidean action would be

$$\hat{I} = -\frac{1}{16\pi} \int R(g)^{\frac{1}{2}} d^4x \tag{1.8}$$

However this action contains second derivatives of the metric which have to be removed by integration by parts to give an action which is quadratic in first derivatives of the metric, as is required by the path integral approach. One therefore gets a surface term in the action which is often ignored but which turns out to be ver important [2, 3]

$$\hat{I} = -\frac{1}{16\pi} \int R(g)^{\frac{1}{2}} d^4x - \frac{1}{8\pi} \int K(h)^{\frac{1}{2}} d^3x + C \tag{1.9}$$

where h is the induced metric and K is the trace of the second fundamental form of the boundary of the region over which the action is being evaluated. The term C is an arbitrary constant which may depend on the induced metric h on the boundary but not on the metric g in the interior of the region. When the boundary metric h is such that the boundary can be embedded in flat space, it is natural to choose C so that the action is zero when g is the flat metric. However not all boundary metrics h can be embedded even locally in flat space. Also the existence of such a rigid boundary surface is not very physical. I shall therefore be describing ways in which one can eliminate the boundary term and deal only with compact manifolds.

I shall be talking mainly about the Euclidean approach to pure gravity but these ideas can also be extended to supergravity. For instance one could probably regard superspace S as some sort of fibre bundle over the spacetime manifold M with a Grassmannian fibre with coordinates $\Theta_A, \tilde{\Theta}^{A'}$ which represents the Euclidean version of a Majorana spinor. The first, and so far the most successful, application of Euclidean methods was to the thermal properties of black holes. Since much of this work has already been published I shall describe it only briefly. I shall then describe some as yet unpublished work on the gravitational vacuum and shall end up with the volume canonical ensemble which enables one to treat the foamlike structure of spacetime.

2. THERMAL PROPERTIES

The partition function $Z[\beta]$ for the thermal canonical ensemble for a scalar field ϕ in flat spacetime at temperature $T = \beta^{-1}$ is defined as

$$Z[\beta] = \Sigma_n \langle \phi_n \mid \exp(-\beta H) | \phi_n \rangle \tag{2.1}$$

where $|\phi_n\rangle$ is a complete orthonormal basis of states for the field ϕ. One can also represent Z as a path integral

$$Z[\beta] = \int D[\phi] \exp(-\hat{I}[\phi]) \tag{2.2}$$

where the path integral is taken over all field configurations ϕ that are real in Euclidean space and are periodic with period β in the Euclidean time coordinate (i.e. they are periodic in imaginary Minkowski time). One can represent the partition function for higher spin boson fields in a similar way but, in the case of fields with gauge degrees of freedom, one has to include the Fadeev-Popov ghosts to subtract out the unphysical degrees of freedom. One can also treat fermion fields in a similar way though in this case the fields have to be antiperiodic i.e. they reverse sign when the Euclidean time coordinate increases by β.

It would be natural to define the partition function for the thermal canonical ensemble for gravity in an analogous manner.

In order to avoid the infra-red problems caused by having an infinite volume of thermal gravitons with an infinite mass, one would like to enclose the system in (say) a spherical box of radius r_o with perfectly reflecting walls. This of course is rather unphysical becuase one cannot make a rigid perfectly reflecting box even for electromagnetic radiation let alone for gravitational radiation. It leads to problems near the walls of the box but I shall ignore these for the present anyway. The world tube of the spherical box identified with period β in the Euclidean time coordinate defines a 3-dimensional manifold ∂M with topology $S^2 \times S^1$ with a positive definite metric h which is the product of the standard metric of a 2-sphere of radius r_o and 1-dimensional metric on a circle of circumference β. The partition function $Z[\beta]$ for the gravitational field at temperature $T = \beta^{-1}$ in the spherical box is defined by a path integral over all metrics g on all manifolds M which have ∂M as their boundary and which induce the given metric h on ∂M.

One expects, or at least one hopes, that the dominant contribution to the path integral will come from metrics near a background metric g_o that extremises the action i.e. is a solution of the classical field equations with the given boundary conditions.

One can expand the action in a Taylor series about the back ground metric g_o

$$\hat{I}[g] = \hat{I}[g_o] + I_2[g_o, \bar{g}] + \text{higher order terms} \tag{2.3}$$

where $g = g_o + \bar{g}$ and I_2 is quadratic in the perturbations $\bar{g}$. Then

$$\log Z = -\hat{I}[g_o] + \log \int D[\bar{g}] \exp(-I_2) \tag{2.4}$$
$$+ \text{higher order terms}$$

One can regard the first term as the contribution to log Z of the background metric and the second, "one loop" term as the contribution of thermal gravitons on the background. The measure $D[\bar{g}]$ for the one loop term can be expressed as

$$D[\bar{g}] = \Pi_n \mu da_n \pi^{-\frac{1}{2}} \tag{2.5}$$

where μ is a normalization quantity or regulator and a_n are the coefficients in the expansion of the perturbation $\bar{g}$ in terms of eigenfunctions of the second order elliptic operator A which determines I_2 i.e.

$$\bar{g} = \Sigma a_n \phi_n \tag{2.6}$$

where $A\phi_n = \lambda_n \phi_n$ and (2.7)

$$I_2 = \int \bar{g} \, A \, \bar{g} (g_o)^{\frac{1}{2}} d^4x \tag{2.8}$$

The one loop term L is formally [4]

$$L = \frac{\det(\mu^{-2} C)}{(\det(\mu^{-2}(A + B)))^{\frac{1}{2}}} \tag{2.9}$$

where B is the gauge fixing operator and C is the ghost operator. The number $N(\lambda)$ of eigenvalues of an operator has an asymptotic expansion of the form

$$N(\lambda) = \sum_{n=o}^{\infty} P_n \lambda^{2-n} \tag{2.10}$$

P_o is proportional to the 4-volume of the background metric times the number of components of the field on which the operator acts. The higher coefficients P_n are polynomials in the metric, the curvature and its covariant derivatives of degree 2n in derivatives of the metric. The one loop term therefore diverges badly. The various regularization schemes all amount to dividing out by the distribution of eigenvalues given by the P_o and P_1 terms. In general however, except in flat space and a few other special background metrics, the net P_2 term will be non-zero. This means that there will be some finite number (not necessarily an integer) of extra eigenvalues whose dimensionality is not cancelled out. Because each extra eigenvalue has dimensions $(\text{length})^{-2}$, one has to divide it by the normalization quantity μ^2. Thus the one loop term is μ dependent. One cannot absorb this dependence in a redefinition of the coupling constant in the original action but, since it is proportional to the Euler number χ one might absorb it in a topological term $k\chi$ where k was a scale dependent topological coupling constant. I shall return to this later.

One background metric which satisfies the boundary conditions for the canonical ensemble is the flat metric on periodically identified Euclidean space i.e. on the manifold M with topology $S^1 \times R^3$. With an appropriate choice of the constant C, the action for this background metric is zero and so it makes no contribution to log Z. The one loop term representing the quadratic fluctuations around the flat space background metric gives the standard result for thermal radiation with two helicity states

$$\log Z = \frac{16\pi^3 r_o^3}{135\beta^3} \tag{2.11}$$

this can be interpreted as the partition function of thermal gravitons in flat space. Higher loops, if they had any meaning which is doubtful, would represent the effects of interactions between these gravitons.

The Schwarzschild metric is another solution with the given boundary conditions. The metric is normally expressed as

$$ds^2 = -(1 - \frac{2M}{r})\, dt^2 + (1 - \frac{2M}{r})^{-1}\, dr^2 + r^2 d\Omega^2 \tag{2.12}$$

Putting $t = -i\tau$ converts this into a positive definite metric for $r > 2M$. There is an apparent singularity at $r = 2M$ but this is like the apparent singularity at the origin of polar coordinates as can be seen by defining a new radial coordinate $x = 4M(1 - 2Mr^{-1})^{\frac{1}{2}}$. Then the metric becomes

$$ds^2 = (\frac{x}{4M})^2\, d\tau^2 + (\frac{r^2}{4M^2})^2\, dx^2 + r^2 d\Omega^2 \tag{2.13}$$

This will be regular at $x = 0$, $r = 2M$ if τ is regarded as an angular variable and is identified with period $8\pi M$ (I am using units in which $G = 1$). The manifold defined by $x \geq 0$, $0 \leq \tau \leq 8\pi M$ is called the Euclidean section of the Schwarzschild solution. On it the metric is positive definite, asymptotically flat and non-singular (the curvature singularity at $r = 0$ does not lie on the Euclidean section).

The Scwarzschild metric will satisfy the boundary conditions if

$$M = \frac{\beta}{8\pi} .$$

Because it has $R = 0$, the action will come entirely from the surface term. This gives [3]

$$\hat{I} = 4\pi M^2 = \frac{\beta^2}{16\pi} \quad (2.14)$$

Thus the Schwarzschild background metric will give a contribution of

$$\frac{-\beta^2}{16\pi}$$

to log Z. By its definition

$$Z = \Sigma \exp(-\beta E_n) \quad (2.15)$$

where E_n is the energy of the nth state of the gravitational field. Thus

$$\langle E \rangle = \frac{\Sigma E_n \exp(-\beta E_n)}{\Sigma \exp(-\beta E_n)} = -\frac{\partial}{\partial\beta} \log Z \quad (2.16)$$

Inserting the contribution $\frac{-\beta^2}{16\pi}$ to log Z from the action of the background Schwarzschild metric one obtains

$$\langle E \rangle = \frac{\beta}{8\pi} = M \quad (2.17)$$

This is what one would expect. However one can also compute the entropy defined as

$$S = -\Sigma p_n \log p_n = \beta \langle E \rangle + \log Z \quad (2.18)$$

where $p_n = Z^{-1} \exp(-\beta E_n)$ is the probability of being in the nth state. Applied to the Schwarzschild background metric this gives

$$S = 4\pi M^2 = \tfrac{1}{4}A$$

where A is the area of the event horizon. This relation between intrinsic gravitational entropy and the area of event horizons is quite general. It arises from a combination of the facts that the gravitational action is not scale invariant (which is what makes it nonrenormalizable) and the fact that the Euclidean section of the Schwarzschild solution has the topology $S^2 \times R^2$ which is different from periodically identified flat space which has topology $S^1 \times R^3$ [5]. There is no analogue in QCD or other lower spin theories

The one loop term contains a μ dependent term of the form

$$\frac{106\chi}{45} \log (\mu\beta) \text{ where } \chi = 2$$

is the Euler number of the Euclidean section of the Schwarzschild solution. This arises from

$$\frac{106\chi}{45}$$

extra eigenvalues whose dimensionality is not cancelled by the regularization procedure. Three of these extra eigenvalues are zero and correspond to translations of the black hole inside the box. They give a contribution of $\log (\mu^3\beta^3 V)$ to $\log Z$. On the other hand a non relativistic particle of mass M in a box of volume V at temperature β^{-1} would have a partition function proportional to $V(M\beta^{-1})^{3/2}$. Thus it seems that to reproduce this one should take μ proportional to β^{-1}.

The eigenfunctions of the operator A + B can be divided into two classes [4]. First there are the eigenfunctions corresponding to metric perturbations $\bar{g}$ which are proportional to the background metric g_0. These represent conformal perturbations. The eigenvalues are all negative for perturbations around flat space or around any metric with R = 0. One therefore has to rotate the contour of integration of the conformal factor to lie parallel to the imaginary axis. The remaining eigenfunctions are traceless and correspond to nonconformal perturbations of the metric. At flat space they are all positive and therefore one adopts the rule that one integrates over real conformal equivalence classes of metrics. However for perturbations around the Schwarzschild background metric, one of these nonconformal eigenvalues is negative [6]. Since it appears in the one loop term under a square root sign, it contributes a factor i to the partition function Z which is therefore imaginary. One might have expected some such pathology in Z because the canonical ensemble for gravity breaks down due to its attractive nature.

One way of saying this is to express Z as the Laplace transform of the density of states N(E) where N(E)dE is the number of states of the gravitational field in the box with energies between E and E + dE:

$$Z[\beta] = \int_0^\infty N(E) \exp(-\beta E) dE \tag{2.19}$$

Thus N(E) is the inverse Laplace transform

$$N(E) = \frac{1}{2\pi i} \int_{-i\infty}^{i\infty} Z[\beta] \exp(\beta E) d\beta \tag{2.20}$$

However if one substitutes for Z the estimate

$$\exp\left(\frac{-\beta^2}{16\pi}\right)$$

arising from the action of the background metric, the integral does not converge. To make it converge one has to rotate the contour of the β integration in 2.20 from the imaginary axis. The fact that Z is imaginary then leads to a real N(E) as one would expect because the micro canonical ensemble for gravity is well-defined: a black hole in a box can be in stable equilibrium with thermal radiation if the total energy in the box is fixed but it is not stable if the temperature of the box is fixed because black holes have negative specific heat [7].

One can also consider ensembles where, in addition to temperature, one has chemical potentials for the angular momentum about some axis and the electric charge. The partition function would be given by a path integral over all fields which on some boundary surface were the same at the points (t, r, Θ, ϕ) and (t + iβ, r, Θ, ϕ + iΩβ) in a gauge in which $A_o = \Phi$ where Ω is the angular velocity and Φ is the electrostatic potential. In keeping with the Euclidean approach Ω and Φ should be taken to be imaginary and then the partition function should be analytically continued back to real values of Ω and Φ.

The stationary phase metric in the path integral for the partition function will be periodically identified Euclidean space in rotating coordinates and the Euclidean section of the Kerr-Newman solution with imaginary angular momentum and electric charge (imaginary electric charge produces a real field F_{ab} in the Euclidean regime). A similar treatment to that for the Schwarzschild solution shows that the Kerr-Newman solution also has intrinsic quantum gravitational entropy equal to one quarter of the area of the event horizon.

3. THE GRAVITATIONAL VACUUM

In Yang-Mills theory one can describe the vacuum state with $F_{ab} = 0$ on a spacelike surface by potentials A_a which are pure gauge i.e. they are of the form

$$A_a = \Lambda^{-1}\partial_a\Lambda \tag{3.1}$$

where Λ is an element of the gauge group G. If one fixes Λ to be the unit element at infinity or equivalently if one compactifies the spacelike surface by adding a point at infinity to make it a 3-sphere, one has a map from S^3 into G. In the simple case that G is SU(2), it is also a 3-sphere topologically and so these maps can be divided into homotopy equivalence classes characterized by an integer n which is the degree of the mapping. One therefore has a degenerate family of vacuum states described by an integer n [8]. The transition amplitude to tunnelling from an initial vacuum n_1 to a final vacuum n_2 is given by a path integral over all Yang-Mills fields on Euclidean space which die away at large distance and which have a Pontryagin number $n_1 - n_2$. The action of such fields is bounded below by

$$\frac{1}{8\pi^2} n_1 - n_2$$

so that the tunnelling is suppressed.

One can perform a similar analysis for the gravitational vacuum. The zero field configuration, flat space, on a spacelike surface can be described by different tetrad fields which can be regarded as maps of the 3-sphere (the spacelike surface compactified by the addition of a point at infinity) into SO(4), the group of tetrad rotations. In fact since one wants to consider fermion fields as well one should consider spin frames rather than just ordinary tetrads. These correspond to maps of the 3-sphere into SU(2) x SU(2), the covering group of SO(4). Since each SU(2) is a 3-sphere topologically, these maps are divided into homtopy classes characterized by two integers (n, m) which are the degrees of the mapping of the 3-sphere on the two factors. One thus has a double infinity of degenerate gravitational vacua. The transition amplitude to tunnel from an initial vacuum (n_1, m_1) to a final vacuum (n_2, m_2) will be given by a path integral over all asymptotically Euclidean metrics on parallelizable manifolds with Euler number $\chi = (n_2 - n_1) + (m_2 - m_1) + 1$
and Hirzebruch signature $\tau = \frac{2}{3}\left[(n_2 - n_1) - (m_2 - m_1)\right]$

Asymptotically Euclidean means that the metric approaches the standard Euclidean metric on R^4 outside some compact region in which the topology will differ from that of R^4. Parallelizable means that one can define continuous tetrad and spin frame fields on the manifold and the conditions on χ and τ ensure that these fields can interpolate between the fields on the initial and final surfaces. In fact the condition that the manifolds admit a spin structure i.e. allow spinors to be defined consistently, implies that the manifold is parallelizable [9, 10]. In the asymptotically Euclidean case it also implies that τ is a multiple of 16. Thus the gravitational vacua are divided into classes which cannot tunnel into each other but live in completely separate universes.

The Euler number χ and the signature τ can be expressed as integrals of the curvature

$$\chi = \frac{1}{128\pi^2} \int R_{abcd} R^{efgh} \varepsilon^{ab}{}_{ef} \varepsilon^{cd}{}_{gh} (g)^{\frac{1}{2}} d^4x + \text{surface terms} \qquad (3.2)$$

$$\tau = \frac{1}{96\pi^2} \int R_{abcd} R^{ab}{}_{ef} \varepsilon^{cdef} (g)^{\frac{1}{2}} d^4x + \text{surface terms} \qquad (3.3)$$

The surface terms are rather complicated in general but for an asymptotically Euclidean metric there is a contribution of one to χ for each asymptotically Euclidean region and a zero contribution to τ.

The Euler number χ is given by

$$\chi = B_o - B_1 + B_2 - B_3 + B_4 \qquad (3.4)$$

The pth Betti number B_p is the number of independent closed p-surfaces that are not boundaries of some p + 1 surface. In a compact manifold B_p is equal to the number of square intergrable harmonic p forms. For a compact manifold, $B_p = B_{4-p}$ and $B_o = B_4 = 1$. If the manifold is simply connected $B_1 = B_3 = 0$ so $\chi \geq 2$.

It seems to me that one should probably restrict attention to simply connected manifolds. In this case χ and τ classify compact manifolds that admit spinor structure up to homotopy. It is conjectured (by Poincaré) that they classify the manifolds up to homeomorphisms. It can be shown that there is no possible classification scheme for non-simply connected 4 manifolds.

The B_2 harmonic 2 forms (Maxwell fields) can be divided into B_2^+ self-dual and B_2^- anti-self-dual 2 forms. Then $\tau = B_2^+ - B_2^-$. It is also equal to $8(n^+ - n^-)$ where n^+ and n^- are the numbers of zero modes of the massless Dirac equations with positive helicidies respectively.

In Yang-Mills theory it is convenient to compactify Euclidean space R^4 by adding a point at infinity to convert the manifold into S^4. The original flat metric can be recovered from the metric on S^4 by a conformal transformation that sends the added point to infinity. A similar procedure can be adopted in the gravitational case: one can conformally compactify an asymptotically Euclidean manifold by adding a point at infinity though in this case the resultant compact manifold need not be S^4 but could be any manifold that admitted a spinor structure. In fact one could regard this as a precise definition of what is meant by asymptotically Euclidean analogous to Penrose's definition of asymptotic flatness for Lorentzian manifolds [11]. What one does is start with a compact manifold with a smooth positive definite metric, pick a point z and send it to infinity by a conformal transformation. One obtains an asymptotically Euclidean metric with Euler umber one less than that of the compact manifold but with the same signature.

The transition amplitude from the initial to the final vacuum can be expressed as a path integral over all asymptotically Euclidean metrics on the (simply connected) manifold M with the given values of χ and τ

$$Z = \langle 0_- | 0_+ \rangle = \int D[e] \exp(-\hat{I}[g]) \tag{3.5}$$

The path integral can be decomposed into an integral over conformal factors Ω followed by an integral over conformal equivalence classes $\{e^m_a\}$ of tetrads. Under a conformal transformation

$$\tilde{e}^m_a = \Omega e^m_a, \quad \tilde{g}_{ab} = \Omega^2 g_{ab} \tag{3.6}$$

the action becomes

$$\hat{I}[\Omega e] = -\frac{1}{16\pi} \int (\Omega^2 R + 6\Omega_{;a}\Omega_{;b} g^{ab})(g)^{\frac{1}{2}} d^4x$$

$$- \frac{1}{8\pi} \int \Omega^2 K(h)^{\frac{1}{2}} d^3x + C \tag{3.7}$$

The surface term is evaluated over a large sphere in asymptotically Euclidean space and the constant C is chosen to subtract out the flat space value of the surface term. To perform the path integral over the conformal factor one first finds a conformal factor ω which is equal to one on the boundary surface and which is such that the metric $g^*_{ab} = \omega^2 g_{ab}$ has $R^* = 0$ everywhere. This can be thought of as a choice of conformal gauge. The action of the metric g^* is given entirely by the surface term. One then integrates over conformal factors Ω of the form $1 + y$ about the metric g^* where y vanishes on the boundary. Because the kinetic term $(\nabla y)^2$ appears with a minus sign, one has to integrate over imaginary y [4]. This gives

$$(\det(\mu^{-2}\Delta))^{-\frac{1}{2}} \exp(-\hat{I}[g^*]) = Y[\{e^m_a\}] \tag{3.8}$$

where $\Delta = -\Box + \frac{1}{6}R$ is the conformally invariant scalar operator. The transition amplitude Z is then given by an integral over all conformal equivalence classes of tetrads.

$$Z = \int D[\{e\}]\, Y[\{e\}] \tag{3.9}$$

One can express this procedure in terms of the conformally compactified manifold $\tilde{M}$ and metric $\tilde{g}$. Pick a point z in $\tilde{M}$ and send it to infinity by the conformal transformation

$$\omega(x) = 4\pi^2\, \Delta^{-1}(x,z) \tag{3.10}$$

where $\Delta^{-1}_{\sim}$ is the Green's function on the compact manifold in the metric $\tilde{g}$. The metric $g^* = \omega^2 \tilde{g}$ then is asymptotically Euclidean with $R^* = 0$.

Let (λ_n, ϕ_n) be the eigenvalues and eigenfunctions of Δ on $(\tilde{M},\tilde{g})$. The eigenfunction ϕ_0 corresponding to the lowest eigenvalue λ_0 will have the same sign everywhere. One can therefore use it as the conformal factor in a regular conformal transformation $\tilde{g} \to \tilde{g}'$ that makes $\tilde{R}'$ have everywhere the same sign as λ_0. From this it follows that if the lowest eigenvalue λ_0 is positive, the Green's function $\Delta^{-1}(x,z)$ does not pass through zero anywhere. Therefore the metric g^* is non-singular. On the other hand, if λ_0 is negative, $\Delta^{-1}(x,z)$ passes through zero and g^* is singular. I shall give an interpretation of this shortly.

The action of the metric g^* (omitting the C term) is

$$\hat{I}[g^*] = 6\pi^3\Delta^{-1}(z,z) \tag{3.11}$$

This of course is infinite because the Green's function diverges at z but so is the action if the surface term

$$-\frac{1}{8\pi}\int K(h)^{\frac{1}{2}}d^3x$$

is evaluated on a 3-sphere of infinite radius. To obtain a finite value for the action one has to make an (infinite) subtraction of the flat space value of the surface term. In the conformally compactified procedure this corresponds to regularizing $\Delta^{-1}(z,z)$. One obtains the correct result by a conformally invariant dimensional regularization procedure of adding extra flat dimensions. This is equivalent to zeta function regularization [12, 13] plus a correction term $\tilde{R}/288\pi^2$.

<u>The Positive Action Conjecture</u> [4, 5, 6] asserts that the action of any regular asymptotically Euclidean metric with R = 0 is positive or zero, being zero if and only if the metric is flat. In conformally compactified terms, the conjecture is that of $\Delta^{-1}(z,z)$ is greater than or equal to zero for all $(\tilde{M},\tilde{g})$ that do not have negative or zero eigen values for Δ. It is the analogue in one higher dimension of the Positive Energy Conjecture in ordinary general relativity (this has now been proved) [14].

One can represent

$$\Delta^{-1}(z,z) = \Sigma\lambda_n^{-1}\,\phi_n(z)\phi_n(z) \tag{3.12}$$

The regularization makes it zero for the standard metric on S^4. As one deforms the metric away from the spherical metric, the lowest eigenvalue λ_0 decreases and $\Delta^{-1}(z,z)$ becomes positive. When the metric is deformed so far that λ_0 passes through zero. $\Delta^{-1}(z,z)$ passes through infinity and becomes negative and the conformal factor $\Delta^{-1}(x,z)$ passes through zero.

To interpret this physically it is helpful to compare it with the initial value problem for classical general relativity which can be formulated in a similar way. One starts with a compact 3-dimensional manifold $\tilde{S}$ with a smooth metric $\tilde{h}$. One picks a point z and sends it to infinity with a conformal factor $\omega(x) = 4\pi\Delta^{-1}(x,z)$ where Δ^{-1} is now the Green's function for the 3-dimensional conformally invariant operator. The metric $h^* = \omega^4\tilde{h}$ then is asymptotically flat and has $R^* = 0$. It therefore satisfies the constraint equation for the time symmetric initial value problem (one can also deal with the non-symmetric initial value problem in a similar way). The ADM mass is given by the regularized value of $8\pi\Delta^{-1}(z,z)$.

The standard metric on the 3-sphere gives the initial data for flat space. As one deforms the metric $\tilde{h}$ away from the spherical metric one obtains initial data for a time symmetric imploding-exploding gravitational wave wave which will have positive energy or mass. As one increases the strength of the wave one reaches a critical state at which the energy in the wave is so great that it curls up the initial surface and cuts itself off from infinity. This corresponds to an eigenvalue of Δ passing through zero. One then passes to a new class of initial data, that for black holes with an apparent event horizon [15]. Such data can be obtained from a pair $(\tilde{S}, \tilde{h})$ with no negative eigenvalues by sending two points, z_1 and z_2, to infinity by the conformal factor $\omega(x) = 4\pi\Delta^{-1}(x,z_1) + 4\pi\Delta^{-1}(x,z_2)$.

It might seem reasonable to adopt a similar procedure in quantum gravity. Instead of simply sending one point z of the compact manifold $\tilde{M}$ to infinity to obtain an asymptotically Euclidean metric, one might send n points $z_1, z_2 \ldots z_n$ to infinity to obtain a metric with n asymptotically Euclidean regions. One can do this using the conformal factor

$$\omega(x) = 4\pi^2 \int \Delta^{-1}(x,y)\, J(y)(g)^{\frac{1}{2}} d^4y \tag{3.13}$$

where the "infinity current" J(y) is defined by

$$J(y) = \Sigma\delta(y,z_n) \tag{3.14}$$

The action of the metric $g^* = \omega^2\tilde{g}$ which has $R^* = 0$ is

$$\begin{aligned}\hat{I}[g^*] &= 6\pi^3 \int J(x)\Delta^{-1}(x,y)J(y)(g)^{\frac{1}{2}}(g)^{\frac{1}{2}}d^4x d^4y \\ &= 6\pi^3 J\Delta^{-1}J\end{aligned} \tag{3.15}$$

The vacuum to vacuum amplitude, or rather the vacua to vacua amplitude, corresponding to the manifold $\tilde{M}$ with n points removed is then

$$Z_n = \int D[\{e\}]\, (\det\Delta)^{-\frac{1}{2}} \exp(-6\pi^3 J\Delta^{-1}J) \tag{3.16}$$

integrated over all conformal equivalence classes {e} of tetrad fields. One can now sum this amplitude over all numbers, strengths and positions of the infinity points or, equivalently, over all infinity currents J. This produces a factor of $(\det \Delta^{-1})^{-\frac{1}{2}}$ which exactly cancels the factor of $(\det \Delta)^{-\frac{1}{2}}$ arising from the integral over conformal factors.

There remains the integral over conformal equivalence classes of tetrad fields. The components e^m_a of the tetrad at a point x form a 16 dimensional vector space. Identifying the tetrad with components e^m_a with the tetrad Ωe^m_a for any nonzero Ω, i.e. taking conformal equivalence classes, reduces one to a compact 15 dimensional projective space of conformal tetrads which is factored to a 9 dimensional compact space of conformal metrics by the SO(4) group of tetrad rotations. Because the space of conformal tetrads at each point is compact, one can adopt a measure in which it has unit volume. The path integral over conformal equivalence classes of tetrad fields then gives one, so one has

$$\Sigma Z_n = 1 \tag{3.17}$$

One might think that one should also sum over all possible topologies of the compact manifold $\tilde{M}$. However I think that one is already doing this effectively by integrating over all conformal tetrad fields. There will in general be 3 surfaces where $\det(e^m_a) = 0$ and the metric is degenerate. In general the null vector of the matrix e^m_a will not lie in the surface $\det(e) = 0$. This allows one to introduce new coordinates or, equivalently, to go to another manifold with a different topology on which the tetrad field and the metric are nondegenerate. There will in general be 2 surfaces on which the null vector of e^m_a lies in the surface $\det(e) = 0$. The scalar curvature R maybe singular there but it will still be intergrable so the action can still be defined. Thus integrating over all conformal tetrad fields or, equivalently, over all conformal positive semi-definite metrics effectively incorporates a sum over all manifold topologies.

This remarkable result was obtained without any regularization because the divergences in the path integration over conformal factors exactly cancelled the divergences in the action of the asymptotically Euclidean metrics which are normally removed by subtracting out the flat space value of the surface term. One might regard equation 3.17 as a statement of unitarity for gravity when one includes all topological possibilities. However, it is a bit formal. What one would like to do is to use this approach to calculate physical amplitudes and probabilities. I must confess that I have not yet found a way of doing this, though I do have some vague ideas. It seems that most conformal metrics (and all metrics with spinor structure and nonzero τ) have negative or zero eigenvalues of Δ. This means that g*, the stationary phase metric under conformal transformations, will contain regions which are closed off from infinity. However, there will be other metrics in the conformal equivalence class in which these regions are not closed off. Thus they should have some physical effects though with reduced probabilities because they are not connected in the stationary phase metric. What these effects are could be is a matter of speculation but they might correspond to virtual black holes which could appear, swallow a particle, spit out another particle of a different species with the same charge, momentum and angular momentum and then disappear.

4. ASYMPTOTICALLY LOCALLY EUCLIDEAN METRICS

Under a scale transformation $g \to k^2 g$ (k constant), the action transforms as $I \to k^2 I$. This implies that the action of any asymtotically Euclidean metric which was a solution of the Einstein equations would have to be zero because it would have to be an extremum of the action under all perturbations including dilations. However the Positive Action Conjecture asserts that any asymptotically Euclidean metric with $R = 0$ has positive or zero action, the action being zero if and only if the metric is flat. Thus, if this conjecture holds, there can be no nontrivial asymptotically Euclidean vacuum gravitational instantons i.e. complete non-singular solutions of the vacuum field equations. However the positive action conjecture does not exclude the possibility of vacuum instantons which are asymptotically locally Euclidean (I shall abbreviate this by ALE). In other words, outside some compact region they approach the Euclidean metric on flat space identified under some discrete subgroup G of SO(4). The first ALE instanton was found by Eguchi and Hanson [16]. It is asymptotic to Euclidean space with the point x^a identified with the point $- x^a$. It represents the transition between an initial state and its image under TP. Further ALE instantons have been found explicitly by Gibbons and myself [17] and implicitly by Hitchin [18] using Penrose's twistor technique [19]. These correspond to larger discrete subgroups G and their physical interpretation is not yet clear.

All these metrics have self-dual or anti self-dual curvature. This suggests a Generalized Positive Action Conjecture: any asymptotically locally Euclidean metric with $R = 0$ has positive or zero action, the action being zero if and only if the curvature is self-dual or anti self-dual. This conjecture is supported by the fact that one can show that self-dual or anti self-dual metrics are local minima of the action among metrics with $R = 0$. If the generalized conjecture holds, they are global minima of the action among ALE metrics with $R = 0$ just as the self-dual or anti self-dual Yang-Mills instantons are global minima of the Yang-Mills action. One would therefore expect that metrics near them would make the dominant contribution to transition amplitudes determined by these boundary conditions.

A particularly interesting application is to supergravity theories [20]. In a self-dual metric the curvature seen by the primed spinors is zero so there are two independent covariantly constant primed spinors $\iota_{A'}$ and $o_{A'}$ which can be chosen so that

$$\iota_{A'} o^{A'} = 1 \qquad \bar{\iota}_{A'} = o^{A'} \tag{4.1}$$

In an ALE self-dual metric, $B_2^+ = \tau$ and $B_2^- = 0$ so there are τ self-dual harmonic two-forms (Maxwell fields) which are normalizable i.e. L^2 and no L^2 anti self-dual harmonic two-forms. The self-dual harmonic two-forms can be represented by τ symmetric spinors which satisfy the spin-1 equation

$$\nabla^{AA'} \phi^{j}_{AB} = 0 \tag{4.2}$$

Multiplying by one or other of the two covariantly constant spinors one gets 2τ zero modes of the Majorana spin 3/2 equation in the gauge $\gamma^a \psi_a = 0$

$$\nabla^{AA'} \phi^{ij}_{ABA'} = 0 \tag{4.3}$$

where $\phi^{ij}_{ABA'} = \phi^{j}_{AB}\, \alpha^{i}_{A'}$. None of these modes is a pure gauge transformation i.e. of the form $\nabla_{AA'}\phi_B$. Multiplying by another covariantly constant spinor and symmetrising on the primed indices one obtains 3τ transverse traceless zero modes of the metric perturbation equation in the harmonic gauge

$$\nabla^{AA'} \nabla_{AA'} \phi^{jh}_{BCB'C'} - 2\Psi_{BC}{}^{DE} \phi^{jh}_{DEB'C'} = 0 \tag{4.4}$$

where $\phi^{jh}_{BCB'C'} = \phi^{j}_{BC}\, \alpha^{h}_{B'C'}$.

Up to three of these modes (depending on the discrete group G) may be pure gauge transformations corresponding to the global rotations of the instanton in asymptotically locally Euclidean space. The relations between the zero modes of different spins are actually global supersymmetry transformations with covariantly constant spinor parameters.

The action of the background self-dual ALE instanton is zero so it makes no contribution to log Z. The one loop term in simple supergravity can be expressed

$$Z = \frac{\det B(\det C)^{\frac{1}{2}}}{(\det E)^{\frac{1}{2}}\,(\det F)^{3/2}} \tag{4.5}$$

where B is the vector ghost operator corresponding to diffeomorphism gauge fixing, C is the spin 3/2 operator (it appears to the power $\frac{1}{2}$ because it is a Majorana field), E is the metric perturbation operator equation 4.4 and F is the spin $\frac{1}{2}$ supersymmetry ghost operator (it appears to the power 3/2 because it is a Majorana field and because when one averages over supersymmetry gauges one introduces an extra factor of $(\det F)^{-\frac{1}{2}}$ [21]). There are also six non-propagating components of the tetrad and six non-propagating components of the auxillary fields [22, 23] but these cancel with the twelve non-propagating tetrad rotation ghosts.

One can use global supersymmetry transformations to relate the non zero modes of the operators B, C, E, F, [20]. The multiplicities are such that the non zero eigenvalues cancel completely in the one loop term 4.5. Thus the infinite dimensional one loop path integral reduces to a finite dimensional integral over zero modes. This cancellation between boson and fermion operators is one of the most attractive features of supergravity theories and raises the hope that one might be able to make some reasonable mathematical sense out of them. It should be pointed out however that in the usual language of Feynman diagrams, the one loop term 4.5 has a logarithmic divergence because there are different numbers of zero modes in the numerator and denominator (the proof that all one loop terms are finite in supergravity applies only in topologically trivial metrics). The one loop term introduces a parameter μ which reflects the measure on the zero modes about which there is, as yet, no agreement.

The 3τ gravitational zero modes in the operator E correspond to global rotation, dilation and other self-dual perturbations of the instanton metric. To deal with them one introduces collective coordinates for the orientation, scale etc. of the instanton. They give rise to a factor of the form

$$\mu^{3\tau}\,\rho^{6\tau-1}d\rho$$

where ρ is some length scale of the instanton. The 2τ spin 3/2 zero modes in the operator C occur in the numerator in equation 4.5 and make the vacuum to vacuum amplitude Z zero in the absence of sources. To deal with this one adds a term

$$\int (\phi_{ABA'}\, \eta^{ABA'} + \tilde{\phi}_{AA'B'}\, \tilde{\eta}^{AA'B'})(g)^{\frac{1}{2}} d^4x \qquad (4.6)$$

where the Majorana spin 3/2 field is represented by the Euclidean spinors $\phi_{ABA'}$ and $\tilde{\phi}_{AA'B'}$ and the source current is represented by η and $\tilde{\eta}$. The vacuum to vacuum amplitude $Z[\eta,\tilde{\eta}]$ in the prescence of the sources is then proportional to

$$\prod_n \int \phi^n_{ABA'}\, \eta^{ABA'} (g)^{\frac{1}{2}} d^4x \qquad (4.7)$$

where $\phi^n_{ABA'}$

are the 2τ positive helicity spin 3/2 zero modes. To the Z given by equation 4.7 one has to add the value of Z of order unity arising from the one loop about flat Euclidean space. One then functionally differentiates with respect to η at 2τ points to get a 2τ point function which violates helicity conservation i.e. it con erts τ particles of positive helicity into the same number of negative helicity particles.

The simplest self-dual ALE instanton, the Eguchi-Hanson [16], has $\tau = 1$. One therefore obtains a helicity changing two point function which could be regarded as a spin 3/2 mass term in this metric. However, because the instanton is asymptotic to Euclidean space with the points x^a identified with $-x^a$, one can also interpret this as a 4 point function in Euclidean space which converts two spin 3/2 particles with positive helicity into their TP counterparts, two particles of negative helicity but the same momenta.

The situation will be similar for extended supergravity theories except that there will now be $2\tau N$ spin 3/2 zero modes where N is the number of spin - $\frac{3}{2}$ fields. One will therefore get $2\tau N$ point helicity changing amplitudes. These theories will also contain spin 1 fields and may contain spin $\frac{1}{2}$ and spin 0 fields. There will be no zero modes of the spin $\frac{1}{2}$ and spin 0 fields in a self-dual ALE instanton background metric. There will be τ spin 1 zero modes but these do not appear in the one loop term because they do not arise from a potential. In fact if they are gauge fields (abelian or otherwise) the zero modes will be quantized by the analogue of the Dirac condition for magnetic monopoles.

5. SPACETIME FOAM

I now return to the question of how to treat conformal equivalence classes of metrics on a compact manifold $\tilde{M}$ that possess negative or zero eigenvalues of the conformally invariant operator Δ. In these cases the asymptotically Euclidean member g^* of the conformal equivalence class which has $R^* = 0$ and which is the stationary phase point under conformal transformations is singular and contains regions which are cut off by zero conformal factor. On the other hand, one can find a compact metric g' in the conformal equivalence class which has $R' = -4h$ where h is a positive constant. The metric g' can be normalized by requiring that its 4-volume is unity.

The significance of the metrics g' is that they are the stationary phase points under conformal transformations in the volume canonical ensemble that I shall define below. The physical idea is that in quantum gravity one has to sum over metrics of very complicated topology. This "foamlike structure" [24] will be everywhere so that one cannot really think of a spacetime as being asymptotically Euclidean although this is a convenient viewpoint when interpreting amplitudes as S-matrix elements.

To keep things finite it is convenient to consider compact metrics with some given large value of the 4-volume V. This is not to say that spacetime necessarily is compact. It is merely a normalization device like periodic boundary conditions in ordinary quantum mechanics. One would compute the density of certain quantities per unit volume and then take the limit as V tends to infinity.

In order to constrain the 4-volume one adds a term $\Lambda V/8\pi$ to the gravitational action where Λ is a Lagrange multiplier. This is of the same form as the cosmological term but the motivation for its introduction and its actual value are very different: observational evidence indicates that any cosmological term must be so small as to be practically negligible whereas the Lagrange multiplier will turn out to give the dominant contribution when it is of order one in Planck units. One forms the partition function $Z[\Lambda]$ for what I call the volume canonical ensemble [25] by performing a path integral over all compact positive semi-definite metrics

$$Z[\Lambda] = \int D[e] \exp(-\hat{I}[e])$$

$$= \Sigma_n \langle g_n| \exp(-\Lambda V/8\pi) |g_n\rangle \quad (5.1)$$

where the action includes the Λ term. The partition function is the Laplace transform of N(V)dV, the number of states of the gravitational field between V and V + dV,

$$Z[\Lambda] = \int_0^\infty N(V) \exp(-\Lambda V/8\pi) dV \qquad (5.2)$$

Thus N(V) is the inverse Laplace transform

$$N(V) = (1/16\pi^2 i) \int Z[\Lambda] \exp(\Lambda V/8\pi) d\Lambda \qquad (5.3)$$

where the contour is taken parallel to the imaginary Λ axis and to the right of the singularity in $Z[\Lambda]$ at $\Lambda = 0$. This ensures that $N(V) = 0$ for $V \leq 0$.

The path integral 5.1 for Z can be broken up into a path integral over conformal factors Ω in one conformal equivalence class $\{e\}$ of tetrad fields

$$Y[\{e\}, \Lambda] = \int D[\Omega] \exp(-\hat{I}[\Omega e]) \qquad (5.4)$$

followed by an integral over conformal equivalence classes

$$Z[\Lambda] = \int D[\{e\}]\, Y[\{e\}\, \Lambda] \qquad (5.5)$$

Under a conformal transformation $\tilde{e} = \Omega e$, the action (including the Λ term) becomes

$$\hat{I}[\tilde{e}] = -\frac{1}{8\pi} \int (3\Omega\Delta\Omega - \Lambda\Omega^4)(e) d^4x \qquad (5.6)$$

Thus $Z[\Lambda]$ can be regarded as the average of $\lambda\phi^4$ theory over all conformal equivalence classes of metrics. The stationary phase point in the integral 5.4 over conformal factors will occur at the metric $-h\Lambda^{-1}g'$ for which $R = 4\Lambda$, $V = h^2\Lambda^{-2}$ and $I = -h^2/8\pi\Lambda$. The quantity h is a function of the conformal equivalence class and will be stationary at those equivalence classes for which g' is a solution of the Einstein equations with a Λ term

$$R_{ab} - \tfrac{1}{2} R g_{ab} + \Lambda g_{ab} = 0 \qquad (5.7)$$

A few solutions are known explicitly on compact manifolds such as S^4, CP^2 and $S^2 \times S^2$. One is really interested however in solutions on very complicated simply-connected manifolds with high Euler number and signature. Although one cannot hope to find such metrics explicitly, one can nevertheless estimate their action. For a solution of 5.7 one has

$$\chi = \frac{1}{32\pi^2} \int (C_{abcd} C^{abcd} + \frac{8}{3}\Lambda^2)(e)d^4x \tag{5.8}$$

$$\tau = \frac{1}{48\pi^2} \int (C_{abcd} *C^{abcd})(e)d^4x \tag{5.9}$$

Combining equations 5.8 and 5.9 one has

$$2\chi - 3|\tau| \geq h^2/6\pi^2 \tag{5.10}$$

the equality holding if and only if the Weyl tensor is self dual or anti self-dual.

For solutions of 5.7, h has the lower bound of $-(24)^{\frac{1}{2}}\pi$, the value for the standard metric on S^4. If $\tau \neq 0$ and if the manifold admits a spinor structure, $h \geq 0$. From 5.8 and 5.9 one would expect that for large Euler number $h \sim d\chi^{\frac{1}{2}}$ where $d \lesssim 2(3)^{\frac{1}{2}}\pi$. This is supported by a number of examples which have been supplied by Nigel Hitchin which indicate that most solutions lie in the range

$$2^{\frac{1}{2}}\pi < d < 2^2 \pi \tag{5.11}$$

One would hope that one could obtain a value for the partition function $Z[\Lambda]$ by expanding the metric or tetrad in a perturbation series about each solution of the classical field equations. When Λ is small, the action of these solutions will be widely separated an and one should have a good "dilute gas approximation". However when Λ is large, fluctuations around one solution may change the topology and so overlap with fluctuations around another solution, without significantly increasing the action and so being damped. One way of avoiding such an overlap might be to add a gauge fixing term $\alpha \int R^2(e)d^4\chi$ to the action to fix the conformal gauge in which the path integral 5.4 over conformal factors was evaluated. The path integral for Z will then become

$$Z[\Lambda] = \int D[\Omega]\, D[e] \det \alpha\Delta \exp - \tilde{I}[e,\Omega]) \tag{5.12}$$

where

$$\tilde{I}[e,\Omega] = -\frac{1}{8\pi}\int (3\Omega\Delta\Omega - \Lambda\Omega^4 - 8\pi\alpha R^2)(e)d^4x$$

$$- 16\,\alpha\, h^2[e_o] \qquad (5.13)$$

The det $\alpha\Delta$ term in 5.12 is the Fadeev-Popov ghost and the last term in 5.13 is designed to make the gauge-fixing term zero at the solution e_o. The gauge-fixing term $\int R^2(e)d^4x$, unlike the action $\int R(e)d^4x$ would probably not remain bounded if the metric changed topology by passing through a degenerate metric. It might thus confine the perturbation expansion to the topology of the solution about which it was made, and prevent it from overlapping with perturbation expansions centred on other solutions. It would also give rise to a perturbation expansion that was at least formally renormalizable. The fact that it was only a gauge-fixing term and not a part of the action itself might avoid some of the pathologies associated with such terms.

So far no progress has been made with this approach. However one can estimate the ordinary one loop term L about a solution with Euler number χ.

$$L = \left(\frac{\Lambda}{\Lambda_o}\right)^{-\gamma} \qquad (5.14)$$

where γ is the integrated trace anomaly for gravity [26]

$$\gamma = \frac{106\chi}{45} + \frac{73\,h^2}{120\pi^2} \qquad (5.15)$$

and Λ_o is related to the normalization constant μ. Thus metrics with Euler number χ make a contribution to Z of the form

$$\left(\frac{\Lambda}{\Lambda_o}\right)^{-\gamma} \exp\,(b\chi\Lambda^{-1}) \qquad (5.16)$$

where $b = d^2/8\pi$. This gives a stationary phase point in the inverse Laplace transform 5.3 for N(V) at $\Lambda = \Lambda_s$ where

$$\Lambda_s = 4\pi V^{-1}\left[\gamma \pm \left(\gamma^2 + \frac{Vb\chi}{2\pi}\right)^{1/2}\right] \tag{5.17}$$

Comparing the contributions from metrics of different χ, one finds that the dominant one comes when

$$-\gamma\chi^{-1} \log \left(\frac{\Lambda_s}{\Lambda_o}\right) + b\,\Lambda_s^{-1} = 0 \tag{5.18}$$

For $\Lambda_o \sim 1$, this is satisfied when $\chi \sim V$ i.e. there is one unit of topology or one gravitational instanton per Planck volume.

REFERENCES

1. M.J. Perry, Nucl.Phys.B., to be published.
2. J. York, Phys.Rev.Lett, 28, 1082, 1972.
3. G.W. Gibbons and S.W. Hawking, Phys.Rev.D15, 2752, 1977.
4. G.W. Gibbons, S.W. Hawking and M.J. Perry, Nucl.Phys.B, to be published.
5. S.W. Hawking, Phys.Rev.D, to be published.
6. D.N.Page, Phys.Rev.D. to be published.
7. S.W. Hawking, Phys.Rev.D13, 191, 1976.
8. R.Jackiw and C. Rebbi, Phys.Lett 67B, 189, 1977.
9. R.P. Geroch, J.Math.Phys 9, 1739, 1968; J.Math.Phys.11, 343, 1970.
10. C.J. Isham, Spinor Fields in Four Dimensional Spacetime, Imperial College preprint, 1978.
11. R. Penrose, Proc.Roy.Soc. A284, 159, 1965.
12. J.S. Dowker and R. Critchley, Phys.Rev.D13, 3224, 1976.
13. S.W. Hawking, Comm.Math.Phys. 55, 133, 1977.
14. S.T. Yau and R. Schoen, "Incompressible Minimal Surfaces, Three Dimensional Manifolds with Non-Negative Scalar Curvature, and the Positive Mass Conjecture in General Relativity.
15. S.W. Hawking, "The Event Horizon" in "Les Astres Occlus" ed. B.S. deWitt and C.M. deWitt, Gordon and Breach, 1973.
16. T. Eguchi and A.J. Hanson, Phys.Lett 74B, 249, 1978.
17. G.W. Gibbons and S.W. Hawking, Gravitational Multi-Instantons, D.A.M.T.P. preprint.
18. N. Hitchin, in preparation.
19. R. Penrose, J.Gen.Rel. and Gravitation, 7, 31, 1976.
20. C.N. Pope and S.W. Hawking, "Symmetry Breaking by Instantons

in Supergravity" D.A.M.T.P. preprint.

21. M.J. Perry, Nucl.Phys.B., to be published.

22. S. Ferrara and P. van Nieuwenhuizen, "The Auxiliary Fields of Supergravity, CERN preprint.

23. K. Stelle and P. West, "Minimal Auxiliary Fields for Supergravity", Imperial College preprint.

24. J.A. Wheeler in "Relativity Groups and Topology", proceedings of the Les Houches Summer School, 1963, ed by B.S. deWitt and C.M. deWitt, Gordon and Breach, New York, 1964.

25. S.W. Hawking, "Spacetime Foam" D.A.M.T.P. preprint.

26. G.W. Gibbons and M.J. Perry "Quantizing Gravitational Instantons, D.A.M.T.P. preprint.

27. M.J. Duff, Abstracts of Contributed Papers for GR VIII Conferences, Waterloo, Ontario (1977)

Nuclear Physics B138 (1978) 141–150

PATH INTEGRALS AND THE INDEFINITENESS OF THE GRAVITATIONAL ACTION

G.W. GIBBONS
Max Planck Institut für Physik und Astrophysik, 8 München 40, Postfach 401212, FDR

and

Department of Applied Mathematics and Theoretical Physics, University of Cambridge, Silver Street, Cambridge CB3 9EW, UK

S.W. HAWKING and M.J. PERRY
Department of Applied Mathematics and Theoretical Physics, University of Cambridge, Silver Street, Cambridge CB3 9EW, UK

Received 3 January 1978

The Euclidean action for gravity is not positive definite unlike those of scalar and Yang-Mills fields. Indefiniteness arises because conformal transformations can make the action arbitrarily negative. In order to make the path integral converge one has to take the contour of integration for the conformal factor to be parallel to the imaginary axis. The path integral will then converge at least in the one-loop approximation if a certain positive action conjecture holds. We perform a zeta function regularization of the one-loop term for gravity and obtain a non-trivial scaling behaviour in cases in which the background metric has non-zero curvature tensor, and hence non-trivial topologies.

1. Introduction

The path integral approach appears to be the best method of quantizing gauge fields. It has been applied with considerable success to Yang-Mills fields. In this approach, one considers quantities of the form

$$Z = \int D[\phi] \exp(iI[\phi]) , \tag{1.1}$$

where $D[\phi]$ is some measure on the space of all fields ϕ. $I[\phi]$ is the action of the fields ϕ, and the integral is taken over all fields that satisfy some boundary or periodicity condition. In order to give this oscillating integral a well-defined meaning one makes a Wick rotation of the time axis and passes to Euclidean space. For real fields

ϕ on Euclidean space the action $I[\phi]$ is $i\hat{I}[\phi]$, where $\hat{I}[\phi]$ is real. For Yang-Mills fields $\hat{I}[\phi]$ is positive semi-definite, so one can expect the path integral to converge. The Euclidean action $\hat{I}[\phi]$ is also positive semi-definite for scalar fields. The action $\hat{I}[\phi]$ for spin-$\frac{1}{2}$ fields is not positive definite if the fields are regarded as commuting quantities, but this difficulty can be overcome by adopting formal rules for the evaluation of path integrals over anticommuting quantities.

One would like to extend the path integral approach to include gravitation. The gravitational action is

$$I[g] = \frac{1}{16\pi G}\int_{\mathrm{M}} R(-g)^{1/2}\,\mathrm{d}^4x + \frac{1}{8\pi G}\int_{\partial\mathrm{M}} [K](-h)^{1/2}\,\mathrm{d}^3x\,, \tag{1.2}$$

where the volume integral of the Ricci scalar R is taken over M with boundary ∂M, and $[K]$ is the difference in the trace of the second fundamental form of the boundary in the metric g and the flat space metric, [1]. The boundary term is present because R contains second derivatives of the metric which have to be removed by integration. In analogy with Yang-Mills theory, one might hope to define the path integral for gravity by Wick rotating the boundary so that its induced metric h becomes positive definite everywhere, and then integrating over all positive definite metrics g which induce the given metric h on the boundary. Unfortunately this procedure will not lead to a convergent integral because the Euclidean action

$$\hat{I}[g] = \frac{-1}{16\pi G}\int R(g)^{1/2}\,\mathrm{d}^4x - \frac{1}{8\pi G}\int [K](h)^{1/2}\,\mathrm{d}^3x \tag{1.3}$$

is not positive semi-definite. (The minus sign comes from the direction of the Wick rotation, which has to be chosen to be consistent with that for the matter fields.) Under conformal transformations of the metric $\tilde{g}_{ab} = \Omega^2 g_{ab}$, R transforms as

$$\tilde{R} = \Omega^{-2}R - 6\Omega^{-3}\Box\Omega\,, \tag{1.4}$$

and

$$\tilde{K} = \Omega^{-1}K + 3\Omega^{-2}\Omega_{,a}n^a\,, \tag{1.5}$$

where n^a is the unit outward normal to the boundary ∂M. Thus

$$\hat{I}[\tilde{g}] = \frac{-1}{16\pi G}\int_{\mathrm{M}} \Omega^2 R + 6\Omega_{,a}\Omega^{,a}(g)^{1/2}\,\mathrm{d}^4x - \frac{1}{8\pi G}\int_{\partial\mathrm{M}} [\Omega^2 K](h)^{1/2}\,\mathrm{d}^4x\,. \tag{1.6}$$

One sees that $\hat{I}$ may be as negative, as one wants, by choosing a rapidly varying conformal factor Ω.

It is perhaps not surprising that one does not obtain convergence for the integral over all metrics g which induce a given fixed metric h on the boundary. For if the integral did converge, one could give a well-defined meaning to the canonical ensemble for the quantum gravitational field in a box by integrating over all metrics which fill in a boundary which is the product of the box and a periodically identified imaginary time axis [1]. However, the attractive nature of gravity and the possibility of having a black hole cause the canonical ensemble to break down [2]. One would expect that this problem could be overcome by the use of the microcanonical ensemble since this should still be well-defined. It is, however, cumbersome to work with; for actual calculations it would be convenient if one could give at least a formal meaning to the integral over all metrics spanning a given fixed boundary. In sect. 2, we shall propose a method of doing this based on a rotation of the contour of integration of the conformal factor at each point. In sect. 3, we shall use these methods to give a zeta function regularization [3,4] of the single closed loop for gravity. We derive the scaling behaviour and the trace of the effective energy-momentum tensor. This agrees with the calculation by Tsao [5] and Critchley [6].

2. The conformal factor

Let $\mathcal{H}$ denote the space of all positive definite metrics g on M which induce the given metric h on the boundary ∂M. The metrics $\mathcal{H}$ can be divided into equivalence classes under conformal transformations with a conformal factor Ω which is positive and equal to one on ∂M. Let $\{\lambda_n, \phi_n\}$ be the eigenvalues and eigenfunctions of the conformally invariant scalar wave operator ($\Delta = -\Box + \frac{1}{6}R$) on M with Dirichlet boundary conditions i.e. $\Delta\phi_n = \lambda_n\phi_n$ and $\phi_n = 0$ on ∂M. If there is an eigenfunction ϕ_a with $\lambda_a = 0$, then $\widetilde{\phi}_a = \Omega^{-1}\phi_a$ will be an eigenfunction in the metric $\widetilde{g}_{ab} = \Omega^2 g_{ab}$ with $\widetilde{\lambda}_a = 0$. The eigenfunctions with non-zero eigenvalues do not have any simple behaviour under conformal transformations. However, they and their eigenvalues will change smoothly under smooth changes in Ω. The eigenvalues cannot change sign under such smooth changes in Ω because the zero eigenvalues are conformally invariant. Thus the number of negative eigenvalues (which will be finite) is the same for all metrics in a conformal equivalence class. The number of zero eigenvalues is conformally invariant.

Let $\Omega = 1 + y$ where y is zero on ∂M. Then,

$$\hat{I}[\widetilde{g}] = \hat{I}[g] - \frac{1}{16\pi G}\int (y\Delta y + 2Ry)(g)^{1/2}\,\mathrm{d}^4x$$

$$= \hat{I}[g] - \frac{1}{16\pi G}\int \{(y - \Delta^{-1}R)\,\Delta(y - \Delta^{-1}R) - R\Delta^{-1}R\}(g)^{1/2}\,\mathrm{d}^4x\,, \quad (2.1)$$

where Δ^{-1} is the inverse of the operator Δ which vanishes on ∂M, and $\Delta^{-1}R$ denotes

$$\int \Delta^{-1}(x, x')\,R(x')(g(x'))^{1/2}\,\mathrm{d}^4x'\,.$$

Thus $\hat{I}[\widetilde{g}] = \hat{I}^1 + \hat{I}^2$, where

$$\hat{I}^1 = I[g] + \frac{1}{16\pi G}\int R\Delta^{-1}R(g)^{1/2}\,\mathrm{d}^4x\,, \quad (2.2)$$

$$\hat{I}^2 = \frac{-1}{16\pi G}\int z\Delta z(g)^{1/2}\,\mathrm{d}^4x\,, \tag{2.3}$$

where

$$z = y - \Delta^{-1}R\,. \tag{2.4}$$

Under conformal transformations $g'_{ab} = \omega^2 g_{ab}$, $\hat{I}^1$ is unchanged. If Δ has no negative or zero eigenvalues in the metric g, one can choose $\omega > 0$ and $\omega = 1$ on ∂M such that $R' = 0$ everywhere. In this case $\hat{I}[g] = \hat{I}^1[g']$ which is given entirely by the surface terms, K. We make what we call the *positive action conjecture*: $\hat{I} \geqslant 0$ for all asymptotically Euclidean positive definite metrics g' for which $R' = 0$. By asymptotically Euclidean we mean that outside some compact region the manifold is R^4 and the metric tends to the flat Euclidean metric on R^4. In other words, the manifold is flat in both space and time directions.

There is a close connection between the positive action conjecture and the positive energy conjecture which states that the mass, or energy, of a solution of the classical Einstein equations with Lorentzian signature is greater than, or equal to zero, if the solution evolves from a non-singular asymptotically flat initial surface. Although the positive energy conjecture can be proved only in restricted cases, or under certain assumptions [7–10], it is generally believed. If it held also in five dimensions, it would imply the positive action conjecture because an asymptotically Euclidean four-dimensional manifold with a positive definite metric for which $R = 0$ could be regarded as a time-symmetric initial surface for a five-dimensional spacetime with a metric of Lorentzian signature. In that situation, the mass of the five-dimensional metric would be equal to the action of the four-dimensional metric. Page [11] has obtained some results which support the positive action conjecture. He has, however, shown that it is false for metrics like the Schwarzschild solution which are asymptotically flat in space but which contain horizons and are therefore periodic in the Euclidean time variable [1].

If the operator Δ has negative eigenvalues but no zero eigenvalues, its inverse Δ^{-1} and $\hat{I}^1$ will still be defined but the conformal factor ω needed to make R' zero will pass through zero so that metric g' will be singular although its action will still be well-defined. Consider the one-parameter family $g[v]$ of metrics where $g[0] = g_0$, a solution of the Einstein equations. For small values of v, Δ will have only positive eigenvalues and, subject to the discussion above, $\hat{I}^1[g[v]]$ will be positive. At larger values of v, Δ may develop negative eigenvalues. This will mean that in the complex v plane $\hat{I}^1[g[v]]$ will have poles at $v = v_1, v_2, \ldots$, where a positive eigenvalue passes through zero and becomes negative. Presumably, one should take one's contour of integration in the space of all conformal equivalence classes of metrics to lie just above or below these poles in the real axis.

Under conformal transformations $g' = \omega^2 g$, $\hat{I}^1$ is unchanged. Thus one can integrate $\exp(-\hat{I}^2)$ over each conformal equivalence class by metrics and then integrate the result multiplied by $\exp(-I^1)$ over conformal equivalence classes. The operator

Δ appears in $\hat{I}^2$ with a minus sign. This means that for conformal equivalence classes for w ich Δ has only positive eigenvalues, and in particular, for conformal equivalence classes near solutions of the field equations, one has to integrate over values of z w ich are purely imaginary. In fact it seems that path integrals for quantum gravity are defined only in the one-loop approximation [12], i.e. for metrics which are near a solution of the field e uations. One prescription for evaluating the one-loop terms if to divide the space of all metrics into conformal equivalence classes. In each conformal equivalence class, choose the metric g' for which $R' = 0$. This will be a reg lar metric for equivalence classes near a solution of the field equations). Integrate over conformal factors of the form $\Omega = 1 + i\xi$ (ξ real) about each metric g' and then integrate over conformal equivalence classes. If $\hat{I}^1$ is a local minimum on the conformal equivalence classes, and if at the equivalence class of our solution g_0 of the field equations, then the path integral over conformal equivalence classes will converge inthe one-loop approximation if the non-conformal perturbations are taken to be purely real. On the other hand, there may be an n-dimensional (n finite) space of perturbations which reduce $\hat{I}^1$. In this case, one has to take the perturbations lying in this space to be purely imaginary in order to make the path integral converge in the one-loop approximation. This will introduce a factor of i^n in Z. In the case of the chwarzschild solution, Page [11] has sho n that n is at least one. It seems plausible that n is in fact equal to one in this case. This would mean that the partition function for the Schwarzschild solution is imaginary. However, the partition function would be expected to be pathological in this case because the canonical ensemble for black holes breaks down on account of the fact that black holes have negative specific heat [13,14]. However, one can obtain the density of states of the gravitational field in a given energy interval in the microcanonical ensemble by an inverse Laplace transform of the partition function. In order to make this transform converge, one has to rotate the contour of integration through 90° and this introduces another factor of i so that one obtains a real value for the density of states.

3. Single closed loops

One would expect that the dominant contribution to the path integral would come from metrics near the metric g_0 that minimizes I on $\mathcal{H}$. This metric will be a solution of the classical Einstein equations with the given boundary conditions. Particular examples are flat space, and the Euclidean Schwarzschild [1] and Taub-NUT [15] solutions. One can expand the action in a Taylor series about the background metric g_0:

$$\hat{I}[g] = \hat{I}[g_.] + I_2[g_0, \phi] + \text{higher terms}\,, \tag{3.1}$$

where

$$I_2[g_0, \phi] = \frac{1}{32\pi G} \int \phi^{ab} A_{abcd} \phi^{cd} (g_0)^{1/2}\, d^4x\,, \tag{3.2}$$

$$g^{ab} = g_0^{ab} + \phi^{ab}\,, \tag{3.3}$$

$$A_{abcd} = \tfrac{1}{4} g_{cd} \nabla_a \nabla_b - \tfrac{1}{4} g_{ac} \nabla_d \nabla_b + \tfrac{1}{8}(g_{ac} g_{bd} + g_{ab} g_{cd}) \nabla_e \nabla^e + \tfrac{1}{2} R_{ad} g_{bc}$$
$$- \tfrac{1}{4} R_{ab} g_{cd} + \tfrac{1}{16} R_{ad} g_{bc} - \tfrac{1}{8} R g_{ac} g_{bd} + (a \leftrightarrow b) + (c \leftrightarrow d) + (a \leftrightarrow b, c \leftrightarrow d)\,. \tag{3.4}$$

If one neglects the higher-order terms, the path integral becomes

$$Z = \exp(-\hat{I}[g_0]) \int D[\phi] \exp(-\hat{I}_2[g_0, \phi])\,. \tag{3.5}$$

To evaluate the Gaussian path integral over the metric perturbations ϕ, one expands ϕ in terms of eigenfunctions of the operator A.

$$\phi^{ab} = \sum_n a_n \phi_n^{ab}\,, \tag{3.6}$$

where

$$A_{abcd} g_0^{ac} g_0^{bf} \phi_n^{cd} = \lambda_n \phi_n^{ef}\,. \tag{3.7}$$

The measure $D[\phi]$ is taken to be $\Pi_n \mu \mathrm{d}a_n/(32\pi G)^{1/2}$, where μ is a normalisation factor. This gives

$$\log Z = -\hat{I}[g_0] - \tfrac{1}{2}\log \det(\mu^{-2} A)\,, \tag{3.8}$$

$$\det A = \prod_n \lambda_n\,. \tag{3.9}$$

The trouble with this is that A has a large number of zero eigenvalues corresponding to the fact that the action is unchanged under an infinitesimal diffeomorphism (gauge transformation)

$$x^a \to x^a + \epsilon \xi^a\,,$$
$$g_{ab} \to g_{ab} + 2\epsilon \xi_{(a;b)}\,. \tag{3.10}$$

One would like to factor out the gauge freedom by integrating only over gauge-inequivalent ϕ^{ab}. One would then obtain an answer which depended on the determinant of A on the quotient of all fields ϕ^{ab} modulo infinitesimal gauge transformations. The way to do this has been given by Feynman [16], DeWitt [17], Fadeev and Popov [18], and 't Hooft [19]. One adds a gauge-fixing term

$$\hat{I}_g = \frac{1}{32\pi G} \int \phi^{ab} B_{abcd} \phi^{cd} (g_0)^{1/2}\, \mathrm{d}^4x \tag{3.11}$$

to the action. The operator B is chosen so that for any ϕ which satisfies the appropriate boundary condition there is a unique transformation ξ^a which vanishes on the boundary such that $B_{abcd}(\phi^{cd} + 2\xi^{(c;d)}) = 0$. We shall choose the harmonic gauge in the background metric

$$B_{abcd} = \tfrac{1}{4} g_{bd} \nabla_a \nabla_c - \tfrac{1}{8} g_{cd} \nabla_a \nabla_b - \tfrac{1}{8} g_{ab} \nabla_c \nabla_d + \tfrac{1}{16} g_{ab} g_{cd} \nabla_e \nabla^e$$
$$+ (a \leftrightarrow b) + (c \leftrightarrow d) + (a \leftrightarrow b, c \leftrightarrow d)\,. \tag{3.12}$$

The operator $A + B$ will in general have no zero eigenvalues. However, $\det(A + B)$ contains the eigenvalues of the arbitrarily chosen operator B. To cancel them out one has to divide by the determinant of B on the subspace of all ϕ^{ab} which are pure gauge transformations, i.e. of the form $\phi^{ab} = 2\xi^{(a;b)}$ for some ξ^a which vanish on the boundary. The determinant of B on this subspace is equal to the square of the determinant of the operator C on the space of all vector fields which vanish on the boundary, where

$$C_{ab}\xi^b = -g_{0ad}(\xi^{d;b} + \xi^{b;d} - g_0^{db}\xi^c{}_{;c})_{;b}$$
$$= -g_{0ad}(\xi^{d;b}{}_{;b} + R^d_b\xi^b) . \qquad (3.13)$$

Thus one obtains

$$\log Z = -\hat{I}[g_0] - \tfrac{1}{2}\log\det(\mu^{-2}(A + B)) + \log\det(\mu^{-2}C) . \qquad (3.14)$$

The last term is the so-called ghost contribution.

The operator $A + B$ is

$$\tfrac{1}{4}g_{bd}\nabla_a\nabla_c - \tfrac{1}{4}g_{ad}\nabla_c\nabla_b + \tfrac{1}{8}(g_{ac}g_{bd} - \tfrac{1}{2}g_{ab}g_{cd})\,\nabla_e\nabla^e + \tfrac{1}{2}R_{ad}g_{bc} - \tfrac{1}{4}R_{ab}g_{cd}$$
$$+ \tfrac{1}{16}Rg_{ad}g_{bc} - \tfrac{1}{8}Rg_{ac}g_{bd} + (a \leftrightarrow b) + (c \leftrightarrow d) + (a \leftrightarrow b, c \leftrightarrow d) , \qquad (3.15)$$

and will in general have only non-zero eigenvalues. It will have negative eigenvalues corresponding to the conformal transformations $\phi^{ab} \propto g_0^{ab}$. This means that one cannot apply the zeta function technique [4] directly to evaluate $\det(A + B)$. However, one sees from eq. (3.15) that if g_0 is a solution of the vacuum Einstein equations, $A + B$ can be expressed as $-F + G$; where

$$F = \tfrac{1}{16}\nabla_a\nabla^a \qquad (3.16)$$

operates on the trace of ϕ^{ab}, $\phi = \phi^{ab}g_{0ab}$, and

$$G_{abcd} = \tfrac{1}{8}(g_{ac}g_{bd} + g_{bc}g_{ad})\,\nabla_e\nabla^e + \tfrac{1}{4}(R_{acbd} + R_{adbc}) \qquad (3.17)$$

operates on the tracefree part of ϕ^{ab}, $\tilde{\phi}^{ab} = \phi^{ab} - \tfrac{1}{4}g_0^{ab}\phi$.

The operator F is positive definite. In order to make the path integral over the conformal factor ϕ converge, one has to turn the contour of integration for the coefficients of ϕ in terms of the eigenfunctions of F to lie along the imaginary axis. Because one has to do this at all solutions of the Einstein equations, including flat space, we adopt this as the general form. The operator G is positive at flat space. We therefore adopt the prescription that the coefficients of $\tilde{\phi}_{ab}$ in terms of the eigenfunctions of G should be integrated along the real axis. However, at other solutions of the Einstein equations, such as the Schwarzschild solution, G may have a finite number, n, of negative eigenvalues because of the Riemann tensor term. For each of these negative eigenvalues λ_i, one will have to rotate the contour of integration of the corresponding coefficient a_i to lie along the imaginary axis. This will introduce a factor of $(+i)^n$ into Z, where n is the number of such eigenvalues.

Let $\tilde{G} = G - P$, where P denotes projection on any eigenfunctions with negative or zero eigenvalues of G. Then F, G and C are all positive definite in a metric g_0 which is a solution of the Einstein equations. Thus, one can evaluate their determinants by the zeta function technique [4]. One forms a zeta function from the eigenvalues $\{\lambda_n(L)\}$ of an operator L

$$\zeta(s, L) = \sum_n \lambda_n^{-s}(L) . \tag{3.18}$$

This will converge for Re $s > 2$ and can be analytically continued to a meromorphic function analytic at $s = 0$. The derivative of ζ at $s = 0$ is formally equal to $-\Sigma_n \log \lambda_n$, so det L is defined to be $\exp(-\zeta'(0))$. Thus the path integral gives

$$\begin{aligned}\log Z = & -\hat{I}[g_0] + \tfrac{1}{2}\zeta'(0, F) + \tfrac{1}{2}\zeta'(0, \widetilde{G}) - \zeta'(0, C) \\ & -\tfrac{1}{2}\sum_i \log \lambda_i \\ & + \tfrac{1}{2}\log(\mu^2)(\zeta(0, F) + \zeta(0, \widetilde{G}) + n - 2\zeta(0, C)) \\ & + \tfrac{1}{2} in\pi .\end{aligned} \tag{3.19}$$

Using asymptotic expansion of the heat kernel one can evaluate the zeta functions at $s = 0$. From the results of Gilkey [20] one has

$$\begin{aligned}\zeta(0, F) &= \frac{1}{2880\pi^2}\int R_{abcd}R^{abcd}(g_0)^{1/2}\, d^4x , \\ \zeta(0, \widetilde{G}) + n &= \frac{21}{320\pi^2}\int R_{abcd}R^{abcd}(g_0)^{1/2}\, d^4x , \\ \zeta(0, C) &= \frac{-11}{2880\pi^2}\int R_{abcd}R^{abcd}(g_0)^{1/2}\, d^4x .\end{aligned} \tag{3.20}$$

In each case there is a possible boundary term which may vanish in the asymptotically flat case. These values determine the behaviour of $\log Z$ under scale transformations of the boundary metric, $\widetilde{h} = kh$, where k is a constant. The new background metric will then be $\widetilde{g}_0 = kg_0$ and $\hat{I}[\widetilde{g}_0] = k\hat{I}[g_0]$. The eigenvalues of the operators F, G and C will be multiplied by k^{-1}. Since the normalization quantity μ has been assumed to be independent of position, it would seem reasonable to assume that it also is independent of the scale of the background metric. Thus, the new value of Z is given by

$$\log \widetilde{Z} = \log Z + (1 - k)\,\hat{I}[g_0] + \tfrac{1}{2}\log k(\zeta(0, F) + \zeta(0, \widetilde{G}) + n - \zeta(0, C)) . \tag{3.21}$$

The last term in brackets is equal to

$$\frac{53}{720\pi^2}\int R_{abcd}R^{abcd}(g_0)^{1/2}\, d^4x .$$

The fact that this is positive implies that Z will be very small for small scales ($k \ll 1$). This same quantity appears as the trace of the effective energy-momentum tensor of the single closed loop defined by

$$T_{ab} = -2(g_0)^{-1/2} \frac{\delta \log Z}{\delta g_0^{ab}} . \tag{3.22}$$

Then

$$T_a{}^a = \frac{53}{720\pi^2} R_{abcd} R^{abcd} \tag{3.23}$$

For a compact space which is a solution of the vacuum Einstein equations, the Euler number is given by

$$\chi = \frac{1}{32\pi^2} \int R_{abcd} R^{abcd} (g_0)^{1/2} \, d^4x . \tag{3.24}$$

Thus the scaling behaviour is determined by the topology. In the non-compact case there are surface terms both in the scaling behaviour and in the Euler number. These terms will both be zero for spatially asymptotically flat metrics like the Schwarzschild solution.

Note added in proof

The boundary term in (1.2) has also been discussed by York [21]. A discussion of trace anomalies together with further references is given by Duff [22]. R. Schoen and S.-T. Yau have recently announced a rigorous proof of the positive energy conjecture for data on maximal slices which are homeomorphic to R^3.

References

[1] G.W. Gibbons and S.W. Hawking, Phys. Rev. D15 (1977) 2752.
[2] S.W. Hawking, Phys. Rev. D13 (1976) 191.
[3] J.S. Dowker and R. Critchley, Phys. Rev. D13 (1976) 3224.
[4] S.W. Hawking, Comm. Math. Phys. 55 (1977) 133.
[5] H-S. Tsao, Phys. Lett. B68 (1977) 79.
[6] R. Critchley, The trace anomaly for gravitons, University of Manchester preprint (1977).
[7] D. Brill, Ann. of Phys. 7 (1959) 466.
[8] D. Brill and S. Deser, Ann of Phys. 50 (1968) 548.
[9] R. Geroch, J. Math. Phys. 13 (1972) 956; Ann N.Y. Acad. Sci. 224 (1973) 108.
[10] P.S. Jang and R. Wald, J. Math. Phys. 18 (1977) 41.
[11] D.N. Page, The positive action conjecture, DAMTP Cambridge preprint (1977), Phys. Rev. D., to appear.
[12] S.W. Hawking, Quantum gravity, Talk given at GR8, Waterloo, Ontario, 1977, Phys. Rev. D., to appear.
[13] S.W. Hawking, Phys. Rev. D13 (1976) 191.
[14] G.W. Gibbons and M.J. Perry, Proc. Roy. Soc. A358 (1978) 467.

[15] S.W. Hawking, Phys. Lett. A60 (1977) 81.
[16] R.P. Feynmann, Quantizing the gravititional and Yang-Mills fields, in Magic without magic, ed. J.R. Klauder (W.H. Freeman, San Francisco, 1972).
[17] B.S. DeWitt, The dynamical theory of groups and fields, in Relativity, groups and topology, ed. C. and B.S. DeWitt (Gordon and Breach, New York, 1964).
[18] L. Fadeev and V.N. Popov, Phys. Lett. B25 (1967) 29.
[19] G. 't Hooft, Nucl. Phys. B33 (1971) 173.
[20] P. Gilkey, J. Diff. Geom. 10 (1975) 601.
[21] J.W. York, Phys. Rev. Lett. 28 (1972) 1082.
[22] M.J. Duff, Nucl. Phys. B125 (1977) 334.

PHYSICAL REVIEW LETTERS

VOLUME 42 26 FEBRUARY 1979 NUMBER 9

Proof of the Positive-Action Conjecture in Quantum Relativity

Richard M. Schoen
Department of Mathematics, University of California, Berkeley, California 94720

and

Shing-Tung Yau
Department of Mathematics, Stanford University, Stanford, California 94305
(Received 27 November 1978)

We extend our previous method of proving the positive-mass conjecture to prove the positive-action conjecture of Hawking for asymptotically Euclidean metric. This result is crucial in proving the path integral convergent in the Euclidean quantum gravity theory.

In Hawking's[1] Euclidean approach to quantum gravity, the space of all four-dimensional Riemannian metrics is divided into conformal classes, each of which has a representative of zero scalar curvature. Then an important conjecture in this theory is the positivity of the action for any asymptotically Euclidean metric with zero scalar curvature. In this Letter, we settle this conjecture in the affirmative by extending our previous method of proving the positive-mass conjecture.[2-4] However, we emphasize here that a direct extension of our previous method will not work because of dimensional reasons. (Later on we will indicate how counter examples arise at several delicate points of our previous approach. These examples show that one cannot simply wave hands on these points without going through more delicate arguments.)

Let M be a four-dimensional asymptotically Euclidean manifold with m asymptotic regions. Then in each asymptotic region there is a coordinate chart such that the metric is given by

$$(1+4I/3\pi r^2)(dx^2+dy^2+dz^2+dt^2)+O(r^{-3}), \qquad (1)$$

where $r^2=x^2+y^2+z^2+t^2$. Our method shows that if the Ricci scalar curvature of M is nonnegative everywhere, then I is positive unless M is the flat Euclidean metric. This, of course, settles the positive-action conjecture in the affirmative.

Below we outline the procedure of the demonstration. The delicate points will be discussed here and will be presented in detail elsewhere.

As in Ref. 4, we divide the proof into two parts. The first step is to prove that the action is nonnegative. The second step is to prove that if the action is zero, then the metric is flat. Both of these two parts are much more subtle than the previous paper because of the dimension reason.

As in Ref. 4, in the first step, we assume that the action is negative and then derive a contradiction. In this case, we conformally deform our metric as in Ref. 4 to make the scalar curvature $R>0$ outside a compact set. At this point one would think that one can argue as in Ref. 4 to establish the existence of a complete minimal hypersurface lying between two hyperplanes by solving the boundary-value problem and taking the limit. This turns out to be a *serious* mistake because of dimensional reasons. In Ref. 4, the existence of such a hypersurface comes from the negativity of the mass. However, in the case under consideration here, we can prove the existence of such a hypersurface whether the action is negative or not. Hence, the existence of such an object will not create a contradiction to the negativity of the action in contrast to the argument in Ref. 4.

In order to overcome this difficulty, we solve the *free* boundary-value problem for the minimal hypersurface in an expanding sequence of cylinders in the asymptotically Euclidean region. We will present the existence proof elsewhere (which is by no means trivial). Then also by a rather *nontrivial* analysis, we show that the complete minimal hypersurface H so constructed is still asymptotically flat in the sense of Ref. 4 and has vanishing Arnowitt-Deser-Misner[5] mass. [Asymptotically flat in the sense of Ref. 4 means that H is diffeomorphic to R^3 and the metric has the form $(1+m/2r)^4\delta_{ij}+O(r^{-2})$.]

Since H is obtained by a minimizing procedure, we know that the second variation of it must be nonnegative. By a similar manipulation as in Ref. 4 and using the fact that the scalar curvature of M is positive outside a compact set, we know that

$$\int_H(|\nabla f|^2+\bar{R}f^2)\geq 0, \qquad (2)$$

where $\bar{R}$ is the scalar curvature of H and f belongs to the class of smooth functions which are asymptotically constant at infinity of H. Furthermore, when f is nonzero outside a compact set of H, we can assume (2) to be a strict inequality. (The advantage of our way of producing H is that we have a wider class of f whereas in the former approach, f must approach zero at infinity of H.)

Clearly, (2) implies

$$\int_H(8|\nabla f|^2+\bar{R}f^2)\geq 0. \qquad (3)$$

Now, we multiply the metric of H by u^4, where u is a function solving the equation (with boundary values asymptotic to 1 at infinity)

$$-\nabla^2 u+\tfrac{1}{8}\bar{R}u=0. \qquad (4)$$

The resulting metric has zero scalar curvature, and by studying Eq. (4) *carefully*, we can prove that the Arnowitt-Deser-Misner mass of H is given by

$$-(4\pi)^{-1}\int_H(|\nabla u|^2+\tfrac{1}{8}\bar{R}u^2), \qquad (5)$$

which is negative by (3) and the fact that u is not zero outside a compact set of H. This contradicts our theorem on the positivity of the mass.

In step 2, we can argue, as in Ref. 4, that when the action is zero the Ricci tensor of M must be identically zero. In three dimensions, this implies that M is flat. In our case, the argument is not trivial. If H is compact, one may try to argue by some variation formula of H. However, when H is not compact, this is not possible and we proceed as follows. Using the assumption that M is asymptotically Euclidean, we construct a geodesic line in M so that each segment of this geodesic minimizes length. Then one can use the Ricci flat assumption to prove that M is flat.

Some of our arguments have worked for locally asymptotically flat manifolds and we shall discuss this case elsewhere.

This research was supported in part by the National Science Foundation under Grant No. MCS78-04872.

[1]S. W. Hawking, Phys. Rev. D 18, 1747 (1978).

[2]R. Schoen and S.-T. Yau, Proc. Nat. Acad. Sci. U. S. A. 75, 2567 (1978).

[3]R. Schoen and S.-T. Yau, "Existence of Incompressible Minimal Surfaces and the Topology of Three Dimensional Manifolds with Non-negative Scalar Curvature" (to be published).

[4]R. Schoen and S.-T. Yau, "On the Proof of the Positive Energy Conjecture in General Relativity" (to be published).

[5]R. Arnowitt, S. Deser, and C. W. Misner, Phys. Rev. 118, 1100 (1960).

Commun. math. Phys. 55, 133—148 (1977)

Communications in
Mathematical
Physics

Zeta Function Regularization of Path Integrals in Curved Spacetime

S. W. Hawking

Department of Applied Mathematics and Theoretical Physics, University of Cambridge, Cambridge CB3 9EW, England

Abstract. This paper describes a technique for regularizing quadratic path integrals on a curved background spacetime. One forms a generalized zeta function from the eigenvalues of the differential operator that appears in the action integral. The zeta function is a meromorphic function and its gradient at the origin is defined to be the determinant of the operator. This technique agrees with dimensional regularization where one generalises to n dimensions by adding extra flat dimensions. The generalized zeta function can be expressed as a Mellin transform of the kernel of the heat equation which describes diffusion over the four dimensional spacetime manifold in a fith dimension of parameter time. Using the asymptotic expansion for the heat kernel, one can deduce the behaviour of the path integral under scale transformations of the background metric. This suggests that there may be a natural cut off in the integral over all black hole background metrics. By functionally differentiating the path integral one obtains an energy momentum tensor which is finite even on the horizon of a black hole. This energy momentum tensor has an anomalous trace.

1. Introduction

The purpose of this paper is to describe a technique for obtaining finite values to path integrals for fields (including the gravitational field) on a curved spacetime background or, equivalently, for evaluating the determinants of differential operators such as the four-dimensional Laplacian or D'Alembertian. One forms a gemeralised zeta function from the eigenvalues λ_n of the operator

$$\zeta(s) = \sum_n \lambda_n^{-s} . \tag{1.1}$$

In four dimensions this converges for $\mathrm{Re}(s) > 2$ and can be analytically extended to a meromorphic function with poles only at $s=2$ and $s=1$. It is regular at $s=0$. The derivative at $s=0$ is formally equal to $-\sum_n \log \lambda_n$. Thus one can define the determinant of the operator to be $\exp(-d\zeta/ds)|_{s=0}$.

In situations in which one knows the eigenvalues explicitly one can calculate the zeta function directly. This will be done in Section 3, for the examples of thermal radiation or the Casimir effect in flat spacetime. In more complicated situations one can use the fact that the zeta function is related by an inverse Mellin transform to the trace of the kernel of the heat equation, the equation that describes the diffusion of heat (or ink) over the four dimensional spacetime manifold in a fifth dimension of parameter time t. Asymptotic expansions for the heat kernel in terms of invariants of the metric have been given by a number of authors [1–4].

In the language of perturbation theory the determinant of an operator is expressed as a single closed loop graph. The most commonly used method for obtaining a finite value for such a graph in flat spacetime is dimensional regularization in which one evaluates the graph in n spacetime dimensions, treats n as a complex variable and subtracts out the pole that occurs when n tends to four. However it is not clear how one should apply this procedure to closed loops in a curved spacetime. For instance, if one was dealing with the four sphere, the Euclidean version of de Sitter space, it would be natural to generalize that S^4 to S^n [5, 6]. On the other hand if one was dealing with the Schwarzschild solution, which has topology $R^2 \times S^2$, one might generalize to $R^2 \times S^{n-2}$. Alternatively one might add on extra dimensions to the R^2. These additional dimensions might be either flat or curved. The value that one would obtain for a closed loop graph, would be different in these different extensions to n dimensions so that dimensional regularization is ambiguous in curved spacetime. In fact it will be shown in Section 5 that the answer given by the zeta function technique agrees up to a multiple of the undetermined renormalization parameter with that given by dimensional regularization where the generalization to n dimensions is given by adding on extra flat dimensions.

The zeta function technique can be applied to calculate the partition functions for thermal gravitons and matter quanta on black hole and de Sitter backgrounds. It gives finite values for these despite the infinite blueshift of the local temperature on the event horizons. Using the asymptotic expansion for the heat kernel, one can relate the behaviour of the partition function under changes of scale of the background spacetime to an integral of a quadratic expression in the curvature tensor. In the case of de Sitter space this completely determines the partition function up to a multiple of the renormalization parameter while in the Schwarzschild solution it determines the partition function up to a function of r_0/M where r_0 is the radius of the box containing a black hole of mass M in equilibrium with thermal radiation. The scaling behaviour of the partition function suggests that there may be a natural cut off at small masses when one integrates over all masses of the black hole background.

By functional differentiating the partition function with respect to the background metric one obtains the energy momentum tensor of the thermal radiation. This can be expressed in terms of derivatives of the heat kernel and is finite even on the event horizon of a black hole background. The trace of the energy momentum is related to the behaviour of the partition function under scale transformations. It is given by a quadratic expression in the curvature and is non zero even for conformally invariant fields [7–12].

The effect of the higher order terms in the path integrals is discussed in Section 9. They are shewn to make an insignificant contribution to the partition function for thermal radiation in a black hole background that is significantly bigger than the Planck mass. Generalised zeta functions have also been used by Dowker and Critchley [11] to regularize one-loop graphs. Their approach is rather different from that which will be given here.

2. Path Integrals

In the Feynmann sum over histories approach to quantum theory one considers expressions of the form

$$Z=\int d[g]d[\phi]\exp\{iI[g,\phi]\}\,, \tag{2.1}$$

where $d[g]$ is a measure on the space of metrics g, $d[\phi]$ is a measure on the space of matter fields ϕ and $I[g,\phi]$ is the action. The integral is taken over all fields g and ϕ that satisfy certain boundary or periodicity conditions. A situation which is of particular interest is that in which the fields are periodic in imaginary time on some boundary at larg distance with period β [13]. In this case Z is the partition function for a canonical ensemble at the temperature $T=\frac{1}{\beta}$.

The dominant contribution to the path integral (2.1) will come from fields that are near background fields g_0 and ϕ_0 which satisfy the boundary or periodicity conditions and which extremise the action i.e. they satisfy the classical field equations. One can expand the action in a Taylor series about the background fields:

$$I[g,\phi]=I[g_0,\phi_0]+I_2[\tilde{g}]+I_2[\tilde{\phi}]+\text{higher order terms}, \tag{2.2}$$

where

$$g=g_0+\tilde{g}\,,\qquad \phi=\phi_0+\tilde{\phi}$$

and $I_2[\tilde{g}]$ and $I_2[\tilde{\phi}]$ are quadratic in the fluctuations $\tilde{g}$ and $\tilde{\phi}$. Substituting (2.2) into (2.1) and neglecting the higher order terms one has

$$\begin{aligned}\log Z=&\,iI[g_0,\phi_0]+\log\int d[\tilde{g}]\exp iI_2[\tilde{g}]\\&+\log\int d[\tilde{\phi}]\exp iI_2[\tilde{\phi}]\,.\end{aligned} \tag{2.3}$$

The background metric g_0 will depend on the situation under consideration but in general it will not be a real Lorentz metric. For example in de Sitter space one complexifies the spacetime and goes to a section (the Euclidean section) on which the metric is the real positive definite metric on a four sphere. Because the imaginary time coordinate is periodic on this four sphere, Z will be the partition function for a canonical ensemble. The action $I[g_0,\phi_0]$ of the background de Sitter metric gives the contribution of the background metric to the partition function while the second and third terms in Equation (2.3) give the contributions of thermal gravitons and matter quanta respectively on this background. In the case of the canonical ensemble for a spherical box with perfectly reflecting walls the background metric can either be that of a Euclidean space or it can be that of a section (the Euclidean section) of the complexified Schwarzschild solution on which the metric is real positive definite. Again the action of the background metric gives the contribution of the background metric to the partition function. This corresponds to an entropy equal to one quarter of the area of the event horizon in units in which $G=c=\hbar=k=1$. The second and third terms in Equation (2.3) give the contributions of thermal gravitons and matter quanta on a Schwarzschild background. In the case of the grand canonical ensemble for a box with temperature $T=\beta^{-1}$ and angular velocity Ω one considers fields which, on the walls of the box, have the same value at the point (t, r, θ, ϕ) and at the point $(t+i\beta, r, \theta, \phi+i\beta\Omega)$. This boundary cannot be filled in with any real metric but it can be filled in with a complex flat metric or with a complex section (the quasi Euclidean section [13]) of the Kerr solution. In both cases the metric is strongly elliptic (I am grateful to Dr. Y. Manor for this point) [14] if the rotational velocity of the boundary is less than that of light. A metric g is said to be strongly elliptic if there is a function f such that $\mathrm{Re}(fg)$ is positive definite. It seems necessary to use such strongly elliptic background metrics to make the path integrals well defined. One could take this to be one of the basic postulates of quantum gravity.

The quadratic term $I_2[\phi]$ will have the form

$$I_2[\tilde{\phi}] = -\int \tfrac{1}{2}\tilde{\phi} A \tilde{\phi} (-g_0)^{1/2} d^4x\,, \tag{2.4}$$

where A is a second order differential operator constructed out of the background fields g_0, ϕ_0. (In the case of the fermion fields the operator A is first order. For simplicity I shall deal only with boson fields but the results can easily be extended to fermions.) The quadratic term $I_2[\tilde{g}]$ in the metric fluctuations can be expressed similarly. Here however, the second order differential operator is degenerate i.e. it does not have an inverse. This is because of the gauge freedom to make coordinate transformations. One deals with this by taking the path integral only over metrics that satisfy some gauge condition which picks out one metric from each equivalence class under coordinate transformations. The Jacobian from the space of all metrics to the space of those satisfying the gauge condition can be regarded in perturbation theory as introducing fictitious particles known as Feynmann-de Witt [15, 16] or Fadeev-Popov ghosts [17]. The path integral over the gravitational fluctuations will be treated in another paper by methods similar to those used here for matter fields without gauge degrees of freedom.

In the case when the background metric g_0 is Euclidean i.e. real and positive definite the operator A in the quadratic term $I_2[\phi]$ will be real, elliptic and self-adjoint. This means that it will have a complete spectrum of eigenvectors ϕ_n with real eigenvalues λ_n:

$$A\phi_n = \lambda_n \phi_n\,. \tag{2.5}$$

The eigenvectors can be normalized so that

$$\int \phi_n \phi_m (g_0)^{1/2} d^4x = \delta_{nm}\,. \tag{2.6}$$

Note that the volume element which appears in the (2.6) is $(g_0)^{1/2}$ because g_0 is positive definite. On the other hand the volume element that appears in the action I is $(-g)^{1/2} = -i(g)^{1/2}$ where the minus sign corresponds to a choice of the direction of Wick rotation of the time axis into the complex plane.

If the background metric g_0 is not Euclidean, the operator A will not be self-adjoint. However I shall assume that the eigen functions ϕ_n are still complete. If this is so, one can express the fluctuation $\tilde{\phi}$ in terms of the eigen functions.

$$\tilde{\phi} = \sum a_n \phi_n \,. \tag{2.7}$$

The measure $d[\phi]$ on the space of all fields $\tilde{\phi}$ can then be expressed in terms of the coefficients a_n:

$$\mathrm{d}[\phi] = \prod_n \mu da_n \,, \tag{2.8}$$

where μ is some normalization constant with dimensions of mass or inverse length. From (2.5)–(2.8) it follows that

$$\begin{aligned} Z[\tilde{\phi}] &= \int d[\phi] \exp i I_2[\tilde{\phi}] \\ &= \prod_n \int \tfrac{1}{2}\mu da_n \exp(-\lambda_n a_n^2) \\ &= \prod_n \tfrac{1}{2}\mu\pi^{1/2}\lambda_n^{-1/2} \\ &= (\det(4\mu^{-2}\pi^{-1}A))^{-1/2} \,. \end{aligned} \tag{2.9}$$

3. The Zeta Function

The determinant of the operator A clearly diverges because the eigenvalues λ_n increase without bound. One therefore has to adopt some regularization procedure. The technique that will be used in this paper will be called the zeta function method. One forms a generalized zeta function from the eigenvalues of the operator A:

$$\zeta(s) = \sum_n \lambda_n^{-s} \,. \tag{3.1}$$

In four dimensions this will converge for $\mathrm{Re}(s) > 2$. It can be analytically extended to a merophorphic function of s with poles only at $s=2$ and $s=1$ [18]. In particular it is regular at $s=0$. The gradient of zeta at $s=0$ is formally equal to $-\sum_n \log\lambda_n$. One can therefore *define* $\det A$ to be $\exp(-d\zeta/ds|_{s=0})$ [19]. Thus the partition function

$$\log Z[\tilde{\phi}] = \tfrac{1}{2}\zeta'(0) + \tfrac{1}{2}\log(\tfrac{1}{4}\pi\mu^2)\zeta(0) \,. \tag{3.2}$$

In situations in which the eigenvalues are known, the zeta function can be computed explicitly. To illustrate the method, I shall treat the case of a zero rest mass scalar field ϕ contained in a box of volume V in flat spacetime at the temperature $T = \beta^{-1}$. The partition function will be defined by a path integral over all fields ϕ on the Euclidean space obtained by putting $\tau = it$ which are zero on the walls of the box and which are periodic in τ with period β. The operator A in the action is the negative of the four dimensional Laplacian on the Euclidean space. If the dimensions of the box are large compared to the characteristic wavelength β, one can approximate the spatial dependence of the eigenfunctions by plane waves with periodic boundary conditions. The eigenvalues are then

$$\lambda_n=(2\pi\beta^{-1}n)^2+k^2 \tag{3.3}$$

and the density of eigenvalues in the continuum limit is

$$\frac{2V}{(2\pi)^3}\int d^3k \tag{3.4}$$

when $n>0$ and half that when $n=0$. The zeta function is therefore

$$\zeta(s)=\frac{4\pi V}{(2\pi)^3}\left\{\int dkk^{2-2s}+2\sum_{n=1}^{\infty}\int dkk^2(4\pi^2\beta^{-2}n^2+k^2)^{-s}\right\}. \tag{3.5}$$

The second term can be integrated by parts to give

$$-\frac{8\pi V}{(2\pi)^3}\sum_{n=1}^{\infty}\int dk(4\pi^2\beta^{-2}n^2+k^2)^{-s+1}(2-2s)^{-1}. \tag{3.6}$$

Put $k=2\pi n\beta^{-1}\sinh y$. This gives

$$\begin{aligned}&-\frac{8\pi V}{(2\pi)^3}\sum_{n=1}^{\infty}\int dy(2\pi\beta^{-1}n)^{-2s+3}(2-2s)^{-1}(\cosh y)^{-2s+3}\\ &=-\frac{8\pi V}{(2\pi)^3}(2\pi\beta^{-1})^{3-2s}\times\zeta_R(2s-3)\\ &\quad(2-2s)^{-1}\times\frac{1}{2}\,\frac{\Gamma(1/2)\Gamma(s-3/2)}{\Gamma(s-1)},\end{aligned} \tag{3.7}$$

where ζ_R is the usual Riemann zeta function $\sum_n n^{-s}$. The first term in (3.5) seems to diverge at $k=0$ when s is large and positive. This infra red divergence can be removed if one assumes that the box containing the radiation is large but finite. In this case the k integration has a lower cut off at some small value ε. If s is large, the k integration then gives a term proportional to ε^{3-2s}. When analytically continued to $s=0$, this can be neglected in the limit $\varepsilon\to 0$, corresponding to a large box.

The gamma function $\Gamma(s-1)$ has a pole at $s=0$ with residue -1. Thus the generalised zeta function is zero at $s=0$ and

$$\begin{aligned}\zeta'(0)&=2\pi V\beta^{-3}\zeta_R(-3)\Gamma(1/2)\Gamma(-3/2)\\ &=\frac{\pi^2}{45}VT^3\end{aligned} \tag{3.8}$$

thus the partition function for scalar thermal radiation at temperature T in a box of volume V is given by

$$\log Z=\frac{\pi^2VT^3}{90}. \tag{3.9}$$

Note that because $\zeta(0)=0$, the partition function does not depend on the undetermined normalization parameter μ. However, this will not in general be the case in a curved space background.

From the partition function one can calculate the energy, entropy and pressure of the radiation.

$$E=-\frac{d}{d\beta}\log Z=\frac{\pi^2}{30}VT^4\,, \tag{3.10}$$

$$S=\beta E+\log Z=\frac{2\pi^2}{45}VT^3\,, \tag{3.11}$$

$$P=\beta^{-1}\frac{d}{dV}\log Z=\frac{\pi^2}{90}T^4\,. \tag{3.12}$$

One can calculate the partition functions for other fields in flat space in a similar manner. For a charged scalar field there are twice the number of eigenfunctions so that $\log Z$ is twice the value given by Equation (3.9). In the case of the electromagnetic field the operator A in the action integral is degenerate because of the freedom to make electromagnetic gauge transformations. One therefore has, as in the gravitational case, to take the path integral only over fields which satisfy some gauge condition and to take into account the Jacobian from the space of all fields satisfying the gauge condition. When this is done one again obtains a value $\log Z$ which is twice that of Equation (3.9). This corresponds to the fact that the electromagnetic field has two polarization states.

One can also use the zeta function technique to calculate the Casimir effect between two parallel reflecting planes. In this case instead of summing over all field configurations which are periodic in imaginary time, one sums over fields which are zero on the plates. Defining Z to be the path integral over all such fields over an interval of imaginary time τ one has

$$\log Z=\frac{\pi^2 A\tau b^{-3}}{720}\,, \tag{3.13}$$

where b is the separation and A the area of the plates. Thus the force between the plates is

$$F=\tau^{-1}\frac{d}{db}\log Z=-\frac{\pi^2 Ab^{-4}}{240}\,. \tag{3.14}$$

4. The Heat Equation

In situations in which one does not know the eigenvalues of the operator A, one can obtain some information about the generalized zeta function by studying the heat equation.

$$\frac{d}{dt}F(x,y,t)+AF(x,y,t)=0 \tag{4.1}$$

here x and y represent points in the four dimensional spacetime manifold, t is a fifth dimension of parameter time and the operator A is taken to act on the first argument of F. With the initial conditions

$$F(x,y,0)=\delta(x,y) \tag{4.2}$$

the heat kernel F represents the diffusion over the spacetime manifold in parameter time t of a unit quantity of heat (or ink) placed at the point y at $t=0$. One can express F in terms of the eigenvalues and eigenfunctions of A:

$$F(x,y,t)=\sum_n \exp(-\lambda_n t)\phi_n(x)\phi_n(y)\,. \tag{4.3}$$

In the case of a field ϕ with tensor or spinor indices, the eigenfunctions will carry a set of indices at the point x and a set at the point y. If one puts $x=y$, contracts over the indices at x and y and integrates over all the manifold one obtains

$$Y(t)\equiv\int \mathrm{Tr}F(x,x,t)(g_0)^{1/2}d^4x=\sum_n \exp(-\lambda_n t)\,. \tag{4.4}$$

The generalized zeta function is related to $Y(t)$ by a Mellin transform:

$$\zeta(s)=\sum_n{}' \lambda_n^{-s}=\frac{1}{\Gamma(s)}\int_0^\infty t^{s-1}Y(t)dt\,. \tag{4.5}$$

A number of authors e.g. [1–4] have obtained asymptotic expansions for F and Y valid as $t\to 0^+$. In the case that the operator A is a second order Laplacian type operator on a four dimensional compact manifold.

$$Y(t)\sim\sum_n B_n t^{n-2}\,, \tag{4.6}$$

where the coefficients B_n are integrals over the manifold of scalar polynomials in the metric, the curvature tensor and its covariant derivatives, which are of order $2n$ in the derivatives of the metric

i.e. $$B_n=\int b_n(g_0)^{1/2}d^4x\,. \tag{4.7}$$

DeWitt [1,2] has calculated the b_n for the operator $-\Box+\xi R$ acting on scalars,

$$\begin{aligned} b_0&=(4\pi)^{-2}\\ b_1&=(4\pi)^{-2}(\tfrac{1}{6}-\xi)R\\ b_2&=(2880\pi^2)^{-1}\\ &\quad\cdot[R^{abcd}R_{abcd}-R^{ab}R_{ab}+30(1-6\xi)^2R^2+(6-30\xi)\Box R]\,. \end{aligned} \tag{4.8}$$

Note that b_1 is zero when $\xi=\frac{1}{6}$ which corresponds to a conformally invariant scalar field.

In the case of a non-compact spacetime manifold one has to impose boundary conditions on the heat equation and on the eigenfunctions of the operator A. This can be done by adding a boundary to the manifold and requiring the field or its normal derivative to be zero on the boundary. An example is the case of a black hole metric such as the Euclidean section of the Schwarzschild solution in which one adds a boundary at some radius $r=r_0$. This boundary represents the walls of a perfectly reflecting box enclosing the black hole. For a manifold with boundary the asymptotic expansion for Y takes the form [20].

$$Y(t)=\sum_n (B_n+C_n)t^{n-2}\,, \tag{4.9}$$

where, as before, B_n has the form (4.7) and

$$C_n=\int c_n(h)^{1/2}d^3x\,,$$

where c_n is a scalar polynomial in the metric, the normal to the boundary and the curvature and their covariant derivatives of order $2n-1$ in the derivatives of the metric and h is the induced metric on the boundary. The first coefficient c_0 is zero because their is no polynomial of order -1. McKean and Singer [3] showed that $c_1=\frac{-1}{48\pi^2}K$ when $\xi=0$ where K is the trace of the second fundamental form of the boundary. In the case of a Schwarzschild black hole in a spherical box of radius r_0, c_2 must be zero in the limit of large r_0 because all polynomials of degree 3 in the derivatives of the metric go down faster than r_0^{-2}.

In a compact manifold with or without boundary with a strongly elliptical metric g_0 the eigenvalues of a Laplacian type operator A will be discrete. If there are any zero eigenvalues they have to be omitted from the definition of the generalized zeta function and dealt with separately. This can be done by defining a new operator $\tilde{A}=A-P$ where P denotes projection on the zero eigenfunctions. Zero eigenvalues have important physical effects such as the anomaly in the axial vector current conservation [21, 22]. Let $\varepsilon>0$ be the lowest eigenvalue of $\tilde{A}$ (from now on I shall simply use A and assume that any zero eigenfunctions have been projected out). Then

$$\zeta(s)=\frac{1}{\Gamma(s)}\left[\int_0^1 t^{s-1}Y(t)dt+\int_1^\infty t^{s-1}Y(t)dt\right]. \tag{4.10}$$

As $t\to\infty$, $Y\to e^{-\varepsilon t}$. Thus the second integral in Equation (4.10) converges for all s. In the first integral one can use the asymptotic expression (4.9). This gives

$$\sum_n \frac{B_n+C_n}{n+s-2}\,. \tag{4.11}$$

Thus ζ has a pole at $s=2$ with residue B_0 and a pole at $s=1$ with residue B_1+C_1. There would be a pole at $s=0$ but it is cancelled out by the pole in $\Gamma(s)$. Thus $\zeta(0)=B_2+C_2$. Similarly the values of ζ at negative integer values of s are given by (4.11) and (4.10).

5. Other Methods of Regularization

A commonly used method to evaluate the determinant of the operator A is to start with the integrated heat kernel

$$Y(t)=\sum_n \exp(-\lambda_n t)\,. \tag{5.1}$$

Multiply by $\exp(-m^2 t)$ and integrate from $t=0$ to $t=\infty$

$$\int_0^\infty Y(t)\exp(-m^2 t)dt = \sum_n{}' (\lambda_n + m^2)^{-1} \tag{5.2}$$

then integrate over m^2 from $m^2=0$ to $m^2=\infty$ and interchange the orders of integration to obtain

$$\int_0^\infty t^{-1} Y(t)dt = \left[\sum_n{}' \log(\lambda_n + m^2)\right]_0^\infty . \tag{5.3}$$

One then throws away the value of the righthand side of (5.3) at the upper limit and claims that

$$\begin{aligned}\log \det A &= \sum \log\lambda_n \\ &= -\int_0^\infty t^{-1} Y(t)dt .\end{aligned} \tag{5.4}$$

This is obviously a very dubious procedure. One can obtain the same result from the zeta function method in the following way. One has

$$\begin{aligned}\log \det A &= -\zeta'(0) \\ &= -\frac{d}{ds}\left[\frac{1}{\Gamma(s)}\int_0^\infty t^{s-1} Y(t)dt\right] .\end{aligned} \tag{5.5}$$

Near $s=0$

$$\frac{1}{\Gamma(s)} = s + \gamma s^2 + O(s^3) , \tag{5.6}$$

where γ is Euler's constant.

Thus

$$\begin{aligned}\log \det A = -\lim_{s\to 0}\Big[&(1+2\gamma s)\int_0^\infty t^{s-1} Y(t)dt \\ &+(s+\gamma s^2)\int_0^\infty t^{s-1} \log t\, Y(t)dt\Big] .\end{aligned} \tag{5.7}$$

If one ignores the fact that the two integrals in Equation (5.7) diverged when $s=0$, one would obtain Equation (5.4). Using the asymptotic expansion for Y, one sees that the integral in Equation (5.4) has a t^{-2}, t^{-1}, and a $\log t$ divergence at the lower limit with coefficients $\frac{1}{2}B_0$, B_1, and B_2 respectively. The first of these is often subtraced out by adding an infinite cosmological constant to the action while the second is cancelled by adding an infinite multiple of the scalar curvature which is interpreted as a renormalization of the gravitational constant. The logarithmic term requires an infinite counter term of a new type which is quadratic in the curvature.

To obtain a finite answer from Equation (5.4) dimensional regularization is often used. One generalizes the heat equation from 4+1 dimensions to $2\omega+1$ dimensions and then subtracts out the pole that occurs in (5.4) at $2\omega=4$. As mentioned in the introduction, this is ambiguous because there are many ways that one could generalize a curved spacetime to 2ω dimensions. The simplest generalization would be to take the product of the four dimensional spacetime manifold with $2\omega-4$ flat dimensions. In this case the integrated heat kernel Y would be multiplied by $(4\pi t)^{2-\omega}$. Then (5.4) would become

$$\log \det A = -\int_0^\infty t^{1-\omega}(4\pi)^{2-\omega}Y(t)dt\,. \tag{5.8}$$

This has a pole at $2\omega=4$ with residue $\zeta(0)$ and finite part $-\zeta'(0)+(2\gamma+\log 4\pi)\times\zeta(0)$. Thus, the value of the $\log Z$ derived by the dimensional regularization using flat dimensions agrees with the value obtained by the zeta function method up to a multiple of $\zeta(0)$ which can be absorbed in the normalization constant. However, if one extended to $2\omega+1$ dimensions in some more general way than merely adding flat dimensions, the integrated heat kernel would have the form

$$Y(t_0\omega)=\sum_n B_n(\omega)t^{n-\omega}\,, \tag{5.9}$$

where the coefficients $B_n(\omega)$ depend on the dimensions 2ω. The finite part at $\omega=2$ would then acquire an extra term $B'_2(2)$. This could not be absorbed in the normalization constant μ. One therefore sees that the zeta function method has the conceptual advantages that it avoids the dubious procedures used to obtain Equation (5.4), it does not require the subtraction of any pole term or the addition of infinite counter terms, and it is unambiguous unlike dimensional regularization which depends on how one generalizes to 2ω dimensions.

6. Scaling

In this Section I shall consider the behaviour of the partition function Z under a constant scale transformation of the metric

$$\tilde{g}_{ab}=kg_{ab}\,. \tag{6.1}$$

If A is a Laplacian type operator for a zero rest mass field, the eigenvalues transform as

$$\lambda_n=k^{-1}\lambda_n\,. \tag{6.2}$$

Thus the new generalized zeta function is

$$\tilde{\zeta}(s)=k^s\zeta(s) \tag{6.3}$$

and

$$\log\det\tilde{A}=\log\det A-\log k\zeta(0)\,. \tag{6.4}$$

Thus

$$\log\tilde{Z}=\log Z+\tfrac{1}{2}\log k\zeta(0) + (\log\tilde{\mu}-\log\mu)\zeta(0)\,. \tag{6.5}$$

If one assumed that the normalization constant μ remained unchanged under a scale transformation, the last term would vanish. This assumption is equivalent to assuming that the measure in the path integral over all configurations of the field ϕ is defined not on a scalar field but on a scalar density of weight $\frac{1}{2}$. This is because the eigenfunctions of the operator A would have to transform according to

$$\tilde{\phi}_n = k^{-1}\phi_n \tag{6.6}$$

in order to maintain the normalization condition (2.6). The coefficients a_n of the expansion of a given scalar field ϕ would therefore transform according to

$$\tilde{a}_n = k a_n \tag{6.7}$$

and the normalization constant μ would transform according to

$$\tilde{\mu} = k^{-1}\mu \tag{6.8}$$

if the measure is defined on the scalar field itself, i.e. if

$$d[\phi] = \Pi_x d\phi(x)\,. \tag{6.9}$$

However if the measure is defined on densities of weight $\frac{1}{2}$, i.e.

$$d[\phi] = \Pi_x (g(x))^{1/4} d\phi(x) \tag{6.10}$$

then the normalization parameter μ is unchanged.

The weight of the measure can be deduced from considerations of unitarity. In the case of a scalar field one can use the manifestly unitary formalism of summing over all particle paths. This gives the conformally invariant scalar wave equation if the fields are taken to be densities of weight $\frac{1}{2}$ [23]. By contrast, the "minimally coupled" wave equation $\Box\phi = 0$ will be obtained if the weight is 1. In the case of a gravitational field itself one can use the unitary Hamiltonian formalism. From this Fadeev and Popov [17] deduce that the measure is defined on densities of weight $\frac{1}{2}$ and is scale invariant. Similar procedures could be used to find the weight of the measure for other fields. One would expect it to be $\frac{1}{2}$ for massless fields.

These scaling arguments give one certain amounts of information about the partition function. In DeSitter space they determine it up to the arbitrariness of the normalization parameter μ because DeSitter space is completely determined by the scale. Thus

$$\log Z = B_2 \log r/r_0\,, \tag{6.11}$$

where r is the radius of the space and r_0 is related to μ. In the case of a Schwarzschild black hole of mass M in a large spherical box of radius r_0,

$$\log Z = B_2 \log M/M_0 + f(r_0 M^{-1})\,, \tag{6.12}$$

where again M_0 is related to μ. If the radius of the box is large compared to M, one would expect that the partition function should approach that for thermal radiation at temperature $T = (8\pi M)^{-1}$ in flat space. Thus one would expect

$$f = \frac{r_0^3}{34560 M^3} + O\left(\frac{r_0^2}{M^2}\right). \tag{6.13}$$

It should be possible to verify this and to calculate the lower order terms by developing suitable approximations to the eigenvalues of the radial equation in the Schwarzschild solution. In particular f and $\log Z$ will be finite. This contrasts with the result that one would obtain if one naively assumed that the thermal radiation could be described as a fluid with a density of $\log Z$ equal to $\pi^2/_{90}\bar{T}^3$ where $\bar{T}=T(1-2Mr^{-1})^{-1/2}$ is the local temperature. Near the horizon $\bar{T}$ would get very large because of a blueshift effect and so $\log Z$ would diverge.

For a conformally invariant scalar field $B_2=-\frac{2}{45}$ for DeSitter space and $\frac{1}{45}$ for the Schwarzschild solution. The fact that B_2 is positive in the latter case may provide a natural cut off in the path integral when one integrates over background metrics will all masses M. If the measure on the space of gravitational fields is scale invariant then the action of the background fields will give an integral of the form

$$\int_0^\infty \exp(-4\pi M^2)M^{-1}dM\,. \tag{6.14}$$

This converges nicely at large M but has a logarithmic divergence at $M=0$. However if one includes a contribution of the thermal radiation the integral is modified to

$$\int_0^\infty \exp(-4\pi M^2)M^{-1+B_2}dM\,. \tag{6.15}$$

This converges if B_2 is positive. Such a cut off can however be regarded as suggestive only because it ignores the contributions of high order terms which will be important near $M=0$. One might hope that these terms might in turn be represented by further black hole background metrics.

7. Energy-Momentum Tensor

By functionally differentiating the partition function one obtains the energy momentum tensor of the thermal radiation

$$T_{ab}=-2(g_0)^{-1/2}\frac{\delta\log Z}{\delta g_0^{ab}}\,. \tag{7.1}$$

The energy momentum tensor will be finite even on the event horizon of a black hole background metric despite the fact that the blueshifted temperature $\bar{T}$ diverges there. This shows that the energy momentum tensor cannot be that of a perfect fluid with pressure equal to one third the energy density.

One can express the energy momentum tensor in terms of derivatives of the heat kernel F:

$$\delta\log Z=\tfrac{1}{2}\delta\zeta'(0)-\mu^{-1}\delta\mu\zeta(0)-\tfrac{1}{2}\log(\tfrac{1}{4}\pi\mu^2)\delta\zeta(0)\,. \tag{7.2}$$

The second term on the right of (7.2) will vanish if one assumes that μ does not change under variations of the metric. This will be the case if the measure is defined on densities of weight $\frac{1}{2}$. The third term can be expressed as the variation of an integral quadratic in the curvature tensor and can be evaluated directly. To calculate the first term one writes

$$\zeta'(0) = \frac{d}{ds}\left[\frac{1}{\Gamma(s)}\int\int_0^\infty t^{s-1}F(x,x,t)(g_0)^{1/2}d^4x\,dt\right]\Bigg|_{s=0} . \tag{7.3}$$

Therefore

$$\delta\zeta'(0) = \frac{d}{ds}\left[\frac{1}{\Gamma(s)}\int\int_0^\infty t^{s-1}\delta[F(x,x,t)(g_0)^{1/2}d^4x\,dt\right]\Bigg|_{s=0} . \tag{7.4}$$

To calculate δF one uses the varied heat equation

$$\left(A + \frac{\partial}{\partial t}\right)\delta F(x,y,t) + \delta A F(x,y,t) = 0 \tag{7.5}$$

with $\delta[(g_0(y))^{1/2}F(x,y,0)] = 0$. The solution is

$$\delta[(g_0(y))^{1/2}F(x,y,t)] = -\int\int_0^t F(x,z,t-t')\delta A F(z,y,t')g_0(y)g_0(z)^{1/2}d^4z\,dt' . \tag{7.6}$$

Therefore

$$\delta\int F(x,x,t)(g_0)^{1/2}d^4x = -t\int \delta A F(z,z,t)(g_0)^{1/2}d^4z . \tag{7.7}$$

Where the operator δA acts on the first argument of F.

The operator δA involves δg^{ab} and its covariant derivatives in the background metric. Integrating by parts, one obtains an expression for T^{ab} in terms of F and its covariant derivatives. For a conformally invariant scalar field.

$$\begin{aligned} T_{ab} = \frac{d}{ds}&\left[\frac{1}{\Gamma(s)}\int_0^\infty t^s\left(\tfrac{2}{3}{}_aF_b - \tfrac{1}{6}g_{ab\,c}F^c - \tfrac{1}{3}F_{ab}\right.\right.\\ &\left.\left. + \tfrac{1}{3}g_{ab}F^c_c + \tfrac{1}{6}R_{ab}F - \tfrac{1}{12}g_{ab}RF\right)dt\right]\\ &- \log\left(\tfrac{1}{4}\pi\mu^2\right)\frac{\delta B_2}{\delta g_{ab}}(g_0)^{-1/2} . \end{aligned} \tag{7.8}$$

Where indices placed before or after F indicates differentiation with respect to the first or second arguments respectively and the two arguments are taken at the point x at which the energy momentum tensor is to be evaluated. In an empty spacetime the quantity B_2 is the integral of a pure divergence so B_2 vanishes.

8. The Trace Anomaly

Naively one would expect T^a_a, the trace of the energy momentum tensor, would be zero for a zero rest mass field. However this is not the case as can be seen either directly from (7.8) or by the following simple argument. Consider a scale transformation in which the metric is multiplied by a factor $k=1+\varepsilon$. Then $\delta g_{ab}=\varepsilon g_{ab}$ and

$$\begin{aligned}\int T^a_a(g_0)^{1/2}d^4x &= 2\frac{d\log Z}{dk}\\ &= B_2(1+\tfrac{1}{2}\mu^{-1}d\mu/dk)\\ &= B_2 \qquad (8.1)\end{aligned}$$

if the measure is defined on densities of weight $\frac{1}{2}$. Thus for the case of a conformally invariant scalar field

$$T^a_a = \frac{1}{2880\pi^2}[R_{abcd}R^{abcd} - R_{ab}R^{ab} + \Box R]\ . \qquad (8.2)$$

The trace anomalies for other zero rest mass fields can be calculated in a similar manner.

These results for the trace anomaly agree with those of a number of other authors [7–12]. However, they disagree with some calculations by the point separation method [24] which do not obtain any anomaly. The trace anomaly for DeSitter completely determines the energy momentum because it must be a multiple of the metric by the symmetry. In a two dimensional black hole in a box the trace anomaly also determines the energy momentum tensor and in the four dimensional case it determines it up to one function of position [25].

9. Higher Order Terms

The path integral over the terms in the action which are quadratic in the fluctuations about the background fields are usually represented in perturbation theory by a single closed loop without any vertices. Functionally differentiating with respect to the background metric to obtain the energy momentum tensor corresponds to introducing a vertex coupling the field to the gravitational field. If one then feeds this energy momentum tensor as a perturbation back into the Einstein equations for the background field, the change in the $\log Z$ would be described by a diagram containing two closed loops each with a gravitational vertex and with the two vertices joined by a gravitational propagator. Under a scale transformation in which the metric was multiplied by a constant factor k, such a diagram would be multiplied by k^{-2}. Another diagram which would have the same scaling behaviour could be obtained by functionally differentiating $\log Z$ with respect to the background metric at two different points and then connecting these points by a gravitational propagator. In fact all the higher order terms have scaling behaviour k^{-n} where $n \geq 2$. Thus one would expect to make a negligible contribution to the partition function for black holes of significantly more than the Planck mass. The higher order terms will however be important near the Planck mass and will cause the scaling argument in Section 6 to break down. One might nevertheless hope that just as a black hole background metric corresponds to an infinite sequence of higher order terms in a perturbation expansion around flat space, so the higher order terms in expansion about a black hole background might in turn be represented by more black holes.

Acknowledgement. I am grateful for discussions with a number of colleagues including G. W. Gibbons, A. S. Lapedes, Y. Manor, R. Penrose, M. J. Perry, and I. M. Singer.

References

1. DeWitt, B.S.: Dynamical theory of groups and fields in relativity, groups and topology (eds. C. M. and B. S. DeWitt). New York: Gordon and Breach 1964
2. DeWitt, B.S.: Phys. Rep. **19** C, 295 (1975)
3. McKean, H.P., Singer, J.M.: J. Diff. Geo. **5**, 233—249 (1971)
4. Gilkey, P.B.: The index theorem and the heat equation. Boston: Publish or Perish 1974
5. Candelas, P., Raine, D.J.: Phys. Rev. D **12**, 965—974 (1975)
6. Drummond, I.T.: Nucl. Phys. **94** B, 115—144 (1975)
7. Capper, D., Duff, M.: Nuovo Cimento **23** A, 173 (1974)
8. Duff, M., Deser, S., Isham, C.J.: Nucl. Phys. **111** B, 45 (1976)
9. Brown, L.S.: Stress tensor trace anomaly in a gravitational metric: scalar field. University of Washington, Preprint (1976)
10. Brown, L.S., Cassidy, J.P.: Stress tensor trace anomaly in a gravitational metric: General theory, Maxwell field. University of Washington, Preprint (1976)
11. Dowker, J.S., Critchley, R.: Phys. Rev. D **13**, 3224 (1976)
12. Dowker, J.S., Critchley, R.: The stress tensor conformal anomaly for scalar and spinor fields. University of Manchester, Preprint (1976)
13. Gibbons, G.W., Hawking, S.W.: Action integrals and partition functions in quantum gravity. Phys. Rev. D (to be published)
14. Manor, Y.: Complex Riemannian sections. University of Cambridge, Preprint (1977)
15. Feynman, R. P.: Magic without magic, (eds. J. A. Wheeler and J. Klaunder). San Francisco: W. H. Freeman 1972.
16. DeWitt, B.S.: Phys. Rev. **162**, 1195—1239 (1967)
17. Fadeev, L.D., Popov, V.N.: Usp. Fiz. Nauk **111**, 427—450 (1973) [English translation in Sov. Phys. Usp. **16**, 777—788 (1974)]
18. Seeley, R.T.: Amer. Math. Soc. Proc. Symp. Pure Math. **10**, 288—307 (1967)
19. Ray, D.B., Singer, I.M.: Advances in Math. **7**, 145—210 (1971)
20. Gilkey, P.B.: Advanc. Math. **15**, 334—360 (1975)
21. 't Hooft, G.: Phys. Rev. Letters **37**, 8—11 (1976)
22. 't Hooft, G.: Computation of the quantum effects due to a four dimensional pseudoparticle. Harvard University, Preprint
23. Hartle, J.B., Hawking, S.W.: Phys. Rev. D **13**, 2188—2203 (1976)
24. Adler, S., Lieverman, J., Ng, N.J.: Regularization of the stress-energy tensor for vector and scalar particles. Propagating in a general background metric. IAS Preprint (1976)
25. Fulling, S. A., Christensen, S.: Trace anomalies and the Hawking effect. Kings College London, Preprint (1976)

Communicated by R. Geroch

Received February 10, 1977

PHYSICAL REVIEW D VOLUME 36, NUMBER 8 15 OCTOBER 1987

Conformal rotation in perturbative gravity

Kristin Schleich
Department of Physics, University of Chicago, Chicago, Illinois 60637
and Department of Physics, University of California, Santa Barbara, California 93106
(Received 25 June 1987)

The classical Euclidean action for general relativity is unbounded below; therefore Euclidean functional integrals weighted by this action are manifestly divergent. However, as a consequence of the positive-energy theorem, physical amplitudes for asymptotically flat spacetimes can indeed be expressed as manifestly convergent Euclidean functional integrals formed in terms of the physical degrees of freedom. From these integrals, we derive expressions for these same physical quantities as Euclidean integrals over the full set of variables for gravity computed as metric perturbations off a flat background. These parametrized Euclidean functional integrals are weighted by manifestly positive actions with rotated conformal factors. They are similar in form to Euclidean functional integrals obtained by the Gibbons-Hawking-Perry prescription of contour rotation.

I. INTRODUCTION

When searching for a quantum theory to describe a physical system, one in general starts by attempting to quantize its classical counterpart. In the case of gravity, such attempts have met with many difficulties of both a conceptual and technical nature. This has led to proposals for alternate quantum theories for gravity, such as string theories,[1] that hold promise for alleviating at least some of these difficulties. However, although Einstein gravity may not provide the correct theory of quantum gravity, it is the correct theory in the low-energy limit. This means that an understanding of its quantum properties is interesting and useful as a qualitative guide to the low-energy limit of proposed quantum theories for gravity. In addition, quantized Einstein gravity is a useful tool in investigation of quantum cosmology and in minisuperspace models, where one is interested in the qualitative behavior of the theory below the Planck scale.

In order to study the quantum mechanics of a theory, one needs to construct quantities such as wave functions describing the possible states of the system and transition amplitudes between these states. There are two parts to this; first one needs to develop formal expressions for these quantities that incorporate the kinematics of the theory. In general, these expressions will contain ultraviolet divergences. Therefore, next one needs to regulate and renormalize these formal expressions in order to get physical answers. In the case of Einstein gravity, the theory calculated perturbatively is nonrenormalizable.[2] However, quantities can still be computed to a given order in perturbation theory by including the appropriate counterterms; the problem is that in order to compute to all orders, an infinite number of different counterterms must be included. However, in order to begin to regulate and renormalize, one must have the results of the first step; formal expressions for the physical quantities.

One productive method of formulating the kinematics of a theory is to express such quantities as functional integrals, an approach that is especially convenient for further formal manipulation. Functional integrals directly implement the sum over histories formulation of quantum mechanics which connects the quantum amplitudes to the classical action. They are especially useful in the case of theories with local invariances because functional integrals for quantum amplitudes can be formulated to manifestly display these invariances. Lorentzian functional integrals,

$$\int d\phi(x)\exp\{iS[\phi(x)]\} , \tag{1.1}$$

give quantities such as transition amplitudes by summing over the appropriate class of field configurations weighted by the classical action $S[\phi]$. Euclidean functional integrals, which involve sums over field configurations weighted by the classical Euclidean action $I[\phi]$,

$$\int d\phi(x)\exp\{-I[\phi(x)]\} , \tag{1.2}$$

express ground-state wave functions or generating functions in a form useful for actual computations.

One proposed approach to quantizing Einstein gravity is to use Euclidean functional integrals to construct the states of the theory.[3–5] In this approach one forms these integrals using the Euclidean action for general relativity. The appropriate action when the induced three-metric h_{ij} is fixed on the boundary is

$$l^2I[g]=-\int_M d^4x\, g^{1/2}R-2\int_{\partial M} d^3x\, h^{1/2}K , \tag{1.3}$$

where $l=(16\pi G)^{1/2}$ is the Planck length in the units $\hbar=c=1$ and K is the extrinsic curvature of the boundary hypersurface ∂M. Immediately, there is a difficulty; unlike the Euclidean actions for more familiar gauge theories such as electromagnetism, that for gravity is not positive definite. This can be seen by writing the metric $g_{\alpha\beta}$ in terms of a metric $\tilde{g}_{\alpha\beta}$ in the conformal equivalence class of $g_{\alpha\beta}$ and a conformal factor Ω:

$$g_{\alpha\beta}=\Omega^2\tilde{g}_{\alpha\beta} . \tag{1.4}$$

This decomposition is fixed by requiring $\tilde{g}_{\alpha\beta}$ to satisfy a coordinate invariant condition of the form

$$R(\tilde{g})=0 \tag{1.5}$$

and fixing boundary conditions on Ω such as $\Omega=1$ on ∂M. In these variables the action (1.3) becomes

$$l^2 I[\tilde{g},\Omega]=-\int_M d^4x\, \tilde{g}^{1/2}[\Omega^2 R(\tilde{g})+6(\nabla\Omega)^2] -2\int_{\partial M} d^3x\, h^{1/2}K \;. \tag{1.6}$$

It is readily apparent that (1.6) will become arbitrarily negative for Ω that vary rapidly enough. Consequently Euclidean functional integrals for gravity of the form (1.2) weighted by (1.6) will be manifestly divergent.[6] This divergence is one appearing in the kinematical formulation of the theory; it is not related to the ultraviolet divergences. It must be taken care of first before one can regulate and renormalize these quantities.

In order to construct convergent Euclidean functional integrals for the kinematics of the theory, Gibbons, Hawking, and Perry proposed an additional formal manipulation called conformal rotation. First change the variables of integration in the functional integral from $g_{\alpha\beta}$ to Ω and $\tilde{g}_{\alpha\beta}$ which satisfy the condition (1.5). Next distort the contour of the Ω integration to complex values. The action (1.6) then becomes positive definite; the integration over the conformal factor becomes manifestly positive and for asymptotically flat spacetimes the positive action theorem[7] guarantees the positivity of the surface term. Consequently, the resulting Euclidean functional integral is then convergent.

Conformal rotation provides a method of forming convergent Euclidean functional integrals for gravity. However it does so by starting with a divergent Euclidean gravitational integral, a quantity that does not really exist, and manipulating it to produce a convergent one. This manipulation is not needed to construct Euclidean functional integrals for more familiar theories with invariances such as electromagnetism and Yang-Mills theories; the classical Euclidean actions of these theories are manifestly positive. Therefore it would be useful to have a more physically based motivation for Euclidean gravitational integrals in their conformally rotated form. In this paper we will provide such motivation. In doing so we will concentrate on the kinematical formulation of the theory; in the rest of the paper we will not discuss regulating and renormalizing these quantities. Instead we will assume that these procedures can be carried out following the standard methods to handle the ultraviolet divergences in Einstein gravity as needed. Therefore, in this paper the term convergence will refer to the formal properties of the Euclidean functional integrals alone. To begin, we will first construct physical quantities as manifestly convergent Euclidean functional integrals from the fundamental formulation of the quantum theory in terms of its physical degrees of freedom. We expect that we can do so for asymptotically flat spacetimes by virtue of the positive-energy theorem.[8,9] Then starting from these integrals in the physical variables we derive convergent parametrized Euclidean functional integrals with rotated conformal factors for the same physical quantities. First we will do so for the theory of linearized gravity, reviewing previous work done in collaboration with Hartle.[10] We will then discuss how to extend this to perturbative gravity, by which we mean Einstein gravity in asymptotically flat spacetimes when the metric configurations are treated as perturbations on a flat background.

Classically, theories with local invariances such as gauge or parametrized theories are usually given in a form in which not all field configurations are physically distinct. In electromagnetism for example, fields A_μ that differ by a gauge transformation, $A_\mu \rightarrow A_\mu + \partial_\mu \Lambda$, are physically equivalent. The classical dynamics of such theories is summarized in an action that is a local functional of the field variables and this is also manifestly invariant under both the local invariance and Lorentz transformations. However, the initial data cannot be freely specified for theories with local invariances because it has to be compatible with the invariance for the evolution of the system to be consistent. This means that the full set of variables contains redundant fields as well as the physical degrees of freedom. The physical components are invariant; their initial data can be freely given. The rest are redundant variables needed to display the symmetry of the theory in a local set of fields. In electromagnetism, the physical fields are the transverse components of the vector potential A_i^T; the longitudinal and time components, A_i^L and A_0, are redundant variables which change under gauge transformations. As the action is manifestly invariant, its physical content can be expressed in terms of the physical variables alone. However, in general this form of the action will be nonlocal in the original potentials and will not display all of the invariances of the theory. For example, the action for electromagnetism can be written in terms of A_i^T alone; however this form is neither local nor manifestly Lorentz invariant.

Although a theory with invariances is most elegantly presented classically using redundant variables, its quantum mechanics is really based on the dynamics as given in terms of the physical degrees of freedom. Physical amplitudes can be constructed as sums over histories in the physical degrees of freedom weighted by the physical action. For example, the ground-state wave functional for electromagnetism can be given as a Euclidean functional integral over A_i^T weighted by the action of the theory expressed in terms of A_i^T. When the physical variables can be explicitly solved for, as in electromagnetism, these integrals are similar in form to those of scalar field theory. These integrals for quantum amplitudes contain the correct physical content of the theory; however, they usually do not display all its invariances in the most transparent form. They also typically do not present the theory in a tractable form for calculation when the physical fields cannot be explicitly isolated. Therefore it is useful to have expressions for these same quantities as functional integrals over the full set of variables. Starting with the integrals in terms of the physical degrees of freedom, one inserts additional integrals over the redundant variables to recover expressions displaying the original invariance and locality of the classical theory. These redundant integrals must be added in a way that leaves the value unchanged, so that the resulting parametrized integrals give the same amplitudes as those in the physical variables. This process can be demonstrated explicitly when the physical degrees of freedom can be explicitly isolated; it can be carried through formally if they cannot. This procedure is most familiar for Lorentzian functional integrals; it provides the correct form and measure for Lagrangian integrals in the full set of variables from the Hamiltonian form given in the physical fields.[11,12] It can be carried out for Euclidean functional integrals as well.

Euclidean functional integrals for both linearized gravity and perturbative gravity are convergent when given in the physical variables. This is manifestly so for linearized gravity; it is implied by the positive-energy theorem for asymptotically flat spacetimes in the interacting case. We shall therefore study how to relate these convergent integrals in the physical degrees of freedom to those over the full set of variables in order to gain a more physically based understanding of conformal rotation. We will show that indeed parametrized Euclidean functional integrals weighted by positive actions for physical quantities can be derived from those given in terms of the physical degrees of freedom.

Before we proceed to treat gravity itself, it is useful to introduce some of the basic techniques in a simpler context. In Sec. II we shall use a simple model to illustrate some basic issues in adding redundant variables to Euclidean functional integrals. Next, in Sec. III we shall derive convergent Euclidean functional integrals for the ground state given in terms of the physical variables for both linearized gravity and perturbative gravity. Because the physical variables can be explicitly solved for in the linearized case, the connection between the parametrized Euclidean functional integral and that in the physical variables can be carried out explicitly. In Sec. IV we shall make this connection and derive the conformally rotated Euclidean functional integral for the ground state of linearized gravity. As the physical variables cannot be explicitly solved for in the case of the interacting theory, it is not practical to parametrize the Euclidean functional integral for the ground state directly. In Sec. V we will discuss how to proceed in this case and derive conformally rotated Euclidean functional integrals for gravity when the metric configurations can be treated as interacting perturbations on a flat background.

II. FUNCTIONAL INTEGRALS FOR THEORIES WITH INVARIANCES

We will first discuss quantizing theories with invariances using a simple model which is a generalization of one discussed by Hartle and Kuchař.[13,14] It allows us to outline the basic techniques in the more familiar context of single-particle quantum mechanics. (1) Beginning with the classical theory expressed in its manifestly gauge-invariant form, one first isolates the physical degrees of freedom and expresses the dynamics in terms of them. (2) One next formulates the quantum theory as functional integrals in the physical variables weighted by the appropriate physical action (the one that gives the dynamics in the physical variables). (3) Finally, one adds in integrations over the redundant variables to recover the manifest invariance expressed in the full set of variables. One does this in such a way that the resulting parametrized functional integrals equal those given in the physical variables. The model will be too simple to adequately illustrate all of the issues, but will be a useful conceptual guide when we turn to the case of gravity.

Let us consider a system consisting of n variables $q^a(t)$ which are the physical degrees of freedom and two variables $\phi(t)$ and $\lambda(t)$ which are the redundant variables. The Lagrangian is

$$L=l(q^a,\dot{q}^a)+l^g(\phi,\dot{\phi},\lambda)\ , \tag{2.1a}$$

where

$$l(q^a,\dot{q}^a)=\tfrac{1}{2}\delta_{ab}\dot{q}^a\dot{q}^b-V(q^a)\ , \tag{2.1b}$$

$$l^g(\phi,\dot{\phi},\lambda)=\tfrac{1}{2}\mu(\dot{\phi}-\lambda)^2-\kappa(\dot{\phi}-\lambda)F(q^a)\ . \tag{2.1c}$$

L is invariant under the transformation

$$\phi(t)=\phi(t)+\Lambda(t)\ , \tag{2.2a}$$

$$\lambda(t)=\lambda(t)+\dot{\Lambda}(t)\ , \tag{2.2b}$$

which serves as the analog of a gauge transformation in the model. Because of the invariance, there will be a constraint on the initial data for the model; it has to be consistent with (2.2) so that the system will evolve preserving it. Of course, gauge theories are not usually written in a set of variables in which the physical degrees of freedom are obvious. More typically they are described in a set of fields in which the gauge-invariant components are not already isolated but in which the invariance is manifest. However, the model does display the basic content of a gauge theory, albeit in a very simple form.

A useful way of displaying the constraint on the initial data and showing how it is preserved by the dynamics is to study the system in its Hamiltonian form. In the following we will briefly discuss constrained Hamiltonian dynamics; we refer the reader to the literature[15-17] for a more thorough and elegant presentation.

One first solves for the momenta conjugate to the variables in terms of the velocities using $p_a=\partial L/\partial\dot{q}^a$. Immediately, one finds that this Legendre transformation is singular; λ has no conjugate momenta as it occurs in (2.1a) without time differentiation. This is a primary constraint on the system:

$$p_\lambda=\frac{\partial L}{\partial\dot{\lambda}}=0\ . \tag{2.3}$$

It reflects the fact that λ is not a dynamical variable; it is a Lagrange multiplier. The rest of the momenta can be solved for in terms of the velocities. What this means is that the classical phase space of the system is smaller than what one would naively guess; it consists of the variables q^a and ϕ and their respective conjugate momenta p_a and π. One finds that the Hamiltonian corresponding to (2.1a) is

$$H=h(q^a,p_a)+h^g(\lambda,\pi,\phi)+\lambda\pi\ , \tag{2.4a}$$

$$h(q^a,p_a)=\tfrac{1}{2}\delta^{ab}p_a p_b+V(q^a)\,, \tag{2.4b}$$

$$h^g(\lambda,\pi,\phi,q^a)=\frac{1}{2\mu}[\pi+\kappa F(q^a)]^2\,. \tag{2.4c}$$

Infinitesimal canonical transformations of the phase-space variables are implemented by taking their Poisson brackets with the generator of the transformation. The brackets are defined by

$$\{A,B\}=\sum_\alpha\left[\frac{\partial A}{\partial p_\alpha}\frac{\partial B}{\partial q^\alpha}-\frac{\partial A}{\partial q^\alpha}\frac{\partial B}{\partial p_\alpha}\right], \tag{2.5a}$$

where α labels the n variables q^a and ϕ. The fundamental brackets between the canonical variables is

$$\{p_\alpha,q^\beta\}=-\delta^\beta_\alpha\,. \tag{2.5b}$$

Of special interest are canonical transformations that give the invariances and dynamics of the system. In particular, Hamilton's equations of motion, $\dot p_\alpha=-\partial H/\partial q^\alpha$ and $\dot q^\alpha=\partial H/\partial p_\alpha$, follow from the brackets of the variables with H, which is the generator of time evolution.

The initial conditions for solving Hamilton's equations are given by a point in the phase space of the system (p_α,q^α). However, this point cannot be freely specified; in order for the system to evolve consistently it must stay in the phase space, so (2.3) must be preserved. Consequently, the equations of motion imply that the momentum conjugate to ϕ vanishes:

$$\{p_\lambda,H\}=\pi=0\,. \tag{2.6}$$

How is this constraint related to the invariance (2.2) of the model? It generates the infinitesimal gauge transformations of the canonical variables. The variable ϕ transforms as in (2.2a):

$$\delta_\Lambda\phi=\{\phi,\pi\}\Lambda=\Lambda\,. \tag{2.7}$$

The variables p_a and q^a are unchanged; they are gauge invariant. Again for consistency, Eq. (2.6), called a secondary or dynamical constraint, must also be preserved in time. It is, because $\{\pi,H\}=0$. Therefore we have found all the constraints needed for consistent evolution of the model.

What (2.6) tells us is that not all regions of phase space are allowed by the dynamics. The allowed region consists of configurations of p_a, q^a, and ϕ with $\pi=0$. When restricted to this region, the Hamiltonian

$$h^P=\tfrac{1}{2}\delta^{ab}p_a p_b+V(q^a)+\frac{\kappa^2}{2\mu}F^2(q^a) \tag{2.8}$$

is a function of p_a and q^a only. The variable ϕ does not enter into the dynamics; it is a redundant variable whose value is arbitrary. Therefore the physical phase space is just (p_a,q^a); the physical content of the theory is described in terms of these variables. However, note that the presence of the interaction terms between the redundant and physical variables in the Lagrangian (2.1) has resulted in an added term of $[\kappa^2/(2\mu)]F^2$ to the potential in (2.8). It is the analog in the model of a nonlocal interaction induced when the redundant fields are eliminated in a self-interacting gauge theory. In summary, this analysis tells us that the physical degrees of freedom are p_a and q^a and their evolution is given by the physical Hamiltonian (2.8).

Having isolated the physical degrees of freedom, we see that the Hamiltonian when written as a function of the full set of variables is not unique.[17] We could, for example, add any function of the constraint to it, $\tilde H=H+g(p_\alpha,q^\alpha)\pi$, and it would still generate the same dynamics. This means that the action in the full set of variables, $S=\int dt(p_\alpha\dot q^\alpha-H)$, is also not unique if the requirements leading to its form in the original set of variables are relaxed. There are many alternate actions that will generate the same dynamics as that corresponding to (2.1). The constraints themselves can be similarly generalized; they only have to vanish up to constraints as well. This emphasizes that the physical dynamics is not contained in the canonical form of the full Hamiltonian but only in its value when the constraints are satisfied.

We have introduced a lot of analytical machinery to study the dynamics of a simple model. However, the power of the Hamiltonian formulation of constrained systems is that this kind of analysis can be carried through when the physical degrees of freedom cannot be explicitly isolated. (1) Primary constraints arise in systems with invariances because some variables are Lagrange multipliers. (2) Consistent time evolution then requires secondary constraints on the canonical variables. These constraints generate the gauge transformations of the theory. (3) The variables canonically conjugate to these secondary constraints are arbitrary parameters specified by choice of gauge. The dynamics of the system is independent of this choice as it is gauge invariant. (4) The physical phase space is the subspace of the canonical phase space that is orthogonal to both constraints and their canonically conjugate variables. The physical content of the theory is determined by the Hamiltonian on this space. We shall use the extension of this formalism to field theory to isolate the physical degrees of freedom for gravity in Sec. III.

Having explicitly reduced the model to its physical degrees of freedom we can proceed to construct quantum amplitudes as sums over histories in terms of them. We shall take the states of the model to be labeled by $|q^a,t\rangle$. The transition amplitude or propagator is then

$$\langle q'^a,t'|q^a,t\rangle=\int dp_a dq^a\exp\left[i\int_t^{t'}dt[p_a\dot q^a-h^P(p_a,q^a)]\right]. \tag{2.9}$$

The sum is over phase-space paths which begin at q^a at t and end at q'^a at t'. The measure (in which we do not display constant factors) is the canonically invariant Liouville measure on paths in the physical phase space: $dp\,dq/2\pi\hbar$. The action in the exponent is the classical action for the physical theory in Hamiltonian form. The integrand of (2.9) as it stands is purely oscillatory; in order to define the functional integral one needs to make it convergent.[18] This can be done by inserting factors of the form $\exp(-\delta\int dt\,p_a{}^2)$ where δ is a positive real constant for all the variables and taking the limit $\delta=0$

after evaluation. This prescription defines these integrals independent of the phase of the integrand. We shall assume that this procedure is implied when we write Lorentzian functional integrals in the rest of the paper.

Equation (2.9) is a formal expression for the transition amplitude; in order to make it concrete we need to spell out how to compute the sum over paths. There are various different methods for doing this. One standard way is by time slicing;[19,14] the interval $t'-t$ is divided up into N discrete time steps of length ϵ where $N\epsilon=t'-t$. By giving the value of the physical phase-space coordinates $(p_a(i),q^a(i))$ at every time step, paths can be described by straight lines connecting $(p_a(i),q^a(i))$ to $(p_a(i+1),q^a(i+1))$ for all N steps. The boundary conditions fix $q^a(0)=q^a$ and $q^a(N)=q'^a$. The functional integral (2.9) is then evaluated for N steps and the transition amplitude results in the limit as $N\to\infty$ while keeping $N\epsilon=t'-t$:

$$\langle q'^a,t'\,|\,q^a,t\rangle=\lim_{N\to\infty}\int\frac{dp_a(0)}{(2\pi)^n}\prod_{i=1}^{N-1}\left[\frac{dp_a(i)dq^a(i)}{(2\pi)^n}\right]\exp\{iS^P[p_a(i),q^a(i)]\}\ , \tag{2.10a}$$

$$S^P=\epsilon\sum_{i=0}^{N-1}\left[p_a(i)\left[\frac{q^a(i+1)-q^a(i)}{\epsilon}\right]-h^P(p_a(i),q^a(i))\right]\ . \tag{2.10b}$$

The initial $p_a(0)$ is summed over, but the final $p_a(N)$ is not because (2.10b) is independent of it. Another useful way to implement the integral is to expand the variables in a Fourier decomposition and write it as a product of integrations over mode amplitudes. We shall use this method to evaluate functional integrals for linearized gravity. The precise form of the measure in (2.9) will depend on how the sum over paths is specified; different methods contribute different factors to the measure in the same way that changing coordinates in an ordinary functional integral introduces factors of the Jacobian. From now on we will assume that some such prescription for the path integra s is supplied and concentrate instead on the main issue of constructing functional integrals for physical quantities.

Because the physical Hamiltonian of our model is quadratic in the momenta, the momentum integrals can be carried out explicitly. This results in the familiar Lagrangian form of the functional integral for the transition function

$$\langle q'^a,t'\,|\,q^a,t\rangle=\int dq^a\exp\left[i\int_t^{t'}dt\,l^P(q^a,\dot q^a)\right]\ , \tag{2.11a}$$

$$l^P(q^a,\dot q^a)=\tfrac{1}{2}\delta_{ab}\dot q^a\dot q^b-V(q^a)-\frac{\kappa^2}{2\mu}F^2(q^a)\ . \tag{2.11b}$$

The transition from (2.9) to (2.11) provides the correct form of the measure from the canonically invariant one of Hamiltonian quantum mechanics. It is especially useful in field theories where the correct form of the Lorentzian measure is not obvious.

Euclidean functional integrals provide expressions for certain states of the system; an important example is the ground-state wave function. The Euclidean functional integral for the ground state can be derived from the Feynman-Kac formula[19] and the transition amplitude (2.9) or (2.11). Given a complete set of eigenstates $\Psi_m(q^a)$ with energy E_m of the physical Hamiltonian, the transition amplitude can be written as

$$\langle q^a,0\,|\,q'^a,t\rangle=\sum_m\Psi_m(q^a)\Psi_m^*(q'^a)\exp(iE_mt)\ . \tag{2.12}$$

Next take q'^a to be at a minimum of $V(q)$, which for concreteness we take at $q'^a=0$, and rotate $t\to-i\tau$. If the energy spectrum of the physical Hamiltonian is bounded below then the dominant contribution as $\tau\to-\infty$ will be proportional to the ground state; if E_0 is renormalized to zero then one has

$$\lim_{\tau\to-\infty}\langle q^a,0\,|\,0,-i\tau\rangle\sim\Psi_0(q^a)\Psi_0^*(0) \tag{2.13}$$

as all other terms fall off exponentially. If we carry out the same procedure resulting in the Feynman-Kac formula using the path-integral form of the transition amplitude (2.11) instead of (2.12), the result is a Euclidean functional integral for the ground-state wave function up to a normalization:

$$\Psi_0(q^a)=\mathcal{N}\int dq^a\exp(-i^P[q^a])\ , \tag{2.14}$$

where i^P is the Euclidean action

$$i^P=\int_{-\infty}^{0}d\tau\left[\tfrac{1}{2}\delta_{ab}\dot q^a\dot q^b+V(q^a)+\frac{\kappa^2}{2\mu}F^2(q^a)\right]\ . \tag{2.15}$$

The class of paths summed over in (2.14) is all those matching q^a at $\tau=0$ that go to 0 in the infinite past. $\mathcal{N}$ is the normalizing constant which includes the factors needed to renormalize the ground-state energy to zero. We also have, using (2.9),

$$\Psi_0(q^a)=\mathcal{N}\int dp_a dq^a\exp\left[\int_{-\infty}^{0}d\tau[ip_a\dot q^a-h^P(p_a,q^a)]\right]\ . \tag{2.16}$$

[Note that the momenta are not rotated in passing from (2.9) to (2.14) and a divergent expression would result if they were.] The usual configuration space integral (2.14) can be derived from (2.16) by integrating over the momenta. The functional integral over the phase-space variables for the ground state (2.16) is less familiar but it is useful especially when the physical degrees of freedom cannot be isolated explicitly. One sees that if the Hamiltonian in terms of the physical variables is bounded below, then the Euclidean functional integrals of the theory will be convergent. This will be the case in the model if $V+[\kappa^2/(2\mu)]F^2$ is bounded below.

By adding integrations over the redundant variables, the functional integrals can be expressed as integrals over the full set of extended variables. It is easy to see that there are many different ways to do so, corresponding to different ways of forming identities out of these variables. Therefore, what do we want to achieve by this procedure? As stated at the beginning of this section, one would like to recover the manifest gauge invariance in the full set of variables. If possible, one would like to recover a functional integral weighted by an action that is a local functional of the original variables as well. Both of these goals can be achieved for Lorentzian functional integrals by aiming to recover an integral weighted by the classical action. Whether or not this is the case for Euclidean functional integrals will depend on the properties of the classical Euclidean action for the theory.

For our simple model, the parametrization can be carried out directly because the action we want to recover is quadratic in the redundant variables. In order to add in the redundant variables to the transition amplitude (2.11) we will use the following two integrals. First, suppose $f(x)$ is 0 for a unique value of x. Then one has the identity

$$1=\int dx\left|\frac{df}{dx}\right|\delta[f(x)]\ . \qquad (2.17)$$

The second integral is a Gaussian:

$$\exp(-ia^2)=\int_{-\infty}^{\infty}\frac{dx}{\sqrt{i\pi}}\exp(ix^2+2iax)\ . \qquad (2.18)$$

It can be verified by completing the square in the exponent. Now consider for a moment the form of the physical Lagrangian; it consists of two parts. One is $l(q^a,\dot q^a)$, which is the first of the two terms in the original Lagrangian (2.1). The other, $[\kappa^2/(2\mu)]F^2$, comes from the term coupling F to the redundant variables. This suggests combining integrals of the form (2.17) and (2.18) so that we form this extra term. Specifically, what we want is

$$\exp\left[-i\int_t^{t'}dt\frac{\kappa^2}{2\mu}F^2(q^a)\right]=\int d\phi\, d\lambda\det\left|\frac{\delta\Phi}{\delta\phi}\right|\delta[\Phi(\phi)]\exp\left[i\int_t^{t'}dt\, l^g(\phi,\dot\phi,\lambda)\right]\ , \qquad (2.19)$$

where $\Phi(\phi)$ is a function that vanishes for only one value of ϕ. Equation (2.19) can be verified by doing the integration over λ (it is a Gaussian of the form (2.18) and then doing the integration over ϕ using the δ function. It is manifestly gauge invariant; l^g (2.1c) is manifestly invariant and the determinant in (2.19) is the product of factors needed to make the integral over the δ function unity for any Φ. It is the Faddeev-Popov determinant for the analog of the gauge-fixing condition $\Phi(\phi)=0$ in the model. This formula (2.19) can be constructed explicitly using some method of summing over the paths. For example, using time slicing, one implements the determinant and δ function by using the identity (2.17) on each time slice.

If we now substitute the right-hand side of (2.19) for the left in (2.11) we arrive at the manifestly invariant functional integral for the transition amplitude

$$\langle q'^a,t'|q^a,t\rangle =\int dq^a d\phi\, d\lambda\det\left|\frac{\delta\Phi}{\delta\phi}\right|\delta[\Phi(\phi)]\exp(iS[q^a,\phi,\lambda])\ , \qquad (2.20)$$

where S is the manifestly gauge-invariant action made from the sum of l and l^g. Thus we have recovered the familiar Faddeev-Popov prescription for the functional integral for the transition amplitude in a gauge theory.[11,12]

The above analysis is not a very general or powerful way to look at adding redundant variables to a theory. A much more general way is to begin with the Hamiltonian form of the functional integral and add in integrations over the redundant phase-space variables using functional δ functions. It is an especially useful approach when the physical degrees of freedom cannot be isolated explicitly. By exponentiating the δ functions of the constraints with the Lagrange multipliers of the theory and then performing the momentum integrations, one can produce the familiar Faddeev-Popov form of functional integrals for gauge theories.[20,21] We will not discuss these methods further here because we will be doing so in the case of gravity in later sections. However it should be emphasized that the fundamental idea behind all these methods is that the quantum amplitudes are given as functional integrals in the physical variables; integrations over the redundant variables are added to recover the manifest invariance of the theory.

The ideas used to find parametrized Euclidean functional integrals from those in the physical variables are directly analogous to those in the Lorentzian case. However there is an additional restriction on the process; in order for Euclidean functional integrals for physical quantities to be well defined they must be convergent. Therefore, beginning with convergent Euclidean integrals in the physical variables, we add in convergent integrals in the redundant variables to recover well-defined parametrized Euclidean integrals that display the manifest invariance of the theory. Again we may add quantities using δ functions as in (2.17) but (2.18) becomes

$$\exp(-a^2)=\int_{-\infty}^{\infty}\frac{dx}{\sqrt{\pi}}\exp[-(x^2-2iax)]\ . \qquad (2.21)$$

Note that the integrand is complex on the right-hand side, but the integral results in a real quantity. As an example, we will consider how to derive a manifestly invariant form of the Euclidean functional integral for the ground state (2.14) using these identities. First of all, the classical form of the manifestly gauge-invariant Euclidean action corresponding to (2.1) is

$$I[q^a,\phi,\lambda]=i+I^g\ , \tag{2.22a}$$

$$i=\int d\tau \tfrac{1}{2}\delta_{ab}\dot{q}^a\dot{q}^b+V(q^a)\ , \tag{2.22b}$$

$$I^g=\int d\tau \tfrac{1}{2}\mu(\dot{\phi}-\lambda)^2-2i\kappa(\dot{\phi}-\lambda)F(q^a)\ . \tag{2.22c}$$

It is obtained from the Lorentzian action by rotating both $t\to -i\tau$ and $\lambda\to i\lambda$; the rotation of λ is required in order to preserve gauge invariance. This Euclidean action is uniquely determined by requiring gauge invariance and the same form as the Lorentzian action in the original set of variables. Note that this prescription has resulted in a complex term. Using (2.22c) as a guide, we can construct the analog of (2.19) for Euclidean functional integrals using (2.17) and (2.21):

$$\exp\left[-\int_{-\infty}^{0}\frac{\kappa^2}{2\mu}F^2(q^a)\right]=\int d\phi\, d\lambda \det\left|\frac{\delta\Phi}{\delta\phi}\right|\delta[\Phi(\phi)]\exp(-I^g[\phi,\lambda])\ . \tag{2.23}$$

This identity is true when μ is positive. However when μ is negative the Gaussian integration over the redundant variables becomes manifestly divergent. Therefore (2.23) does not exist for $\mu<0$. In that case we cannot recover a parametrized integral for the ground state weighted by the local Euclidean action (2.22a). This does not mean that Euclidean functional integrals do not exist for the theory; as we have already seen, they do in terms of the physical variables if $V+[\kappa^2/(2\mu)]F^2$ is bounded below. Nor does it mean that a manifestly invariant parametrized integral cannot be found.

Instead of adding redundant variables by using I^g we can form an identity which is a manifestly convergent Gaussian integral when $\mu<0$. It is

$$\exp\left[-\int_{-\infty}^{0}\frac{\kappa^2}{2\mu}F^2(q^a)\right]=\int d\phi\, d\lambda \det\left|\frac{\delta\Phi}{\delta\phi}\right|\delta[\Phi(\phi)]\exp(-\hat{I}^g[q^a,\lambda,\phi])\ , \tag{2.24a}$$

$$\hat{I}^g=\int_{-\infty}^{0}d\tau[-\tfrac{1}{2}\mu(\dot{\phi}-\lambda)^2+\kappa(\dot{\phi}-\lambda)F(q^a)]\ . \tag{2.24b}$$

Not only has the sign in front of μ been changed but the interaction term has also been modified from its local Euclidean form (2.22c); it is now real. This is necessary in order to get the sign of the F^2 term to be the same as that in the physical Euclidean action. Using this identity in (2.14) we find a manifestly convergent integral for the ground state:

$$\Psi_0(q^a)=\mathcal{N}\int dq^a d\phi\, d\lambda \exp(-\hat{I}[q^a,\phi,\lambda])\ , \tag{2.25a}$$

$$\hat{I}=i[q^a]+\hat{I}^g[q^a,\lambda,\phi]\ . \tag{2.25b}$$

The action that this integral is weighted by is positive, but is no longer of the same form as the Lorentzian action in the original set of variables, as reflected in the sign of μ and the form of the interaction. This change in the form of the action will correspond in the case of gravity to the Euclidean action being nonlocal in the original set of variables. It could be obtained from (2.22a) formally by performing an additional rotation on the variables $\phi\to i\phi$ and $\lambda\to i\lambda$, the analog of conformal rotation.

The action (2.25b) we obtained is not unique; there are many other positive actions that we could have used instead. For example, we could have constructed the manifestly gauge-invariant identity

$$1=\int d\phi\, d\lambda \det\left|\frac{\delta\Phi}{\delta\phi}\right|\delta[\Phi(\phi)]\times\exp\left[-\int_{-\infty}^{0}d\tau[-\tfrac{1}{2}\mu(\dot{\phi}-\lambda)^2]\right]\ . \tag{2.26}$$

Using this identity one could recover a manifestly gauge-invariant ground state of the form (2.24a) weighted instead by the action

$$\tilde{I}=i^p[q^a]+\int_{-\infty}^{0}d\tau[-\tfrac{1}{2}\mu(\dot{\phi}-\lambda)^2]\ , \tag{2.27}$$

where i^p is (2.15), the action in terms of the physical variables.

Because the interaction between the physical and redundant variables is simple in the model, appropriate convergent identities are easy to guess. For a more complicated interacting theory, the appropriate modification of the interaction terms may not be so obvious, as is the case in the full theory of gravity. The fundamental requirement is that it be chosen so that the parametrized integral equal that in terms of the physical degrees of freedom when evaluated. However, the model does demonstrate the basic principle; convergent, manifestly invariant Euclidean functional integrals can be found for the theory that equal those given in terms of the physical variables. The quantum mechanics of a system with redundant variables is really given in terms of the physical degrees of freedom. If the Hamiltonian in terms of the physical variables is bounded below, the quantum theory is well behaved. In that case convergent Euclidean functional integrals for the states of the theory can be formulated in terms of the physical variables. We can also find convergent Euclidean integrals for these states that reflect the invariances of the theory, even if the classical Euclidean action is unbounded below. However, if it is unbounded, then we cannot recover integrals that are weighted by a Euclidean action that is a local functional of the variables. They are weighted by nonlocal actions; moreover the form of the nonlocal action is not unique. Which nonlocal action one wants to use depends on what is most convenient.

III. THE GROUND STATE

Physical amplitudes for gravity such as ground-state wave functionals can be expressed as convergent Euclidean functional integrals in terms of the physical degrees of freedom for asymptotically flat spacetimes. This will be possible because the Hamiltonian is positive definite for such metrics by virtue of the positive-energy theorem. To illustrate this we will construct the Euclidean functional integral that describes the ground state for asymptotically flat metrics in the physical fields. We shall give a precise definition of this integral by treating gravity as interacting metric perturbations on a flat background. This will be the starting point for deriving a convergent parametrized Euclidean functional integral for the same state.

The appropriate action for Einstein's theory of relativity when the induced three-metric h_{ij} is fixed on the boundary is

$$l^2 S[g] = \int_M d^4x\, g^{1/2} R(g) + 2\int_{\partial M} d^3x\, h^{1/2} K\ , \quad (3.1)$$

where K is the extrinsic curvature of the boundary of the manifold ∂M. It is invariant under general coordinate transformations, $x^\alpha \to \bar{x}^\alpha$, which change the metric $g_{\alpha\beta}(x) \to \bar{g}_{\alpha\beta}(\bar{x})$. The invariance of the action means that some metric components are not physical as the form depends on the coordinate system. Thus, as in the model problem, there will be constraints. In order to display these constraints and isolate the physical degrees of freedom, we will write the theory in Hamiltonian form.[22–24]

It is convenient to begin by dividing the spacetime into a family of spacelike hypersurfaces labeled by t, a function constant on each hypersurface. The four metric is decomposed with respect to these surfaces:

$$ds^2 = -(N^2 - N_i N^i)dt^2 + 2N_i dx^i dt + h_{ij} dx^i dx^j\ . \quad (3.2)$$

For the spacetime to be asymptotically flat, the metric components must satisfy certain falloff conditions.[23,22] A sufficient behavior is that in an appropriate set of coordinates the metric components fall off like the Schwarzchild metric at spatial infinity:

$$ds^2 \sim -\left[1 - \frac{2M}{r}\right]dt^2 + \left[\delta_{ij} + \frac{2Mx^i x^j}{r^3}\right]dx^i dx^j + O\left[\frac{1}{r^2}\right]. \quad (3.3)$$

Thus, also choosing a coordinate system in which the x^i are asymptotically Euclidean, the conditions on the metric components and their derivatives are

$$\begin{aligned} &g_{ij} - \delta_{ij} \sim 1/r,\quad \partial_k g_{ij} \sim 1/r^2\ ,\\ &N - 1 \sim 1/r,\quad \partial_k N \sim 1/r^2\ ,\\ &N^i \sim 1/r,\quad \partial_k N^i \sim 1/r^2\ , \end{aligned} \quad (3.4)$$

as r approaches infinity.

The Hamiltonian corresponding to (3.1) is

$$H = \int d^3x\, N^\mu \mathcal{H}_\mu + E\ , \quad (3.5a)$$

$$\mathcal{H}_0 = l^2 G_{ijkl}\pi^{ij}\pi^{kl} - \frac{1}{l^2} hR(h)\ , \quad (3.5b)$$

$$\mathcal{H}_i = -2D_j \pi_i^j\ , \quad (3.5c)$$

$$E = \frac{1}{l^2}\int d^2s_i(\partial_j h_{ij} - \partial_i h_{jj})\ , \quad (3.5d)$$

where π^{ij} is the momenta conjugate to h_{ij}, the metric on the constant t hypersurfaces, D_j is the corresponding covariant derivative, and $G_{ijkl} = \frac{1}{2}(h_{ik}h_{jl} + h_{il}h_{jk} - h_{ij}h_{kl})$. Indices in (3.5) are lowered using h_{ij}. E is the surface integral needed in order to get the correct equations of motion for asymptotically flat spacetimes from Hamilton's principle.[25] N^0 and N^i have no conjugate momenta. This is because of the general coordinate invariance of the theory; not all components of the metric are dynamical as metrics that differ by coordinate transformations describe the same spacetime. Consequently N^0 and N^i are Lagrange multipliers enforcing the constraints $\mathcal{H}_0$ and $\mathcal{H}_i$. $\mathcal{H}_0$ is a factor of $h^{1/2}$ different from its usual form; this choice simplifies subsequent calculations. Therefore N^0 also differs from the conventional lapse N as given in (3.2); $N = h^{1/2}N^0$. This results in the Hamiltonian density having an overall weight of one as required by the density of the original action (3.1). The action expressed in terms of the phase-space variables is

$$S = \int_t^{t'} dt \left[\int d^3x\, \pi^{ij}\dot{h}_{ij} - H\right]\ , \quad (3.6)$$

where t and t' label the boundary hypersurfaces of the manifold. The overdot means the derivative with respect to t.

As in the model problem, the invariances and dynamics of the theory can be conveniently displayed using Poisson brackets. The constraints are the generators of the infinitesimal gauge transformations of the canonical variables:

$$\delta_F h_{ij}(x) = \left\{h_{ij}(x), \int d^3x' F^\mu(x')\mathcal{H}_\mu(x')\right\}\ , \quad (3.7a)$$

$$\delta_F \pi^{ij}(x) = \left\{\pi^{ij}(x), \int d^3x' F^\mu(x')\mathcal{H}_\mu(x')\right\}\ , \quad (3.7b)$$

computed using the fundamental brackets

$$\begin{aligned} &\{\pi^{ij}(x), h_{kl}(x')\} = -\delta_{kl}^{ij}(x,x')\ ,\\ &\delta_{kl}^{ij}(x,x') = \tfrac{1}{2}(\delta_k^i\delta_l^j + \delta_l^i\delta_k^j)\delta(x,x')\ . \end{aligned} \quad (3.7c)$$

$\mathcal{H}_i$ generates infinitesimal diffeomorphisms by F^i in the spacelike hypersurface and $\mathcal{H}$ generates the infinitesimal transformations of the variables caused by deformations of the hypersurface by F^0 in the direction of its normal. The algebra of the constraints closes which means the constraints are first class:

$$\{\mathcal{H}_\mu(x), \mathcal{H}_\nu(x')\} = \int d^3x''\, U_{\mu\nu}^\rho(x,x';x'')\mathcal{H}_\rho(x'')\ . \quad (3.8)$$

$U_{\mu\nu}^\rho$ are the structure functions of the theory which depend on the three-metric; the nonvanishing ones are[24]

$$U_{00}^i(x,x';x'') = hh^{ij}(x'')\delta_{,j}(x,x')[\delta(x,x'') + \delta(x',x'')]\ , \quad (3.9a)$$

$$U_{i0}^0(x,x';x'') = \delta_{,i}(x,x')[\delta(x,x'') + \delta(x',x'')]\ , \quad (3.9b)$$

$$U_{ij}^k(x,x';x'') = \delta_i^k\delta_{,j}(x,x')\delta(x',x'') + \delta_j^k\delta_{,i}(x,x')\delta(x,x'')\ , \quad (3.9c)$$

where the spatial derivative is with respect to the first coordinate in the δ function. The action (3.6) is invariant under the gauge transformation (3.7) provided that N^μ transforms as

$$\delta_F N^\mu(x) = F^\mu(x) - \int d^3x' d^3x'' U^\mu_{\nu\rho}(x',x'';x) N^\nu(x') F^\rho(x'') \quad (3.10)$$

and the initial and final hypersurfaces are not deformed, $F^0(x,t)=F^0(x',t')=0$. Therefore the classical action is in completely parametrized form; it is manifestly diffeomorphism invariant.

What does this mean? The physical content of the theory is independent of the parametrization of the family of hypersurfaces or the choice of coordinates in them. Because the constraints generate these gauge transformations the physical degrees of freedom must have vanishing Poisson brackets with them. Given h_{ij} and π^{ij} that commute with the constraints on an initial hypersurface, the closure of the algebra and the form of the Hamiltonian (3.5) ensures that they evolve such that they do on future slices too. Therefore it is sufficient to isolate the physical degrees of freedom on the initial data.

The physical variables are identified as those canonical variables (h_{ij},π^{ij}) that identically satisfy the constraints on the initial constant t hypersurface. The constraints can be viewed as fixing four conditions on the six metric components and their conjugate momenta. Suppose we could make a change of variables to a set of canonical variables for gravity in which the constraints were four of the momenta. Then as in the model problem, the fields conjugate to these momenta are also no longer dynamical. They are arbitrary parameters in the theory corresponding to a choice of coordinates. Therefore, four gauge conditions can be given to fix this choice. The same holds true when the constraints are written in terms of (h_{ij},π^{ij}). Four functions fixing gauge must also be specified on the canonical variables in addition to the constraints to reduce the system to its physical degrees of freedom. Counting these degrees of freedom, the twelve canonical fields have been reduced to four: two metric components and their conjugate momenta. This is what is expected for a massless spin-2 field.

The identification of the physical variables can be carried out explicitly for the theory of linearized gravity;[26] they are the transverse trace-free components of the metric perturbation and its conjugate momenta. The Hamiltonian for linearized gravity is obtained from that of the full theory by expanding the metric components in perturbations around flat space, $g_{ij}=\delta_{ij}+l\gamma_{ij}$, $N^0=1+l(n^0-\frac{1}{2}\gamma^i_i)$, $N^i=ln^i$, and truncating the result at quadratic order. The γ^i_i contribution to the expansion of N^0 is from it being a density of weight -1. The leading term in the expansion of $N^0\mathcal{H}_0$ cancels the surface term E; the result is

$$H_2=\tfrac{1}{4}\int d^3x\,[\,4(\pi^{ij})^2-2(\pi^i_i)^2+(\partial_k\gamma_{ij})^2-(\partial_k\gamma^i_i)^2 -2\partial_k\gamma^{ki}(\partial_j\gamma^j_i-\partial_i\gamma^j_j)+n^\mu\mathcal{H}^{(1)}_\mu]\,, \quad (3.11a)$$

$$\mathcal{H}^{(1)}_0=\partial_i\partial_j\gamma^{ij}-\partial^2\gamma^i_i\,, \quad (3.11b)$$

$$\mathcal{H}^{(1)}_i=-2\partial_j\pi^j_i\,, \quad (3.11c)$$

where we have used the convention that $(\pi^{ij})^2=\delta_{ik}\delta_{jl}\pi^{ij}\pi^{kl}$ with similar contractions holding for any other tensor written in this notation. Also, in expressions for linearized gravity, indices are raised and lowered using the flat metric δ_{ij}. The physical degrees of freedom can be isolated using the Fourier decomposition of the perturbations on an initial hypersurface. Labeling the Fourier transform of the spatial dependence of the tensor fields by the wave vector k^i, the perturbations can be written as Fourier components on the tensor space. There are six different types of components for each value of k^i: three which are parallel to k^i, two that are transverse to k^i and trace-free, and one that is the trace of the transverse part of the tensor. The constraints (3.11b) and (3.11c) fix the longitudinal components of the momenta and the trace of the transverse component of the metric perturbation. The variables conjugate to these will no longer be dynamical and therefore depend on the choice of gauge. Thus the physical part of the three-metric and its conjugate momenta are the two independent Fourier components of each that lie in the tensor subspace that is both transverse to k^i and trace-free, that is, γ^{TT}_{ij} and π^{ij}_{TT}.

Formally, the physical variables for general relativity are isolated using the same method illustrated above for the linearized theory; however, unlike linearized gravity, these variables cannot be explicitly constructed because of the full theory's nonlinear interactions. This means that it is very useful and necessary to be able to express quantities in terms of the full set of fields that are equivalent to those in the physical variables. An example of how this can be done will be discussed for the case of the transition functional for the interacting theory.

Having identified the physical degrees of freedom, the quantum mechanics of the theory can be formulated in terms of them. A basic unit in this is the transition functional which gives the evolution of the metric configuration specified on an initial hypersurface to that fixed on the final hypersurface; it is the generalization of the transition amplitude or propagator in quantum mechanics. When the physical configurations can be explicitly found then this integral is the direct analog of (2.9). This is the case for the theory of linearized gravity:

$$G[h'^{TT}_{ij},t';h^{TT}_{ij},t]=\int d\gamma^{TT}_{ij}d\pi^{ij}_{TT}\exp\left[i\int dt\left[\int d^3x\,\pi^{ij}_{TT}\dot\gamma^{TT}_{ij}-h_2[\gamma^{TT},\pi_{TT}]\right]\right], \quad (3.12a)$$

$$h_2=\int d^3x\,[(\pi^{ij}_{TT})^2+\tfrac{1}{4}(\partial_k\gamma^{TT}_{ij})^2]\,. \quad (3.12b)$$

The sum is over all paths in the physical phase space which consists of the field configurations $(\pi^{ij}_{TT},\gamma^{TT}_{ij})$ where the transverse traceless metric perturbations take on the values h'^{TT}_{ij} and h^{TT}_{ij} fixed on the constant t' and t hypersurfaces, respectively. The measure is the functional generalization of the Liouville measure of the model problem. Again constant factors will not be displayed in the measure in general, although we will when providing an explicit formulation of the functional integral for linearized gravity in the Appendix.

When the physical degrees of freedom cannot be found explicitly, functional integrals that are equivalent to those in terms of the physical variables can be formulated in terms of the full set of canonical variables. This is done by using functional δ functions of the constraints and gauge degrees of freedom in the measure to eliminate all but the physical variables from the sum. Let us consider this first for linearized gravity, where we can add back in the other phase-space variables explicitly. In this case, π^{ij}_L the longitudinal components of the momenta are fixed by the momentum constraints (3.11c). Their conjugate variables γ^L_{ij} are fixed by specifying three gauge-fixing conditions. Similarly, γ^T_{ij}, the trace of the transverse component of the metric is fixed by the linearized Hamiltonian constraint (3.11b); its conjugate variable π^{ij}_T is also fixed by a choice of gauge. Writing the gauge-fixing functions as G^ν, the following identity is true:

$$1=\int d\mathcal{H}^{(1)}_\mu dG^\nu\delta[\mathcal{H}^{(1)}_\mu(\pi_L,\gamma^T)]\delta[G^\nu(\gamma^L,\pi_T)]\ . \qquad (3.13)$$

Next change the variables of functional integration from $\mathcal{H}^{(1)}_\mu$ and G^ν to the redundant phase space variables; the Jacobian of this transformation is simply the determinant of the Poisson brackets of the constraints with the gauge-fixing functions: $|\{\mathcal{H}^{(1)}_\mu,G^\nu\}|$. If we add the resulting form of the identity into the transition functional (3.12), we arrive at the following functional integral:

$$G[h'^{TT}_{ij},t';h^{TT}_{ij},t]=\int d\gamma_{ij}d\pi^{ij}\delta[\mathcal{H}^{(1)}_\mu]\delta[G^\nu]\,|\{\mathcal{H}^{(1)}_\mu,G^\nu\}|\exp\left[i\int_t^{t'}dt\left[\int d^3x\,\pi^{ij}\dot\gamma_{ij}-H_2\right]\right]\ . \qquad (3.14)$$

Because the constraints are enforced by δ function, the action for linearized gravity in terms of the extended variables can be used in (3.14). This integral over the full set of canonical variables is manifestly identical to that written in the physical degrees of freedom.

This same procedure can be carried out when the physical degrees of freedom cannot be explicitly isolated. The resulting transition functional for the full theory of gravity is

$$G[h'_{ij},t';h_{ij},t]=\int dh_{ij}d\pi^{ij}\delta[\mathcal{H}_\mu]\delta[G^\nu]\,|\{\mathcal{H}_\mu,G^\nu\}|\exp\left[i\int_t^{t'}dt\left[\int d^3x\,\pi^{ij}\dot h_{ij}-E\right]\right]\ . \qquad (3.15)$$

This integral is weighted by the classical action (3.6) with the constraints set to zero. $G^\nu(\pi^{ij},h_{ij})$ are four functions of the canonical variables fixing the gauge. In order to do this, they must have nonzero Poisson brackets with the constraints. The sum is over all metric configurations matching the initial and final data on the boundary hypersurfaces t and t'. Note that the gauge choice must be consistent with this data. The determinant $|\{\mathcal{H}_\mu,G^\nu\}|$ included in the measure is precisely the factor needed to make it equal to the Liouville measure in the physical degrees of freedom. In order to make this path integral meaningful, the class of metric configurations summed over must be specified. In this paper we shall make this integral concrete by defining it to be over metric perturbations. One first selects a one-parameter family of constant t hypersurfaces and chooses coordinates on those surfaces that are asymptotically Euclidean. The metric of each hypersurface is then taken to be of the form $h_{ij}=\delta_{ij}+l\gamma_{ij}$ where γ_{ij} is a bounded function with the falloff behavior (3.4). Its corresponding momentum π^{ij} is also bounded with falloff behavior $\pi^{ij}\sim 1/r^2$ at spatial infinity. The sum over geometries matching the boundary data is then done by summing over all fields with this behavior on each intermediate hypersurface between the initial and final hypersurface. This sum over geometries does not include those geometries that cannot be written in this form in this coordinate system or those that develop singularities. However the transition functional will be useful for cases where the initial and final metrics are almost flat because then one expects the major contribution to its value to come from nearby configurations that are also almost flat.

Both sides of Eq. (3.15) really depend on the physical degrees of freedom only; however, the specification of the path integral depends on the gauge chosen. This is because we cannot express the true (gauge-invariant) degrees of freedom explicitly but only in a particular gauge. The two components of h_{ij} that we choose as physical are free data on the $t=$const boundary hypersurface; the other components of h_{ij} must agree with the gauge choice.

From transition functionals in the physical degrees of freedom we can derive Euclidean functional integrals for the ground state by continuing $t\to -i\tau$ and then taking the leading behavior as $\tau\to-\infty$ in analogy to the Feynman-Kac prescription in quantum mechanics. These integrals will be convergent and therefore this procedure is allowed if the physical Hamiltonian is positive. In linearized gravity, the Hamiltonian (3.12b) in terms of the physical variables is manifestly positive; implementing the construction procedure used in Sec. II on the transition functional (3.12a) one obtains the Euclidean functional integral for the ground-state wave functional for linearized gravity:

$$\Psi_0[h^{TT}_{ij}]=\mathcal{N}\int d\gamma^{TT}_{ij}d\pi^{ij}_{TT}\times\exp\left[\int_{-\infty}^{0}d\tau\left[\int d^3x\,i\pi^{ij}_{TT}\dot\gamma^{TT}_{ij}-h_2\right]\right]. \qquad (3.16)$$

The sum is over all transverse traceless tensors matching the data h_{ij}^{TT} given at the boundary $\tau=0$ that fall off fast enough at Euclidean infinity so that the action is finite. $\mathcal{N}$ is a normalization factor independent of the perturbations. The equivalent configuration-space functional integral can be derived by doing the quadratic momentum integrals:

$$\Psi_0[h_{ij}^{TT}]=\mathcal{N}\int d\gamma_{ij}^{TT}\exp(-i_2)\ , \tag{3.17a}$$

$$i_2=\frac{1}{4}\int_{-\infty}^{0} d^4x[(\dot{\gamma}_{ij}^{TT})^2+(\partial_k\gamma_{ij}^{TT})^2]\ . \tag{3.17b}$$

When evaluated, this integral results in a wave functional composed of the product of harmonic-oscillator ground states whose arguments are the two independent amplitudes of the transverse traceless Fourier components for each wave vector k^i (Refs. 27 and 28). Therefore the Feynman-Kac procedure is explicitly seen to give the ground-state wave functional in this case.

For asymptotically flat spacetimes, the positive-energy theorem[8,9] states that E is positive for nonsingular vacuum spacetimes where h_{ij} and π^{ij} satisfy the constraints. This means that we can formally construct convergent Euclidean functional integrals over the physical variables. However, the analog of the ground-state wave functional (3.16) for linearized gravity cannot be written explicitly because the physical degrees of freedom cannot be explicitly isolated. Formally this state can be derived for asymptotically flat spacetimes by first computing the transition functional and then using the Feynman-Kac prescription. By the positive-energy theorem, the minimum energy metric configuration is flat space; this means the analog of setting q^a at the minimum of the potential is to fix the initial geometry to be flat. The final geometry is taken to be the argument of the wave functional. For the transition functional (3.15) one has

$$\Psi_0[\gamma_{ij}]=\lim_{\tau\to-\infty}(\mathcal{N}G[\delta_{ij}+l\gamma_{ij},0;\delta_{ij},-i\tau])\ , \tag{3.18}$$

where t is continued to $-i\tau$ in (3.15) and the data given on the $\tau=0$ boundary is almost flat. In (3.18) we have written the argument of the wave functional in terms of the perturbation from flat space γ_{ij} for convenience. Because the functional integration is carried out before the analytic continuation, the constraints are enforced and the Hamiltonian is positive. Thus (3.18) will be a well-defined procedure for obtaining the ground state; however it has several disadvantages. It does not exploit the simplicity of the boundary conditions as (3.17) does because it cannot be explicitly written in terms of the physical variables. It cannot be easily written as a configuration-space path integral because it is necessary to exponentiate the constraints to make it quadratic in the momenta before performing the momentum integrations. The transition functional itself is not parametrized and thus does not display the manifest invariances of the theory. Equation (3.18) is our starting point; our aim is to derive a corresponding convergent parametrized Euclidean functional integral for this state. Before proceeding to do this for perturbative gravity, it is instructive to study how the ground-state wave functional for the linearized theory (3.15) is parametrized. We shall do so in Sec. IV and then treat the interacting case in Sec. V.

IV. LINEARIZED GRAVITY

The transition from Euclidean functional integrals over the physical variables to those over the extended variables can be explicitly worked out for the linearized version of Einstein's theory. This is because the physical degrees of freedom can be explicitly identified and the action in terms of them (3.17) is a quadratic functional. In this section we shall review work done in a previous paper[10] showing how this connection is made for the ground-state wave functional for linearized gravity. Boulware has also obtained this result in the linearized theory.[29] The results of this section will be a useful guide to carrying out the construction of the parametrized Euclidean integral for the ground-state wave functional for asymptotically flat spacetimes calculated perturbatively.

Our starting point is the ground-state wave functional for the linearized theory (3.17). What we want to do is to add integrations over the redundant variables to this expression until we arrive at an expression for Ψ_0 that is manifestly gauge invariant under the linearized diffeomorphisms of the full theory:

$$\gamma_{\alpha\beta}=\gamma_{\alpha\beta}+2\partial_{(\alpha}f_{\beta)} \tag{4.1}$$

and O(4) invariant. What form of the parametrized action can we expect to get? An O(4)-invariant, gauge-invariant Euclidean action that is also local in the perturbation $\gamma_{\alpha\beta}$ is the linearized version of (1.3):[27]

$$I_2=-\tfrac{1}{2}\int_M d^4x\,\gamma^{\alpha\beta}\dot{G}_{\alpha\beta}-\tfrac{1}{2}\int_{\partial M} d^3x\,\gamma^{ij}(\dot{K}_{ij}-\delta_{ij}\dot{K})\ , \tag{4.2a}$$

$$\dot{G}_{\alpha\beta}=-\tfrac{1}{2}(-\partial^2\bar{\gamma}_{\alpha\beta}-\delta_{\alpha\beta}\partial_\gamma\partial_\eta\bar{\gamma}^{\gamma\eta}+\partial_\alpha\partial_\gamma\bar{\gamma}^{\gamma}_{\beta}+\partial_\beta\partial_\gamma\bar{\gamma}^{\gamma}_{\alpha})\ , \tag{4.2b}$$

$$\dot{K}_{ij}=-\tfrac{1}{2}(\dot{\gamma}_{ij}-\partial_i\gamma_{0j}-\partial_j\gamma_{0i})\ , \tag{4.2c}$$

where $\bar{\gamma}_{\alpha\beta}$ is the trace reversed metric perturbation

$$\bar{\gamma}_{\alpha\beta}=\gamma_{\alpha\beta}-\tfrac{1}{2}\delta_{\alpha\beta}\gamma^{\delta}_{\delta}\ . \tag{4.3}$$

We cannot end up with a functional integral for Ψ_0 involving this action because, as in the full theory, it is not positive definite. In particular, on perturbations of the form $\gamma_{\alpha\beta}=2\chi\delta_{\alpha\beta}$,

$$I_2=-6\int_M d^4x\,(\partial_\alpha\chi)^2\ . \tag{4.4}$$

However (4.2a) is not the only gauge- and O(4)-invariant action for linearized gravity. As in the model, there are many others if the action is not required to be local in the original set of variables. We shall find such an action by adding in integrations over the redundant variables to the physical functional integral; our aim will be to construct one as close as possible in form to (4.2a).

To add back the redundant integrations it is useful to decompose $\gamma_{\alpha\beta}$ into pieces corresponding to the physical variables and pieces corresponding to the redundant ones. As the result (4.4) suggests, it is convenient to begin by decomposing $\gamma_{\alpha\beta}$ into conformal equivalence classes

$$\gamma_{\alpha\beta}=\phi_{\alpha\beta}+2\chi\delta_{\alpha\beta} \tag{4.5}$$

This decomposition is fixed by the O(4)-invariant, gauge-invariant condition

$$R^{(1)}(\phi)=\partial_\alpha\partial_\beta\phi^{\alpha\beta}-\partial^2\phi^\beta_\beta=0 . \tag{4.6}$$

χ is now related to $\gamma_{\alpha\beta}$ by the second-order equation

$$R^{(1)}(\gamma)=-6\partial^2\chi \tag{4.7}$$

subject to the boundary conditions that $\chi=0$ on the $\tau=0$ hypersurface and at Euclidean infinity.

The perturbation $\phi_{\alpha\beta}$ is then decomposed as

$$\phi_{\alpha\beta}=t_{\alpha\beta}+l_{\alpha\beta}+\phi^T_{\alpha\beta}+\phi^L_{\alpha\beta} , \tag{4.8}$$

where the components are defined as follows: let n^α be the unit vector orthogonal to the surfaces of constant τ. Consider the subspaces of the space of tensor functions whose elements $t_{\alpha\beta}$, $l_{\alpha\beta}$, $\phi^T_{\alpha\beta}$, and $\phi^L_{\alpha\beta}$ satisfy the conditions

$$\partial^\alpha t_{\alpha\beta}=0,\quad n^\alpha t_{\alpha\beta}=0,\quad t^\alpha_\alpha=0 , \tag{4.9a}$$

$$\partial^\alpha l_{\alpha\beta}=0,\quad l^\alpha_\alpha=0,\quad \int_M d^4x\, t^{\alpha\beta}l_{\alpha\beta}=0 , \tag{4.9b}$$

$$\partial^\alpha\phi^T_{\alpha\beta}=0,\quad n^\alpha\phi^T_{\alpha\beta}=0,\quad \int_M d^4x\, t^{\alpha\beta}\phi^T_{\alpha\beta}=0 , \tag{4.9c}$$

$$\int_M d^4x\, t^{\alpha\beta}\phi^L_{\alpha\beta}=0,\quad \int_M d^4x\, l^{\alpha\beta}\phi^L_{\alpha\beta}=0 ,$$
$$\int_M d^4x\, \phi^{T\alpha\beta}\phi^L_{\alpha\beta}=0 . \tag{4.9d}$$

If the orthogonality conditions are required to hold for all tensors in the subspaces, then the decomposition of $\phi_{\alpha\beta}$ (4.8) is unique. A more explicit version of the decomposition will be given in the Appendix. The condition (4.6) is seen to fix $\phi^T_{\alpha\beta}=0$. The tensors $t_{\alpha\beta}$ correspond to the physical variables γ^{TT}_{ij}; the rest are redundant.

The utility of this decomposition is that under gauge transformations (4.1) $t_{\alpha\beta}$, $l_{\alpha\beta}$, and χ are unchanged. Since the action (4.2) is a gauge invariant, it can be expressed in terms of these variables:

$$I_2=\tfrac{1}{4}\int_M d^4x\,[(\partial_\gamma t_{\alpha\beta})^2+(\partial_\gamma l_{\alpha\beta})^2-24(\partial_\gamma\chi)^2]$$
$$-\tfrac{1}{4}\int_{\partial M} d^3x\, n^\alpha\partial_\alpha[2(n^\beta l_{\beta\gamma})^2-\tfrac{3}{2}(n^\beta n^\gamma l_{\beta\gamma})^2] . \tag{4.10}$$

Using this decomposition of the metric we can proceed to add in the redundant degrees of freedom by inserting in (3.17) identities composed of Gaussian integrals over the gauge-invariant quantities and gauge-fixing δ functions over the noninvariant ones. Although the final answer is independent of the gauge choice, the following arguments will be clearer if a particular gauge is fixed. The choice

$$\Phi^\alpha(\phi)=\partial_\beta\bar\phi^{\alpha\beta}=0 \tag{4.11a}$$

when combined with the condition (4.5) fixes the $\phi^L_{\alpha\beta}$ components up to a gauge transformation (4.1) where f_α also satisfies

$$-\partial^2 f_\alpha=0 . \tag{4.11b}$$

This remaining gauge freedom can be eliminated by fixing the value of these components on the boundaries. Boundary conditions on the other redundant variables must also be specified in order to define the class of tensor field configurations that will be summed over. A simple and convenient set of boundary conditions that corresponds to those given for ground state in the physical variables (3.17) and satisfies the above requirements on the redundant ones is to (1) take $t_{\alpha\beta}$ to match the argument h^{TT}_{ij} of the wave function at $\tau=0$, (2) require that $\chi=0$ and the spatial part γ_{ij} of the remaining components vanish there, (3) require that the gauge condition (4.11a) be satisfied on the boundary, and (4) take all components of $\phi_{\alpha\beta}$ and χ to vanish rapidly enough at Euclidean infinity so that the action is finite. On configurations satisfying these four conditions, the surface term in the action (4.10) vanishes.

In terms of the decomposition, the action i_2 (3.17b) in the physical degrees of freedom is

$$i_2=\tfrac{1}{4}\int_M d^4x\,(\partial_\gamma t_{\alpha\beta})^2 . \tag{4.12}$$

Next, we want a positive-definite action quadratic in the redundant variables that is gauge invariant and O(4) invariant to use in forming Gaussian integrals. If, in addition, we require this integral to be at most quadratic in the derivatives, the most general form of the action that satisfies these requirements is

$$I^g_2=\tfrac{1}{4}\int_M d^4x[(\partial_\gamma l_{\alpha\beta})^2+a(\partial_\gamma\chi)^2] , \tag{4.13}$$

where a is an arbitrary positive constant. O(4) rotations that mix n^α with the spatial unit vectors $\hat{\mathbf{x}}^i$ will mix the components $t_{\alpha\beta}$ and $l_{\alpha\beta}$; therefore the coefficients of these terms must be the same if O(4) invariance is to be preserved in the full action. This does not similarly restrict the coefficient of $(\partial_\alpha\chi)^2$ as it transforms as a scalar. The constant a must be positive, however, for the action to be positive definite.

As in the model problem, integrals over the action (4.13) and the gauge-fixing condition (4.11a) can be added to the Euclidean functional integral for the ground-state wave function by forming the appropriate combinations of Gaussians and δ functions. In this case we want the identities

$$1=\int dl_{\alpha\beta}d\phi^L_{\alpha\beta}d\chi\,\delta[\Phi^\alpha(\phi)]\left|\frac{\delta\Phi^\alpha}{\delta f^\beta}\right|\exp(-I^g_2[l,\chi]) , \tag{4.14a}$$

$$1=\int d\phi^{T}_{\alpha\beta}\delta[R^{(1)}(\phi)]\left|\frac{\delta R^{(1)}}{\delta\omega}\right| . \tag{4.14b}$$

In (4.14) the functional integrations are over the configurations that we have specified by the decomposition and the boundary conditions. A specific measure is required for these identities to be true. This will be given explicitly in the Appendix for calculating these integrals using mode amplitudes. The determinant in (4.14a) is the Faddeev-Popov determinant of the operator constructed by varying the gauge-fixing term Φ^α (4.11a) with respect to the gauge parameter f^α (4.1). The boundary conditions for this operator are determined by those on the gauge parameter. In order to keep the boundary conditions fixed, $f^\alpha=0$ at $\tau=0$ and at Euclidean infinity. The determinant is calculated using the spectrum of this operator subject to these same boundary conditions. The determinant in (4.14b) is of the operator constructed by varying the condition fixing conformal equivalence classes (4.6) by an infinitesimal conformal transformation

$$\phi_{\alpha\beta}=\phi_{\alpha\beta}+2\delta_{\alpha\beta}\omega . \tag{4.15}$$

As for the Faddeev-Popov determinant, this one is computed using the spectrum of the operator determined for the boundary conditions that ω vanishes at $\tau=0$ and at Euclidean infinity.

These determinants can be conveniently represented using functional integration over Grassmann variables.[18] Let $-i\overline{\mathcal{C}}_\mu$ and $\mathcal{C}^\mu$ be eight real anticommuting Grassmann fields. Then the Faddeev-Popov determinant in (4.14a) becomes

$$\left|\frac{\delta\Phi^\mu}{\delta f^\nu}\right|=\int d\overline{\mathcal{C}}_\mu d\mathcal{C}^\mu \exp(-I_2^{\text{gh}}[\overline{\mathcal{C}},\mathcal{C}]) , \tag{4.16a}$$

$$I_2^{\text{gh}}=\int_M d^4x(-i\overline{\mathcal{C}}_\mu\delta_c^{(2)}\Phi^\mu) , \tag{4.16b}$$

where $\delta_c^{(2)}\Phi^\mu$ is the value of the gauge-fixing function on the linearized gauge transformation (4.1) with $\mathcal{C}^\mu$ as the gauge parameter. The boundary conditions needed to compute the spectrum of the operator are implemented by requiring that $\overline{\mathcal{C}}_\mu$ and $\mathcal{C}^\mu$ vanish at $\tau=0$ and Euclidean infinity. Again, a particular measure is needed for this identity to hold. For the gauge-fixing choice (4.11a),

$$\delta_c^{(2)}\Phi^\mu=-\partial^2\mathcal{C}^\mu . \tag{4.17}$$

The determinant in (4.14b) can also be exponentiated in a similar manner; however, we will choose to leave it in the measure.

Inserting these identities into the Euclidean functional integral and exponentiating the Faddeev-Popov determinant using (4.16) we arrive at the following expression for the ground-state wave function:

$$\Psi_0[h_{ij}^{TT}]=\int d\phi_{\alpha\beta}d\chi\, d\overline{\mathcal{C}}_\mu d\mathcal{C}^\mu\delta[\Phi^\alpha(\phi)]\delta[R^{(1)}(\phi)]\left|\frac{\delta R^{(1)}}{\delta\omega}\right|\exp\{-(\hat{I}_2[\phi,\chi]+I_2^{\text{gh}}[\overline{\mathcal{C}},\mathcal{C}])\} . \tag{4.18}$$

Here $\hat{I}_2$ is the sum of i_2 and I_2^{g}

$$\hat{I}_2=\tfrac{1}{4}\int_M d^4x[(\partial_\gamma t_{\alpha\beta})^2+(\partial_\gamma l_{\alpha\beta})^2+a(\partial_\gamma\chi)^2] , \tag{4.19}$$

where a is any positive constant. The integral is over all ten components of $\phi_{\alpha\beta}$ and over the linearized piece of the conformal factor χ in the class of configurations described previously. Thus the integral is of the form of one over all gauge inequivalent metrics in a conformal equivalence class fixed by $R^{(1)}=0$ plus an integration over the conformal factor.

The action (4.19) is gauge invariant, O(4) invariant and, for positive a it is positive definite so that the integral (4.18) converges. This action cannot be made to agree with the action I_2 (4.10) because this would require that a be negative and lead to a divergent functional integral. The action $\hat{I}_2$ when a is set to 24 is exactly what would be obtained formally from I_2 by conformal rotation; that is, rotating $\chi\to i\chi$.

As suggested before, the action $\hat{I}_2$ is not local in the original metric perturbation $\gamma_{\alpha\beta}$; however, it can be expressed in terms of it

$$\hat{I}_2[\gamma]=I_2[\gamma]+\frac{a+24}{144}\int d^4x\, R^{(1)}(\gamma)\frac{1}{-\partial^2}R^{(1)}(\gamma) . \tag{4.20}$$

It is physically equivalent to I_2, gauge and O(4) invariant, and positive definite for $a>0$. Thus at the expense of locality, one can construct convergent Euclidean functional integrals for gravity that manifestly display the invariances of the theory. There are many different forms of these convergent Euclidean functional integrals as the nonlocal action needed is not unique; a is not fixed by the process. These integrals are very similar to those obtained by the Gibbons, Hawking, and Perry prescription of conformal rotation applied to the linearized theory. However, the Gibbons, Hawking, and Perry prescription omits the Faddeev-Popov determinant $|\delta R^{(1)}/\delta\omega|$ of the decomposition fixing δ function in the measure. Our analysis shows that this factor is needed for the manifestly invariant Euclidean integral to equal that given in terms of the physical fields.

We have derived (4.18) directly from the Euclidean functional integral for the ground state given in terms of its physical variables. This is not the only way to arrive at this integral. Alternatively, we could have begun by first parametrizing the Lorentzian transition functional (3.12a); however, instead of aiming for its usual form weighted by the local action S_2 we construct one weighted by the Lorentzian version of $\hat{I}_2$. This parametrized Lorentzian functional integral can be derived along the same lines as the Euclidean one; the result is

$$G[h'^{TT}_{ij},t';h^{TT}_{ij},t]=\int d\phi_{\alpha\beta}d\mathcal{X}\,d\overline{\mathcal{C}}_\mu d\mathcal{C}^\mu\delta[\Phi^\alpha(\phi)]\delta[R^{(1)}(\phi)]\left|\frac{\delta R^{(1)}}{\delta\omega}\right|\exp(i\hat{S}_2[\phi,\mathcal{X}]+iS_2^{\rm gh}[\overline{\mathcal{C}},\mathcal{C}])\ , \tag{4.21}$$

where the class of configurations summed over is specified by the Lorentzian analogs of the decompositions (4.6) and (4.7), the gauge-fixing condition (4.11), and boundary conditions on both the initial and final hypersurface. The action $\hat{S}_2$ is the Lorentzian version of (4.19)

$$\hat{S}_2=-\tfrac{1}{4}\int_M d^4x\,[(\partial_\gamma t_{\alpha\beta})^2+(\partial_\gamma l_{\alpha\beta})^2+a(\partial_\gamma\mathcal{X})^2]\ . \tag{4.22}$$

The ghost term $S_2^{\rm gh}$ is the analog of (4.16b) where the gauge transformation is done using the Lorentzian signature. The usual form of the Lorentzian functional integral weighted by the action S_2 can be recovered by setting $a=-24$ and doing the integral over $\mathcal{X}$ using the δ function of $R^{(1)}$. By choosing a positive, the analytic continuation of $t\to-i\tau$ can be carried out on (4.21) because the corresponding Euclidean action is positive. Taking the initial data to vanish and the limit $\tau\to-\infty$ results in precisely the same Euclidean functional integral for the ground state (4.18) obtained directly.

V. PERTURBATIVE GRAVITY

In the previous section we showed how convergent Euclidean functional integrals for linearized gravity could be derived by appropriately adding integrals over the redundant variables to the Euclidean functional integral in terms of the physical variables. We found that in order to maintain the convergence of the integral we were led to a parametrized action that was not local in the original metric perturbation. This motivates us to look for a convergent Euclidean functional integral weighted by a nonlocal action for the ground-state functional (3.18) of the full theory, where it is not possible to parametrize the physical functional integral directly. We will derive this in three steps. (1) We will parametrize the Lorentzian transition functional (3.15) in phase space and perform the momentum integrations to get the parametrized local Lorentzian transition functional in configuration space. This will be carried out using Becchi-Rouet-Stora (BRS) invariance following the method of Fradkin and Vilkovisky. (2) Then, as suggested at the end of the previous section, we will find an alternate Lorentzian functional with a nonlocal action that equals the first term by term in perturbation theory. This new Lorentzian functional integral will be constructed so as to have manifestly positive Euclidean action. (3) We shall then use this alternate transition functional in the definition of the ground state (3.18). Consequently the rotation of $t\to-i\tau$ can be carried out term by term. Doing this will result in a manifestly convergent parametrized Euclidean functional integral for the ground state.

The basic idea is to parametrize the Lorentzian functional integral (3.15) in phase space[30] by introducing extra fields to exponentiate the functional δ functions and determinants. The gauge conditions used in (3.15) are given on the canonical variables; one would like to generalize these to include the additional variables so that calculations can be done in other gauges. For this generalization to be correct, the parametrized integral must be manifestly independent of gauge choice and must equal the original transition functional in a canonical gauge. Fradkin and Vilkovisky proved that this is indeed the case for gauge theories and gravity in a series of papers.[21,31,32] It is useful to review how this connection is made for asymptotically flat spaces and to discuss the boundary conditions needed in the parametrized transition functional.

The canonical phase space (h_{ij},π^{ij}) is extended by adding the lapse and shift N^μ and their conjugate momenta P_μ. In addition, eight real anticommuting Grassmann fields $\mathcal{C}^\mu$ and $-i\overline{\mathcal{C}}_\mu$ and their conjugate momenta, $\overline{\mathcal{P}}_\mu$ and $i\mathcal{P}^\mu$ are also added. The Grassmann parity, σ, of the anticommuting variables is odd and that of the commuting variables is even. There is a new structure on this extended phase space, the ghost number. $\mathcal{P}^\mu$ and $\mathcal{C}^\mu$ have ghost number 1 and $\overline{\mathcal{P}}_\mu$ and $\overline{\mathcal{C}}_\mu$ have ghost number -1. The rest of the variables h_{ij}, π^{ij}, N^μ, and P_μ have ghost number 0. Quantities composed of products of these fields have ghost number equal to the sum of that of their components. The corresponding Poisson brackets are

$$\begin{aligned}\{\overline{\mathcal{P}}_\mu(x),\mathcal{C}^\nu(x')\}&=-\delta^\nu_\mu(x,x')\ ,\\ \{\mathcal{P}^\nu(x),\overline{\mathcal{C}}_\mu(x')\}&=-\delta^\nu_\mu(x,x')\ ,\\ \{P_\mu(x),N^\nu(x')\}&=-\delta^\nu_\mu(x,x')\ ,\end{aligned} \tag{5.1}$$

where the Poisson brackets of the Grassmann fields are anticommuting, $\{\overline{\mathcal{P}}_\mu,\mathcal{C}^\nu\}=\{\mathcal{C}^\nu,\overline{\mathcal{P}}_\mu\}$. The Poisson brackets of quantities containing both commuting and anticommuting variables can be computed using the relation

$$\{A,BC\}=\{A,B\}C+(-1)^{\sigma_A\sigma_B}B\{A,C\}\ , \tag{5.2}$$

where the parity of a quantity, for example σ_A of A, is odd if it contains an odd number of Grassmann variables and even otherwise.

The next step in constructing the transition functional on the extended phase space is to define the BRS transformation on the extended variables. It generalizes the local gauge transformations of general relativity to a global transformation that mixes the commuting and anticommuting variables. The generator of this transformation is

$$\Omega=\int[-i\mathcal{P}^\mu(x)P_\mu(x)+\mathcal{C}^\mu(x)\mathcal{H}_\mu(x)-\tfrac{1}{2}\mathcal{C}^\mu(x)\mathcal{C}^\nu(x')U^\rho_{\nu\mu}(x',x;x'')\overline{\mathcal{P}}_\rho(x'')]\ , \tag{5.3}$$

where we have introduced the convention to be used in this section that all spatial variables x, x', etc, *repeated* under the integral sign are to be integrated over. The $U^{\alpha}_{\nu\mu}$ are the first-order structure functions (3.9). Ω has ghost number 1 and it follows by the algebra of the constraints and (5.2) that it is nilpotent, $\{\Omega,\Omega\}=0$. The BRS transformations of the variables are given by their Poisson brackets with Ω:

$$\{h_{ij}(x),\Omega\epsilon\}=\int \mathcal{C}^{\mu}(x')\{h_{ij}(x),\mathcal{H}_{\mu}(x')\}\epsilon \ , \tag{5.4a}$$

$$\{\pi^{ij}(x),\Omega\epsilon\}=\int [\mathcal{C}^{\mu}(x')\{\pi^{ij}(x),\mathcal{H}_{\mu}(x')\}-\tfrac{1}{2}\mathcal{C}^{\mu}(x')\mathcal{C}^{\nu}(x'')\{\pi^{ij}(x),U^{\rho}_{\nu\mu}(x'',x';\bar{x})\}\bar{\mathcal{P}}_{\rho}(\bar{x})]\epsilon \ ,$$

$$\{N^{\mu}(x),\Omega\epsilon\}=i\mathcal{P}^{\mu}(x)\epsilon, \quad \{P_{\mu}(x),\Omega\epsilon\}=0 \ , \tag{5.4c}$$

$$\{\mathcal{C}^{\mu}(x),\Omega\epsilon\}=\int \tfrac{1}{2}\mathcal{C}^{\nu}(x')\mathcal{C}^{\rho}(x'')U^{\mu}_{\rho\nu}(x'',x';x)\epsilon \ , \tag{5.4d}$$

$$\{\bar{\mathcal{P}}_{\mu},\Omega\epsilon\}=-\mathcal{H}_{\mu}(x)\epsilon+\int \mathcal{C}^{\nu}(x')U^{\rho}_{\nu\mu}(x',x;x'')\bar{\mathcal{P}}_{\rho}(x'')\epsilon \ , \tag{5.4e}$$

$$\{\bar{\mathcal{C}}_{\mu}(x),\Omega\epsilon\}=iP_{\mu}(x)\epsilon, \quad \{\mathcal{P}^{\mu}(x),\Omega\epsilon\}=0 \ . \tag{5.4f}$$

The constant anticommuting parameter ϵ is introduced so that the transformation preserves the Grassmann parity of the variables. From its form, one sees that these transformations will contain those of the canonical variables (3.7) with $\mathcal{C}^{\mu}\epsilon$ as the transformation parameter.

The action needed to form the transition functional on the extended phase space is the generalization of the action (3.6) to the extended variables. It is found by requiring it to be BRS invariant and have ghost number 0. The general form of this action is

$$S_{\mathrm{FV}}=\int_{t}^{t'} dt([\pi^{ij}(x)\dot{h}_{ij}(x)+P_{\mu}(x)\dot{N}^{\mu}(x)+\dot{\mathcal{P}}^{\mu}(x)\bar{C}_{\mu}(x)+C^{\mu}(x)\bar{\mathcal{P}}_{\mu}(x)]-E+\{\Phi,\Omega\}) \ . \tag{5.5}$$

Φ is an arbitrary functional of any of the variables on the extended phase space such that it has ghost number -1. It is the analog of the gauge choice in the usual Hamiltonian form. This is more apparent if we take a specific form for Φ,

$$\Phi=\int [i\bar{\mathcal{C}}_{\mu}(x)\varphi^{\mu}(x)+\bar{\mathcal{P}}_{\mu}(x)N^{\mu}(x)] \ , \tag{5.6}$$

where φ^{μ} is an arbitrary function with ghost number 0 of the extended set of variables. It is convenient to restrict it to be independent of the momenta $\mathcal{P}^{\mu}$ and $\bar{\mathcal{P}}_{\mu}$. The Poisson brackets of Φ with Ω is then

$$\begin{aligned}\{\Phi,\Omega\}=\int (&-N^{\mu}(x)[\mathcal{H}_{\mu}(x)-\mathcal{C}^{\nu}(x')U^{\rho}_{\nu\mu}(x',x;x'')\bar{\mathcal{P}}_{\rho}(x'')]-i\bar{\mathcal{P}}_{\mu}(x)\mathcal{P}^{\mu}(x)-P_{\mu}(x)\varphi^{\mu}(x)\\ &+i\bar{\mathcal{C}}_{\mu}(x)\{\varphi^{\mu}(x),\mathcal{H}_{\nu}(x')\}\mathcal{C}^{\nu}(x')+\bar{\mathcal{C}}_{\mu}(x)\{\varphi^{\mu}(x),P_{\nu}(x')\}\mathcal{P}^{\nu}(x')\\ &-\tfrac{1}{2}i\bar{\mathcal{C}}_{\mu}(x)\mathcal{C}^{\nu}(x')\mathcal{C}^{\rho}(x'')\{\varphi^{\mu}(x),U^{\lambda}_{\rho\nu}(x'',x';x''')\}\bar{\mathcal{P}}_{\lambda}(x''')) \ .\end{aligned} \tag{5.7}$$

With this choice, the action (5.5) contains terms corresponding to the functional δ functions of the constraints and gauge-fixing condition and measure as well as additional terms that give dynamics to the extra fields. The integrand of (5.5) will transform into itself under BRS transformations. This is because (1) the factor of the form pdq is canonically invariant, (2) E commutes with Ω, and (3) $\{\{\Phi,\Omega\},\Omega\}=0$ by the Jacobi identity because Ω is nilpotent.

The BRS transformation will also act on the values of the variables fixed on the t and t' constant boundary hypersurfaces. In general, initial and final data for half of the variables is specified to determine the classical evolution of a Hamiltonian system. For the corresponding classical action (5.5) to be BRS invariant, a consistent BRS-invariant set of such data must be selected. We shall choose a set of boundary conditions that generalize those used in the physical transition functional (3.15). The three-metric h_{ij} equals its values given on the t and t' constant hypersurfaces as before. Next, note that under an infinitesimal BRS transformation (5.4a), components proportional to $\mathcal{C}^{\mu}$ and $\bar{\mathcal{C}}_{\mu}$ are added to h_{ij}. Therefore $\mathcal{C}^{\mu}$ and $\bar{\mathcal{C}}_{\mu}$ must vanish on the boundary to preserve the initial and final values of the three-metric. In order that $\bar{\mathcal{C}}_{\mu}=0$ be unchanged by BRS transformation, the condition that $P_{\mu}=0$ on the boundaries is also required as seen by (5.4f). The remaining variables π^{ij}, $\mathcal{P}^{\mu}$, $\bar{\mathcal{P}}_{\mu}$, and N^{μ} are not fixed on the boundaries and will be integrated over on the initial hypersurface.

The transition functional on the extended phase space is

$$G[h'_{ij},t';h_{ij},t]=\int D\mu \exp\left\{i\int dt\, S_{\mathrm{FV}}\right\} , \tag{5.8a}$$

where the measure is

$$D\mu=d\pi^{ij}dh_{ij}dN^{\mu}dP_{\mu}d\mathcal{C}^{\mu}d\bar{\mathcal{C}}_{\mu}d\bar{\mathcal{P}}^{\mu}d\mathcal{P}_{\mu} \tag{5.8b}$$

and the sum is over all phase-space paths subject to the boundary conditions already discussed previously. The measure in (5.8b) is the usual canonically invariant Liouville measure on paths in the extended phase space. Because all the variables are dynamical there are no divergences arising from summing over equivalent field configurations. This is because the BRS transformation is not a local gauge transformation but a global one. The transition functional (5.8) is BRS invariant because both the action and measure are invariant. As seen in the case of the physical transition functional (3.15), the gauge choice φ^μ must be compatible with the data fixed on the boundaries.

How is the transition functional on the extended set of variables related to (3.15)? Fradkin and Vilkovisky proved a general theorem[31] that for theories with first-class constraints such as Yang-Mills theories and Einstein gravity, the value of this BRS-invariant transition functional is independent of the choice of Φ. In addition, by appropriately choosing φ^μ (5.8) reduces to (3.15) when the extra phase-space variables are integrated over. Specifically, one takes $\varphi^\mu = 1/\beta G^\mu$ where β is an arbitrary constant and rescales $\mathcal{C}^\mu = \beta\mathcal{C}^\mu$, $P_\mu = \beta P_\mu$. The measure of (5.8) is unchanged by this scaling because one variable is Grassmann and the other is not. Because of the Fradkin-Vilkovisky theorem, the functional integral is independent of the value of β. This allows one to set $\beta=0$. Doing the integrations over the extended variables, one now obtains the expression for the transition functional in the physical variables (3.15). Therefore the transition functional (5.8) is equal to the physical transition functional for arbitrary choices of gauge including those involving the redundant variables. Thus, (5.8) displays the equivalence of canonical and covariant gauge choices for gravity.

We have obtained a general parametrized path integral for the physical transition amplitude. Because it is quadratic in the momenta, the integrations over these variables can be performed to arrive at the equivalent parametrized configuration-space transition functional. This is most easily discussed by restricting φ^μ to be independent of the momenta π^{ij} although it can be shown for more general cases.[21,31] One shifts the momenta π^{ij}, $\mathcal{P}^\mu$, and $\overline{\mathcal{P}}_\mu$ in the action (5.6) by the appropriate combination of variables to bring it into quadratic form:

$$\pi^{ij}(x)=\pi^{ij}(x)+h^{1/2}[K^{ij}(x)-h^{ij}K(x)]-\frac{i}{N^0(x)}\overline{\mathcal{C}}_\mu(x)\int\frac{\delta\varphi^\mu(x)}{\delta h_{ij}(x')}\mathcal{C}^0(x')\,, \tag{5.9a}$$

$$\mathcal{P}^\mu(x)=\mathcal{P}^\mu(x)-i\dot{\mathcal{C}}^\mu(x)+\int[-\tfrac{1}{2}\overline{\mathcal{C}}_\lambda(\bar{x})\mathcal{C}^\nu(x')\mathcal{C}^\rho(x'')\{\varphi^\lambda(\bar{x}),U^\mu_{\rho\nu}(x'',x';x)\}-iN^\nu(x'')\overline{\mathcal{C}}^\rho(x')U^\mu_{\rho\nu}(x',x'';x)]\,, \tag{5.9b}$$

$$\overline{\mathcal{P}}_\mu(x)=\overline{\mathcal{P}}_\mu(x)+i\dot{\overline{\mathcal{C}}}_\mu(x)+\int\overline{\mathcal{C}}_\nu(x')\{\varphi^\nu(x'),P_\mu(x)\}\,, \tag{5.9c}$$

where K_{ij} is the extrinsic curvature

$$K_{ij}=-\frac{1}{2N^0h^{1/2}}(\dot{h}_{ij}-2D_{(i}N_{j)})\,. \tag{5.9d}$$

(K_{ij} differs from the usual form because N^0 is a density of weight -1.) The additional term in (5.9a) arises because the Poisson brackets $\{\varphi^\nu,\mathcal{H}_\mu\}$ in (5.7) is linear in the momenta. After this shift the action becomes

$$S_{\mathrm{FV}}=\int_t^{t'}dt\,[-i\overline{\mathcal{P}}_\mu(x)\mathcal{P}^\mu(x)-N^0G_{ijkl}\pi^{ij}(x)\pi^{kl}(x)]+S_L\,, \tag{5.10a}$$

$$\begin{aligned}S_L=&\int_M d^4x\,N^0h\,[K^{ij}K_{ij}-K^2+R(h)]-\int_t^{t'}dt\,E\\&+\int dt(\,P_\mu(x)\Phi^\mu(x)-i\overline{\mathcal{C}}_\mu(x)\{\Phi^\mu(x),\mathcal{H}_\nu(x')\}\,|_{\mathrm{cl}}\mathcal{C}^\nu(x')\\&\qquad-i\overline{\mathcal{C}}_\mu(x)\{\Phi^\mu(x),P_\nu(x')\}[\dot{\mathcal{C}}^\nu(x')-N^\rho(x'')\mathcal{C}^\lambda(\bar{x})U^\nu_{\lambda\rho}(x'',\bar{x};x')])\,.\end{aligned} \tag{5.10b}$$

The notation $\{\Phi^\mu(x),\mathcal{H}_\nu(x')\}\,|_{\mathrm{cl}}$ means to evaluate the Poisson brackets at the classical value of π^{ij}, $\pi^{ij}_{\mathrm{cl}}=-h^{1/2}(K^{ij}-h^{ij}K)$. S_L is the sum of the classical Lorentzian action for Einstein's theory, the integral over the Arnowitt-Deser-Misner (ADM) energy and contributions from ghost terms. The gauge choice

$$\Phi^\mu=\dot{N}^\mu-\varphi^\mu \tag{5.11}$$

has been redefined to simplify the form of the ghost term. Performing the integrations over π^{ij}, $\mathcal{P}^\mu$, and $\overline{\mathcal{P}}_\mu$ in (5.8) using the action (5.10a) results in a contribution of the determinant $|N^0G_{ijkl}|^{-1/2}$ to the measure.

The transition functional is now over the configuration space variables. However the part of (5.10b) that gives the Faddeev-Popov determinant still appears in the Hamiltonian form; the infinitesimal gauge transformation of the gauge-fixing term is implemented by the Poisson brackets with the constraints. In order to convert this term into Lorentzian form, where infinitesimal gauge transformations can be implemented by Lie derivatives, we need to make another change of variables. This change can be found by comparing the result of performing an infinitesimal Lorentzian gauge transformation parametrized by the vector f^μ,

$$\delta_f g_{\alpha\beta}=f^\gamma\partial_\gamma g_{\alpha\beta}+g_{\alpha\gamma}\partial_\beta f^\gamma+g_{\beta\gamma}\partial_\alpha f^\gamma\,, \tag{5.12}$$

to the result of performing a Hamiltonian one (3.7) parametrized by the vector F^μ. What choice of F^μ will give the same transformation of the metric using (3.7) as f^μ does using (5.12)? The answer is

$$F^i=f^i+N^if^0,\quad F^0=N^0f^0\,. \tag{5.13}$$

Therefore, in order to rewrite the Poisson brackets in (5.10b) as a Lorentzian gauge transformation the variable $\mathcal{C}^\mu$ has to be changed as indicated in (5.13); this change of variables will introduce a factor of N^0 in the measure. The final result for the parametrized transition functional is

$$G[h'_{ij},t';h_{ij},t]=\int D\bar{\mu}\,\delta[\Phi^{\mu}]\exp(iS_L)\ , \qquad (5.14a)$$

where

$$l^2S_L=S[g]-\int_t^{t'}dt\,E+S^{gh}[g,\bar{\mathcal{C}},\mathcal{C}]\ , \qquad (5.14b)$$

$$S^{gh}=\int dt[-i\bar{\mathcal{C}}_{\mu}(x)\delta_{\mathcal{C}}\Phi^{\mu}(x)]\ . \qquad (5.14c)$$

E is given by (3.5d) and $S(g)$ is the Einstein action (3.1). The measure is

$$D\bar{\mu}=\frac{1}{N^0}\,|\,N^0G_{ijkl}\,|^{-1/2}dh_{ij}dN^{\mu}d\mathcal{C}^{\mu}d\bar{\mathcal{C}}_{\mu}$$

$$=g^{00}(g)^{-3/2}dg_{\alpha\beta}d\mathcal{C}^{\mu}d\bar{\mathcal{C}}_{\mu}\ . \qquad (5.14d)$$

The notation $\delta_{\mathcal{C}}\Phi^{\mu}$ means to perform the infinitesimal transformation (5.12) on the gauge-fixing term with $\mathcal{C}^{\mu}$ as the vector field. The action and measure have been rewritten in terms of the metric $g_{\alpha\beta}$. The integration over P^{μ} has resulted in a gauge-fixing δ function in the measure of (5.14a). The boundary conditions on the remaining variables are $\mathcal{C}^{\mu}=\bar{\mathcal{C}}_{\mu}=0$ and h_{ij} matches the values on the t and t' constant hypersurfaces; the N^{μ} are integrated over on the initial hypersurface.

By the series of steps sketched above we have derived the Lorentzian transition functional in its fully parametrized form. It is local in the metric variables $g_{\alpha\beta}$ and manifestly invariant. To make this path integral definite, we shall take the class of tensor field configurations summed over to be defined in terms of metric perturbations as discussed in Sec. III for the transition functional (3.15). Choosing an asymptotically flat coordinate system, the metric is written as $g_{\alpha\beta}=\eta_{\alpha\beta}+l\gamma_{\alpha\beta}$. The initial and final data are taken to be of the form $h_{ij}=\delta_{ij}+l\gamma_{ij}$ and the sum is taken over all bounded perturbations $\gamma_{\alpha\beta}$ with the fall-off behavior (3.4) that match this data at t and t'.

The transition functional (5.14a) defined on this class of metric configurations cannot be computed exactly but we can compute its asymptotic expansion in powers of l using perturbation theory. To do this, the action (5.14b) is expanded in the metric perturbation $l\gamma_{\alpha\beta}$ and separated into a quadratic piece and an interaction piece which contains the higher-order terms. One finds that

$$S_L=S_2[\gamma]+S_2^{gh}[\bar{\mathcal{C}},\mathcal{C}]+S_I[\gamma,\bar{\mathcal{C}},\mathcal{C}]\ , \qquad (5.15a)$$

$$S_I=\sum_{k=3}^{\infty}S_k[\gamma]+S_k^{gh}[\gamma,\bar{\mathcal{C}},\mathcal{C}]\ , \qquad (5.15b)$$

where S_2 is the Lorentzian action for linearized gravity corresponding to the linearized Euclidean action (4.1) and S_2^{gh} is the quadratic Lorentzian ghost term corresponding to (4.16b). The S_k and S_k^{gh} are the contributions to S_I of order $k-2$ in l that come from the expansion of the curvature and ghost terms in (5.14b) to the appropriate order in the metric perturbation. The exponential of the interaction term is then written as its power series

$$G[\gamma'_{ij},t';\gamma_{ij},t]$$
$$=\int D\bar{\mu}\,\delta[\Phi^{\mu}]\exp[i(S_2+S_2^{gh})]\left[\sum_{j=0}^{\infty}\frac{1}{j!}(S_I)^j\right] \qquad (5.16)$$

and the order of functional integration and integration over spacetime points is interchanged. The terms involving $g_{\alpha\beta}$ in the measure $D\bar{\mu}$ (5.14d) are also expanded in powers of l. For simplicity we will assume that Φ^{μ} is linear in $g_{\alpha\beta}$. The δ function of the gauge-fixing condition will then be linear in $\gamma_{\alpha\beta}$. If Φ^{μ} is not linear in the metric, then its δ function will be more complicated to evaluate; basically additional terms will enter into the measure of the functional integral from the change of variables needed to make it linear.

The leading-order contribution to this Lorentzian path integral for the transition functional is simply that for the linearized theory. The changes to the transition functional introduced by the interactions are computed perturbatively to the desired order in l by including the appropriate contributions to the transition functional for the linearized theory from Gaussian integrations over $S_2+S_2^{gh}$ weighted by the interaction terms. In order to explicitly carry out these computations to arrive at physical quantities, the theory needs to be regulated and renormalization counterterms need to be introduced. However, as stated in the Introduction, we are concentrating on finding formal integrals expressing the kinematics of the theory and therefore will assume that the standard procedures for handling the divergences in these quantities can be implemented as needed.

Given the transition functional as computed in perturbation theory the next step is to construct an alternate transition functional that is (1) identical to (5.16) order by order in perturbation theory and (2) convergent when $t\to -i\tau$. The observation that the leading-order term in (5.16) is linearized gravity and the discussion of Sec. IV suggest the following approach to finding a nonlocal action for the full theory that is physically equivalent to the local invariant one (5.14b). First decompose the metric perturbation into the equivalence classes of the linearized theory (4.5), $\gamma_{\alpha\beta}=\phi_{\alpha\beta}+2\chi\eta_{\alpha\beta}$ where $\chi=0$ on the boundaries. Using this set of variables (5.16) becomes

$$G[\gamma'_{ij},t';\gamma_{ij},t]=\int D\mu'\delta[\Phi^{\mu}]\delta[R^{(1)}(\phi)]\exp(iS_2[\phi,\chi]+iS_2^{gh}[\bar{\mathcal{C}},\mathcal{C}])\left[\sum_{j=0}^{\infty}\frac{1}{j!}(S_I[\phi,\chi,\bar{\mathcal{C}},\mathcal{C}])^j\right]\ , \qquad (5.17)$$

where S_2 and S_I are now written in terms of the decomposition (4.5) and

$$D\mu'=[g^{00}(g)^{-3/2}]\left|\frac{\delta R^{(1)}(\phi)}{\delta\omega}\right|d\phi_{\alpha\beta}d\chi\,d\mathcal{C}^{\mu}d\bar{\mathcal{C}}_{\mu}\ . \qquad (5.18)$$

The factors of g in (5.18) are also written in terms of the decomposition (4.5). This form of the transition amplitude is identical to (5.16) as it differs only by a change of variables. To lowest order in l, (5.18) is the transition functional for linearized gravity weighted by the local action S_2. In Sec. IV we argued that the alternate transition functional (4.21) weighted by the nonlocal action $\hat{S}_2$ (4.22) is physically equivalent. Moreover, when a is positive, the Feynman-Kac procedure implemented using this alternate functional produces a manifestly convergent Euclidean functional integral for the ground state. Therefore we will construct the alternate transition functional by using $\hat{S}_2(\phi,\mathcal{X})$ instead of $S_2(\phi,\mathcal{X})$ in the exponent of (5.17) and then appropriately modifying the interaction terms so as to obtain the same result. It is convenient to set $a=24$ in $\hat{S}_2$ to match the absolute value of the coefficient of the corresponding term in S_2. Then we will demonstrate the equivalence.

The next step is to find the correct interaction terms for the alternate transition functional. This can be done by studying the form of the Gaussian integrations over $\mathcal{X}$ in (5.17). An arbitrary interaction term in (5.17) can be written as

$$(S_I[\phi,\mathcal{X},\bar{\mathcal{C}},\mathcal{C}])^j=\int \mathcal{X}(x_1)\mathcal{X}(x_2)\cdots\mathcal{X}(x_n)\times F(\phi,\bar{\mathcal{C}},\mathcal{C};x_1,x_2,\ldots,x_n)\,, \quad (5.19)$$

where F is a function of the other variables and will in general contain δ functions of the coordinates and partial derivatives. The integrations over $\mathcal{X}$ in (5.17) are then of the form

$$F[\phi,\bar{\mathcal{C}},\mathcal{C}]=\int d\mathcal{X}\exp\left[+6i\int_M d^4x(\partial_\alpha\mathcal{X})^2\right]\times\int[\mathcal{X}(x_1)\mathcal{X}(x_2)\cdots\mathcal{X}(x_n)\times F(\phi,\bar{\mathcal{C}},\mathcal{C};x_1,x_2,\ldots,x_n)]. \quad (5.20)$$

The sign of the $\mathcal{X}$ term in S_2 has opposite sign from that of $\hat{S}_2$. Because $\mathcal{X}=0$ on the boundaries, the classical contribution to this integral vanishes. The only contributions to this integral will come from the fluctuations around the classical path; this means that only interaction terms even in $\mathcal{X}$ will contribute as Gaussian integrals over an odd number of variables vanishes. If n is even then the integration over $\mathcal{X}$ in (5.20) will give a factor of $(-1)^{n/2}$ relative to integration over the same interaction term weighted by an exponential with opposite sign. If we now modify the interaction terms by taking $\mathcal{X}$ to $i\mathcal{X}$, as well as changing the sign in the exponent then the added factor of $(i)^n$ will give the same contribution; consequently

$$F[\phi,\bar{\mathcal{C}},\mathcal{C}]=\int d\mathcal{X}\exp\left[-6i\int_M d^4x(\partial_\alpha\mathcal{X})^2\right]\int(i)^n\mathcal{X}(x_1)\mathcal{X}(x_2)\cdots\mathcal{X}(x_n)F(\phi,\bar{\mathcal{C}},\mathcal{C};x_1,x_2,\ldots,x_n)\,. \quad (5.21)$$

A transition functional weighted by a nonlocal action that is physically equivalent to (5.17) in perturbation theory is thus

$$G[\gamma'_{ij},t';\gamma_{ij},t]=\int D\mu'\delta[\Phi^\mu]\delta[R^{(1)}(\phi)]\exp(i\hat{S}_2[\phi,\mathcal{X}]+iS_2^{\mathrm{gh}})\left[\sum_{j=0}^{\infty}\frac{1}{j!}(S_I[\phi,i\mathcal{X},\bar{\mathcal{C}},\mathcal{C}])^j\right], \quad (5.22)$$

where the factors of $\mathcal{X}$ that appear in the measure (5.18) are also taken to $i\mathcal{X}$. The modified action in (5.22) is complex; however, the resulting transition functional is the same as that weighted by the real action because only terms even in $\mathcal{X}$ contribute to it.

Now the rotation of the time coordinate can be carried out before functional integration order by order in l because the resulting quadratic action $\hat{I}_2$ is manifestly positive definite. The resulting parametrized Euclidean functional integral for the ground-state wave functional (3.18) is

$$\Psi_0[\gamma_{ij}]=\int D\mu'\delta[\Phi^\mu]\delta[R^{(1)}(\phi)]\exp(-\hat{I}_2-I_2^{\mathrm{gh}})\left[\sum_{j=0}^{\infty}\frac{1}{j!}(I_I[\phi,i\mathcal{X},\bar{\mathcal{C}},\mathcal{C}])^j\right], \quad (5.23)$$

where I_I are the Euclideanized interaction terms corresponding to (5.15b). This integral is convergent and O(4) invariant; it does not manifestly display the coordinate invariance of the theory because the conformal factor was isolated using the linearized scalar curvature. However, its action is simply related to the manifestly gauge invariant one of the full theory.

Can we recover a parametrized, convergent Euclidean functional integral that is manifestly diffeomorphism invariant? The answer is a qualified yes. One can do so by using an alternate Lorentzian perturbation theory to evaluate (5.14); however, because this new theory is very nonlocal in terms of the original perturbations, it does not obviously reproduce the same physical integral. To get this alternate Lorentzian perturbation theory one first forms the identity

$$1=\int d\Omega\left|\frac{\delta R(\Omega^{-2}g)}{\delta\Omega}\right|\delta[R(\Omega^{-2}g)]\,, \quad (5.24)$$

where the scalar curvature is evaluated on $\Omega^{-2}g_{\alpha\beta}$. The integration over Ω is defined to be over configurations of the form $\Omega=1+l\mathcal{X}$ where $\mathcal{X}\sim 1/r$ at spatial infinity and vanishes on the boundaries. The metric is assumed to be perturbative, $g_{\alpha\beta}=\eta_{\alpha\beta}+lh_{\alpha\beta}$, with appropriate falloff behavior as before. This identity is true if the scalar curvature of $g_{\alpha\beta}$ is sufficiently small. One then inserts (5.24) into (5.14) and then changing variables to $g_{\alpha\beta}=\Omega^2\bar{g}_{\alpha\beta}$ one obtains

$$G[h'_{ij},t';h_{ij},t]=\int D\bar{\mu}\,\delta[\Phi^\mu]\delta[R(\bar{g})]\exp(iS_L)\,, \quad (5.25a)$$

where

$$D\bar{\mu}=\bar{g}^{00}(\bar{g})^{-3/2}\Omega^4\left|\frac{\delta R}{\delta\Omega}\right|d\bar{g}_{\alpha\beta}d\chi\, d\mathscr{C}^{\mu}d\bar{\mathscr{C}}_{\mu}\ . \qquad (5.25b)$$

S_L is (5.14b) evaluated with $g_{\alpha\beta}=\Omega^2\bar{g}_{\alpha\beta}$. One can in principle define this integral in a perturbation expansion in l using $\bar{g}_{\alpha\beta}=\eta_{\alpha\beta}+l\phi_{\alpha\beta}$, $\Omega=1+l\chi$; however, in practice evaluating it is difficult because one has to solve a nonlinear equation for $\phi_{\alpha\beta}$ in order to integrate over the decomposition fixing δ function. If one proceeds by solving this equation for one component of $\phi_{\alpha\beta}$ in terms of the others as a power series in l, one can then similarly expand the action and measure by performing this substitution. The leading order of (2.25a) is found to be the transition amplitude for linearized gravity weighted by its action $S_2[\phi,\chi]$. One again can make the same arguments of (5.18)–(5.22) to derive the corresponding nonlocal action. The resulting Euclidean functional integral in this case is then

$$\Psi_0[\gamma_{ij}]=\int D\bar{\mu}\,\delta[\Phi^{\mu}]\delta[R(\bar{g})]\exp(-\hat{I})\ , \qquad (5.26a)$$

where

$$\hat{I}=I[\bar{g},\hat{\Omega}]-\int d\tau\, E[\phi,i\chi]+I^{\mathrm{gh}}[\phi,i\chi,\bar{\mathscr{C}},\mathscr{C}]\ . \qquad (5.26b)$$

$I[\bar{g},\Omega]$ is the Euclidean action for Einstein gravity (1.6) evaluated for $\Omega=1+il\chi$. E and I^{gh} are the appropriate forms of (3.5d) and (5.14c) and again, factors of χ are taken to $i\chi$ in the measure. The above functional integral is to be evaluated perturbatively in the same manner as (5.23). The difficulty with this form is checking that it equals (5.23) order by order in perturbation theory; that is, that the interactions obtained by solving the decomposition-fixing δ function perturbatively in $\phi_{\alpha\beta}$ give the same results as the local Lorentzian perturbation theory. That it does can be verified to the next order in l after linearized gravity explicitly; the equivalence was checked with the help of the MACSYMA ITENSR package. However, it is difficult to carry this out to higher orders. This Euclidean functional integral is manifestly diffeomorphism invariant and convergent. It is weighted by a nonlocal action different than that used in (2.23) and it is plausible, though not explicitly verified to all orders, that it equals the functional integral in the physical fields. We thus have obtained two prescriptions for Euclidean functional integrals with manifestly positive actions. This emphasizes that the form of the convergent parametrized Euclidean functional integral is not unique when its action is nonlocal in the original set of variables.

VI. CONCLUSION

When the Euclidean action for a theory with invariances is unbounded below, Euclidean functional integrals for its quantum states may or may not exist. Whether or not they do is determined by the action for the theory expressed in the physical degrees of freedom. In the case of linearized gravity, the physical degrees of freedom can be explicitly identified and the Hamiltonian in terms of them is positive definite. One can proceed to quantize the theory in its physical variables and indeed can form convergent Euclidean functional integrals in terms of the transverse traceless metric perturbations. Although the physical variables cannot be explicitly isolated for Einstein's theory in asymptotically flat spacetimes, the Hamiltonian is positive for metric configurations that satisfy the constraints by the positive-energy theorem. Again this means that convergent Euclidean functional integrals for the theory can be given in terms of the physical variables. However, in either theory, because of the unboundedness of the Euclidean action, one cannot find expressions equal to those given in the physical fields involving an action that are manifestly invariant and local in the full set of variables. Still, one can find useful parametrized forms of the Euclidean integrals that come close to achieving these goals.

For linearized gravity, we can recover functional integrals weighted by a manifestly gauge- and O(4)-invariant Euclidean action. It is even local in the set of variables $\phi_{\alpha\beta}$ and χ. However it is nonlocal when expressed in the original metric perturbation $\gamma_{\alpha\beta}$. The action weighting these convergent Euclidean functional integrals is the same as would be obtained by conformally rotating χ in the classical Euclidean action (3.1) for linearized gravity.

For the case of asymptotically flat spacetimes, the task of finding convergent parametrized Euclidean functional integrals is more difficult because the physical variables cannot be isolated explicitly. However, as we showed, one can proceed by using the manifestly invariant form of the Lorentzian transition functional as a guide. One looks for a parametrized transition functional weighted by an alternate nonlocal action that equals the first when both are evaluated in perturbation theory. There are many such nonlocal actions which will have this property. In addition we require that this action be chosen so that the Euclidean functional integrals corresponding to the alternate transition functional will be manifestly convergent. In Sec. V we demonstrated that one such choice resulted in manifestly convergent, O(4)-invariant Euclidean functional integrals. This action was not manifestly diffeomorphism invariant; however it was simply related to the manifestly invariant one of the full theory. This particular form of the Euclidean functional integral is local in the variables $\phi_{\alpha\beta}$ and χ of the linearized theory used to derive it; in this sense it is almost local in the metric perturbations $\gamma_{\alpha\beta}$. This property makes it an especially convenient form for calculation.

Can convergent Euclidean functional integrals for physical quantities be weighted by manifestly diffeomorphism invariant actions? The procedure by which (5.24) was derived suggests that such integrals correspond to another choice of a nonlocal action. However, there is a price to be paid; such integrals are highly nonlocal in the metric perturbations. This makes explicitly verifying their equivalence to those expressed in the physical degrees of freedom via perturbation theory difficult. However, as discussed at the end of Sec. V, the equivalence can be verified to hold through the first order in the interaction for the conformally rotated action of the full theory.

How are these Euclidean functional integrals for both linearized and perturbative gravity related to the conformally rotated ones of Gibbons, Hawking, and Perry? Their form suggests that these functional integrals could be obtained formally by an appropriate distortion of the contour of integration over χ to $i\chi$. However, this prescription of conformal rotation begins with an Euclidean functional integral that is manifestly divergent and consequently not well defined. In addition, it is hard to get the correct Jacobian factor in the measure by this method. Moreover, we showed that there are many more prescriptions for convergent Euclidean functional integrals than this one as there are many possible positive nonlocal actions. A more satisfactory way to view these integrals is the one presented in this paper; they arise naturally in the course of quantizing a theory with invariances. First one isolates the physical degrees of freedom. Then one constructs functional integrals for physical quantities in terms of them. Finally one adds in integrations over the redundant variables to recover manifest invariance. When doing this for quantities expressed as Euclidean functional integrals, there is an additional restriction on the parametrization process; one is only allowed to add back in manifestly convergent quantities. However, the result of this process is mostly determined by the form that is desired for the final answer; the content of the theory is contained in the functional integrals given in the physical variables.

ACKNOWLEDGMENTS

I would like to thank J. Hartle for suggesting this problem and many valuable discussions during the course of this work. I would also like to thank K. Kuchař for his helpful comments on this work in progress. This work was supported in part by NSF Grant No. PHY 85-06686. The use of the MACSYMA ITENSR package was supported by an Institute of Geophysics and Planetary Physics grant from Lawrence Livermore National Laboratory. I most sincerely thank AT&T Bell Laboratories for their generous and longstanding financial support of my graduate research. This paper was presented as a thesis to the Department of Physics, The University of Chicago, in partial fulfillment of the requirements for the Ph.D degree.

APPENDIX: THE MEASURE

To show that the functional integrals in (4.15) and (4.16) have a definite and concrete meaning, we shall evaluate the factors making up the measure $d\phi_{\alpha\beta}d\chi$, etc., in a particular set of "coordinates on the function spaces" with the specific gauge choice (4.11a). To make the mode sums discrete, we will take our spacetime to be a finite box of volume L^4 interior to the planes $\tau=0$, $\tau=-L$, and $x^i=\pm L/2$. We shall then use the coefficients of the Fourier expansion of the integration variables in this box as our coordinates on the function space. The expansion is

$$
\begin{aligned}
t_{\alpha\beta}(x)&=t^{\rm cl}_{\alpha\beta}(x)+\sum_{\nu=1}^{2}\sum_k{}' t^{(\nu)}(k)t^{(\nu)}_{\alpha\beta}(k,x)\,,\\
l_{\alpha\beta}(x)&=\sum_{\nu=1}^{3}\sum_k{}' l^{(\nu)}(k)l^{(\nu)}_{\alpha\beta}(k,x)\,,\\
\phi^L_{\alpha\beta}(x)&=\sum_{\nu=1}^{4}\sum_k{}' \phi^{L(\nu)}(k)\phi^{L(\nu)}_{\alpha\beta}(k,x)\,,\\
\phi^T_{\alpha\beta}(x)&=\sum_k{}' \phi^{T}(k)\phi^{T}_{\alpha\beta}(k,x)\,,\\
\chi(x)&=\sum_k{}' \chi(k)s(k,x)\,.
\end{aligned}
\tag{A1}
$$

This expansion is done by expanding around the classical solution of the linearized Einstein equations which satisfies the boundary conditions as fixed in Sec. IV. $t^{\rm cl}_{\alpha\beta}(x)$ is the classical solution which matches the argument of the wave function h^{TT}_{ij} on the $\tau=0$ boundary and vanishes on the other boundary surfaces; with our boundary conditions, the classical solutions of the other metric components vanish. (We assume that the finite volume box is chosen to be large enough so that the compact support of the initial data at $\tau=0$ is interior to it.)

The class of configurations that is summed over for the fluctuations around the classical solution is specified by the boundary conditions that the spatial components of the fields vanish at $\tau=0$ and $\tau=-L$ and are periodic in the spatial directions. The gauge choice we shall use is (4.11a). The modes on the right-hand side of (A1) will be constructed to satisfy these conditions. The notation $\sum_k'$ in (A1) means the sum over all $k^0>0$ and k such that $k^i\neq 0$. Modes with $k^i=0$ will have infinite action in the infinite-volume limit and thus will not contribute to the functional integrals; we omit them for convenience in defining the tensor modes. To explicitly construct these modes it is useful to first define, for a given k^α satisfying the above restrictions,

$$
\begin{aligned}
s(k,x)&=\frac{2}{L^2}\sin(k_0\tau)\sin(k_ix^i),\quad k^3>0\\
&=\frac{2}{L^2}\sin(k_0\tau)\cos(k_ix^i),\quad k^3<0\,,\\
p_\alpha(k,x)&=\frac{1}{kk_0}s^\gamma\partial_\gamma\dot{s}(k,x)n_\alpha+\frac{k_0}{k}s(k,x)s_\alpha\,,\\
p_{\alpha\beta}(k,x)&=\frac{1}{k^2}[\,(s^\gamma\partial_\gamma)^2s(k,x)n_\alpha n_\beta\\
&\qquad-2s^\gamma\partial_\gamma\dot{s}n_{(\alpha}s_{\beta)}+\ddot{s}s_\alpha s_\beta]\,,
\end{aligned}
\tag{A2}
$$

where $s_\alpha(k)$ is the unit vector in the direction of the projection of k^α onto the space orthogonal to n^α and $k=(k_\alpha k^\alpha)^{1/2}$. Using $\epsilon^{(\nu)}_\alpha$, two orthonormal vectors transverse to both k_α and s_α, we then construct the unit tensors

$$
\begin{aligned}
t^{(1)}_{\alpha\beta}(k)&=\sqrt{2}\,\epsilon^{(1)}_{(\alpha}\epsilon^{(2)}_{\beta)}\,,\\
t^{(2)}_{\alpha\beta}(k)&=\frac{1}{\sqrt{2}}(\epsilon^{(1)}_\alpha\epsilon^{(1)}_\beta-\epsilon^{(2)}_\alpha\epsilon^{(2)}_\beta)\,,\\
\phi^T_{\alpha\beta}(k)&=\frac{1}{\sqrt{2}}(\epsilon^{(1)}_\alpha\epsilon^{(1)}_\beta+\epsilon^{(2)}_\alpha\epsilon^{(2)}_\beta)\,.
\end{aligned}
\tag{A3}
$$

Then the tensor modes are

$$
\begin{aligned}
t^{(\nu)}_{\alpha\beta}(k,x) &= s(k,x)\, t^{(\nu)}_{\alpha\beta}(k)\ , \\
\phi^{T}_{\alpha\beta}(k,x) &= s(k,x)\, \phi^{T}_{\alpha\beta}(k)\ , \\
\phi^{L(\nu)}_{\alpha\beta}(k,x) &= \frac{\sqrt{2}}{k}\epsilon^{(\nu)}_{(\alpha}\partial_{\beta)}s(k,x), \quad \nu=1,2\ , \\
\phi^{L(3)}_{\alpha\beta}(k,x) &= \frac{\sqrt{2}}{k}\partial_{(\alpha}p_{\beta)}(k,x)\ , \\
\phi^{L(4)}_{\alpha\beta}(k,x) &= \frac{1}{k^2}\partial_{\alpha}\partial_{\beta}s(k,x)\ , \\
l^{(\nu)}_{\alpha\beta}(k,x) &= \sqrt{2}\epsilon^{(\nu)}_{(\alpha}p_{\beta)}(k,x), \quad \nu=1,2\ , \\
l^{(3)}_{\alpha\beta}(k,x) &= (\tfrac{1}{3})^{1/2}\phi^{T}_{\alpha\beta}(k,x)-(\tfrac{2}{3})^{1/2}p_{\alpha\beta}(k,x)\ .
\end{aligned}
\tag{A4}
$$

These tensor modes are real and normalized to 1 on the finite volume. The real functions $t^{(\nu)}(k)$, $\phi^{T(\nu)}(k)$, $\phi^{L(\nu)}(k)$, and $\chi(k)$ become the coordinates on the space of functions over which we integrate and the actions may be expressed in terms of them. For example, the action I^{g}_{2} (4.12) is

$$
\begin{aligned}
I^{g}_{2} &= \sum_{k}{}' I^{g}_{2}(k)\ , \\
I^{g}_{2}(k) &= \frac{k^2}{4}\sum_{\nu=1}^{3}\{[l^{(\nu)}(k)]^2+a[\chi(k)]^2\}\ ,
\end{aligned}
\tag{A5}
$$

and the action i_2 is

$$
\begin{aligned}
i_2 &= i^{\rm cl}_{2}[h^{TT}_{ij}]+\sum_{k}{}' i_2(k)\ , \\
i_2(k) &= \frac{k^2}{4}\sum_{\nu=1}^{2}[t^{(\nu)}(k)]^2\ ,
\end{aligned}
\tag{A6}
$$

and $i^{\rm cl}_{2}[h^{TT}_{ij}]$ is the classical action evaluated for the appropriate classical solution. The functional δ functions and Faddeev-Popov determinants are, in the gauge (4.11a),

$$
\delta[\partial_\alpha\phi^{\alpha\beta}]\left|\frac{\delta G^{\alpha}}{\delta f^{\beta}}\right| = \prod_{k}{}' D_1 \left[\prod_{\nu=1}^{4}\delta[k\phi^{L(\nu)}(k)]\right][k^2]^4\ , \tag{A7a}
$$

$$
\delta[R^{(1)}(\phi^T)]\left|\frac{\delta R^{(1)}}{\delta\omega}\right| = \prod_{k}{}' D_2\delta[k^2\phi^{T}(k)][k^2]\ , \tag{A7b}
$$

where D_1 and D_2 are numerical constants determined by the normalization of the fields and implementation of the δ functions and determinants. The products of factors $[k^2]^4$ and $[k^2]$ in Eqs. (A7) are the Faddeev-Popov determinants represented as modes. Defining the measure (4.13) as

$$
dl_{\alpha\beta}d\phi^{L}_{\alpha\beta}d\chi = \prod_{k}{}' \frac{N}{D_1}\left[\prod_{\nu=1}^{3}dl^{(\nu)}(k)\right]\left[\prod_{\mu=1}^{4}d\phi^{L(\mu)}(k)\right]d\chi(k)\ , \tag{A8a}
$$

$$
d\phi^{T}_{\alpha\beta} = \prod_{k}{}' \frac{1}{D_2}d\phi^{T}(k)\ , \tag{A8b}
$$

and using (A5) and (A7) Eqs. (4.15) become (suppressing the label k on the mode amplitudes)

$$
\begin{aligned}
1 &= \int \prod_{k}{}' N\left[\prod_{\nu=1}^{3}dl^{(\nu)}\right]d\chi\left[\prod_{\mu=1}^{4}d\phi^{L(\mu)}\delta[k\phi^{L(\mu)}]\right][k^2]^4\exp[-I^{g}_{2}(k)]\ , \\
1 &= \int \prod_{k}{}' d\phi^{T}\delta[k^2\phi^{T}][k^2]\ ,
\end{aligned}
\tag{A9}
$$

where N is a constant needed to set (A9) to unity including a factor of $a^{1/2}$ as well as other numerical constants.

Using the Fourier modes (A1) the path integral over the physical degrees of freedom can also be made concrete. Defining the measure over the fluctuations around the classical solution to be

$$
dt_{\alpha\beta} = \prod_{k}{}' \frac{\pi}{4}\left[\prod_{\nu=1}^{2}dt^{(\nu)}(k)\right] \tag{A10}
$$

and using (A6), the wave functional is

$$
\Psi_0[h^{TT}_{ij}] = \mathcal{N}\exp(-i^{\rm cl}_{2}[h^{TT}_{ij}])\int\prod_{k}{}' \frac{\pi}{4}\left[\prod_{\nu=1}^{2}dt^{(\nu)}(k)\right]\exp[-i_2(k)]\ , \tag{A11}
$$

where $\mathcal{N}$ is a normalization parameter that can be computed explicitly in the following way. First evaluate the path integral (A11) over the fluctuations in the finite volume L^4. Then fix $\mathcal{N}$ by requiring that the resulting wave functional be the normalized product of the ground-state harmonic-oscillator wave functions whose arguments are the amplitudes of the Fourier transform of $h^{TT}_{ij}(x)$ in the appropriate measure as $L\to\infty$. One finds that $\mathcal{N}$ is the properly normalized ground-state wave functional for 0 initial data on the boundary surface at $\tau=-L$.

Finally for completeness we give the form of the parametrized wave functional (4.18) in the Fourier coordinates after integration over the Grassmann fields:

$$\Psi_0[h_{ij}^{TT}]=\mathcal{N}\int \prod_k{}' \frac{\pi N}{4} d^{10}\phi \, d\chi \left(\prod_{\mu=1}^{4} \delta[k\phi^{L(\mu)}]\right)\delta[k^2\phi^T][k^2]^5 \exp(-\hat{I}_2[h_{ij}^{TT},t^{(\nu)},l^{(\nu)},\chi]) \,, \tag{A12}$$

where $\prod'_k [\pi N/(4D_1 D_2)]d^{10}\phi \, d\chi$ is the product of (A8a), (A8b), and (A10) and $\hat{I}_2$ is the sum of (A5) and (A6).

[1]See, for example, M. B. Green, J. H. Schwarz, and E. Witten, *Superstring Theory* (Cambridge University Press, Cambridge, England, 1987), Vols. 1 and 2.

[2]See, for example, S. Weinberg, in *General Relativity: An Einstein Centenary Survey,* edited by S. W. Hawking and W. Israel (Cambridge University Press, Cambridge, 1979).

[3]S. W. Hawking, in *General Relativity: An Einstein Centenary Survey* (Ref. 2).

[4]S. W. Hawking, Nucl. Phys. **B239**, 257 (1984).

[5]J. B. Hartle and S. W. Hawking, Phys. Rev. D **28**, 2960 (1983).

[6]G. W. Gibbons, S. W. Hawking, and M. J. Perry, Nucl. Phys. **B138**, 141 (1978).

[7]R. Schoen and S. T. Yau, Phys. Rev. Lett. **42**, 547 (1979).

[8]R. Schoen and S. T. Yau, Phys. Rev. Lett. **43**, 1457 (1979).

[9]E. Witten, Commun. Math. Phys. **80**, 381 (1981).

[10]J. Hartle and K. Schleich, in *Quantum Field Theory and Quantum Statistics,* edited by T. A. Batalin, C. J. Isham, and G. A. Vilkovisky (Hilger, Bristol, to be published).

[11]L. D. Faddeev, Teor. Mat. Fiz. **1**, 3 (1969) [Theor. Math. Phys. **1**, 1 (1970)].

[12]L. D. Faddeev and V. N. Popov, Phys. Lett. **25B**, 30 (1967).

[13]J. B. Hartle and K. Kuchař, J. Math. Phys. **25**, 57 (1984).

[14]J. B. Hartle and K. Kuchař, in *Quantum Theory of Gravity,* edited by S. Christensen (Hilger, Bristol, 1984).

[15]P. A. M. Dirac, Can. J. Math. **2**, 129 (1950).

[16]P. A. M. Dirac, Proc. R. Soc. London **A246**, 326 (1958).

[17]A. Hanson, T. Regge, and C. Teitelboim, *Constrained Hamiltonian Systems* (Academia Nazionale dei Lincei, Rome, 1976).

[18]See, for example, P. Ramond, *Field Theory: A Modern Primer* (Benjamin Cummings, Reading, MA, 1981).

[19]R. P. Feynman and A. R. Hibbs, *Quantum Mechanics and Path Integrals* (McGraw-Hill, New York, 1965).

[20]L. D. Faddeev and V. N. Popov, Usp. Fiz. Nauk **111**, 427 (1973) [Sov. Phys. Usp. **16**, 777 (1974)].

[21]E. S. Fradkin and G. A. Vilkovisky, CERN Report No. TH-2332, 1977 (unpublished).

[22]Robert Wald, *General Relativity* (University of Chicago Press, Chicago, 1984).

[23]C. W. Misner, K. S. Thorne, and J. A. Wheeler, *Gravitation* (Freeman, San Francisco, 1973).

[24]B. S. DeWitt, Phys. Rev. **160**, 1113 (1967).

[25]C. Teitelboim and T. Regge, Ann. Phys. (N.Y.) **88**, 286 (1974).

[26]R. Arnowitt and S. Deser, Phys. Rev. **113**, 745 (1959).

[27]J. B. Hartle, Phys. Rev. D **29**, 2730 (1984).

[28]K. Kuchař, J. Math. Phys. **11**, 3322 (1970).

[29]D. Boulware (private communication).

[30]For a review, see M. Henneaux, Phys. Rep. **126**, 1 (1985).

[31]E. S. Fradkin and G. A. Vilkovisky, Lett. Nuovo Cimento **13**, 5 (1975).

[32]E. S. Fradkin and G. A. Vilkovisky, Phys. Lett. **55B**, 6224 (1975).

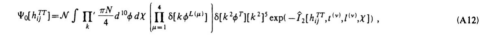

Nuclear Physics B298 (1988) 178–186
North-Holland, Amsterdam

QUANTUM TUNNELING AND NEGATIVE EIGENVALUES

Sidney COLEMAN*

Lyman Laboratory of Physics, Harvard University, Cambridge, Massachusetts 02138, USA

Received 31 August 1987

In the path-integral approach to the decay of a metastable state by quantum tunneling, the tunneling process is dominated by a solution to the imaginary-time equations of motion, called the bounce. In all known cases, the second variational derivative of the euclidean action at the bounce has one and only one negative eigenvalue. This note explains this phenomenon by showing it is an inevitable feature of the bounce for a wide class of systems. This class includes a set of particles interacting through potentials obeying some mild technical restrictions, and also theories of interacting scalar and gauge fields. There may exist solutions in other ways like bounces and which have more than one negative eigenvalue, but, even if they do exist, they have nothing to do with tunneling.

In the semiclassical approximation to quantum dynamics, we frequently study false ground states, time-independent states that are classically stable but decay through quantum tunneling. The decay probability per unit time of such a state is of the form

$$\Gamma = A\,e^{-B/\hbar}(1 + O(\hbar))\,. \tag{1}$$

For field-theory applications, it's useful to have a manifestly Lorentz-invariant method of computing A and B. The usual one is based on the euclidean (imaginary-time) version of Feynman's sum over histories [1]. One begins by finding a bounce, a time-reversal invariant solution of the imaginary-time equations of motion that approaches the false ground state at infinity. (The trivial constant solution is excluded.) The coefficient B is the euclidean action evaluated at the bounce. The coefficient A is the product of certain collective-coordinate factors and the square root of the absolute value of the determinant of the second variation of the action evaluated at the bounce (with zero eigenvalues from collective coordinates omitted).

In all known cases the second variation at the bounce has one and only one negative eigenvalue. There is a hand-waving argument for this. The bounce shifts the energy of the false ground state in the same way an instanton shifts the energy

* Work supported in part by the National Science Foundation under grant no. PHY82-15249.

0550-3213/88/$03.50

of a true ground state. The formula for the instanton energy shift *is* the formula described in the preceding paragraph, without the phrase "absolute value". Thus the negative eigenvalue is necessary to insure that the energy shift is imaginary, that is to say, that the state becomes unstable.

Even if we're gullible enough to swallow this argument, we're still left with an unanswered question. What are we to do if we find a solution for which the second variation has more than one negative eigenvalue? Is any number of negative eigenvalues satisfactory? Or must the number of negative eigenvalues be odd? Or perhaps equal to one modulo four?

There exist somewhat more plausible and considerably more elaborate arguments connecting the bounce to tunneling, but they are no more helpful on this point. The purpose of this note is to answer the question by showing that, for a wide class of dynamical systems, the bounce has one and only one negative eigenvalue. If we find a solution with more than one negative eigenvalue, we should throw it away; it's the wrong solution and has nothing to do with the tunneling process.

Let q^a be the generalized coordinates of our system, where a runs over a finite or infinite set. I shall establish the stated result when the euclidean lagrangian is of the form

$$L = \tfrac{1}{2} m_{ab}(q) \dot{q}^a \dot{q}^b + V(q) . \tag{2}$$

Here the sum over repeated indices is implied, m_{ab} is some positive-definite symmetric matrix function of the q's, V is some function of the q's, and the overdot denotes differentiation with respect to imaginary time.

I will choose coordinates such that the false ground state is at the origin, $q^a = 0$, and add a constant to V such that $V(0) = 0$. Because the false ground state is classically stable, the origin must be a local minimum of V. Because tunneling occurs, V must be negative somewhere. Thus we have a situation like that sketched in fig. 1: The origin is surrounded by a region of positive V, separated by a surface, Σ, on which V vanishes, from an exterior region of negative V. I shall have to make one technical assumption* about V, that ∇V vanishes nowhere on Σ. This ensures that Σ has everywhere a well-defined normal.

Some comments:

(i) As advertised, this is a wide class of systems. It includes both the theory of an arbitrary number of particles interacting through arbitrary velocity-independent potentials, and the theory of a set of scalar fields interacting through nonderivative interactions, coupled to abelian or nonabelian gauge fields (in temporal gauge). Unfortunately, it does not include one very interesting problem, quantum tunneling from de Sitter space in the theory of scalar fields coupled to einsteinian gravity. (At least, I have not been able to put this problem into appropriate form, or into any tractable generalization of it.)

* This assumption can be weakened considerably; see the appendix.

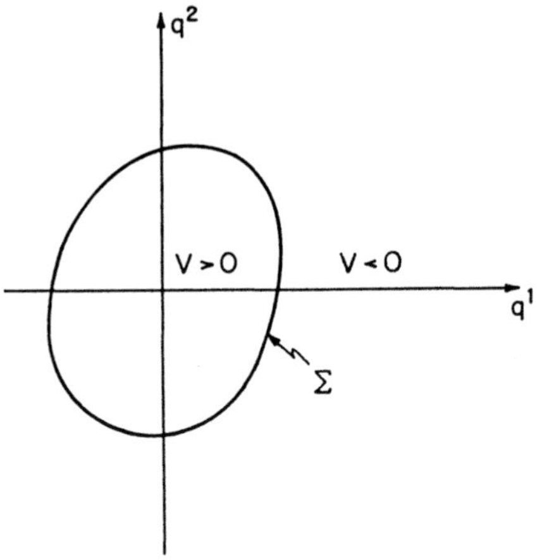

Fig. 1. The shape of the potential, V, for a tunneling problem. The system tunnels from the origin, where $V=0$, through a region of positive V, to the surface Σ, where V vanishes, and from which it can escape to the region of negative V.

(ii) Fig. 1 is drawn for the simplest case. However, it would not affect the argument if things were more complicated. For example, there could be additional regions of positive V, disconnected from the origin. Likewise, the region of positive V connected to the origin could extend to infinity in some directions, which would make Σ noncompact or even disconnected.

(iii) Although I'll deal with infinite numbers of degrees of freedom, I'll make no attempt at rigorous functional analysis; I'll blithely assume that all sums are as convergent and all functionals as smooth as I need.

Now for the argument:

For the systems under study, there is a well-developed WKB theory of barrier penetration [2], which can be summarized by the statement that the system tunnels through the barrier along the path of least resistance. In more detail: We define a metric in configuration space by

$$\mathrm{d}l^2 = m_{ab}\,\mathrm{d}q^a\,\mathrm{d}q^b\,, \tag{3}$$

and define a functional of paths that begin at the origin and terminate somewhere on Σ,

$$S_{\mathrm{B}}[q(l)] = \int \mathrm{d}l\sqrt{2V(q(l))}\,. \tag{4}$$

To avoid worrying about paths that penetrate the region of negative V, by convention we terminate all paths on their first intersection with Σ. (The suffix B denotes barrier penetration. This is the first of three closely related functionals we shall encounter.) A minimal barrier penetration path, $q_0(l)$, is one that minimizes S_B. The coefficient B in eq. (1) is given by

$$B = 2S_B[q_0(l)]. \tag{5}$$

For many problems, only the absolute minimum of S_B is important, but in other cases, even local minima are of interest. For example, penetration of Σ at different points might lead to escape into different valleys in V, and we might be interested in the probability of tunneling into each of these valleys, even if the probabilities for some valleys are exponentially small compared to those for others. In any case, local saddle points or maxima are of no interest. Thus, the second variation at the stationary point must be positive semidefinite operator,

$$\delta^2 S_B\Big|_{q_0(l)} \geqslant 0. \tag{6}$$

This equation will eventually be connected to the paucity of negative eigenvalues.

Eq. (6) must be interpreted with a small amount of care. The reason is that the domain of S_B, the set of paths on which S_B is defined, is not an open set (in any of the usual topologies for path space). In particular, there exist smooth curves in path space, passing through $q_0(l)$, such that every path on one side of $q_0(l)$ is in the domain of S_B, while every path on the other side penetrates the region of negative V. I can say nothing, and need to say nothing, about the derivatives of S_B along such a curve. I only assert that, for twice-differentiable curves which pass through $q_0(l)$ and which lie completely in the domain of S_B, at $q_0(l)$ the derivative of S_B along the curve vanishes and the second derivative is positive.

Let us define a functional

$$S'[q(t)] = \int \mathrm{d}t\, L(\dot{q}(t), q(t)), \tag{7}$$

for motions that begin at the origin and end on Σ. Just as for S_B, by convention all motions are terminated at their first intersection with Σ; thus V is never negative in the region of integration. Note that although the integral is like that in Hamilton's principle, the boundary conditions are non-hamiltonian; the final endpoint is not completely fixed, nor is the time interval. However, a stationary point of S' is *a fortiori* stationary under variations that do leave the final endpoint and the time fixed, and thus is a solution of the Euler-Lagrange equations,

$$\frac{\mathrm{d}}{\mathrm{d}t} m_{ab} \frac{\mathrm{d}q^b}{\mathrm{d}t} = \frac{\partial L}{\partial q^a}. \tag{8}$$

This implies that the euclidean energy,

$$E=\frac{1}{2}\left(\frac{\mathrm{d}l}{\mathrm{d}t}\right)^2-V, \tag{9}$$

is a constant of the motion.

Every motion $q(t)$ defines a unique path $q(l)$. (Of course, the mapping is not invertible; there are many motions that traverse the same path at different speeds.) S' may be written in terms of $q(l)$ and $\mathrm{d}l/\mathrm{d}t$:

$$S'=\int \mathrm{d}l\left[\frac{1}{2}\frac{\mathrm{d}l}{\mathrm{d}t}+V\frac{\mathrm{d}t}{\mathrm{d}l}\right]. \tag{10}$$

Because the time interval is not fixed, to stationarize this for fixed $q(l)$ requires just ordinary calculus, not calculus of variations. We see that the stationary point is a minimum (because V is positive) and obeys

$$\frac{\mathrm{d}l}{\mathrm{d}t}=\sqrt{2V}\,. \tag{11}$$

If we insert this in eq. (10), S' becomes S_{B}. In equations,

$$S'[q(t)]\geqslant S_{\mathrm{B}}[q(l)], \tag{12}$$

with equality obtained for that $q(t)$ which obeys eq. (11).

Thus there is a one-to-one correspondence between the stationary points of S' and those of S_{B}.* In particular, for $q_0(t)$, the motion that corresponds to $q_0(l)$,

$$S'[q_0(t)]=S_{\mathrm{B}}[q_0(l)]=\tfrac{1}{2}B, \tag{13}$$

and, from eqs. (6) and (12),

$$\delta^2 S'\Big|_{q_0(t)}\geqslant 0. \tag{14}$$

Of course, this equation must be interpreted with the same care as eq. (6).

By eq. (11), $q_0(t)$ is a zero-energy solution of the equations of motion. Thus it approaches the origin in infinite time. Because ∇V nowhere vanishes on Σ, $q_0(t)$ reaches Σ in finite time. Thus, with no loss of generality we can take t to occupy the range $[-\infty,0]$. At $t=0$, again by eq. (11), $\dot{q}_0^a=0$. Thus we can extend $q_0(t)$ to a

* This correspondence is very close to that which connects Hamilton's principle to Jacobi's principle of least action [3]. I have used this correspondence before in discussing the connection between the path-integral and WKB formulations of tunneling [4]. However, I have chosen to give the argument from first principles here, rather than referring to the literature, both because my boundary conditions are not quite the usual ones and because I need somewhat more information about the correspondence than I can find in the literature.

solution of the equations of motion for all time, by reflection:

$$q_0(-t) = q_0(t). \tag{15}$$

The extended $q_0(t)$ is the bounce. It stationarizes the usual euclidean action,

$$S = \int_{-\infty}^{\infty} L(\dot{q}(t), q(t))\,\mathrm{d}t, \tag{16}$$

restricted to functions that approach the origin at plus and minus infinity.

We now turn to a study of the eigenvalues and eigenfunctions of $\delta^2 S$ evaluated at $q_0(t)$.

(i) Because q_0 is an even function of t, we can always choose the eigenfunctions to be either even or odd functions of t.

(ii) Because of the time-translation invariance of the equations of motion, $\dot{q}_0$ is an eigenfunction with eigenvalue zero. It is an odd function of t.

(iii) The eigenfunction(s) with lowest eigenvalue must be even.

Proof: Let $\psi(t) = -\psi(-t)$ be an eigenfunction with lowest eigenvalue. Define

$$\tilde{\psi} = \begin{cases} \psi, & t > 0 \\ -\psi, & t < 0. \end{cases} \tag{17}$$

Because $\psi(0) = 0$, this is continuous and piecewise continuously differentiable. If ψ is normalized, so is $\tilde{\psi}$, and the expectation value of $\delta^2 S$ for $\tilde{\psi}$ is the same as that for ψ. Since this is the minimum expectation value, the lowest eigenvalue, $\tilde{\psi}$ must also be an eigenfunction with lowest eigenvalue. So therefore is $\psi + \tilde{\psi}$. But this is preposterous, since this function vanishes on the entire negative axis, and can hardly be a solution of the eigenvalue equation, an ordinary differential equation.

Thus there must be at least one even eigenfunction with a negative eigenvalue, which I will call ψ_1. If there is a second eigenfunction with a negative eigenvalue, ψ_2, it may be either even or odd. I shall show that in either case there is a contradiction.

If ψ_2 is even, we can find coefficients a and b such that

$$\left(a\psi_1(0) + b\psi_2(0)\right) \cdot \nabla V(q_0(0)) = 0. \tag{18}$$

Thus we can construct a one-parameter family of motions,

$$q_\lambda(t) = q_0(t) + \lambda\left(a\psi_1(t) + b\psi_2(t)\right) + \mathrm{O}(\lambda^2), \tag{19}$$

such that for sufficiently small λ, $q_\lambda(0)$ is in Σ,

$$V(q_\lambda(0)) = 0. \tag{20}$$

I wish to evaluate S' for $q_\lambda(t)$, restricted to $t \leqslant 0$. However, before I can do this, I must check that this function is among those for which S' is defined, that is to say, that q_λ lies in the region of positive V for all negative t. I'll do this by computing the time derivatives of $V(q_\lambda(t))$ at $t=0$. Because $q_\lambda(t)$ is an even function of t,

$$\left.\frac{\mathrm{d}V(q_\lambda(t))}{\mathrm{d}t}\right|_{t=0}=0, \tag{21}$$

while

$$\left.\frac{\mathrm{d}^2V(q_\lambda(t))}{\mathrm{d}t^2}\right|_{t=0}=\left.\frac{\mathrm{d}^2V(q_0(t))}{\mathrm{d}t^2}\right|_{t=0}+\mathrm{O}(\lambda)=\nabla V\cdot m^{-1}\nabla V+\mathrm{O}(\lambda), \tag{22}$$

by the Euler-Lagrange equations. Thus V is positive for sufficiently small λ and for t in some neighborhood of zero. But for t outside this neighborhood, $q_0(t)$ is strictly in the interior of the region of positive V, and thus so is $q_\lambda(t)$ for sufficiently small λ. Hence, for sufficiently small λ, it is legitimate to compute $S'[q_\lambda(t)]$.

Because q_λ is an even function of t,

$$S[q_\lambda(t)]=2S'[q_\lambda(t)]. \tag{23}$$

Thus,

$$\left.\frac{\mathrm{d}^2S[q_\lambda(t)]}{\mathrm{d}\lambda^2}\right|_{\lambda=0}=2\left.\frac{\mathrm{d}^2S'[q_\lambda(t)]}{\mathrm{d}\lambda^2}\right|_{\lambda=0}. \tag{24}$$

The left side of this equat on is negative, because ψ_1 and ψ_2 are eigenfunctions with negative eigenvalues. But the right side is non-negative by eq. (14).

If ψ_2 is odd, we can find coefficients a and b such that

$$\left(a\dot\psi_2(0)+b\ddot q_0(0)\right)\cdot\nabla V(q_0(0))=0. \tag{25}$$

If we construct the one-parameter family of motions,

$$q_\lambda(t)=q_0(t)+\lambda\left(a\psi_2(t)+b\dot q_0(t)\right), \tag{26}$$

$q_\lambda(0)=q_0(0)$ is in Σ for all λ. Furthermore, for sufficiently small λ, $q_\lambda(t)$, for $t\leqslant 0$, is in that class of functions for which S' is defined. The reasoning here is the same as before; note that eq. (25) is necessary to establish (21). Because the coefficient of λ in eq. (26) is an odd function of t,

$$S[q_\lambda(t)]=S'[q_\lambda(t)]+S'[q_{-\lambda}(t)]. \tag{27}$$

Differentiating this twice, we again obtain eq. (24). The left side of this equation is negative, because ψ_2 is an eigenfunct on with negative eigenvalue and $\dot q_0$ is an eigenfunction with eigenvalue zero. But the right side is positive, by eq. (14).

This co plete the argument. It has been long and niggling; perhaps a short sloppy summary will be useful. The WKB formulation of tunneling tells us to search for a minimal barrier penetration path; the path-integral formulation tells us to search for a bounce. These two prescriptions are equivalent, in much the same way that Hamilton's principle and Jacobi's principle are equivalent. But there is an apparent paradox. The minimal barrier penetration path is a true minimum; all small deviations from it increase the action. However, the problem of small vibrations about the bounce has an eigenfunction with negative eigenvalue; small deviations in this direction decrease the action. The resolution of the paradox is the observation that not all small deviations from the bounce map into possible barrier penetration paths; they may over- or undershoot the escape surface, Σ. However, if there are two eigenfunctions with negative eigenvalues, we can always build a deviation that lands dead on Σ, and we have a true contradiction.

This work was completed while I was a visitor to the Theory Division of CERN. I would like to thank CERN for its hospitality.

Appendix

WHAT HAPPENS IF ∇V VANISHES?

In the body of this paper, I made a technical assumption, that ∇V vanishes nowhere on Σ. This can be replaced by a considerably weaker assumption, that at every point on Σ for which ∇V vanishes, the matrix of second derivatives of V is invertible. This can in turn be replaced by an even weaker (though somewhat less natural) assumption, that the matrix has at least one negative eigenvalue. (This follows from invertibility because Σ is the boundary of a region of negative V.)

I shall show that, under the stated assumption, the minimal barrier penetration path can not intersect Σ at a point where ∇V vanishes. Thus, in the neighborhood of the minimal path, we have nonvanishing ∇V, and the rest of the proof follows as before.

The argument proceeds by contradiction. I shall assume that $q_0(t)$, the minimum of S', intersects Σ at a point where ∇V vanishes. I shall then construct a path arbitrarily close to $q_0(l)$ with a smaller value of S'.

Because ∇V vanishes at the assumed intersection point, it takes an infinite time for $q_0(t)$ to reach Σ. Thus t occupies the range $[-\infty, \infty]$. By our assumption, there exists a vector of unit length, e, such that at the intersection point,

$$e^a \frac{\partial^2 V}{\partial q^a \partial q^b} e^b \equiv V' < 0 . \tag{28}$$

(Note that this equation is true even if there are only continuous negative eigenvalues, as could be the case for an infinite number of variables.)

I shall define a new motion, $q_1(t)$, by

$$\begin{aligned} q_1(t) &= q_0(t)\,, & t &\leqslant T\,, \\ &= q_0(t) + Ae\sin[\omega(t-T)]\,, & t &\geqslant T\,, \end{aligned} \tag{29}$$

where A, T, and ω are real parameters which I shall choose shortly.

The parameter A is fixed by demanding that q_1 intersect Σ at $T' = T + \pi/(2\omega)$. Let us expand $V(T')$ in a series in powers of A, neglecting terms of higher than second order. (I shall justify this neglect shortly.)

$$V(T') = V_0 + AV_1 + A^2V_2\,. \tag{30}$$

For sufficiently large T, $V_0 = V(q_0(T'))$ is arbitrarily small. So is V_1, which is proportional to $\nabla V(q_0(T'))$. V_2 is the first coefficient that does not go to zero for large T; it is arbitrarily close to the negative constant, V''. Because V_0 is positive, the quadratic equation for A, $V(T')=0$, has two real roots, one positive and one negative. For both roots, A is arbitrarily small for sufficiently large T. This justifies neglecting the higher terms in the series. Because A can be made arbitrarily small, $q_1(l)$ can be made arbitrarily close to $q_0(l)$.

Now let us expand $S'[q_1(t)]$ in a series in powers of A,

$$S'[q_1(t)] = S'_0 + AS'_1 + A^2S'_2 + \cdots\,. \tag{31}$$

I shall analyze the terms in this expression one by one: (i) S'_0 is an integral over the same positive integrand as that which defines $S'[q(t)]$, but the range of integration is smaller, $[-\infty, T']$ rather than the whole line. Thus S'_0 is strictly less than $S'[q_0(t)]$. (ii) AS'_1 can always be made less than or equal to zero by choosing the sign of A appropriately. (iii) For T sufficiently large,

$$\begin{aligned} A^2S'_2 &= \tfrac{1}{2}A^2\int_T^{T'} \mathrm{d}t\left[\omega^2\cos^2\omega(t-T) + V''\sin^2\omega(t-T)\right] \\ &= \tfrac{1}{8}\pi A^2(\omega^2 + V'')\,. \end{aligned} \tag{32}$$

If we choose ω^2 to be less than $-V''$, this is strictly negative. (iv) Thus the first three terms in the series all make $S'[q_1(t)]$ less than $S'[q_0(t)]$. But since the term proportional to A^2 has a nonzero coefficient, the higher terms can not change this inequality, for sufficiently small A. This completes the proof.

References

[1] S. Coleman *in:* The whys of subnuclear physics, ed. A. Zichichi (Plenum, 1979) pp. 805–916, and references therein

[2] T. Banks, C. Bender and T.T. Wu, Phys. Rev. D8 (1973) 3346, 3366

[3] L.D. Landau and E.M. Lifshitz, Mechanics (Pergamon, 1960) p. 140

[4] S. Coleman, Phys. Rev. D15 (1977) 2929

63T-196. R. L. BISHOP, University of Illinois, Urbana, Illinois. *A relation between volume, mean curvature, and diameter.*

Let M be a compact Riemannian manifold having volume V, mean curvature, bounded below by B, and diameter D. Let V_1 be the volume of a ball of radius D in the simply-connected Riemannian manifold having constant sectional curvature, mean curvature B, and the same dimension as M. Then $V \leq V_1$. Equality holds only if M has constant curvature. The proof consists in getting bounds on the Jacobians of an exponential map of M, using the minimizing properties of Jacobi fields with respect to 2nd variation, and integrating the bound on the ball of radius D in the tangent space, the domain of exp. The details for a similar result on Kaehler manifolds will appear in a paper by the author and S. I. Goldberg. Myers' theorem that when $B > 0$, $D \leq \pi/\sqrt{B/(n-1)}$ is a corollary to the proof. (Received March 20, 1963.)

[The proofs of this and related results were published in R. L Bishop and R. J. Crittenden, Geometry of Manifolds, Academic Press, 1964, pp. 253-257.]

THE COSMOLOGICAL CONSTANT IS PROBABLY ZERO

S.W. HAWKING
Department of Applied Mathematics and Theoretical Physics,
Silver Street, Cambridge, CB3 9EW, England

Received 12 August 1983
Revised manuscript received 24 October 1983

It is suggested that the apparent cosmological constant is not necessarily zero but that zero is by far the most probable value. One requires some mechanism like a three-index antisymmetric tensor field or topological fluctuations of the metric which can give rise to an effective cosmological constant of arbitrary magnitude. The action of solutions of the euclidean field equations is most negative, and the probability is therefore highest, when this effective cosmological constant is very small.

The cosmological constant is probably the quantity in physics that is most accurately measured to be zero: observations of departures from the Hubble Law for distant galaxies place an upper limit of the order of

$$|\Lambda|/m_p^2 < 10^{-120} , \qquad (1)$$

where m_p is the Planck mass. On the other hand, one might expect that the zero point energies of quantum fluctuations would produce an effective or induced Λm_p^{-2} of order one if the quantum fluctuations were cut off at the Planck mass. Even if this were renormalized exactly to zero, one would still get a change in the effective Λ of order $\mu^4 m_p^{-2}$ whenever a symmetry in the theory was spontaneously broken, where μ is the energy at which the symmetry was broken. There are a large number of symmetries which seem to be broken in the present epoch of the universe, including chiral symmetry, electroweak symmetry and possibly, supersymmetry. Each of these would give a contribution to Λ that would exceed the upper limit (1) by at least forty orders of magnitude.

It is very difficult to believe that the bare value of Λ is fine tuned so that after all the symmetry breakings, the effective Λ satisfies the inequality (1). What one would like to find is some mechanism by which the effective value of Λ could relax to zero. Although there have been a number of attempts to find such a mechanism (see e.g. refs. [1,2]), I think it is fair to say that no satisfactory scheme has been suggested. In this paper, I want to propose instead a very simple idea: the cosmological constant can have any value but it is much more probable for it to have a value very near zero. A preliminary version of this argument was given in ref. [3].

My proposal requires that a variable effective cosmological constant be generated in some manner and that the path integral includes all, or some range, of values of this effective cosmological constant. One possibility would be to include the value of the cosmological constant in the variables that are integrated over in the path integral. A more attractive way would be to introduce a three-index antisymmetric tensor field $A_{\mu\nu\rho}$. This would have gauge transformations of the form

$$A_{\mu\nu\rho} \to A_{\mu\nu\rho} + \nabla_{[\mu} C_{\nu\rho]} , \qquad (2)$$

The action of the field is F^2 where F is the field strength formed from A:

$$F_{\mu\nu\rho\sigma} = \nabla_{[\mu} A_{\nu\rho\sigma]} . \qquad (3)$$

Such a field has no dynamics: the field equations imply that F is a constant multiple of the four-index antisymmetric tensor $\epsilon_{\mu\nu\rho\sigma}$. However, the F^2 term in the action behaves like an effective cosmological

0.370-2693/84/$ 03.00 © Elsevier Science Publishers B.V.
(North-Holland Physics Publishing Division)

constant [4]. Its value is not determined by field equations. Three-index antisymmetric tensor fields arise naturally in the dimensional reduction of $N = 1$ supergravity in eleven dimensions to $N = 8$ supergravity in four dimensions. Other mechanisms that would give an effective cosmological constant of arbitrary magnitude include topological fluctuations of the metric [5] and a scalar field ϕ with a potential term $V(\phi)$ but no kinetic term. In this last case, the gravitational field equations could be satisfied only if ϕ was constant. The potential $V(\phi)$ then acts as an effective cosmological constant.

In the path integral formulation of quantum theory, the amplitude to go from a field configuration $\phi_1(x)$ on the surface $t = t_1$ to a configuration $\phi_2(x)$ on $t = t_2$ is

$$\langle \phi_2, t_2 | \phi_1, t_1 \rangle = \int \mathrm{d}[\phi] \, \exp(\mathrm{i}I[\phi]) , \tag{4}$$

where $\mathrm{d}[\phi]$ is a measure on the space of all field configurations $\phi(x, t)$, $I[\phi]$ is the action of the field configuration and the integral is over all field configurations which agree with ϕ_1 and ϕ_2 at $t = t_1$ and $t = t_2$ respectively. The integral (4) oscillates and does not converge. One can improve the situation by making a rotation to euclidean space by defining a new coordinate $\tau = \mathrm{i}t$. The transition amplitude then becomes

$$\langle \phi_2, \tau_2 | \phi_1, \tau_1 \rangle = \int \mathrm{d}[\phi] \, \exp(-\tilde{I}[\phi]) , \tag{5}$$

where $\tilde{I} = -\mathrm{i}I$ is the euclidean action which is bounded below for well behaved field theories in flat space. One can interpret $\exp(-\tilde{I}[\phi])$ as being proportional to the probability of the euclidean field configuration $\phi(x, \tau)$. One calculates amplitudes like (5) in euclidean space and then analytically continues them in $\tau_2 - \tau_1$ back to real time separations.

One can adopt a similar euclidean approach in the case of gravity [6,7]. There is a difficulty because the euclidean gravitational action is not bounded below. This can be overcome by dividing the space of all positive definite metrics up into equivalence classes under conformal transformations. In each equivalence class one integrates over the conformal factor on a contour which is parallel to the imaginary axis [8,3]. The dominant contribution to the path integral comes from metrics which are near to solutions of the field equations. Of particular interest are solutions in which the dynamical matter fields, i.e. the matter fields apart from $A_{\mu\nu\rho}$ or ϕ are near their ground state values over a large region. This would be a reasonable approximation to the universe at the present time. The ground state of the matter fields plus the contribution of the $A_{\mu\nu\rho}$ or ϕ fields will generate an effective cosmological constant Λ_e. If the effective value Λ_e is positive, the solutions are necessarily compact and their four-volume is bounded by that of the solution of greatest symmetry, the four-sphere of radius $(3\Lambda_e^{-1})^{1/2}$. The euclidean action $\tilde{I}$ will be negative and will be bounded below by

$$-3\pi m_p^2/\Lambda_e . \tag{6}$$

If Λ_e is negative, the solutions can be either compact or non-compact [5]. If they are compact, the action $\tilde{I}$ will be finite and positive. If they are non-compact, $\tilde{I}$ will be infinite and positive.

The probability of a given field configuration will be proportional to

$$\exp(-\tilde{I}) . \tag{7}$$

If Λ_e is negative, $\tilde{I}$ will be positive and the probability will be exponentially small. If Λ_e is positive, the probability will be of the order of

$$\exp(3\pi m_p^2/\Lambda_e) . \tag{8}$$

Clearly, the most probable configurations will be those with very small values of Λ_e. This does not imply that the effective cosmological constant will be small everywhere in these configurations. In regions in which the dynamical fields differ from the ground state values there can be an apparent cosmological constant as in the inflationary model of the universe.

[1] F. Wilczek, in: The very early universe, eds. G.W. Gibbons, S.W. Hawking and S.T.C. Siklos (Cambridge U.P., Cambridge, 1983).
[2] A.D. Dolgov, in: The very early universe, eds. G.W. Gibbons, S.W. Hawking and S.T.C. Siklos (Cambridge U.P., Cambridge, 1983).
[3] S.W. Hawking, The cosmological constant, Phil. Trans. Roy. Soc. A., to be published.
[4] A. Aurilia, H. Nicolai and P.K. Townsend, Nucl. Phys. B176 (1980) 509.
[5] S.W. Hawking, Nucl. Phys. B144 (1978) 349.
[6] S.W. Hawking, The path integral approach to quantum gravity, in: General relativity: an Einstein centenary survey, eds. S.W. Hawking and W. Isreal (Cambridge U.P., Cambridge, 1979).
[7] S.W. Hawking, Euclidean quantum gravity, in: Recent developments in gravitation, Cargese Lectures, eds. M. Levy and S. Deser (1978).
[8] G.W. Gibbons, S.W. Hawking and M.J. Perry, Nucl. Phys. B138 (1978) 141.

II. BLACK HOLES

Commun. math. Phys. 43, 199—220 (1975)

Particle Creation by Black Holes

S. W. Hawking

Department of Applied Mathematics and Theoretical Physics, University of Cambridge, Cambridge, England

Received April 12, 1975

Abstract. In the classical theory black holes can only absorb and not emit particles. However it is shown that quantum mechanical effects cause black holes to create and emit particles as if they were hot bodies with temperature $\frac{\hbar\kappa}{2\pi k} \approx 10^{-6}\left(\frac{M_\odot}{M}\right)$ °K where κ is the surface gravity of the black hole. This thermal emission leads to a slow decrease in the mass of the black hole and to its eventual disappearance: any primordial black hole of mass less than about 10^{15} g would have evaporated by now. Although these quantum effects violate the classical law that the area of the event horizon of a black hole cannot decrease, there remains a Generalized Second Law: $S+\frac{1}{4}A$ never decreases where S is the entropy of matter outside black holes and A is the sum of the surface areas of the event horizons. This shows that gravitational collapse converts the baryons and leptons in the collapsing body into entropy. It is tempting to speculate that this might be the reason why the Universe contains so much entropy per baryon.

1.

Although there has been a lot of work in the last fifteen years (see [1, 2] for recent reviews), I think it would be fair to say that we do not yet have a fully satisfactory and consistent quantum theory of gravity. At the moment classical General Relativity still provides the most successful description of gravity. In classical General Relativity one has a classical metric which obeys the Einstein equations, the right hand side of which is supposed to be the energy momentum tensor of the classical matter fields. However, although it may be reasonable to ignore quantum gravitational effects on the grounds that these are likely to be small, we know that quantum mechanics plays a vital role in the behaviour of the matter fields. One therefore has the problem of defining a consistent scheme in which the space-time metric is treated classically but is coupled to the matter fields which are treated quantum mechanically. Presumably such a scheme would be only an approximation to a deeper theory (still to be found) in which space-time itself was quantized. However one would hope that it would be a very good approximation for most purposes except near space-time singularities.

The approximation I shall use in this paper is that the matter fields, such as scalar, electro-magnetic, or neutrino fields, obey the usual wave equations with the Minkowski metric replaced by a classical space-time metric g_{ab}. This metric satisfies the Einstein equations where the source on the right hand side is taken to be the expectation value of some suitably defined energy momentum operator for the matter fields. In this theory of quantum mechanics in curved space-time there is a problem in interpreting the field operators in terms of annihilation and creation operators. In flat space-time the standard procedure is to decompose the field into positive and negative frequency components. For example, if ϕ is a massless Hermitian scalar field obeying the equation $\phi_{;ab}\eta^{ab}=0$ one expresses ϕ as

$$\phi=\sum_i\{f_i a_i+\bar{f}_i a_i^\dagger\} \tag{1.1}$$

where the $\{f_i\}$ are a complete orthonormal family of complex valued solutions of the wave equation $f_{i;ab}\eta^{ab}=0$ which contain only positive frequencies with respect to the usual Minkowski time coordinate. The operators a_i and $a_i^\dagger$ are interpreted as the annihilation and creation operators respectively for particles in the ith state. The vacuum state $|0\rangle$ is defined to be the state from which one cannot annihilate any particles, i.e.

$$a_i|0\rangle=0 \quad \text{for all } i\,.$$

In curved space-time one can also consider a Hermitian scalar field operator ϕ which obeys the covariant wave equation $\phi_{;ab}g^{ab}=0$. However one cannot decompose into its positive and negative frequency parts as positive and negative frequencies have no invariant meaning in curved space-time. One could still require that the $\{f_i\}$ and the $\{\bar{f}_i\}$ together formed a complete basis for solutions of the wave equations with

$$\tfrac{1}{2}i\int_S(f_i\bar{f}_{j;a}-\bar{f}_j f_{i;a})d\Sigma^a=\delta_{ij} \tag{1.2}$$

where S is a suitable surface. However condition (1.2) does not uniquely fix the subspace of the space of all solutions which is spanned by the $\{f_i\}$ and therefore does not determine the splitting of the operator ϕ into annihilation and creation parts. In a region of space-time which was flat or asymptotically flat, the appropriate criterion for choosing the $\{f_i\}$ is that they should contain only positive frequencies with respect to the Minkowski time coordinate. However if one has a space-time which contains an initial flat region (1) followed by a region of curvature (2) then a final flat region (3), the basis $\{f_{1i}\}$ which contains only positive frequencies on region (1) will not be the same as the basis $\{f_{3i}\}$ which contains only positive frequencies on region (3). This means that the initial vacuum state $|0_1\rangle$, the state which satisfies $a_{1i}|0_1\rangle=0$ for each initial annihilation operator a_{1i}, will not be the same as the final vacuum state $|0_3\rangle$ i.e. $a_{3i}|0_1\rangle\neq 0$. One can interpret this as implying that the time dependent metric or gravitational field has caused the creation of a certain number of particles of the scalar field.

Although it is obvious what the subspace spanned by the $\{f_i\}$ is for an asymptotically flat region, it is not uniquely defined for a general point of a curved space-time. Consider an observer with velocity vector v^a at a point p. Let B be the least upper bound $|R_{abcd}|$ in any orthonormal tetrad whose timelike vector coincides with v^a. In a neighbourhood U of p the observer can set up a local inertial coordinate system (such as normal coordinates) with coordinate radius of the order of $B^{-\frac{1}{2}}$. He can then choose a family $\{f_i\}$ which satisfy equation (1.2) and which in the neighbourhood U are approximately positive frequency with respect to the time coordinate in U. For modes f_i whose characteristic frequency ω is high compared to $B^{\frac{1}{2}}$, this leaves an indeterminacy between f_i and its complex conjugate $\bar{f}_i$ of the order of the exponential of some multiple of $-\omega B^{-\frac{1}{2}}$. The indeterminacy between the annihilation operator a_i and the creation operator $a_i^\dagger$ for the mode is thus exponentially small. However, the ambiguity between the a_i and the $a_i^\dagger$ is virtually complete for modes for which $\omega < B^{\frac{1}{2}}$. This ambiguity introduces an uncertainty of $\pm\frac{1}{2}$ in the number operator $a_i^\dagger a_i$ for the mode. The density of modes per unit volume in the frequency interval ω to $\omega + d\omega$ is of the order of $\omega^2 d\omega$ for ω greater than the rest mass m of the field in question. Thus the uncertainty in the local energy density caused by the ambiguity in defining modes of wavelength longer than the local radius of curvature $B^{-\frac{1}{2}}$, is of order B^2 in units in which $G = c = \hbar = 1$. Because the ambiguity is exponentially small for wavelengths short compared to the radius of curvature $B^{-\frac{1}{2}}$, the total uncertainty in the local energy density is of order B^2. This uncertainty can be thought of as corresponding to the local energy density of particles created by the gravitational field. The uncertainty in the curvature produced via the Einstein equations by this uncertainty in the energy density is small compared to the total curvature of space-time provided that B is small compared to one, i.e. the radius of curvature $B^{-\frac{1}{2}}$ is large compared to the Planck length 10^{-33} cm. One would therefore expect that the scheme of treating the matter fields quantum mechanically on a classical curved space-time background would be a good approximation, except in regions where the curvature was comparable to the Planck value of 10^{66} cm^{-2}. From the classical singularity theorems [3–6], one would expect such high curvatures to occur in collapsing stars and, in the past, at the beginning of the present expansion phase of the universe. In the former case, one would expect the regions of high curvature to be hidden from us by an event horizon [7]. Thus, as far as we are concerned, the classical geometry–quantum matter treatment should be valid apart from the first 10^{-43} s of the universe. The view is sometimes expressed that this treatment will break down when the radius of curvature is comparable to the Compton wavelength $\sim 10^{-13}$ cm of an elementary particle such as a proton. However the Compton wavelength of a zero rest mass particle such as a photon or a neutrino is infinite, but we do not have any problem in dealing with electromagnetic or neutrino radiation in curved space-time. All that happens when the radius of curvature of space-time is smaller than the Compton wavelength of a given species of particle is that one gets an indeterminacy in the particle number or, in other words, particle creation. However, as was shown above, the energy density of the created particles is small locally compared to the curvature which created them.

Even though the effects of particle creation may be negligible locally, I shall show in this paper that they can add up to have a significant influence on black holes over the lifetime of the universe $\sim 10^{17}$ s or 10^{60} units of Planck time. It seems that the gravitational field of a black hole will create particles and emit them to infinity at just the rate that one would expect if the black hole were an ordinary body with a temperature in geometric units of $\kappa/2\pi$, where κ is the "surface gravity" of the black hole [8]. In ordinary units this temperature is of the order of $10^{26} M^{-1}$ °K, where M is the mass, in grams of the black hole. For a black hole of solar mass (10^{33} g) this temperature is much lower than the 3 °K temperature of the cosmic microwave background. Thus black holes of this size would be absorbing radiation faster than they emitted it and would be increasing in mass. However, in addition to black holes formed by stellar collapse, there might also be much smaller black holes which were formed by density fluctuations in the early universe [9, 10]. These small black holes, being at a higher temperature, would radiate more than they absorbed. They would therefore presumably decrease in mass. As they got smaller, they would get hotter and so would radiate faster. As the temperature rose, it would exceed the rest mass of particles such as the electron and the muon and the black hole would begin to emit them also. When the temperature got up to about 10^{12} °K or when the mass got down to about 10^{14} g the number of different species of particles being emitted might be so great [11] that the black hole radiated away all its remaining rest mass on a strong interaction time scale of the order of 10^{-23} s. This would produce an explosion with an energy of 10^{35} ergs. Even if the number of species of particle emitted did not increase very much, the black hole would radiate away all its mass in the order of $10^{-28} M^3$ s. In the last tenth of a second the energy released would be of the order of 10^{30} ergs.

As the mass of the black hole decreased, the area of the event horizon would have to go down, thus violating the law that, classically, the area cannot decrease [7, 12]. This violation must, presumably, be caused by a flux of negative energy across the event horizon which balances the positive energy flux emitted to infinity. One might picture this negative energy flux in the following way. Just outside the event horizon there will be virtual pairs of particles, one with negative energy and one with positive energy. The negative parti le is in a region which is classically forbidden but it can tunnel through the event horizon to the region inside the black hole where the Killing vector which represents time translations is spacelike. In this region the particle can exist as a real parti le with a timelike momentum vector even though its energy relative to infinity as measured by the time translation Killing vector is negative. The other particle of the pair, having a positive energy, can escape to infinity where it constitutes a part of the thermal emission described above. The probability of the negative energy particle tunnelling through the horizon is governed by the surface gravity κ since this quantity measures the gradient of the magnitude of the Killing vector or, in other words, how fast the Killing vector is becoming spacelike. Instead of thinking of negative energy particles tunnelling through the horizon in the positive sense of time one could regard them as positive energy particles crossing the horizon on past-directed world-lines and then being scattered on to future-directed world-lines by the gravitational field. It should be emphasized that these pictures of the mechanism responsible for the thermal emission and area decrease are heuristic only and should not be taken too literally. It should not be thought unreasonable that a black hole, which is an excited state of the gravitational field, should decay quantum mechanically and that, because of quantum fluctuation of the metric, energy should be able to tunnel out of the potential well of a black hole. This particle creation is directly analogous to that caused by a deep potential well in flat space-time [18]. The real justification of the thermal emission is the mathematical derivation given in Section (2) for the case of an uncharged non-rotating black hole. The effects of angular momentum and charge are considered in Section (3). In Section (4) it is shown that any renormalization of the energy-momentum tensor with suitable properties must give a negative energy flow down the black hole and consequent decrease in the area of the event horizon. This negative energy flow is non-observable locally.

The decrease in area of the event horizon is caused by a violation of the weak energy condition [5–7, 12] which arises from the indeterminacy of particle number and energy density in a curved space-time. However, as was shown above, this indeterminacy is small, being of the order of B^2 where B is the magnitude of the curvature tensor. Thus it can have a diverging effection a null surface like the event horizon which has very small convergence or divergence but it can not untrap a strongly converging trapped surface until B becomes of the order of one. Therefore one would not expect the negative energy density to cause a breakdown of the classical singularity theorems until the radius of curvature of space-time became 10^{-33} cm.

Perhaps the strongest reason for believing that black holes can create and emit particles at a steady rate is that the predicted rate is just that of the thermal emission of a body with the temperature $\kappa/2\pi$. There are independent, thermodynamic, grounds for regarding some multiple of the surface gravity as having a close relation to temperature. There is an obvious analogy with the second law of thermodynamics in the law that, classically, the area of the event horizon can never decrease and that when two black holes collide and merge together, the area of the final event horizon is greater than the sum of the areas of the two original horizons [7, 12]. There is also an analogy to the first law of thermodynamics in the result that two neighbouring black hole equilibrium states are related by [8]

$$dM = \frac{\kappa}{8\pi} dA + \Omega dJ$$

where M, Ω, and J are respectively the mass, angular velocity and angular momentum of the black hole and A is the area of the event horizon. Comparing this to

$$dU = TdS + pdV$$

one sees that if some multiple of A is regarded as being analogous to entropy, then some multiple of κ is analogous to temperature. The surface gravity is also analogous to temperature in that it is constant over the event horizon in equilibrium. Beckenstein [19] suggested that A and κ were not merely analogous to entropy and temperature respectively but that, in some sense, they actually were the entropy and temperature of the black hole. Although the ordinary second law of thermodynamics is transcended in that entropy can be lost down black holes, the flow of entropy across the event horizon would always cause some increase in the area of the horizon. Beckenstein therefore suggested [20] a Generalized Second Law: Entropy + some multiple (unspecified) of A never decreases. However he did not suggest that a black hole could emit particles as well as absorb them. Without such emission the Generalized Second Law would be violated by for example, a black hole immersed in black body radiation at a lower temperature than that of the black hole. On the other hand, if one accepts that black holes do emit particles at a steady rate, the identification of $\kappa/2\pi$ wi h temperature and $\frac{1}{4}A$ with entropy is established and a Generalized Second Law confirmed.

2. Gravitational Collapse

It is now generally believed that, according to classical theory, a gravitational collapse will produce a black hole which will settle down rapidly to a stationary axisymmetric equilibrium state characterized by its mass, angular momentum and electric charge [7, 13]. The Kerr-Newman solution represent one such family of black hole equilibrium states and it seems unlikely that there are any others. It has therefore become a common practice to ignore the collapse phase and to represent a black hole simply by one of these solutions. Because these solutions are stationary there will not be any mixing of positive and negative frequencies and so one would not expect to obtain any particle creation. However there is a classical phenomenon called superradiance [14–17] in which waves incident in certain modes on a rotating or charged black hole are scattered with increased amplitude [see Section (3)]. On a particle description this amplification must correspond to an increase in the number of particles and therefore to stimulated emission of particles. One would therefore expect on general grounds that there would also be a steady rate of spontaneous emission in these superradiant modes which would tend to carry away the angular momentum or charge of the black hole [16]. To understand how the particle creation can arise from mixing of positive and negative frequencies, it is essential to consider not only the quasi-stationary final state of the black hole but also the time-dependent formation phase. One would hope that, in the spirit of the "no hair" theorems, the rate of emission would not depend on details of the collapse process except through the mass, angular momentum and charge of the resulting black hole. I shall show that this is indeed the case but that, in addition to the emission in the super-radiant modes, there is a steady rate of emission in all modes at the rate one would expect if the black hole were an ordinary body with temperature $\kappa/2\pi$.

I shall consider first of all the simplest case of a non-rotating uncharged black hole. The final stationary state for such a black hole is represented by the Schwarzschild solution with metric

$$ds^2 = -\left(1-\frac{2M}{r}\right)dt^2 - \left(1-\frac{2M}{r}\right)^{-1} dr^2 + r^2(d\theta^2 + \sin^2\theta d\phi^2)\,. \tag{2.1}$$

As is now well known, the apparent singularities at $r=2M$ are fictitious, arising merely from a bad choice of coordinates. The global structure of the analytically extended Schwarzschild solution can be described in a simple manner by a Penrose diagram of the r-t plane (Fig. 1) [6, 13]. In this diagram null geodesics in the r-t plane are at $\pm 45°$ to the vertical. Each point of the diagram represents a 2-sphere of area $4\pi r^2$. A conformal transformation has been applied to bring infinity to a finite distance: infinity is represented by the two diagonal lines (really null surfaces) labelled $\mathscr{I}^-$ and $\mathscr{I}^-$, and the points I^+, I^-, and I^0. The two horizontal lines $r=0$ are curvature singularities and the two diagonal lines $r=2M$ (really null surfaces) are the future and past event horizons which divide the solution up into regions from which one cannot escape to $\mathscr{I}^+$ and $\mathscr{I}^-$. On the left of the diagram there is another infinity and asymptotically flat region.

Most of the Penrose diagram is not in fact relevant to a black hole formed by gravitational collapse since the metric is that of the Schwarzschild solution

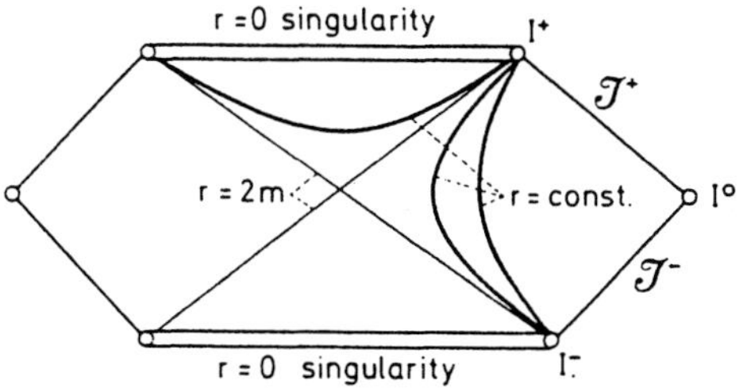

Fig. 1. The Penrose diagram for the analytically extended Schwarzschild solution

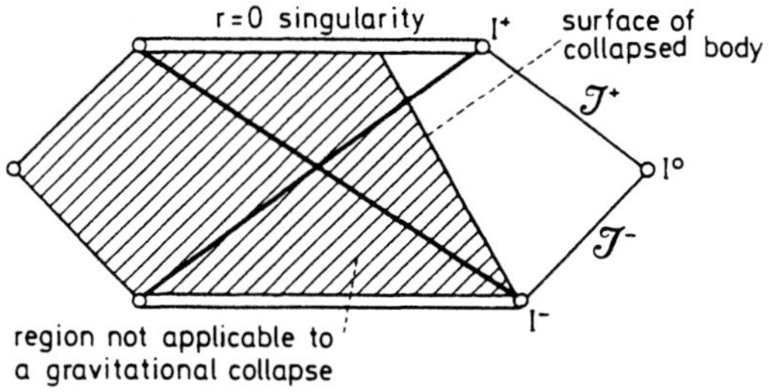

Fig. 2. Only the region of the Schwarzschild solution outside the collapsing body is relevant for a black hole formed by gravitational collapse. Inside the body the solution is completely different

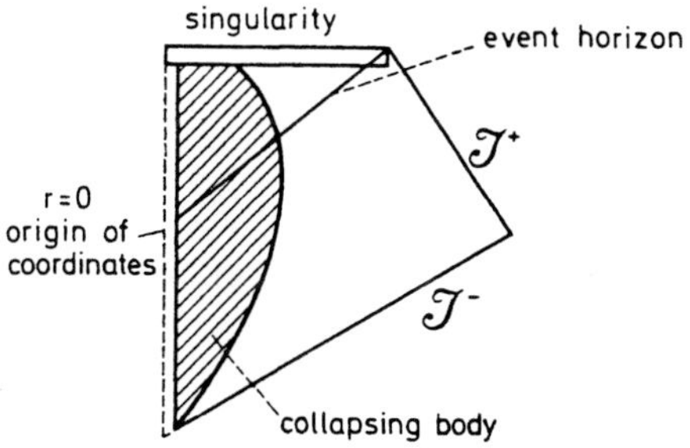

Fig. 3. The Penrose diagram of a spherically symmetric collapsing body producing a black hole. The vertical dotted line on the left represents the non-singular centre of the body

only in the region outside the collapsing matter and only in the asymptotic future. In the case of exactly spherical collapse, which I shall consider for simplicity, the metric is exactly the Schwarzchild metric everywhere outside the surface of the collapsing object which is represented by a timelike geodesic in the Penrose diagram (Fig. 2). Inside the object the metric is completely different, the past event horizon, the past $r=0$ singularity and the other asymptotically flat region do not exist and are replaced by a time-like curve representing the origin of polar coordinates. The appropriate Penrose diagram is shown in Fig. 3 where the conformal freedom has been used to make the origin of polar coordinates into a vertical line.

In this space-time consider (again for simplicity) a massless Hermitian scalar field operator ϕ obeying the wave equation

$$\phi_{;ab}g^{ab}=0\,. \tag{2.2}$$

(The results obtained would be the same if one used the conformally invariant wave equation:

$$\phi_{;ab}g^{ab}+\tfrac{1}{6}R\phi=0\,.)$$

The operator ϕ can be expressed as

$$\phi=\sum_i\{f_i\boldsymbol{a}_i+\bar{f}_i\boldsymbol{a}_i^\dagger\}\,. \tag{2.3}$$

The solutions $\{f_i\}$ of the wave equation $f_{i;ab}g^{ab}=0$ can be chosen so that on past null infinity $\mathscr{I}^-$ they form a complete family satisfying the orthonormality conditions (1.2) where the surface S is $\mathscr{I}^-$ and so that they contain only positive frequencies with respect to the canonical affine parameter on $\mathscr{I}^-$. (This last condition of positive frequency can be uniquely defined despite the existence of "supertranslations" in the Bondi-Metzner-Sachs asymptotic symmetry group [21, 22].) The operators $\boldsymbol{a}_i$ and $\boldsymbol{a}_i^\dagger$ have the natural interpretation as the annihilation and creation operators for ingoing particles i.e. for particles at past null infinity $\mathscr{I}^-$. Because massless fields are completely determined by their data on $\mathscr{I}^-$, the operator ϕ can be expressed in the form (2.3) everywhere. In the region outside the event horizon one can also determine massless fields by their data on the event horizon and on future null infinity $\mathscr{I}^+$. Thus one can also express ϕ in the form

$$\phi=\sum_i\{p_i\boldsymbol{b}_i+\bar{p}_i\boldsymbol{b}_i^\dagger+q_i\boldsymbol{c}_i+\bar{q}_i\boldsymbol{c}_i^\dagger\}\,. \tag{2.4}$$

Here the $\{p_i\}$ are solutions of the wave equation which are purely outgoing, i.e. they have zero Cauchy data on the event horizon and the $\{q_i\}$ are solutions which contain no outgoing component, i.e. they have zero Cauchy data on $\mathscr{I}^+$. The $\{p_i\}$ and $\{q_i\}$ are required to be complete families satisfying the orthonormality conditions (1.2) where the surface S is taken to be $\mathscr{I}^+$ and the event horizon respectively. In addition the $\{p_i\}$ are required to contain only positive frequencies with respect to the canonical affine parameter along the null geodesic generators of $\mathscr{I}^+$. With the positive frequency condition on $\{p_i\}$, the operators $\{\boldsymbol{b}_i\}$ and $\{\boldsymbol{b}_i^\dagger\}$ can be interpreted as the annihilation and creation operators for outgoing particles, i.e. for particles on $\mathscr{I}^+$. It is not clear whether one should impose some positive frequency condition on the $\{q_i\}$ and if so with respect to what. The choice of the $\{q_i\}$ does not affect the calculation of the emission of particles to $\mathscr{I}^+$. I shall return to the question in Section (4).

Because massless fields are completely determined by their data on $\mathscr{I}^-$ one can express $\{p_i\}$ and $\{q_i\}$ as linear combinations of the $\{f_i\}$ and $\{\bar{f}_i\}$:

$$p_i=\sum_j(\alpha_{ij}f_j+\beta_{ij}\bar{f}_j)\,, \tag{2.5}$$

$$q_i=\sum_j(\gamma_{ij}f_j+\eta_{ij}\bar{f}_j)\,. \tag{2.6}$$

These relations lead to corresponding relations between the operators

$$\boldsymbol{b}_i=\sum_j(\bar{\alpha}_{ij}\boldsymbol{a}_j-\bar{\beta}_{ij}\boldsymbol{a}_j^\dagger)\,, \tag{2.7}$$

$$\boldsymbol{c}_i=\sum_j(\bar{\gamma}_{ij}\boldsymbol{a}_j-\bar{\eta}_{ij}\boldsymbol{a}_j^\dagger)\,. \tag{2.8}$$

The initial vacuum state $|0\rangle$, the state containing no incoming particles, i.e. no particles on $\mathscr{I}^-$, is defined by

$$a_i|0\rangle = 0 \quad \text{for all } i. \tag{2.9}$$

However, because the coefficients β_{ij} will not be zero in general, the initial vacuum state will not appear to be a vacuum state to an observer at $\mathscr{I}^+$. Instead he will find that the expectation value of the number operator for the ith outgoing mode is

$$\langle 0_-|b_i^\dagger b_i|0_-\rangle = \textstyle\sum_j |\beta_{ij}|^2 . \tag{2.10}$$

Thus in order to determine the number of particles created by the gravitational field and emitted to infinity one simply has to calculate the coefficients β_{ij}. One would expect this calculation to be very messy and to depend on the detailed nature of the gravitational collapse. However, as I shall show, one can derive an asymptotic form for the β_{ij} which depends only on the surface gravity of the resulting black hole. There will be a certain finite amount of particle creation which depends on the details of the collapse. These particles will disperse and at late retarded times on $\mathscr{I}^+$ there will be a steady flux of particles determined by the asymptotic form of β_{ij}.

In order to calculate this asymptotic form it is more convenient to decompose the ingoing and outgoing solutions of the wave equation into their Fourier components with respect to advanced or retarded time and use the continuum normalization. The finite normalization solutions can then be recovered by adding Fourier components to form wave packets. Because the space-time is spherically symmetric, one can also decompose the incoming and outgoing solutions into spherical harmonics. Thus, in the region outside the collapsing body, one can write the incoming and outgoing solutions as

$$f_{\omega' lm} = (2\pi)^{-\frac{1}{2}} r^{-1} (\omega')^{-\frac{1}{2}} F_{\omega'}(r) e^{i\omega' v} Y_{lm}(\theta, \phi) , \tag{2.11}$$

$$p_{\omega lm} = (2\pi)^{-\frac{1}{2}} r^{-1} \omega^{-\frac{1}{2}} P_\omega(r) e^{i\omega u} Y_{lm}(\theta, \phi) , \tag{2.12}$$

where v and u are the usual advanced and retarded coordinates defined by

$$v = t + r + 2M \log\left|\frac{r}{2M} - 1\right| , \tag{2.13}$$

$$u = t - r - 2M \log\left|\frac{r}{2M} - 1\right| . \tag{2.14}$$

Each solution $p_{\omega lm}$ can be expressed as an integral with respect to ω' over solutions $f_{\omega' lm}$ and $\bar{f}_{\omega' lm}$ with the same values of l and $|m|$ (from now on I shall drop the suffices l, m):

$$p_\omega = \textstyle\int_0^\infty (\alpha_{\omega\omega'} f_{\omega'} + \beta_{\omega\omega'} \bar{f}_{\omega'}) d\omega' . \tag{2.15}$$

To calculate the coefficients $\alpha_{\omega\omega'}$ and $\beta_{\omega\omega'}$, consider a solution p_ω propagating backwards from $\mathscr{I}^+$ with zero Cauchy data on the event horizon. A part $p_\omega^{(1)}$ of the solution p_ω will be scattered by the static Schwarzchild field outside the collapsing body and will end up on $\mathscr{I}^-$ with the same frequency ω. This will give a $\delta(\omega' - \omega)$ term in $\alpha_{\omega\omega'}$. The remainder $p_\omega^{(2)}$ of p_ω will enter the collapsing body

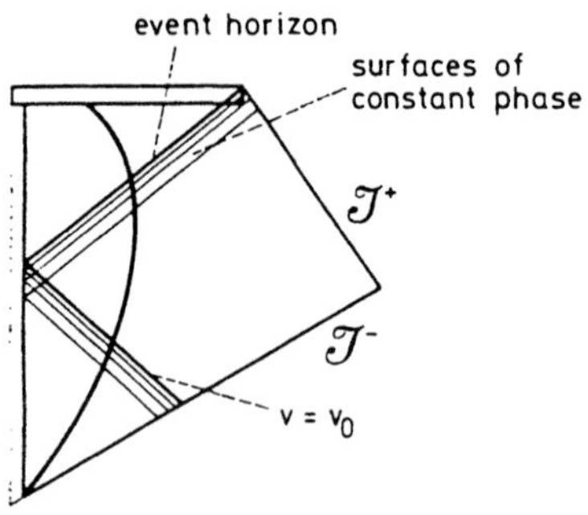

Fig. 4. The solution p_ω of the wave equation has an infinite number of cycles near the event horizon and near the surface $v = v_0$

where it will be partly scattered and partly reflected through the centre, eventually emerging to $\mathscr{I}^-$. It is this part $p_\omega^{(2)}$ which produces the interesting effects. Because the retarded time coordinate u goes to infinity on the event horizon, the surfaces of constant phase of the solution p_ω will pile up near the event horizon (Fig. 4). To an observer on the collapsing body the wave would seem to have a very large blue-shift. Because its effective frequency was very high, the wave would propagate by geometric optics through the centre of the body and out on $\mathscr{I}^-$. On $\mathscr{I}^-$ $p_\omega^{(2)}$ would have an infinite number of cycles just before the advanced time $v = v_0$ where v_0 is the latest time that a null geodesic could leave $\mathscr{I}^-$, pass through the centre of the body and escape to $\mathscr{I}^+$ before being trapped by the event horizon. One can estimate the form of $p_\omega^{(2)}$ on $\mathscr{I}^-$ near $v = v_0$ in the following way. Let x be a point on the event horizon outside the matter and let l^a be a null vector tangent to the horizon. Let n^a be the future-directed null vector at x which is directed radially inwards and normalized so that $l^a n_a = -1$. The vector $-\varepsilon n^a$ (ε small and positive) will connect the point x on the event horizon with a nearby null surface of constant retarded time u and therefore with a surface of constant phase of the solution $p_\omega^{(2)}$. If the vectors l^a and n^a are parallelly transported along the null geodesic γ through x which generates the horizon, the vector $-\varepsilon n^a$ will always connect the event horizon with the same surface of constant phase of $p_\omega^{(2)}$. To see what the relation between ε and the phase of $p_\omega^{(2)}$ is, imagine in Fig. 2 that the collapsing body did not exist but one analytically continued the empty space Schwarzchild solution back to cover the whole Penrose diagram. One could then transport the pair (l^a, n^a) back along to the point where future and past event horizons intersected. The vector $-\varepsilon n^a$ would then lie along the past event horizon. Let λ be the affine parameter along the past event horizon which is such that at the point of intersection of the two horizons, $\lambda = 0$ and $\dfrac{dx^a}{d\lambda} = n^a$. The affine parameter λ is related to the retarded time u on the past horizon by

$$\lambda = -Ce^{-\kappa u} \tag{2.16}$$

where C is constant and κ is the surface gravity of the black hole defined by $K^a_{;b} K^b = -\kappa K^a$ on the horizon where K^a is the time translation Killing vector.

$\left(\text{For a Schwarzchild black hole } \kappa = \frac{1}{4M}\right)$. It follows from this that the vector $-\varepsilon n^a$ connects the future event horizon with the surface of constant phase $-\frac{\omega}{\kappa}(\log\varepsilon - \log C)$ of the solution $p_\omega^{(2)}$. This result will also hold in the real space-time (including the collapsing body) in the region outside the body. Near the event horizon the solution $p_\omega^{(2)}$ will obey the geometric optics approximation as it passes through the body because its effective frequency will be very high. This means that if one extends the null geodesic γ back past the end-point of the event horizon and out onto $\mathscr{I}^-$ at $v = v_0$ and parallelly transports n^a along γ, the vector $-\varepsilon n^a$ will still connect γ to a surface of constant phase of the solution $p_\omega^{(2)}$. On $\mathscr{I}^- n^a$ will be parallel to the Killing vector K^a which is tangent to the null geodesic generators of $\mathscr{I}^-$:

$$n^a = DK^a\,.$$

Thus on $\mathscr{I}^-$ for $v_0 - v$ small and positive, the phase of the solution will be

$$-\frac{\omega}{\kappa}(\log(v_0 - v) - \log D - \log C)\,. \tag{2.17}$$

Thus on $\mathscr{I}^-$ $p_\omega^{(2)}$ will be zero for $v > v_0$ and for $v < v_0$

$$p_\omega^{(2)} \sim (2\pi)^{-\frac{1}{2}}\omega^{-\frac{1}{2}}r^{-1}P_\omega^- \exp\left(-i\frac{\omega}{\kappa}\left(\log\left(\frac{v_0 - v}{CD}\right)\right)\right) \tag{2.18}$$

where $P_\omega^- \equiv P_\omega(2M)$ is the value of the radial function for P_ω on the past event horizon in the analytically continued Schwarzchild solution. The expression (2.18) for $p_\omega^{(2)}$ is valid only for $v_0 - v$ small and positive. At earlier advanced times the amplitude will be different and the frequency measured with respect to v, will approach the original frequency ω.

By Fourier transforming $p_\omega^{(2)}$ one can evaluate its contributions to $\alpha_{\omega\omega'}$ and $\beta_{\omega\omega'}$. For large values of ω' these will be determined by the asymptotic form (2.18). Thus for large ω'

$$\alpha_{\omega\omega'}^{(2)} \approx (2\pi)^{-1}P_\omega^-(CD)^{\frac{i\omega}{\kappa}} \exp(i(\omega - \omega')v_0)\left(\frac{\omega'}{\omega}\right)^{\frac{1}{2}} \Gamma\left(1 - \frac{i\omega}{\kappa}\right)(-i\omega')^{-1 + \frac{i\omega}{\kappa}}, \tag{2.19}$$

$$\beta_{\omega\omega'}^{(2)} \approx -i\alpha_{\omega(-\omega')}^{(2)}\,. \tag{2.20}$$

The solution $p_\omega^{(2)}$ is zero on $\mathscr{I}^-$ for large values of v. This means that its Fourier transform is analytic in the upper half ω' plane and that $p_\omega^{(2)}$ will be correctly represented by a Fourier integral in which the contour has been displaced into the upper half ω' plane. The Fourier transform of $p_\omega^{(2)}$ contains a factor $(-i\omega')^{-1 + \frac{i\omega}{\kappa}}$ which has a logarithmic singularity at $\omega' = 0$. To obtain $\beta_{\omega\omega'}^{(2)}$ from $\alpha_{\omega\omega'}^{(2)}$ by (2.20) one has to analytically continue $\alpha_{\omega}^{(2)}{}_{\omega'}$ anticlockwise round this singularity. This means that

$$|\alpha_{\omega\omega'}^{(2)}| = \exp\left(\frac{\pi\omega}{\kappa}\right)|\beta_{\omega\omega'}^{(2)}|\,. \tag{2.21}$$

Actually, the fact that $p_{\omega}^{(2)}$ is not given by (2.18) at early advanced times means that the singularity in $\alpha_{\omega\omega'}$ occurs at $\omega'=\omega$ and not at $\omega'=0$. However the relation (2.21) is still valid for large ω'.

The expectation value of the total number of created particles at $\mathscr{I}^+$ in the frequency range ω to $\omega+d\omega$ is $d\omega\int_0^\infty|\beta_{\omega\omega'}|^2d\omega'$. Because $|\beta_{\omega\omega'}|$ goes like $(\omega')^{-\frac{1}{2}}$ at large ω' this integral diverges. This infinite total number of created particles corresponds to a finite steady rate of emission continuing for an infinite time as can be seen by building up a complete orthonormal family of wave packets from the Fourier components p_ω. Let

$$p_{jn}=\varepsilon^{-\frac{1}{2}}\int_{j\varepsilon}^{(j+1)\varepsilon}e^{-2\pi in\varepsilon^{-1}\omega}p_\omega d\omega \tag{2.22}$$

where j and n are integers, $j\geqq 0$, $\varepsilon>0$. For ε small these wave packets will have frequency $j\varepsilon$ and will be peaked around retarded time $u=2\pi n\varepsilon^{-1}$ with width ε^{-1}. One can expand $\{p_{jn}\}$ in terms of the $\{f_\omega\}$

$$p_{jn}=\int_0^\infty(\alpha_{jn\omega'}f_{\omega'}+\beta_{jn\omega'}\bar{f}_{\omega'})d\omega' \tag{2.23}$$

where

$$\alpha_{jn\omega'}=\varepsilon^{-\frac{1}{2}}\int_{j\varepsilon}^{(j+1)\varepsilon}e^{-2\pi in\varepsilon^{-1}\omega}\alpha_{\omega\omega'}d\omega \qquad \text{etc.} \tag{2.24}$$

For $j\gg\varepsilon$, $n\gg\varepsilon$

$$\begin{aligned}|\alpha_{jn\omega'}|&=\left|(2\pi)^{-1}P_\omega^-\omega^{-\frac{1}{2}}\Gamma\left(1-\frac{i\omega}{\kappa}\right)\varepsilon^{-\frac{1}{2}}(\omega')^{-\frac{1}{2}}\right.\\&\qquad\left.\cdot\int_{j\varepsilon}^{(j+1)\varepsilon}\exp i\omega''(-2\pi n\varepsilon^{-1}+\kappa^{-1}\log\omega')d\omega''\right|\\&=\left|\pi^{-1}P_\omega^-\omega^{-\frac{1}{2}}\Gamma\left(1-\frac{i\omega}{\kappa}\right)\varepsilon^{-\frac{1}{2}}(\omega')^{-\frac{1}{2}}z^{-1}\sin\tfrac{1}{2}\varepsilon z\right|\end{aligned} \tag{2.25}$$

where $\omega=j\varepsilon$ and $z=\kappa^{-1}\log\omega'-2\pi n\varepsilon^{-1}$. For wave-packets which reach $\mathscr{I}^+$ at late retarded times, i.e. those with large values of n, the main contribution to $\alpha_{jn\omega'}$ and $\beta_{jn\omega'}$ come from very high frequencies ω' of the order of $\exp(2\pi n\kappa\varepsilon^{-1})$. This means that these coefficients are governed only by the asymptotic forms (2.19, 2.20) for high ω' which are independent of the details of the collapse.

The expectation value of the number of particles created and emitted to infinity $\mathscr{I}^+$ in the wave-packet mode p_{jn} is

$$\int_0^\infty|\beta_{jn\omega'}|^2d\omega'\,. \tag{2.26}$$

One can evaluate this as follows. Consider the wave-packet p_{jn} propagating backwards from $\mathscr{I}^+$. A fraction $1-\Gamma_{jn}$ of the wave-packet will be scattered by the static Schwarzchild field and a fraction Γ_{jn} will enter the collapsing body.

$$\Gamma_{jn}=\int_0^\infty(|\alpha_{jn\omega'}^{(2)}|^2-|\beta_{jn\omega'}^{(2)}|^2)d\omega' \tag{2.27}$$

where $\alpha_{jn\omega'}^{(2)}$ and $\beta_{jn\omega'}^{(2)}$, are calculated using (2.19, 2.20) from the part $p_{jn}^{(2)}$ of the wave-packet which enters the star. The minus sign in front of the second term on the right of (2.27) occurs because the negative frequency components of $p_{jn}^{(2)}$ make a negative cont ibution to the flux into the collapsing body. By (2.21)

$$|\alpha_{jn\omega'}^{(2)}|=\exp(\pi\omega\kappa^{-1})|\beta_{jn\omega'}^{(2)}|\,. \tag{2.28}$$

Thus the total number of particles created in the mode p_{jn} is

$$\Gamma_{jn}(\exp(2\pi\omega\kappa^{-1})-1)^{-1}. \tag{2.29}$$

But for wave-packets at late retarded times, the fraction Γ_{jn} which enters the collapsing body is almost the same as the fraction of the wave-packet that would have crossed the past event horizon had the collapsing body not been there but the exterior Schwarzchild solution had been analytically continued. Thus this factor Γ_{jn} is also the same as the fraction of a similar wave-packet coming from $\mathscr{I}^-$ which would have crossed the future event horizon and have been absorbed by the black hole. The relation between emission and absorption cross-section is therefore exactly that for a body with a temperature, in geometric units, of $\kappa/2\pi$.

Similar results hold for the electromagnetic and linearised gravitational fields. The fields produced on $\mathscr{I}^-$ by positive frequency waves from $\mathscr{I}^+$ have the same asymptotic form as (2.18) but with an extra blue shift factor in the amplitude. This extra factor cancels out in the definition of the scalar product so that the asymptotic forms of the coefficients α and β are the same as in the Eqs. (2.19) and (2.20). Thus one would expect the black hole also to radiate photons and gravitons thermally. For massless fermions such as neutrinos one again gets similar results except that the negative frequency components given by the coefficients β now make a positive contribution to the probability flux into the collapsing body. This means that the term $|\beta|^2$ in (2.27) now has the opposite sign. From this it follows that the number of particles emitted in any outgoing wave packet mode is $(\exp(2\pi\omega\kappa^{-1})+1)^{-1}$ times the fraction of that wave packet that would have been absorbed by the black hole had it been incident from $\mathscr{I}^-$. This is again exactly what one would expect for thermal emission of particles obeying Fermi-Dirac statistics.

Fields of non-zero rest mass do not reach $\mathscr{I}^-$ and $\mathscr{I}^+$. One therefore has to describe ingoing and outgoing states for these fields in terms of some concept such as the projective infinity of Eardley and Sachs [23] and Schmidt [24]. However, if the initial and final states are asymptotically Schwarzchild or Kerr solutions, one can describe the ingoing and outgoing states in a simple manner by separation of variables and one can define positive frequencies with respect to the time translation Killing vectors of these initial and final asymptotic space-times. In the asymptotic future there will be no bound states: any particle will either fall through the event horizon or escape to infinity. Thus the unbound outgoing states and the event horizon states together form a complete basis for solutions of the wave equation in the region outside the event horizon. In the asymptotic past there could be bound states if the body that collapses had had a bounded radius for an infinite time. However one could equally well assume that the body had collapsed from an infinite radius in which case there would be no bound states. The possible existence of bound states in the past does not affect the rate of particle emission in the asymptotic future which will again be that of a body with temperature $\kappa/2\pi$. The only difference from the zero rest mass case is that the frequency ω in the thermal factor $(\exp(2\pi\omega\kappa^{-1})\mp 1)^{-1}$ now includes the rest mass energy of the particle. Thus there will not be much emission of particles of rest mass m unless the temperature $\kappa/2\pi$ is greater than m.

One can show that these results on thermal emission do not depend on spherical symmetry. Consider an asymmetric collapse which produced a black hole which settled to a non-rotating uncharged Schwarzchild solution (angular momentum and charge will be considered in the next section). The fact that the final state is asymptotically quasi-stationary means that there is a preferred Bondi coordinate system [25] on $\mathscr{I}^+$ with respect to which one can decompose the Cauchy data for the outgoing states into positive frequencies and spherical harmonics. On $\mathscr{I}^-$ there may or may not be a preferred coordinate system but if there is not one can pick an arbitrary Bondi coordinate system and decompose the Cauchy data for the ingoing states in a similar manner. Now consider one of the $\mathscr{I}^-$ states $p_{\omega lm}$ propagating backwards through this space-time into the collapsing body and out again onto $\mathscr{I}^-$. Take a null geodesic generator γ of the event horizon and extend it backwards beyond its past end-point to intersect $\mathscr{I}^-$ at a point y on a null geodesic generator λ of $\mathscr{I}^-$. Choose a pair of null vectors $(l^a, \hat{n}^a)$ at y with l^a tangent to γ and $\hat{n}^a$ tangent to λ. Parallelly propagate $l^a, \hat{n}^a$ along γ to a point x in the region of space-time where the metric is almost that of the final Schwarzchild solution. At x $\hat{n}^a$ will be some linear combination of l^a and the radial inward directed null vector n^a. This means that the vector $-\varepsilon\hat{n}^a$ will connect x to a surface of phase $-\omega/\kappa$ $(\log\varepsilon - \log E)$ of the solution $p_{\omega lm}$ where E is some constant. As before, by the geometric optics approximation, the vector $-\varepsilon\hat{n}^a$ at y will connect y to a surface of phase $-\omega/\kappa$ $(\log\varepsilon - \log E)$ of $p^{(2)}_{\omega lm}$ where $p^{(2)}_{\omega lm}$ is the part of $p_{\omega lm}$ which enters the collapsing body. Thus on the null geodesic generator λ of $\mathscr{I}^-$, the phase of $p^{(2)}_{\omega lm}$ will be

$$-\frac{i\omega}{\kappa}(\log(v_0 - v) - \log H) \tag{2.30}$$

where v is an affine parameter on λ with value v_0 at y and H is a constant. By the geometrical optics approximation, the value of $p^{(2)}_{\omega lm}$ on λ will be

$$L\exp\left\{-\frac{i\omega}{\kappa}[\log(v_0 - v) - \log H]\right\} \tag{2.31}$$

for $v_0 - v$, small and positive and zero for $v > v_0$ where L is a constant. On each null geodesic generator of $\mathscr{I}^-$ $p^{(2)}_{\omega lm}$ will have the form (2.31) with different values of L, v_0, and H. The lack of spherical symmetry during the collapse will cause $p^{(2)}_{\omega lm}$ on $\mathscr{I}^-$ to contain components of spherical harmonics with indices (l', m') different from (l, m). This means that one now has to express $p^{(2)}_{\omega lm}$ in the form

$$p^{(2)}_{\omega lm} = \sum_{l'm'} \int_0^\infty \{\alpha^{(2)}_{\omega lm\omega' l'm'} f_{\omega' l'm'} + \beta^{(2)}_{\omega lm\omega' l'm'} \bar{f}_{\omega' l'm'}\} d\omega' . \tag{2.32}$$

Because of (2.31), the coefficients $\alpha^{(2)}$ and $\beta^{(2)}$ will have the same ω' dependence as in (2.19) and (2.20). Thus one still has the same relation as (2.21):

$$|\alpha^{(2)}_{\omega lm\omega' l'm'}| = \exp(\pi\omega\kappa^{-1})\,|\beta^{(2)}_{\omega lm\omega' l'm'}| . \tag{2.33}$$

As before, for each (l, m), one can make up wave packets p_{jnlm}. The number of particles emitted in such a wave packet mode is

$$\sum_{l',m'} \int_0^\infty |\beta_{jnlm\omega' l'm'}|^2 d\omega' . \tag{2.34}$$

Similarly, the fraction Γ_{jnlm} of the wave packet that enters the collapsing body is

$$\Gamma_{jnlm}=\sum_{l',m'}\int_0^\infty \{|\alpha^{(2)}_{jnlm\omega' l' m'}|^2-|\beta^{(2)}_{jnlm\omega' l' m'}|^2\}d\omega' . \tag{2.35}$$

Again, Γ_{jnlm} is equal to the fraction of a similar wave packet coming from $\mathscr{I}^-$ that would have been absorbed by the black hole. Thus, using (2.33), one finds that the emission is just that of a body of temperature $\kappa/2\pi$: the emission at late retarded times depends only on the final quasi-stationary state of the black hole and not on the details of the gravitational collapse.

3. Angular Momentum and Charge

If the collapsing body was rotating or electrically charged, the resulting black hole would settle down to a stationary state which was described, not by the Schwarzchild solution, but by a charged Kerr solution characterised by the mass M, the angular momentum J, and the charge Q. As these solutions are stationary and axisymmetric, one can separate solutions of the wave equations in them into a factor $e^{i\omega u}$ or $e^{i\omega v}$ times $e^{-im\phi}$ times a function of r and θ. In the case of the scalar wave equation one can separate this last expression into a function of r times a function of θ [26]. One can also completely separate any wave equation in the non-rotating charged case and Teukolsky [27] has obtained completely separable wave equations for neutrino, electromagnetic and linearised gravitational fields in the uncharged rotating case.

Consider a wave packet of a classical field of charge e with frequency ω and axial quantum number m incident from infinity on a Kerr black hole. The change in mass dM of the black hole caused by the partial absorption of the wave packet will be related to the change in area, angular momentum and charge by the classical first law of black holes:

$$dM=\frac{\kappa}{8\pi}dA+\Omega dJ+\Phi dQ \tag{3.1}$$

where Ω and Φ are the angular frequency and electrostatic potential respectively of the black hole [13]. The fluxes of energy, angular momentum and charge in the wave packet will be in the ratio $\omega:m:e$. Thus the changes in the mass, angular momentum and charge of the black hole will also be in this ratio. Therefore

$$dM(1-\Omega m\omega^{-1}-e\Phi\omega^{-1})=\frac{\kappa}{8\pi}dA . \tag{3.2}$$

A wave packet of a classical Boson field will obey the weak energy condition: the local energy density for any observer is non-negative. It follows from this [7, 12] that the change in area dA induced by the wave-packet will be non-negative. Thus if

$$\omega<m\Omega+e\Phi \tag{3.3}$$

the change in mass dM of the black hole must be negative. In other words, the black hole will lose energy to the wave packet which will therefore be scattered with the same frequency but increased amplitude. This is the phenomenon known as "superradiance".

For classical fields of half-integer spin, detailed calculations [28] show that there is no superradiance. The reason for this is that the scalar product for half-integer spin fields is positive definite unlike that for integer spins. This means that the probability flux across the event horizon is positive and therefore, by conservation of probability, the probability flux in the scattered wave packet must be less than that in the incident wave packet. The reason that the above argument based on the first law breaks down is that the energy-momentum tensor for a classical half-integer spin field does not obey the weak energy condition. On a quantum, particle level one can understand the absence of superradiance for fermion fields as a consequence of the fact that the Exclusion Principle does not allow more than one particle in each outgoing wave packet mode and therefore does not allow the scattered wave-packet to be stronger than the incident wave-packet.

Passing now to the quantum theory, consider first the case of an unchanged, rotating black hole. One can as b fore pick an arbitrary Bondi coordinate frame on $\mathscr{I}^-$ and decompose the operator ϕ in terms of a family $\{f_{\omega lm}\}$ of incoming solutions where the indices ω, l, and m refer to the advanced time and angular dependence of f on $\mathscr{I}^-$ in the given coordinate system. On $\mathscr{I}^+$ the final quasi-stationary state of the black hole defines a preferred Bondi coordinate system using which one can define a family $\{p_{\omega lm}\}$ of outgoing solutions. The index l in this case labels the spheroidal harmonics in terms of which the wave equation is separable. One proceeds as before to calculate the asymptotic form of $p^{(2)}_{\omega lm}$ on $\mathscr{I}^-$. The only difference is that because the horizon is rotating with angular velocity Ω with respect to $\mathscr{I}^+$, the effective frequency near a generator of the event horizon is not ω but $\omega - m\Omega$. This means that the number of particles emitted in the wave-packet mode p_{jnlm} is

$$\{\exp(2\pi\kappa^{-1}(\omega - m\Omega)) \mp 1\}^{-1}\Gamma_{jnlm}\,. \tag{3.4}$$

The effect of this is to cause the rate of emission of particles with positive angular momentum m to be higher than that of particles with the same frequency ω and quantum number l but with negative angular momentum $-m$. Thus the particle emission tends to carry away the angular momentum. For Boson fields, the factor in curly brackets in (3.4) is negative for $\omega < m\Omega$. However the fraction Γ_{jnlm} of the wave-packet that would have been absorbed by the black hole is also negative in this case because $\omega < m\Omega$ is the condition for superradiance. In the limit that the temperature $\kappa/2\pi$ is very low, the only particle emission occurs is an amount $\mp\Gamma_{jnlm}$ in the modes for which $\omega < m\Omega$. This amount of particle creation is equal to that calculated by Starobinski [16] and Unruh [29], who considered only the final stationary Kerr solution and ignored the gravitational collapse.

One can treat a charged non-rotating black hole in a rather similar way. The behaviour of fields like the electromagnetic or gravitational fields which do not carry an electric charge will be the same as before except that the charge on the black will reduce the surface gravity k and hence the temperature of the black hole. Consider now the simple case of a massless charged scalar field ϕ which obeys the minimally coupled wave equation

$$g^{ab}(\nabla_a - ieA_a)(\nabla_b - ieA_b)\phi = 0\,. \tag{3.5}$$

The phase of a solution p_ω of the wave equation (3.5) is not gauge-invariant but the propagation vector $ik_a = \nabla_a(\log p_\omega) - ieA_a$ is. In the geometric optics or WKB limit the vector k_a is null and propagates according to

$$k_{a;b}k^b = -eF_{ab}k^b. \tag{3.6}$$

An infinitessimal vector z^a will connect points with a "guage invariant" phase difference of $ik_a z^a$. If z^a is propagated along the integral curves of k^a according to

$$z^a_{;b}k^b = -eF^a_b z^b \tag{3.7}$$

z^a will connect surfaces of constant guage invariant phase difference.

In the final stationary region one can choose a guage such that the electromagnetic potential A_a is stationary and vanishes on $\mathscr{I}^+$. In this guage the field equation (3.5) is separable and has solutions p_ω with retarded time dependence $e^{i\omega u}$. Let x be a point on the event horizon in the final stationary region and let l^a and n^a be a pair of null vectors at x. As before, the vector $-\varepsilon n^a$ will connect the event horizon with the surface of actual phase $-\omega/\kappa$ $(\log\varepsilon - \log C)$ of the solution p_ω. However the guage invariant phase will be $-\kappa^{-1}(\omega - e\Phi)(\log\varepsilon - \log C)$ where $\Phi = K^a A_a$ is the electrostatic potential on the horizon and K^a is the time-translation Killing vector. Now propagate l^a like k^a in Eq. (3.6) back until it intersects a generator λ of $\mathscr{I}^-$ at a point y and propagate n^a like z^a in Eq. (3.7) along the integral curve of l^a. With this propagation law, the vector $-\varepsilon n^a$ will connect surfaces of constant guage invariant phase. Near $\mathscr{I}^-$ one can use a different electromagnetic guage such that A^a is zero on $\mathscr{I}^-$. In this guage the phase of $p^{(2)}_\omega$ along each generator of $\mathscr{I}^-$ will have the form

$$-(\omega - e\phi)\kappa^{-1}\{\log(v_0 - v) - \log H\} \tag{3.8}$$

where H is a constant along each generator. This phase dependence gives the same thermal emission as before but with ω replaced by $\omega - e\Phi$. Similar remarks apply about charge loss and superradiance. In the case that the black hole is both rotating and charged one can simply combine the above results.

4. The Back-Reaction on the Metric

I now come to the difficult problem of the back-reaction of the particle creation on the metric and the consequent slow decrease of the mass of the black hole. At first sight it might seem that since all the time dependence of the metric in Fig. 4 is in the collapsing phase, all the particle creation must take place in the collapsing body just before the formation of the event horizon, and that an infinite number of created particles would hover just outside the event horizon, escaping to $\mathscr{I}^+$ at a steady rate. This does not seem reasonable because it would involve the collapsing body knowing just when it was about to fall through the event horizon whereas the position of the event horizon is determined by the whole future history of the black hole and may be someway outside the apparent horizon, which is the only thing that can be determined locally [7].

Consider an observer falling through the horizon at some time after the collapse. He can set up a local inertial coordinate patch of radius $\sim M$ centred on the point where he crosses the horizon. He can pick a complete family $\{h_\omega\}$ of solutions of the wave equations which obey the condition:

$$\tfrac{1}{2}i\int_S (h_{\omega_1}\bar{h}_{\omega_2;a}-\bar{h}_{\omega_2}h_{\omega_1;a})d\Sigma^a=\delta(\omega_1-\omega_2) \tag{4.1}$$

(where S is a Cauchy surface) and which have the approximate coordinate dependence $e^{i\omega t}$ in the coordinate patch. This last condition determines the splitting into positive and negative frequencies and hence the annihilation and creation operators fairly well for modes h_ω with $\omega>M$ but not for those with $\omega<M$. Because the $\{h_\omega\}$, unlike the $\{p_\omega\}$, are continuous across the event horizon, they will also be continuous on $\mathscr{I}^-$. It is the discontinuity in the $\{p_\omega\}$ on $\mathscr{I}^-$ at $v=v_0$ which is responsible for creating an infinite total number of particles in each mode. p_ω by producing an $(\omega')^{-1}$ tail in the Fourier transforms of the $\{p_\omega\}$ at large negative frequencies ω'. On the other hand, the $\{h_\omega\}$ for $\omega>M$ will have very small negative frequency components on $\mathscr{I}^-$. This means that the observer at the event horizon will see few particles with $\omega>M$. He will not be able to detect particles with $\omega<M$ because they will have a wavelength bigger than his particle detector which must be smaller than M. As described in the introduction, there will be an indeterminacy in the energy density of order M^{-4} corresponding to the indeterminacy in the particle number for these modes.

The above discussion shows that the particle creation is really a global process and is not localised in the collapse: an observer falling through the event horizon would not see an infinite number of particles coming out from the collapsing body. Because it is a non-local process, it is probably not reasonable to expect to be able to form a local energy-momentum tensor to describe the back-reaction of the particle creation on the metric. Rather, the negative energy density needed to account for the decrease in the area of the horizon, should be thought of as arising from the indeterminacy of order of M^{-4} of the local energy density at the horizon. Equivalently, one can think of the area decrease as resulting from the fact that quantum fluctuations of the metric will cause the position and the very concept of the event horizon to be somewhat indeterminate.

Although it is probably not meaningful to talk about the local energy-momentum of the created particles, one may still be able to define the total energy flux over a suitably large surface. The problem is rather analogous to that of defining gravitational energy in classical general relativity: there are a number of different energy-momentum pseudo-tensors, none of which have any invariant local significance, but which all agree when integrated over a sufficiently large surface. In the particle case there are similarly a number of different expressions one can use for the renormalised energy-momentum tensor. The energy-momentum tensor for a classical field ϕ is

$$T_{ab}=\phi_{;a}\phi_{;b}-\tfrac{1}{2}g_{ab}g^{cd}\phi_{;c}\phi_{;d}\,. \tag{4.2}$$

If one takes this expression over int the quantum theory and regards the ϕ's as operators one obtains a divergent result because there is a creation operator for each mode to the right of an annihilation operator. One therefore has to subtract out the divergence in some way. Various methods have been proposed for this (e.g. [30]) but they all seem a bit ad hoc. However, on the analogy of the pseudo-tensor, one would hope that the different renormalisations would all give the

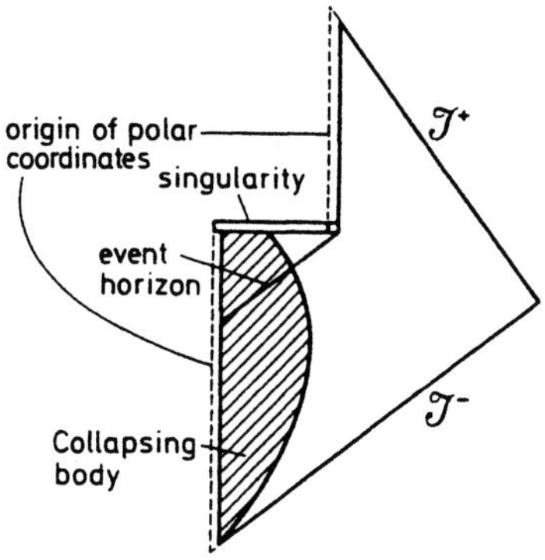

Fig. 5. The Penrose diagram for a gravitational collapse followed by the slow evaporation and eventual disappearance of the black hole, leaving empty space with no singularity at the origin

operator which agrees with the normal ordered operator near $\mathscr{I}^+$, which obeys the conservation equations, and which is stationary in the final quasi-stationary region will give the same energy flux over any surface of constant r. Thus it will give positive energy flux out across the event horizon or, equivalently, a negative energy flux in across the event horizon.

This negative energy flux will cause the area of the event horizon to decrease and so the black hole will not, in fact, be in a stationary state. However, as long as the mass of the black hole is large compared to the Planck mass 10^{-5} g, the rate of evolution of the black hole will be very slow compared to the characteristic time for light to cross the Schwarzchild radius. Thus it is a reasonable approximation to describe the black hole by a sequence of stationary solutions and to calculate the rate of particle emission in each solution. Eventually, when the mass of the black hole is reduced to 10^{-5} g, the quasi-stationary approximation will break down. At this point, one cannot continue to use the concept of a classical metric. However, the total mass or energy remaining in the system is very small. Thus, provided the black hole does not evolve into a negative mass naked singularity there is not much it can do except disappear altogether. The baryons or leptons that formed the original collapsing body cannot reappear because all their rest mass energy has been carried away by the thermal radiation. It is tempting to speculate that this might be the reason why the universe now contains so few baryons compared to photons: the universe might have started out with baryons only, and no radiation. Most of the baryons might have fallen into small black holes which then evaporated giving back the rest mass energy of baryons in the form of radiation, but not the baryons themselves.

The Penrose diagram of a black hole which evaporates and leaves only empty space is shown in Fig. 5. The horizontal line marked "singularity" is really a region where the radius of curvature is of the order the Planck length. The matter that runs into this region might reemerge in another universe or it might even reemerge in our universe through the upper vertical line thus creating a naked singularity of negative mass.

References

1. Isham, C. J.: Preprint (1973)
2. Ashtekar, A., Geroch, R. P.: Quantum theory of gravity (preprint 1973)
3. Penrose, R.: Phys. Rev. Lett. **14**, 57—59 (1965)
4. Hawking, S. W.: Proc. Roy. Soc. Lond. **A 300**, 187—20 (1967)
5. Hawking, S. W., Penrose, R.: Proc. Roy. Soc. Lond. **A 314**, 529—548 (1970)
6. Hawking, S. W., Ellis, G. F. R.: The large scale structure of space-time. London: Cambridge University Press 1973
7. Hawking, S. W.: The event horizon. In: Black holes. Ed. C. M. DeWitt, B. S. DeWitt. New York: Gordon and Breach 1973
8. Bardeen, J. M., Carter, B., Hawking, S. W.: Commun. math. Phys. **31**, 161—170 (1973)
9. Hawking, S. W.: Mon, Not. Roy. astr. Soc. **152**, 75—78 (1971)
10. Carr, B. J., Hawking, S. W.: Monthly Notices Roy. Astron. Soc. **168**, 399—415 (1974)
11. Hagedorn, R.: Astron. Astrophys. **5**, 184 (1970)
12. Hawking, S. W.: Commun. math. Phys. **25**, 152—166 (1972)
13. Carter, B.: Black hole equilibrium states. In: Black holes. Ed. C. M. DeWitt, B. S. DeWitt. New York: Gordon and Breach 1973
14. Misner, C. W.: Bull. Amer. Phys. Soc. **17**, 472 (1972)
15. Press, W. M., Teukolsky, S. A.: Nature **238**, 211 (1972)
16. Starobinsky, A. A.: Zh. E.T.F. **64**, 48 (1973)
17. Starobinsky, A. A., Churilov, S. M.: Zh. E.T.F. **65**, 3 (1973)
18. Bjorken, T. D., Drell, S. D.: Relativistic quantum mechanics. New York: McGraw Hill 1965
19. Beckenstein, J. D.: Phys. Rev. D. **7**, 2333—2346 (1973)
20. Beckenstein, J. D.: Phys. Rev. D. **9**,
21. Penrose, R.: Phys. Rev. Lett. **10**, 66—68 (1963)
22. Sachs, R. K.: Proc. Roy. Soc. Lond. **A 270**, 103 (1962)
23. Eardley, D., Sachs, R. K.: J. Math. Phys. **14** (1973)
24. Schmidt, B. G.: Commun. Math. Phys. **36**, 73—90 (1974)
25. Bondi, H., van der Burg, M. G. J., Metzner, A. W. K.: Proc. Roy. Soc. Lond. **A 269**, 21 (1962)
26. Carter, B.: Commun. math. Phys. **10**, 280—310 (1968)
27. Teukolsky, S. A.: Ap. J. **185**, 635—647 (1973)
28. Unruh, W.: Phys. Rev. Lett. **31**, 1265 (1973)
29. Unruh, W.: Phys. Rev. D. **10**, 3194—3205 (1974)
30. Zeldovich, Ya. B., Starobinsky, A. A.: Zh. E.T.F. **61**, 2161 (1971), JETP **34**, 1159 (1972)

Communicated by J. Ehlers

S. W. Hawking
California Institute of Technology
W. K. Kellogg Radiation Lab. 106-38
Pasadena, California 91125, USA

PHYSICAL REVIEW D VOLUME 13, NUMBER 8 15 APRIL 1976

Path-integral derivation of black-hole radiance*

J. B. Hartle
Department of Physics, University of California, Santa Barbara, California 93106
and California Institute of Technology, Pasadena, California 91125

S. W. Hawking†
Department of Applied Mathematics and Theoretical Physics, University of Cambridge, Cambridge, England
and California Institute of Technology, Pasadena, California 91125

(Received 17 November 1975)

The Feynman path-integral method is applied to the quantum mechanics of a scalar particle moving in the background geometry of a Schwarzschild black hole. The amplitude for the black hole to emit a scalar particle in a particular mode is expressed as a sum over paths connecting the future singularity and infinity. By analytic continuation in the complexified Schwarzschild space this amplitude is related to that for a particle to propagate from the past singularity to infinity and hence by time reversal to the amplitude for the black hole to absorb a particle in the same mode. The form of the connection between the emission and absorption probabilities shows that a Schwarzschild black hole will emit scalar particles with a thermal spectrum characterized by a temperature which is related to its mass, M, by $T = \hbar c^3/8\pi GMk$. Thereby a conceptually simple derivation of black-hole radiance is obtained. The extension of this result to other spin fields and other black-hole geometries is discussed.

I. INTRODUCTION AND SYNOPSIS

The Feynman path-integral method[1] is a natural way to formulate the quantum mechanics of matter fields moving in curved background spacetimes.[2-4] In this method the amplitude $K(x,x')$ for a particle to propagate from one space time point x' to another x is expressed as an integral over all the paths connecting the two points. The integral has the form

$$K(x,x') \sim \sum_{\text{paths}} e^{iS(x,x')/\hbar}\,, \tag{1.1}$$

where $S(x,x')$ is the classical action for a particular path connecting x' and x. The amplitude K is called the propagator.

This formulation of quantum mechanics has several advantages. Because the sum is over paths in the four-dimensional space time and because S is a four-dimensional scalar the expression for K is manifestly covariant. Expressing the propagator as a sum over paths gives it a direct physical interpretation. Since the propagator is expressed directly as a functional integral a problem of finding an approximate form for it reduces immediately to a problem of approximating the functional integral. For these reasons the Feynman path-integral method is an attractive way to do quantum mechanics in curved space times. In this paper we shall use the path-integral method to derive the thermal radiation emitted by black holes.[5] In the following we shall give a qualitative outline of our methods and results. The details and proofs will be presented in the subsequent sections.

Figure 1 shows the Penrose diagram for the Schwarzschild geometry. The unshaded part of this diagram represents the geometry outside a spherically symmetric collapsing body. The shaded part should be replaced by the geometry inside. Let us now consider the probability that a particle is emitted by the black hole and detected by an observer a constant distance away in a positive-frequency mode peaked about some point A when there are no incoming particles in the distant past. This probability can be related to the amplitude to propagate from some point B on the future singularity to the observation point A. This in turn can be represented as a sum over paths of the form in Eq. (1.1), where the paths summed over are those which begin at the point B on the future singularity and end at the observation point A. A typical such path (BCA) is shown in Fig. 1. These are exactly the paths which correspond to a pair of particles being created (near C), one of which falls into the black hole and the other of which propagates out to the observer. We do not sum over paths which start on $\mathscr{I}^-$ since they would represent the propagation of incoming particles in the distant past. We do not sum over paths which pass through the shaded region since that should properly be replaced by the interior of the collapsing star and will not contribute to the particle production at late times (as we shall show in more detail subsequently). We consider only propagation from the future singularity.

If one attempts to evaluate the production probability by applying the method of stationary phase to the integrals involved, then it is readily seen that there are no real stationary paths which connect the future singularity to a positive-frequency mode for a stationary exterior observer. Such paths would be permissible classical paths with the positive energy connecting the two surfaces and there are none (although these are classical paths with negative energy). However, if the coordinates of the point B on the future singularity are displaced to complex values then stationary-phase paths can be found. These paths are, in fact, in the real manifold and connect the *past* singularity to the external observer. Thus, the amplitude to propagate to the external observer from a point B on the future singularity can be related to the amplitude to propagate from a corresponding point D on the past singularity. In turn, by symmetry under time reversal, the modulus of the latter amplitude is the same as that of the amplitude to propagate from the time-reversed point of A into the future singularity at the time-reversed point of D. This shows that by appropriately distorting the contours of integration into the complex coordinate plane the amplitude for a black hole to emit particles can be related to the amplitude for it to absorb. If we consider the amplitude for the black hole to emit particles in a definite mode with energy E as measured by a distant observer then this connection is

$$(\text{probability to emit a particle with energy } E) = e^{-2\pi E/\kappa} \times (\text{probability to absorb a particle with energy } E)\,, \tag{1.2}$$

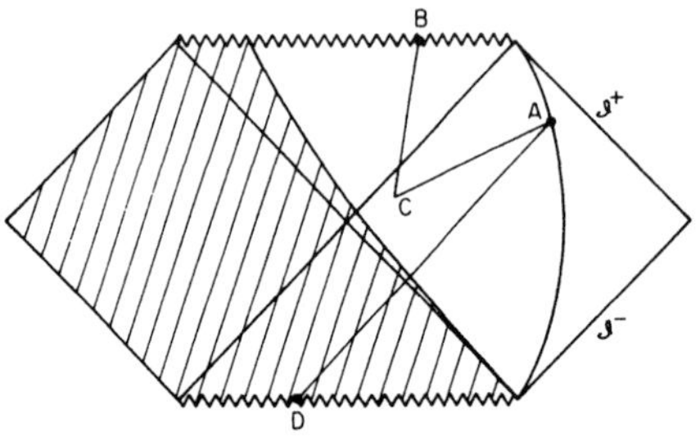

FIG. 1. The unshaded part of the diagram represents the Schwarzschild geometry outside a spherically symmetric collapse. T e world line O is that of an observer, who remains outside the black hole at a fixed radius. The nonstationary path BCA corres nds to particle production by the black hole. A pair of particles is created near C. One falls into the future singularity at B while the other propagates out to t e observer and is detected at A. The amplitude that a particle is produced by the black hole and detected by the observer in a given mode at late times can be expressed as an integral over the amplitude to propagate between a point B on the future singularity and a point A on the observer's world line. In turn the propagator can be expressed as a sum over the paths which connect these points. By analytically continuing the point B into the complexified Schwarzschild space, the amplitude to propagate to A from a real point B on the future singularity can be related to the amplitude to propagate to A from a reflected point D on the past singularity. This latter process is just the time-reversed proces of absorption of a particle by the black hole. In this way the probability for a black hole to emit a scalar particle can be related to the probability for it to absorb one. The relation implies the thermal radiance.

where κ is the surface gravity of the black hole; in the case of a Schwarzschild black hole of mass M, $\kappa = 1/4M$. (Here as in the following we are using units where $\hbar = c = G = 1$.) This connection between emission and absorption is exactly that necessary to establish the result of Ref. 5 that a Schwarzschild black hole emits particles with a thermal spectrum corresponding to a temperature $T = \kappa/2\pi k$. To see this, imagine surrounding the black hole by a thermal cavity and adjusting the temperature, T, until the whole system is in equilibrium. The temperature of the radiation is then the temperature of the black hole. In equilibrium the rate of emission particles by the black hole must exactly equal the rate of absorption. S nce the ratio of the probability of having N photons in a particular mode in the cavity with energy E to the probability of having $N-1$ photons in the same mode is $\exp(-E/kT)$ this equilibrium condition will be true when $T = \kappa/2\pi k$.

In the following we shall give th details of this simple argument. In Sec. II the quantum mechanics of scalar particles moving in a Schwarzschild background geometry is formulated in terms of path integrals. Section III contains a derivation of the necessary analytic properties of the propagator on the complexified manifold, and in Sec. IV the thermal radiation from a Schwarzschild black hole is deduced. Section V discusses the generalization to Reissner-Nordström and Kerr black h les and higher-spin particles.

II. PATH-INTEGRAL QUANTUM MECHANICS OF A SCALAR PARTICLE IN A SCHWARZSCHILD BACKGROUND

In this section we shall formulate the quantum mechanics of a scalar particle moving in a curved background geometry in terms of Feynman path integrals. Our work here is a generalization and interpretation of that of Feynman for a scalar field in a flat-space background,[7] and as far as the general formulation of path integrals in curved backgrounds goes closely parallels the previous work of the DeWitts.[2-3] The considerations of this section are intended to motivate the definition of the propagator in Sec. III as a solution of the inhomogeneous scalar wave equation with certain boundary conditions. As a consequence in the present section we shall not be uniformly mathematically precise, but this is a familiar situation when working with path integrals.

The path of a scalar particle through spacetime may be specified by giving the four coordinates x^α as a function of a parameter time w. Letting x stand for all four coordinates we write this as $x=x(w)$. Suppose we consider the motion of a particle which starts at a spacetime point x' at $w=0$ and arrives at x at $w=W$. An action functional which describes the classical motion of such a particle is

$$S[x(w)]=\frac{1}{4}\int_0^W dw\, g(\dot{x},\dot{x}), \tag{2.1}$$

where g is the metric on the curved spacetime and $\dot{x}$ represents the tangent vector whose components are dx^α/dw. The path which extremizes S satisfies the geodesic equation with w as an affine parameter. Thus, for timelike paths w may be taken to be a constant multiple of the proper time while for spacelike paths it could be taken to be the same multiple of the proper distance.

The action functional of Eq. (2.1) is not the usual one in which the integrand is $[-g(\dot{x},\dot{x})]^{1/2}$. However, (2.1) is a perfectly valid classical action which has obvious advantages for a path-integral formulation in that it is quadratic in the four-velocities (see the Appendix). In contrast to the usual form, the action in Eq. (2.1) continues analytically from timelike to spacelike paths. In addition it gives correctly the relativistic quantum mechanics of a scalar particle in flat spacetime and is therefore a natural generalization to curved backgrounds. We shall not discuss other choices of the action further here.

The basic assumption of the Feynman path-integral method is that the amplitude for a particle to travel a particular path in spacetime is proportional to $\exp\{iS[x(w)]\}$. To have a clearer idea of what this means imagine dividing the parameter time w into many small intervals at values w_i. The amplitude for observations of spacetime position at *each* parameter time w_i to yield the set of values $\{x_i=x(w_i)\}$ is proportional to $\exp\{iS[x(w)]\}$ in the limit as the intervals become infinitesimally separated. Amplitudes for more restricted sets of observations may be constructed by summing this amplitude over the unobserved positions. For example, the amplitude that an observation of spacetime position at one parameter time yields the value x' and a second observation a parameter time W later yields the value x is

$$F(W,x,x')=\int \delta x[w]\exp\left[\frac{1}{4}i\int_0^W g(\dot{x},\dot{x})dw\right], \tag{2.2}$$

where the integral is over all the unobserved positions at parameter times between 0 and W. In other words, the integral is a functional integral over all paths which have $x(0)=x'$ and $x(W)=x$.

The parameter w has been introduced as an observable which plays a role analogous to ordinary time in nonrelativistic quantum mechanics.[8] However, there is no experiment in which it is directly observed (since particles do not carry clocks). All physical observations can be obtained from the amplitude $K(x,x')$ for a particle to be localized at two spacetime points x' and x. $K(x,x')$ is called the propagator. This amplitude can be constructed in two steps: first by summing over all paths which connect x' to x in a given parameter time W and then by summing[9] over the unobserved value, W. The first sum is just F in Eq. (2.2). In the second sum an appropriate weight must be assigned to each elapsed parameter time W. In flat space if the scalar particle has a rest mass m this weight is[7] $i\exp(-im^2W)$. It is natural to adopt this also for the curved-space case. The expression for the propagator then takes the form[10]

$$K(x,x')=i\int_0^\infty dW \exp(-im^2W)F(W,x,x'), \tag{2.3}$$

where F is given by Eq. (2.2). The restriction of the integral to positive W is the requirement that the particles always propagate forward in parameter time.

From this definition it easily follows that $K(x,x')$ is symmetric in x and x'. Letting $w=W-w'$ leaves the action in Eq. (2.1) unchanged but interchanges x and x' in the sense that as functions of w', $x(0)=x$ and $x(W)=x'$. Thus $F(W,x,x')$ is symmetric in x and x' and it follows immediately from Eq. (2.3) that $K(x,x')$ is also.

Equations (2.2) and (2.3) are the basic relations which are needed to define the quantum mechanics of a free scalar particle moving in a curved spacetime. Before this definition is complete the integrals in Eqs. (2.2) and (2.3) need to be given meaning. We now turn to this question but for simplicity and definiteness restrict our attention to scalar particles propagating in the Schwarzschild geometry.

The first problem is the definition of the path integral in Eq. (2.2). To solve this we analytically continue the variables in this formal expression to values where the integral is well defined. In particular w and W are continued to negative imaginary values $-i\omega$ and $-i\Omega$, respectively, and the coordinates are continued to a domain where the metric has signature +4. In the case of the Schwarzschild geometry which in Kruskal coordinates z, y, θ, φ has the form

$$ds^2 = (32M^3e^{-r/2M}/r)(-dz^2+dy^2)+r^2d\Omega^2, \qquad (2.4)$$

with $d\Omega^2 = d\theta^2+\sin^2\theta d\varphi^2$ and $r(y,z)$ defined by

$$-z^2+y^2 = (r/2M-1)e^{r/2M}, \qquad (2.5)$$

this can be accomplished by letting $z = i\zeta$ and keeping ζ real. Then the analytically continued metric γ is given by the line element

$$d\sigma^2 = (32M^3e^{-r/2M}/r)(d\zeta^2+dy^2)+r^2d\Omega^2, \qquad (2.6)$$

with r now defined by

$$\zeta^2+y^2 = (r/2M-1)e^{r/2M}. \qquad (2.7)$$

The analytically continued expression for F becomes

$$F(\Omega,x,x') = \int \delta x[\omega]\exp\left[-\tfrac{1}{4}\int_0^{\Omega}\gamma(\dot{x},\dot{x})d\omega\right], \qquad (2.8)$$

where x is to be understood as $x^\alpha = (\zeta, y, \theta, \varphi)$ and $\dot{x}$ as $dx^\alpha/d\omega$. The space covered by the coordinate ranges $-\infty<\zeta<\infty$, $-\infty<y<\infty$, $0\le\theta\le\pi$, $0\le\varphi<2\pi$ is complete, has the topology $R^2\times S^2$, and, since r ranges only over values greater than $2M$, the metric γ is regular with the exception of the trivial polar singularities at $\theta=0$ and π. The integration now extends over all paths in this space which start with $\omega=0$ at x' and end with $\omega=\Omega$ at x (see Fig. 2). This path integral can be precisely defined.[11] Our basic assumption[12] is that the function F when defined in this way and analytically continued back to real values of the coordinates and parameter time variable gives the correct propagator defined heuristically by Eq. (2.2).

This procedure not only gives definition to the integral in Eq. (2.2) but also identifies the class of paths over which integration is done. Imagine taking a particular path in the space with positive-definite metric and continuing both coordinates and parameter to complex values. The complexified path is now a two-dimensional sheet in the space of complex coordinates given by the four complex analytic functions $x(\omega)$ of the complex parameter ω. The analytic functions are completely fixed by their real values for real ω. What does the class of paths defined above look like when the analytic continuation reaches real values of W and the contours of the path integration are deformed to real coordinate values? It seems clear that the resulting class of paths will not be confined by any finite boundaries in the Kruskal coordinates. In particular, they will cross and recross the singularities at $r=0$. These singularities are poles in the metric considered as functions of the complex coordinates. The path integral across the singularity is defined by giving a prescription for which way the contour of integration goes around the pole. In turn this is determined by the analytic continuation of the path from the positive-definite section and the deformation of the contours to real values everywhere except near $r=0$. If the paths cross $r=0$ then they extend into the Schwarzschild geometries with negative mass. This is illustrated schematically in Fig. 2(b).

The actual computation of $F(\Omega,x,x')$ is greatly facilitated by noticing that it satisfies a parabolic partial differential equation

$$\frac{\partial F}{\partial\Omega} = \bar{\Box}^2F, \qquad (2.9)$$

where $\bar{\Box}^2 = \gamma^{\alpha\beta}\bar{\nabla}_\alpha\bar{\nabla}_\beta$ and $\bar{\nabla}_\alpha$ indicates covariant differentiation with respect to the metric γ. The derivation of this result is reviewed in the Appendix. The boundary conditions on Eq. (2.9) which yield

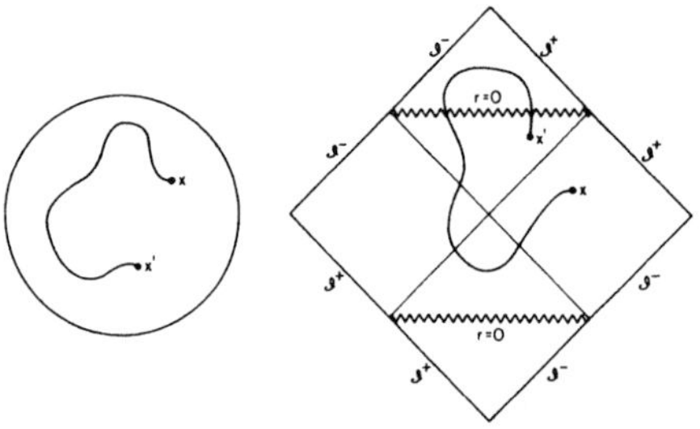

(a) (b)

FIG. 2. (a) A compactified representation of a constant θ, constant φ slice of the positive-definite spacetime whose metric is given in Eq. (2.6). The heavy circle represents infinity. There are no sin ularities. A typical path connecting two points x' and x is shown. (b) A Penrose diagram for the Schwarzschild geometry showing in addition the regions of negative mass (or $r\le 0$) above and below the sin ularities. A typical member of the class of paths continued analytically to this real section from the positive-definite spacetime represented in (a) is shown. Such paths may cross and recross the singularities at $r=0$. Inte rations over paths which cross the sin ularities are specified by choosing contours of inte ration which are the analytic continuations of those in the positive-definite section.

F as defined by the path integral are first that

$$F(0,x,x') = \delta(x,x'), \qquad (2.10)$$

where $\delta(x,x')$ is the four-dimensional δ function in the space with positive-definite metric [equal to $\delta^{(4)}(x-x')\gamma^{-1/2}$]. Second, F must vanish as x approaches the infinity of the space with positive-definite metric. The first boundary condition is simply the requirement that at $\Omega = 0$ the particle be localized at x'. The second follows from the exponential damping of the integral as x tends to infinity.

The solution to Eq. (2.8) and associated boundary conditions will be analytic in x where the metric is and analytic in Ω except at the origin.[13] The solution can thus be continued back to physical values of the coordinates and parameter time. There it will satisfy the analytically continued equation

$$i\frac{\partial F}{\partial W} = -\Box^2 F, \qquad (2.11)$$

where $\Box^2 = g^{\alpha\beta}\nabla_\alpha\nabla_\beta$ and ∇_α is covariant differentiation with respect to the metric g. Equation (2.11) can be thought of as the Schrödinger equation for propagation in the parameter time W.

The second problem involved in giving meaning to Eq. (2.3) for the propagator is the integral over W. To resolve this we need the asymptotic form of $F(W,x,x')$ for both small and large W. The behavior of F for small W is essentially given by the definition of the path integral itself. For small W and fixed x and x' the action [Eq. (2.1)] for a typical path will become large. The only paths which contribute significantly to the path integral are the stationary paths, i.e., the geodesics connecting x and x'. For these paths the action is

$$S(W,x,x') = \tfrac{1}{4}s(x,x')/W, \qquad (2.12)$$

where $s(x,x')$ is the square of the geodesic distance between x and x'. If there were a single geodesic connecting x and x' then for small W we could write

$$F(W,x,x') = \exp[is(x,x')/(4W)]N(W,x,x'),$$

where $N(W,x,x')$ is a real normalization factor. Equation (2.11) can then be solved for small-W behavior of N. When note is taken of the identity

$$(\nabla_\alpha s)(\nabla^\alpha s) = 4s, \qquad (2.13)$$

one finds that

$$N(W,x,x') = D(x,x')W^{-2} + \cdots, \qquad (2.14)$$

where D is independent of W; its exact form will not concern us. The reader will recognize this as essentially the WKB approximation to the solution of Eq. (2.11). Considerably more detail on the derivation can be found in Refs. 2 and 8(d).

In general there will be several geodesics connecting x and x'. In that case the small-W behavior of F will be

$$F(W,x,x') = \sum_c e^{is_c(x,x')/(4W)} \times W^{-2}[D_c(x,x') + O(W^{-1})], \qquad (2.15)$$

where the sum is over each class of geodesics which connect x to x'. This small-W behavior will not be uniformly valid over the whole range of values of x. In particular where neighboring geodesics which start at x' intersect (caustics) we expect the approximation to break down. For example, $s(x,x')$ will have a branch point at such an intersection, but we know from general considerations that F has none.

If Eq. (2.15) is integrated over a smooth function of x then as W tends to zero there will be a significant contribution to the integral only for values of x for which $s_c(x,x')$ is nearly stationary keeping x' fixed. This will happen only for x close to x'. In other words, $F(0,x,x')$ is proportional to a δ function. The proportionality factor is just unity because the normalization factors in Eq. (2.15) must also give rise to Eq. (2.10). Thus,

$$\lim_{W\to 0} F(W,x,x') = \delta(x,x'). \qquad (2.16)$$

For large values of Ω standard estimates[13] for the solution of the parabolic equation (2.9) defined by the boundary condition in Eq. (2.10) show that $F(\Omega,x,x')$ decreases at least as fast as Ω^{-2} for x and x'. Physically this is nothing more than the spreading of an initially localized wave packet with increasing Ω. Thus, at large Ω, F can be expressed as

$$F(\Omega,x,x') = \Omega^{-2}[F_0(x,x') + F_1(x,x')\Omega^{-1} + \cdots]. \qquad (2.17)$$

We will assume that expansion can be continued back to real values of W and x by term. Thus, in particular we have for large W

$$F(W,x,x') = O(W^{-2}). \qquad (2.18)$$

The large-W behavior of F shows that the integral in Eq. (2.3) always converges at the upper limit. This is not the case at the lower limit where F diverges as W^{-2} [Eq. (2.15)]. To make the integral finite we shall insert a convergence factor $\exp(-\epsilon/W)$, where ϵ is a small positive constant. Physical quantities are to be computed with ϵ finite and then the limit $\epsilon \to 0$ is to be taken. Thus,

$$K(x,x') = i\int_0^\infty dW \exp(-im^2W - \epsilon/W)F(W,x,x'). \qquad (2.19)$$

This method of regularization corresponds physically to the requirement that the particle propagate forward in the parameter time W. It correctly gives the usual Feynman propagator for a scalar particle in flat spacetime as we shall show in the next section. With Eq. (2.19) our definition of the propagator is now essentially precise. We shall now examine the consequences.

III. ANALYTICITY PROPERTIES OF THE PROPAGATOR

In the preceding section an integral representation for the Feynman propagator $K(x,x')$ was derived in terms of the propagator for a definite parameter time W, $F(W,x,x')$. It follows immediately from Eq. (2.9), the parameter time Schrödinger equation, Eq. (2.11), and the boundary conditions, Eqs. (2.15) and (2.17), that $K(x,x')$ is a solution of the inhomogeneous wave equation in the Schwarzschild background

$$(\Box^2 - m^2)K(x,x') = -\delta(x,x')\,. \tag{3.1}$$

As an alternative to the path integral, $K(x,x')$ could be defined as a solution to Eq. (3.1) with suitable boundary conditions. This approach is a useful one because some properties of $K(x,x')$ can be deduced directly from the differential equation and because it is more easily generalizable to the propagation of particles with higher spin. In this section we shall derive the boundary conditions for $K(x,x')$ from its path-integral definition.

First, we illustrate the procedure with the example of a massless scalar particle in flat space. The solution of the parameter time Schrödinger equation for F which satisfies the boundary condition of Eq. (2.10) is

$$F(W,x,x') = i(4\pi W)^{-2} e^{is(x,x')/(4W)}\,, \tag{3.2}$$

where $s(x,x')$ denotes the square of the Minkowski interval between x and x'. From Eq. (2.19) the propagator is then

$$K(x,x') = -\frac{i}{4\pi^2}\frac{1}{s(x,x')+i\epsilon}\,, \tag{3.3}$$

which is the correct Feynman propagator. In any coordinates in which the Minkowski metric is analytic, $s(x,x')$ will be an analytic function of the coordinates and $K(x,x')$ will also be an lytic except at the poles where $s(x,x') = -i\epsilon$. These poles correspond to the null geodesics connecting x and x'. It is regularity at infinity plus the location of these poles in the complex coordinate plane that uniquely fixes $K(x,x')$ as a solution of the inhomogeneous wave equation. More concretely, in the usual rectangular Minkowski coordinates with $x = (t, \vec{x})$ and $x' = (t', \vec{x}')$, $K(x,x')$ has poles at $t - t' = \pm(|\vec{x} - \vec{x}'| - i\epsilon)$. Thus $K(x,x')$ is that solution of the inhomogeneous wave equation which is regular at infinity and for which the singularities corresponding to propagation along future-directed null geodesics lie below the real t axis, while those corresponding to propagation along the past-directed null geodesics lie above the real t axis. Elsewhere in the complex t plane $K(x,x')$ is analytic. We shall now consider the analogous boundary conditions for Eq. (3.1) in the Schwarzschild geometry.

To begin with let us consider the case in which x' is exterior to the black hole and x lies on the horizon. It is convenient to use null Kruskal coordinates U and V in which the Schwarzschild metric takes the form

$$ds^2 = -(32M^3 e^{-r/2M}/r)dU dV + r^2 d\Omega^2\,, \tag{3.4}$$

with r defined by

$$UV = (1 - r/2M)e^{r/2M}\,. \tag{3.5}$$

On the horizon $r = 2M$ and either $U=0$ or $V=0$. We analytically continue the nonzero member of the pair (U, V) to complex values and refer to the surface thus obtained as the complexified horizon. Since the metric is analytic in the Kruskal coordinates on the complexified horizon, the function $F(W,x,x')$ will also be analytic there. Any singularities in $K(x,x')$ will therefore come from the end points of the integration over W. From the asymptotic expression [Eq. (2.17)] for $F(W,x,x')$ at large values of W one easily sees that the integral converges for large W for all complex values of x. Any singularities of $K(x,x')$ must therefore come from the $W=0$ end point. To analyze these, divide the interval $[0,\infty]$ in W into two pieces $[0,W_0]$ and $[W_0,\infty]$, where W_0 is small. The integral from W_0 to infinity gives a contribution $K_0(x,x')$ to $K(x,x')$ which is analytic in x. In the part from 0 to W_0 our expression for the small-W behavior of $F(W,x,x')$ may be used wherever it is valid. The result for $K(x,x')$ when $m=0$ is

$$K(x,x') = K_0(x,x') - i\sum_c \frac{e^{is_c(x,x')/4W_0}}{s_c(x,x')+i\epsilon}D_c(x,x')\,. \tag{3.6}$$

This expression can be used to continue $K(x,x')$ to values of x off the complexified horizon. With $m \neq 0$, $K(x,x')$ also has singularities whenever $s_c(x,x') = -i\epsilon$, i.e., slightly displaced from wherever there is a null geodesic connecting x' to the complexified horizon. We shall now locate these points.

To start with we shall show that all null geodesics which start from real v lues of x' intersect the complexified horizon on the real sectio , i.e., at real values of U and V. For definiteness let us consider first the geodesics which connect a real x' exterior to the hole to the future horizon. Instead of the affine parameter V on the horizon it is convenient to use the Killing time v related to it by $V=\exp(\kappa v)$. This is just the familiar advanced time of the Eddington-Finkelstein coordinates which cover the region $V \geq 0$. Complex null geodesics may be represented by giving the four coordinates as functions of a complex affine parameter which we shall call λ. The geodesic is thus really a two-dimensional sheet in the complex coordinate space.

From the relation $V=\exp(\kappa v)$, a given value of v and one displaced from it by $\mathrm{Im} v = 2\pi/\kappa$ correspond to the same value of V. A consequence of this is that by studying the null geodesics whose v coordinates are confined to a strip of width $2\pi/\kappa$ in the complex plane one learns about the null geodesics for all other values of v. It is convenient to choose a strip which includes the real v axis. We may then suppose that at $\lambda=0$ the coordinates assume the real starting values $v', r', \theta', \varphi'$. Our question is what complex values of v in this strip with real values of θ and φ and $r=2M$ lie on the two-dimensional sheet which represents a complex null geodesic?

Null geodesics in this stationary spherically symmetric spacetime may without loss in generality be taken to be in the equatorial plane, $\theta=\pi/2$. They are characterized by two constants of the motion e and l which may take complex values. The definitions of these constants are

$$e=\left(1-\frac{2M}{r}\right)\frac{dv}{d\lambda}, \quad l=r^2\frac{d\varphi}{d\lambda}. \tag{3.7}$$

Then from the null condition $g(\dot{x},\dot{x})=0$ we derive the familiar expressions

$$v-v'=\int_{r'}^{2M}\frac{dr}{1-2M/r}\left\{1-\frac{1}{[1-b^2r^{-2}(1-2M/r)]^{1/2}}\right\}, \tag{3.8}$$

and

$$\varphi-\varphi'=\int_{r'}^{2M}\frac{b\,dr/r^2}{[1-b^2r^{-2}(1-2M/r)]^{1/2}}. \tag{3.9}$$

Here we have written b for the impact parameter l/e. The multiplicative arbitrariness in the affine parameter λ implies that the invariant ratio b is sufficient to completely characterize a particular null geodesic.

The purely real geodesics connecting x' to the future horizon correspond to real values of b between 0 and $3\sqrt{3}M$. For these values, $r=r'$ and $r=2M$ can be connected by a purely real contour. For this reason for real b between 0 and $3\sqrt{3}M$ it is convenient to take the cuts of $f(r)=[1-b^2r^{-2}(1-2M/r)]^{1/2}$ to avoid the positive real r axis and to have $r=r'$ and $r=2M$ on the same sheet of the Riemann surface of f. The complex analytic structure of f for other values of b is then fixed by analytic continuation in that variable.

A given complex value of b and a given contour in the r plane connecting $r=r'$ with $r=2M$ should determine a unique complex null geodesic. For example, for every real b between 0 and $3\sqrt{3}M$ and a purely real contour between $r=r'$ and $r=2M$ there is a unique real null geodesic. For the same value of b a second contour which could not be obtained from the first by a smooth distortion would determine a different complex null geodesic.

However, it is easily verified from Eqs. (3.8) and (3.9) that there are no poles in the integrands of these equations to prevent one contour which connects $r=r'$ and $r=2M$ on the given sheet from being distorted into any other. The most convenient choice for the contour, therefore, is simply to take it to lie along the real axis provided that it does not intersect a cut of $f(r)$.

For a given contour the integral in Eq. (3.9) defines a function φ/b which is a multivalued function of b^2. In order to obtain a unique connection between φ and b it will be necessary to restrict b^2 to a given sheet of the Riemann surface of φ/b. We shall call this the physical sheet. This sheet must include the real axis between $b^2=0$ and $b^2=27M^2$ which corresponds to the physical real null geodesics. There is a branch point of the function φ/b at $b^2=27M^2$. It is therefore convenient to define the physical sheet as the plane cut along the positive real axis from $27M^2$ to infinity and containing the physical real values from $b^2=0$ to $27M^2$. It is then easily verified that the contour in Eq. (3.9) can always be chosen to lie along the real axis. With b^2 on the physical sheet the integral in Eq. (3.9) then defines a unique connection between φ and b.

Of interest in the present instance are null geodesics which have real values of φ at the end point. An elementary analysis of the integral in Eq. (3.9) shows that since the contour can be chosen real for b^2 on the physical sheet, the imaginary part of the integrand is always of one sign and the imaginary part of φ does not vanish unless b^2 is real and between 0 and $27M^2$. However, these values of b mean that v at the end point will also be real. Thus, complex null geodesics starting from x' intersect the complexified future horizon only for real values of V. A similar conclusion clearly holds for the past horizon. There are then singularities of K on the complexified horizon slightly displaced from the real values of U and V at which null geodesics from x' intersect the horizon according to the relation $s(x,x')=-i\epsilon$. We shall now determine the direction of these displacements.

Suppose that x_0 is the end point on the future horizon of a real null geodesic which starts at x'. If V_0 represents the value of V associated with x_0, then for values of V near to V_0, s will behave as

$$s(x,x') = \left(\frac{\partial s}{\partial V}\right)_{x_0} (V - V_0) + \cdots . \tag{3.10}$$

A positive value of $(\partial s/\partial V)_{x_0}$ means that the solution of $s(x,x') = -i\epsilon$ is in the lower half plane while a negative value means that it is in the upper half plane. To determine the correct sign let $k\delta V$ be the displacement vector from the null geodesic to a neighboring geodesic which starts at x' but ends on the future horizon a small affine parameter distance δV to the future of V_0. Thus on the horizon $k = \partial/\partial V$. If l is the tangent vector to the null geodesic then on the horizon $l \cdot k < 0$. The equation of geodesic deviation implies that

$$d^2(l\cdot k)/d\lambda^2 = 0 , \tag{3.11}$$

and this relation can be used to propagate $l\cdot k$ back along the null geodesic to $\lambda = 0$, where both geodesics originate. One finds

$$l \cdot k = c\lambda , \tag{3.12}$$

where c is a negative constant. Since the tangent vector along the neighboring geodesic is the sum of l and $k\delta V$, Eq. (3.12) implies that the neighboring geodesic is timelike. Thus $s(x,x')$ is negative as x runs along the neighboring curve and $(\partial s/\partial V)_{x_0} < 0$. The singularities of the propagator corresponding to $s(x,x') = -i\epsilon$ therefore lie in the upper half plane and the propagator will be analytic in the lower half V plane on the complexified horizon.

In a similar manner the analytic properties of the propagator on the past complexified horizon can be deduced. For x' located at a real point outside the black hole, and x on the complexified past horizon, $K(x,x')$ will be analytic in the upper half U plane.

The analytic properties which we have deduced from the path integral for the propagator on the complexified horizon may now be considered as boundary conditions which *define* the propagator as a particular solution of the inhomogeneous wave equation, Eq. (3.1). For all of our subsequent results we could have started from this definition of the propagator in terms of its analytic properties on the complexified horizon, but such a definition would lack the physical motivation which our definition in terms of the path integral gives.

The inhomogeneous scalar wave equation together with the boundary conditions just deduced may be used to derive the analytic properties of the propagator for regions other than the complexified horizon. To complete the program outlined in the Introduction we shall be concerned in particular with the analytic properties in the Schwarzschild coordinate t. This is connected to the null coordinates U and V through the relations

$$\left.\begin{aligned} U &= (1 - r/2M)^{1/2} e^{(r-t)/4M} \\ V &= (1 - r/2M)^{1/2} e^{(r+t)/4M} \end{aligned}\right\} U>0, \quad V>0 \quad \text{(region II)} \tag{3.13a}$$

$$\left.\begin{aligned} U &= -(r/2M - 1)^{1/2} e^{(r-t)/4M} \\ V &= (r/2M - 1)^{1/2} e^{(r+t)/4M} \end{aligned}\right\} U<0, \quad V>0 \quad \text{(region I)} \tag{3.13b}$$

and similar relations with the signs of U and V changed in the quadrants reflected in the origin of the U-V plane. These relations are indicated schematically on a Penrose diagram in Fig. 3.

For definiteness let us first consider the case when x' is exterior to the black hole and x is in region II. The portion of the future horizon with

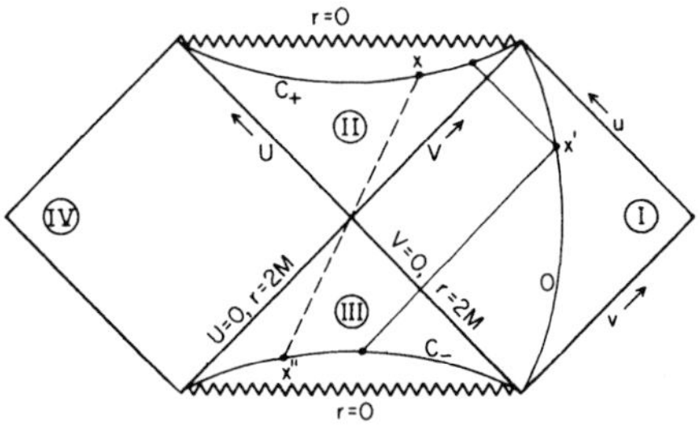

FIG. 3. A Penrose diagram for the Schwarzschild geometry. The amplitude for a black hole to emit particles which are detected in a mode of energy E by an observer in region I may be related [Eq. (4.4)] to the integral of $\exp(-iEt)$ times the propagator to go from a point x on a surface C_+ of constant r in region II to a point x' on the detector's world line in region I. The integral is over the coordinate t on C_+. The propagator is analytic in the coordinate t except for singularities at those values where a null geodesic from x intersects the complexified surface C_+. One of these values is the real value of t corresponding to the intersection with C_+ of the radial real future-directed null geodesic from x. As shown in the text, if x has a time coordinate with an imaginary part $-4\pi M$, it actually corresponds to the point x'' in region III, which is x reflected in the origin. There is thus another singularity in t with imaginary part $-4\pi M$ corresponding to the radial real past-directed null geodesic from x intersecting the surface C_- which is the reflection of C_+. The two singularities just discussed are repeated at intervals of $8\pi M$ in Imt. The location of the singularity is illustrated in Fig. 4(a).

$V \geq 0$ together with the part of the past horizon with $U \geq 0$ are a complete characteristic Cauchy surface for region II. The propagator in the interior region is uniquely determined through the differential equation by the initial data on this Cauchy surface. These are just the values of the propagator on the relevant parts of the horizon.

Complex values of t with r, θ, and φ fixed correspond to certain complex values of U and V according to Eq. (3.13a). If we let $t = \tau + i\sigma$ then in particular

$$U = |U| e^{-i\sigma/4M}, \quad V = |V| e^{i\sigma/4M}. \tag{3.14}$$

The problem of determining the propagator at a certain complex value of t may be considered as a problem of solving the wave equation, Eq. (3.1), for a fixed value of σ in the real coordinates $|U|$ and $|V|$. Since the metric is independent of t the equation is hyperbolic for any value of σ and the characteristic initial-value problem is well posed. The analyticity of the propagator on the complexified horizon in the upper half U plane and in the lower half V plane implies that the Cauchy data for this real problem are regular provided $-4\pi M < \sigma < 0$. The standard existence and uniqueness theorem for the hyperbolic characteristic initial value problem guarantees that there will be a solution for the propagator for this range of σ. To determine whether the resulting solution is analytic in t we need only verify that the Cauchy-Riemann condition is satisfied. This is (a bar denotes complex conjugation)

$$\left(\frac{\partial K}{\partial \bar{t}}\right)_r = \frac{1}{4M}\left(\bar{V}\frac{\partial K}{\partial \bar{V}} - \bar{U}\frac{\partial K}{\partial \bar{U}}\right) = 0, \tag{3.15}$$

where the derivative with respect to $\bar{t}$ is being taken at constant r. Evidently this condition is satisfied by the data for $-4\pi M < \sigma < 0$ since K is an analytic function of U and V in the appropriate half planes on the complexified horizon. Furthermore $(\partial/\partial\bar{t})_r$ commutes with $\Box^2$ so that determining $(\partial K/\partial\bar{t})_r$ may be regarded as a problem of solving the wave equation with zero data on the characteristic Cauchy surfaces. The unique answer to this problem is $(\partial K/\partial\bar{t})_r = 0$. One concludes, therefore, that for x' in the exterior region and fixed r, θ, φ in the region $U > 0$, $V > 0$, $K(x, x')$ is analytic in t in a strip of width $4\pi M$ below the real axis.

The strip of analyticity cannot be extended above the real axis because immediately above it there are singularities corresponding to the real null geodesics which connect a value of t on the surface of given r to x' (see Fig. 3). The strip of analyticity cannot be pushed below $\sigma = -4\pi M$ either. From Eq. (3.14), this value of σ corresponds to a U and V which are again in the real section but reversed in sign. The propagator $K(x, x')$ with x in region II when continued in t to $t - 4\pi M i$ then equals the propagator from a point x'' in region III to x' in the exterior region. The point x'' is just x reflected in the origin of the U-V plane. This identity will be the basis of our derivation of black-hole radiance in the next section, but for the present we simply note that it implies that immediately below the line $\sigma = -4\pi M$ there are singularities corresponding to the real null geodesic which connect a point on the surface of constant r in region III to x'. The singularities in this case lie below the real axis because x'' lies in region III and from the relations analogous to Eq. (3.14) the small positive imaginary value of U and a negative value of V which locate the pole correspond to a negative imaginary value of t.

In this way the analytic properties of the propagator $K(x, x')$ in the variable t become apparent. For fixed θ, φ, r in region II with x' located in region I they are illustrated in Fig. 4(a). The propagator is periodic in $\sigma = \mathrm{Im}(t)$ with period $8\pi M$. The regions of analyticity corresponding to the upper half U plane and lower half V plane on the complexified horizon are the shaded strips of width $4\pi M$. There are periodic singularities corresponding to the real null geodesics which connect x' to a point on the curve of given θ, φ, r either in the past or in the future.

If x and x' are both located in region I the propagator is still periodic in σ with period $8\pi M$ as a consequence of Eq. (3.13b). Now, however, there are real values of t both in the past and in the future of x' for which there are real null geodesics connecting it to the fixed r, θ, φ curve. For the values to the future of x' the corresponding singularities are displaced slightly above the real t axis. For the values of t to the past, the corresponding singularities are displaced slightly below the real t axis. These singularities are shown in Fig. 4(b). There are no singularities corresponding to real null geodesics near $\sigma = \pm 4\pi M$ since the corresponding real points lie in region IV, every point of which is separated by a spacelike interval from x'.

The propagator $K(x, x')$ is periodic in imaginary t because it follows from the path-integral definition that the propagator is an analytic function of the Kruskal coordinates U and V except at the singularities we have described. However, the Schwarzschild coordinate, t, has a logarithmic singularity as a function of U and V and is multivalued; it is defined only modulo $8\pi i M$. Thus if the propagator has a singularity at some value of U and V, it will have periodic singularities at intervals of $8\pi i M$ when expressed as a function of t. The propagator is similar to that suggested by Unruh.[15] By contrast, the propagator proposed by

Boulware[14] is not periodic in t because it is not analytic on the two horizons.

The existence of periodic singularities in t implies that observers moving on lines of constant r, θ, φ in the extended Schwarzschild solution will detect particles. This is very similar to the fact that, as Unruh has pointed out, observers moving on world lines of uniform acceleration in Minkowski space will also detect particles. On the other hand, an observer moving along either of the two horizons will not see any particles. This is an illustration of the fact that the concept of particles is observer-dependent.[16] In the next section we shall show that the propagator constructed here will give for observers at a constant distance from the black hole the same rate of particle production as was obtained in Ref. 5 through a study of the mixing of positive and negative frequencies in a gravitational collapse.

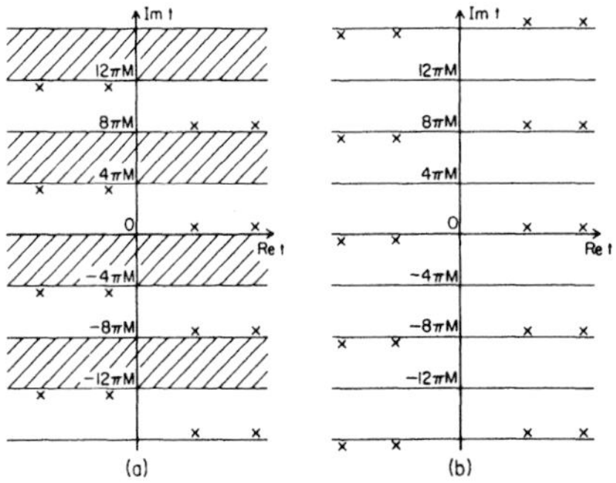

FIG. 4. The analytic structure of the propagator in the complex t plane. (a) shows the analytic structure of $K(x, x')$ for x' fixed outside the black hole (region I) and fixed values of r, θ, φ inside the future horizon (region II). The propagator is periodic in Im(σ) with period $8\pi M$ as a consequence of the relation of t [Eq. (3.14)] to the coordinates U and V in which the metric is analytic everywhere except at the physical singularity. The shaded regions are the regions of analyticity in t which are deduced from the analyticity of the propagator in the upper half U plane and lower half V plane on the complexified horizon. The crosses locate the singularities which correspond to the real null geodesics which connect x' to the curve of constant r, θ, φ. A typical situation is illustrated in Fig. 3. There are singularities immediately above the real axis corresponding to the null geodesics which connect x' to the fixed r, θ, φ curve in region II. There are singularities below the $\mathrm{Im}\,t=-4\pi M$ line corresponding to the real geodesics which connect x' to the fixed r, θ, φ curve in region III. In each case there is an infinite sequence of singularities (only a few of which are shown) which arises because there are null geodesics which spiral an arbitrary number of times near $r=3M$ and thus can connect the fixed r, θ, φ curve to x' at increasingly large values of $|t-t'|$. The singularities at other values of Imt are duplicates of these as a consequence of the periodicity of the propagator in Imt with period $8\pi M$. (b) shows the similar analytic structure when both x and x' are in Region I. The propagator remains periodic in Imt with period $8\pi M$. There are now infinite sequences of null geodesics which connect x' to a curve of fixed r, θ, φ in the future and in the past. Correspondingly there are singularities above and below the real t axis. These are periodically repeated in Imt with period $8\pi M$.

IV. BLACK-HOLE RADIANCE

In the preceding section we demonstrated the analyticity of the propagator $K(x, x')$ in a strip in the t plane when x is in region II and x' is in region I of the Schwarzschild geometry. We shall now use this analyticity to derive the thermal radiation from a Schwarzschild black hole.

Suppose we surround a Schwarzschild black hole by particle detectors at some large constant radius R. These detectors measure particles coming out from inside the surface in modes $f_j(t', r', \theta', \varphi')$ which are purely positive-frequency (with respect to t') solutions of the scalar wave equation. The amplitude that a particle is detected in a mode $f_i(x')$ having started in a mode $h_j(x)$ on some surface which bounds a region interior to R is

$$-\int d\sigma^{\mu}(x')\int d\sigma^{\nu}(x)\bar{f}_i(x')\,\overleftrightarrow{\partial}_{\mu}K(x', x)\,\overleftrightarrow{\partial}_{\nu}h_j(x), \tag{4.1}$$

where the integral over x' is taken over the surface $r'=R$ and the integral over x is over the interior bounding surface. The notation $a\overleftrightarrow{\partial}_{\mu}b$ means $ab_{,\mu}-a_{,\mu}b$.

Suppose for the moment that the particle detectors are confined to a time interval $t'\in(-t_1', t_1')$, where t_1' is very large. Eventually the limit $t_1'\to\infty$ will be taken. The interior bounding surface mentioned above can then be taken to be a spacelike surface through the precollapse star at $-t_1'$, a complete spacelike surface inside the future horizon a d a timelike surface connecting this to the $r'=R$ surface outside at $t'=t_1'$. The spacelike surface inside the future horizon will be taken to be part of a constant-r surface C_+ outside the matter and a spacelike extension inside it. The complete spacelike surface inside the future horizon could have been chosen to be the future singularity were it not convenient to avoid mathematical complications associated with the singularity in the metric at $r=0$ by keeping it away from those points.

We now calculate the total probability that a particle is measured by a detector in a mode $f_i(x')$ which is peaked in time about some late time t_0'. This probability will be the sum of the square of the amplitude in Eq. (4.1) over all modes $h_j(x)$ which are consistent with our knowledge of the bounding surface. There will be no contribution from modes which are localized on the spacelike part at $t' = -t_1'$ because we are assuming that there are no scalar particles in the initial state. The contribution of the other two surfaces is significantly restricted by the fact that the propagator $K(x, x')$ in the exterior geometry of a collapsing star at late times will be a function only of the difference $t' - t$ as a consequence of the time transition invariance of the Schwarzschild geometry. Thus if $f_i(x')$ is peaked about a late time t_0' only times t comparable to t_0' will contribute in Eq. (4.1). In particular there will be no contribution from the timelike surface which starts at t_1' since by choosing t_1' large enough this only intersects values of t much larger than t_0'. Furthermore, the only part of the spacelike surface inside the horizon which contributes is the part with $t \approx t_0'$ and for sufficiently late times this is well outside the matter. The conclusion of this is that if one is interested in the probability of production of particles at late times the details of the collapse may be ignored. The only part of the amplitude in Eq. (4.1) which contributes to the probability of particle production is that in which the integral over x is taken over the complete spacelike surface inside the horizon, and this can be idealized as a surface C_+ of constant r between 0 and $2M$ in the exact Schwarzschild geometry. We are explicitly assuming here that the propagator between a point interior to the future horizon and a point outside the black hole in the geometry of a collapsing star is well approximated at late times by the propagator we have obtained in the analytically extended Schwarzschild metric.

To compute the total probability that a particle is detected we next note that there is no information on the state of the particle on the future singularity. The total probability is obtained by summing the modulus squared of Eq. (4.1) over a complete set of states on C_+. It is not necessary to carry this sum out in detail to derive the black-hole radiance as we shall now show.

Of chief interest is the amplitude for a black hole to emit a mode with a definite positive energy E. The time dependence of such a mode is $f \sim \exp(-iEt')$. Because of the time translation invariance the modes $h_i(x)$ in the complete set on C_+ may also be classified into modes with the time dependence $\exp(-iEt)$, although since t is a spacelike coordinate inside the horizon, E is not to be interpreted as a local energy. The fact that K is a function only of the difference $t' - t$ means that the integral over t and t' in Eq. (4.1) will lead to an energy-conservation δ function. When probabilities are computed the formal square of the δ function will be replaced by a density-of-states factor in the usual way. Factoring out this energy-conservation δ function there remains of Eq. (4.1).

$$-\int d\sigma(\vec{R}')\int d\sigma(\vec{R})\left[\bar{f}_i(\vec{R}')\frac{\overleftrightarrow{\partial}}{\partial r}\mathcal{E}_E(\vec{R}',\vec{R})\frac{\overleftrightarrow{\partial}}{\partial r}h_j(\vec{R})\right], \tag{4.2}$$

where $\vec{R}$ and $\vec{R}'$ denote the coordinates r, θ, φ and r', θ', φ' respectively, $f_i(\vec{R})$ and $h_j(\vec{R})$ denote the angular parts of the respective modes, and $d\sigma(\vec{R})$ and $d\sigma(\vec{R}')$ are appropriately weighted angular integrals. The crucial information about the emission is contained in the amplitude $\mathcal{E}_E$ defined by

$$\mathcal{E}_E(\vec{R}',\vec{R}) = \int_{-\infty}^{+\infty} dt\, e^{-iEt}K(0,\vec{R}';t,\vec{R}). \tag{4.3}$$

Making use of the symmetry of $K(x, x')$ under interchange of x and x' this can also be written in what will be the more convenient form

$$\mathcal{E}_E(\mathrm{R}',\mathrm{R}) = \int_{-\infty}^{+\infty} dt\, e^{-iEt}K(t,\vec{R};0,\vec{R}'). \tag{4.4}$$

The amplitude $\mathcal{E}_E$ is the component with energy E of the amplitude to propagate from the surface C_+ to a point $(0, \vec{R}')$ outside the black hole.

Following the program outlined in the Introduction we now wish to relate the amplitude for emission as contained in Eqs. (4.2) and (4.4) to an amplitude for the black hole to absorb a particle in the same mode by distorting the contours of integration into the complex plane. To do this it is enough to concentrate on the amplitude $\mathcal{E}_E(\vec{R}',\vec{R})$ and distort the contour of the t integration in Eq. (4.4) downward by an amount $-4\pi M i$. This distortion is permissible since the main result of the preceding section is that $K(t, \vec{R}; 0, \vec{R})$ is analytic in a strip of width $4\pi M$ below the real axis. Equation (4.4) becomes

$$\mathcal{E}_E(\vec{R}',\vec{R}) = e^{-\pi E/\kappa}\int_{-\infty}^{+\infty} dt\, e^{-iEt}K(t-i\pi/\kappa,\vec{R};0,\vec{R}'), \tag{4.5}$$

where we have written the surface gravity of the black hole κ instead of $1/4M$. Equation (3.14) shows that translating t by an amount $-i\pi/\kappa$ is equivalent to reflecting the Kruskal coordinates U and V in the origin. The integral in Eq. (4.5) can thus be interpreted as the component with energy E of the amplitude to propagate to a point $(0, \vec{R}')$ outside the black hole from the surface C_- in region III, which is C_+ reflected in the origin of the U-V plane. If this integral is inserted in

Eq. (4.2) in place of $\mathcal{E}_E$ we obtain minus the amplitude for a particle to be detected in a mode f_i of definite energy E having started on C_- in region III in a mode h_j with the same energy. The minus sign occurs because the appropriate normal to the surface C_- is reversed. By time-reversal invariance the modulus squared of this amplitude is exactly equal to the modulus squared of the amplitude for the black hole to *absorb* a particle which starts at $(0, \vec{R}')$ in a mode f_i with energy E and arrives at C_+ in a mode h_j with the same energy. When the sum over the complete set of states h_j is performed we have the general relation

$$\begin{aligned}&(\text{probability for a Schwarzschild black hole to emit a particle in a mode with energy } E)\\ &\quad = e^{-2\pi E/\kappa}\times(\text{probability for a Schwarzschild black hole to absorb a particle in the same mode}).\end{aligned} \tag{4.6}$$

This is the fundamental connection between emission and absorption stated in the Introduction. This connection shows that a black hole will emit particles with a thermal spectrum characterized by a temperature $T = \kappa/2\pi k$. Thus we recover[17] the result of Ref. 5.

V. ROTATION, CHARGE, AND SPIN

In this section we shall comment on the generalization of our results to particles with higher spin and to black holes with rotation and charge.

For fields of spin greater than zero it is difficult to express the propagator in terms of an integral over paths. However, we shall assume that the analytic properties of the higher-spin propagators are the same of those we have derived for the scalar field. Namely, we shall assume that if U and V are affine parameters along the past and future horizons, respectively, both increasing toward the future, the propagator from a point outside the black hole to a point on the horizon is analytic in the upper half U plane and in the lower half V plane. If this assumption is taken then the generalization of our results to higher-spin particle is immediate since the derivation of black-hole radiance given in the preceding section for Schwarzschild black holes and that to be given below for more general black holes depend only on this analyticity property and the correct construction of the emission and absorption amplitudes. We now proceed to the generalization of our argument to the Kerr and Reissner-Nordström black holes. For simplicity we treat these two cases separately, leaving it to the reader to join the arguments together for the general rotating charged black hole.

A. The Kerr black hole

Figure 5(a) shows the familiar Penrose diagram for the axis of a Kerr black hole. Following our argument for the Schwarzschild case, the amplitude for the black hole to emit a particle can be related to an integral of the propagator $K(x, x')$ in which x' is fixed outside the black hole and x is integrated over a spacelike surface interior to the horizon which divides the past and future. It is convenient to call this surface C_+ and take for it a surface of constant r in the usual Boyer-Lindquist coordinates such that $r_- < r < r_+$. In order to generalize our result to the Kerr geometry we shall thus need a set of coordinates in which the metric is regular over at least regions I, II, and III shown in Fig. 5(a). Fortunately Carter[18] has given such a set of coordinates. To avoid lengthy redefinition we shall use his notation wherever it does not differ from that used elsewhere in this paper. The reader is referred to Carter's paper for symbols not defined here.

Region I in the Kerr geometry can be covered by Boyer-Lindquist coordinates (t, r, θ, φ) with $r > r_+$ (Carter uses $\hat{t}, \hat{\varphi}$ for our t, φ). Regions II and III can be covered by a similar patch with $r_- < r < r_+$. Regions I and II can be covered by a coordinate patch of the Kerr-Newman form involving an advanced time v. Similarly regions I and III can be covered by a patch involving a retarded time u. (Carter uses u for our v and $-w$ for our u.) In region I these coordinates are connected by rela-

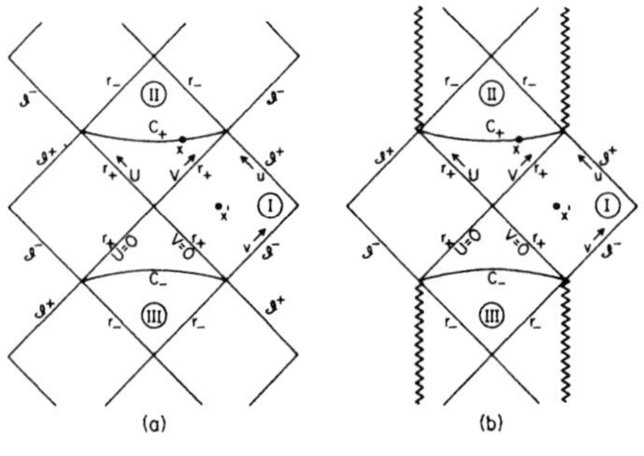

FIG. 5. (a) The Penrose diagram for the axis of a Kerr black hole. (b) The Penrose diagram for a Reissner-Nordström black hole.

tions of the form

$$t = u + f(r), \qquad (5.1)$$

$$t = v - f(r), \qquad (5.2)$$

where $df/dr = (r^2 + a^2)/\Delta$. Carter then introduces a new azimuthal Killing angle φ^+ defined by

$$\varphi^+ = \varphi - \omega_+ t, \qquad (5.3)$$

where $\omega_+ = a/(r_+^2 + a^2)$ is the angular frequency of the horizon. He also introduces two new null coordinates x and y which correspond to our U and V. These are defined as

$$U = -e^{-\kappa u}, \quad V = e^{\kappa v}, \quad \text{region I}, \qquad (5.4a)$$

$$U = e^{-\kappa u}, \quad V = e^{\kappa v}, \quad \text{region II}, \qquad (5.4b)$$

$$U = -e^{-\kappa u}, \quad V = -e^{\kappa v}, \quad \text{region III}, \qquad (5.4c)$$

where $\kappa = \frac{1}{2}(r_+ - r_-)/(r_+^2 + a^2)$ is the surface gravity of the black hole. Thus V is Carter's x while U is the negative of Carter's y. In this new (U, V, r, φ^+) patch the metric is analytic in regions I, II, and III [see Carter's Eq. (26)]. The future horizon is located at $U = 0$ and V is an affine coordinate along it. The past horizon is at $V = 0$ and U is an affine coordinate along it. The coordinate r may be considered to be defined in terms of U and V by Eqs. (5.1), (5.2), and (5.4). Figure 5(a) gives a schematic representation of the various definitions.

In the Schwarzschild case the amplitude for a black hole to emit a particle with energy E was ultimately related to an integral $\mathcal{E}_E(\vec{R}', \vec{R})$ over a surface of constant r inside the future horizon of $\exp(-iEt)$ times the propagator to go from a point x on that surface to a point x' outside the hole. In the Kerr case we will be interested in the amplitude for the emission of a particle of energy E and an angular momentum along the axis of rotation m (m is not to be confused with the rest mass). In a similar fashion this amplitude can be related to an integral of the form

$$\mathcal{E}_{Em}(\vec{R}', \vec{R}) = \int_{-\infty}^{+\infty} dt \int_0^{2\pi} d\varphi\, e^{-i(Et - m\varphi)} K(t, \varphi, \vec{R}; 0, 0, \vec{R}'). \qquad (5.5)$$

Here K is the propagator in the Kerr geometry which, because of the time-translation invariance and axial symmetry, depends only on the difference in the t and φ coordinates of x and x'. The quantities $\vec{R}$ and $\vec{R}'$ stand for the r, θ coordinates of x and x'. The integral is over the surface C_+. Arguments similar to those given in the Schwarzschild case will show that, for fixed r, θ, φ^+, K is analytic with U in the upper half plane and V in the lower half plane. We can therefore distort the contour of the t integration downward by an amount $-i\pi/\kappa$ keeping r, θ, φ^+ fixed since that amounts to rotating U by an angle π and V by an angle $-\pi$. Keeping φ^+ fixed means from Eq. (5.3) that $\varphi \to \varphi - i\pi\omega_+/\kappa$ in the process. Thus,

$$\mathcal{E}_{Em}(\vec{R}', \vec{R}) = e^{-\pi(E - m\omega_+)/\kappa} \int_{-\infty}^{+\infty} dt \int_0^{2\pi} d\varphi K(t - i\pi/\kappa, \varphi - i\pi\omega_+/\kappa, \vec{R}; 0, 0, \vec{R}'). \qquad (5.6)$$

Since this displacement of t is equivalent to [see Eq. (5.4)] $U \to -U$ and $V \to -V$ the integration is now over the reflected surface of constant r which is in region III and which we have shown as C_- in Fig. 5(a). The remaining integral can be related to the amplitude for the black hole to absorb a particle of energy E and angular momentum m. Thus we have

(probability for a Kerr black hole to emit a mode with energy E and angular momentum m)

$$= e^{-2\pi(E - m\omega_+)/\kappa} \times \text{(probability for a Kerr black hole to absorb a mode with } E \text{ and } m). \qquad (5.7)$$

This is exactly the relation necessary to establish that a rotating black hole will emit particles with an expected number per mode proportional to $\{\exp[(E - m\omega_+)2\pi/\kappa] - 1\}$ so that $\kappa/2\pi k$ may be interpreted as the black hole's temperature.[5,6]

B. Reissner-Nordström black hole

The situation with the Reissner-Nordström black hole is similar. Here, however, we shall investigate the amplitude for the black hole to emit particles of charge q. Figure 5(b) shows the Penrose diagram for the Reissner-Nordström geometry. The propagator for a particle of charge q is a solution of the wave equation

$$g^{\alpha\beta}(\nabla_\alpha - iqA_\alpha)(\nabla_\beta - iqA_\beta)K(x, x') = -\delta(x, x'), \qquad (5.8)$$

in the Reissner-Nordström background geometry. In the usual gauge the only nonvanishing component of A_μ is

$$A_t(x) = e/r, \qquad (5.9)$$

where e is the charge on the black hole. However, such a gauge K cannot be expected to be analytic in the upper half U plane or lower half V plane as is required by our boundary conditions because the components of $A_\mu(x)$ will not be analytic on the horizon in coordinates which are analytic there. For example, if we use the (u, r, θ, φ) coordinates which are analytic on the future horizon, $A_r(x) = (e/r)(1 - 2M/r + e^2/r^2)^{-1}$, which diverges at $r = r_+$. A gauge in which A_μ is stationary and regular on both horizons can be found by making the transformation

$$A_\mu \to A_\mu + \Lambda_{,\mu}, \tag{5.10}$$

where

$$\Lambda = \Phi t, \tag{5.11}$$

and Φ is the potential on the horizon equal to e/r_+. In the new gauge A_μ will be analytic in the same domain as the metric but it will not vanish at infinity. This means that the time dependence of a mode of energy E at large distance from the hole will not be $\exp(-iEt)$ but rather $\exp[-i(E - q\Phi)t]$.

As in the Schwarzschild and Kerr cases the amplitude for the black hole to emit a particle with energy E and charge q can be expressed in terms of an integral of the propagator over a constant-r surface C_+ which lies between the future horizon at r_+ and the inner horizon at r_-. By the time invariance of the propagator the amplitude for emission depends on the integral

$$\mathcal{E}_E(\vec{R}', \vec{R}) = \int_{-\infty}^{+\infty} dt\, e^{-i(E-q\Phi)t}\, K(t, \vec{R}; 0, \vec{R}'), \tag{5.12}$$

where $\vec{R}$ stands for the coordinates r, θ, φ.

In the new gauge K will be analytic in the upper half U plane and the lower half V plane. The coordinates U and V here are related to those of Carter as described in Eq. (5.4) and above. As in the Kerr case this analyticity implies that the contour of the t integration in Eq. (5.12) may be distorted downward in the complex t plane by an amount $-i\pi/\kappa$ keeping r fixed since this amounts to rotating U by π and V by π. Thus,

$$\mathcal{E}_E(\vec{R}', \vec{R}) = e^{\pi(E-q\Phi)/\kappa} \int_{-\infty}^{+\infty} dt\, e^{-i(E-q\Phi)t} \times K(t - i\pi/\kappa, \vec{R}; 0, \vec{R}'). \tag{5.13}$$

Since displacing t by $-i\pi/\kappa$ is equivalent to the reflection $U \to -U$ and $V \to -V$ this integral may be written

$$\mathcal{E}_E(\vec{R}', \vec{R}) = e^{\pi(E-q\Phi)/\kappa} \int_{-\infty}^{+\infty} dt\, e^{-iEt} K(t, \vec{R}; 0, \vec{R}'). \tag{5.14}$$

where the integral is now over the reflected surface C_- illustrated in Fig. 5(b). The integral in Eq. (5.14) can be related, as before, to the amplitude for the black hole to absorb a particle of charge q. Following through the arguments which led to Eq. (4.6) we have

(probability for a Reissner-Nordström black hole to emit a particle of charge q and energy E

$= e^{-2\pi(E-q\Phi)/\kappa} \times$ (probability for a Reissner-Nordström black hole to absorb a particle of charge q and energy E). (5.15)

This is exactly the relation necessary to establish that a rotating black hole will emit scalar charged particles with an expected number per mode proportional to $\{\exp[(E - q\Phi)/kT] - 1\}^{-1}$, where $kT = \kappa/2\pi$. (See Refs. 5 and 19.)

APPENDIX: DERIVATION OF THE DIFFUSION EQUATION FOR $F(\Omega, x, x')$

The amplitude $F(\Omega, x, x')$ is defined by the path integral

$$F(\Omega, x, x') = \int \delta x[\omega] \exp\left[-\tfrac{1}{4}\int_0^\Omega \gamma(\dot{x}, \dot{x}) d\omega\right]. \tag{A1}$$

where γ is the positive-definite metric of Eq. (2.6), $\dot{x} = dx/d\omega$, and the sum ranges over all paths with $x(0) = x'$ and $x(\Omega) = x$. We shall now derive the diffusion equation [Eq. (2.8)] for F and in the process discuss the interpretation of the differential $\delta x[\omega]$.

Divide the interval $[0, \Omega]$ up into $N+1$ intervals each ϵ long. With a natural assignment of the weight to the integrals over spacetime the path integral in Eq. (2.1) may be interpreted as

$$F(\Omega, x, x') = \lim_{N\to\infty} \int \frac{d^4x_N}{A}[\gamma(x_N)]^{1/2} \int \frac{d^4x_{N-1}}{A} \cdots \int \frac{d^4x_1}{A}[\gamma(x_1)]^{1/2} \exp\left[\sum_{i=0}^{N} S(\epsilon, x_{i+1}, x_i)\right], \tag{A2}$$

where $x_0 = x'$, $x_{N+1} = x$, and S is given by integral

$$S(\epsilon, x_{i+1}, x_i) = \tfrac{1}{4}\int_0^{\epsilon} d\omega\, \gamma(\dot{x}, \dot{x}), \tag{A3}$$

evaluated along the geodesic path which connects x_i at $\omega = 0$ with x_{i+1} at $\omega = \epsilon$. The constant A is a normalization fixed by the requirement that the amplitude for a particle to propagate from one point to *any* other point in the spacetime is unity. This is equivalent to

$$A = \int d^4x [\gamma(x)]^{1/2} \exp[-S(\epsilon, x, x')]. \tag{A4}$$

The reason that Eq. (A2) is correct is that as ϵ becomes smaller and smaller the action for the paths which connect fixed x_i to x_{i+1} will become larger and larger and hence their contribution to the integral will be exponentially damped. The dominant contribution will come from the stationary path for which S is a minimum. This is a geodesic between the points.

The diffusion equation can be derived by considering the relation between $F(\Omega + \epsilon, x, x')$ and $F(\Omega, x, x')$. From Eq. (A2) this is

$$F(\Omega + \epsilon, x, x') = \int d^4y [\gamma(y)]^{1/2} \exp[-S(\epsilon, x, y)] \times F(\Omega, y, x')/A. \tag{A5}$$

Write $y = x + z$ and let the integration be over z. On the right expand S and F in powers of z. On the left expand F in powers of ϵ. Analysis of the integral shows that only the first few terms of the expansion on the right contribute to the part of F linear in ϵ giving an expression for $\partial F/\partial\Omega$. This analysis has been carried out by Cheng in a general coordinate system.[4] However, the calculations are considerably simplified if a Riemann normal coordinate system is introduced at x. In such a coordinate system

$$\gamma_{\alpha\beta}(y) = \delta_{\alpha\beta} - \tfrac{1}{3}R_{\alpha\gamma\beta\delta}z^{\gamma}z^{\delta} + O(z^3). \tag{A6}$$

where $\delta_{\alpha\beta}$ denotes the Kronecker δ. From the definition of normal coordinates it follows that the geodesics from the origin are straight lines so that $\dot{x}^{\alpha} = z^{\alpha}/\epsilon$ and

$$S(\epsilon, x, x+z) = \tfrac{1}{4}\delta_{\alpha\beta}z^{\alpha}z^{\beta}/\epsilon. \tag{A7}$$

From the form of S we see that the only significant part of the integral as $\epsilon \to 0$ comes from the region where $z^{\alpha} \sim \epsilon^{1/2}$. Thus only terms $O(z^2)$ in the expansion of $(\gamma)^{1/2}$ and F need to kept under the integral and the limits may be extended to infinity. Then, the zeroth-order term in an expansion of Eq. (A5) gives the normalization condition

$$A = \int d^4z \exp(-\tfrac{1}{4}\delta_{\alpha\beta}z^{\alpha}z^{\beta}/\epsilon) = (4\pi\epsilon)^2. \tag{A8}$$

Because the odd integrations in z vanish, the first-order term in ϵ is

$$\epsilon\frac{\partial F}{\partial\Omega} = B^{\alpha\beta}(\tfrac{1}{2}F_{,\alpha\beta} - \tfrac{1}{6}R_{\alpha\beta}F). \tag{A9}$$

where the curvature term comes from the expansion

$$\gamma^{1/2} = 1 - \tfrac{1}{6}R_{\alpha\beta}z^{\alpha}z^{\beta} + O(z^3), \tag{A10}$$

and $B^{\alpha\beta}$ is the integral

$$B^{\alpha\beta} = \int d^4z (z^{\alpha}z^{\beta}/A) \exp(-\tfrac{1}{4}\delta_{\mu\nu}z^{\mu}z^{\nu}/\epsilon) = 2\epsilon\delta^{\alpha\beta}. \tag{A11}$$

Using this expression we finally arrive at the diffusion equation for F by replacing the partial derivatives in the normal coordinates by covariant derivatives $\bar{\nabla}_{\alpha}$ with respect to the metric γ. The equation is

$$\frac{\partial F}{\partial\Omega} = (\gamma^{\alpha\beta}\bar{\nabla}_{\alpha}\bar{\nabla}_{\beta} - \tfrac{1}{3}R)F. \tag{A12}$$

This is exactly Eq. (2.8) taking into account the fact that R vanishes for Schwarzschild geometry.

The factor $-R/3$ in Eq. (A12) is a consequence of our particular choice of weight in the coordinate integrals in Eq. (A2). If we had replaced $\exp[-S(\epsilon, x_{i+1}, x_i)]$ in the integrals with

$$[\gamma(x_{i+1})/\gamma(x_i)]^{s/2} \exp[-S(\epsilon, x_{i+1}, x_i)],$$

then the resulting equation would have been

$$\frac{\partial F}{\partial\Omega} = \left[\gamma^{\alpha\beta}\bar{\nabla}_{\alpha}\bar{\nabla}_{\beta} + \tfrac{1}{3}(s-1)R\right]F. \tag{A13}$$

Thus any amount of scalar curvature can be had both here and in the equation for K by the appropriate choice of the action. For the vacuum black-hole solutions we are considering R vanishes and these equations are all identical and we will not consider this issue further. However, see the remarks in Ref. 3.

ACKNOWLEDGMENTS

We are grateful for the hospitality of Kip Thorne and the relativity group at Caltech. We would also like to thank Werner Israel and Richard Feynman for helpful conversations.

*Work supported in part by the National Science Foundation under Grants Nos. MPS-75-01398 and GP-43905.
†Sherman Fairchild Distinguished Scholar at the California Institute of Technology.

[1]R. P. Feynman, Rev. Mod. Phys. 20, 327 (1948).
[2]C. Morette, Phys. Rev. 81, 848 (1951).
[3]B. S. DeWitt, Rev. Mod. Phys. 29, 377 (1957).
[4]K. S. Cheng, J. Math. Phys. 13, 1723 (1972).
[5]S. W. Hawking, Nature 248, 30 (1974); Commun. Math. Phys. 43, 199 (1975).
[6]S. W. Hawking (unpublished).
[7]R. P. Feynman, Phys. Rev. 76, 769 (1949).
[8]For other discussions of proper time in connection with relativistic quantum mechanics see (a) V. Fock, Phys. Zeit. Sowet. Un. 12, 404 (1937); (b) Y. Nambu, Prog. Theor. Phys. 5, 82 (1950); (c) J. Schwinger, Phys. Rev. 82, 64 (1951); (d) B. S. DeWitt, Phys. Rep. 19C, 295 (1975).
[9]One might have though that the parameter time W, which in the case of timelike paths is a multiple of the proper time, is determined completely by x, x' and the path and not an independently specifiable quantity. However, that is the case only for differentiable paths, and nondifferentiable paths are being summed over here. The situation is analogous to that in nonrelativistic quantum mechanics where one can ask for the amplitude to propagate from a space point $\vec{x}$ to a space point $\vec{x}'$ along a given path in a definite time t. If the path were differentiable t would be determined, but the paths summed over in the Feynman integral for the propagator are not differentiable.
[10]The definition of $K(x, x')$ corresponds to the I_+ of Ref. 7 and the Δ_F of J. D. Bjorken and S. D. Drell, *Relativistic Quantum Fields* (McGraw-Hill, New York, 1965).
[11]See, for example, E. Nelson, J. Math. Phys. 5, 332 (1964), and M. Kac, *Probability and Related Topics in the Physical Sciences* (Interscience, New York, 1959).
[12]This is analogous to the "Euclidicity postulate" in Euclidean quantum field theory. See, for example, K. Symanzik, in *Local Quantum Theory*, Proceedings of the International School of Physics "Enrico Fermi" Course 45, edited by R. Jost (Academic, New York, 1969), and R. F. Streater, Rep. Prog. Phys. 38, 771 (1975).
[13]See, for example, S. D. Eidel'man, *Parabolic Systems* (North-Holland, Amsterdam, 1969), p. 178.
[14]D. Boulware, Phys. Rev. D 11, 1404 (1975).
[15]W. G. Unruh (unpublished).
[16]A. Ashtekar and A. Magnon (unpublished).
[17]A remark is probably in order here comparing the conventions used here and in Ref. 5. We have chosen the action in Eq. (2.1) so that, taking account of our signature convention (+2), the variable part of it for nonrelativistic paths reduces to $+\int dt(d\vec{x}/dt)^2$. This is the usual choice for the nonrelativistic action and leads to a Schrödinger equation of the usual form and a time dependence for solutions with energy E of the form $\exp(-iEt)$. In Ref. 5 wave functions with definite frequency had time dependence $\exp(i\omega t)$. Exactly the same physical results would be obtained with this convention provided one started with an action of the opposite sign.
[18]B. Carter, Phys. Rev. 174, 1559 (1968).
[19]G. W. Gibbons, Commun. Math. Phys. 44, 245 (1975).

Proc. R. Soc. Lond. A. 358, 467–494 (1978)
Printed in Great Britain

Black holes and thermal Green functions

BY G. W. GIBBONS AND M. J. PERRY
Department of Applied Mathematics and Theoretical Physics, University of Cambridge, Silver Street, Cambridge, U.K.

(*Communicated by S. W. Hawking, F.R.S. – Received* 12 *July* 1976)

This paper concerns itself with the possibility of thermal equilibrium between a black hole and a heat bath implied by Hawking's discovery of black hole emission. We argue that in an isolated box of radiation, for sufficiently high energy density a black hole will condense out. We introduce thermal Green functions to discuss this equilibrium and are able to extend the original arguments, that the equilibrium is possible based on fields interacting solely with the external gravitational field, to the case when mutual and self interactions are included.

1. INTRODUCTION

One of the most remarkable recent developments in gravitation theory has been the recognition of the close connection between event horizons, thermodynamics and quantum field theory. In particular the discovery by Hawking (1974, 1975) that black holes emit particles in a thermal fashion has completely changed one's view of quantum gravity.

The origin of this train of ideas began with the discovery by Christodoulou (1970) that one could express the mass M of a Kerr black hole in terms of a part due to the rotational kinetic energy and an irreducible part M_0 such that only the rotational part could be extracted by means of idealized Penrose (1969) processes, and the simultaneous discovery by Hawking (1971) of the result that the area A of a general black hole could never decrease. It was seen immediately that the area and irreducible mass were related by the simple formula

$$A = 16\pi M_0^2 \tag{1.1}$$

(we use units in which $G = c = \hbar = k = 1$ throughout) and where, if one takes into account the possible charge, Q of a black hole, one has for the Kerr–Newman family the relation (Christodoulou & Ruffini 1971)

$$M^2 = (M_0 + Q^2/4M_0)^2 + J^2/4M_0^2, \tag{1.2}$$

where J is the angular momentum of the hole.

This led to the observation by Smarr (1973) that again for the Kerr–Newman family one might express the mass, or its differential by the formulae

$$\tfrac{1}{2}M = \kappa A/8\pi + \tfrac{1}{2}\Phi Q + \Omega J, \tag{1.3}$$

$$\mathrm{d}M = \kappa\,\mathrm{d}A/8\pi + \Phi\,\mathrm{d}Q + \Omega\,\mathrm{d}J, \tag{1.4}$$

where for the Kerr–Newman family, Ω, the angular velocity Φ, the electric potential and κ the surface gravity have the particular forms

$$\kappa = \frac{r_+ - r_-}{2r_0^2}, \tag{1.5}$$

$$\Omega = J/Mr_0^2, \tag{1.6}$$

$$\Phi = Qr_+/r_0^2, \tag{1.7}$$

where
$$r_\pm = M \pm \sqrt{(M^2 - Q^2 - J^2/M^2)} \tag{1.8}$$

and
$$r_0 = 2M_0 = [M^2 - \tfrac{1}{2}Q^2 - (J^2 + \tfrac{1}{4}Q^4)^{\frac{1}{2}}]^{\frac{1}{2}} + [M^2 - \tfrac{1}{2}Q^2 + (J^2 + \tfrac{1}{4}Q^4)^{\frac{1}{2}}]^{\frac{1}{2}}. \tag{1.9}$$

In a discussion of black hole perturbations, Hartle & Hawking (1972) produced a version of (1.4) valid for a general black hole (with $Q = 0$) with suitable definitions of κ and Ω. Given the No-hair theorem that the solution depends on 3 parameters (Carter 1971; Robinson 1974) which may be taken as A, J and Q^2 one sees by a simple scaling argument that (1.3) must also be true irrespective of the detailed form of the Kerr–Newman metrics. Doubling the linear dimension of the hole keeping the geometry similar to itself will scale up A, J and Q^2 by a factor 4, but will merely double M. M is thus a homogeneous function of A, J and Q^2 of degree $\frac{1}{2}$ and (1.3) follows by Eulers Theorem. The explicit form of the function for the Kerr family is of course just Christodoulou's formula (1.1).

The obvious analogy of these formulae with the laws of thermodynamics for a homogeneous system with internal energy U, temperature T, entropy S and a set of extensive and intensive state variables x_i and X_i:

$$\mathrm{d}U = T\,\mathrm{d}S + \sum_i X_i\,\mathrm{d}x_i \tag{1.10}$$

$$U = TS + \sum_i X_i x_i, \tag{1.11}$$

$$\mathrm{d}S \geqslant 0 \tag{1.12}$$

led to the formulation of the 'Four laws of black hole mechanics' by Bardeen, Carter & Hawking (1973).

The analogy was carried further by the suggestion of Bekenstein (1973, 1974, 1975) that some multiple of the area A of a black hole represent the information lost down the hole during gravit tional collapse and that a 'generalized second law' holds.

The problem with this suggestion was that classically a black hole cannot remain in thermal equilibrium with a bath of black body radiation. The hole must inevitably grow until it has absorbed all the radiation present. That is it acts as a perfect sink at absolute zero whereas equation (1.4) implies a non-zero temperature proportional to κ. The missing link in the argument has been supplied by Hawking (1974, 1975) who has shown that if one considers quantized matter fields propagating in classical black hole geometries, with no mutual or self interactions, one discovers that a black hole will emit particles exactly as a hot body with temperature T

$$T = \kappa/2\pi \tag{1.13}$$

and thus
$$S = \tfrac{1}{4}A. \tag{1.14}$$

That is the differential emission rate in a mode with energy E_i, angular momentum n_i, charge e_i is given by

$$R_{nEe} = \frac{\Gamma_{nEe}}{\exp\left(E_i - n_i\Omega_{\mathrm{H}} - e_i\Phi_{\mathrm{H}}\right)2\pi/\kappa \pm 1}\,\mathrm{d}E. \tag{1.15}$$

This implies, by the principle of detailed balance, that a black hole can be in equilibrium with a heat bath.

When interactions are taken into account it is inappropriate to consider the emission rate, whose precise form it is impossible to specify without a detailed calculation based on the particular interactions considered. Rather one should consider the equilibrium configuration itself. It is the purpose of the paper to do just that.

We begin in § 2 by considering this equilibrium from a purely heuristic point of view using the thermodynamic formulae (1.13)–(1.15) together with the usual expressions for a black body gas. This simple consideration indicates (as has already been pointed out by Hawking (1976)) that for sufficiently large energy densities an isolated box, volume V, of heat radiation will exhibit a black hole condensation. That is as one fills up such a box with energy its temperature will at first rise, until a critical temperature $T_c = (\alpha V)^{-\frac{1}{5}} \times 1.784 \times 10^{-2}$ is reached, where α is Stefan's constant. The temperature will then suddenly drop by a factor 2.568 because a black hole of mass $M = 0.97702E$, where E is the total energy, has condensed out. Thereafter the temperature will fall and the mass of the condensed hole will rise (see figure 1).

In order to investigate the equilibrium in greater detail we follow Hartle & Hawking's (1977) path integral derivation of black hole radiance, in concentrating attention upon Green functions (Gibbons & Perry 1976). These propagators introduced by Hartle & Hawking (1977) have, for non-rotating black holes and scalar particles the striking property of being periodic in the complex Schwarzschild time coordinate with a period $2\pi i\kappa^{-1}$. This periodicity is typical of 'thermal Green functions' which have long been used in conventional many body theory to describe grand canonical ensembles of particles. Accordingly we (in § 3) introduce thermal Green functions for scalar particles which are suitable for describing a system in thermal equilibrium in a static gravitational field. We outline how one constructs interacting Green functions in a perturbation theory, starting from the interacting Green functions.

In § 4 we show that the Hartle–Hawking non-interacting propagator for scalar particles is the same as the non-interacting thermal Green function of temperature $T = \kappa/2\pi$ which describes thermal equilibrium from the point of view of observers at rest with respect to a Schwarzschild hole. We then argue that if one includes interactions this correspondence will still hold since one would construct interacting thermal or Hartle–Hawking propagators, starting from the non-interacting propagators by identical methods.

A further advantage of the technique is that it enables one to establish the completely thermal nature of the Hawking emission. The previous demonstrations of the fact (Wald 1975; Parker 1975; Hawking 1976) are rather lengthy and are restricted to free fields.

The generalization of these results to fields of with spin (excluding gravitation) and to charged and rotating black hole is given in § 5. Section 6 comprises conclusions and some speculations about possible quantum gravity effects. We also include an appendix showing how a simple extension of the work of § 3 provides a means of describing the cosmological black body background by using thermal Green functions.

2. The thermodynamics of a black hole in an isolated box

In this section, we shall apply the usual thermodynamic arguments relating to hot bodies in equilibrium with black body radiation to the case of a black hole in a box. This is done by assuming the expressions for the entropy S_H and temperature T_H of the black hole given in § 1 are valid. In §§ 4 and 5, we shall provide a justification of this rather formal procedure by a discussion of the process of particle creation.

We consider an isolated system consisting of a box of volume V, containing a total energy E in the form of various particle species. The walls of the box are chosen to be rigid and adiathermal. That is, they do not permit the passage of energy in any form, including gravitational waves. Mathematically, we can represent these boundary conditions by making the walls of the box perfectly reflecting: hence on the walls, a scalar field ϕ vanishes, the transverse part of the electromagnetic field vanishes, and for gravitational waves the appropriate components of the Weyl tensor vanish. These boundary conditions cannot be relaxed, since the consequence of this would be that no stable equilibrium could exist. From a statistical mechanical point of view one must therefore use a microcanonical ensemble rather than a grand canonical ensemble, as has been pointed out by Hawking (1976). In the absence of gravitation, the answer to this problem is well-known. For simplicity we consider a system with no net electric charge or angular momentum.

The energy E_g, entropy S_g and temperature T are related by

$$E_g = aVT^4, \tag{2.1}$$

$$S_g = \tfrac{4}{3}aVT^3. \tag{2.2}$$

a is Stefan's constant, and is equal to $\frac{1}{15}\pi^2(n_b + \frac{7}{8}n_f + \frac{1}{2}n_s)$ where n_b, n_f and n_s are the number of zero mass boson fields (excluding scalars), the number of zero mass fermion fields, and the number of zero mass scalar fields. In practice these numbers are temperature dependent: at temperatures less than the rest mass of the particle, the contribution to the energy density is small, whereas at temperatures greater than the rest mass, the particles behave as if their rest mass were effectively zero.

We can now consider the effect of including gravity into this picture. The 'volume' of the box is taken to be the same as before. However, it is possible that a black hole can condense out of the radiation gas, rather in the way a liquid drop can condense out of a saturated vapour. The entropy S is now given by $S = S_g + S_H$ and energy by $E = E_g + M$. With the specified boundary conditions, equilibria are determined by the maxima of S, subject to the constraint that the energy is constant. Thus

$$S = 4\pi M^2 + \tfrac{4}{3} a V T^3, \tag{2.3}$$

$$E = M + a V T^4. \tag{2.4}$$

Setting $x = M/E$ and $y = (1/3\pi)(aV/E^5)^{\frac{1}{4}}$, this is equivalent to maximizing

$$f(x) = x^2 + y(1-x)^{\frac{3}{4}}$$

with respect to x on the closed interval $[0, 1]$.

For $y > 2^5 3^{-1} 5^{-\frac{5}{4}} = 1.4266$, there are no turning points, and the maximum value of f is at $x = 0$. The equilibrium configuration is then pure radiation. Below this value, but for $y > y_c = 1.0144$, there are two turning points; a local minimum for $x < \frac{4}{5}$ and a local maximum for $x > \frac{4}{5}$. However, the global maximum of f is still given by $x = 0$. For $y < y_c$ there is a global maximum of f with $x > x_c = 0.97702$. This represents a stable equilibrium configuration consisting of a black hole and black-body radiation, the mass of which is then related to the temperature of the radiation by $8\pi M T = 1$.

The relation between T and E is shown schematically in figure 1. Thus one could imagine filling the enclosure with energy and, at first, raising the temperature. However, at a sufficiently high energy density, a black hole will condense out, after which the temperature will fall. That is, the system has a negative specific heat. For a given volume, there is an upper bound to the temperature $T_m = x_c^{-\frac{1}{5}} T_c$, where

$$T_c = \frac{1}{8\pi x_c} (3\pi y_c)^{\frac{4}{5}} (aV)^{-\frac{1}{5}}. \tag{2.5}$$

The dotted lines represent unstable 'superheated' or 'supercooled' configurations. This first order phase transition is rather analogous to that occurring when a liquid drop condenses out of a vapour. Indeed one can define a black hole 'vapour pressure'

$$P = \tfrac{1}{3} a T^4 \tag{2.6}$$

where $T = \kappa/2\pi$. Its qualitative dependence on the size of the black hole, its charge, and its angular momentum are precisely those of a liquid drop (Frenkel 1948). The smaller the black hole, the easier it is for a particle to escape by the quantum tunnelling process. Similarly, it is easier for a molecule to escape from the surface of a small drop, because the surface tension increases the vapour pressure for small drops. A neutral particle will less readily evaporate from a charged drop than from a neutral drop because to do so will decrease the drop size, and hence increase the electrical energy. In the same way a particle with zero angular momentum will escape less readily from a rotating drop than from a non-rotating one, because its escape would increase the kinetic energy of the drop. The qualitative dependence of the surface gravity – and hence 'vapour pressure' of a black hole – upon its charge and angular momentum corresponds precisely with this analogy.

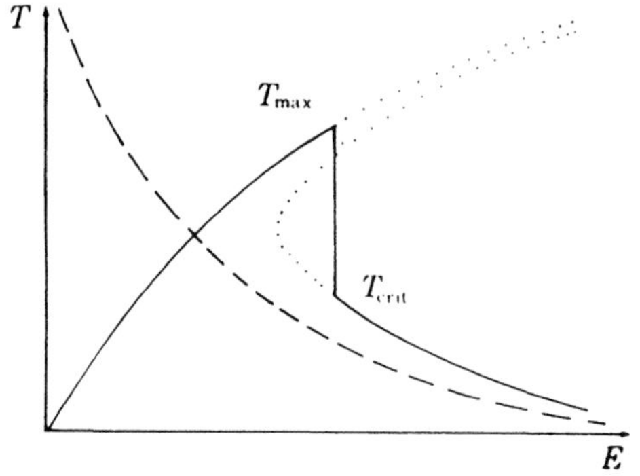

FIGURE 1. The relation between the temperature of an enclosure of volume V and the total energy E. The solid line is the stable equilibrium configuration. At small energy densities one is on the pure radiation curve. For high enough energy densities the configuration jumps to that consisting of a large black hole and a small amount of cooler radiation. The dotted line represents unstable configurations. The dashed line gives the relation between total energy and temperature for a black hole on its own.

It is perhaps worth remarking here that the size of the box, $L \sim M^{\frac{5}{3}}$ at the point when a black hole condenses out. Thus, there is no danger of the box being crushed by gravitational forces before the condensation has occurred. This is because these forces are only significant if $M \sim L$, and one expects a black hole always to have a mass greater than the Planck Mass.

If, instead of considering an adiathermal box with rigid walls, we had adopted some different boundary conditions, we should have had to consider extrema of some different thermodynamic function. The various possibilities are considered in table 1. We see that the only case in which an equilibrium between radiation and a black hole is possible is the one we have considered. If we consider a box with adiathermal walls, but allow work to be done on the gas, we find no true equilibrium at all. This is because one may always force more radiation into a more and more massive hole. There is however, an unstable equilibrium when $8\pi MT = 1$. If on the other hand we allow the free passage of heat keeping the termperature T, and hence the pressure constant, one finds that the gas is stable against forming a small black hole which would be hotter than the surroundings, since this would immediately evaporate. However, if a large black hole were formed, it would gobble up more and more radiation since it would be cooler than its surroundings. This situation can, in fact, only be stab lized by keeping the total energy available finite; that is by restricting oneself to constant energy and volume.

TABLE 1. THE EQUILIBRIUM CONFIGURATION OF A SELF GRAVITATING MASS OF NEUTRAL MATTER

variables held constant	thermodynamic function minimized	minimum configuration
S, P, T	enthalpy, $H = M + T(S - 4\pi M^2)$	no true minimum ($M \to \infty$) local maximum at $8\pi MT = 1$
P, T	Gibbs energy $G = \frac{1}{2}M$	no black hole. Pure radiation
V, T, P	Helmholtz energy $F = \frac{1}{2}M - PV$	no black hole. Pure radiation
S, V	total energy E	see figure 1

3. THERMAL GREEN FUNCTIONS IN GLOBALLY STATIC SPACETIMES

In this section, we shall discuss the treatment of system in thermal equilibrium in a gravitational field from the point of view of statistical mechanics, and quantum field theory. Since a system in equilibrium is by definition one which is independent of time as judged by observers at rest with respect to the system, we shall demand that the spacetime be stationary. However, although it cannot be said to be a true equilibrium configuration, it is possible for a gas of zero rest mass particles to be described by using equilibrium thermodynamic ideas if the spacetime is merely conformally stationary (Israel 1972). Since this situation prevails in a Robertson–Walker universe, we shall discuss this in an appendix.

To begin with, we restrict ourselves to considering a finite system in a static spacetime. That is, in a compact spatial region V, the metric may be expressed in the form

$$ds^2 = g_{00}(x^\alpha)\,dt^2 + h_{\alpha\beta}\,dx^\alpha dx^\beta \tag{3.1}$$

where $\alpha, \beta, \ldots = 1, 2, 3$, and $g_{00} < 0$. t, which is proportional to the proper time along the worldlines of the stationary observers is undetermined up to a multiple which is fixed by demanding that it coincide with the proper time of one particular observer. $(-g_{00}(x^a))^{\frac{1}{2}}$ then gives the gravitational redshift factor between this observer and one at x^a V is bounded by walls isolating the system from the exterior. Since the detailed boundary conditions will not affect our answers, we shall proceed as if the spatial region were compact, with no boundary. For instance, using periodic boundary conditions in a cube in flat space is entirely equivalent to working on a flat compact 3-torus. For definiteness, we shall begin by considering a massive, neutral non self-interacting scalar field described by a Hermitian, Heisenberg, operator-valued distribution ϕ. ϕ obeys a covariant generalization of the Klein–Gordon equation.

$$\phi^{a}{}_{;a} - m^2\phi + \xi R\phi = 0. \tag{3.2}$$

R is the Ricci scalar, and ξ a free parameter. If one wanted (3.2) to be conformally invariant when $m^2 = 0$, one would set $\xi = \frac{1}{6}$. This equation of motion may be derived from the Lagrangian density

$$\mathscr{L} = \tfrac{1}{2}(-g)^{\frac{1}{2}}\{\phi_{,a}\phi^{,a} + m^2\phi^2 - \xi R\phi^2\} \tag{3.3}$$

One may construct from (3.3) a Hamiltonian H which represents the total energy, or equivalently the time translation operator relative to static observers.

$$H = -\frac{1}{2}\int h^{\frac{1}{2}}\{-g_{00}\dot{\phi}^2 + \phi_{,\alpha}\phi_{,\beta}h^{\alpha\beta} + m^2\phi^2 - \xi R\phi^2\}\,\mathrm{d}^3x, \tag{3.4}$$

where
$$\dot{\phi} = \partial\phi/\partial t = \mathrm{i}[H, \phi]. \tag{3.5}$$

In order to find the energy eigenstates of the field, we resolve ϕ into normal modes

$$\phi = \sum_n a_n \frac{\psi_n(x^\alpha)}{\sqrt{(2\omega_n)}}\,\mathrm{e}^{-\mathrm{i}\omega_n t} + a_n^\dagger \frac{\overline{\psi}_n(x^\alpha)}{\sqrt{(2\omega_n)}}\,e^{\mathrm{i}\omega_n t} \tag{3.6}$$

where a_n, $a_n^\dagger$ are the annihilation and creation operators respectively, and $\psi_n(x)$ obeys

$$\nabla_f^2\psi_n - (g_{00})_{,\alpha}h^{\alpha\beta}\psi_{n,\beta} + m^2(-g_{00})^{\frac{1}{2}}\psi_n - \xi R(-g_{00})^{\frac{1}{2}}\psi_n = \mathscr{L}\psi_n = -\omega_n^2\psi_n. \tag{3.7}$$

The first term in (3.7) is the Laplace–Beltrami operator with respect to the 3-metric $f_{\alpha\beta} = h_{\alpha\beta}/-g_{00}$, which we call the optical metric. The projection of null geodesics into the spatial sections are geodesics of this metric. The normal modes are normalized by the condition

$$\int \psi_n(x)\,\psi_m(x)\sqrt{(h/-g_{00})}\,\mathrm{d}^3x = \delta_{nm}. \tag{3.8}$$

The elliptic differential operator $\mathscr{L}$ is self-adjoint with respect to this scalar product. The eigenvalue problem of this type has been quite extensively studied. (McKean & Singer 1969). The basic technique is to study the elementary solution $e(x, x'; \beta)$ of the associated diffusion equation

$$\partial\psi/\partial\beta = \mathscr{L}\psi. \tag{3.9}$$

In terms of the normal modes, this is

$$e(x, x', \beta) = \sum_n \psi_n(x)\,\psi_n(x')\,\mathrm{e}^{-\omega_n^2\beta}. \tag{3.10}$$

A careful analysis near $\beta = 0$ shows it to be given asymptotically by, for x near x'

$$e(x, x'; \beta) = \frac{1}{(4\pi\beta)^{\frac{3}{2}}}\exp\left[-\frac{f(x, x')}{4\beta}\right], \tag{3.11}$$

where $f(x, x')$ is the square of the optical distance between x and x'. Letting $x \to x'$ and integrating over V, one has

$$Z_1 = \sum_n \exp[-\omega_n^2\beta] \approx \frac{V_0}{(4\pi\beta)^{\frac{3}{2}}} \tag{3.12}$$

where V_0 is the volume of V measured by the optical metric. The application of theorems relating the behaviour of a function for small values of its argument

(Tauberian and Abelion theorems) enables one to compute the number of eigenvalues between ω and $\omega + \mathrm{d}\omega$, $(\mathrm{d}N/\mathrm{d}\omega)\,\mathrm{d}\omega$, for large ω

$$\frac{\mathrm{d}N}{\mathrm{d}\omega}\,\mathrm{d}\omega = \frac{V_0\,\omega^2}{2\pi^2}\,\mathrm{d}\omega \tag{3.13}$$

hence

$$\omega_n^2 \approx 2^{-\frac{1}{2}}3^{\frac{3}{2}}\pi^{\frac{7}{4}}(n/V_0)^{\frac{3}{2}}. \tag{3.14}$$

One can also find the small β behaviour of

$$Z_2 = \sum_n \exp(-\omega_n\beta). \tag{3.15}$$

It is

$$Z_2 \approx \pi^2 V_0/2\beta^3. \tag{3.16}$$

It is interesting to note that Z_1 and Z_2 are respectively the non-relativistic, and relativistic single particle canonical partition functions, if we interpret $\beta = T^{-1}$ as an inverse temperature. The high temperature forms (3.12) and (3.16) are just the standard ones except that it is the optical volume which appears rather than the proper volume. This is because of the red-shifting of temperature. That is, for any system in thermal equilibrium, the local temperature multiplied by $(-g_{00})^{\frac{1}{2}}$ is constant. One may now proceed to construct n-particle partition functions from these forms in the standard way (Feynman 1974). Thus, if one calculates the total entropy of the system at high temperatures, one obtains

$$S = (4\pi^2/45)\,V_0 T^3 = (4\pi^2/45)\int \sqrt{h}\,(T/\sqrt{-g_{00}})^3\,\mathrm{d}^3x \tag{3.17}$$

which is just the entropy of a collection of sub-systems of proper volume $\sqrt{h}\,\mathrm{d}^3x$, and local temperature $T/(-g_{00})^{\frac{1}{2}}$.

When discussing systems with a small number of particles, one is concerned with deviations from the ground state, $|0\rangle$ from the point of view of the static observers. A suitable tool for this is the Feynman propagator.

For a general spacetime, there are various inequivalent definitions of this object, some of which are discussed by Gibbons (1975). The common feature of all definitions is that the Feynman propagator is a symmetric solution of the inhomogeneous equation

$$G(x,x')^{;a}{}_{;a} + (\xi R - m^2)\,G(x,x') = -\delta(x,x'). \tag{3.18}$$

If one has a notion of 'in' and 'out' vacua $|0_-\rangle$ and $|0_+\rangle$ respectively, one such definition is

$$G_\infty(x,x') = \mathrm{i}\,\frac{\langle 0_+|\,T\phi(x)\,\phi(x')\,|0_-\rangle}{\langle 0_+|0_-\rangle}. \tag{3.19}$$

T is the Wick time-ordering operator. This will of course depend on the definition of $|0_+\rangle$ and $|0_-\rangle$, that is it will not be invariant under Bogoliubov transformations of the bases defining $|0_+\rangle$ and $|0_-\rangle$. In the present, globally static case, it seems reasonable to define $|0_+\rangle$ and $|0_-\rangle$ using the same bases $\{p_n\} = \psi_n(x)\,e^{-i\omega_n t}/\sqrt{(2\omega_n)}$. This results in the propagator, referred to as G_F and is given explicitly by

$$G_F(x^\alpha, t;\, x'^\alpha, t') = \begin{cases} i\sum\limits_n \dfrac{\psi_n(x^\alpha)\,\psi_n(x^{\alpha'})}{2\omega_n}\exp[-i\omega_n(t-t')] & (t > t') \\[2ex] i\sum\limits_n \dfrac{\psi_n(x^\alpha)\,\psi_n(x^{\alpha'})}{2\omega_n}\exp[i\omega_n(t-t')] & (t' > t). \end{cases} \tag{3.20}$$

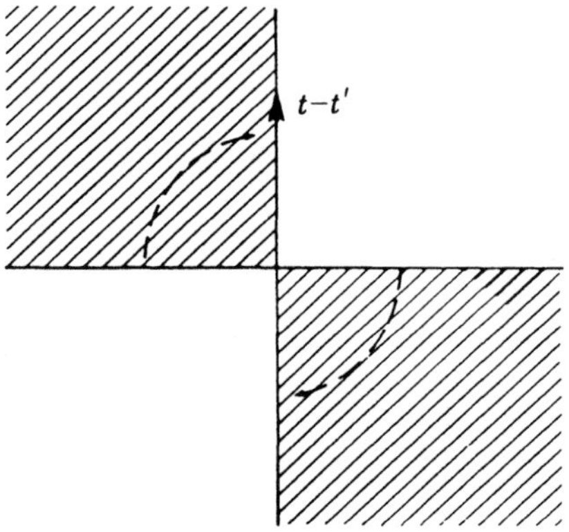

FIGURE 2. The complex $t-t'$ plane. G_F is analytic in the shaded regions G_E is obtained by analytically continuing along the dotted paths.

The ambiguities in (3.20) are resolved by giving ω_n a small imaginary part. This shows that G_F propagates positive frequencies to the future, and negative frequencies to the past. We can extend the definition of G_F to complex values of $t-t'$, by adopting the convention that T acts on the real part of its argument. This implies that G_F, which is a function of $(t-t')^2$, will be analytic in the second and fourth quadrants of the $t-t'$ plane. Thus, if we wish to consider purely imaginary values of $t-t' = i(\tau-\tau')$ we approach the negative $(\tau-\tau')$-axis from the second quadrant, and the positive $(\tau-\tau')$-axis from the fourth quadrant, (see figure 2). This results in the function

$$\begin{aligned} G_E(x^\alpha, t;\, x^{\alpha'}, t') &= \sum_n \frac{\psi_n(x^\alpha)\,\psi_n(x^{\alpha'})}{2\omega_n}\exp[-\omega_n(\tau-\tau')] \quad (\tau > \tau') \\ &= \sum_n \frac{\psi_n(x^\alpha)\,\psi_n(x^{\alpha'})}{2\omega_n}\exp[\omega_n(\tau-\tau')] \qquad (\tau' > \tau). \end{aligned} \tag{3.21}$$

This is the Green function for the elliptic equations corresponding to (3.19) on the same manifold as before, but with the Euclideanized (positive definite) metric:

$$ds^2 = -g_{00}\,d\tau^2 + h_{\alpha\beta}\,dx^\alpha\,dx^\beta. \tag{3.22}$$

This Green function is unique given the boundary conditions that it should die away for large values of $|\tau-\tau'|$.

An alternative approach is based on the proper time formalism of Schwinger and de Witt, or equivalently the path integral formulation of Feynman. In this approach, one defines $G(x, x')$ by the equations

$$G(x,x') = \mathrm{i}\int_0^\infty \mathrm{d}w\, \mathrm{e}^{-\mathrm{i}m^2w-\epsilon/w} F(x.x';\, w), \tag{3.23}$$

$$F(x,x';\, w)^{;a}{}_{;a} + \xi R(x)\, F(x,x';\, w) = \mathrm{i}\,\partial F/\partial w, \tag{3.24}$$

$$F(x,x';\, 0) = \delta(x,x'). \tag{3.25}$$

F may be expressed as a path integral

$$F(x,x';\, w) = \int \delta x[w] \exp\left[-\int_0^w \tfrac{1}{4} g\left(\frac{\mathrm{d}x}{\mathrm{d}s}, \frac{\mathrm{d}x'}{\mathrm{d}s}\right) \mathrm{d}s\right] \tag{3.26}$$

provided that a suitable weighting is chosen for the 'sum over histories', expressed by $\int \delta x[w]$ (Hartle & Hawking 1976). In order to give a precise meaning to (3.26), and hence find the appropriate solution of (3.24), one is led to complexify the spacetime, analytically continuing into a region with positive definite metric, and solving the analytically continued version of (3.24).

$$\square F + \xi R F = \partial F/\partial \Omega \tag{3.27}$$

where $\square$ denotes the four-dimensional Laplace–Beltrami operator in the Euclidean region. Under suitable boundary conditions, F may be determined by using the methods described by McKean & Singer (1969). In the present situation, it is clear that the propagator so determined is identical to G_{E} found above. This procedure is manifestly independent of any basis. It does of course depend on the coordinate chart in which one performs the calculations.

The importance of G_∞ is that it permits the

(i) calculation of amplitudes

(ii) evaluation of the expectation values of operators bilinear in the fields

(iii) construction of a perturbation theory when interactions are included.

1. Consider an initial state $|f_{\mathrm{i}}\rangle$, defined by a solution of the Klein–Gordon equation $f(x)$ of the form

$$|f_{\mathrm{i}}\rangle = \int \overline{f(x)}\, \overleftrightarrow{\partial a}\, \phi(x)\, |0_-\rangle\, \mathrm{d}\Sigma^a \tag{3.28}$$

where the integral is over an initial Cauchy surface Σ. Similarly, we may construct a final state $|g_{\mathrm{f}}\rangle$ defined by a solution of the Klein–Gordon equation $g(x)$, and the integral being taken over a final Cauchy surface Σ', which is to the future of Σ. Then one finds the amplitude for going from $|f_{\mathrm{i}}\rangle$ to $|g_{\mathrm{f}}\rangle$ to be

$$\frac{\langle g_{\mathrm{f}}|f_{\mathrm{i}}\rangle}{\langle 0_+|0_-\rangle} = \iint \mathrm{d}\Sigma^a \mathrm{d}\Sigma'^b g(x')\, \overleftrightarrow{\partial}{}^a_{x'}\, G_\infty(x',x)\, \overleftrightarrow{\partial}{}^b_x \bar{f}(x). \tag{3.29}$$

One may therefore interpret $G_\infty(x, x')$ as the amplitude for a particle to travel from x to x', since G_∞ has the property of propagating 'positive frequency' functions in the future basis to the future, and 'negative frequency' functions in the past basis to the past. One conventionally regards segments of the paths from x to x' which make up the sum in (3.26) as being those of 'particles' if future-directed, and 'antiparticles' if past-directed. (Stueckelberg 1942; Feynman 1949). Of course, for the Hermitian field there is no distinction between particles and antiparticles. Furthermore, since particles cannot be localized $G_\infty(x, x')$ will always appear together with a Klein–Gordon scalar product in any physically measurable amplitude.

2. By taking certain coincidence limits of derivatives of $G(x, x')$ one may obtain quantities like the expectation value of the stress tensor

$$\frac{\langle 0_+| T_{ab} |0_-\rangle}{\langle 0_+|0_-\rangle}$$

(Schwinger 1951; Gibbons 1975).

As is well-known, in order to obtain sensible results, formal manipulations of divergent quantities must be made, but this does seem to provide sensible results in some situations (see for example, the Casimir Effect, Brown & Maclay (1969)). Some attempts have been made for the case of a black hole radiating into empty space both in two spacetime dimensions (Fulling & Davies 1976; Unruh 1976) and the physical case of four dimensions (Fulling & Christensen, 1976).

3. If one includes self-interactions, for example by adding an extra term like $\lambda\phi^4$ to the Lagrangian, one may proceed to construct the interaction picture operators, and expand the S-matrix in powers of λ. This is entirely analogous to the usual flat space procedure (Drummond 1975). Alternatively, one may use the 'functional' methods (Lurié 1969) to the same end.

We turn now to discussing fluctuations about a system in thermal equilibrium at a temperature $T = \beta^{-1}$. We are now interested in the expectation values over a grand canonical ensemble. This expectation value is defined for any operator $\boldsymbol{A}$ by

$$\langle A\rangle_\beta = \frac{\mathrm{Tr}\,[\exp(-\beta H)\,A]}{\mathrm{Tr}\,[\exp(-\beta H)]} \tag{3.30}$$

where Tr denotes the operation of taking the trace over a complete set of states.

As in the case of fluctuations about the vacuum state, we can construct a thermal Green function $G_T(x, x')$ (Abrikosov, Gorkov & Dzyaloskinskii 1965; Kadanoff & Baym 1962; Fetter & Walecka 1971) by the formula

$$G_\mathrm{T}(x, x') = \mathrm{i}\langle T\phi(x)\,\phi(x')\rangle_\beta \tag{3.31}$$

which describes fluctuations about thermal equilibrium. From its definition, it follows that G_T is a solution of the inhomogeneous Klein–Gordon equation. Explicitly we have, in the basis of states used before

$$G_\mathrm{T}(x^\alpha, t; x'^\alpha, t') = \mathrm{i}\sum_n \frac{\psi_n(x^\alpha)\,\psi_n(x^{\alpha'})}{2\omega_n}\{(1+n_\mathrm{B})\exp[-\mathrm{i}\omega_n(t-t')] + n_\mathrm{B}\exp[\mathrm{i}\omega_n(t-t')]\} \quad (t > t'),$$

$$\mathrm{i}\sum_n \frac{\psi_n(x^\alpha)\,\psi_n(x^{\alpha'})}{2\omega_n}\{(1+n_\mathrm{B})\exp[\mathrm{i}\omega_n(t-t')] + n_\mathrm{B}\exp[-\mathrm{i}\omega_n(t-t')]\} \quad (t < t'), \tag{3.32}$$

where
$$n_\mathrm{B} = \frac{1}{\exp(\omega_n\beta) - 1}. \tag{3.33}$$

Just as with G_F, the first term corresponds to adding a particle to the system. The new second term describes the effect of subtracting a particle from the equilibrium configuration. As discussed by Bowers & Zimmerman (1973), we may describe this last process as the propagation of a 'hole' – not to be confused with either a black or white hole, or an antiparticle – to the past. If our field were not hermitian, we should have both 'holes' and 'anti-holes'. The latter would be propagated to the future. This is an extension of the more familiar ideas of a *single* sea of electrons, and holes, in a semiconductor.

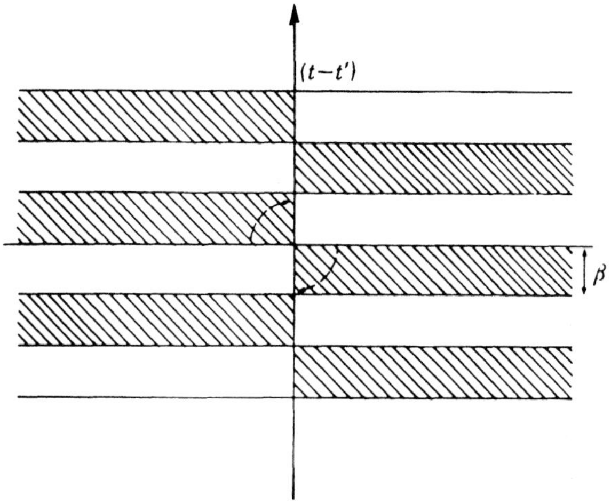

FIGURE 3. The complex $t-t'$ plane. G_T is analytic in the shaded regions. G_{ET} is obtained by analytically continuing along the dotted paths. It is a periodic function of period $\beta = 1/T$.

In principle, we could deduce the usual relation for the energy density and pressure of the distribution from G_T using the coincidence limit techniques mentioned earlier. If we wished to consider self-interactions, we could carry out a perturbation calculation just as one does for the vacuum case. The methods for doing this are entirely analogous to those described in Fetter & Walecka (1971). We shall not discuss them in detail. One may also use 'functional methods' (Dolan & Jackiw 1974; Bernard 1974).

Just as with the Feynman propagator $G_F(x, x')$, one may consider $G_T(x, x')$ as a function of the complex variable $(t-t')$. An examination of the power series in (3.32) shows that $G_T(x, x')$ will be analytic in the second and fourth quadrant of the $(t-t')$ plane provided that $|\mathrm{Im}\,(t-t')| < \beta$. Furthermore, its values at $\mathrm{Im}\,(t-t') = \pm\beta$ are the same as on the real axis. Thus $G_T(x, x')$ may be analytically continued throughout the complex plane as a periodic function of $(t-t')$, with period $i\beta$. This may also be shown directly from the Heisenberg field operators (see § 5). In particular, on the imaginary axis, $t-t' = i(\tau-\tau')$, τ, τ' real, G_T is a periodic function of $(\tau-\tau')$. The function so obtained is called G_{ET}.

This Euclidean thermal Green function could also be obtained by the path integral method described previously by solving the elliptic differential equation equation (3.27), on a manifold with the positive definite metric (3.22) provided one compactifies it by identifying points whose τ coordinates differ by β. Thus, the condition for thermal equilibrium is completely described by the cyclic boundary condition.

Having obtained these two-point Green functions in the non-interacting case, one may construct the n-point Green functions, (that is the expectation values of time-ordered products of n field operators), using Wick's theorem. If interactions are present, one would presumably need to have recourse to perturbation theory in order to construct the interacting 2 and n-point functions. (Abrikosov *et al.* 1965; Kadanoff & Baym 1962; Fetter & Walecka 1971.) Of course, the procedure will only gain sensible results if the theory is renormalizable in flat space, and remains so in a spacetime with curvature. Without detailed calculations it is impossible to be dogmatic on this point, however we are encouraged by the results of Utiyama (1962) for the case of quantum electrodynamics and Drummond (1975) for $\lambda\phi^4$ theory in de Sitter space to expect no obstacles to this programme in the cases we are considering.

The essential point is that all physical information is contained in principle in these n-point functions and these can be built up from the non-interacting two point functions. The standard argument using Heisenberg operators, given in § 5, shows that the interacting Green functions will also be periodic. Considered as a function of a local static observers proper time $t(-g_{00})^{\frac{1}{2}}$ it will have a period of $i\beta(-g_{00})^{\frac{1}{2}}$, corresponding to a local red-shifted temperature of $T(-g_{00})^{\frac{1}{2}}$.

4. Schwarzschild black holes and thermal Green functions

When discussing particle creation by a collapsing black hole, one should (as was first discussed by Hawking 1975), use a spacetime which is truly dynamic. However, because in any collapse the geometry is believed to settle down rapidly to a stationary state described by the Kerr–Newman solution, and because the details of the collapse do not affect the emission rate at late times, it is more convenient to carry out the discussion using the exact stationary solutions. To do this, one must set up boundary conditions appropriate to a collapse, rather than some hypothetical and physically unrealistic 'eternal black hole' situation. Furthermore, when discussing the thermodynamics of self-gravitating objects, one ought to use the micro-canonical ensemble (Lynden-Bell & Wood 1968; Hawking 1976), rather than grand canonical ensemble. If however the gravitational field is not allowed to respond to the effects of the created particles, as is true in this semi-classical approximation, then the answers obtained will be equivalent no matter which ensemble is used.

A system emitting particles to infinity is not in an equilibrium state. It is necessary to isolate the system by confining the particles to some finite region – a 'box'. Thus, we imagine placing around the black hole at late times a box having the same properties as that described in § 2 (see figure 4).

In this section, we shall consider only the simplest case, the Schwarzschild black hole, and a massive hermitean scalar field. The metric of this spacetime, in Kruskal coordinates is

$$ds^2 = -\frac{r}{32M^3}\exp\left(-\frac{r}{2M}\right)dU\,dV + r^2(d\theta^2 + \sin^2\theta\, d\phi^2), \tag{4.1}$$

$$UV = \left(1-\frac{r}{2M}\right)\exp\left(\frac{r}{2M}\right). \tag{4.2}$$

U and V are the retarded and advanced null coordinates respectively, and range over values such that $-\infty < UV < 1$.

The Penrose–Carter diagram of this space is shown in figure 5. Regions I and IV are invariant under a one parameter group of time translations generated by the timelike Killing vector $k = \partial/\partial t$, where t is given by

$$\exp(t/2M) = -U/V. \tag{4.3}$$

$\partial/\partial t$ is future directed in I and past directed in IV. Regions II and III are invariant under a one parameter group of space translations generated by the spacelike Killing vector $k = \partial/\partial t$ where

$$\exp(t/2M) = U/V. \tag{4.4}$$

The point $U = V = 0$ is a fixed point under this group. That is, it is the analogue for Lorentz boosts, of the axis of symmetry, or fixed point, for ordinary rotations (Boyer 1969). This can be seen immediately if one introduces the coordinates.

$$T = \tfrac{1}{2}(U+V), \tag{4.5}$$

$$Z = \tfrac{1}{2}(U-V), \tag{4.6}$$

when the metric becomes

$$ds^2 = -\frac{32M^3}{r}\exp\left(-\frac{r}{2M}\right)(dT^2 - dZ^2) + r^2(d\theta^2 + \sin^2\theta\, d\phi^2). \tag{4.7}$$

All geometrical quantities are functions of $Z^2 - T^2$ only. That is, the geometry is invariant under translation along the orbits of the Killing vector

$$k = T\,\partial/\partial Z + Z\,\partial/\partial T. \tag{4.8}$$

k vanishes at $T = Z = 0$. In the more familiar rotation case, one would be considering a geometry depending only upon $X^2 + Y^2$ for some suitable choice of coordinates. The corresponding Killing vector would be

$$\tilde{k} = X\,\partial/\partial Y - Y\,\partial/\partial X \tag{4.9}$$

whose orbits are the circles $X^2 + Y^2 = \text{const}$. The fixed point is, of course $X = Y = 0$. The surfaces $V = 0$, $U < 0$ and $U = 0$, $V > 0$ are respectively the past and future

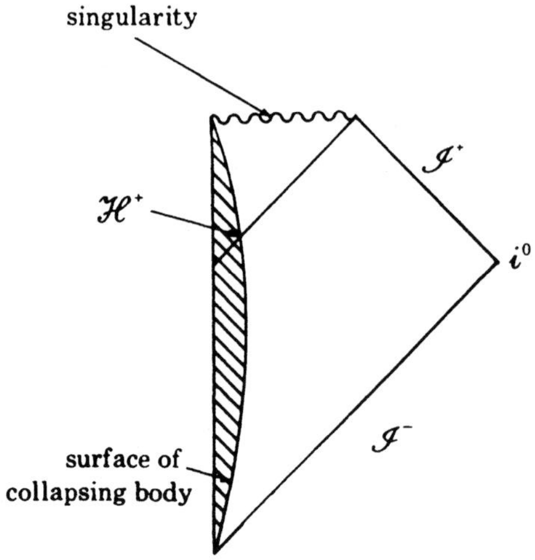

FIGURE 4. The Penrose–Carter diagram of a black hole collapsing inside a box and coming into thermal equilibrium. $\mathscr{H}^+$ is the event horizon. The shaded area is the interior of the collapsing body. The dashed line represents the walls of the box.

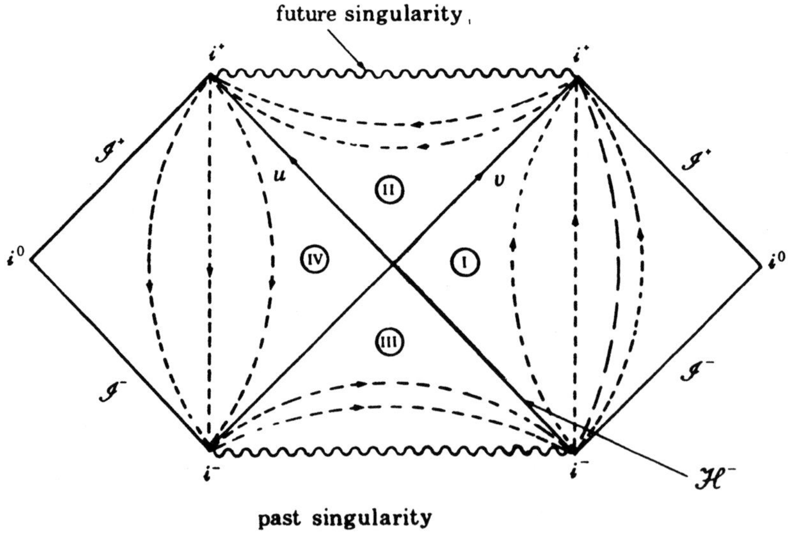

FIGURE 5. Penrose–Carter diagram of the Kruskal manifold. The thin lines represent the orbits of the Killing vector $k = \partial/\partial t$ together with its sense. $\mathscr{I}^+$ and $\mathscr{I}^-$ are the future and past null infinities, i^+ and i^- are the future and past timelike infinities, and $\mathscr{H}^+$ and $\mathscr{H}^-$ the future and past event horizons.

horizons $\mathscr{H}^-$ and $\mathscr{H}^+$, of observers who remain in region I. $\mathscr{I}^-$ and $\mathscr{I}^+$ are past and future null infinity. i^- and i^+ represent past and future timelike infinity.

Observers who remain at constant r in region I can describe their observations in terms of Schwarzschild coordinates (t, r, θ, ϕ) in which the metric takes the manifestly time invariant form

$$ds^2 = -\left(1-\frac{2M}{r}\right)dt^2 + \frac{dr^2}{1-2M/r} + r^2(d\theta^2 + \sin^2\theta\, d\phi^2). \tag{4.10}$$

For discussing the horizons $\mathscr{H}^-$ and $\mathscr{H}^+$ it is also convenient to introduce retarded and advanced null coordinates u, v respectively defined by

$$du = -4M\frac{dU}{U} = dt - \frac{dr}{1-2M/r}, \tag{4.11}$$

$$dv = +4M\frac{dV}{V} = dt + \frac{dr}{1-2M/r}. \tag{4.12}$$

If one expresses the metric in terms of (u, r, θ, ϕ) or (v, r, θ, ϕ) it is manifestly independent of u and v, and regular throughout regions I and IV, or I and II respectively.

When discussing quantum field theory in this space, we are primarily concerned with region I external to the black hole. One may take as past Cauchy surfaces a sequence of space-like surfaces which as they are moved into the past tend to $\mathscr{I}^- \cup i^- \cup \mathscr{H}^-$. There are then two fairly natural choices of bases for solutions of the Klein–Gordon equation. The first we shall call B_S^-. This is constructed such that the functions with positive Klein–Gordon norm in this basis have positive frequencies with respect to the Schwarzschild time coordinate t. On the past horizon it is more convenient to use the retarded null coordinate u. Positive frequency is then interpreted as meaning analytic in the upper half u-plane on $\mathscr{H}^-$. One may define a corresponding basis in the future, B_S^+, using the advanced null coordinate v. These bases are those with which a static observer would most naturally choose to describe his observations.

It is also convenient to decompose these bases into two disjoint sets, one describing horizon states, and the other describing states on $\mathscr{I}$. For example

$$B_S^- = B_S^-(\mathscr{H}^-) \cup B_S^-(\mathscr{I}^- \cup i^-).$$

$B_S^-(\mathscr{H}^-)$ consists of functions with zero Cauchy data on $\mathscr{I}^- \cup i^-$ and $B_S^-(\mathscr{I}^- \cup i^-)$ of functions with zero Cauchy data on $\mathscr{H}^-$.

Another useful basis may be constructed using the Kruskal coordinates U and V. We shall call this basis B_K^-. It consists of two disjoint sets of functions $B_K^-(\mathscr{H}^-)$ and $B_K^-(\mathscr{I}^- \cup i^-)$. The first set of functions are chosen to have zero Cauchy data at past infinity. They consist of wave-packets which enter region I through past horizon $\mathscr{H}^-$ but have no incoming component from large distances from the hole. The second set of functions is complementary to the first set in that they have zero data on $\mathscr{H}^-$ and consist of incoming waves from the past infinity. In order to qualify as a positive frequency function in this basis, a member of $B_{\mathrm{K}}^{-}(\mathscr{H}^{-})$ must have data on $\mathscr{H}^{-}$ which is holomorphic in the lower U-plane. That is, it is positive frequency with respect to an affine parameter along the past horizon. On the other hand, the incoming waves, which are members of $B_{\mathrm{K}}^{-}(\mathscr{I}^{-} \cup i^{-})$ are made up of wave-packets, which at some large fixed value of r are holomorphic in the lower half V-plane. On interchanging past and future, and U and V, one may similarly define a dual basis B_{K}^{+}.

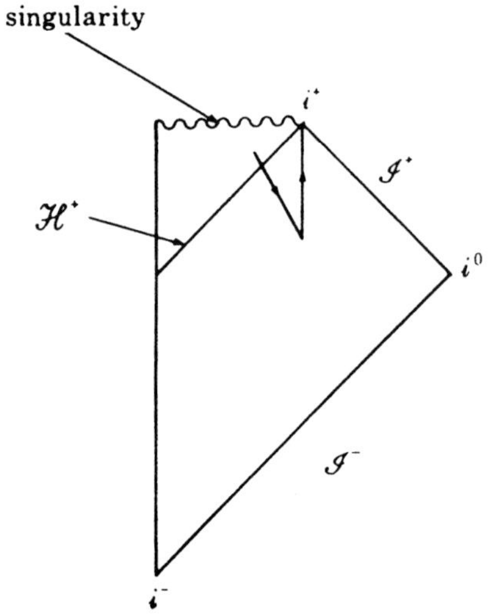

Figure 6. Carter–Penrose–Feynman diagram of particle creation by gravitational collapse to a black hole final state.

The crucial point of Hawking's original derivation of black hole radiance was to show that one could indeed use the Kruskal manifold to describe the production of particles by the gravitational collapse of a body if one chose as one's initial state that corresponding to the vacuum state of the basis $B_{\mathrm{K}}^{-}(\mathscr{H}^{-}) \cup B_{\mathrm{S}}^{-}(\mathscr{I}^{-} \cup i^{-})$. That is, it would contain no particles coming from infinity, and no excitations corresponding to $B_{\mathrm{K}}^{-}(\mathscr{H}^{-})$. This point has been elaborated by Unruh (1976). If one decomposes this initial state in terms of the basis B_{S}^{+} one finds that it consists of a sum of states each of which consists of one or more pairs of particles. Each pair onsists of a particle described by one of the functions of $B_{\mathrm{K}}^{+}(\mathscr{H}^{+})$ and one particle described by a function in $B_{\mathrm{S}}^{+}(\mathscr{I}^{+} \cup i^{+})$. The amplitude for each of these processes – e.g. creation of a pair, may be associated with a spacetime diagram showing two future-directed world lines beginning from a point outside the horizon, one leading into the hole, and one escaping to infinity. Adopting the Feynman view that a particle moving to the future is the same as an antiparticle moving into the past, one can regard such a diagram as consisting of a single world line, one portion of which moves backwards in time (see figure 6). Since there is no absolute notion of a particle in curved space, one should perhaps use a different sort of world line for each basis. If one now averages over the states $B_S^+(\mathscr{H}^+)$ which are not measurable by external observers, one obtains a thermal density matrix (Wald 1975; Parker 1975; Hawking 1976). These calculations which are rather long, establish the thermal nature of the emission in the non-interacting case. We believe a thermal Green function approach is both easier and susceptible of a straight-forward generalization to the interacting case.

We begin by considering a spherical box placed around the hole. Its walls are shown as a dashed line in the Penrose–Carter diagram (figure 5). The field ϕ is to satisfy the condition that it vanish on the surface of the box. One may now omit the states describing particles to and from infinity from our bases. The initial state corresponding to the collapse in the past will contain no excitations in the B_K^- basis. We denote it by $|0_K^-\rangle$. Static observers will describe their observations in terms of $|0_S^+\rangle$, which because of the staticity of the metric differs from $|0_S^-\rangle$ only by a phase factor. The corresponding propagator

$$G_S(x, x') = \mathrm{i}\,\frac{\langle 0_S^+|\,T\phi(x)\,\phi(x')\,|0_S^-\rangle}{\langle 0_S^+|0_S^-\rangle} \tag{4.13}$$

will be analytic in the second and fourth quadrants on the complex $t-t^-$ plane. It may be constructed in an entirely analogous fashion to that adopted by Boulware (1975). The normal modes used would take the form

$$f_l(r)\,Y_{lm}(\theta, \phi)\exp(-\mathrm{i}\omega_n t), \tag{4.14}$$

where f_l satisfies the radial wave equation

$$\frac{1}{r^2}\frac{\mathrm{d}}{\mathrm{d}r}\left(r^2\left(1-\frac{2M}{r}\right)\frac{\mathrm{d}f_l}{\mathrm{d}r}\right)+\left[\omega_n^2\left(1-\frac{2M}{r}\right)^{-1}-\left(m^2+\frac{l(l+1)}{r^2}\right)\right]f_l(r)=0. \tag{4.15}$$

f_l is to vanish on the box at $r = R$ say, and will behave like

$$n(\omega)\exp(\mathrm{i}\omega r^*)+b(\omega)\exp(-\mathrm{i}\omega r^*)$$

as $r^* \to -\infty$, where

$$r^* = r - 2M\ln\left|\frac{r-2M}{2M}\right|. \tag{4.16}$$

$a(\omega)$ and $b(\omega)$ will differ only by a phase and their moduli are determined by the condition that the modes have the correct Klein–Gordon norm. Strictly speaking, we should use wave packets peaked at about ω as described by Hawking (1975). When this is done, these solutions represent modes which enter region I through the past horizon, bounce off the walls of the box, and leave region I through the future horizon.

Not only can we construct a vacuum propagator, but we can construct a thermal Green function of temperature $T = \beta^{-1}$, G_{TS}, used by the external observers in perfect analogy to that constructed in § 3. It will have the same analyticity properties as shown in figure 3, and described in § 3.

Just as there is no 'frequency mixing' between the bases B_S^- and B_S^+, so there is none between the bases B_K^- and B_K^+. This can be seen as follows. Consider a positive frequency member of B_K^-, i.e. a solution of the Klein–Gordon equation (3.2) that is analytic in the lower half U-plane on $\mathscr{H}^-$ and zero on the walls of the box $r = R$. The behaviour of a member of B_K^- on the future horizon $\mathscr{H}^+$ is determined by solving the Cauchy problem for the wave equation given the boundary conditions on the past horizon and the walls of the box. We may consider this problem not only for real values of U and V, but for complex ones also. That is, we may consider solving this problem on a succession of surfaces in which the real part of the variable t is allowed to take all values, but the imaginary part is held constant at σ say. For each value of σ we have a real manifold, and a real metric. The complexified Cauchy surfaces will consist of the surface $V = 0$, and the half-line in the complex U-plane running from the origin at an angle $\pi + \sigma\kappa$ (the complexified past horizon) and the surface $U = 0$ and the half-line in the complex V-plane starting at the origin making an angle $-\sigma\kappa$ with the real axis, (the complexified future horizon). If for each value of σ the initial data are regular then by standard theorems for the Cauchy problem, they will be correspondingly regular on the complexified future horizon. As we vary σ from zero to π/κ, the initial surface runs over the entire lower half V-plane, whilst the final surface runs over the entire lower half U-plane. Therefore any member of B_K^- holomorphic in the lower half U-plane on H^- will be holomorphic on the lower half V-plane on $\mathscr{H}^+$.

Since the state we began in is $|0_K^-\rangle$, and $|0_K^+\rangle$ differs from this merely by a phase, the appropriate propagator to describe fluctuations about this state is G_K defined by

$$G_K(x, x') = \mathrm{i}\frac{\langle 0_K^+|\, T\phi(x)\,\phi(x')|\, 0_K^-\rangle}{\langle 0_K^+|0_K^-\rangle}. \tag{4.17}$$

This is an analytic function of x' in the upper half U-plane on $\mathscr{H}^-$, $V = 0$, and in the lower half V-plane on $\mathscr{H}^+$, $U = 0$, if x lies in region I. It thus enjoys the same properties as the Hawking–Hartle propagator constructed by the path integral method. Regarded as a function of the coordinates (t, r, θ, ϕ) it will be a function of $(t-t')$, which by equations (4.2) and (4.3) will be analytic in half-strips of width $8\pi M$ which are alternately to the left and right of the imaginary axis, as shown in figure 3. For the choice $\beta = 8\pi M$, the analytic structure of G_K is identical to that of G_{ST} – the thermal Green function according to static observers. Therefore, any observation made in the state $|0_K^-\rangle$, which can be calculated from G_K will be the same as one measured in a thermal equilibrium system of temperature $1/8\pi M$, which can be calculated from $G_S(8\pi M)^{-1}$. Since the measurements in the presence of interactions can be calculated entirely from G_K, or $G_S(8\pi M)^{-1}$, this equivalence also holds in this case.

It is interesting to look at these results from the point of view of the path integral methods of constructing propagators. If we consider only region I with the standard Schwarzschild metric (4.10), then a natural way of Euclideanizing is to replace t by $\mathrm{i}\tau$. This results in the positive definite metric

$$\mathrm{d}s^2 = \left(1 - \frac{2M}{r}\right)\mathrm{d}\tau^2 + \frac{\mathrm{d}r^2}{1-2M/r} + r^2(\mathrm{d}\theta^2 + \sin^2\theta\,\mathrm{d}\phi^2). \tag{4.18}$$

This suffers from a coordinate singularity at $r = 2M$. It may be removed by identifying points whose τ co-ordinates differ by $8\pi M$. The point $r = 2M$ then appears as a symmetry axis. That is, the analytically continued Killing vector $\partial/\partial t$ becomes an axial Killing vector $\partial/\partial\tau$. The path integral method then naturally yields a periodic propagator. This is in effect the method adopted by Hartle & Hawking. If one wished to obtain the Schwarzschild propagator at some temperature T, one would identify after time intervals of T^{-1} as explained in § 3. One would also impose the condition that F (3.24) diffused outwards through $r = 2M$. Clearly, if $T = 1/8\pi M$, this second condition is unnecessary since the singularities of the Euclidean manifold have been eliminated. This example shows that the path integral method depends crucially upon the global structure of the Euclidean manifold to which one continues.

The same arguments may be applied to the creation of particles in de Sitter space (Gibbons & Hawking 1976). It is interesting to note that for reasons unconnected with thermal particle creation, various authors have considered interacting field theories on the Euclidean version of de Sitter space and were able to carry out perturbation and renormalization procedures without any difficulty (Drummond 1975).

5. The effects of rotation, charge and spin

In this section, we shall consider the generalization of these results to higher spin fields, and to rotating and charged black holes.

If one wishes to consider higher spin fields, it seems necessary to restrict oneself to spins less than $\frac{3}{2}$ (Gibbons 1976), except for gravitational perturbations. For spin one, the treatment is similar to that outlined before, merely more complicated. One merely replaces the Klein–Gordon equation by the Proca or Maxwell equations. When considering fermions, which are described by using spinors, it is necessary to generalize the ordinary treatment of spinors in curved space, to the case where spacetime is complexified. This has been described in outline by Penrose (1976). Normally one introduces a spin space V_2 of two complex dimensions at each spacetime point. Each member of V_2 is called a spinor. One may also associate with V_2 a complex conjugate space $\bar{V}_2$ whose members are dashed spinors. When the complexifies one retains these two spaces, but drops the requirement that there is any relation between them. The second space is now denoted by $\tilde{V}_2$. In each space, we introduce a symplectic form ϵ and $\tilde{\epsilon}$. One can now define spinor dyads $\{o, \iota\}$ and $\{o', \iota'\}$ normalized by $\epsilon(o, \iota) = 1 = \tilde{\epsilon}(\tilde{o}, \tilde{\iota})$. The space $T = V_2 \otimes \tilde{V}_2$ is spanned by the basis l, m, $\tilde{m}$, n defined by

$$l = o\tilde{o} \quad m = o\tilde{\iota} \quad \tilde{m} = \iota\tilde{o} \quad n = \iota\tilde{\iota}. \tag{5.1}$$

Using ϵ and $\tilde{\epsilon}$, one may construct a symmetric metric g on T such that

$$g(l, n) = 1 \qquad g(m, \tilde{m}) = -1 \tag{5.2}$$

and all other scalar products vanish. T and its associated scalar product, have the same structure as the tangent space of a complex Riemannian manifold, M. One gives a spin structure on such a manifold at each point on M by giving an explicit isomorphism between T and the tangent space at M. In our case, the tangent space is spanned by the four null vectors $\partial/\partial U$, $\partial/\partial V$, $\partial/\partial\xi$, and $\partial/\partial\tilde{\xi}$ where

$$\xi = e^{i\phi}\cot\tfrac{1}{2}\theta \tag{5.3}$$

$$\tilde{\xi} = e^{-i\phi}\cot\tfrac{1}{2}\theta \tag{5.4}$$

and where θ and ϕ are complex coordinates on the sphere. We choose to associate l, n, m and $\tilde{m}$ with appropriate multiples of $\partial/\partial U$, $\partial/\partial V$, $\partial/\partial\xi$, and $\partial/\partial\tilde{\xi}$. Thus, the spinors o and ι are associated with the null vectors $\partial/\partial U$ and $\partial/\partial V$ respectively. Now the vector $U\partial/\partial U$ is parallely propagated along both the orbits of $V\partial/\partial V$ and $U\partial/\partial U$. Similarly, the vector $V\partial/\partial V$ is parallely propagated along $V\partial/\partial V$ and $U\partial/\partial U$. It follows that the spinors $\sqrt{V}\,\iota$ and $\sqrt{U}\,o$ are also parallely propagated over the entire complex $U-V$ plane.

Any spinor propagator will be a spinorial function of two spacetime points x and x'. It will have two sets of spinor indices one defined at x and the other defined at x', in the sense that if one alters the spinor dyad at x, then the propagator will undergo a corresponding transformation with respect to its spinorial indices at that point. More about bispinors can be found in Lichnerowicz (1964). The direction of the spinor at x' is determined by the rule that it be parallely transported along a geodesic from x to x'. In the present case, it is clear that in order to specify the geodesic parallel propagator, we simply demand that the ratio of the components of the propagator in the basis o, ι are proportional to $\sqrt{U}$ and $\sqrt{V}$.

In region I, U and V are related to Schwarzschild coordinates by

$$U = -\exp(\kappa(r-t))\,(r/(2M)-1)^{\frac{1}{2}}, \tag{5.5}$$

$$V = \exp(\kappa(r+t))\,(r/(2M)-1)^{\frac{1}{2}}. \tag{5.6}$$

Thus it is clear that if we move x' an amount $2\pi i/\kappa$ in the complex t-plane, the spinor function will undergo a change of sign. On the other hand, bitensor propagators will be continuously rotated into themselves.

This is entirely consistent with the Thermal character of these propagators. To generalize them to the fermion case, one has of course to consider anticommuting field operators. The Wick time-ordering operation now includes a sign change for field interchange. In addition, one must use the Fermi–Dirac distribution function, which will then appear in the analogue of equation (3.32). The effect of this is to render the thermal propagator antiperiodic. That is the propagator changes sign when $t-t'$ is shifted through an amount $i\beta$. Thus the geometrical ideas relating thermodynamics and black holes also provide an unusually vivid illustration of the connection between the connection between the geometrical properties of spinors, and the anticommutativity of the corresponding field operators.

We turn now to the case of charged non-rotating black holes, and a charged scalar field ϕ. The presence of a conserved quantum number, the net charge of the system, requires the introduction of the charge operator N

$$N = -i\int \phi^{\dagger}\overleftrightarrow{D}_a\phi\, d\Sigma^a, \tag{5.7}$$

where $D_a = \partial_a - \mathrm{i}eA_a$, and an associated chemical potential, μ. The expectation value of any operator in this ensemble is now defined to be

$$\langle A\rangle_p = \frac{\mathrm{Tr}\,[\exp(-\beta(H-\mu N))\,A]}{\mathrm{Tr}\,[\exp(-\beta(H-\mu N))]}. \tag{5.8}$$

The relevant propagator for our problem is of the form

$$\langle T\,\phi(x)\,\phi(x')^{\dagger}\rangle \tag{5.9}$$

and will of course be gauge-dependent. The spacetime is given by the Reissner–Nordstrom solution

$$\mathrm{d}s^2 = -\left(1-\frac{2M}{r}+\frac{Q^2}{r^2}\right)\mathrm{d}t^2 + \frac{\mathrm{d}r^2}{1-2M/r+Q^2/r^2} + r^2(\mathrm{d}\theta^2 r\sin^2\theta\,\mathrm{d}\phi^2), \tag{5.10}$$

where Q is the charge on the hole.

One possible vector potential is

$$A = (Q/r)\,\mathrm{d}t \tag{5.11}$$

which dies away suitably at infinity, but is singular on the past and future event horizons $r = r_+ = M+\surd(M^2-Q^2)$. In order to find a well-behaved vector potential, one can perform a gauge transformation of the form

$$A = A + \mathrm{d}\chi \tag{5.12}$$

with
$$\chi = -Qt/r_+.$$

Q/r_+ is the electric potential of the black hole. Just as in the Schwarzschild case, one may introduce Kruskal-type coordinates U and V on the two horizons. We shall not describe these in detail but note that U and V are related to t in the exterior region by

$$\exp(2\kappa t) = -U/V, \tag{5.13}$$

where κ is the surface gravity of the hole defined by

$$\kappa = \frac{\surd(M^2-Q^2)}{(M+\surd(M^2-Q^2))^2}. \tag{5.14}$$

Exterior observers will conventionally describe their observations using the coordinates (t, r, θ, ϕ) and the first gauge. However, the path integral method of Hartle & Hawking will provide a propagator suitably analytic in U and V in a gauge that is well behaved on the horizons, for instance that of (5.12). To change gauge, one merely multiplies the propagator by

$$\exp(\mathrm{i}e\chi(x))\exp(-\mathrm{i}e\chi(x')) = \exp\frac{-\mathrm{i}eQ}{r_+}(t-t'). \tag{5.15}$$

The result of this is that on shifting $t-t'$ through an amount $\mathrm{i}\beta$, G takes on a factor of $\exp(eQ\beta/r_+)$. This gauge gives G manifestly the property of quasi-periodicity, which is precisely the behaviour of a thermal Green function for a system with chemical potential $\mu = eQ/r_+$, and temperature $\kappa/2\pi$. This can be seen most readily from the following argument. Let $\phi(t)$, $\phi(t')$ be two Heisenberg field operators. Their time evolution is governed by

$$\phi(t) = e^{iHt}\phi(0)\,e^{-iHt}, \tag{5.16}$$

where H is the Hamiltonian. We are concerned with quantities of the form

$$\mathrm{Tr}\,(\exp(-\beta(H-\mu N))\,\phi(t)\,\phi(t')^{\dagger}). \tag{5.17}$$

One has

$$[N,\phi] = -\phi^{\dagger} \tag{5.18}$$

$$[N,\phi^{\dagger}] = \phi. \tag{5.19}$$

Furthermore, ϕ and $\phi^{\dagger}$ commute at spacelike separations. If one now considers the continuation into the Euclidean region keeping the spatial coordinates fixed, they will also commute. Thus one has

$$\begin{aligned} \mathrm{Tr}\,(e^{-\beta(H-\mu N)}\phi^{\dagger}(t)\,\phi(t')) &= \mathrm{Tr}\,(e^{-\beta(H-\mu N)}\phi^{\dagger}(t)\,e^{\beta H}e^{-\beta H}\,\phi(t')) \\ &= \mathrm{Tr}\,(e^{\mu N\beta}\,\phi^{\dagger}(t+i\beta)\,e^{-\beta H}\,\phi(t')) \\ &= \mathrm{Tr}\,(e^{-\beta H}\,\phi(t')\,e^{\mu N\beta}\,\phi^{\dagger}(t+i\beta)) \\ &= e^{-\mu\beta}\,\mathrm{Tr}\,(e^{-\beta(H-\mu N)}\phi(t')\,\phi^{\dagger}(t+i\beta)) \\ &= e^{-\mu\beta}\,\mathrm{Tr}\,(e^{-\beta(H-\mu N)}\phi^{\dagger}(t+i\beta)\,\phi(t')). \end{aligned} \tag{5.20}$$

Thus

$$G_{\mathrm{T}}(x^{\alpha},x'^{\alpha};t-t') = e^{\mu\beta}\,G_{\mathrm{T}}(x^{\alpha},x^{\alpha'};t-t'+i\beta) \tag{5.21}$$

since G_{T} is a function only of $t-t'$, and the trace is invariant under cyclic permutation.

We shall now discuss rotating black holes. In order to discuss the horizons, one must work in a suitably co-rotating coordinate system. In the case of the Kerr solution, this means changing from the Boyer–Lindquist system (t,r,θ,ϕ) which is appropriate for observers outside the black hole, to the system introduced by Carter (1968). The main feature is a change to a new angular coordinate ϕ' by

$$\phi' = \phi - \Omega_{\mathrm{H}}t, \tag{5.22}$$

where Ω_{H} is the angular velocity of the hole. Since the solution is axisymmetric, the orbital angular momentum of the particle is restricted to be integral. That is the modes are proportional to $\exp(in\phi)$, n integral. It is convenient – since angular momentum is conserved to treat modes of different angular momenta separately. Each such mode has its own propagator, changing the coordinates ϕ is then analogous to a gauge transformation. That is the modes are multiplied by $\exp(in\Omega_{\mathrm{H}}t)$. Now the Hartle–Hawking propagator is a suitably analytic function of Kruskal co-ordinates. On changing back to Boyer–Lindquist coordinates, it is multiplied by a factor of $\exp(-in\Omega_{\mathrm{H}}(t-t'))$. If one displaces $t-t'$ an amount $i\beta$, $\beta = 2\pi/\kappa$, κ is the surface gravity of the hole, one sees that the propagator picks up a factor of $\exp(n\Omega_{\mathrm{H}}\beta)$. That is the emission is as if there were a non-zero chemical potential equal to $n\Omega_{\mathrm{H}}$, and at a temperature $\kappa/2\pi$.

It is worth noting that we have not used any detailed information about the particular spacetimes. Similar results will hold for any stationary spacetime with an horizon in it. In particular, it holds by a simple extension of these arguments to all Kerr–Newman solutions, and to cosmological event horizons discussed by Gibbons & Hawking (1976) (de Sitter space) and Lapedes (1976) (Taub-NUT type spaces).

6. Conclusions and speculations

In this paper, we have considered the result of putting a certain amount of energy E into an isolated box, volume V, with reflecting walls, which is then allowed to reach an equilibrium state. We have argued, used both simple heuristic ideas based upon the concept of black hole entropy, and by considering quantized matter fields on fixed, classical black hole geometries, that for suitably large energy densities an equilibrium configuration is possible consisting of a black hole and thermal radiation. Our principle innovation is the introduction of thermal Green functions, which we feel will have many applications both to the discussion of the thermodynamics of event horizons, and thermal equilibrium in gravitational fields.

Our main new result is the demonstration that mutual- and self-interactions between the matter fields will not change Hawking's (1975) original picture of the relation between black holes and thermodynamics, which was based on the rate of emission of non-interacting particles from a black hole.

It is interesting to consider quantum effects due not only to ordinary matter but also due to the gravitational field itself. A complete discussion of this must, of course, await a suitably formulated quantum theory of gravity. However, there are some remarks that one can make on the basis of the theory developed so far.

To begin with, it is clearly possible to consider quantizing deviations h_{ab} of the metric g_{ab}, from some background classical metric, $\mathring{g}_{ab}$. That is, to write

$$g_{ab} = \mathring{g}_{ab} + h_{ab} \tag{6.1}$$

and regard h_{ab} as a quantum field. The theory so obtained (Lichnerowicz 1964; de Witt 1964) looks rather like theories we have considered alone. h_{ab} appears as a spin 2 field with complicated self-interactions. One may construct the corresponding non-interacting Feynman Green functions on a black hole back-ground, which will of course enjoy the thermal periodicity properties. One could then proceed to build up the interacting Green functions by perturbation theory. Since this type of theory is non-renormalizable, (Deser & van Nieuwenhuizen 1975), the validity of this procedure is somewhat dubious.

Another approach would be to start with the flat space metric $\mathring{g}_{ab} = \eta_{ab}$, and build up the interacting Green functions. That is we first lay down a preferred coordinate system (T, X, Y, Z) the Einstein action $(-g^{\frac{1}{2}})R$, does not depend explicitly on T, and hence a conserved Hamiltonian H will exist. We construct the quantity

$$\mathrm{G}_{ab,cd} = \mathrm{i}\langle T h_{ab}(x)\, h_{cd}(x')\rangle_E \tag{6.2}$$

where $\langle\ \rangle_E$ now denotes an average over a microcanonical ensemble of total energy E. The use of the microcanonical ensemble is necessary because of the fact that self-gravitating systems can have negative specific heats (Lynden-Bell & Wood 1968; Hawking 1976). One could attempt to use this object to find the temperature of the enclosure as a function of the total energy E. One might hope that this would give results which are qualitatively the same as those obtained in § 2. This is evidently not a practicable procedure, and as before, the validity of the process is rather doubtful because of the non-renormalizability of gravity.

In addition, one has the traditional difficulties in dealing with two light cones – that is in deciding with respect to which causal structure the time ordering operator in (6.2) should be defined. This difficulty will also arise in a less severe form in perturbation calculations about black hole geometries.

It is interesting to note that Green functions for the full nonlinear Einstein equations have been considered in the classical theory (Sciama, Waylen & Gilman 1969; Lynden-Bell 1967; Altshuler 1967).

We wish to thank S. W. Hawking for stimulating discussions.

Appendix. Thermal Green functions in Robertson–Walker spacetimes

As we noted in the beginning of § 3, thermal equilibrium usually requires the spacetime to be stationary (Israel 1972). However in the case of zero rest-mass particles – for instance neutrinos and photons – one can using relativistic kinetic theory, find equilibrium distribution functions when the spacetime is merely conformally stationary. In particular the physically important 3 K microwave background can be dealt with in this way. In this section we shall provide the quantum field theoretic analogue of this description.

All three types of Friedman model, with metrics

$$ds^2 = -dt^2 + R^2(t)\,(dr^2 + \sin^2 r\, d\Omega^2), \tag{A 1}$$

$$ds^2 = -dt^2 + R^2(t)\,(dr^2 + r^2\, d\Omega^2), \tag{A 2}$$

$$ds^2 = -dt^2 + R^2(t)\,(dr^2 + \sinh^2 r\, d\Omega^2) \tag{A 3}$$

are clearly conformal to a static spacetime with a time like Killing vector $\partial/\partial\eta$, where

$$d\eta = dt/R(t). \tag{A 4}$$

One can now consider the thermal propagator for the Weyl neutrino, and Maxwell fields in the static spacetimes, at some temperature $T = \beta^{-1}$

$$ds^2 = -d\eta^2 + dr^2 + \sin^2 r\, d\Omega^2, \tag{A 5}$$

$$ds^2 = -d\eta^2 + dr^2 + r^2\, d\Omega^2, \tag{A 6}$$

$$ds^2 = -d\eta^2 + dr^2 + \sinh^2 r\, d\Omega^2. \tag{A 7}$$

These will be periodic in η with period $i\beta$. Since the Weyl and Maxwell equations are conformally invariant (Penrose 1969), suitably rescaled they will serve as Green functions in the physical metrics (A 1–3). According to the comoving observers with 4-velocities $\partial/\partial t$ they will be approximately periodic, with period $i\beta R(t)$, provided the Hubble time $R/\dot{R}$ is much larger than β. That is, the propagators will seem to be thermal with a temperature $T = \beta^{-1} R(t)^{-1}$. This is the standard adiabatic cooling of a radiation gas by a slow expansion.

One thus has a phenomenological description of the 3 K background. It does not, of course, explain *why* the propagator is periodic in the background static space, nor does it provide the value of β.

References

Abrikosov, A. A., Gorkov, L. P. & Dzyaloskinskii, I. Ye. 1965 *Quantum field theoretical methods in statistical physics*. Oxford: Pergamon.

Altshuler, B. L. 1967 *Soviet Phys. JETP*, **24**, 766.

Bardeen, J. M., Carter, B. & Hawking, S. W. 1973 *Commun. Math. Phys.* **31**, 162.

Bekenstein, J. D. 1973 *Phys. Rev.* D **7**, 2333.

Bekenstein, J. D. 1974 *Phys. Rev.* D **9**, 3292.

Bekenstein, J. D. 1975 *Phys. Rev.* D **12**, 3077.

Bernard, C. 1974 *Phys. Rev.* D **9**, 3312.

Boulware, D. 1975 *Phys. Rev.* D **11**, 1406.

Bowers, R. L. & Zimmerman, R. L. 1973 *Phys. Rev.* D **7**, 296.

Boyer, R. H. 1969 *Proc. R. Soc. Lond.* A **311**, 245.

Brown, L. S. & Maclay, G. J. 1969 *Phys. Rev.* **184**, 1272.

Carter, B. 1968 *Phys. Rev.* **174**, 1559.

Carter, B. 1971 *Phys. Rev. Lett.* **25**, 1596.

Christodoulou, D. 1970 *Phys. Rev. Lett.* **25**, 1596.

Christodoulou, D. & Ruffini, R. 1971 *Phys. Rev.* D **4**, 2552.

De Witt, B. S. 1964 The dynamical theory of groups and fields. In *Relativity groups and topology* (ed. B. S. de Witt & C. de Witt). London: Gordon and Breach.

Deser, S & van Nieuwenhuizen, P. 1975 *Phys. Rev.* D **10**, 401, 411.

Dolan, L. & Jackiw, R. 1974 *Phys. Rev.* D **9**, 3320.

Drummond, I. T. 1975 *Nucl. Phys.* B **94**, 115.

Fetter, A. L. & Walecka, J. D. 1971 *Quantum theory of many-particle systems*, New York: McGraw-Hill.

Feynman, R. P. 1949 *Phys. Rev.* **76**, 749.

Feynman, R. P. 1974 *Statistical Mechanics*, Menlo Park, California: Benjamin and Co.

Frenkel, J. 1948 *Kinetic theory of liquids*, London: Oxford University Press.

Fulling, S. A. & Christensen, S. 1976 *Trace anomalies and the Hawking effect*, King's College, London, preprint.

Fulling, S. A. & Davies, P. C. W. 1976 *Proc. R. Soc. Lond.* A **348**, 393.

Gibbons, G. W. 1975 *Commun. Math. Phys.* **45**, 191.

Gibbons, G. W. 1976 *J. Phys.* A **9**, 145.

Gibbons, G. W. & Hawking, S. W. 1976 Cosmological event horizons, thermodynamics, and particle creation, preprint.

Gibbons, G. W. & Perry, M. J. 1976 *Phys. Rev. Lett.* **36**, 985.

Hartle, J. B. & Hawking, S. W. 1972 *Commun. Math. Phys.* **27**, 283.

Hartle, J. B. & Hawking, S. W. 1977 *Phys. Rev.* (to be published).

Hawking. S. W. 1971 *Phys. Rev. Lett.* **26**, 1344.

Hawking, S. W. 1974 *Nature, Lond.* **248**, 30.

Hawking, S. W. 1975 *Commun. Math. Phys.* **43**, 199.

Hawking, S. W. 1976 *Phys. Rev.* D **13**, 191.
Israel, W. 1072 The relativistic Boltzmann equation. In *General relativity* (ed. L. O'Raifeartaigh). London: Oxford University Press.
Kadanoff, L. P. & Baym, G. 1962 *Quantum statistical mechanics.* Menlo Park, California: Benjamin.
Lapedes, A. 1976 DAMTP Cambridge preprint.
Lichnerowicz, A. 1964 Commutateurs et anticommutateurs. In *Relativity groups and topology* (ed. B. S. de Witt & C. de Witt). London: Gordon and Breach.
Lurié, D. 1969 *Particles and fields.* New York: Interscience.
Lynden-Bell, D. 1967 *Mon. Not. R. Ast. Soc.* **135**, 413.
Lynden-Bell, D. & Wood, R. 1968 *Mon. Not. R. Ast. Soc.* **138**, 495
McKean, H. P. & Singer, I. M. 1969 *J. Diff. Geom.* **1**, 43.
Parker, L. 1975 *Phys. Rev.* D **12**, 1590.
Penrose, R. 1969 *Riv. Nuovo Cimento* **1**, 252.
Penrose, R. 1976 *J. Gen. Relativity Gravity* **7**, 31.
Robinson, D. C. 1974 *Phys. Rev.* D **10**, 458.
Sciama, D. W., Waylen, P. C. & Gilman, R. C. 1969 *Phys. Rev.* **187**, 1762.
Schwinger, J. 1951 *Phys. Rev.* **82**, 664.
Smarr, L. 1973 *Phys. Rev. Lett.* **30**, 71. Erratum: *Phys. Rev. Lett.* **30**, 521.
Stueckelberg, E. 1942 *Helv. phys. Acta.* **15**, 23.
Unruh, W. 1976 *Phys. Rev.* D **14**, 870.
Utiyama, R. 1962 *Phys. Rev.* **125**, 1727.
Wald, R. 1975 *Commun. Math. Phys.* **45**, 9.

PHYSICAL REVIEW D VOLUME 15, NUMBER 10 15 MAY 1977

Action integrals and partition functions in quantum gravity

G. W. Gibbons* and S. W. Hawking
Department of Applied Mathematics and Theoretical Physics, University of Cambridge, England
(Received 4 October 1976)

One can evaluate the action for a gravitational field on a section of the complexified spacetime which avoids the singularities. In this manner we obtain finite, purely imaginary values for the actions of the Kerr-Newman solutions and de Sitter space. One interpretation of these values is that they give the probabilities for finding such metrics in the vacuum state. Another interpretation is that they give the contribution of that metric to the partition function for a grand canonical ensemble at a certain temperature, angular momentum, and charge. We use this approach to evaluate the entropy of these metrics and find that it is always equal to one quarter the area of the event horizon in fundamental units. This agrees with previous derivations by completely different methods. In the case of a stationary system such as a star with no event horizon, the gravitational field has no entropy.

I. INTRODUCTION

In the path-integral approach to the quantization of gravity one considers expressions of the form

$$Z=\int d[g]d[\phi]\exp\{iI[g,\phi]\}, \tag{1.1}$$

where $d[g]$ is a measure on the space of metrics g, $d[\phi]$ is a measure on the space of matter fields ϕ, and $I[g,\phi]$ is the action. In this integral one must include not only metrics which can be continuously deformed into the flat-space metric but also homotopically disconnected metrics such as those of black holes; the formation and evaporation of macroscopic black holes gives rise to effects such as baryon nonconservation and entropy production.[1-4] One would therefore expect similar phenomena to occur on the elementary-particle level. However, there is a problem in evaluating the action I for a black-hole metric because of the spacetime singularities that it necessarily contains.[5-7] In this paper we shall show how one can overcome this difficulty by complexifying the metric and evaluating the action on a real four-dimensional section (really a contour) which avoids the singularities. In Sec. II we apply this procedure to evaluating the action for a number of stationary exact solutions of the Einstein equations. For a black hole of mass M, angular momentum J, and charge Q we obtain

$$I=i\pi\kappa^{-1}(M-Q\Phi), \tag{1.2}$$

where

$$\kappa=(r_+-r_-)\,2^{-1}(r_+^{\,2}+J^2M^{-2})^{-1},$$

$$\Phi=Qr_+(r_+^{\,2}+J^2M^{-2})^{-1},$$

$$r_\pm=M\pm(M^2-J^2M^{-2}-Q^2)^{1/2}$$

in units such that

$$G=c=\hbar=k=1.$$

One interpretation of this result is that it gives a probability, in an appropriate sense, of the occurrence in the vacuum state of a black hole with these parameters. This aspect will be discussed further in another paper. Another interpretation which will be discussed in Sec. III of this paper is that the action gives the contribution of the gravitational field to the logarithm of the partition function for a system at a certain temperature and angular velocity. From the partition function one can calculate the entropy by standard thermodynamic arguments. It turns out that this entropy is zero for stationary gravitational fields such as those of stars which contain no event horizons. However, both for black holes and de Sitter space[8] it turns out that the entropy is equal to one quarter of the area of the event horizon. This is in agreement with results obtained by completely different methods.[1,4,8]

II. THE ACTION

The action for the gravitational field is usually taken to be

$$(16\pi)^{-1}\int R(-g)^{1/2}\,d^4x.$$

However, the curvature scalar R contains terms which are linear in second derivatives of the metric. In order to obtain an action which depends only on the first derivatives of the metric, as is required by the path-integral approach, the second derivatives have to be removed by integration by parts. The action for the metric g over a region Y with boundary ∂Y has the form

$$I=(16\pi)^{-1}\int_Y R(-g)^{1/2}d^4x+\int_{\partial Y} B(-h)^{1/2}\,d^3x. \tag{2.1}$$

The surface term B is to be chosen so that for metrics g which satisfy the Einstein equations the action I is an extremum under variations of the metric which vanish on the boundary ∂Y but which may have nonzero normal derivatives. This will be satisfied if $B=(8\pi)^{-1}K+C$, where K is the trace of the second fundamental form of the boundary ∂Y in the metric g and C is a term which depends only on the induced metric h, on ∂Y. The term C gives rise to a term in the action which is independent of the metric g. This can be absorbed into the normalization of the measure on the space of all metrics. However, in the case of asymptotically flat metrics, where the boundary ∂Y can be taken to be the product of the time axis with a two-sphere of large radius, it is natural to choose C so that $I=0$ for the flat-space metric η. Then $B=(8\pi)^{-1}[K]$, where $[K]$ is the difference in the trace of the second fundamental form of ∂Y in the metric g and the metric η.

We shall illustrate the procedure for evaluating the action on a nonsingular section of a complexified spacetime by the example of the Schwarzschild solution. This is normally given in the form

$$ds^2=-(1-2Mr^{-1})dt^2+(1-2Mr^{-1})^{-1}dr^2+r^2d\Omega^2. \tag{2.2}$$

This has singularities at $r=0$ and at $r=2M$. As is now well known, the singularity at $r=2M$ can be removed by transforming to Kruskal coordinates in which the metric has the form

$$ds^2=32M^3r^{-1}\exp[-r(2M)^{-1}](-dz^2+dy^2)+r^2d\Omega^2, \tag{2.3}$$

where

$$-z^2+y^2=[r(2M)^{-1}-1]\exp[r(2M)^{-1}], \tag{2.4}$$

$$(y+z)(y-z)^{-1}=\exp[t(2M)^{-1}]. \tag{2.5}$$

The singularity at $r=0$ now lies on the surface $z^2-y^2=1$. It is a curvature singularity and cannot be removed by coordinate changes. However, it can be avoided by defining a new coordinate $\zeta=iz$. The metric now takes the positive-definite or Euclidean form

$$ds^2=32M^3r^{-1}\exp[-r(2M)^{-1}](d\zeta^2+dy^2)+r^2d\Omega^2, \tag{2.6}$$

where r is now defined by

$$\zeta^2+y^2=[r(2M)^{-1}-1]\exp[r(2M)^{-1}]. \tag{2.7}$$

On the section on which ζ and y are real (the Euclidean section), r will be real and greater than or equal to $2M$. Define the imaginary time by $\tau=it$. It follows from Eq. (2.5) that τ is periodic with period $8\pi M$. On the Euclidean section τ has the character of an angular coordinate about the "axis" $r=2M$. Since the Euclidean section is nonsingular we can evaluate the action (2.1) on a region Y of it bounded by the surface $r=r_0$. The boundary ∂Y has topology $S^1\times S^2$ and so is compact.

The scalar curvature R vanishes so the action is given by the surface term

$$I=(8\pi)^{-1}\int[K]d\Sigma. \tag{2.8}$$

But

$$\int K\,d\Sigma=\frac{\partial}{\partial n}\int d\Sigma, \tag{2.9}$$

where $(\partial/\partial n)\int d\Sigma$ is the derivative of the area $\int d\Sigma$ of ∂Y as each point of ∂Y is moved an equal distance along the outward unit normal n. Thus in the Schwarzschild solution

$$\begin{aligned}\int K\,d\Sigma&=-32\pi^2M(1-2Mr^{-1})^{1/2}\\&\quad\times\frac{d}{dr}[ir^2(1-2Mr^{-1})^{1/2}]\\&=-32\pi^2iM(2r-3M).\end{aligned} \tag{2.10}$$

The factor $-i$ arises from the $(-h)^{1/2}$ in the surface element $d\Sigma$. For flat space $K=2r^{-1}$. Thus

$$\int K d\Sigma=-32\pi^2iM(1-2Mr^{-1})^{1/2}\,2r. \tag{2.11}$$

Therefore

$$\begin{aligned}I&=(8\pi)^{-1}\int[K]d\Sigma\\&=4\pi iM^2+O(M^2r_0^{-1})\\&=\pi iM\kappa^{-1}+O(M^2r_0^{-1}),\end{aligned} \tag{2.12}$$

where $\kappa=(4M)^{-1}$ is the surface gravity of the Schwarzschild solution.

The procedure is similar for the Reissner-Nordström solution except that now one has to add on the action for the electromagnetic field F_{ab}. This is

$$-(16\pi)^{-1}\int F_{ab}F^{ab}(-g)^{1/2}d^4x. \tag{2.13}$$

For a solution of the Maxwell equations, $F^{ab}{}_{;b}=0$ so the integrand of (2.13) can be written as a divergence

$$F_{ab}F_{cd}g^{ac}g^{bd}=(2F^{ab}A_a)_{;b}. \tag{2.14}$$

Thus the value of the action is

$$-(8\pi)^{-1}\int F^{ab}A_a\,d\Sigma_b. \tag{2.15}$$

The electromagnetic vector potential A_a for the Reissner-Nordström solution is normally taken to be

$$A_a = Qr^{-1} t_{;a}. \tag{2.16}$$

However, this is singular on the horizon as t is not defined there. To obtain a regular potential one has to make a gauge transformation

$$A'_a = (Qr^{-1} - \Phi) t_{;a}, \tag{2.17}$$

where $\Phi = Q(r_+)^{-1}$ is the potential of the horizon of the black hole. The combined gravitational and electromagnetic actions are

$$I = i\pi\kappa^{-1}(M - Q\Phi). \tag{2.18}$$

We have evaluated the action on a section in the complexified spacetime on which the induced metric is real and positive-definite. However, because R, F_{ab}, and K are holomorphic functions on the complexified spacetime except at the singularities, the action integral is really a contour integral and will have the same value on any section of the complexified spacetime which is homologous to the Euclidean section even though the induced metric on this section may be complex. This allows us to extend the procedure to other spacetimes which do not necessarily have a real Euclidean section. A particularly important example of such a metric is that of the Kerr-Newman solution. In this one can introduce Kruskal coordinates y and z and, by setting $\zeta = iz$, one can define a nonsingular section as in the Schwarzschild case. We shall call this the "quasi-Euclidean section." The metric on this section is complex and it is asymptotically flat in a coordinate system rotating with angular velocity Ω, where $\Omega = JM^{-1}(r_+^2 + J^2M^{-2})^{-1}$ is the angular velocity of the black hole. The regularity of the metric at the horizon requires that the point (t, r, θ, ϕ) be identified with the point $(t + i2\pi\kappa^{-1}, r, \theta, \phi + i2\pi\Omega\kappa^{-1})$. The rotation does not affect the evaluation of the $\int [K] d\Sigma$ so the action is still given by Eq. (2.18). One can also evaluate the gravitational contribution to the action for a stationary axisymmetric solution containing a black hole surrounded by a perfect fluid rigidly rotating at some different angular velocity. The action is

$$I = i2\pi\kappa^{-1}\left[(16\pi)^{-1}\int_\Sigma R K^a d\Sigma_a + 2^{-1}M\right], \tag{2.19}$$

where $K^a\partial/\partial x_a = \partial/\partial t$ is the time-translation Killing vector and Σ is a surface in the quasi-Euclidean section which connects the boundary at $r = r_0$ with the "axis" or bifurcation surface of the horizon $r = r_+$. The total mass, M, can be expressed as

$$M = M_H + 2\int_\Sigma (T_{ab} - \tfrac{1}{2} g_{ab} T) K^a d\Sigma^b, \tag{2.20}$$

where

$$M_H = (4\pi)^{-1}\kappa A + 2\Omega_H J_H. \tag{2.21}$$

M_H is the mass of the black hole, A is the area of the event horizon, and Ω_H and J_H are respectively the angular velocity and angular momentum of the black hole.[9] The energy-momentum tensor of the fluid has the form

$$T_{ab} = (p + \rho) u_a u_b + p g_{ab}, \tag{2.22}$$

where ρ is the energy density and p is the pressure of the fluid. The 4-velocity u_a can be expressed as

$$\lambda u^a = K^a + \Omega_m m^a, \tag{2.23}$$

where Ω_m is the angular velocity of the fluid, m^a is the axial Killing vector, and λ is a normalization factor. Substituting (2.21) and (2.22) in (2.20) one finds that

$$M = (4\pi)^{-1}\kappa A + 2\Omega_H J_H + 2\Omega_m J_m - \int (\rho + 3p) K^a d\Sigma_a, \tag{2.24}$$

where

$$J_m = -\int T_{ab} m^a d\Sigma^b \tag{2.25}$$

is the angular momentum of the fluid. By the field equations, $R = 8\pi(\rho - 3p)$, so this action is

$$I = 2\pi i\kappa^{-1}\left[M - \Omega_H J_H - \Omega_m J_m - \kappa A(8\pi)^{-1} + \int \rho K^a d\Sigma_a\right]. \tag{2.26}$$

One can also apply (2.26) to a situation such as a rotating star where there is no black hole present. In this case the regularity of the metric does not require any particular periodicity of the time coordinate and $2\pi\kappa^{-1}$ can be replaced by an arbitrary periodicity β. The significance of such a periodicity will be discussed in the next section.

We conclude this section by evaluating the action for de Sitter space. This is given by

$$I = (16\pi)^{-1}\int_Y (R - 2\Lambda)(-g)^{1/2} d^4x + (8\pi)^{-1}\int_{\partial Y} [K] d\Sigma, \tag{2.27}$$

where Λ is the cosmological constant. By the field equations $R = 4\Lambda$. If one were to take Y to be the ordinary real de Sitter space, i.e., the section on which the metric was real and Lorentzian, the volume integral in (2.27) would be infinite. However, the complexified de Sitter space contains a section on which the metric is the real positive-definite metric of a 4-sphere of radius $3^{1/2}\Lambda^{-1/2}$. This Euclidean section has no boundary so that the value of this action on it is

$$I=-12\pi i\Lambda^{-1}, \tag{2.28}$$

where the factor of $-i$ comes from the $(-g)^{1/2}$.

III. THE PARTITION FUNCTION

In the path-integral approach to the quantization of a field ϕ one expresses the amplitude to go from a field configuration ϕ_1 at a time t_1 to a field configuration ϕ_2 at time t_2 as

$$\langle \phi_2, t_2 | \phi_1, t_1\rangle = \int d[\phi]\exp(iI[\phi]), \tag{3.1}$$

where the path integral is over all field configurations ϕ which take the values ϕ_1 at time t_1 and ϕ_2 at time t_2. But

$$\langle \phi_2, t_2 | \phi_1, b_1\rangle = \langle \phi_2 | \exp[-iH(t_2-t_1)] | \phi_1\rangle, \tag{3.2}$$

where H is the Hamiltonian. If one sets $t_2-t_1=-i\beta$ and $\phi_1=\phi_2$ and the sums over all ϕ_1 one obtains

$$\operatorname{Tr}\exp(-\beta H)=\int d[\phi]\exp(iI[\phi]), \tag{3.3}$$

where the path integral is now taken over all fields which are periodic with period β in imaginary time. The left-hand side of (3.3) is just the partition function Z for the canonical ensemble consisting of the field ϕ at temperature $T=\beta^{-1}$. Thus one can express the partition function for the system in terms of a path integral over periodic fields.[10] When there are gauge fields, such as the electromagnetic or gravitational fields, one must include the Faddeev-Popov ghost contributions to the path integral.[11-13]

One can also consider grand canonical ensembles in which one has chemical potentials μ_i associated with conserved quantities C_i. In this case the partition function is

$$Z=\operatorname{Tr}\exp\left[-\beta\left(H-\sum_i \mu_i C_i\right)\right]. \tag{3.4}$$

For example, one could consider a system at a temperature $T=\beta^{-1}$ with a given angular momentum J and electric charge Q. The corresponding chemical potentials are then Ω, the angular velocity, and Φ, the electrostatic potential. The partition function will be given by a path integral over all fields ϕ whose value at the point $(t+i\beta, r, \theta, \phi+i\beta\Omega)$ is $\exp(q\beta\Phi)$ times the value at (t, r, θ, ϕ), where q is the charge on the field.

The dominant contribution to the path integral will come from metrics g and matter fields ϕ which are near background fields g_0 and ϕ_0 which have the correct periodicities and which extremize the action, i.e., are solutions of the classical field equations. One can express g and ϕ as

$$g=g_0+\bar{g}, \quad \phi=\phi_0+\bar{\phi} \tag{3.5}$$

and expand the action in a Taylor series about the background fields

$$I[g,\phi]=I[g_0,\phi_0]+I_2[\bar{g}]+I_2[\bar{\phi}] + \text{higher-order terms}, \tag{3.6}$$

where $I_2[\bar{g}]$ and $I_2[\bar{\phi}]$ are quadratic in the fluctuations $\bar{g}$ and $\bar{\phi}$. If one neglects higher-order terms, the partition function is given by

$$\ln Z = iI[g_0,\phi_0]+\ln\int d[\bar{g}]\exp(iI_2[\bar{g}]) + \ln\int d[\bar{\phi}]\exp(iI_2[\bar{\phi}]). \tag{3.7}$$

But the normal thermodynamic argument

$$\ln Z=-WT^{-1}, \tag{3.8}$$

where $W=M-TS-\sum_i \mu_i C_i$ is the "thermodynamic potential" of the system. One can therefore regard $iI[g_0,\phi_0]$ as the contribution of the background to $-WT^{-1}$ and the second and third terms in (3.7) as the contributions arising from thermal gravitons and matter quanta with the appropriate chemical potentials. A method for evaluating these latter terms will be given in another paper.

One can apply the above analysis to the Kerr-Newman solutions because in them the points (t, r, θ, ϕ) and $(t+2\pi i\kappa^{-1}, r, \theta, \phi+2\pi i\Omega\kappa^{-1})$ are identified (the charge q of the graviton and photon are zero). It follows that the temperature T of the background field is $\kappa(2\pi)^{-1}$ and the thermodynamic potential is

$$W=\tfrac{1}{2}(M-\Phi Q), \tag{3.9}$$

but

$$W=M-TS-\Phi Q-\Omega J. \tag{3.10}$$

Therefore

$$\tfrac{1}{2}M=TS+\tfrac{1}{2}\Phi Q+\Omega J, \tag{3.11}$$

but by the generalized Smarr formula[9,14]

$$\tfrac{1}{2}M=\kappa(8\pi)^{-1}A+\tfrac{1}{2}\Phi Q+\Omega J. \tag{3.12}$$

Therefore

$$S=\tfrac{1}{4}A, \tag{3.13}$$

in complete agreement with previous results.

For de Sitter space

$$WT^{-1}=-12\pi\Lambda^{-1}, \tag{3.14}$$

but in this case $W=-TS$, since $M=J=Q=0$ because this space is closed. Therefore

$$S=12\pi\Lambda^{-1}, \tag{3.15}$$

which again agrees with previous results. Note that the temperature T of de Sitter space cancels out the period. This is what one would expect since the temperature is observer dependent and related to the normalization of the timelike Killing vector.

Finally we consider the case of a rotating star in equilibrium at some temperature T with no event horizons. In this case we must include the contribution from the path integral over the matter fields as it is these which are producing the gravitational field. For matter quanta in thermal equilibrium at a temperature T volume $V \gg T^{-3}$ of flat space the thermodynamic potential is given by

$$WT^{-1}=-i\int p(-\eta)^{1/2}d^4x=-pVT^{-1}. \tag{3.16}$$

In situations in which the characteristic wavelengths, T^{-1}, are small compared to the gravitational length scales it is reasonable to use this fluid approximation for the density of thermodynamic potential; thus the matter contributing to the thermodynamic potential will be given by

$$W_m T^{-1}=-i\int p(-g)^{1/2}d^4x=T^{-1}\int pK^a d\Sigma_a \tag{3.17}$$

(because of the signature of our metric $K^a d\Sigma_a$ is negative), but by Eq. (2.26) the gravitational contribution to the total thermodynamic potential is

$$W_g=M-\Omega_m J_m+\int_\Sigma \rho K^a d\Sigma_a . \tag{3.18}$$

Therefore the total thermodynamic potential is

$$W=M-\Omega_m J_m+\int_\Sigma (p+\rho)K^a d\Sigma_a , \tag{3.19}$$

but

$$p+\rho=\bar{T}s+\sum_i \bar{\mu}_i n_i , \tag{3.20}$$

where $\bar{T}$ is the local temperature, s is the entropy density of the fluid, $\bar{\mu}_i$ is the local chemical potentials, and n_i is the number densities of the ith species of particles making up the fluid. Therefore

$$W=M-\Omega_m J_m+\int_\Sigma \left(\bar{T}s+\sum_i \bar{\mu}_i n_i\right)K^a d\Sigma_a . \tag{3.21}$$

In thermal equilibrium

$$\bar{T}=T\lambda^{-1}, \tag{3.22}$$

$$\bar{\mu}_i=\mu_i\lambda^{-1}, \tag{3.23}$$

where T and μ_i are the values of $\bar{T}$ and $\bar{\mu}_i$ at infinity.[9] Thus the entropy is

$$S=-\int su^a d\Sigma_a . \tag{3.24}$$

This is just the entropy of the matter. In the absence of the event horizon the gravitational field has no entropy.

---

*Present address: Max-Planck-Institute für Physik und Astrophysik, 8 München 40, Postfach 401212, West Germany. Telephone: 327001.

[1]S. W. Hawking, Commun. Math. Phys. 43, 199 (1975).
[2]R. M. Wald, Commun. Math. Phys. 45, 9 (1975).
[3]S. W. Hawking, Phys. Rev. D 14, 2460 (1976).
[4]S. W. Hawking, Phys. Rev. D 13, 191 (1976).
[5]R. Penrose, Phys. Rev. Lett. 14, 57 (1965).
[6]S. W. Hawking and R. Penrose, Proc. R. Soc. London A314, 529 (1970).
[7]S. W. Hawking and G. F. R. Ellis, *The Large Scale Structure of Spacetime* (Cambridge Univ. Press, Cambridge, England, 1973).
[8]G. W. Gibbons and S. W. Hawking, preceding paper, Phys. Rev. D 15, 2738 (1977).
[9]J. Bardeen, B. Carter, and S. W. Hawking, Commun. Math. Phys. 31, 161 (1973).
[10]R. P. Feynman and Hibbs, *Quantum Mechanics and Path Integrals* (McGraw-Hill, New York, 1965).
[11]C. W. Bernard, Phys. Rev. D 9, 3312 (1974).
[12]L. Dolan and R. Jackiw, Phys. Rev. D 9, 3320 (1974).
[13]L. D. Faddeev and V. N. Popov, Phys. Lett. 25B, 29 (1967).
[14]L. Smarr, Phys. Rev. Lett. 30, 71 (1973); 30, 521(E) (1973).

Instability of flat space at finite temperature

David J. Gross and Malcolm J. Perry
Department of Physics, Princeton University, Princeton, New Jersey 08544

Laurence G. Yaffe
Department of Physics, California Institute of Technology, Pasadena, California 91109
(Received 29 June 1981)

The instabilities of quantum gravity are investigated using the path-integral formulation of Einstein's theory. A brief review is given of the classical gravitational instabilities, as well as the stability of flat space. The Euclidean path-integral representation of the partition function is employed to discuss the instability of flat space at finite temperature. Semiclassical, or saddle-point, approximations are utilized. We show how the Jeans instability arises as a tachyon in the graviton propagator when small perturbations about hot flat space are considered. The effect due to the Schwarzschild instanton is studied. The small fluctuations about this instanton are analyzed and a negative mode is discovered. This produces, in the semiclassical approximation, an imaginary part of the free energy. This is interpreted as being due to the metastability of hot flat space to nucleate black holes. These then evolve by evaporation or by accretion of thermal gravitons, leading to the instability of hot flat space. The nucleation rate of black holes is calculated as a function of temperature.

I. INTRODUCTION

Gravity, unlike the other fundamental forces of nature, is universally attractive and cannot be screened. This property of gravity, to which we owe our ability to detect this incredibly weak interaction, is the source of many instabilities.

The instability of gravity already appears in classical Newtonian theory. As Jeans[1] showed, a static, homogeneous nonrelativistic fluid is unstable under long-wavelength gravitational perturbations.[2] Consider a nonviscous fluid of mass density ρ, pressure p, and velocity $\vec{v}$ that satisfies the equation of continuity and the Navier-Stokes equation

$$\frac{\partial\rho}{\partial t}+\vec{\nabla}\cdot(\rho\vec{v})=0\,,\qquad \frac{\partial\vec{v}}{\partial t}+(\vec{v}\cdot\vec{\nabla})\vec{v}=-\frac{1}{\rho}\vec{\nabla}p-\vec{g}\,, \tag{1.1}$$

where $\vec{g}$ is the gravitational field, given by

$$\vec{\nabla}\times\vec{g}=0\,,\quad \vec{\nabla}\cdot\vec{g}=-4\pi G\rho\,. \tag{1.2}$$

One now considers small perturbations $\delta\rho$, δp, $\delta\vec{v}$, and $\vec{g}$ about the static, homogeneous nongravitating fluid of constant density ρ and pressure p. One finds that they are governed by the equation

$$\frac{\partial^2\delta\rho}{\partial t^2}-V_s^{\,2}\nabla^2\delta\rho=4\pi G\rho\delta\rho\,, \tag{1.3}$$

where V_s is the speed of sound in the fluid ($V_s^{\,2}=\partial P/\partial\rho$). Note that the right-hand side of Eq. (1.3) has the form of a "mass term," but with the wrong sign. Therefore the solutions of this equation

$$\delta\rho=C\exp(i\vec{k}\cdot\vec{x}-i\omega t)\,,\qquad \omega=(\vec{k}^2V_s^{\,2}-4\pi G\rho)^{1/2} \tag{1.4}$$

will grow exponentially if the wave number k is less than k_J, where

$$k_J=(4\pi G\rho/V_s^{\,2})^{1/2}\,. \tag{1.5}$$

This instability is due to the attractive nature of gravity, which, in contrast to the damping of charge density fluctuations in a plasma, "antiscreens" mass density fluctuations thus leading to their amplification.

The same instability occurs if we consider a gas in isothermal equilibrium in a finite volume. Suppose that we have a spherical ball of perfect gas that is in isothermal equilibrium under its own gravitational field.[3,4] At each point, the equation of state is

$$p=\rho T/m \,, \tag{1.6}$$

where p is the pressure, ρ the mass density, T the temperature, and m is the mass of each molecule. The equation of hydrostatic equilibrium for such a system is

$$\frac{1}{r^2}\frac{d}{dr}\left[\frac{r^2}{\rho}\frac{dp}{dr}\right]=-4\pi G\rho \,, \tag{1.7}$$

where r is the distance from the center of the cloud. At the origin, $r=0$, $\rho'(r)=0$. The final boundary condition is the value of the pressure at the outermost edge of the cloud which must be specified. These conditions determine the solution to (1.6) and (1.7). Suppose we surround the cloud by a spherical box of variable radius R, with external pressure P, containing N molecules, and whose walls are fixed at temperature T. Then it follows that

$$\left[\frac{\partial P}{\partial R}\right]_{N,T}=-\frac{P(8\pi PR^4-Gm^2N^2)}{R^2(4\pi PR^2-NT)} \tag{1.8}$$

along the curve of equilibrium.

For R sufficiently large, then $(\partial P/\partial R)<0$ as is usually expected. However, if R is decreased, then the pressure increases to a maximum of

$$P=\frac{Gm^2N^2}{8\pi R^4} \,. \tag{1.9}$$

Thereafter an instability sets in since $(\partial p/\partial R)>0$. Any small fluctuation in the gas that decreases the pressure will decrease the volume of the box. This will decrease the pressure further, and so the system is unstable.

It is essentially this mechanism that causes stars to condense out of clouds of interstellar gas. This instability is terminated when a cloud becomes sufficiently hot that nuclear reactions are initiated and support the cloud against further collapse. Again the basic reason for this instability is due to attractive nature of gravitation.

In the general theory of relativity, as formulated by Einstein, gravitational instabilities are even more severe, since the spacetime manifold is warped by the presence of matter. Gravitational collapse can give rise to singularities in the fabric of spacetime. Thus, for example, a star with a mass greater than the Oppenheimer-Volkoff limit[5] (currently estimated to be about 1.4 $M_\odot$)[6] cannot support itself against gravitational collapse. After collapsing to a white dwarf (a degenerate gas of electrons) and then to a neutron star (a degenerate gas of neutrons), it will continue to collapse.

It seems that this collapse will lead first to the formation of an event horizon and then to a spacetime singularity.[7] Thus, a black hole is formed as a direct result of this type of instability. (It could be that a spacetime singularity is formed without an event horizon: a naked singularity. Such behavior is ruled out by the cosmic censorship hypothesis.[8] However, this conjecture has not been proved.) The type of singularity associated with a black hole is relatively mild. If however we assume that the universe is well described by a Friedmann-Robertson-Walker model with a density larger than the critical density of 2×10^{-29} g cm^{-3}, which has not been observationally ruled out, then the universe itself is unstable in the sense that it too must undergo collapse leading to a spacelike singularity to the future of all observers.

Given the inevitable instabilities of gravity one might worry about the stability of the ground state of quantum gravity. One is accustomed to regarding Minkowski space as the ground state, or vacuum, of quantum gravity. Small perturbations about this vacuum are certainly stable; however one might find that flat space is quantum mechanically unstable. This would occur if the "potential" for gravity had the form given in Fig. 1, so that the metastable vacuum A would decay by tunneling through a barrier to some configuration B. This concern is nontrivial since there is no way to define a local energy density in gravity whose positivity ensures stability. Moreover the Einstein action is not positive definite, even when continued

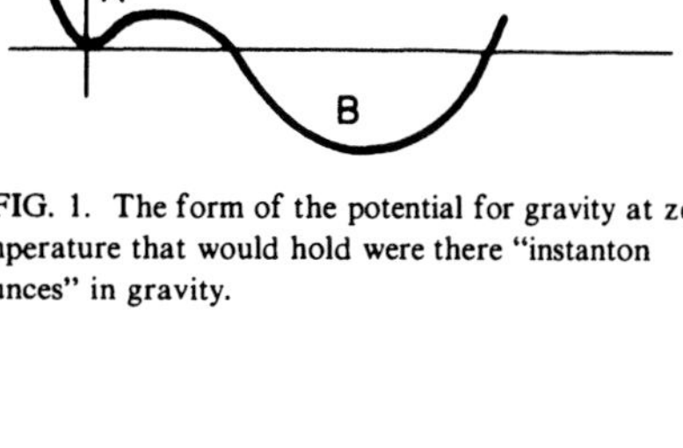

FIG. 1. The form of the potential for gravity at zero temperature that would hold were there "instanton bounces" in gravity.

to Euclidean space (i.e., the curvature can be arbitrarily large and positive or negative). Thus the Euclidean functional integral formalism of quantum gravity[9] is not well defined. One might be worried that this implies the nonexistence of a ground state.

It is remarkable that the issue of the stability of flat space was only settled recently. First Schoen and Yau[10] proved that there are no instanton bounces, i.e., solutions of Einstein's equations with Euclidean metric that asymptotically approach flat space, which would represent the tunneling through a barrier. (We shall return to discuss the meaning of such instantons in Sec. II.) This implies that flat space is stable under processes that can be treated by semiclassical (or WKB) methods. Second, Schoen and Yau[11] proved the long-outstanding conjecture that the total energy (defined in terms of the asymptotic behavior of the gravitational field) of asymptotically flat manifolds (which satisfy Einstein's equations in the presence of matter that itself has positive energy in any frame) is positive semidefinite, and that only Minkowski space has zero energy.[12] This is the positive-energy theorem, which since energy is conserved seems to preclude the possibility of flat space decaying by any mechanism.

In this paper we explore the instabilities of Minkowski space at a finite temperature. Here, unlike the case of zero temperature, we find two distinct sources of instability. These imply that hot flat space is unstable and will decay. This result obtains for any finite, however small, temperature.

One source of instability is expected on the basis of the classical Jeans analysis. Hot flat space is not empty. It contains a gas of gravitons in thermal equilibrium, which are a source for the gravitational field. For nonvanishing temperature, flat space contains "matter." Thus one expects a Jeans instability to occur for large-wavelength density fluctuations of the thermal gravitons. We shall show how this instability arises when one uses functional integral techniques to examine the partition function for quantum gravity.

The second source of instability is the nucleation of black holes. This is a quantum effect, which cannot be understood on the classical level. We discovered this effect by investigating the contribution of Euclidean instantons (i.e., the Euclidean section of the Schwarzschild metric) to the functional integral. These instantons have been discussed by many authors, notably Hawking who attempted to use them to deduce the thermodynamic properties of black holes.[13] Our interpretation of their meaning is somewhat different. We found that there exist small fluctuations about these instantons that decrease the action.[14] These give rise to an imaginary part in the free energy of flat space, which is to be interpreted as a finite lifetime for decay. The decay proceeds by thermal (= quantum) fluctuations nucleating black holes of radius $R=\hbar c/4\pi kT$ and mass $=\hbar c^3/8\pi GkT$ (where T is the temperature). These then either expand or contract by absorption or emission of gravitational radiation.

We shall be discussing throughout the paper the quantum version of Einstein's theory. As is well known, this theory is problematical at high energies or small distances due to its nonrenormalizability. However we believe that our considerations should be insensitive to possible short-distance modifications of the theory (e.g., R^2 or $R^{ab}R_{ab}$ terms in the action, supergravity, etc.) as long as we restrict the temperature T to be small (in natural units, $kT << m_P c^2$). We shall use units in which $\hbar=c=k=1$, and often will express masses, inverse lengths, and temperatures in units of the Planck mass $m_P \equiv (\hbar c/G)^{1/2}$ or $m_P = 2.18\times 10^{-5}$ g. When discussing spacetime, we shall use the signature $(-+++)$.

The main purpose of this paper is to use the Euclidean functional integral formalism to explore the instabilities of gravity. It is evident that gravity must exhibit instabilities by the arguments of this section. The Jeans instability, of course, has been known since 1902. Also, it is clear that a second instability should exist. If we imagine a box whose walls are kept at constant temperature, then it will certainly be filled with thermal gravitons. Now if we assume that this system is in some sense ergodic, then all points in phase space will eventually be reached from any reasonable initial state, for example, from a configuration of thermal gravitons. In particular the spontaneous formation of a black hole (as opposed to collapse via the Jeans mechanism) would appear to be a possible process. If this process can happen, it will happen. However, this argument does not yield the rate for such events to occur. One of the main results of our work is the calculation of this rate from the Euclidean formulation of gravity.

Black-hole formation, via nucleation or via the Jeans instability, renders the canonical ensemble ill defined. This is because, even classically, some trajectories run into the boundaries of the allowed region of phase space. Such a catastrophe happens when a black hole starts to grow and engulfs the box in which the system is contained. One way out of this impasse would be to use the microcanonical ensemble to discuss the thermodynamic properties of gravity. We do not know how to carry out such a program.

Although this paper deals with the instabilities of self-gravitating systems its conclusions bear on an issue of much greater importance. In the functional integral formulation of quantum gravity the following question arises. Should one, when summing over path histories of spacetime, include the contributions of all possible metrics, independent of their topology? In ordinary field theories (e.g., gauge theories) the issue of which topological classes of field configurations must be included in the path integral is resolved by energy considerations. Thus for example instantons must be included in the Euclidean path integral (for Yang-Mills theories) in order to construct the correct ground state. Here one can show that this must be so by continuously deforming the naive vacuum into a widely separated instanton-antiinstanton pair with a finite cost in action. Thus one can smoothly construct a configuration that has a nonzero topological charge in any given finite volume.

In gravity, however, energy considerations are notoriously problematic. Furthermore the topology of spacetime is not additive—one can add handles to a manifold (with a positive increase of the Euler character) but there are no corresponding "antihandles." In fact there is no convincing argument that one *must* include anything but continuous deformations of flat space in the path integral. It would be very desirable to investigate a set of physical situations in which this issue arises. Given that one has little hope, with present techniques, of going beyond semiclassical approximations this means that one must come up with very special situations. These must have the property that they give rise to boundary conditions for the Euclidean functional integral that allow for the existence of finite-action gravitational saddle points (instantons) with nontrivial topology. The only case we know of where this condition is met is the example of hot flat space discussed below.

In the case of the canonical ensemble the boundary conditions for the Euclidean functional integral for the partition function do indeed allow two kinds of saddle points. There is, of course, the topologically trivial flat-space saddle point, but also the Schwarzschild instanton with nonzero Euler character. Thus we can address the important issue of whether one should or should not include strange topologies in the path integral in the context of a well-defined calculation.

As described below, we find that the effect of summing over Schwarzschild instantons is to produce a totally reasonable physical process, namely, the nucleation at finite temperature of black holes. If we were to ignore these topologically nontrivial manifolds there would be no mechanism, in the semiclassical limit, of producing this expected process. We therefore present this calculation as evidence that one *should* include nontrivial topologies in the path integral, and that no strange effects need emerge.

The remainder of the article is arranged as follows. In Sec. II we discuss various quantum-mechanical systems at finite temperatures, by exploring the functional integral representation of the partition function. We discuss the stability of various systems. We show how many of the features of the finite-temperature behavior can be calculated using semiclassical approximations. We then examine the issue of stability in gravitation. In Sec. III, we show that the functional integral can be used to explore the gravitational vacuum at zero temperature. Perturbation theory is discussed, and we conclude that, at least semiclassically, the theory is stable. In Sec. IV, we extend our treatment to perturbations of flat space at finite temperature. We show how the Jeans instability emerges in the language of quantum field theory. In Sec. V, we discuss the role of gravitational instantons in the finite-temperature case. We discover that there is a further instability associated with the nucleation of black holes. This process cannot be described by perturbations of flat spacetime. Finally, we discuss some of the physical consequences of this instability in Sec. VI.

II. TOY MODELS

In this section we consider quantum theories describing a single degree of freedom. In particular, we discuss how certain features of the finite-temperature behavior of a system may be extracted from the Euclidean function integral formalism. Most of this material is, or should be, well known.[15] Our intention is to remind the reader of various points which will prove useful when we apply familiar techniques in an unfamiliar context.

The finite-temperature theory is defined as that given by the canonical ensemble. Thus, the density matrix, $\exp(-\beta\hat{H})$ ($\beta \equiv 1/T$, T is the temperature) represents the equilibrium behavior of the system that is weakly coupled to an external heat bath. All thermodynamic quantities may be extracted from the partition function

$$Z \equiv \mathrm{Tr}[\exp(-\beta\hat{H})] . \tag{2.1}$$

For example, the free energy $\mathcal{F} \equiv -\beta^{-1}\ln Z$. The expected value of any observable $\mathcal{O}$ is given by

$$\langle \mathcal{O} \rangle = \mathrm{Tr}[\mathcal{O}\exp(-\beta\hat{H})]/Z . \tag{2.2}$$

Functional integral representations may be derived for these quantities by repeatedly inserting a complete set of states. If the Hamiltonian has the standard form $\hat{H} = \frac{1}{2}\hat{p}^2 + V(\hat{x})$, then the partition function is given by

$$Z = \lim_{N\to\infty} \int \frac{dp_1}{2\pi\hbar} \cdots \frac{dp_N}{2\pi\hbar} dx_1 \cdots dx_N \left\langle p_N \left| \left[1 - \frac{\epsilon}{\hbar}\hat{H} \right] \right| x_N \right\rangle \langle x_N | p_{N-1} \rangle$$
$$\times \left\langle p_{N-1} \left| \left[1 - \frac{\epsilon}{\hbar}\hat{H} \right] \right| x_{N-1} \right\rangle \cdots \left\langle p_1 \left| \left[1 - \frac{\epsilon}{\hbar}\hat{H} \right] \right| x_1 \right\rangle \langle x_1 | p_N \rangle$$
$$= \lim_{N\to\infty} \int \left[\frac{dp_i}{2\pi\hbar} \right] (dx_i) \exp\left[-\frac{\epsilon}{\hbar} \sum_i \left[\tfrac{1}{2}p_i^2 + V(x_i) + ip_i(x_{i+1} - x_i)/\epsilon \right] \right]$$
$$= \lim_{N\to\infty} \int \frac{(dx_i)}{(2\pi\hbar\epsilon)^{1/2}} \exp\left\{ -\frac{\epsilon}{\hbar} \sum_i \left[\frac{1}{2} \left[\frac{x_{i+1} - x_i}{\epsilon} \right]^2 + V(x_i) \right] \right\}$$
$$\equiv \int_{x(0)=x(\beta\hbar)} \mathcal{D}X(t) \exp\left[-\frac{1}{\hbar} \int_0^{\beta\hbar} dt [\tfrac{1}{2}\dot{x}(t)^2 + V(x(t))] \right] . \tag{2.3}$$

Here $\epsilon \equiv \beta\hbar/N$ and $x_{N+1} \equiv x_1$. The argument of the exponential is $(-1/\hbar)$ times the Euclidean action, $S_E \equiv \int_0^{\beta\hbar} dt [\frac{1}{2}\dot{x}^2 + V(x)]$, and the integral is over all periodic trajectories with period $\beta\hbar$.

If we rescale time $t \to \beta\hbar\tau$, then the action becomes

$$\frac{1}{\hbar} S_{\rm cl} = \beta \int_0^1 d\tau \{ \tfrac{1}{2}[x(\tau)/\partial\tau]^2/(\beta\hbar)^2 + V(x(\tau)) \} .$$

This shows that the action of any nonstatic trajectory becomes arbitrarily large as $(\beta\hbar) \to 0$. Therefore, for small $(\beta\hbar)$ the integral is highly peaked about static trajectories, and in the $(\beta\hbar) \to 0$ limit the quantum partition function reduces to the classical result

$$Z_{\rm cl} = \int dx \exp[-\beta V(x)] .$$

This limit may be regarded as either the classical ($\hbar \to 0$, β fixed) limit, or the high-temperature ($\beta \to 0$, $\hbar$ fixed) limit; the important fact is that the temperature becomes arbitrarily large compared to the spacing between quantum levels.

Typically of more interest is the semiclassical limit ($\hbar \to 0$, $\beta\hbar$ fixed). This is equivalent to the weak-coupling, $g^2 \to 0$, limit if we replace $V(x)$ by the rescaled potential $(1/g^2)V(gx)$. In this limit the integrand is highly peaked about trajectories which minimize the classical action. These are periodic trajectories obeying the Euclidean equations of motion

$$-\ddot{x}(t) + V'(x(t)) = 0 . \tag{2.4}$$

Suppose that the potential has the simple form shown in Fig. 2, with $V(x)$ having a single minimum with positive curvature:

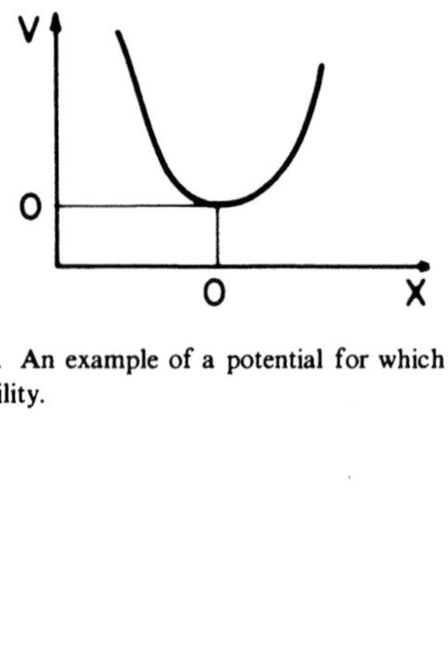

FIG. 2. An example of a potential for which there is no instability.

$$V(x) \approx \tfrac{1}{2}\omega^2 x^2 + O(x^3) .$$

Then the only classical trajectory obeying (2.4) is simply $x(t)=0$. Expanding about this trajectory yields

$$Z = \int_{\text{periodic}} \mathscr{D}x(t) \exp\left\{-\frac{1}{\hbar}\int_0^{\beta\hbar} dt[\tfrac{1}{2}(\dot{x}^2+\omega^2x^2) + O(x^3)]\right\}$$

$$= [\det{}_+(-\partial_t{}^2+\omega^2)]^{-1/2}[1+O(\hbar)]$$

$$= \left[\frac{e^{-\beta\hbar\omega/2}}{1-e^{-\beta\hbar\omega}}\right][1+O(\hbar)] . \qquad (2.5)$$

(Here, $\det_+(-\partial_t{}^2+\omega^2)$ indicates the determinant of $(-\partial_t{}^2+\omega^2)$ on the interval $[0,\beta h]$ with periodic boundary conditions. See Appendix A for the evaluation of such functional determinants.) This, of course, is the standard result for a harmonic oscillator. Higher-order terms in the saddle-point expansion around $x(t)=0$ yield an asymptotic series in powers of $\hbar$.

Now consider a potential of the form shown in Fig. 3. $x=0$ is now only a local minimum of the potential. The genuine equilibrium thermodynamic behavior of the theory obviously depends crucially on the behavior of the potential for $x>b$. A particle initially placed to the left of the barrier will have a finite probability of escaping. However, if the associated lifetime is very long, then it makes sense to speak of the metastable quasiequilibrium state describing particles confined within the potential well. Two different effects contribute to the finite lifetime of this state, quantum-mechanical tunneling through the barrier, and classical thermal excitation over the barrier. At sufficiently low temperatures tunneling will dominate, while at high temperatures thermal excitation will dominate.

To calculate the thermodynamic properties of this metastable state, one may begin by expanding the functional integral (2.3) about the local minimum $x(t)=0$. This yields a free energy

$$\mathscr{F} = \tfrac{1}{2}\hbar\omega\{1+(2/\beta\hbar\omega)\ln[1-\exp(-\beta\hbar\omega)]\}+O(\hbar^2) . \qquad (2.6)$$

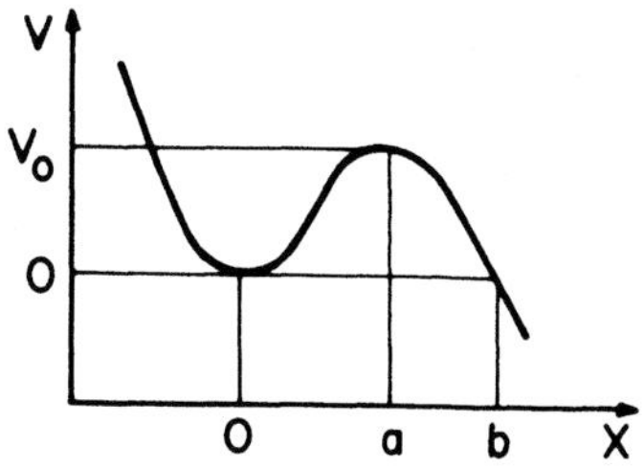

FIG. 3. An example of a potential that exhibits an instability. A particle with zero energy at $x=0$ can tunnel into region $x>b$.

$\mathscr{F}$ is real to all orders in the perturbative expansion around $x=0$: no sign of any instability is found. However, $x(t)=0$ is no longer the only periodic solution of the Euclidean equations of motion (2.4). Since these equations are identical to the usual classical equations of a point particle moving in the potential $-V(x)$, it is trivial to see that other periodic solutions exist which describe a particle oscillating back and forth under the barrier. The period of these trajectories is given by

$$\tau(E) = 2\int_{x_1}^{x_2} dx\{2[V(x)-E]\}^{1/2} , \qquad (2.7)$$

where $E \equiv V(x) - \tfrac{1}{2}\dot{x}^2$ is the conserved energy. If $V(x) \sim V_0 - \tfrac{1}{2}\omega_0 x^2$ near $x=a$, then $\tau(E)$ varies from $+\infty$ down to $2\pi/\omega_0$ as E varies from 0 to V_0. Since the only trajectories which contribute to the functional integral are those with period $\beta\hbar$, one finds that for temperatures $0 \le kT \le \hbar\omega_0/2\pi \equiv \beta_0{}^{-1}$ there is another extrema of the functional integral given by the periodic trajectory $x=\bar{x}(t)$ for which $\tau(E)=\beta\hbar$. These solutions are commonly called "bounces" (see Fig. 4).

At temperatures above the critical temperature $\beta_0{}^{-1}$ the periodic bounce degenerates to the static solution, $\bar{x}(t)=a$, which simply sits at the top of the barrier.

One may now try to evaluate the contribution to the functional integral coming from the neighborhood of the bounce. Expanding in $\delta \equiv x - \bar{x}(t)$, one finds

$$\delta Z = \exp\{-S_E[\bar{x}(t)]/\hbar\} \int [\mathscr{D}\delta(t)] \exp\left\{-\frac{1}{2\hbar}\int_0^{\beta\hbar} dt\, \delta[-\partial_t{}^2 + V''(\bar{x}(t))]\delta\right\} + O(\delta^3)$$

$$= \exp\{-S_E[\bar{x}(t)]/\hbar - \tfrac{1}{2}\ln\det{}_+[-\partial_t{}^2 + V''(\bar{x}(t)) + O(\hbar)]\} , \qquad (2.8)$$

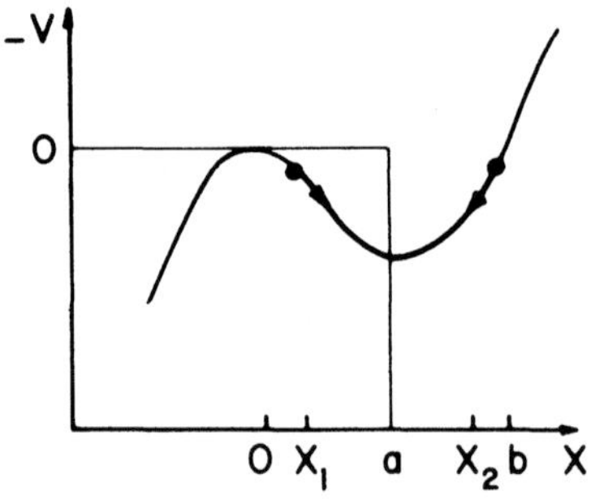

FIG. 4. A plot of $-V$, V being defined as in Fig. 3. A particle can oscillate between x_1 and x_2, the so-called "bounce" solution.

where

$$S_E[\bar{x}(t)]=\begin{cases}\beta\hbar E+2\int_{x_1}^{x_2}dx\{2[V(x)-E]\}^{1/2}, & \beta>\beta_0\\ \beta\hbar V_0\,, & \beta<\beta_0\,.\end{cases}$$

Although formally correct, this result is ill defined. The quadratic operator $\hat{\mathscr{M}}\equiv-\partial_t{}^2+V''(\bar{x}(t))$ is not positive definite. For $\beta<\beta_0$, $\hat{\mathscr{M}}$ equals $-\partial_t{}^2-\omega_0{}^2$ and the lowest eigenvector, $\delta(t)$ = const, has a negative eigenvalue $-\omega_0{}^2$. For $\beta>\beta_0$, $\delta(t)=\dot{\bar{x}}(t)$ is a zero mode of $\hat{\mathscr{M}}$, $\hat{\mathscr{M}}\dot{\bar{x}}=\partial_t[-\ddot{\bar{x}}+V'(\bar{x}(t))]=0$. Since $\dot{\bar{x}}(t)$ changes sign there must be a lower eigenvector with a negative eigenvalue. This shows that $\bar{x}(t)$ is only a saddle point, not a minimum, of $S_E[x(t)]$ and consequently the Gaussian integral leading to (2.8) was actually divergent.

These negative modes could have been predicted from the outset if we had been more precise about the meaning of the metastable state. A careful definition always requires a process of analytic continuation. Suppose we began with a potential of the form shown in Fig. 5, where $x=0$ is the global minimum. The functional integral is perfectly well defined and $\mathscr{F}$ is obviously real. If we now analytically change the potential back to the desired form (Fig. 3), then certain contours of integration in function space will have to be rotated in order for the integral to remain convergent. Examining this in slightly greater detail allows one to show that half of the contour of integration away from the bounce in the direction of the zero mode must be rotated to the imaginary direction.[16,17] Therefore for $\beta<\beta_0$ one finds

$$\begin{aligned}\mathrm{Im}\mathscr{F}&=Z^{-1}(2\beta)^{-1}\,|\det{}_+(-\partial_t{}^2-\omega_0{}^2)\,|^{-1/2}\\&\quad\times\exp(-\beta V_0)\\&=[(\sinh\beta\hbar\omega/2)/(2\beta\sin\beta\hbar\omega_0/2)]\\&\quad\times\exp(-\beta V_0)\,.\end{aligned}\tag{2.9}$$

For $\beta>\beta_0$, one must also deal with the zero mode of $\hat{\mathscr{M}}$. However, this is a simple consequence of the fact that $\bar{x}(t)$ is not time translation invariant. To preserve the symmetry one must simultaneously expand about all time translates of the bounce, $\bar{x}(t+t_0)$, $0\le t_0<\beta h$. This is accomplished by standard collective-coordinate techniques, which automatically remove the zero mode of $\hat{\mathscr{M}}$.[17] One finds

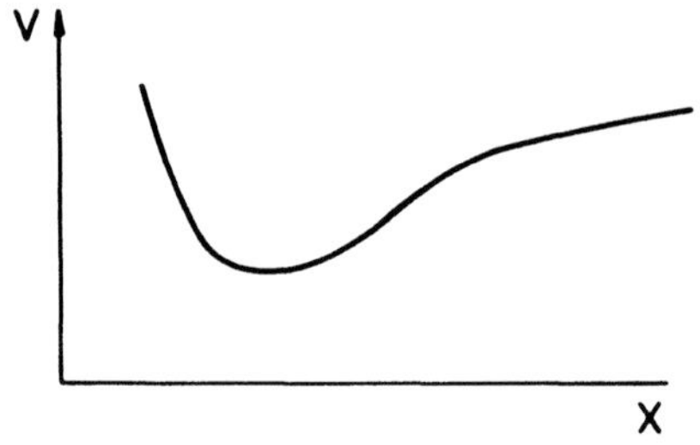

FIG. 5. A potential from which the potential shown in Fig. 3 might be obtained by some suitable analytic continuation.

$$\begin{aligned}\mathrm{Im}\mathscr{F}&=\tfrac{1}{2}\hbar Z^{-1}(W/2\pi\hbar)^{1/2}\,|\det{}'_+[-\partial_t{}^2+V''(\bar{x}(t))]\,|^{-1/2}\exp\{-S_E[\bar{x}(t)]/\hbar\}\\&=[(\sinh\beta h\omega/2)/(2\pi\tau'/\hbar)^{1/2}]\exp\{-S_E[\bar{x}(t)]/\hbar\}\,.\end{aligned}\tag{2.10}$$

Here $W\equiv\int_0^{\beta h}dt\,[\dot{\bar{x}}(t)]^2=2\int_{x_1}^{x_2}dx\{2[V(x)-E]\}^{1/2}$, and $\tau'=d\tau(E)/dE$. ($\det'_+$ indicates the determinant with zero modes removed. See Appendix A for its evaluation.)

This shows how the instability of the metastable state is reflected in the presence of nontrivial saddle points in the functional integral. Note that the energy of the bounce with the correct period $\beta\hbar$ is simply the energy for which the probability of escape is largest, i.e., it maximizes the product of the Boltzmann factor, $\exp(-\beta E)$, times the tunneling probability

$$\exp\left[-\frac{2}{\hbar}\int_{x_1}^{x_2}dx\{2[V(x)-E]\}^{1/2}\right].$$

The critical temperature $T_0=\beta_0{}^{-1}=\hbar\omega_0/2\pi$ is the temperature above which thermal excitation over the top of the barrier is the dominant decay mechanism. The fact that the periodic bounce $\bar{x}(t)$ becomes static above this temperature reflects the fact that this is a classical process.

Finally, one may ask how to relate the imaginary part of the free energy to the actual decay rate Γ. This requires a further WKB calculation. For temperatures $T<T_0$, one finds[18]

$$\Gamma=(2/\hbar)\text{Im}\mathcal{F}\ , \tag{2.11}$$

which reduces to the familiar relation $\Gamma=(2/\hbar)\times\text{Im}E$, when $T\to 0$. For $T>T_0$, one finds[18]

$$\Gamma=(\omega_0\beta/\pi)\text{Im}\mathcal{F}\ , \tag{2.12}$$

which reduces to the classical result $\Gamma=(\omega/2\pi)\times\exp(-\beta V_0)$ as $T\to\infty$. [These formulas are correct outside a narrow transition region where $T-T_0=O(h^{3/2})$. They are also correct in multidimensional systems. See Ref. 18 for further details.]

Once a tunneling process has occurred to a new classically allowed region, the subsequent evolution will be governed by classical mechanics until a new equilibrium state is neared.[19] For further discussion of this point see Sec. VI.

III. PERTURBATION THEORY, ZERO TEMPERATURE, AND STABILITY

The action from which we begin our construction of quantum gravity is the Einstein-Hilbert action. This is defined for a Lorentzian metric g_{ab} on a manifold M as

$$I=\frac{1}{16\pi G}\int_M R(-g)^{1/2}d^4x\ . \tag{3.1}$$

R is the Ricci scalar of the metric g_{ab}. In cases of interest, M will have some boundary ∂M, typically a boundary at infinity. If we consider variations of the metric in M whose normal derivatives vanish on ∂M, then extrema of I yield the vacuum Einstein equations

$$R_{ab}=0\ . \tag{3.2}$$

However, in a functional integral, we wish to have an action which reproduces the vacuum Einstein equations under all variations of the metric that vanish on ∂M. To find an action with these properties, we add a surface term to the action,[20] giving

$$I=\frac{1}{16\pi G}\int_M R(-g)^{1/2}d^4x+\frac{1}{8\pi G}\int_{\partial M}K(\pm h)^{1/2}d^3x+C\ . \tag{3.3}$$

h_{ab} is the induced metric on the boundary, the $+$ $(-)$ being taken depending on whether the boundary is spacelike (timelike). K is the trace of the second fundamental form. C is a term which depends only on the metric h_{ab}. It could be absorbed into the measure in the functional integral. However, for spacetimes which admit a single asymptotically flat region, so that ∂M is a timelike tube at infinity, C can be written as

$$C=\frac{-1}{8\pi G}\int K^0(h)^{1/2}d^3x\ . \tag{3.4}$$

K^0 is the second fundamental form of ∂M embedded in flat space. With this choice, the action for Minkowski spacetime is zero.

In classical general relativity, one can ask about the stability of flat spacetime. One would expect that stability would be guaranteed by the positivity of the energy. The total energy of an asymptotically flat spacetime is the Arnowitt-Deser-Misner (ADM) mass.[21] This is

$$E_{\text{ADM}}=\int_{\partial\Sigma}(\hat{h}^k_{l;k}-\hat{h}^k_{k;l})d^2S^l\ , \tag{3.5}$$

where $\hat{h}_{ab}$ is the metric induced on a spacelike hypersurface Σ which has a boundary at infinity which is a large S^2. In vacuum gravitation, $E_{\text{ADM}}\geq 0$.[11,12] This result holds even if Σ has a complicated topology resulting from the presence of black holes. If we couple matter with an energy-momentum tensor T_{ab} to the gravitational field, the Einstein equations become

$$R_{ab}-\tfrac{1}{2}Rg_{ab}=8\pi GT_{ab}\ . \tag{3.6}$$

If the energy-momentum density of matter is nonnegative in all frames, i.e., if $T^{00}\geq|T^{ab}|$ for any orthonormal tetrad, then $E_{\text{ADM}}\geq 0$.[11,12] Furthermore, if $E_{\text{ADM}}=0$ then the spacetime is Minkowski space. This indicates that Minkowski space is classically stable in the presence of fixed matter sources, since there is nothing into which it could decay.

We wish to explore the structure of quantum gravity rather than classical gravity at both zero and nonzero temperature. In order to do this we follow the methods outlined in Sec. II.

We wish to construct a functional integral representation for the partition function of quantum gravity. A formal construction is not much more difficult here than in the better known case of an Abelian or non-Abelian gauge theory.[22,23] For zero temperature we recover the vacuum-to-vacuum amplitude discussed by Faddeev and Popov[24] and others.[25] We shall therefore give a brief outline of the construction.

Start with the canonical variables[21,26]: $\hat{h}_{ij}(x)$, the components of the three metric, and the canonically conjugate momenta $\pi_{ij}(x)$ $(i,j,k,\ldots=1,2,3)$. The canonical commutation relations are

$$[\pi_{ij}(x),\hat{h}_{kl}(x')]=-i\delta_{i(k}\delta_{l)j}\delta(x,x') .$$

We work within an unphysical Hilbert space spanned by all $h_{ij}(x)$. Physical states must be invariant under gauge (general coordinate) transformations. This requires that they be annihilated by four constraints

$$\mathscr{H}_i=-2\pi_i{}^j{}_{;j} , \tag{3.7}$$
$$\mathscr{H}_0=\mathscr{H}_\perp=\hat{h}^{-1/2}(\pi_{ij}\pi^{ij}-\tfrac{1}{2}\pi^2)-\hat{h}^{1/2}R .$$

(A semicolon denotes the covariant derivative with respect to the metric $\hat{h}_{ij}$.) One can construct a projection operator $\hat{\Lambda}$ that projects out the physical subspace as

$$\hat{\Lambda}\equiv\int\mathscr{D}N_a(\vec{x})\exp\left[i\int d\vec{x}N_b(\vec{x})\mathscr{H}^b(\vec{x})\right] . \tag{3.8}$$

The Hamiltonian is simply

$$\hat{H}=\int d^3x\,\mathscr{H}_\perp-P_\perp , \quad P_\perp\equiv\oint_{r\to\infty}d^2S^l(\hat{h}_{lr,r}-\hat{h}_{rr,l}) . \tag{3.9}$$

Now we follow the steps of Sec. II, and derive a function integral for

$$Z=\mathrm{tr}(\hat{\Lambda}e^{-\beta H})=\lim_{N\to\infty}\mathrm{tr}(\hat{\Lambda}e^{-\epsilon\hat{H}})^N \quad (\epsilon=\beta/N)$$
$$=\int\mathscr{D}N_a(\vec{x},t)\mathscr{D}\pi_{ij}(\vec{x},t)\mathscr{D}\hat{h}_{ij}(\vec{x},t)\exp\left[\int_0^\beta dt\left[\int d^3x(N_a\mathscr{H}^a+i\pi_{ij}\dot{\hat{h}}^{ij})-P_\perp\right]\right] , \tag{3.10}$$

where the integration is over metrics $\hat{h}_{ij}(\vec{x},t)$ and $N_a(\vec{x},t)$ which are strictly periodic in Euclidean time t: $\hat{h}_{ij}(\vec{x},0)=h_{ij}(\vec{x},\beta)$. The integration over π_{ij}, which only appears quadratically, can now be done explicitly. ly. This can then be written in the standard, Euclidean form of the functional integral for gravity, first derived by Faddeev and Popov[24] (N_a has been rewritten in terms of g_{0a}):

$$Z=\int\mathscr{D}[g_{ab}(x)]\exp[-\hat{I}(g)+\text{gauge-fixing terms}] , \tag{3.11}$$

where the Euclidean action is given by

$$\hat{I}[g]=\frac{-1}{16\pi G}\int Rg^{1/2}d^4x -\frac{1}{8\pi G}\int Kh^{1/2}d^3x+C . \tag{3.12}$$

Below, we shall specify the form of the gauge-fixing terms.

The functional integral is evaluated by integrating over all metrics which are positive definite and obey appropriate boundary conditions. We are interested in two types of boundary condition which correspond to the zero-temperature vacuum and to the canonical ensemble at temperature $T=1/\beta$.

The boundary conditions appropriate to the vacuum are termed asymptotically Euclidean (AE).[27] An AE metric is one in which the metric approaches the flat metric on R^4 outside some compact set. For the action to be finite, the metric must then look like

$$ds^2=\left[1+\frac{\alpha}{r^2}\right]\delta_{ab}dx^a dx^b+O(r^{-3}) , \tag{3.13}$$

where r is a four-dimensional radial coordinate and α a function of the coordinates, but independent of r. The boundary at infinity is topologically S^3.

The boundary conditions for the canonical ensemble at temperature $T=\beta^{-1}$ are termed asymptotically flat (AF).[27] An AF metric is one in which the metric approaches the flat metric on $R^3\times S^1$ outside some compact set. Finite action requires the metric to be asymptotically

$$ds^2=d\tau^2+\left[1+\frac{\alpha}{r^2}\right]\delta_{ij}dx^i dx^j$$
$$+\text{terms which fall off faster} \quad (i,j=1,2,3) . \tag{3.14}$$

Here r is a three-dimensional radial coordinate. α can be a function of the coordinates, but is independent of r; τ is a coordinate which is periodic with period β. The boundary of infinity is topologically $S^2 \times S^1$. This case will be discussed in detail in Secs. IV and V.

This functional integral construction of quantum gravity is poorly understood. Are we to integrate over all manifolds or perhaps only those topologically equivalent to flat space? How can one render the functional integral well defined when R can be arbitrarily large? These and other questions are the subject of much recent research and speculation.[28] At the moment the only feasible way to treat the functional integral is by saddle-point methods. This is adequate for a treatment of the small perturbations about Minkowski space and for a semiclassical analysis of vacuum stability.

The saddle-point evaluation starts by constructing stationary points of the action, namely, solutions of the Euclidean Einstein equations. Expansions about these saddle points are performed by writing

$$g_{ab} = g_{ab}^{(\text{saddle point})} + \phi_{ab} \; . \tag{3.15}$$

Treating ϕ_{ab} as a quantum field and $g_{ab}^{(\text{saddle point})}$ as a c-number background field will generate the usual perturbation expansion, which can be expressed in terms of Feynman diagrams. The saddle-point metric g_{ab} (normally assumed to be a nonsingular geodesically complete four-manifold) is colloquially termed a gravitational instanton. Different instantons will have various physical interpretations. But first we must find all gravitational instantons consistent with our AE boundary conditions. The positive-action theorem, first proved by Schoen and Yau,[10] states that for any AE metric with $R=0$, the action I is non-negative and $I=0$ if and only if g_{ab} is flat. However the action for any AE instanton must be zero. This follows from the fact that any AE instanton will be a solution of $R_{ab}=0$. Such a metric will always admit a uniform dilatation $g_{ab} \to \lambda g_{ab}$, which will map the old solution into a new solution. However such a dilatation will map the action $I \to \lambda I$. But such a dilatation could be produced by a coordinate transformation $x^a \to \lambda x^a$, which must leave the action invariant. Thus, the action for any AE instanton can only be zero. The positive-action theorem then guarantees that such a metric is flat. For zero temperature we need only examine the perturbations about flat space. Since the action is not positive definite for metrics that do not satisfy $R=0$, there is cause for concern that these perturbations might be unstable.

Let us, therefore, examine the pertubations about an arbitrary saddle-point metric, $\hat{g}_{ab}$. It is useful at this stage to specify the gauge that we will employ. It is simplest to consider the covariant Lorentz gauge

$$\nabla_a (\phi^{ab} - \tfrac{1}{2} \hat{g}^{ab} \phi) = X^b \; , \tag{3.16}$$

where X^b is some specified function of the coordinates. (∇_a is the covariant derivative with respect to the background metric g_{ab}.) We then employ the 't Hooft averaging procedure. The net result is an effective action that contains the field ϕ_{ab} and a set of anticommuting Faddeev-Popov vector fields η^a. This effective action is rather complicated. To simplify it, we introduce the following decompositions[29]:

$$\phi_{ab} = \phi_{ab}^{TT} + \tfrac{1}{4} \hat{g}_{ab} \phi + (\nabla_a \xi_b + \nabla_b \xi_a - \tfrac{1}{2} \hat{g}_{ab} \nabla_c \xi^c) \; . \tag{3.17}$$

ϕ_{ab}^{TT} is the transverse tracefree ($\nabla_a \phi^{abTT} = 0$, $\hat{g}^{ab} \phi_{ab}^{TT} = 0$) part of ϕ_{ab}; ϕ is the trace part; and ξ_b is the longitudinal tracefree part. The two vector fields ξ_b and η_b are then decomposed according to the Hodge-de Rham decomposition

$$\begin{aligned} \eta_b &= \nabla_b \chi + \eta_b^c + \eta_b^H \; , \\ \xi_b &= \nabla_b \psi + \xi_b^c + \xi_b^H \; . \end{aligned} \tag{3.18}$$

η_b^c, ξ_b^c are the coexact parts of η and ξ and are hence divergence free. η_b^H, ξ_b^H are the harmonic parts of η and ξ. The number of square-integrable harmonic vectors is equal to the dimensionality of the first cohomology group $H^1(M,R)$ on the manifold associated with metric $\hat{g}_{ab}$. Since we are almost always dealing with simply connected manifolds, we will ignore the harmonic sector in what follows.

We now expand the effective action in terms of ϕ and η. There are no terms linear in ϕ since the vacuum Einstein equations are satisfied. The quadratic terms are

$$\frac{1}{16\pi G} [\tfrac{1}{4} \phi_{ab}^{TT} G^{abcd} \phi_{cd}^{TT}] \; , \tag{3.19a}$$

$$\frac{1}{16\pi G} [-\tfrac{1}{2} \phi F \phi] \; , \tag{3.19b}$$

$$\frac{1}{16\pi G} [2 X_{ab}(\psi) X^{ab}(F\psi)] \; , \tag{3.19c}$$

$$\frac{1}{16\pi G} [Y_{ab}(\xi_e^c) Y^{ab}(C_f^e \xi^{cf})] \; , \tag{3.19d}$$

$$\frac{1}{16\pi G}[X_{ab}(\chi)X^{ab}(F\chi)] , \tag{3.19e}$$

$$\frac{1}{16\pi G}[Y_{ab}(\eta_e^c)Y^{ab}(C_f^e\eta^{cf})] , \tag{3.19f}$$

where

$$\begin{aligned} G_{abcd} &= -g_{ac}g_{bd}\Box - 2R_{acbd} , \\ X_{ab}(\psi) &= (\nabla_a\nabla_b + \nabla_b\nabla_a - \tfrac{1}{2}\hat{g}_{ab}\Box)\psi , \\ F &= -\tfrac{1}{8}\Box , \\ Y_{ab}(\eta_e) &= \nabla_a(\eta_e\delta_b^e) + \nabla_e(\eta_e\delta_a^e) , \\ G_f^e &= -\delta_f^e\Box . \end{aligned} \tag{3.20}$$

These quadratic terms determine the propagators of the fields in the theory. Further terms can be obtained by expanding the effective action to higher order and would include the vertices of the theory. The metric perturbation ϕ_{ab} has ten degrees of freedom, of which five correspond to a spin-2 piece (ϕ_{ab}^{TT}), three to a spin-1 piece (ξ_b^c), and two to spin-0 pieces (ϕ,ψ). The Faddeev-Popov terms contain a spin-1 piece (η_b^c), and a spin-0 piece (χ). If we consider perturbations about flat space ($\hat{g}_{ab}=\eta_{ab}$) then the operators G, F, and C are all manifestly positive definite. However, the term $\phi F\phi$ in the effective action is negative definite. This is the perturbation-theory remnant of the fact that certain conformal transformations can be made on a given metric such as to make the Euclidean action arbitrarily negative.[9] Thus we might think that Z, which is proportional to $[\det(F)]^{-1/2}$, contains a factor of $i^{(\dim F)}$ and that the perturbations about flat space are unstable. However, this negative-metric piece is not significant. First, a detailed examination of the propagator shows that this spin-zero piece does not couple to the conserved energy-momentum tensor of other fields. Thus it cannot represent a physical instability of the system. Second, a more sophisticated analysis, which is performed in a family of covariant gauges, shows that the analog of operator F is a gauge-dependent operator, in contrast to the operators G and C which are explicitly gauge independent.[29] In fact, if we choose to work in the gauge

$$\nabla_a(\phi^{ab} - \hat{g}^{ab}\phi) = X^b \tag{3.21}$$

rather than the Lorentz gauge, we would discover that the operator F vanishes identically. Since all physical observables are gauge invariant, they must be independent of F and so we cannot be troubled by such terms. Third, even if we were to ignore the above arguments and proceed to calculate physical observables in perturbation theory we would find an explicit cancellation of $\det F$. This is because a factor of $\det F$ occurs three times in the evaluation of Z. Two factors of $(\det F)^{-1/2}$ arise from the integrations over ϕ and ψ, and a factor of $\det F$ comes from integrating the Faddeev-Popov ghosts.

In curved space, the situation is slightly different. If the operator F is positive definite then the terms (3.19c) and (3.19e) are also positive definite, when $R_{ab}=0$ as is the case for a classical solution. Suppose that a normalizable eigenfunction of F is ϕ_n with eigenvalue λ_n. Then

$$\begin{aligned} \int_M \phi_n F\phi_n g^{1/2}d^4x &= \int_M (\nabla_a\phi_n)(\nabla^a\phi_n)g^{1/2}d^4x \\ &\quad + \int_{\partial M}\phi_n(\nabla_a\phi_n)d\Sigma^a \\ &= \lambda_n\int_M \phi_n{}^2 g^{1/2}d^4x . \end{aligned} \tag{3.22}$$

Since ϕ_n must vanish on the boundary ∂M of the manifold M, it follows that $\lambda_n > 0$ for all square-integrable eigenfunctions.

The operator C acts on divergence-free vectors. Suppose that it has an eigenfunction ζ_a with eigenvalue λ_n. Then an identity due to Yano and Nagano[30]

$$\begin{aligned} \int_M [(\nabla_a\zeta^a)^2 - \tfrac{1}{2}(\nabla_a\zeta_b + \nabla_b\zeta_a)(\nabla^a\zeta^b + \nabla^b\zeta^a) \\ + \zeta^a\Box\zeta_a - 2\zeta^a R_{ab}\zeta^b]g^{1/2}d^4x = 0 \end{aligned} \tag{3.23}$$

shows that C is positive semidefinite, the zeros being associated only with Killing vectors. However, deformations constructed from Killing vectors are not included in functional integrals as this would overcount field configurations. This is because a Killing vector refers to a continuous symmetry of a background metric and maps a space into the identical space. Therefore (3.19d) and (3.19f) are positive for $R_{ab}=0$.

The situation is rather different for the operator G. One cannot prove that it is positive definite, and indeed, in Sec. V, where we discuss finite-temperature instantons, we will encounter a space in which G has both zero and negative eigenvalues.[31] The zero modes are associated with transformations between distinct solutions of the Einstein equations with the same action. There is one such mode for each degree of freedom of the instanton, and these modes will be handled by the collective-coordinate method. The negative eigenvalues of G do have physical significance and will lead to the instability of hot flat space.

We therefore see that flat space at zero temperature is stable quantum mechanically as well as classically. Of course we can only verify this for small perturbations about flat space, but it seems unlikely, in view of the positive-action theorem, that nonperturbative instabilities could arise. However, one must note that quantum gravity is probably nonrenormalizable beyond the one-loop level, and that any statement regarding quantum gravity to all orders in perturbation theory is dangerous.

IV. FINITE-TEMPERATURE PERTURBATION THEORY

We shall now investigate quantum gravity at finite temperature. We wish to describe the properties of a system placed in an arbitrarily large spatial volume which is kept at some fixed temperature $T=1/\beta$. The equilibrium state of such a system will be described by the canonical partition function

$$Z \equiv e^{\beta\mathcal{F}} = \mathrm{Tr}\, e^{-\beta H}\Big|_{\text{physical degrees of freedom}} \tag{4.1}$$

for which we derived a functional integral representation in the previous section [Eq. (3.11)]. This system describes the purest vacuum of all; in the absence of matter only fluctuations of the metric field are present. It is these that we must sum over in the functional integral, integrating over all asymptotically flat, Euclidean four-metrics, which are periodic in Euclidean time with period β: $g_{ab}(\tau,x)=g_{ab}(\tau+\beta,x)$.

Once again we only know how to treat this problem in the semiclassical approximation. To this end we must first find all saddle points of the classical action, i.e., periodic Euclidean finite-action solutions of Einstein's equations, and expand about each one. There is one trivial periodic solution, namely, flat space. The metric $g_{ab}(x)=\delta_{ab}$ is clearly periodic and has zero action. The contribution of this saddle point to $\mathcal{F}$, to lowest order in perturbation theory, will simply be the free energy of an ideal gas of gravitons at temperature T. In higher orders the interaction free energy of the gravitons will appear and will produce an (Jeans) instability. In the following section we shall consider the contribution of other (instanton) saddle points.

Perturbation theory about flat space at finite temperature proceeds much the same way as at zero temperature, the only difference being that the fields $g_{ab}(x)$ are periodic in t with period β. Thus the Euclidean frequencies p_0 are quantized in units of $2\pi n/\beta=\omega_n$ and integrals over p_0 are replaced by discrete sums

$$\int dp_0 \to \frac{2\pi}{\beta}\sum_{\omega_n} .$$

Otherwise the zero-temperature Feynman rules are unchanged. Note that if fermionic matter fields are included in the theory, they must be antiperiodic in t, and therefore their frequencies are quantized in units of $\omega_n=(2n+1)\pi/\beta$. On the other hand Faddeev-Popov ghosts, although fermionic, serve to represent a bosonic determinant and thus must be periodic.[23,32]

The free energy is given by the sum of all vacuum graphs, evaluated with finite-temperature propagators. To lowest order

$$\begin{aligned} Z &= e^{\beta\mathcal{F}} \\ &= [\det(-\Box)]^{-1/2}_{T,\text{physical degrees of freedom}} . \end{aligned} \tag{4.2}$$

$\Box$ is the appropriate second-order differential operator. At zero temperature, $\mathcal{F}$ is (quartically) divergent; however the temperature-dependent part of $\mathcal{F}$ is ultraviolet finite. The calculation is straightforward, yielding the standard result for the free energy of a relativistic gas of massless particles with two (helicity) degrees of freedom

$$\frac{\mathcal{F}_0(T)}{V} = -\frac{\pi^2 T^4}{45} = \frac{2}{\beta}\int \frac{d^3k}{(2\pi)^3}\ln(1-e^{-\beta k}) . \tag{4.3}$$

One can also readily calculate higher-order [in $(16\pi G)^{1/2}$] corrections to the free energy, arising from the self-interactions of the gas of gravitons. These will be given by vacuum Feynman graphs that contain cubic (or higher-order) vertices. There are two sources of trouble that appear in this perturbative expansion. First, since pure gravity is unrenormalizable, the ultraviolet divergences will be uncontrollable. We have nothing to say about this problem. However, in addition we expect infrared instabilities to show up once we allow for graviton-graviton interactions. As discussed above, a homogeneous matter distribution develops a (Jeans) instability for time-dependent fluctuations of wavelength bigger than $1/k_J$, where $k_J=(4\pi G\rho)^{1/2}$ $(V_s=1)$. We expect such an instability to appear in our gas of gravitons at finite temperature since it does contain "matter", namely, the thermally excited gravitation modes themselves. These carry energy and thus are a source for the gravitational field, leading to the Jeans instability.

How does this instability appear in the partition function? The standard analysis of the Jeans instability, as performed in the Introduction, considers *time-dependent* fluctuations of a homogeneous medium. The Jeans wave number k_J is the maximum wave number for which the frequency of these fluctuations is still real. One can interpret this result by saying that due to the antiscreening effects of gravity in a relativistic gas of density ρ (velocity of sound = 1), the graviton acquires an imaginary "mass," $m_J{}^2=-4\pi G\rho$. This is the analog of the usual plasmon.[33] In our Euclidean calculation, however, we are interested in the static equilibrium properties of the system, in particular in the response of the gravitational field to spatial, time-independent fluctuations of the medium. Here too we expect gravitational antiscreening, which will produce an instability; however the static graviton "mass" need not have the Jeans value. This too is familiar from the analogous case of a plasma.[23] There one finds zero-momentum excitations of the photon have a mass equal to $m_{\rm el}/\sqrt{3}$. On the other hand, the inverse screening length of the plasma, which governs the long-range correlations of the charge density $\rho(x)$, is given by $m_{\rm el}{}^2=\frac{1}{3}g^2T^2$, namely,

$$\langle \rho(\vec{x})\rho(0)\rangle \underset{|\vec{x}|\to\infty}{\sim} e^{-m_{\rm el}|\vec{x}|} .$$

If we calculate the electric screening of a plasma by evaluating Euclidean functional integrals for a charged gas at finite temperature, it is the latter mass that is generated. To evaluate the plasmon frequency one must analytically continue the resulting phonon propagator back to Minkowski space.

We shall now present a very simple argument that the graviton mass which appears in Euclidean propagators is in fact twice the Jeans value, namely,

$$m_g{}^2=-16\pi G\rho(T) , \tag{4.4}$$

where ρ is the thermal density of gravitons ($\rho=\pi^2T^4/15$). In fact the above result holds for any kind of massless "matter."

We wish to consider the static correlation of the gravitational field with itself in the presence of a thermal gas of massless particles. Imagine placing a very small test mass M into the system at the origin. At large distances we can use the weak-field (Newtonian) approximation, whence the gravitation potential ϕ, defined by $g_{00}=-1+2\phi$, satisfies

$$-\nabla^2\phi=4\pi GM\delta^3(\vec{x})+4\pi G\delta\rho(\phi,T) . \tag{4.5}$$

$\delta\rho(\phi,T)=\rho(\phi,T)-\rho(0,T)$ is the change in the energy density of the thermal gas at temperature T due to the gravitation potential ϕ. We can evaluate the energy density far away from the origin, where ϕ can be regarded as small and uniform, by applying the equivalence principle. Thus the energy density of the gas at $\vec{x}$ is the same as if the temperature were $T[1+\phi(x)]$ and there was no gravitational field [since if T is the temperature of the heat bath at infinity, at $\vec{x}$ all energies are redshifted by an amount $1+\phi(x)$]. Therefore,

$$\rho(\phi,T)=N\int\frac{d^3k}{(2\pi)^3}\frac{k}{\exp[\beta k/(1+\phi)]-1}$$
$$=(1+\phi)^4\rho(T) . \tag{4.6}$$

(N = number of degrees of freedom of the massless particles.) Therefore far away from the source, where $\phi << 1$, we have

$$[-\nabla^2-16\pi G\rho(T)]\phi(\vec{x})=4\pi GM\delta^3(\vec{x}) . \tag{4.7}$$

Thus for static weak fields the graviton acts as if it had an imaginary mass given by Eq. (4.4).

According to this argument the graviton should develop an imaginary mass when it couples to thermally excited matter of any kind. We shall illustrate how this emerges in perturbation theory in the simple case of gravitons coupled to massless fermions. We wish to calculate the propagator of the graviton field $h_{ab}=g_{ab}-\delta_{ab}$ to one-loop order. The full propagator may be expressed in terms of the one-particle irreducible self-energy, $\Pi_{ab,cd}(\omega_n,k)$,

$$G_{ab,cd}(\omega_n,\vec{k})=[(\omega_n{}^2+\vec{k}^2)\delta_{c(a}\delta_{b)d}+\Pi_{ab,cd}(\omega_n,\vec{k})]^{-1}+\text{gauge terms} , \tag{4.8}$$

which in turn is given by the one-loop Feynman diagram in Fig. 6. The vertex is given by the energy-momentum tensor of a massless fermion. At $T=0$ gauge invariance, plus Euclidean invariance, implies that $k^a\Pi_{ab,cd}(k)=0$, and this in turn requires that at $T=0$, $\Pi_{ab,cd}(0)=0$. However at finite temperature the energy k_0 is quantized, and there is no such constraint on the longitudinal self-energy $\Pi_{00,00}$. In fact, one can easily show that if we first set $k_0=0$, and then examine $\Pi_{ab,cd}(0,\vec{k})$ as $\vec{k}\to 0$ that only $\Pi_{00,00}(0,\vec{k})$ can be nonvanishing. To evaluate this term we simply evaluate the contribution of the diagram exhibited in Fig. 6 to $\Pi_{00,00}(0,0)$:

$$\Pi_{00,00}(0,0)=-16\pi G\int\frac{d^3p}{(2\pi)^3}\frac{1}{\beta}\sum_n \mathrm{Tr}\left[\gamma_0 p_0\frac{1}{p}\gamma_0 p_0\frac{1}{p}\right], \tag{4.9}$$

where we recall that the fermion energy p_0 take values of $(2n+1)\pi/\beta=\omega_n$. This can be calculated by standard contour techniques[23] and yields the result

$$\Pi_{ab,cd}(0)=\delta_{a0}\delta_{b0}\delta_{c0}\delta_{d0}(-\tfrac{14}{15}\pi^3GT^4)\,. \tag{4.10}$$

This shows indeed that the longitudinal graviton h_{00} develops a one-loop "mass" due to thermal fluctuations, of magnitude

$$m_g{}^2=-\tfrac{14}{15}\pi^3GT^4=-16\pi G\rho_f\,, \tag{4.11}$$

in accord with our expectation (recall that the density of a massless fermi gas is $\rho_f=\frac{7}{120}\pi^2T^4$).

The same effect will be produced even if there are no explicit matter fields present. According to our previous argument the one-loop contribution of thermal gravitons to $\Pi_{00,00}(0)$ will produce a "mass" of[34]

$$m_g{}^2=-16\pi G\rho_g=-\tfrac{16}{15}\pi^3GT^4\,. \tag{4.12}$$

The fact that the graviton acquires an imaginary "mass" at finite temperature means that flat space is unstable. Flat space, which was an absolute minimum of our classical action (with, of course, AE boundary conditions), becomes merely a saddle point of the effective action once we take into account the interactions of the thermal gravitons. (This is illustrated in Fig. 7.) The mechanism for instability is clear, large-scale density fluctuations of the gas of gravitons tend to grow owing to the attractive (antiscreening) gravitational forces. Presumably, these eventually collapse to form black holes. Indeed if we were to calculate the higher-order contributions to the free energy, we would encounter the increasing infrared-divergent "ring" diagrams of Fig. 8. The sum of these yields a contribution to the free energy of

$$\frac{1}{2\beta}\sum_n\int\frac{d^3k}{(2\pi)^3}\mathrm{tr}\ln[1+\Pi(k)/k^2]\,, \tag{4.13}$$

which is complex [since $\Pi_{00,00}(0)<0$].

This Jeans instability of hot flat space suffices to call into question the ability to treat hot gravity by semiclassical methods. In fact unless there is some stabilizing mechanism in the theory it is questionable whether there exist *any* fixed-temperature equilibrium states. If someone provides an indefinite amount of energy in order to keep the walls of our container at a finite temperature, it might be that gravitational collapse continues to occur until the resulting black hole engulfs the walls themselves.

Nevertheless we shall continue to employ the semiclassical approximation in the following section, where we investigate the contribution of instanton saddle points. This is not only because of the interest in elucidating the significance of the Euclidean Schwarzschild solution (the instanton), but also since the instability generated by this mechanism is totally different from the Jeans in-

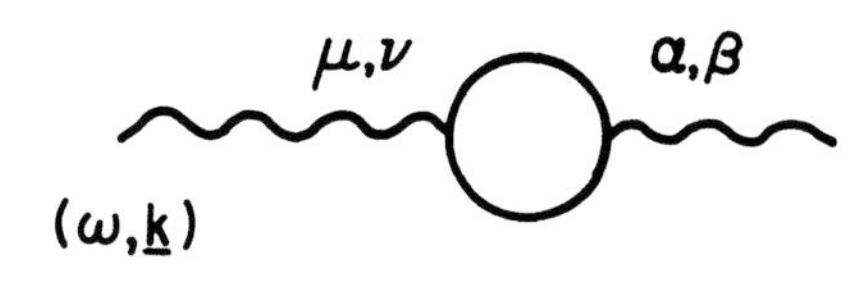

FIG. 6. The diagram which gives a contribution to the polarization tensor.

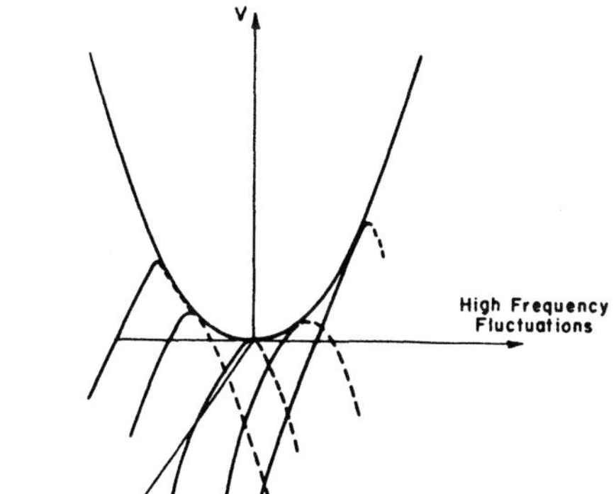

FIG. 7. The effective one-loop potential surface close to flat space-time at nonzero temperature. The potential increases for short-wavelength fluctuations but decreases for long-wavelength fluctuations.

stability. It corresponds, not to long-wavelength fluctuations about flat space, but to the spo tane-ous nucleation of black holes. Furthermore, the radius of these black holes is smaller than the Jeans length by a factor of T/m_P (m_P=Planck mass). Thus one could imagine enclosing the system in a finite volume of size less than the Jeans length, which would eliminate all but the black-hole nucleation instability.

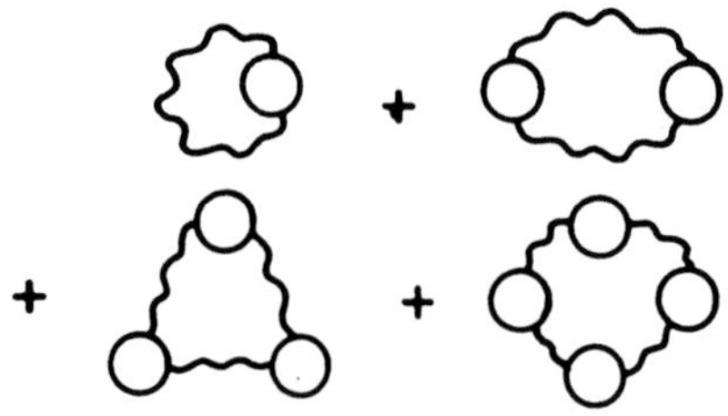

FIG. 8. The ring diagrams that contribute to the imaginary part of F.

V. GRAVITATIONAL INSTANTONS

In the previous section, we discussed a particular AF gravitational instanton, namely, flat space, with the topology of $R^3 \times S^1$. However, unlike the AE, zero-temperature case, flat space is not the unique instanton; there exist other periodic solutions of the Euclidean equations of motion. A familiar instanton is the Euclidean section of the Schwarzschild solution,[35] which is a special case of the Kerr instanton.[36] In Boyer-Lindquist coordinates (t,r,θ,ϕ) this metric takes the form

$$ds^2=(r^2-a^2\cos^2\theta)\left[\frac{dr^2}{\Delta}+d\theta^2\right]+\frac{1}{r^2-a^2\cos^2\theta}(\Delta(dt+a\sin^2\theta\,d\phi)^2+\sin^2\theta[(r^2-a^2)d\phi-a\,dt]^2)\,, \tag{5.1}$$

where

$$\Delta=r^2-2GMr-a^2\,. \tag{5.2}$$

This describes a two-parameter family of solutions. In addition to the translational degrees of freedom, there is a rotation parameter a and a mass parameter M. This metric is positive definite as long as the radial coordinate r is restricted to the region $GM+(G^2M^2+a^2)^{1/2}\leq r<\infty$. The region $r<GM+(G^2M^2+a^2)^{1/2}$ is not part of the space that we are interested in. The locus $r=GM+(G^2M^2+a^2)^{1/2}$ will be a conical singularity unless we identify the point (t,r,θ,ϕ) with the point $(t+2\pi\kappa^{-1},\ r,\ \theta,\ \phi+2\pi\Omega\kappa^{-1})$, where

$$\kappa=\frac{(G^2M^2+a^2)^{1/2}}{2GM[GM+(G^2M^2+a^2)^{1/2}]} \tag{5.3}$$

and

$$\Omega=\frac{a}{2GM[GM+(G^2M^2+a^2)^{1/2}]}\,. \tag{5.4}$$

Furthermore, $\theta=0,\pi$ define symmetry axes with rotations about them generated by the Killing vector $\partial/\partial\phi$. For these axes to be nonsingular, ϕ must be an angular variable with period 2π.

Since our fields must be periodic in t, with period β, it is clear that κ must be fixed to equal

$$\kappa=2\pi T\,. \tag{5.5}$$

Also Ω must equal the chemical potential associated with the angular momentum. The instanton is also characterized by its topological properties. Consider the two-surface $r=GM+(G^2M^2+a^2)^{1/2}$, which is the fixed point of the orbit of the Killing vector $\partial/\partial t$. Such a fixed-point set is sometimes called a "bolt."[37] One can use the G-index theorems to find the Euler character and Hirzebruch signature of the manifold. This bolt has zero self-intersection number and an Euler number of two; thus they equal two and zero, respectively. Finally the action of the instanton is

$$I=\frac{\beta M}{2}\,. \tag{5.6}$$

It has been widely conjectured that the Kerr instanton is the unique AF instanton other than flat space.[38] In any case it is the only one we shall consider.

The above form for the action looks very much like the contribution of a particle, or soliton, of mass $M/2$. Does the instanton represent the co - tribution of black holes to the partition function?

We do not believe so. Recall that M is not a fixed mass, but rather determined in terms of Ω and T by Eqs. (5.3)–(5.5). If we consider the Schwarzschild case, where $a=\Omega=0$, and thus $\kappa=2\pi T=1/4GM$, the action is given by

$$I=\frac{\beta^2}{16\pi G}=\frac{m_P{}^2}{16\pi T^2}\ , \tag{5.7}$$

which is unlike the contribution of any fixed-mass particle to the partition function. In fact we shall argue that the instanton is not a soliton, but rather provides a mechanism for the nucleation of black holes.

There are also, however, other configurations which are arbitrarily close to solutions of the vacuum Einstein equations. These will be needed when we actually calculate the instanton contributions to Z. These configurations are in a distinct topological class to the Kerr solution. We will discuss in detail the sector which has Euler character $\chi=4$ and zero angular momentum. Part of the $\chi=2N$ sector is discussed by Gibbons and Perry,[39] and the region where the rotation is not zero can be obtained by analytic continuation of the solutions of Hauser and Ernst,[40] Kinnersley *et al.*,[41] and Kramer and Neugebauer.[42] These are all very complicated but qualitatively similar to the case discussed below. We discuss the metric

$$ds^2=Vdt^2+V^{-1}[e^{2k}(d\rho^2+dz^2)+\rho^2 d\phi^2]\ . \tag{5.8}$$

ρ,ϕ,z form a cylindrical polar coordinate system, ϕ being identified with period π. The solution is considered to be static and axially symmetric about the $\rho=0$ axis. Hence V and k are functions of ρ and z only:

$$V=\frac{r'_1+r'_1-2GM}{r'_1+r''_1+2GM}\,\frac{r'_2+r''_2-2GM}{r'_2+r''_2+2GM}\ , \tag{5.9}$$

$$k=\tfrac{1}{4}\sum_{n,m=1,2}\ln\frac{r'_n r''_m+(z-z_n-GM)(z-z_m+GM)+\rho^2}{r'_n r'_m+(z-z_n-GM)(z-z_m-GM)+\rho^2}\,\frac{r''_n r'_m+(z-z_n+GM)(z-z_m-GM)+\rho^2}{r''_n r''_m+(z-z_n+GM)(z-z_m-GM)+\rho^2}\ , \tag{5.10}$$

$$r'_n{}^2=\rho^2+(z-z_n-GM)^2\ ,$$
$$r''_n{}^2=\rho^2+(z-z_n+GM)^2\ . \tag{5.11}$$

The Killing vector $\partial/\partial t$ has fixed points at $\rho=0$, $z_1-GM<z<z_1+GM$; and $\rho=0$, $z_2-GM<z<z_2+GM$. These fixed points appear to be rods in the ρ-z plane. However, they are in fact two-surfaces in a four-manifold. For these surfaces to be free of conical singularities, t must be identified with period $8\pi GM$ exactly as in the Kerr case. As ρ, z tend to infinity, $V\to 1$ and $k\to 0$ and thus the metric becomes flat. The periodicity in t means that this metric has AF boundary conditions with a temperature of $T=(8\pi GM)^{-1}$. This metric now has $R_{ab}=0$ everywhere except where $\rho=0$, $z_1+GM<z<z_2-GM$, assuming that $\Delta z=z_2-z_1+2GM>0$ This is the locus corresponding to the gap between the rods in the ρ-z one. There is a conical singularity here which cannot be eliminated by any identification of coordinates compatible with these already performed. However, a conical singularity has the effect of introducing a δ function into the Ricci curvature scalar. Consequently, the contribution to the action from such a singularity is finite. In this case, the action turns out to be

$$I=8\pi GM^2-\beta\frac{GM^2}{(\Delta z+2GM)}\ . \tag{5.12}$$

This is the action we would expect for two "bolts," each of mass M together with a Coulomb interaction between the two masses. In this sense instantons behave as particles; they have normal long-range gravitational interactions. This indicates that we should not expect to be able to find any stationary axisymmetric solutions. But if the masses are separated by arbitrarily large distances, we get arbitrarily close to solutions of the vacuum Einstein equations. In the leading semiclassical approximation we must sum over all these configurations.

We shall consider only the Schwarzschild instanton, with $\Omega=a=0$. To evaluate its contribution to

Z we must integrate over the Gaussian fluctuations about the background Schwarzschild metric. In particular we shall study the stability of such fluctuations using the methods developed in Secs. II and III.

The operators F and C are positive definite. They can be treated by standard methods. The operator G is rather more complex. To determine its eigenvalues we study the solutions of

$$-\Box\phi^{ab}-2R^{acbd}\phi_{cd}=\lambda\phi^{ab}\,, \tag{5.13}$$

where ϕ^{ab} are transverse, tracefree, and normalizable. A variant of this problem has been treated by Regge and Wheeler.[43] They investigated the Lorentz version of this problem with $\lambda=0$. Their methods were refined by subse uent workers, Vishveshwara,[44] Zerilli,[45] Press and Teukolsky,[46] Stewart,[47] and Chandrasekhar.[48] $\lambda=0$ corresponds to a small perturbation of the black hole that remains a classical solution. These authors searched for runaway solutions of the form $\exp(-i\omega t)\times$(function of spatial variables with ω complex). They demonstrated that $\text{Im}\omega=0$ for all solutions of (5.13) and concluded that black holes were classically stable objects.

We, on the other hand, are interested in solutions to (5.13), where g_{ab} is the Euclidean Schwarzschild solution and λ is not necessarily zero. Positive (negative) values of λ will correspond to stable (unstable) Gaussian fluctuations about our instanton.

This e uation can be separated in (t,r,θ,ϕ) coordinates, and is exhibited in Appendix B. We then follow the approach of Regge and Wheeler[43] and divide the space of eigenfunctions into even and odd parity:

$$\phi_{ab}^{(\text{even})}=\begin{bmatrix} H_0(r) & H_1(r) & K_0(r)\partial_\theta & K_0(r)\partial_\phi \\ (\text{sym}) & H_2(r) & K_1(r)\partial_\theta & K_2(r)\partial_\phi \\ (\text{sym}) & (\text{sym}) & r^2[G_1(r)+G_2(r)]\partial_\theta{}^2 & r^2G_2(r)(\partial_\theta\partial_\phi-\cot\theta\partial_\phi) \\ (\text{sym}) & (\text{sym}) & (\text{sym}) & r^2[G_1(r)\sin^2\theta+G_2(r)(\partial_\phi{}^2+\sin\theta\cos\theta\,\partial_\theta)] \end{bmatrix}\exp(i\omega t)Y_{lm}(\theta,\phi)\,, \tag{5.14}$$

$$\phi_{ab}^{(\text{odd})}=\begin{bmatrix} 0 & 0 & \dfrac{-h_0(r)}{\sin\theta}\partial_\phi & h_0(r)\sin\theta\,\partial_\theta \\ 0 & 0 & \dfrac{-h_1(r)}{\sin\theta}\partial_\phi & h_1(r)\sin\theta\,\partial_\theta \\ (\text{sym}) & (\text{sym}) & h_2(r)\left[\dfrac{1}{\sin\theta}\partial_\theta\partial_\phi-\dfrac{\cos\theta}{\sin^2\theta}\partial_\phi\right] & \tfrac{1}{2}h_2(r)\left[\dfrac{1}{\sin\theta}\partial_\phi{}^2+\cos\theta\,\partial_\theta-\sin\theta\,\partial_\theta{}^2\right] \\ (\text{sym}) & (\text{sym}) & (\text{sym}) & -h_2(r)(\sin\theta\partial_\theta\partial_\phi-\cos\theta\,\partial_\phi) \end{bmatrix}$$
$$\times\exp(i\omega t)Y_{lm}(\theta,\phi)\,. \tag{5.15}$$

Substitution of these forms into (5.13), together with the co ditions of tracefree and transversality appli d to ϕ_{ab}, lead to sets of coupled ordinary differential equations in r. By applying Sturm-Lio ville techniques, it is possible to show that for even perturbations with $l\geq 2$, a d for odd p rturbations with $l\geq 1$, that any eigenvalue λ must be positive. The argument fails for $l=0$ or 1 perturbations. If $l=0$ and $\omega>0$, then $\lambda>0$. If $l=1$ and $\omega=0$, then we find three zero modes which are

$$\phi_{ab}^{(i)}=\nabla_a\nabla_b\phi^{(i)}\,, \tag{5.16}$$

$$\phi^{(i)}=(r-GM)\begin{bmatrix}\sin\theta\cos\phi\\ \sin\theta\sin\phi\\ \cos\theta\end{bmatrix}. \tag{5.17}$$

At first sight these zero modes look like gauge transformations. However, the vectors $\xi_a^{(i)}=\nabla_a\phi^{(i)}$ are nonnormalizable. Thus, these modes do not correspond to nonsingular gauge transformations and must be included in the functional integral. They simply represent translations in the x, y, and z directions of the origin of the Schwarzschild instanton. They are dealt with by the standard collective-coordinate method.[49]

For $l=0$, we are forced to resort to numerical methods. Write the perturbations as

$$\phi^{ab}_{(-)}=\text{diag}\left[\left[1-\frac{2m}{r}\right]^{-1}h_0(r),\left[1-\frac{2m}{r}\right]h_1(r),k(r),k(r)\csc^2\theta\right]\exp(-i\omega t)\ . \tag{5.18}$$

Since ϕ^{ab} is tracefree,

$$h_0+h_1+2k=0\ . \tag{5.19}$$

Since it is transverse, it follows that

$$\frac{r-2GM}{r}h_1'+\frac{2r-3GM}{r^2}h_1+2(2GM-r)\frac{k}{r^2}-\frac{GM}{r^2}h_0=0\ , \tag{5.20}$$

where prime$\equiv d/dr$. The radial equations now become

$$\frac{r-2GM}{r}h_1''+\frac{4r^2-22GMr+24G^2M^2}{r^2(r-3GM)}h_1'-\frac{8GM}{r^2(r-3GM)}h_1=-\lambda h_1\ . \tag{5.21}$$

This equation has regular singular points at $r=0$, $r=2GM$, and $r=3GM$ and an irregular singular point at $r=\infty$. Its solutions are not explicitly known. Near $r=2GM$, $h\sim(r-2GM)^\sigma$, $\sigma=0,-1$. For large r, $h\sim\exp(\pm|\lambda|^{1/2}r)$. The solutions for large r are acceptable only for an appropriate choice of sign, and at $r=2GM$, only if $\sigma=0$. The technique for finding eigenfunctions of this type is to integrate out from $r=2GM$ for trial values of λ. Only one value of λ was found to be consistent by this procedure, namely, $\lambda\simeq-0.19\,(GM)^{-2}$. If we start with $h_0(2GM)=h_1(2GM)=-k(2GM)+1$, then we discover that h_0, h_1, and k all tend to zero at infinity monotonically. This is in agreement with the naive expectation that the lowest eigenvalue corresponds to the "smoothest" eigenfunction. Defining the normalization as

$$N^2=\int_M \phi^{ab}_{(-)}\phi_{(-)ab}g^{1/2}d^4x\ , \tag{5.22}$$

we discover that

$$N^2\simeq 112\beta(GM)^{-4}\ . \tag{5.23}$$

Henceforth, we will deal with the normalized eigenfunction $\phi^{ab}_{(-)}N^{-1}=\tilde{\phi}^{ab}$.

We have therefore discovered an unstable mode for small fluctuations about the Schwarzschild instanton. This instanton is therefore not a strict minimum of the action, but rather a saddle point. Its role in the thermodynamics of hot gravity is similar to that of the configuration $x=a$ in our toy model (see Fig. 3). It behooves us to investigate what happens when we roll off the top of the barrier. We therefore consider the effect of an infinitesimal perturbation of this type. It generates the metric

$$g_{ab}=g_{ab}^{(\text{Schwarzschild})}+\epsilon\tilde{\phi}_{ab}\ . \tag{5.24}$$

This space is spherically symmetric, periodic, and static, but it is not a solution of the vacuum Einstein equations. To investigate further, we notice that since

$$g_{ab}=\text{diag}\left[\left[1-\frac{2GM}{r}\right]\left[1+\frac{\epsilon h_0}{N}\right],\left[1-\frac{2GM}{r}\right]^{-1}\left[1+\frac{\epsilon h_1}{N}\right],r^2\left[1+\frac{\epsilon k}{N}\right],r\left[1+\frac{\epsilon k}{N}\right]\sin^2\theta\right]\ , \tag{5.25}$$

it is convenient to define a new circumferential radial coordinate ρ so that spheres $\rho=$const have area $A=4\pi\rho^2$:

$$\rho=r\left[1+\frac{\epsilon k}{N}\right]^{1/2}\ . \tag{5.26}$$

Thus, the area of the fixed point of $\partial/\partial t$, a two-sphere, which used to have area $A=16\pi G^2M^2$ now has area $A=16\pi G^2M^2+0.94\epsilon G$. Note that this surface remains nonsingular without any change in the periodicity of t. Since the "mass" of such a two-sphere is defined to be $m=(A/16\pi G^2)^{1/2}$, this mass of the "bolt" of the new configuration is

$$M+0.0094\epsilon(GM)^{-1}\ . \qquad (5.27)$$

Another measure of mass is the total mass at infinity. This is determined by the trace of the second fundamental form on the boundary at infinity. This is determined by the boundary contribution to the action, and this mass remains unchanged. Accordingly we may think of the transformation as producing a spherically symmetric cloud of material outside the "bolt." This matter will have positive or negative energy density depending on whether ϵ is positive or negative. The action however, will decrease because although the boundary term remains constant, there is a contribution from the volume term:

$$I=4\pi GM^2-0.00094/(GM)^2\epsilon^2\ . \qquad (5.28)$$

Following the discussion of the toy model in Sec. II, we shall interpret the Schwarzschild instanton as indicating a finite probability for black holes, of mass $M=\beta/8\pi G$, to nucleate. The rate of nucleation will be calculated below. The unstable mode will correspond to the subsequent expansion (or collapse) of the black hole as it absorbs (or emits) thermal radiation.

One might be puzzled as to whether we should include such instantons at all in the functional integral. After all they are topologically distinct from flat space (Euler character 2 instead of zero), and perhaps there exists an infinite barrier that prevents such configurations from developing. This objection is clearly fallacious since the instanton action is finite; thus the nucleation rate is nonzero. However we can also show that by singular distortions of the metric (which however never cause the action to diverge), one can "continuously" deform the instanton into flat space. To do this we take the Schwarzschild metric and identify the Euclidean time with period β. However, we now take the mass at infinity m to be arbitrary rather than $M=\beta/8\pi G$. The resultant space will have a conical singularity at $r=2Gm$. The action is

$$I=\tfrac{1}{2}m\beta+2Gm^2\left[\frac{\beta}{4mG}-2\pi\right]$$
$$=m\beta-4\pi Gm^2\ . \qquad (5.29)$$

The action is extremal when $m=\beta/8\pi G$, as expected. This configuration allows for a continuous variation from zero action at $m=0$, flat space, to a maximum at $m=\beta/8\pi G$.

As m increases further, the action decreases without bound. This illustrates the well-known fact that in relativity a topology change can be accomplished continuously without generating a configuration with infinite action. This deformation cannot be achieved in perturbation theory. To generate such a perturbation, we would need to consider the tensor

$$\phi_{ab}=\frac{\partial}{\partial m}(g_{ab})$$
$$=\mathrm{diag}\left[-\frac{2}{r},\frac{2r}{(r-2GM)^2},0,0\right]\ . \qquad (5.30)$$

This is both transverse and tracefree. However, the norm of this mode N is given by

$$N^2=\int \phi^{ab}\phi_{ab}g^{1/2}d^4x \qquad (5.31)$$

and is divergent. It is therefore not included in the perturbation expansion about the Schwarzschild instanton.

The nonnormalizable mode may be regarded as a process whereby mass is directly moved in from infinity to the bolt. This is in contrast to the normalizable mode which corresponds to moving mass from a finite distance. An attempt to picture this is presented in Fig. 9.

Before proceeding to evaluate the contribution of these instantons to the partition functions we must

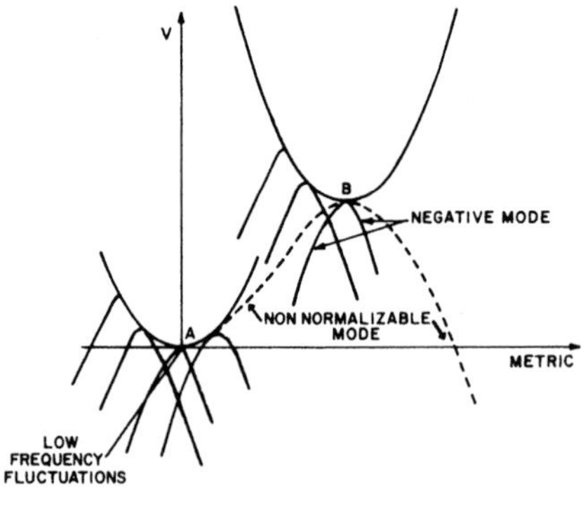

FIG. 9. The one-loop effective potential including the instanton contribution. (A) is flat spacetime. Decrease in the action from (A) represents the Jeans instability. The saddle point at (B) represents the Schwarzschild instanton. The continuous line representing a decrease of the action is the negative mode. The dashed line is the non-normalizable negative mode.

deal with a renormalization problem that does not occur in the flat-space calculation. It turns out that in curved space there is an ultraviolet divergence that occurs even at the one-loop level, which requires a separate counterterm. In a dimensional regularization scheme its contribution to the action is, on shell, (i.e., when g_{ab} is a solution of Einstein's equations),[50]

$$\Delta I = \frac{1}{16\pi^2}\frac{1}{n-4}\frac{53}{45}\int_M R_{abcd}R^{abcd}\sqrt{g}\,d^4 \,. \tag{5.32}$$

This term is proportional to the Euler character of the manifold M on which g_{ab} is defined. For flat space it vanishes, but is nonzero for the instanton. To deal with this one must introduce a regulator mass μ (which we presume should be taken of order the Planck mass m_P). This μ dependence will appear in the contribution of the instanton to the partition function. These remarks apply to all theories of gravity, including supergravity, except for a particular version of $N=8$ supergravity where this counterterm vanishes on shell.[51]

We shall now evaluate (approximately) the contribution to the partition function from the Gaussian fluctuations about each topological sector. This can be done using the standard techniques of ζ-function regularization.[52] The flat space contribution is, as before,

$$Z^{(0)} = \exp\left[\frac{V\pi^2}{45\beta^3}\right], \tag{5.33}$$

i.e., the contribution from thermal gravitons in a box of volume V at temperature $T=1/\beta$. The contribution from the Schwarzschild instanton is

$$Z^{(1)} = \frac{i}{2}\exp(-4\pi GM^2)\left[\exp\left[\frac{V\pi^2}{45\beta^3}\right]\right] \times(\mu\beta)^{212/45}\left[\frac{M}{2\pi\beta}\right]^{3/2} V \,. \tag{5.34}$$

The factor of $i/2$ occurs from the one normalizable negative mode; the first exponential contains the classical action of the Schwarzschild instanton. The second exponential arises from the thermal (= quantum) fluctuations about the instanton. At first sight, it might seem surprising that this term is identical to the flat-space result. However we note that the ultraviolet-finite part of $(1/V)\ln Z^{(1)}$ in the infinite-volume limit can only depend on T, since the factor of GM that appears in the background metric is itself inversely proportional to T. Therefore, $(1/V)\ln Z^{(1)}$ must, on dimensional grounds, be proportional to T^3. The constant of proportionality can be determined to be $\pi^2/45$ in the high-temperature limit. The factor $(\mu\beta)^{212/45}$ arises from the anomalous scaling behavior associated with the counterterm (5.32). The final term, proportional to the spatial volume V, occurs from the integration over the translational degrees of freedom of the instanton. The uncalculated finite part of the determinant, which is a temperature-independent constant, has been absorbed into the definition of μ. Written this way it looks like the single-particle partition function for an object of mass M at a temperature β^{-1} in a box of volume V. However, since $\beta=8\pi GM$, the final factors can be rewritten as $(V/64\pi^3G^{3/2})$, which is independent of M. Similarly, we can estimate the contribution from the N-instanton sector, neglecting the classical interaction between the instantons, as

$$Z^{(N)} = \left[\frac{i}{2}\right]^N \exp(-4\pi GNM^2)\exp\left[\frac{V\pi^2}{45\beta^3}\right](\mu\beta)^{212N/25}\frac{1}{N!}\left[\frac{V}{64\pi^3G^{3/2}}\right]^N . \tag{5.35}$$

Here we have assumed that there exists a negative mode associated with each bolt, thus giving a factor of i^N. The first exponential is the classical action of the N-instanton configuration, and the second the thermal graviton term. The final term results from the collective-coordinate integration. Since each instanton is indistinguishable this produces the familiar factor of $(1/N!)(V/64\pi^3G^{3/2})^N$. The above treatment is the standard dilute-gas approximation, which should be valid for widely separated instantons. If we now sum over all sectors,

$$Z = \sum_{N=0}^{\infty} Z^{(N)} , \tag{5.36}$$

we obtain

$$Z \simeq \exp\left[\frac{V\pi^2}{45\beta^3}\right]\exp\left[\left[\frac{i}{2}\right]\left[\frac{V}{64\pi^3G^{3/2}}\right](\mu\beta)^{212/45}\exp(-4\pi GM^2)\right] . \tag{5.37}$$

This calculation is analogous to the toy model of Sec. II. There are some differences however. In the case of potential theory we saw that for large temperatures the decay of the metastable vacuum was given by the static saddle point at the top of the barrier, and at low temperatures quantum tunneling also occurs. In our case there is only the former mechanism at all temperatures, and the height of the barrier is proportional to $\beta=1/T$. Even with these differences in mind we see no alternative to the conclusion that the instanton represents thermal nucleation of Schwarzschild metrics—black holes, with a nucleation rate per unit volume given by Γ, where

$$\Gamma=\frac{\omega_0\beta}{\pi}\,\mathrm{Im}\left[\frac{F}{V}\right], \tag{5.38}$$

where

$$\omega_0{}^2=-\lambda=\frac{0.19}{(GM)^2}=\left[\frac{1.74}{\beta}\right]^2. \tag{5.39}$$

Therefore the vacuum decay rate (per unit volume) is

$$\Gamma=\frac{0.87}{\beta}(\mu\beta)^{212/45}\frac{m_P{}^3}{64\pi^3}e^{-m_P{}^2/16\pi T^2}. \tag{5.40}$$

VI. INTERPRETATION AND CONCLUSIONS

In a typical low-temperature quantum decay problem, one may regard the Euclidean bounce solution (instanton) as describing the quantum tunneling from one classically allowed region to another. After the tunneling process, the system is in a highly excited state above the true vacuum. Therefore its subsequent evolution is essentially classical. The appropriate initial data are determined by the field of the Euclidean instanton at the moment it r aches its classical turning point, that is, when the mom nta conjugate to the field vanish. Alternatively, if the temperature is sufficiently high so that the rel vant instanton describes a process of thermal activation to the top of the barrier (as opposed to tunneling), then the only difference is that the subsequent classical evolution may be either toward the true vacuum or back to the original metastable state.[16]

In the case of the black-hole nucleation, the situation is slightly different since the instability is absent classically, but only arises due to the presence of thermal (=quantum) fluctuation. Therefore the evolution following black-hole nucleation is not quite classical, but may be consistently computed semiclassically. The necessary ingredients are simply the two effects of absorption of thermal radiation into the hole, plus the emission of thermal radiation due to the Hawking process.[53] The Schwarzschild instanton describes the nucleation of a black hole of critical mass $M=1/8\pi GT$, for which these two effects are in unstable equilibrium. Subsequent fluctuations will, with equal probability, cause the black hole either to grow indefinitely or to evaporate. The reason for this instability is that the black hole has negative specific heat. Far from the unstable equilibrium one can estimate the rates at which the black hole grows or shrinks.

Suppose that the black hole is much hotter than the surrounding matter. Then, if we were only to consider gravitons, the hole will lose mass at a rate of

$$\dot{M}\sim-\frac{1}{G^2M^2}. \tag{6.1}$$

On the other hand, if it were much cooler than the surrounding medium, it would accrete at a rate[54] of

$$\dot{M}\sim T_0{}^4(GM)^2, \tag{6.2}$$

T_0 being the temperature at the walls of the box.

It is amusing to apply our results on finite-temperature instabilities to the standard model of the early universe.[55]

The global expansion in a standard Friedmann-Robertson-Walker model is known to affect the Jeans instability drastically. Instead of exponentially growing large-scale fluctuations, linear perturbations can grow at most as a power of t. The actual spectrum of irregularities one finds at late times is extremely sensitive to assumed initial conditions. In addition to these linearized perturbations, one may try to estimate the probability of direct nucleation of black holes. The nucleation rate for pure gravity is given by Eq. (5.40). It may easily be extended to a general theory containing any number of matter fields. One finds for massless fields

$$\Gamma(T)=0.87T(\mu/T)^{\theta}\frac{m_P{}^3}{64\pi^3}\exp\left[-\frac{m_P{}^2}{16\pi T^2}\right], \tag{6.3}$$

where[56]

$$\theta=\tfrac{1}{45}(212N_2-\tfrac{233}{4}N_{3/2}-13N_1+\tfrac{7}{4}N_{1/2}+N_0). \tag{6.4}$$

N_s is the number of spin-s fields.

Suppose that the universe is given by a Robertson-Walker line element with $k=0$,

$$d^2s = -dt^2 + R^2(t)(dx^2+dy^2+dz^2) , \qquad (6.5)$$

then the number of black holes that have been nucleated in the period from time t_1 to time t_2 per unit comoving volume at time t will be

$$N \sim \frac{1}{R^3(t_2)} \int_{t_1}^{t_2} R^3(t)\Gamma(T(t))dt . \qquad (6.6)$$

The temperature of the universe $T(t)$ is related to the scale factor $R(t)$ by

$$T(t)R(t) = T(t_2)R(t_2) . \qquad (6.7)$$

Assuming the universe to be radiation dominated (as it surely must be when any significant nucleation takes place), then

$$R(t) \sim G^{1/4} t^{1/2} . \qquad (6.8)$$

Hence

$$T(t) \sim G^{-1/4} t^{-1/2} . \qquad (6.9)$$

Thus

$$N \sim \frac{1}{t_2^{3/2}} \int_{t_1}^{t_2} dt\, t^{1+\theta/2} \exp\left[-\frac{m_P{}^2 t}{16\pi}\right] G^{(\theta-1)/4} \times \frac{0.87\mu^\theta m_P{}^3}{64\pi^3} . \qquad (6.10)$$

As $t_1 \to 0$, this integral will only converge if $\theta > -4$. This estimate is presumably invalid for temperatures greater than the Planck temperature, and so for times t_1 less than the Planck time. However, it illustrates that the rate of black-hole production is sensitive to θ. It is impossible to estimate absolute rates without some knowledge of μ, although it is widely supposed that $\mu \sim m_P$. Presumably, this indicates that this process is quite important in the very early universe.

We can now estimate the probability of finding a black hole nucleated by this process. Any black hole will in fact evaporate in this scenario. Suppose a black hole of mass M is nucleated at $t=t_0$ at a temperature of $T_0 \sim G^{-1/3} t_0^{-1/2}$. A black hole, at best, can accrete at a rate of

$$\dot{M}(t) \sim M^2 T^4 G^2 \sim GM^2 t^{-2} . \qquad (6.11)$$

Thus, its mass $M(t)$ is given by

$$M(t) \sim \frac{t}{1+(t_0 G^{-1} - M_0)t/M_0 t_0} G^{-1} . \qquad (6.12)$$

Its temperature $T(t)$ is, as $t \to \infty$,

$$T(t) \sim \frac{t_0 G^{-1} - M_0}{M_0 t_0} , \qquad (6.13)$$

which is constant. Since the blackbody background temperature is cooling like $t^{-1/2}$, there will necessarily come a time when the hole will be hotter than its background, and it will evaporate. (This conclusion also holds in a matter-dominated era.)

From this, we see that any black hole we might observe must have an evaporation time scale longer than 1 Hubble time. This corresponds to an initial mass of around 10^{15} g, which would be nucleated at around $t \sim 10^{-6}$ sec. The number of such black holes per unit comoving volume now is

$$N \sim 10^{-143+37\theta/2} \exp(-10^{36}) \text{ cm}^{-3} \qquad (6.14)$$

assuming $\mu \sim m_P$.

This illustrates the incredible improbability of this phenomenon at any reasonable temperature.

Note added in proof. We understand that B. Allen[57] has confirmed some of the numerical work quoted in Sec. V.

ACKNOWLEDGMENT

D.J.G. and M.J.P. acknowledge support by National Science Foundation Grant No. PHY 80-19754.

APPENDIX A: ONE-DIMENSIONAL DETERMINANTS

Consider the equation

$$-\ddot{u}(t) + V(t)u(t) = \lambda u(t) \qquad (A1)$$

on the interval $0 \le t \le \tau$. Let $u_\lambda^1(t)$ $[u_\lambda^2(t)]$ be the solution with initial data $u_\lambda^1(0)=1$, $\dot{u}_\lambda^1(0)=0$ $[u_\lambda^2(0)=0, \dot{u}_\lambda^2(0)=1]$. Form the matrix

$$M_\lambda(\tau) \equiv \begin{bmatrix} u_\lambda^1(\tau) & u_\lambda^2(\tau) \\ \dot{u}_\lambda^1(\tau) & \dot{u}_\lambda^2(\tau) \end{bmatrix} . \qquad (A2)$$

Note that if $u(t)$ is any solution of (A1), then

$$\begin{bmatrix} u_\lambda(\tau) \\ \dot{u}_\lambda(\tau) \end{bmatrix} = M_\lambda(\tau) \begin{bmatrix} u_\lambda(0) \\ \dot{u}_\lambda(0) \end{bmatrix} . \qquad (A3)$$

Furthermore, $\det M_\lambda(\tau) = u_\lambda^1(\tau)\dot{u}_\lambda^2(\tau) - u_\lambda^2(\tau) \times \dot{u}_\lambda^1(\tau) = 1$ since the Wronskian is constant in time. Consequently, $-\partial_t{}^2 + V(t)$ on the interval $[0,\tau]$ with periodic boundary conditions has an eigenvector with eigenvalue λ if and only if $\text{tr}(M_\lambda(\tau) - 1) = 0$. This implies that

$$\det{}_+[(-\partial_t{}^2+V(t)-\lambda)/(-\partial_t{}^2-\lambda)]$$
$$=\mathrm{tr}(M_\lambda(\tau)-1)/\mathrm{tr}(M_\lambda^0(\tau)-1)\ , \qquad \text{(A4)}$$

where

$$M_\lambda^0(\tau)=\begin{bmatrix}\cos\sqrt{\lambda}\,\tau & (1/\sqrt{\lambda})\sin\sqrt{\lambda}\,\tau\\ -\sqrt{\lambda}\sin\sqrt{\lambda}\,\tau & \cos\sqrt{\lambda}\,\tau\end{bmatrix},$$

since both sides of the equation are meromorphic functions of λ with identical poles and zeros (and both go to one as $\lambda\to\infty$ with $\arg\lambda\neq 0$). Finally, we may define the overall normalization on functional determinants so that

$$\det{}_+[-\partial_t{}^2+V(t)-\lambda]=\mathrm{tr}(M_\lambda(\tau)-1)\ . \qquad \text{(A5)}$$

This agrees with the standard result for a harmonic oscillator,

$$(\det{}_+[-\partial_t{}^2+\omega^2])^{-1/2}$$
$$=\mathrm{Tr}\{\exp[-\beta(\tfrac{1}{2}\hat{p}^2+\tfrac{1}{2}\hat{\omega}^2\hat{x}^2)]\}$$
$$=e^{-\beta\hbar\omega/2}/(1-e^{-\beta\hbar\omega})\ .$$

since $\mathrm{tr}(M_0(\beta\hbar)-1)=2(\cos\beta\hbar\omega-1)$.

Functional determinants with any other choice of boundary conditions may also be computed using the matrix $M_\lambda(\tau)$. For example, determinants with antiperiodic and Dirichlet boundary conditions, respectively, are given by

$$\det{}_-(-\partial_t+V(t))=\mathrm{tr}(M_\lambda(\tau)+1)$$
and
$$\det{}_D(-\partial_t+V(t)-\lambda)=\pi\hbar[M_\lambda(\tau)]_{12}\ . \qquad \text{(A6)}$$

Finally, we must compute $\det'_+[-\partial_t+V''(\bar{x}(t))]$. Since

$$\det{}'_+(-\partial_t+V'')=-(\partial/\partial\lambda)\det(-\partial_t+V''-\lambda)\,|_{\lambda=0}$$
$$=-(\partial/\partial\lambda)\mathrm{tr}(M_\lambda(\tau)-1)\,|_{\lambda=0}$$

we need $M_\lambda(\tau)$ correct to $O(\lambda)$. This may be easily computed using the known zero-mode $\dot{\bar{x}}(t)$, and perturbing in λ. If the bounce is translated in time so that $\ddot{\bar{x}}(0)=0$, then

$$u_0^1(t)=\dot{\bar{x}}(t)/\dot{\bar{x}}(0),\quad u_0^2(t)=\dot{\bar{x}}(t)\dot{\bar{x}}(0)\int_0^t dt'/(\dot{\bar{x}}(t'))^2$$

and

$$u_\lambda^{1,2}(t)=u_0^{1,2}(t)+\lambda\int_0^t dt'[u_0^2(t)u_0^1(t')-u_0^1(t)u_0^2(t')]\times u_0^{1,2}(t')+O(\lambda^2).$$

This yields

$$\mathrm{tr}(M_\lambda(\tau)-1)=\lambda\int_0^\tau dt\int_0^\tau dt'(\dot{\bar{x}}(t)/\dot{\bar{x}}(t'))^2+O(\lambda^2)\ ,$$

so that

$$\det{}'_+[-\partial_t^2+V''(\bar{x}(t))]=-\left[\int_0^\tau dt(\dot{\bar{x}}(t))^2\right]\times\left[\int_0^\tau dt/(\dot{\bar{x}}(t))^2\right]$$
$$=W(E)\tau'(E)\ , \qquad \text{(A7)}$$

where

$$E=V(\bar{x}(t))-\tfrac{1}{2}(\dot{\bar{x}}(t))^2\ ,$$

$$W(E)=2\int_{x_1}^{x_2}dx[2(V(x)-E)]^{1/2}$$

and

$$\tau'(E)=(\partial/\partial E)2\int_{x_1}^{x_2}dx/[2(V(x)-E)]^{1/2}\ .$$

This verifies equation (2.10).

APPENDIX B: THE COMPONENTS OF EQ. (5.13) IN THE SCHWARZSCHILD INSTANTON

We display the components of $-\Box\phi^{ab}-2R^{abcd}\phi_{cd}=\lambda\phi^{ab}$ in Schwarzschild coordinates [we would like to thank Roberta Young for checking Eqs. (B1)–(B10) using CAMAL, an algebra handling program $(G=1)$]:

$$\frac{r}{r-2m}\frac{\partial^2}{\partial t^2}\phi^{00}+\frac{r-2m}{r}\frac{\partial^2}{\partial r^2}\phi^{00}+\frac{1}{r^2}\frac{\partial^2}{\partial\theta^2}\phi^{00}+\frac{1}{r^2\sin^2\theta}\frac{\partial^2}{\partial\phi^2}\phi^{00}+\frac{2(m+r)}{r^2}\frac{\partial}{\partial r}\phi^{00}+\frac{\cot\theta}{r^2}\frac{\partial}{\partial\theta}\phi^{00}$$
$$+\frac{2m^2}{r^3(r-2m)}\phi^{00}+\frac{4m}{(r-2m)^2}\frac{\partial}{\partial t}\phi^{01}+\frac{2m(2r-3m)}{r(r-2m)^3}\phi^{11}-\frac{2m}{(r-2m)}\phi^{22}-\frac{2m\sin^2\theta}{(r-2m)}\phi^{33}=-\lambda\phi^{00}\ , \qquad \text{(B1)}$$

$$\frac{r}{r-2m}\frac{\partial^2}{\partial t^2}\phi^{01}+\frac{r-2m}{r}\frac{\partial^2}{\partial r^2}\phi^{01}+\frac{1}{r^2}\frac{\partial^2}{\partial\theta^2}\phi^{01}+\frac{1}{r^2\sin^2\theta}\frac{\partial^2}{\partial\phi^2}\phi^{01}-\frac{2(m-r)}{r^2}\frac{\partial}{\partial r}\phi^{01}$$
$$+\frac{2m}{(r-2m)}\frac{\partial}{\partial t}\phi^{11}+\frac{\cot\theta}{r^2}\frac{\partial}{\partial\theta}\phi^{01}-\frac{2m}{r^2}\frac{\partial}{\partial t}\phi^{00}+\frac{2(2m-r)}{r^2}\frac{\partial}{\partial\theta}\phi^{02}+\frac{2(2m-r)}{r^2}\frac{\partial}{\partial\theta}\phi^{03}$$
$$-\frac{2r^2-4mr+4m^2}{r^3(r-2m)}\phi^{01}+\frac{2(2m-r)}{r^2}\cot\phi^{02}=-\lambda\phi^{01}\ , \qquad \text{(B2)}$$

$$\frac{r}{r-2m}\frac{\partial^2}{\partial t^2}\phi^{02}+\frac{r-2m}{r}\frac{\partial^2}{\partial r^2}\phi^{02}+\frac{1}{r^2}\frac{\partial^2}{\partial \theta^2}\phi^{02}+\frac{1}{r^2\sin^2\theta}\frac{\partial^2}{\partial \theta^2}\phi^{02}+\frac{4(r-m)}{r^2}\frac{\partial}{\partial r}\phi^{02}$$
$$+\frac{\cot\theta}{r^2}\frac{\partial}{\partial\theta}\phi^{02}+\frac{2m}{(r-2m)^2}\frac{\partial}{\partial t}\phi^{12}+\frac{2}{r^3}\frac{\partial}{\partial\theta}\phi^{01}-\frac{2\cot\theta}{r^2}\frac{\partial}{\partial\theta}\phi^{03}+\left[\frac{4m}{r^3}+\frac{1}{r^2}(1-\cot^2\theta)\right]\phi^{02}=-\lambda\phi^{02}\,, \tag{B3}$$

$$\frac{r}{r-2m}\frac{\partial^2}{\partial t^2}\phi^{03}+\frac{r-2m}{r}\frac{\partial^2}{\partial r^2}\phi^{03}+\frac{1}{r^2}\frac{\partial^2}{\partial \theta^2}\phi^{03}+\frac{1}{r^2\sin^2\theta}\frac{\partial^2}{\partial \phi^2}\phi^{03}+\frac{4(r-m)}{r^2}\frac{\partial}{\partial r}\phi^{03}+\frac{3\cot\theta}{r^2}\frac{\partial}{\partial\theta}\phi^{03}$$
$$+\frac{2m}{(r-2m)^2}\frac{\partial}{\partial t}\phi^{13}+\frac{2}{r^3\sin^2\theta}\frac{\partial}{\partial\phi}\phi^{01}+\frac{2\cot\theta}{r^2\sin^2\theta}\frac{\partial}{\partial\phi}\phi^{02}+\frac{2m}{r^3}\phi^{03}=-\lambda\phi^{03}\,, \tag{B4}$$

$$\frac{r}{r-2m}\frac{\partial^2}{\partial t^2}\phi^{11}+\frac{r-2m}{r}\frac{\partial^2}{\partial r^2}\phi^{11}+\frac{1}{r^2}\frac{\partial^2}{\partial \theta^2}\phi^{11}+\frac{1}{r^2\sin^2\theta}\frac{\partial^2}{\partial \phi^2}\phi^{11}-\frac{2(3m-r)}{r^2}\frac{\partial}{\partial r}\phi^{11}$$
$$+\frac{\cot\theta}{r^2}\frac{\partial}{\partial\theta}\phi^{11}-\frac{4m}{r^2}\frac{\partial}{\partial t}\phi^{01}+\frac{4}{r^2}(2m-r)\frac{\partial}{\partial\theta}\phi^{12}+\frac{4}{r^2}(2m-r)\frac{\partial}{\partial\theta}\phi^{12}+\frac{4}{r^2}(2m-r)\frac{\partial}{\partial\phi}\phi^{13}$$
$$+\frac{2m(2r-3m)(r-2m)}{r^5}\phi^{00}+\frac{16mr-14m^2-4r^2}{r^3(r-2m)}\phi^{11}+\frac{2(r-2m)(r-3m)}{r^2}\phi^{22}$$
$$+\frac{2(r-2m)(r-3m)}{r^2}\phi^{33}\sin^2\theta+\frac{4(2m-r)}{r^2}\cot\theta\phi^{12}=-\lambda\phi^{11}\,, \tag{B5}$$

$$\frac{r}{r-2m}\frac{\partial^2}{\partial t^2}\phi^{12}+\frac{r-2m}{r}\frac{\partial^2}{\partial r^2}\phi^{12}+\frac{1}{r^2}\frac{\partial^2}{\partial \theta^2}\phi^{12}+\frac{1}{r^2\sin^2\theta}\frac{\partial^2}{\partial \phi^2}\phi^{12}+\frac{4(r-2m)}{r^2}\frac{\partial}{\partial r}\phi^{12}+\frac{\cot\theta}{r^2}\frac{\partial}{\partial\theta}\phi^{12}-\frac{2m}{r^2}\frac{\partial}{\partial t}\phi^{02}$$
$$+\frac{2}{r^2}(2m-r)\frac{\partial}{\partial\theta}\phi^{22}+\frac{2}{r^2}(2m-r)\frac{\partial}{\partial\phi}\phi^{23}+\frac{2}{r^3}\frac{\partial}{\partial\theta}\phi^{11}-\frac{2\cot\theta}{r^3}\frac{\partial}{\partial\theta}\phi^{13}+\frac{1}{r^2}(1-\cot^2\theta)\phi^{12}$$
$$+\frac{2(2m-r)}{r^2}\cot\theta\phi^{23}+\frac{2(r-2m)}{r^2}\sin\theta\cos\theta\phi^{33}+\frac{4(2m-r)}{r^3}\phi^{12}=-\lambda\phi^{12}\,, \tag{B6}$$

$$\frac{r}{r-2m}\frac{\partial^2}{\partial t^2}\phi^{13}+\frac{r-2m}{r}\frac{\partial^2}{\partial r^2}\phi^{13}+\frac{1}{r^2}\frac{\partial^2}{\partial \theta^2}\phi^{13}+\frac{1}{r^2\sin^2\theta}\frac{\partial^2}{\partial \phi^2}\phi^{13}+\frac{4(r-2m)}{r^2}\frac{\partial}{\partial r}\phi^{13}$$
$$+\frac{3\cot\theta}{r^2}\frac{\partial}{\partial\theta}\phi^{13}-\frac{2m}{r^2}\frac{\partial}{r^2}\phi^{03}+\frac{2(2m-r)}{r^2}\frac{\partial}{\partial\theta}\phi^{23}+\frac{2}{r^2}(2m-r)\frac{\partial}{\partial\phi}\phi^{33}+\frac{2}{r^3\sin^2\theta}\frac{\partial}{\partial\phi}\phi^{11}$$
$$+\frac{2\cot\theta}{r^2\sin^2\theta}\frac{\partial}{\partial\phi}\phi^{12}+\frac{6(2m-r)\cot\theta}{r^2}\phi^{23}-\frac{4(r-2m)}{r^3}\phi^{13}=-\lambda\phi^{13}\,, \tag{B7}$$

$$\frac{r}{r-2m}\frac{\partial^2}{\partial t^2}\phi^{22}+\frac{r-2m}{r}\frac{\partial^2}{\partial r^2}\phi^{22}+\frac{1}{r^2}\frac{\partial^2}{\partial \phi^2}\phi^{22}+\frac{1}{r^2\sin^2\theta}\frac{\partial^2}{\partial \theta^2}\phi^{22}+\frac{6r-10m}{r^2}\frac{\partial}{\partial r}\phi^{22}+\frac{\cot\theta}{r^2}\frac{\partial}{\partial\theta}\phi^{22}$$
$$+\frac{4}{r^3}\frac{\partial}{\partial\theta}\phi^{12}-\frac{4\cot\theta}{r^2}\frac{\partial}{\partial\phi}\phi^{23}+\frac{2m(2m-r)}{r^6}\phi^{00}+\frac{2(r-3m)}{r^4(r-2m)}\phi^{11}+\frac{4m\sin^2\theta}{r^3}\phi^{33}+\frac{2\cos^2\theta}{r^2}\phi^{33}$$
$$+\frac{2(r-2m)}{r^3}\phi^{22}+\frac{2}{r^2}(1-\cot^2\theta)\phi^{22}=-\lambda\phi^{22}\,, \tag{B8}$$

$$\frac{r}{r-2m}\frac{\partial^2}{\partial t^2}\phi^{23}+\frac{r-2m}{r}\frac{\partial^2}{\partial r^2}\phi^{23}+\frac{1}{r^2}\frac{\partial^2}{\partial \theta^2}\phi^{23}+\frac{1}{r^2\sin^2\phi}\frac{\partial^2}{\partial \phi^2}\phi^{23}+\frac{6r-10m}{r^2}\frac{\partial}{\partial r}\phi^{23}+\frac{3\cot\theta}{r^2}\frac{\partial}{\partial\theta}\phi^{23}$$
$$+\frac{2}{r^3}\frac{\partial}{\partial\theta}\phi^{13}-\frac{2\cot\theta}{r}\frac{\partial}{\partial\phi}\phi^{33}+\frac{2}{r^3\sin^2\theta}\frac{\partial}{\partial\phi}\phi^{12}+\frac{2\cot\theta}{r^2\sin^2\theta}\frac{\partial}{\partial\phi}\phi^{22}+\left[\frac{3}{r^2}(1-\cot^2\theta)-\frac{8m}{r^3}\right]\phi^{23}=-\lambda\phi^{23}\,, \tag{B9}$$

$$\frac{r}{r-2m}\frac{\partial^2}{\partial t^2}\phi^{33}+\frac{r-2m}{r}\frac{\partial^2}{\partial r^2}\phi^{33}+\frac{1}{r^2}\frac{\partial^2}{\partial \theta^2}\phi^{33}+\frac{1}{r^2\sin^2\theta}\frac{\partial^2}{\partial \phi^2}\phi^{33}+\frac{6r-10m}{r_2}\frac{\partial}{\partial r}\phi^{33}$$
$$+\frac{5\cot\theta}{r^2}\frac{\partial}{\partial\theta}\phi^{33}+\frac{4}{r^3\sin^2\theta}\frac{\partial}{\partial\phi}\phi^{13}+\frac{4\cot\theta}{r^2\sin^2\theta}\frac{\partial}{\partial\phi}\phi^{23}+\frac{2m(2m-r)}{r^6\sin^2\theta}\phi^{00}+\frac{2(r-3m)}{r^4\sin^2\theta(r-2m)}\phi^{11}$$
$$+\frac{4\cot\theta}{r^3\sin^3\theta}\phi^{12}+\frac{(4m+2r\cot^2\theta)}{r^3\sin^2\theta}\phi^{22}+\frac{2}{r^3}[(r-2m)+r\cot^2\theta]\phi^{33}=-\lambda\phi^{33}\ . \quad \text{(B10)}$$

These equations are closely related to those of Edelstein and Vishveshwara,[58] who considered perturbations to the Lorentzian Schwarzschild solution, but in a gauge different from ours.

---

[1]J. Jeans, Philos. Trans. Rev. Soc. London A199, 491 (1902).

[2]S. Weinberg, *Gravitation and Cosmology* (Wiley, New York, 1972).

[3]R. Emden, *Gaskugeln* (Teubner, Leipzig, 1907).

[4]S. Chandrasekhar, *An Introduction to the Theory of Stellar Structure* (University of Chicago Press, Chicago, 1939).

[5]J. R. Oppenheimer and G. M. Volkoff, Phys. Rev. 55, 374 (1939).

[6]J. B. Hartle, Phys. Rep. 46, 201 (1978).

[7]S. W. Hawking and R. Penrose, Proc. R. Soc. London A314, 529 (1970).

[8]R. Penrose, Riv. Nuovo Cimento 1, 252 (1969).

[9]G. W. Gibbons, S. W. Hawking, and M. J. Perry, Nucl. Phys. B138, 141 (1978).

[10]R. Schoen and S. T. Yau, Phys. Rev. Lett. 42, 547 (1979).

[11]R. Shoen and S. T. Yau, Commun. Math. Phys. 65, 45 (1979); 79, 231 (1981).

[12]For another proof of these theorems, see E. Witten, Commun. Math. Phys. 80, 381 (1981).

[13]S. W. Hawking, in *General Relativity: An Einstein Centennial Survey,* edited by S. W. Hawking and W. Israel (Cambridge University Press, London, 1979).

[14]An argument for the existence of such modes was previously given by D. N. Page (unpublished).

[15]R. P. Feynman and A. R. Hibbs, *Quantum Mechanics and Path Integrals* (McGraw-Hill, New York, 1965).

[16]J. S. Langer, Ann. Phys. (N.Y.) 41, 108 (1967); 54, 258 (1969); C. Callan and S. Coleman, Phys. Rev. D 16, 1762 (1977).

[17]S. Coleman, in *The Ways of Sub-Nuclear Physics,* edited by A. Zichichi (Plenum, New York, 1979).

[18]I. Affleck, Phys. Rev. Lett. 46, 388 (1981).

[19]S. Coleman, Phys. Rev. D 15, 2929 (1977); M. Stone, Phys. Lett. 67B, 186 (1977).

[20]H. Leutwyler, Nuovo Cimento 42, 159 (1966); J. W. York, Phys. Rev. Lett. 28, 1082 (1972); G. W. Gibbons and S. W. Hawking, Phys. Rev. D 15, 2752 (1977).

[21]R. Arnowitt, S. Deser, and C. W. Misner, Phys. Rev. 116, 1322 (1959).

[22]L. D. Faddeev and V. N. Popov, Phys. Lett. 25B, 29 (1967).

[23]D. Gross, R. Pisarski, and L. G. Yaffe, Rev. Mod. Phys. 53, 43 (1981).

[24]L. D. Faddeev and V. N. Popov, Usp. Fiz. Nauk, 111, 427 (1973) [Sov. Phys. Usp. 16, 777 (1974)].

[25]H. Leutwyler, Phys. Rev. 134, B1155 (1964); E. S. Fradkin and G. Vilkovisky, Phys. Rev. D 8, 4241 (1973); A. Hanson, T. Regge, and C. Teitelboim, Ann. Phys. (N.Y.) 88, 286 (1974).

[26]B. S. DeWitt, Phys. Rev. 160, 1113 (1967).

[27]S. W. Hawking, in *Recent Developments in Gravitation,* edited by M. Levy (Plenum, New York, 1979).

[28]For example, there has been much discussion about questions such as what is the measure for quantum gravity; should all topologies consistent with the boundary conditions be included; and is the Euclidean approach relevant?

[29]G. W. Gibbons and M. J. Perry, Nucl. Phys. B146, 90 (1978).

[30]K. Yano and T. Nogano, Ann. Math. 69, 451 (1959).

[31]M. J. Perry in *Superspace and Supergravity,* edited by S. W. Hawking and M. Rocek (Cambridge University Press, London, 1981).

[32]C. Bernard, Phys. Rev. D 9, 3312 (1974).

[33]D. Pines, *Elementary Excitations in Solids* (Benjamin, New York, 1964).

[34]We have not carried out the rather nasty one-loop computation of m_g for pure gravity. However, we have faith in our general arguments, which was confirmed by the fermion calculations.

[35]G. W. Gibbons and M. J. Perry, Proc. R. Soc. London A358, 467 (1978).

[36]G. W. Gibbons and S. W. Hawking, Phys. Rev. D 15, 2752 (1977).

[37]G. W. Gibbons and S. W. Hawking, Commun. Math. Phys. 66, 291 (1979).

[38]A. S. Lapedes, Phys. Rev. D 22, 1837 (1980).

[39]G. W. Gibbons and M. J. Perry, Phys. Rev. D 22, 313 (1980).

[40]I. Hauser and F. J. Ernst, Phys. Rev. D 20, 362 (1979); 20, 1783 (1979).

[41]C. Hoenselaers, W. Kinnersley, and B. C. Xantho-

poulos, J. Math. Phys. 20, 2530 (1979).
[42]G. Neugebauer, J. Phys. A 13, L19 (1980); D. Kramer and G. Neugebauer, Phys. Lett. 75A, 259 (1980).
[43]T. Regge and J. A. Wheeler, Phys. Rev. 108, 1063 (1957).
[44]C. V. Vishveshwara, Phys. Rev. D 1, 2870 (1970).
[45]F. Zerilli, Phys. Rev. Lett. 24, 737 (1970).
[46]W. Press and S. A. Teukolsky, Astrophys. J. 185, 649 (1973); 193, 443 (1974).
[47]J. M. Stewart, Proc. R. Soc. London A334, 51 (1975).
[48]S. Chandrasekhar, in *General Relativity: An Einstein Centennial Survey*, edited by S. W. Hawking and W. Israel (Cambridge University Press, London, 1979).
[49]G. 't Hooft, Phys. Rev. D 14, 3432 (1976).
[50]H-S Tsao, Phys. Lett. 68B, 79 (1977).
[51]S. M. Christensen, M. J. Duff, G. W. Gibbons, and M. Rocek, Phys. Rev. Lett. 45, 161 (1980).
[52]S. W. Hawking, Commun. Math. Phys. 55, 133 (1977).
[53]S. W. Hawking, Commun. Math. Phys. 43, 199 (1975).
[54]B. Carter, G. W. Gibbons, D. N. C. Lin, and M. J. Perry, Astron. Astrophys. 52, 427 (1976).
[55]P. J. E. Peebles, *Physical Cosmology* (Princeton University Press, Princeton, N. J., 1971).
[56]M. J. Perry, Nucl. Phys. B143, 114 (1978); S. M. Christensen and M. J. Duff, *ibid.* B154, 301 (1979).
[57]B. Allen, personal communication.
[58]L. A. Edelstein and C. V. Vishveshwara, Phys. Rev. D 1, 3514 (1970).

Thermal stress tensors in static Einstein spaces

Don N. Page
Department of Physics, Pennsylvania State University, University Park, Pennsylvania 16802
(Received 30 November 1981)

The Bekenstein-Parker Gaussian path-integral approximation is used to evaluate the thermal propagator for a conformally invariant scalar field in an ultrastatic metric. If the ultrastatic metric is conformal to a static Einstein metric, the trace anomaly vanishes and the Gaussian approximation is especially good. One then gets the ordinary flat-space expressions for the renormalized mean-square field and stress-energy tensor in the ultrastatic metric. Explicit formulas for the changes in $\langle\phi^2\rangle$ and $\langle T_{\mu\nu}\rangle$ resulting from a conformal transformation of an arbitrary metric are found and used to take the Gaussian approximations for these quantities in the ultrastatic metric over to the Einstein metric. The result for $\langle\phi^2\rangle$ is exact for de Sitter space and agrees closely with the numerical calculations of Fawcett and Whiting in the Schwarzschild metric. The result for $\langle T_{\mu\nu}\rangle$ is exact in de Sitter space and the Nariai metric and is close to Candelas's values on the bifurcation two-sphere in the Schwarzschild metric. Thus one gets a good closed-form approximation for the energy density and stresses of a conformal scalar field in the Hartle-Hawking state everywhere outside a static black hole.

I. INTRODUCTION

It is by now well known that a black hole emits thermal radiation. Hawking's original calculation of this effect[1] was in terms of particles that escape to an asymptotically flat region far from the hole where they can be unambiguously defined. In the curved spacetime near the hole, the notion of a particle is ill-defined, so one needs to use other concepts to describe the radiation there. One important such quantity is the renormalized expectation value of the stress-energy tensor[2] $T_{\mu\nu}$, which gives a measure of the back-reaction of the radiation on the geometry via the Einstein equations in the semiclassical approximation. Another important quantity for a real scalar field is the renormalized value of the mean square field $\langle\phi^2\rangle$, which may determine the extent of symmetry restoration near a black hole in theories with spontaneous symmetry breaking.[3]

However, calculations of $T_{\mu\nu}$ and $\langle\phi^2\rangle$ are very difficult and have only been performed in a number of relatively simple spacetimes. The most general class in which exact values are known are the conformally flat metrics, for which the method of Brown and Cassidy[4,5] yields explicit results for $T_{\mu\nu}$ under appropriate boundary conditions.[6,7] But the stress tensor near a black hole is still not known. Christensen and Fulling showed that it depends on the radial energy flux and one unknown function of radius.[8] Candelas expressed the Feynman propagator of a massless scalar field in terms of mode functions which he was able to evaluate numerically near the bifurcation two-sphere to determine $T_{\mu\nu}$ there,[9] but his calculations did not determine it elsewhere. He was also able to evaluate $\langle\phi^2\rangle$ exactly on the bifurcation two-sphere,[9] and Fawcett and Whiting have extended this knowledge to the radial variation of $\langle\phi^2\rangle$ by a numerical calculation.[10]

In this paper the Gaussian path-integral approximation of Bekenstein and Parker[11] is used to construct a thermal propagator for a conformally invariant scalar field in any ultrastatic metric (static with $g_{00}=\text{const}$). A conformal transformation then gives the approximate propagator in any static metric. If this static metric is an Einstein metric ($R_{\mu\nu}=\Lambda g_{\mu\nu}$), the approximation is especially good because the a_2 coefficient[2] vanishes in the corresponding ultrastatic metric. The propagator is then used to get approximate formulas for $\langle\phi^2\rangle$ and $T_{\mu\nu}$. The formula for $\langle\phi^2\rangle$ is exact in de Sitter space and agrees very well with the numerical calculations of Fawcett and Whiting in the Schwarzschild metric. The approximate stress tensor is obtained first in the ultrastatic metric, and then the methods of Brown and Cassidy[4] are extended to give it in the conformally related static

Einstein metric, where it is conserved and has the correct trace. The resulting $T_{\mu\nu}$ is exact in the de Sitter metric and in the Nariai[12] metric. For the Hartle-Hawking[13] thermal state in the Schwarzschild metric, the polynomial expression for $T_{\mu\nu}$ reduces at infinity to that of thermal radiation in flat spacetime and closely reproduces Candelas's expression at the bifurcation two-sphere.

II. PROPAGATOR IN THE OPTICAL METRIC

To consider the thermal properties of a conformally invariant field in a static spacetime, it is helpful to use the optical metric,[14,15] which is the conformally related ultrastatic metric

$$g_{\mu\nu}=\Omega^{-2}\bar{g}_{\mu\nu} . \tag{1}$$

Here Ω^{-2} is a space-dependent conformal factor applied to the physical metric $\bar{g}_{\mu\nu}$ to make $g_{tt}=-1$. It is also convenient to analytically continue $\tau=it$ to real values to obtain a positive-definite (Riemannian rather than Lorentzian) metric in terms of the "Euclidean" time τ, so $g_{00}=1$.

The thermal propagator at temperature T is the Green's function which approaches zero for large spatial separations (for a noncompact metric) and which is periodic[13,14] with period $2\pi\kappa^{-1}=T^{-1}$ in τ. This periodicity may be imposed by identifying τ with $\tau+T^{-1}$ in the Riemannian metric. For a conformally invariant real scalar field ϕ,

$$\begin{aligned}(-\nabla^{\mu}\nabla_{\mu}+\tfrac{1}{6}R)G(x,x')&=\delta(x,x')\\ &\equiv g^{1/2}\delta^{4}(x^{\alpha}-x'^{\alpha}) .\end{aligned} \tag{2}$$

For simplicity, the propagator $G(x,x')$ is defined here without the factor of i that would occur on the right-hand side if one made an analytic continuation from the corresponding real equation in the Lorentzian metric. The propagator in the physical metric is obtained by the conformal transformation

$$\bar{G}(x,x')=\Omega^{-1}(x)G(x,x')\Omega^{-1}(x') . \tag{3}$$

The Schwinger-DeWitt proper-time formalism[16,17,2] with imaginary proper time $s=-iu$ expresses

$$G(x,x')=\int_{0}^{\infty}K(x,x',u)\,du \tag{4}$$

in terms of the heat kernel $K(x,x',u)$, which obeys

$$\left[\frac{\partial}{\partial u}-\nabla^{\mu}\nabla_{\mu}+\tfrac{1}{6}R\right]K(x,x',u)=0 \tag{5}$$

with the boundary condition

$$K(x,x',0)=\delta(x,x') . \tag{6}$$

[Note that my $K(x,x',u)$ is a biscalar and not a bidensity[17] because of my definition (2) of $\delta(x,x')$ as the invariant biscalar Dirac δ function.] Since the ultrastatic optical metric is the product metric

$$ds^{2}=g_{\mu\nu}dx^{\mu}dx^{\nu}=d\tau^{2}+g_{ij}dx^{i}dx^{j} , \tag{7}$$

the heat kernal may be factorized as

$$K(x,x',u)=K_{1}(\tau,\tau',u)K_{3}(\vec{x},\vec{x}\,',u) , \tag{8}$$

where

$$\left[\frac{\partial}{\partial u}-\frac{\partial^{2}}{\partial\tau^{2}}\right]K_{1}(\tau,\tau',u)=0 , \tag{9}$$

$$\left[\frac{\partial}{\partial u}-\nabla^{i}\nabla_{i}+\tfrac{1}{6}R\right]K_{3}(\vec{x},\vec{x}\,',u)=0 \tag{10}$$

and $\vec{x}$ stands for the three spatial variables x^{i}. Equation (9) has the appropriate periodic solution

$$K_{1}(\tau,\tau',u)=\frac{\kappa}{2\pi}\sum_{n=-\infty}^{+\infty}\exp[-\kappa^{2}n^{2}u+i\kappa n(\tau-\tau')] . \tag{11}$$

Equation (10) is difficult to solve exactly, but Bekenstein and Parker[11] have evaluated the Gaussian approximation to the path-integral representation of the heat kernel, which gives

$$K_{3\,\mathrm{Gauss}}(\vec{x},\vec{x}\,',u)=(4\pi u)^{-3/2}\Delta^{1/2}(\vec{x},\vec{x}\,')\exp[-{}^{(3)}\sigma(\vec{x},\vec{x}\,')/2u] . \tag{12}$$

Here ${}^{(3)}\sigma(\vec{x},\vec{x}\,')$ is the three-dimensional geodetic interval biscalar,[17] one-half the square of geodesic distance between $\vec{x}\,'$ and $\vec{x}$ in the optical three-space metric g_{ij}, and

$$\Delta(\vec{x},\vec{x}\,')=g^{-1/2}(\vec{x})\det(-\partial^{2\,(3)}\sigma/\partial x^{i}\partial x'^{j})g^{-1/2}(\vec{x}\,') \tag{13}$$

is a biscalar which obeys[17]

$$\Delta^{-1}(\Delta^{(3)}\sigma^{;i})_{;i}=d=3 \tag{14}$$

in a d-dimensional space and has $\Delta(\vec{x},\vec{x})=1$.

One may compare (12), which applies for all $u\geq 0$, with the Schwinger-DeWitt expansion[17,2]

$$K_3(\vec{x},\vec{x}\,',u)=K_{3\,\mathrm{Gauss}}(\vec{x},\vec{x}\,',u)\sum_{n=0}^{\infty}a_n(\vec{x},\vec{x}\,')u^n\ . \tag{15}$$

Since $a_0(\vec{x},\vec{x}\,')=1$, the Gaussian approximation is equivalent to taking only the first term in the power series, which is also called the WKB approximation.[11] When $\vec{x}$ is near $\vec{x}\,'$, Christensen[18] has given the leading terms in an expansion of $a_1(\vec{x},\vec{x}\,')$ and $a_2(\vec{x},\vec{x}\,')$ in a power series in the separation of $\vec{x}$ and $\vec{x}\,'$. The coefficients are the values and derivatives of these biscalars in the coincidence limit ($\vec{x}=\vec{x}\,'$, denoted by square brackets). $[a_1]=0$ and $[a_{1;\mu}]=0$ automatically for a conformally invariant field ($\xi=\frac{1}{6}$ in his notation), but

$$[a_{1;\nu}^{\ ;\mu}]=\tfrac{1}{180}(2R^{\mu\alpha\beta\gamma}R_{\nu\alpha\beta\gamma}+2R^{\beta}_{\alpha}R^{\alpha\mu}{}_{\beta\nu}-4R^{\mu}_{\alpha}R^{\alpha}_{\nu}+3R^{\mu;\alpha}_{\nu;\alpha}-R^{;\mu}_{;\nu})\ , \tag{16}$$

$$[a_2]=\tfrac{1}{180}(R^{\alpha\beta\gamma\delta}R_{\alpha\beta\gamma\delta}-R^{\alpha\beta}R_{\alpha\beta}+R^{;\alpha}_{;\alpha})\ . \tag{17}$$

Now one can easily show that both of these expressions vanish in the optical metric of any static Einstein metric. This suggests that (12) may be a much better approximation in those metrics than one would otherwise have expected, at least when $\vec{x}$ is near $\vec{x}\,'$. Such a cancellation does not occur in the physical metric, which gives an added motivation besides the factorization (8) for using the optical metric.

By inserting (11) and (12) into (8) and performing the integration (4) one obtains the Gaussian approximation for the propagator,

$$G_{\mathrm{Gauss}}(\tau,\vec{x};0,\vec{x}\,')=\frac{\kappa\Delta^{1/2}\sinh\kappa r}{8\pi^2 r\,(\cosh\kappa r-\cos\kappa\tau)}\ . \tag{18}$$

Here $r=(2^{(3)}\sigma)^{1/2}$ is the spatial distance be ween the two points in the optical metric, and τ is the Euclidean time separation. Δ is given by (13) in terms of the three-dimensional geodetic interval ${}^{(3)}\sigma$, but clearly it would have the same value if ${}^{(3)}\sigma$ were replaced by the four-dimensional geodetic interval $\sigma={}^{(3)}\sigma+\frac{1}{2}(\tau-\tau')^2$ and the four-dimensional Van-Vleck determinant were taken:

$$\Delta=\Delta(x,x')=g^{-1/2}(x)\det(-\partial^2\sigma/\partial x^{\mu}\partial x'^{\nu})g^{-1/2}(x')\ . \tag{19}$$

Actually Bekenstein and Parker say one should sum over all spatial geodesics connecting $\vec{x}$ and $\vec{x}\,'$. [Geodesics in which τ changes by a different multiple of the periodicity are already counted in (11).] However, to get closed-form expressions for the behavior of the propagator when $\vec{x}$ is near $\vec{x}\,'$, and to avoid the effects of caustics, I will consider only the shortest geodesic connecting the two po nts. Then for a calculation of $\langle\phi^2\rangle$ and $T_{\mu\nu}$ only the first few terms of an expansion in Riemann normal coordinates will be needed.[18] The direct geodesic should give the dominant contribution, but by neglecting other possible geodesics I am incurring another error in addition to that of the Bekenstein-Parker Gaussian approximation.

III. RENORMALIZED MEAN-SQUARE FIELD IN THE PHYSICAL METRIC

To obtain the mean-square field and the stress tensor, one takes the coincidence limit ($x\to x'$) of the propagator and the appropriate combination of its derivatives. However, since these formal expressions diverge, one needs to regularize them. I will use the modified point-separation method of Wald[19] and of Adler, Lieberman, and Ng,[20] as corrected by Wald,[21] in which one subtracts from the propagator a locally determined Hadamard elementary solution[22]

$$G^L(x,x')=(4\pi)^{-2}\Delta^{1/2}(2\sigma^{-1}+v\ln\sigma+w)\ . \tag{20}$$

Here

$$v=\sum_{n=0}^{\infty}v_n(x,x')\sigma^n \quad\text{and}\quad w=\sum_{n=0}^{\infty}w_n(x,x')\sigma^n \tag{21}$$

are determ ned recursively[23,24,20] with $w_0(x,x')$, which is arbitrary for a general Hadamard solution, set equal to zero.[20,21] The boundary-condition-dependent remainder

$$G^B(x,x')=G(x,x')-G^L(x,x') \tag{22}$$

then gives finite results for $\langle\phi^2\rangle$ and $T_{\mu\nu}$, although Wald found a trace term must be added to $T_{\mu\nu}$ to correct for the fact that $G^B(x,x')$ is not symmetric.[21] Thus

$$\langle\phi^2(x)\rangle=G^B(x,x)=\lim_{x\to x'}[G(x,x')-G^L(x,x')] \ . \tag{23}$$

(Note that my propagators are normalized at half the value of those in Adler, Lieberman, and Ng[20] and in Wald[21].)

Now one can apply the conformal transformation (3) to (18) to obtain the approximate propagator in the physical metric, write a truncated expansion for it and the Hadamard solution (20) for x near x', and take the coincidence limit (23):

$$\langle\bar\phi^2\rangle_{\text{Gauss}}=\lim_{x\to x'}(4\pi)^{-2}[\Omega^{-1}(x)\Delta^{1/2}\Omega^{-1}(x')(2\sigma^{-1}+\tfrac{1}{3}\kappa^2)-\bar\Delta^{1/2}(2\bar\sigma^{-1})] \ . \tag{24}$$

The barred quantities are the corresponding biscalars in the physical metric $\bar g_{\mu\nu}$ as opposed to the optical metric $g_{\mu\nu}$. By using the expansions[17,18]

$$\Delta^{1/2}=1+\tfrac{1}{12}R_{\alpha\beta}\sigma^\alpha\sigma^\beta+\cdots \ , \tag{25a}$$

$$\bar\Delta^{1/2}=1+\tfrac{1}{12}\bar R_{\alpha\beta}\bar\sigma^\alpha\bar\sigma^\beta+\cdots=1+\tfrac{1}{12}[R_{\alpha\beta}+2\Omega(\Omega^{-1})_{;\alpha\beta}-\tfrac{1}{2}\Omega^{-2}(\Omega^2)_{;\gamma}^{;\gamma}g_{\alpha\beta}]\sigma^\alpha\sigma^\beta+\cdots \ , \tag{25b}$$

in terms of the coordinates

$$\sigma^\alpha=g^{\alpha\beta}(x)\sigma(x,x')_{,\beta}, \quad \bar\sigma^\alpha=\bar g^{\alpha\beta}\bar\sigma_{,\beta}=\sigma^\alpha+\cdots \ , \tag{26}$$

for x with fixed x', and by solving the Hamilton-Jacobi equation

$$\bar\sigma=\tfrac{1}{2}\bar g_{\alpha\beta}\bar\sigma^\alpha\bar\sigma^\beta=\tfrac{1}{2}\bar g^{\alpha\beta}\bar\sigma_{,\alpha}\bar\sigma_{,\beta}=\tfrac{1}{2}\Omega^{-2}g^{\alpha\beta}\bar\sigma_{,\alpha}\bar\sigma_{,\beta} \tag{27}$$

to obtain

$$[\Omega(x)\sigma\Omega(x')]^{-1}\bar\sigma=1+\tfrac{1}{12}[2\Omega(\Omega^{-1})_{;\alpha\beta}-\Omega^{-2}\Omega^{;\gamma}\Omega_{;\gamma}g_{\alpha\beta}]\sigma^\alpha\sigma^\beta+\cdots \ , \tag{28}$$

one finds that

$$\langle\bar\phi^2\rangle_{\text{Gauss}}=(48\pi^2)^{-1}(\kappa^2\Omega^{-2}+\Omega^{-3}\Omega_{;\alpha}^{;\alpha}) =(288\pi^2)^{-1}\Omega^{-2}(6\kappa^2+R-\Omega^2\bar R) \ . \tag{29}$$

This argument also shows that $g^{1/4}(\langle\phi^2\rangle+R/288\pi^2)$ is a conformal invariant.

Now the ultrastatic optical metric has $R_{00}=0$, so if the physical metric is Einstein,

$$\begin{aligned}\bar R_{00}&=\Lambda\bar g_{00}=\Lambda\Omega^2\\ &=R_{00}+2\Omega(\Omega^{-1})_{;00}-\tfrac{1}{2}\Omega^{-2}(\Omega^2)_{;\gamma}^{;\gamma}g_{00}\\ &=-\Omega^{-1}\Omega_{;\alpha}^{;\alpha}-\Omega^{-2}\Omega^{;\alpha}\Omega_{;\alpha} \ .\end{aligned} \tag{30}$$

Then

$$\langle\bar\phi^2\rangle_{\text{Gauss}}=(48\pi^2)^{-1}(\kappa^2\Omega^{-2}-\Omega^{-4}\Omega^{;\alpha}\Omega_{;\alpha}-\Lambda) \ . \tag{31}$$

$2\pi\Omega\kappa^{-1}$ is the length of the orbit of the Euclidean time Killing vector $\partial/\partial\tau$ in the physical metric, so

$$T_{\text{loc}}=T\bar g_{00}^{-1/2}=(\kappa/2\pi)\Omega^{-1} \tag{32}$$

is the local value of the temperature.[25] The acceleration squared of the Killing vector orbit is

$$\begin{aligned}a^2&=\bar g^{\alpha\beta}(\ln\bar g_{00}^{1/2})_{,\alpha}(\ln\bar g_{00}^{1/2})_{,\beta}\\ &=\Omega^{-4}g^{\alpha\beta}\Omega_{,\alpha}\Omega_{,\beta} \ ,\end{aligned} \tag{33}$$

which in flat spacetime would give rise to an Unruh acceleration temperature[26]

$$T_{\text{acc}}=a/2\pi=(\Omega^{-4}\Omega^{;\alpha}\Omega_{;\alpha})^{1/2}/2\pi \ . \tag{34}$$

Thus the mean square field in the Gaussian direct-geodesic approximation can be written directly in terms of physical quantities in a static Einstein metric as

$$\langle\bar\phi^2\rangle_{\text{Gauss}}=\tfrac{1}{12}(T_{\text{loc}}^2-T_{\text{acc}}^2)-\frac{1}{48\pi^2}\Lambda \ . \tag{35}$$

Now consider various examples of this. At temperature T in Minkowski spacetime in Lorentzian coordinates, $T_{\text{loc}}=T$, $T_{\text{acc}}=0$, and $\Lambda=0$, so one gets the standard result for thermal equilibrium in flat spacetime.[27] At $T=0$, the Minkowski vacuum has $\langle\phi^2\rangle=0$, but in Rindler coordinates[28] a static observer is accelerating and sees thermal radiation[26] with $T_{\text{loc}}=T_{\text{acc}}$, so again (35) gives exactly the right answer. For de Sitter space (S^4 when analytically continued to a Riemannian metric) in static coordinates,

$$d\bar{s}^2=(1-\tfrac{1}{3}\Lambda r^2)[d\tau^2+(1-\tfrac{1}{3}\Lambda r^2)^{-2}dr^2+(1-\tfrac{1}{3}\Lambda r^2)^{-1}r^2(d\theta^2+\sin^2\theta\, d\phi^2)]\ , \tag{36}$$

the SO(4,1)-invariant thermal state[29] has $T=(\Lambda/12\pi^2)^{1/2}$, and a static observer at fixed (r,θ,ϕ) sees

$$T_{\rm loc}=T(1-\tfrac{1}{3}\Lambda r^2)^{-1/2}\ , \tag{37}$$

$$T_{\rm acc}=(\Lambda r/6\pi)(1-\tfrac{1}{3}\Lambda r^2)^{-1/2}\ , \tag{38}$$

$$\langle\bar{\phi}^2\rangle_{\rm Gauss}=-\Lambda/72\pi^2\ . \tag{39}$$

This is also exact, because the optical metric [inside the square brackets in (36)] is the open Einstein universe $T\times H^3$ in which the WKB approximation is exact and in which only one spatial geodesic connects any two points,[7] so (18) is exact there. Notice that both $T_{\rm loc}$ and $T_{\rm acc}$ diverge at the static horizon $1-\frac{1}{3}\Lambda r^2=0$, but the divergences cancel and actually leave $\langle\bar{\phi}^2\rangle$ constant. This would not occur if a static state with some other constant temperature T were chosen.

Another symmetric space in which (35) may easily be evaluated is the Nariai metric,[12] which is the limiting form of the Kottler[30] or Schwarzschild-de Sitter metric in which the black-hole and cosmological horizons have the same area. The Riemannian version of the metric is simply the standard metric on $S^2\times S^2$,

$$d\bar{s}^2=(1-\Lambda x^2)[d\tau^2+(1-\Lambda x^2)^{-2}dx^2+(\Lambda-\Lambda^2x^2)^{-1}(d\theta^2+\sin^2\theta\, d\phi^2)]\ . \tag{40}$$

The SO(3)$\times$SO(2,1)-invariant state has $T=\Lambda^{1/2}/2\pi$ and

$$T_{\rm loc}=T(1-\Lambda x^2)^{-1/2}\ , \tag{41}$$

$$T_{\rm acc}=(\Lambda x/2\pi)(1-\Lambda x^2)^{-1/2}\ , \tag{42}$$

$$\langle\bar{\phi}^2\rangle_{\rm Gauss}=0\ . \tag{43}$$

Again the spatial variation of $T_{\rm loc}$ and $T_{\rm acc}$ cancels in $\langle\bar{\phi}^2\rangle_{\rm Gauss}$, which is not *a priori* obvious this time since the optical metric is not homogeneous. The result here is probably not exact, because the WKB approximation on the optical metric is not likely to be exact (though it should be unusually good because of the vanishing of $[a_{1;\mu\nu}]$ and $[a_2]$), and because in this case the optical metric is a wormhole between two open Einstein universes and has an infinite number of spatial geodesics connecting any two points. The contributions from the indirect geodesics are likely to increase $\langle\bar{\phi}^2\rangle$ slightly above zero. However, it is still interesting to use (43) to estimate the gravitational action of the asymptotically Euclidean metric with $R^*=0$ obtained by the conformal transformation

$$g^*_{\mu\nu}(x)=4\pi^2\bar{G}(x,z)\bar{g}_{\mu\nu}(x)\ , \tag{44}$$

which sends the point z to infinity. The action resides entirely in the surface and has the value[31]

$$I[g^*_{\mu\nu}]=6\pi^3[\langle\bar{\phi}^2(z)\rangle+\bar{R}(z)/288\pi^2] \approx\pi\Lambda/12\ . \tag{45}$$

This is positive, consistent with the positive-action theorem,[32] and the corrections mentioned above are only likely to increase it somewhat.

Although they are stationary rather than static, the self-dual multi-Taub-NUT (Newman-Unti-Tamburino) metrics[33] and gravitational multi-instantons[34] are an infinite family of metrics in which (35) gives exactly the right answer. The orbits of the Killing vector $\partial/\partial\tau$ give

$$T_{\rm loc}=(4\pi)^{-1}V^{1/2}\ , \tag{46}$$

$$T_{\rm acc}=(4\pi)^{-1}V^{-3/2}\,|\vec{\nabla}V|\ , \tag{47}$$

$$\langle\bar{\phi}^2\rangle_{\rm Gauss}=(192\pi^2)^{-1}V[1-|\vec{\nabla}(V^{-1})|^2]\ . \tag{48}$$

An exact calculation using the known propagator[35] gives the same answer for $\langle\phi^2\rangle$, which is not too surprising since Huygens's principle[22] applies for these metrics.

Perhaps the most interesting static Einstein metric to consider is the Schwarzschild spacetime surrounding a static black hole,

$$d\bar{s}^2=(1-2M/r)d\tau^2+(1-2M/r)^{-1}dr^2 +r^2(d\theta^2+\sin^2\theta d\phi^2)\ . \tag{49}$$

The Hartle-Hawking thermal state[13] has $T=(8\pi M)^{-1}$, so

$$T_{\rm loc}=T(1-2M/r)^{-1/2}\ , \tag{50}$$

$$T_{\rm acc}=\frac{M}{2\pi r^2}\left[1-\frac{2M}{r}\right]^{-1/2}\ , \tag{51}$$

$$\langle\bar{\phi}^2\rangle_{\text{Gauss}}=\frac{T^2}{12}\left[1+\frac{2M}{r}+\frac{4M^2}{r^2}+\frac{8M^3}{r^3}\right]. \tag{52}$$

This agrees precisely with Candelas's value[9] on the horizon at $r=2M$, and it reduces to the flat-space value at $r=\infty$. Furthermore, it is the same formula that Whiting found[36] that gives excellent agreement with the numerical calculations of Fawcett[10] for all radial values. Thus the renormalized mean-square field is very accurately given in the Schwarzschild metric by the optical-metric Gaussian approximation.

IV. CONFORMAL TRANSFORMATION OF THE STRESS-ENERGY TENSOR

To obtain the renormalized stress-energy tensor $T_{\mu\nu}$, one must compute various second derivatives of the regularized propagator $G^B(x,x')$ and take the coincidence limit. Hence one must know it to distance squared in the point separation. By expanding (18) and (20) to the required order and using the solutions[18,20] of the recursion relations for v_n and w_n for a conformally invariant field, one obtains

$$G_{\text{Gauss}}(x,x')=(4\pi)^{-2}\Delta^{1/2}[2\sigma^{-1}+\tfrac{1}{3}\kappa^2+\tfrac{1}{90}\kappa^4(2(\tau-\tau')^2-\sigma)]+O(\sigma^2)\,, \tag{53}$$

$$G^L(x,x')=(4\pi)^{-2}\Delta^{1/2}[2\sigma^{-1}+\tfrac{1}{360}(3J_{\alpha\beta}-I_{\alpha\beta})\sigma^\alpha\sigma^\beta\ln(\mu^2\sigma)-\tfrac{3}{4}a_2\sigma]+O(\sigma^2)\,, \tag{54}$$

where μ is an undetermined renormalization mass, a_2 is given in the coincidence limit (which is all that matters) by (17), and

$$I_{\mu\nu}=g^{-1/2}\frac{\delta}{\delta g^{\mu\nu}}\int R^2g^{1/2}d^4x=2R_{;\mu\nu}-2RR_{\mu\nu}+(\tfrac{1}{2}R^2-2R_{;\alpha}^{;\alpha})g_{\mu\nu}\,, \tag{55}$$

$$J_{\mu\nu}=g^{-1/2}\frac{\delta}{\delta g^{\mu\nu}}\int R^\alpha_\beta R^\beta_\alpha g^{1/2}d^4x=R_{;\mu\nu}-R_{\mu\nu;\alpha}{}^{\alpha}-2R^\beta_\alpha R^\alpha{}_{\mu\beta\nu}+(\tfrac{1}{2}R^\alpha_\beta R^\beta_\alpha-\tfrac{1}{2}R_{;\alpha}^{;\alpha})g_{\mu\nu}\,. \tag{56}$$

The arbitrariness of μ results in the arbitrariness of adding a multiple of $3J_{\mu\nu}-I_{\mu\nu}$ to the stress tensor,[20,21] but

$$3J_{\mu\nu}-I_{\mu\nu}=g^{-1/2}\frac{\delta}{\delta g^{\mu\nu}}\int\tfrac{3}{2}C_{\alpha\beta\gamma\delta}C^{\alpha\beta\gamma\delta}g^{1/2}d^4x \tag{57}$$

is zero for all metrics conformal to Einstein metrics, so the ambiguity is not present in these cases.

Now if one applies Wald's corrected procedure[21] to $G^B_{\text{Gauss}}=G_{\text{Gauss}}-G^L$, the stress-energy tensor in the optical metric of a static Einstein metric in this Gaussian approximation takes the form

$$T^\mu_\nu=\frac{\pi^2}{90}T^4(\delta^\mu_\nu-4\delta^\mu_0\delta^0_\nu)\,, \tag{58}$$

which is the same as thermal radiation in flat spacetime. It is known that this is exact in the open Einstein universe,[7] which follows from the above-mentioned fact that (18) is exact in $T\times H^3$ (or more precisely $S^1\times H^3$, since τ is given the period $2\pi\kappa^{-1}$).

However, to get the stress-energy tensor in the static Einstein metric itself by this procedure, one must apply the conformal transformation (3) to (53), rewrite (54) in terms of barred quantities in the physical metric, subtract, and differentiate appropriately. It is cumbersome to write $\bar{G}^B$ to sufficient accuracy as a power series in $\bar{\sigma}^\alpha$ to execute this procedure, so it is advantageous to seek a less tedious approach.

Such an indirect approach is provided by the functional-differential scale equation for the change in the stress-energy tensor under a conformal change in the metric,[4]

$$g_{\alpha\beta}(x')\frac{\delta}{\delta g_{\alpha\beta}(x')}g^{1/2}T^\mu_\nu(x)=g_{\nu\gamma}(x)\frac{\delta}{\delta g_{\mu\gamma}(x)}g^{1/2}T^\lambda_\lambda(x')\,. \tag{59}$$

For a conformally invariant quantized field, the right-hand side is given entirely by the trace anomaly[37]

$$T^\lambda_\lambda=\alpha\mathscr{H}+\beta\mathscr{G}+\gamma\Box R\,, \tag{60}$$

where

$$\mathscr{H}=C_{\alpha\beta\gamma\delta}C^{\alpha\beta\gamma\delta}=R_{\alpha\beta\gamma\delta}R^{\alpha\beta\gamma\delta}-2R_{\alpha\beta}R^{\alpha\beta}+\tfrac{1}{3}R^2\,, \tag{61}$$

$$\mathcal{G} = {}^*R_{\alpha\beta\gamma\delta}{}^*R^{\gamma\delta\alpha\beta} = R_{\alpha\beta\gamma\delta}R^{\alpha\beta\gamma\delta} - 4R_{\alpha\beta}R^{\alpha\beta} + R^2 \tag{62}$$

The coefficients α,β,γ depend upon the spin of the field:

$$\alpha = (2^9 45\pi^2)^{-1}[12h(0) + 18h(\tfrac{1}{2}) + 72h(1)] , \tag{63}$$

$$\beta = (2^9 45\pi^2)^{-1}[-4h(0) - 11h(\tfrac{1}{2}) - 124h(1)] , \tag{64}$$

$$\gamma = (2^9 45\pi^2)^{-1}[8h(0) + 12h(\tfrac{1}{2}) + (48 \text{ or } -72)h(1)] , \tag{65}$$

where $h(s)$ is the number of helicity states for the fields of spin s.

There is general agreement by a variety of methods on all of the coefficients except that of $h(1)$, which is 48 in dimensional regularization but -72 in ζ-function regularization.[38] H wever, one can alter this coefficient by adding a counterterm proportional to R^2 to the Lagrangian.[37,38] If one takes the value given by dimensional regularization, then $2\alpha = 2\gamma$. One can easily show that both $\mathcal{G}$ and $3\mathcal{H} + 2\Box R$ vanish in the ultrastatic optical metric conformal to a static Einstein metric, so $2\alpha = 3\gamma$ makes the trace anomaly vanish in such a metric. For the conformally invariant scalar field, this is another consequence of the cancellation of $[a_1]$. It is evidence to support belief in the relative accuracy of the Gaussian approximation (58) for the stress-energy tensor in the optical metric.

Although integrating the functional-differential scale equation (59) between two conformally related metrics is straightforward, it is rather tedious, and the result, when written in terms of derivatives of the conformal factor, is not particularly perspicuous. The term proportional to α is relatively simple to integrate directly, but for the other two terms it is simpler to look for a generalization of the conformally flat solution[4,39,7,5] that gives a conserved stress-energy tensor ith the correct trace in a general sp ime. After some trial and error, the general solution was found which satisfies (59):

$$\bar{T}^\mu_\nu = \Omega^{-4}T^\mu_\nu - 8\alpha\Omega^{-4}[(C^{\alpha\mu}{}_{\beta\nu}\ln\Omega)_{;\alpha}^{;\beta} + \tfrac{1}{2}R^\beta_\alpha C^{\alpha\mu}{}_{\beta\nu}\ln\Omega] + \beta[(4\bar{R}^\beta_\alpha\bar{C}^{\alpha\mu}{}_{\beta\nu} - 2\bar{H}^\mu_\nu) - \Omega^{-4}(4R^\beta_\alpha C^{\alpha\mu}{}_{\beta\nu} - 2H^\mu_\nu)]$$
$$-\tfrac{1}{6}\gamma[\bar{I}^\mu_\nu - \Omega^{-4}I^\mu_\nu] , \tag{66}$$

$$H_{\mu\nu} = \lim_{n\to 4}\frac{1}{n-4}g^{-1/2}\frac{\delta}{\delta g^{\mu\nu}}\int g^{1/2}\mathcal{G}\, d^n x = -R^\alpha_\mu R_{\alpha\nu} + \tfrac{2}{3}RR_{\mu\nu} + (\tfrac{1}{2}R^\alpha_\beta R^\beta_\alpha - \tfrac{1}{4}R^2)g_{\mu\nu} . \tag{67}$$

$I_{\mu\nu}$ is given by (55) above, and barred tensors are evaluated in the metric $\bar{g}_{\mu\nu} = \Omega^2 g_{\mu\nu}$. The covariant derivatives in the α term are with respect to $g_{\mu\nu}$, but this term would be the same if the factor of Ω^{-4} were dropped and the curvature tensors and covariant derivatives inside the square brackets were rewritten as barred quantities in the metric $\bar{g}_{\mu\nu}$. One may directly verify that this $\bar{T}^\mu_\nu$ in the metric $\bar{g}_{\mu\nu}$ is conserved, has the correct trace anomaly, and satisfies the barred version of (59). The solution may be expressed more concisely by stating that

$$g^{1/2}\{T^\mu_\nu + \alpha[(C^{\alpha\mu}{}_{\beta\nu}\ln g)_{;\alpha}^{;\beta} + \tfrac{1}{2}R^\beta_\alpha C^{\alpha\mu}{}_{\beta\nu}\ln g] + \beta[2H^\mu_\nu - 4R^\beta_\alpha C^{\alpha\mu}{}_{\beta\nu}] + \tfrac{1}{6}\gamma I^\mu_\nu\} \tag{68}$$

is conformally invariant, taking the same values in all conformally related metrics for the same conformal state (i.e., boundary conditions such that the propagators are conformally related). (Here g, the determinant of the metric tensor, is to be interpreted as a scalar in a coordinate system which is left unchanged during the conformal transformation.)

The presence of the logarithmic terms in Ω or g means that in general $g^{1/2}T^\mu_\nu$ is not even scale invariant under a constant conformal transformation, but rather changes logarithmically by a multiple of $g^{1/2}(3J^\mu_\nu - I^\mu_\nu)$. This results from the arbitrariness in T^μ_ν noted earlier,[20,21] since changing the scale of the spacetime is equivalent to changing the undetermined renormalization mass μ. However, the ambiguity is absent for metrics conformal to Einstein metrics, since in these metrics $3J^\mu_\nu = I^\mu_\nu$ and (66) only has derivative terms in Ω.

V. RENORMALIZED STRESS-ENERGY TENSOR IN THE PHYSICAL METRIC

If the physical metric $\bar{g}_{\mu\nu}$ is a static Einstein metric with a conformally related ultrastatic (optical) metric $g_{\mu\nu}=\Omega^{-2}\bar{g}_{\mu\nu}$, $\Omega^2=|\bar{g}_{00}|$, then

$$\bar{R}^{\mu}_{\nu}=\Lambda\delta^{\mu}_{\nu}\,, \tag{69}$$

$$\bar{H}^{\mu}_{\nu}=-\tfrac{1}{3}\Lambda^2\delta^{\mu}_{\nu}\,, \tag{70}$$

$$\bar{I}^{\mu}_{\nu}=0\,, \tag{71}$$

$$R^{\mu}_{\nu}=2\Omega^{-1}\Omega_{;\nu}^{;\mu}-4\Omega^{-2}\Omega^{;\mu}\Omega_{;\nu}$$
$$=2\Omega\Omega|^{\mu}_{\nu}-2\Omega|^{\alpha}\Omega|_{\alpha}\delta^{\mu}_{\nu}\,, \tag{72}$$

where the vertical bars represent covariant differentiation with respect to $\bar{g}_{\mu\nu}$, and (66) takes the form

$$\bar{T}^{\mu}_{\nu}=\Omega^{-4}\{T^{\mu}_{\nu}+[8\alpha\Omega^{-2}\Omega_{;\alpha}\Omega^{;\beta}-4(\alpha+\beta)R^{\beta}_{\alpha}]C^{\alpha\mu}{}_{\beta\nu}+2\beta[H^{\mu}_{\nu}+\tfrac{1}{3}\Lambda^2\Omega^4\delta^{\mu}_{\nu}]+\tfrac{1}{6}\gamma I^{\mu}_{\nu}\}\,. \tag{73}$$

Inserting the spin-0 values for α, β, and γ and the Gaussian approximation (58) for T^{μ}_{ν} results in a good approximation for the thermal stress-energy tensor of a conformally invariant scalar field in a static Einstein spacetime.

For example, the thermal state[29] with $T=(\Lambda/12\pi^2)^{1/2}$ in the de Sitter metric gives by this method

$$\bar{T}^{\mu}_{\nu}=-(8640\pi^2)^{-1}\Lambda^2\delta^{\mu}_{\nu}\,, \tag{74}$$

which is exactly the SO(4,1)-invariant result consistent with the trace anomaly. Of course, the Gaussian approximation is exact in this case,[7] but it might not be immediately evident that the Gibbons-Hawking temperature gives an SO(4,1)-invariant state which is not observer dependent.[29]

As a second example, consider the SO(3) $\times$SO(2,1)-invariant state ($T=\Lambda^{1/2}/2\pi$) on the Nariai[12] analytically continued $S^2\times S^2$ metric (40). Again the Gaussian approximation gives the correct homogeneous, isotropic stress-energy tensor

$$\bar{T}^{\mu}_{\nu}=(2880\pi^2)^{-1}\Lambda^2\delta^{\mu}_{\nu}\,, \tag{75}$$

which is somewhat surprising, since the optical metric is not homogeneous and the Gaussian approximation probably does not give the exact propagator. However, it is good enough to give the exact stress-energy tensor.

Now turn to the most interesting static Einstein spacetime, the vacuum ($\Lambda=0$) Schwarzschild metric (49) for the exterior of a nonrotating black hole. The corresponding optical metric is, with $w\equiv 2M/r$,

$$ds^2=d\tau^2+(1-w)^{-2}dr^2$$
$$+(1-w)^{-1}r^2(d\theta^2+\sin^2\theta\, d\phi^2)\,. \tag{76}$$

In the Hartle-Hawking thermal state,[13] the temperature at infinity is $T=\kappa/2\pi$, where $\kappa=1/4M$ is the surface gravity of the black hole. The Gaussian approximation says T^{μ}_{ν} has the thermal form (58) in the optical metric, which has the nature of a static wormhole between an asymptotically flat spacetime for $w<<1$ and an open Einstein universe for $1-w<<1$.

The nonzero curvature components of the optical metric in a coordinate basis with $x^{\alpha}=(\tau, r, \theta, \phi)$ are

$$R^1_1=\kappa^2w^3(-8+6w)\,,$$
$$R^2_2=R^3_3=\kappa^2w^3(4-6w)\,, \tag{77}$$

$$C^{01}{}_{01}=C^{23}{}_{23}=-2C^{02}{}_{02}=-2C^{03}{}_{03}=-2C^{12}{}_{12}$$
$$=-2C^{13}{}_{13}=4\kappa^2w^3(1-w)\,, \tag{78}$$

plus those obtained by $C^{\alpha\mu}{}_{\beta\nu}=C^{[\alpha\mu]}{}_{[\beta\nu]}$. Then one can calculate the nonzero components of H^{μ}_{ν} and I^{μ}_{ν}:

$$H^0_0=\kappa^4w^6(48-96w+45w^2)\,, \tag{79a}$$

$$H^1_1=\kappa^4w^6(-16+32w-15w^2)\,, \tag{79b}$$

$$H^2_2=H^3_3=\kappa^4w^6(32-64w+33w^2)\,, \tag{79c}$$

$$I^0_0=6\kappa^4w^6(96-192w+99w^2)\,, \tag{80a}$$

$$I^1_1=6\kappa^4w^6(-64+144w-81w^2)\,, \tag{80b}$$

$$I^2_2=I^3_3=6\kappa^4w^6(128-264w+135w^2)\,. \tag{80c}$$

The conformal factor to the physical Schwarzschild metric is

$$\Omega^2=1-w\equiv 1-2M/r\,, \tag{81}$$

so one gets

$$\Omega^{-2}\Omega_{;\alpha}\Omega^{;\beta}=\kappa^2w^4\delta^1_{\alpha}\delta^{\beta}_1\,. \tag{82}$$

Inserting all of these expressions back into (73) with the correct values of α, β, and γ gives the approximate stress-energy tensor for a conformally invariant scalar field in the Hartle-Hawking thermal state around a Schwarzschild black hole:

$$\bar{T}^{\mu}_{\nu}=\frac{\pi^2}{90}\left[\frac{1}{8\pi M}\right]^4\left[\frac{1-(4-6M/r)^2(2M/r)^6}{(1-2M/r)^2}(\delta^{\mu}_{\nu}-4\delta^{\mu}_{0}\delta^{0}_{\nu})+24(2M/r)^6(3\delta^{\mu}_{0}\delta^{0}_{\mu}+\delta^{\mu}_{1}\delta^{1}_{\mu})\right]. \tag{83}$$

It may be directly checked that this expression is conserved and has the correct trace. At large distances from the hole ($r>>M$), it has the form of flat-space thermal radiation at the local temperature $T_{\rm loc} = T(1-2M/r)^{-1/2}$, plus correction terms of order M^2/r^6, as expected for the true stress-energy tensor.[8,9] As one moves inward, the calculated energy density $\rho=-\bar{T}^0_0$ goes from the asymptotic value $\rho=aT^4 =(\pi^2/30)\,(8\pi M)^{-4}$ at radial infinity to a maximum of $\rho\simeq3.7730\,aT^4$ at $r\simeq3.1378\,M$ and then decreases through zero at $r\simeq2.3437\,M$ to $\rho=-12\,aT^4$ at $r=2\,M$. Although $T_{\rm loc}$ diverges at the horizon, the correction terms cancel the divergence and give a finite stress-energy tensor

$$\bar{T}^{\mu}_{\nu}(2M)=\frac{\pi^2}{90}T^4[36(\delta^{\mu}_{0}\delta^{0}_{\nu}+\delta^{\mu}_{1}\delta^{1}_{\nu})+12(\delta^{\mu}_{2}\delta^{2}_{\nu}+\delta^{\mu}_{3}\delta^{3}_{\nu})]\,. \tag{84}$$

It is also remarkable that the Gaussian approximation gives $\bar{T}^0_0(2M)=\bar{T}^1_1(2M)$, which is necessary for $\bar{T}^{\mu}_{\nu}$ to be well behaved there.

When one analytically continues from the Riemannian metric (49) to the Lorentzian black-hole metric by replacing real τ with real $t=-i\tau$, the expression (83) has the same form in Lorentzian coordinates $x^{\alpha'} = (t,r,\theta,\phi)$. However, t is singular on the horizon, so it is helpful to go to Kruskal-Szekeres null coordinates[40]

$$U=-(w^{-1}-1)^{1/2}e^{\kappa(r-t)},\quad V=(w^{-1}-1)^{1/2}e^{\kappa(r+t)}\,. \tag{85}$$

Then, following Candelas's notation,[9] the nonzero components of the approximate stress-energy tensor are given in terms of the three functions

$$A(r)=\tfrac{1}{2}(\bar{T}^0_0+\bar{T}^1_1)=-\frac{\pi^2}{90}\left[\frac{1}{8\pi M}\right]^4\left[\frac{1-(8-15M/r)^2(2M/r)^6-192M^8/r^8}{(1-2M/r)^2}\right], \tag{86}$$

$$B(r)=\bar{T}^2_2=\bar{T}^3_3=\frac{\pi^2}{90}\left[\frac{1}{8\pi M}\right]^4\left[\frac{1-(4-6M/r)^2(2M/r)^6}{(1-2M/r)^2}\right], \tag{87}$$

$$D(r)=C(r)(UV)^{-1}=\tfrac{1}{2}(\bar{T}^1_1-\bar{T}^0_0)\left[\frac{-w}{1-w}e^{-2\kappa r}\right]=-\frac{\pi^2}{90}\left[\frac{1}{8\pi M}\right]^4 e^{-r/2M}\sum_{n=1}^{6}n(n+1)\left[\frac{2M}{r}\right]^n, \tag{88}$$

by

$$\bar{T}^U_U=\bar{T}^V_V=A(r)\,, \tag{89}$$

$$\bar{T}^U_V=D(r)U^2,\quad \bar{T}^V_U=D(r)V^2\,, \tag{90}$$

$$\bar{T}^\theta_\theta=\bar{T}^\phi_\phi=B(r)\,. \tag{91}$$

On the horizon ($r=2M$ or $UV=0$), these radial functions have the values

$$A(2M)=36\frac{\pi^2}{90}\left[\frac{1}{8\pi M}\right]^4, \tag{92}$$

$$B(2M)=12\frac{\pi^2}{90}\left[\frac{1}{8\pi M}\right]^4, \tag{93}$$

$$D(2M)=-2Me^{-1}C'(2M)=-112e^{-1}\frac{\pi^2}{90}\left[\frac{1}{8\pi M}\right]^4. \tag{94}$$

One can now compare the Gaussian approximation with Candelas's numerical calculations[9] for the stress-energy tensor on the bifurcation two-sphere ($U=0$, $V=0$), which give

$$A_C(2M)=\frac{\pi^2}{90}\left[\frac{1}{8\pi M}\right]^4\left[-24+480\sum_{l=0}^{\infty}(2l+1)\beta_l\right]\simeq37.71\frac{\pi^2}{90}\left[\frac{1}{8\pi M}\right]^4, \tag{95}$$

$$B_C(2M)=\frac{\pi^2}{90}\left[\frac{1}{8\pi M}\right]^4\left[72-480\sum_{l=0}^{\infty}(2l+1)\beta_l\right]\simeq 10.29\frac{\pi^2}{90}\left[\frac{1}{8\pi M}\right]^4 . \tag{96}$$

Thus the Gaussian values are close to Candelas's numerical values, differing by only 4.5% for $A(2M)$ and 16.6% for $B(2M)$. The Gaussian estimate would predict

$$\sum_{l=0}^{\infty}(2l+1)\beta_l=0.125\,, \tag{97}$$

only 2.9% different from the value Candelas obtained,

$$\sum_{l=0}^{\infty}(2l+1)\beta_l\simeq 0.1286. \tag{98}$$

Candelas's value can also be used to check the derivative of the formula (52) for $\langle\bar{\phi}^2\rangle$ at the horizon, since the stress-energy tensor at the bifurcation two-sphere is determined entirely by Candelas's propagator[9] there and by $\langle\bar{\phi}^2\rangle'\equiv d\langle\bar{\phi}^2\rangle/dr\,|_{r=2M}$. An explicit evaluation by differentiating the propagator and $\langle\bar{\phi}^2\rangle$ in a Riemann normal coordinate system at the bifurcation two-sphere yields

$$\bar{T}^{\mu}_{\nu}(2M)=(\tfrac{1}{6}\kappa\langle\bar{\phi}^2\rangle'+\tfrac{4}{5}\pi^2T^4)(2\delta^{\mu}_{0}\delta^{0}_{\nu}+2\delta^{\mu}_{1}\delta^{1}_{\nu}-\delta^{\mu}_{\nu})+\tfrac{4}{15}\pi^2T^4\delta^{\mu}_{\nu}, \tag{99}$$

so working backward from Candelas's results gives

$$-2^8\pi^2M^3\langle\bar{\phi}^2\rangle'=1.6-2^{11}3\pi^2M^4A_C(2M)=2-8\sum_{l=0}^{\infty}(2l+1)\beta_l\simeq 0.9715. \tag{100}$$

The value 1 obtained from (52) is thus 2.9% larger than Candelas's calculations. The discrepancy is presumably due to the truncation of all indirect geodesics in the optical metric and to the inaccuracy of the Bekenstein-Parker Gaussian approximation. One might seek to reduce the error by evaluating the contributions of the indirect geodesics. In the Schwarzschild optical metric there are an infinite number of these geodesics between any two points, paths that wind around the wormhole an arbitrarily large number of times near the unstable circular orbits at $r=3M$. However, the spherical symmetry implies that the indirect geodesics from x' form a caustic surface at $\theta=\theta'$, $\phi=\phi'$, so in the coincidence limit $x=x'$, $\Delta(x,x')$ has a linear divergence.[17] This means that (18) would have an inverse-square-root singularity in the coincidence limit, so the Gaussian approximation must break down for these indirect geodesics with caustics. One must presumably go to more sophisticated techniques if one wishes to improve upon the accuracy of the truncated Gaussian approximation. However, this method seems to give a reasonably good first approximation to the stress-energy tensor of the thermal state of a conformally invariant scalar field in a static Einstein metric such as the Schwarzschild metric.

Note added in proof. M. R. Brown (unpublished) has independently shown that the trace anomaly vanishes in the optical metric conformal to a static Ricci-flat metric. He also obtained the solution of (59) when $2\alpha=3\gamma$, though in a lengtheir form than my Eq. (66).

ACKNOWLEDGMENTS

This research was stimulated by the formula (52) B. Whiting found as a good fit to the numerical calculations of M. Fawcett (Ref. 10), and I thank these authors for discussing their work with me prior to publication. I also wish to express appreciation to the University of Cambridge and to the Nuffield Quantum Gravity Workshop at the Imperial College of Science and Technology for hospitality which permitted these discussions. Conversations with P. Candelas gave additional information. Financial support was provided in part by NSF Grant PHY-7918430.

[1]S. W. Hawking, Nature 248, 30 (1974); Commun. Math. Phys. 43, 199 (1975).
[2]B. S. DeWitt, Phys. Rep. 19C, 295 (1975).
[3]S. W. Hawking, Commun. Math. Phys. 80, 421 (1981).
[4]L. S. Brown and J. P. Cassidy, Phys. Rev. D 16, 1712 (1977).
[5]B. S. DeWitt, in *General Relativity: An Einstein Centenary Survey*, edited by S. W. Hawking and W. Israel (Cambridge University Press, London, 1979).
[6]T. S. Bunch, Phys. Rev. D 18, 1844 (1978).
[7]P. Candelas and J. S. Dowker, Phys. Rev. D 19, 2902 (1979).
[8]S. M. Christensen and S. A. Fulling, Phys. Rev. D 15, 2088 (1977).
[9]P. Candelas, Phys. Rev. D 21, 2185 (1980).
[10]M. Fawcett and B. Whiting, in *Quantum Theory of Space and Time*, edited by M. J. Duff and C. J. Isham (Cambridge University Press, London, to be published).
[11]J. D. Bekenstein and L. Parker, Phys. Rev. D 23, 2850 (1981).
[12]H. Nariai, Sci. Rep. Tôhoku Univ. 34, 160 (1950); 35, 62 (1951).
[13]J. B. Hartle and S. W. Hawking, Phys. Rev. D 13, 2188 (1976).
[14]G. W. Gibbons and M. J. Perry, Proc. R. Soc. London A358, 467 (1978).
[15]J. S. Dowker and G. Kennedy, J. Phys. A 11, 895 (1978).
[16]J. Schwinger, Phys. Rev. 82, 664 (1951).
[17]B. S. DeWitt, *Dynamical Theory of Groups and Fields* (Gordon and Breach, New York, 1965).
[18]S. M. Christensen, Phys. Rev. D 14, 2490 (1976).
[19]R. M. Wald, Commun. Math. Phys. 54, 1 (1977).
[20]S. L. Adler, J. Lieberman, and Y. J. Ng, Ann. Phys. (N.Y.) 106, 279 (1977).
[21]R. M. Wald, Phys. Rev. D 17, 1477 (1978).
[22]J. Hadamard, *Lectures on Cauchy's Problem in Linear Partial Differential Equations* (Yale University Press, New Haven, 1923).
[23]P. R. Garabedian, *Partial Differential Equations* (Wiley, New York, 1964).
[24]B. S. DeWitt and R. W. Brehme, Ann. Phys. (N.Y.) 9, 220 (1960).
[25]R. C. Tolman, Phys. Rev. 35, 904 (1930); *Relativity, Thermodynamics, and Cosmology* (Clarendon, Oxford, 1934).
[26]W. G. Unruh, Phys. Rev. D 14, 870 (1976).
[27]A. D. Linde, Rep. Prog. Phys. 42, 389 (1979).
[28]W. Rindler, Am. J. Phys. 34, 1174 (1966).
[29]E. A. Tagirov, Ann. Phys. (N.Y.) 76, 561 (1973); P. Candelas and D. J. Raine, Phys. Rev. D 12, 965 (1975); G. W. Gibbons and S. W. Hawking, *ibid.* 15, 2738 (1977).
[30]F. Kottler, Ann. Phys. (Leipzig) 56, 401 (1918).
[31]S. W. Hawking, in *Recent Developments in Gravitation: Cargese, 1978* edited by M. Lévy and S. Deser (Plenum, New York, 1979).
[32]D. N. Page, Phys. Rev. D 18, 2733 (1978); G. W. Gibbons, S. W. Hawking, and M. J. Perry, Nucl. Phys. B138, 141 (1978); R. Schoen and S. -T. Yau, Phys. Rev. Lett. 42, 547 (1979).
[33]S. W. Hawking, Phys. Lett. 60A, 81 (1977).
[34]G. W. Gibbons and S. W. Hawking, Phys. Lett. 78B, 430 (1978); N. J. Hitchin, Math. Proc. Cambridge Philos. Soc. 85, 465 (1979).
[35]D. N. Page, Phys. Lett. 85B, 369 (1979); M. F. Atiyah, Adv. Math Suppl. Studies 7A, 129 (1981).
[36]B. Whiting (private communication).
[37]D. M. Capper and M. J. Duff, Nuovo Cimento 23A, 173 (1974); Phys. Lett. 53A, 361 (1975); S. Deser, M. J. Duff, and C. J. Isham, Nucl. Phys. B111, 45 (1976); M. J. Duff, *ibid.* B125, 334 (1978), and references therein.
[38]L. Parker, in *Recent Developments in Gravitation: Cargèse, 1978*, edited by M. Lévy and S. Deser (Plenum, New York, 1979), and references therein.
[39]T. S. Bunch and P. C. W. Davies, Proc. R. Soc. London A356, 569 (1977).
[40]M. D. Kruskal, Phys. Rev. 119, 1743 (1960); G. Szekeres, Publ. Math. Debreceni 7, 285 (1960).

PHYSICAL REVIEW LETTERS

VOLUME 53 | 30 JULY 1984 | NUMBER 5

Quantum Stress Tensor in Schwarzschild Space-Time

K. W. Howard and P. Candelas
Center for Theoretical Physics, The University of Texas at Austin, Austin, Texas 78712
(Received 13 March 1984)

The vacuum expectation value of the stress-energy tensor for the Hartle-Hawking state in Schwarzschild space-time has been calculated for the conformal scalar field. $\langle T_\mu^\nu \rangle$ separates naturally into the sum of two terms. The first coincides with an approximate expression suggested by Page. The second term is a "remainder" that may be evaluated numerically. The total expression is in good qualitative agreement with Page's approximation. These results are at variance with earlier results given by Fawcett whose error is explained.

PACS numbers: 04.60.+n

The computation of the vacuum expectation value of the stress-energy tensor for Schwarzschild space-time has presented a challenge since the celebrated calculation by Hawking of black-hole radiance. The original motivation, to understand where the particles came from, has to some extent been superceded by a clear understanding of the interplay between "real particles" and "vacuum polarization effects."[1] The picture that has emerged is that, although the density of stress energy associated with the emerging particles is infinite at the horizon, so is the density of stress energy associated with the vacuum polarization. The two effects contribute with opposite sign and the net value of stress-energy density remains finite as the horizon is approached. Although the finiteness of the expectation value of the stress tensor at the horizon is no longer an issue, there remains the problem of understanding its detailed structure in terms of the geometry of the Schwarzschild manifold. As a necessary step in this direction we present here the results of a calculation of $\langle T_\mu^\nu \rangle$ for a conformal scalar field in the Hartle-Hawking vacuum for the region exterior to the horizon.

Suggestions that some sort of semiclassical description of $\langle T_\mu^\nu \rangle$ in terms of the background geometry may be possible are provided by estimates[2,3] of $\langle T_\mu^\nu \rangle$ made on the basis of the properties of the Schwarzschild metric under conformal transformations. An intuitive understanding of the structure of $\langle T_\mu^\nu \rangle$ in terms of the geometry of the background would be of considerable interest with regard to the structure of solutions of the semiclassical Einstein equation

$$R_\mu^\nu - \tfrac{1}{2} R g_\mu^\nu = -8\pi \langle T_\mu^\nu \rangle,$$

and with regard to vacuum energy in Kaluza-Klein theories, the subject of much current work.

The operator expression for the expectation value of the stress-energy operator,

$$\langle T_\mu^\nu \rangle = \langle (\tfrac{2}{3}\phi_{;\mu}\phi_{;}{}^{\nu} - \tfrac{1}{6} g_\mu^\nu \phi_{;\alpha}\phi_{;}{}^{\alpha} - \tfrac{1}{3}\phi\phi_{;\mu}{}^{\nu}) \rangle,$$

was renormalized by means of the covariant point separation procedure of DeWitt[4] and Christensen.[5] In terms of the propagator appropriate to this state

$$G(x,x') = i\langle \phi(x)\phi(x') \rangle$$

that satisfies

$$\Box G(x,x') = -g^{-1/2}\delta(x,x'),$$

the renormalized value of $\langle T_\mu^\nu \rangle$ is given by the expression

$$\langle T_\mu^\nu \rangle_{\rm REN} = \lim_{x' \to x} \{ -i [\tfrac{1}{3} (G_{;\mu\alpha'} g^{\alpha'\nu} + G_{;}{}^{\nu}{}_{\alpha'} g^{\alpha'}{}_{\mu}) - \tfrac{1}{6} G_{;\alpha\beta'} g^{\alpha\beta'} g_\mu^\nu - \tfrac{1}{6} (G_{;\mu}{}^{\nu} + G_{;\alpha'\beta'} g^{\alpha'}{}_{\mu} g^{\beta'\nu})] - \langle T_\mu^\nu \rangle_{\rm subtract} \},$$

where $\langle T_\mu^\nu \rangle_{\rm subtract}$ are the subtraction terms of Christensen.[5]

In principle, the above equation is easy to evaluate. One must only solve the wave equation for the propagator as a sum over mode functions, perform the indicated subtraction, and sum the resultant convergent expression for $\langle T_\mu^\nu \rangle_{\rm REN}$. In practice, a number of difficulties present themselves during this process. For example, some subtlety must be employed in performing the subtraction. The subtraction term must be rewritten as a sum which is compatible with the mode-sum expression for the propagator so that a convergent sum will be obtained when the limit is taken. The summation of the resultant expression for $\langle T_\mu^\nu \rangle_{\rm REN}$ is further complicated by the fact that, as the radial solutions are not expressible in terms of well-known functions, a combination of approximation and numerical analysis must be applied. Our calculation, the details of which will be presented elsewhere, reveals that in Schwarzschild coordinates $\langle T_\mu^\nu \rangle_{\rm REN}$ has the form

$$\langle T_\mu^\nu \rangle_{\rm REN} = \frac{\pi^2}{90(8\pi M)^4} \left\{ \frac{1-(2M/r)^6(4-6M/r)^2}{(1-2M/r)^2} \begin{pmatrix} -3 & 0 & 0 & 0 \\ 0 & 1 & 0 & 0 \\ 0 & 0 & 1 & 0 \\ 0 & 0 & 0 & 1 \end{pmatrix} + 24 \left(\frac{2M}{r} \right)^6 \begin{pmatrix} 3 & 0 & 0 & 0 \\ 0 & 1 & 0 & 0 \\ 0 & 0 & 0 & 0 \\ 0 & 0 & 0 & 0 \end{pmatrix} + 1920 \Delta_\mu^\nu \right\}.$$

The first two terms correspond precisely to the approximation which Page[2] obtained by means of a Gaussian approximation to the proper-time propagator.[6] Being an approximation of WKB type Page's expression might reasonably be expected to furnish a good approximation to the true quantity, particularly in view of the fact that the corresponding approximation[7,8] to $\langle \phi^2 \rangle$ agrees with the true quantity to better than 1%.

The components of the remainder term Δ_μ^ν involve sums over solutions to the radial equation

$$\left\{ \frac{d}{d\xi} (\xi^2 - 1) \frac{d}{d\xi} - l(l+1) - \frac{n^2 (1+\xi)^4}{16(\xi^2 - 1)} \right\} R(\xi) = 0,$$

TABLE I. The sums that appear in Δ_μ^ν.

$$S_1 = \frac{1}{16} \left(\frac{\xi+1}{\xi-1} \right) \sum_{n=1}^{\infty} n^2 \left\{ \sum_{\ell=0}^{\infty} \left[\frac{2\ell+1}{n} p_\ell^n(\xi) q_\ell^n(\xi) - \frac{2}{(\xi^2-1)^{1/2}} \right] + \frac{n}{2} \left(\frac{\xi+1}{\xi-1} \right) \right\}$$

$$S_2 = \left(\frac{\xi-1}{\xi+1} \right) \sum_{n=1}^{\infty} \left\{ \sum_{\ell=0}^{\infty} \left[\frac{2\ell+1}{n} \frac{dp_\ell^n(\xi)}{d\xi} \frac{dq_\ell^n(\xi)}{d\xi} + \frac{2\ell(\ell+1)}{(\xi^2-1)^{3/2}} + \frac{n^2(\xi+1)^4}{16(\xi^2-1)^{5/2}} - \frac{3}{4(\xi^2-1)^{5/2}} \right] - \frac{n^3(\xi+1)^3}{96(\xi-1)^3} + \frac{n}{3(\xi-1)^3} \right\}$$

$$S_3 = \frac{1}{(\xi+1)^2} \sum_{n=1}^{\infty} \left\{ \sum_{\ell=0}^{\infty} \left[\frac{2\ell+1}{n} (\ell+\tfrac{1}{2})^2 p_\ell^n(\xi) q_\ell^n(\xi) - \frac{2(\ell+\frac{1}{2})^2}{(\xi^2-1)^{1/2}} + \frac{n^2(1+\xi)^4}{16(\xi^2-1)^{3/2}} - \frac{1}{4(\xi^2-1)^{3/2}} \right] - \frac{n^3(\xi+1)^4}{48(\xi-1)^2} + \frac{n}{24(\xi-1)^2} (3\xi^2 - 8\xi + 13) \right\}$$

$$S_4 = \frac{1}{(\xi+1)^2} \frac{\partial}{\partial\xi} \sum_{n=1}^{\infty} \left\{ \sum_{\ell=0}^{\infty} \left[\frac{2\ell+1}{n} p_\ell^n(\xi) q_\ell^n(\xi) - \frac{2}{(\xi^2-1)^{1/2}} \right] + \frac{n(\xi+1)}{2(\xi-1)} \right\}$$

$$S_5 = \frac{1}{4(\xi+1)^2} \sum_{n=1}^{\infty} \left\{ \sum_{\ell=0}^{\infty} \left[\frac{2\ell+1}{n} p_\ell^n(\xi) q_\ell^n(\xi) - \frac{2}{(\xi^2-1)^{1/2}} \right] + \frac{n}{2} \left(\frac{\xi+1}{\xi-1} \right) \right\}$$

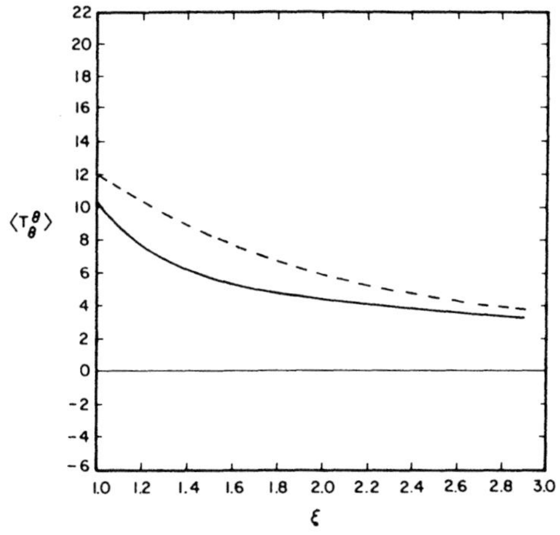

FIG. 1. $[90(8\pi M)^4/\pi^2]\langle T^\theta_\theta\rangle$ as a function of $\xi = r/M - 1$. The dashed line represents Page's approximation.

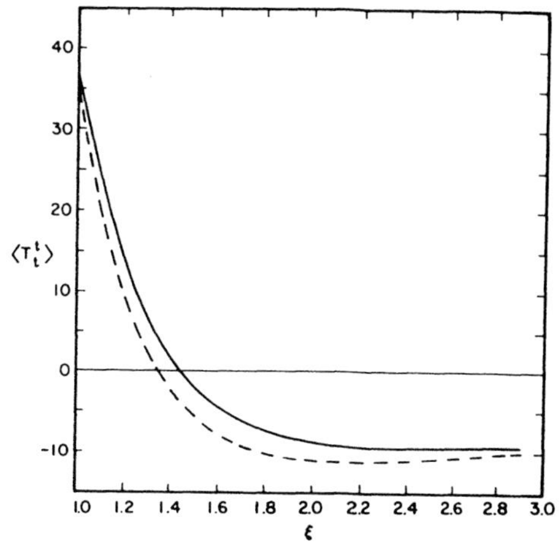

FIG. 2. $[90(8\pi M)^4/\pi^2]\langle T^t_t\rangle$ as a function of $\xi = r/M - 1$. The dashed line represents Page's approximation.

which is expressed in terms of a transformed radial coordinate

$$\xi = r/M - 1.$$

They involve the functions $p_l^n(\xi)$ and $q_l^n(\xi)$ that satisfy the radial equation subject to the boundary conditions, for $n > 0$, that

$$p_l^n(\xi) \sim (\xi - 1)^{n/2} \text{ as } \xi \to 1,$$

while

$$q_l^n(\xi) \sim (\xi - 1)^{-n/2} \text{ as } \xi \to 1$$

and tends to zero as $\xi \to \infty$.

Specifically, we find

$$\Delta_t^t = 5S_1 - S_2 - S_3 - S_4 + S_5,$$

$$\Delta_r^r = -3S_1 + 3S_2 - 3S_3 + (2\xi - 1)S_4 + 3S_5,$$

$$\Delta_\theta^\theta = \Delta_\phi^\phi = -S_1 - S_2 + 2S_3 - (\xi - 1)S_4 - 2S_5,$$

where the sums S_l are exhibited in Table I.

Numerical evaluation of the $S_l(\xi)$ yields the values depicted in Figs. 1–3. We employ the $(-,+,+,+)$ signature and units in which $\hbar = c = G = k = 1$. With this signature $\langle T_t^t\rangle$ has sign opposite to that of the energy density. As $r \to \infty$ all the curves approach the values appropriate to a thermal bath at the Hawking temperature $(8\pi M)^{-1}$.

It is evident from the figures that Δ_μ^ν does not significantly alter the character of the curves expected on the basis of Page's approximation. These results, however, are in definite disagreement with numerical values previously given by Fawcett,[9] which purported to show that the true value of $\langle T_\mu^\nu\rangle$ differed in important respects from Page's approximation.

Fawcett's error occurs prior to performing numerical analysis. At an early stage of calculation

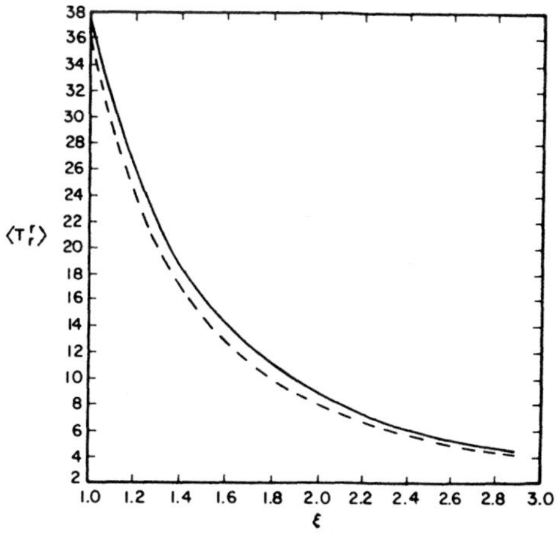

FIG. 3. $[90(8\pi M)^4/\pi^2]\langle T^r_r\rangle$ as a function of $\xi = r/M - 1$. The dashed line represents Page's approximation.

Fawcett[10] effectively writes

$$\langle T_{\mu\nu}\rangle_{\rm REN} = \langle(\phi_{;\mu}\phi_{;\nu} + \tfrac{1}{4}g_{\mu\nu}\phi\Box\phi)\rangle + (6\pi^2)^{-1}g_{\mu\nu}a_2 - \tfrac{1}{12}g_{\mu\nu}\Box\langle\phi^2\rangle - \tfrac{1}{6}\langle\phi^2\rangle_{;\mu\nu},$$

which he employs to evaluate $\langle T_{\theta\theta}\rangle_{\rm REN}$ from which the other components of $\langle T_{\mu\nu}\rangle_{\rm REN}$ are obtained. In his calculation, he assumes that the last term

$$-\tfrac{1}{6}\langle\phi^2\rangle_{;\theta\theta}$$

vanishes on the grounds that $\langle\phi^2\rangle$ is a function of r only. However,

$$\langle\phi^2\rangle_{;\mu\nu} = \langle\phi^2\rangle_{,\mu\nu} - \Gamma_{\mu\nu}{}^{\lambda}\langle\phi^2\rangle_{,\lambda},$$

and therefore

$$\langle\phi^2\rangle_{;\theta\theta} = -\Gamma_{\theta\theta}{}^{r}\frac{\partial}{\partial r}\langle\phi^2\rangle,$$

which is nonzero. If this term is included, then Fawcett's results agree with ours to within numerical error.

It is a pleasure to thank B. Whiting for helpful discussions. This work was supported in part by the National Science Foundation through Grant No. PHY 8205717.

[1]For a review of the physical interpretation of the Hartle-Hawking state and of the other vacua appropriate to Schwarzschild space-time, the reader is referred to D. W. Sciama, P. Candelas, and D. Deutsch, Adv. Phys. **30**, 327 (1981), which contains references to the original literature.

[2]D. N. Page, Phys. Rev. D **25**, 1499 (1982).

[3]M. R. Brown and A. C. Ottewill, Proc. Roy. Soc. London. Ser. A **389**, 379 (1983).

[4]B. DeWitt, *Dynamical Theory of Groups and Fields* (Gordon and Breach, New York, 1965).

[5]S. M. Christensen, Phys. Rev. D **14**, 2490 (1976).

[6]J. D. Bekenstein and L. Parker, Phys. Rev. D **23**, 2850 (1981).

[7]M. S. Fawcett and B. Whiting, "Spontaneous symmetry breaking near a black hole," in Proceedings of the Nuffield Quantum Gravity Conference, London, 1981 (unpublished).

[8]P. Candelas and K. W. Howard, Phys. Rev. D **29**, 1618 (1984).

[9]M. S. Fawcett, Commun. Math. Phys. **89**, 103 (1983).

[10]M. S. Fawcett, Ph.D. thesis, University of Cambridge, Cambridge, England (unpublished).

III. QUANTUM COSMOLOGY

PHYSICAL REVIEW D VOLUME 15, NUMBER 10 15 MAY 1977

Cosmological event horizons, thermodynamics, and particle creation

G. W. Gibbons* and S. W. Hawking
D.A.M.T.P., University of Cambridge, Silver Street, Cambridge, United Kingdom
(Received 4 March 1976)

It is shown that the close connection between event horizons and thermodynamics which has been found in the case of black holes can be extended to cosmological models with a repulsive cosmological constant. An observer in these models will have an event horizon whose area can be interpreted as the entropy or lack of information of the observer about the regions which he cannot see. Associated with the event horizon is a surface gravity κ which enters a classical "first law of event horizons" in a manner similar to that in which temperature occurs in the first law of thermodynamics. It is shown that this similarity is more than an analogy: An observer with a particle detector will indeed observe a background of thermal radiation coming apparently from the cosmological event horizon. If the observer absorbs some of this radiation, he will gain energy and entropy at the expense of the region beyond his ken and the event horizon will shrink. The derivation of these results involves abandoning the idea that particles should be defined in an observer-independent manner. They also suggest that one has to use something like the Everett-Wheeler interpretation of quantum mechanics because the back reaction and hence the spacetime metric itself appear to be observer-dependent, if one assumes, as seems reasonable, that the detection of a particle is accompanied by a change in the gravitational field.

I. INTRODUCTION

The aim of this paper is to extend to cosmological event horizons some of the ideas of thermodynamics and particle creation which have recently been successfully applied to black-hole event horizons. In a black hole the inward-directed gravitational field produced by a collapsing body is so strong that light emitted from the body is dragged back and does not reach an observer at a large distance. There is thus a region of spacetime which is not visible to an external observer. The boundary of the region is called the event horizon of the black hole. Event horizons of a different kind occur in cosmological models with a repulsive Λ term. The effect of this term is to cause the universe to expand so rapidly that for each observer there are regions from which light can never reach him. We shall call the boundary of this region the cosmological event horizon of the observer.

The "no hair" theorems (Israel,[1] Muller zum Hagen *et al.*,[2] Carter,[3] Hawking,[4] Robinson[5,6]) imply that a black hole formed in a gravitational collapse will rapidly settle down to a quasistationary state characterized by only three parameters, the mass M_H, the angular momentum J_H, and the charge Q_H. A black hole of a given M_H, J_H, Q_H therefore has a large number of possible unobservable internal configurations which reflect the different possible initial configurations of the body that collapsed to produce the hole. In purely classical theory this number of internal configurations would be infinite because one could make a given black hole out of an infinitely large number of particles of indefinitely small mass. However, when quantum mechanics is taken into account, one would expect that in order to obtain gravitational collapse the energies of the particle would have to be restricted by the requirement that their wavelength be less than the size of the black hole. It would therefore seem reasonable to postulate that the number of internal configurations is finite. In this case one could associate with the black hole an entropy S_H which would be the logarithm of this number of possible internal configurations.[7,8,9] For this to be consistent the black hole would have to emit thermal radiation like a body with a temperature

$$T_H = G^2\left[\left(\frac{\partial S}{\partial M}\right)_{J,Q}\right]^{-1}.$$

The mechanism by which this thermal radiation arises can be understood in terms of pair creation in the gravitational potential well of the black hole. Inside the black hole there are particle states which have negative energy with respect to an external stationary observer. It is therefore energetically possible for a pair of particles to be spontaneously created near the event horizon. One particle has positive energy and escapes to infinity, the other particle has negative energy and falls into the black hole, thereby reducing its mass. The existence of the event horizon would prevent this happening classically but it is possible quantum-mechanically because one or other of the particles can tunnel through the event horizon. An equivalent way of looking at the pair creation is to regard the positive- and negative-energy particles as being the same particle which tunnels out from the black hole on a spacelike or past-directed timelike world line and is scattered onto a future-directed world line (Hartle and Hawking[10]). When one calculates the rate of particle emission by this process it turns out to be exactly what one would expect from a body with a temperature $T_H = \hbar(2\pi k c)^{-1}\kappa_H$, where κ_H is the surface gravity of the black hole and is related to M_H, J_H, and Q_H by the formulas

$$\kappa_H = (r_+ - r_-)c^2 r_0^{-2},$$

$$r_\pm = c^{-2}[GM \pm (G^2M^2 - J^2M^{-2}c^2 - GQ^2)^{1/2}],$$

$$r_0^2 = r_+^2 + G^{-2}J^2M^{-2}c^2,$$

$$A_H = 4\pi r_0^2.$$

A_H is the area of the event horizon of the black hole.

Combining this quantum-mechanical argument with the thermodynamic argument above, one finds that the total number of internal configurations is indeed finite and that the entropy is given by

$$S_H = (4G\hbar)^{-1}kc^3 A_H.$$

Cosmological models with a repulsive Λ term which expand forever approach de Sitter space asymptotically at large times. In de Sitter space future infinity is spacelike.[11,12] This means that for each observer moving on a timelike world line there is an event horizon separating the region of spacetime which the observer can never see from the region that he can see if he waits long enough. In other words, the event horizon is the boundary of the past of the observer's world line. Such a cosmological event horizon has many formal similarities with a black-hole event horizon. As we shall show in Sec. III it obeys laws very similar to the zeroth, first, and second laws of black-hole mechanics in the classical theory.[13] It also bounds the region in which particles can have negative energy with respect to the observer. One might therefore expect that particle creation with a thermal spectrum would also occur in these cosmological models. In Secs. IV and V we shall show that this is indeed the case: An observer will detect thermal radiation with a characteristic wavelength of the order of the Hubble radius. This would correspond to a temperature of less than 10^{-28} °K so that it is not of much practical significance. It is, however, important conceptually because it shows that thermodynamic arguments can be applied to the universe as a whole and that the close relationship between event horizons, gravitational fields, and thermodynamics that was found for black holes has a wider validity.

One can regard the area of the cosmological event horizon as a measure of one's lack of knowledge about the rest of the universe beyond one's ken. If one absorbs the thermal radiation, one gains energy and entropy at the expense of this region and so, by the first law mentioned above, the area of the horizon will go down. As the area decreases, the temperature of the cosmological radiation goes down (unlike the black-hole case), so the cosmological event horizon is stable. On the other hand, if the observer chooses not to absorb any radiation, there is no change in area of the horizon. This is another illustration of the fact that the concept of particle production and the back reaction associated with it seem not to be uniquely defined but to be dependent upon the measurements that one wishes to consider.[14-16]

The plan of the paper is as follows. In Sec. II we describe the black-hole asymptotically de Sitter solutions found by Carter.[20] In Sec. III we derive the classical laws governing both cosmological and black-hole event horizons. In Sec. IV we discuss particle creation in de Sitter space. We abandon the concept of particles as being observer-independent and consider instead what an observer moving on a timelike geodesic and equipped with a particle detector would actually measure. We find that he would detect an isotropic background of thermal radiation with a temperature $(2\pi)^{-1}\kappa_C$ where $\kappa_C = \Lambda^{1/2}3^{-1/2}$ is the surface gravity of the cosmological event horizon of the observer. Any other observer moving on a timelike geodesic will also see isotropic radiation with the same temperature even though he is moving relative to the first observer. This shows that they are not observing the same particles: Particles are observer-dependent. In Sec. V we extend these results to asymptotically de Sitter spaces containing black holes. The implications are considered in Sec. VI. It seems necessary to adopt something like the Everett-Wheeler interpretation of quantum mechanics because the back reaction and hence the spacetime metric will be observer-dependent, if one assumes, as seems reasonable, that the detection of a particle is accompanied by a change in the gravitational field.

We shall adopt units in which $G = \hbar = k = c = 1$. We shall use a metric with signature +2 and our conventions for the Riemann and the Ricci tensors are

$$v_{a;[b;c]} = \tfrac{1}{2} R^d{}_{abc}\, v_d\,,$$

$$R_{ab} = R_a{}^c{}_{bc}\,.$$

II. EXACT SO UTIONS WITH COSMOLOGICAL EVENT HORIZONS

In this section we shall give some examples of event horizons in exact solutions of the Einstein equations

$$R_{ab}-\tfrac{1}{2}g_{ab}R+\Lambda g_{ab}=8\pi T_{ab}\,. \tag{2.1}$$

We shall consider only the case of Λ positive (corresponding to repulsion). Models with negative Λ do not, in general, have event horizons.

The simplest example is de Sitter space which is a solution of the field equations with $T_{ab}=0$. One can write the metric in the static form

$$ds^2=-(1-\Lambda r^2 3^{-1})dt^2+dr^2(1-\Lambda r^2 3^{-1})^{-1} + r^2(d\theta^2+\sin^2\theta\, d\phi^2)\,. \tag{2.2}$$

This metric has an apparent singularity at $r=3^{1/2}\Lambda^{-1/2}$. This singularity caused considerable discussion when the metric was first discovered.[17,18] However, it was soon realized that it arose simply from a bad choice of coordinates and that there are other coordinate systems in which the metric can be analytically extended to a geodesically complete space of constant curvature with topology $R^1\times S^3$. For a detailed description of these coordinate systems the reader is referred to Refs. 12 and 19. For our purposes it will be convenient to express the de Sitter metric in "Kruskal coordinates":

$$ds^2=3\Lambda^{-1}(UV-1)^{-2}\times[-4dU\,dV+(UV+1)^2(d\theta^2+\sin^2\theta\, d\phi^2)] \tag{2.3}$$

where

$$r=3^{1/2}\Lambda^{-1/2}(UV+1)(1-UV)\,, \tag{2.4}$$

$$\exp(2\Lambda^{1/2}3^{-1/2}t)=-VU^{-1}\,. \tag{2.5}$$

The structure of this space is shown in Fig. 1. In this diagram radial null geodesics are at $\pm 45°$ to the vertical. The dashed curves $UV=-1$ are timelike and represent the origin of polar coordinates and the antipodal point on a three-sphere. The solid curves $UV=+1$ are spacelike and represent past and future infinity $\mathscr{I}^-$ and $\mathscr{I}^+$, respectively.

In region I ($U<0$, $V>0$, $UV>-1$) the Killing vector $K=\partial/\partial t$ is timelike and future-directed. However, in region IV ($U>0$, $V<0$, $UV>-1$, K is still timelike but past-directed, while in regions II and III ($0<UV<1$) K is spacelike. The Killing vector K is null on the two surfaces $U=0$, $V=0$. These are respectively the future and past event horizons for any observer whose world line remains in region I; in particular for any observer moving along a curve of constant r in region I.

By applying a suitable conformal transformation one can make the Kruskal diagram finite and convert it to the Penrose-Carter form (Fig. 2). Radial null geodesics are still $\pm 45°$ to the vertical but the freedom of the conformal factor has been used

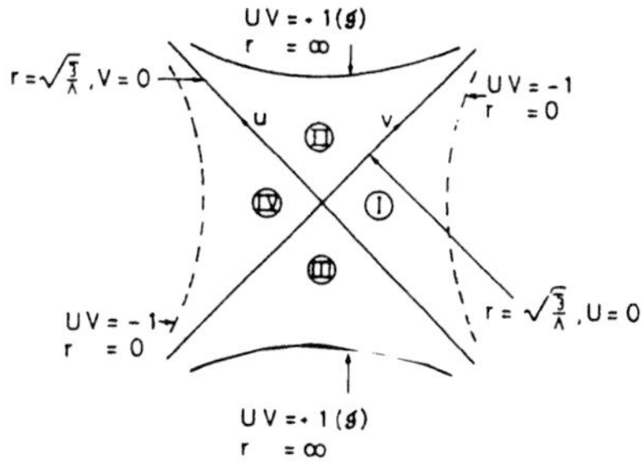

FIG. 1. Kruskal diagram of the (r,t) plane of de Sitter space. In this figure null geodesics are at $\pm 45°$ to the vertical. The dashed curves $r=0$ are the antipodal origins of polar coordinates on a three-sphere. The solid curves $r=\infty$ are past and future infinity $\mathscr{I}^-$ and $\mathscr{I}^+$, respectively. The lines $r=3^{1/2}\Lambda^{-1/2}$ are the past and future event horizons of observers at the origin.

to make the origin of polar coordinates, $r=0$, and future and past infinity, $\mathscr{I}^+$ and $\mathscr{I}^-$, straight lines. Also shown are some orbits of the Killing vector $K=\partial/\partial t$. Because de Sitter space is invariant under the ten-parameter de Sitter group, SO(4, 1), K will not be unique. Any timelike geodesic can be chosen as the origin of polar coordinates and the surfaces $U=0$ and $V=0$ in such coordinates will be the past and future event horizons of an observer moving on this geodesic. If one normalizes K to have unit magnitude at the origin, one can define a "surface gravity" for the horizon by

$$K_{a;b}K^b=\kappa_C K_a \tag{2.6}$$

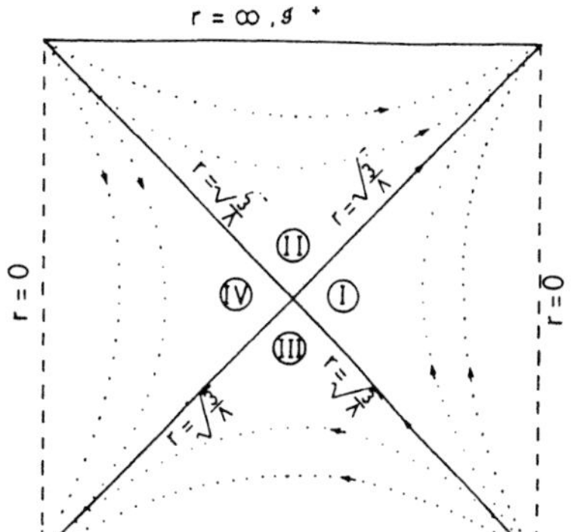

FIG. 2. The Penrose-Carter diagram of de Sitter space. The dotted curves are orbits of the Killing vector.

on the horizon. This gives

$$\kappa_C = \Lambda^{1/2} 3^{-1/2} . \tag{2.7}$$

The area of the cosmological horizon is

$$A_C = 12\pi\Lambda^{-1} . \tag{2.8}$$

One can also construct solutions which generalize the Kerr-Newman family to the case when Λ is nonzero.[20,21] The simplest of these is the Schwarzschild–de Sitter metric. When $\Lambda = 0$ the unique spherically symmetric vacuum spacetime is the Schwarzschild solution. The metric of this can be written in static form:

$$ds^2 = -(1-2Mr^{-1})dt^2 + dr^2(1-2Mr^{-1})^{-1} + r^2(d\theta^2 + \sin^2\theta\, d\phi^2) . \tag{2.9}$$

As is now well known, the apparent singularities at $r = 2M$ correspond to a horizon and can be removed by changing to Kruskal coordinates in which the metric has the form

$$ds^2 = -32M^3 r^{-1} \exp(-2^{-1}M^{-1}r)\, dU\, dV + r^2(d\theta^2 + \sin^2\theta\, d\phi^2) , \tag{2.10}$$

where

$$UV = (1 - 2^{-1}M^{-1}r) \exp(2^{-1}M^{-1}r) \tag{2.11}$$

and

$$UV^{-1} = -\exp(-2^{-1}M^{-1}t) . \tag{2.12}$$

The Penrose-Carter diagram of the Schwarzschild solution is shown in Fig. 3. The wavy lines marked $r = 0$ are the past and future singularities. Region I is asymptotically flat and is bounded on the right by past and future null infinity $\mathscr{I}^-$ and $\mathscr{I}^+$. It is bounded on the left by the surfaces $U = 0$ and $V = 0$, $r = 2M$. These are future and past event horizons for observers who remain outside $r = 2M$. On the left-hand side of the diagram there is another a asymptotically flat region IV. The Killing vector $K = \partial/\partial t$ is now uniquely defined by the condition that it be timelike and of unit magnitude near $\mathscr{I}^+$ and $\mathscr{I}^-$. It is timelike and future-directed in region I, timelike and past-directed in region IV, and spacelike in regions II and III. The Killing vector K is null on the horizons which have area $A_H = 16\pi M^2$. The surface gravity, defined by (2.6), is $\kappa_H = (4M)^{-1}$.

The Schwarzschild solution is usually interpreted as a black hole of mass M in an asymptotically flat space. There is a straightforward generalization to the case of nonzero Λ which represents a black hole in asymptotically de Sitter space. The metric can be written in the static form

$$ds^2 = -(1 - 2Mr^{-1} - \Lambda r^2 3^{-1})dt^2 + dr^2(1 - 2Mr^{-1} - \Lambda r^2 3^{-1})^{-1} + r^2(d\theta^2 + \sin^2\theta\, d\phi^2) . \tag{2.13}$$

If $\Lambda > 0$ and $9\Lambda M^2 < 1$, the factor $(1 - 2Mr^{-1} - \Lambda r^2 3^{-1})$ is zero at two positive values of r. The smaller of of these values, which we shall denote by r_+, can be regarded as the position of the black-hole event horizon, while the larger value r_{++} represents the position of the cosmological event horizon for observers on world lines of constant r between r_+ and r_{++}. By using Kruskal coordinates as above one can remove the apparent singularities in the metric at r_+ and r_{++}. One has to employ separate coordinate patches at r_+ and r_{++}. We shall not give the expressions in full because they are rather messy; however, the general structure can be seen from the Penrose-Carter diagram shown in Fig. 4. Instead of having two regions (I and IV) in which the Killing vector $K = \partial/\partial t$ is timelike, there are now an infinite sequence of such regions, also labeled I and IV depending upon whether K is future- or past-directed. There are also infinite sequences of $r = 0$ singularities and spacelike infinities $\mathscr{I}^+$ and $\mathscr{I}^-$. The surfaces $r = r_+$ and $r = r_{++}$ are black-hole and cosmological event horizons for observers moving on world lines of constant

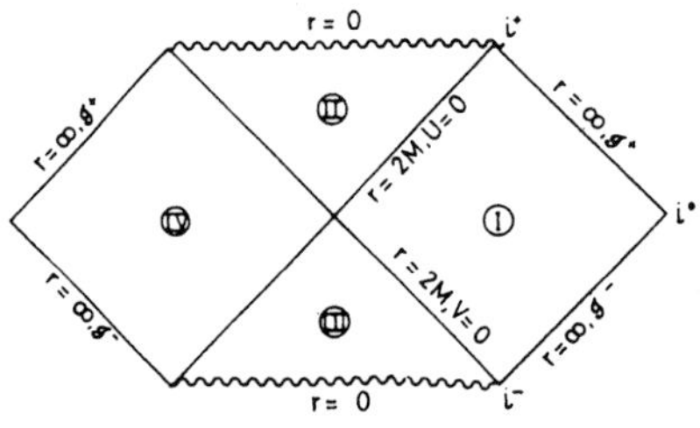

FIG. 3. The Penrose-Carter diagram of the Schwarzschild solution. The wavy lines and the top and bottom are the future and past singularities. The diagonal lines bounding the diagram on the right-hand side are the past and future null infinity of asymptotically flat space. The region IV on the left-hand-side is another asymptotically flat space.

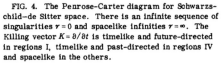

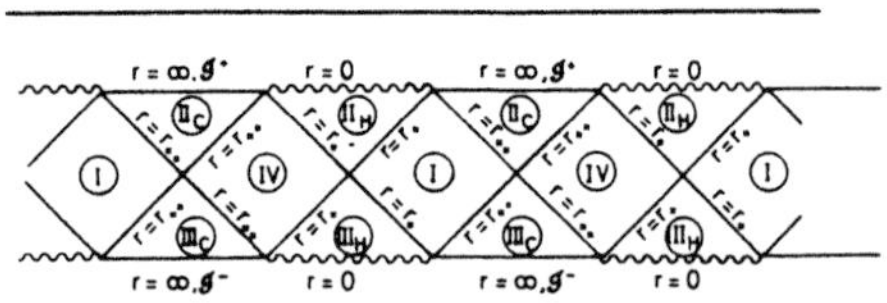

FIG. 4. The Penrose-Carter diagram for Schwarzschild–de Sitter space. There is an infinite sequence of singularities $r = 0$ and spacelike infinities $r = \infty$. The Killing vector $K = \partial/\partial t$ is timelike and future-directed in regions I, timelike and past-directed in regions IV and spacelike in the others.

r between r_+ and r_{++}.

The Killing vector $K = \partial/\partial t$ is uniquely defined by the conditions that it be null on both the black-hole and the cosmological horizons and that its magnitude should tend to $\Lambda^{1/2}3^{-1/2}r$ as r tends to infinity. One can define black-hole and cosmological surface gravities κ_H and κ_C by

$$K_{a;b}K^b = \kappa K_a \tag{2.14}$$

on the horizons. These are given by

$$\kappa_H = \Lambda 6^{-1} r_+^{-1}(r_{++} - r_+)(r_+ - r_{--}), \tag{2.15a}$$

$$\kappa_C = \Lambda 6^{-1} r_{++}^{-1}(r_{++} - r_+)(r_{++} - r_{--}), \tag{2.15b}$$

where $r = r_{--}$ is the negative root of

$$3r - 6M - \Lambda r^3 = 0. \tag{2.16}$$

The areas of the two horizons are

$$A_H = 4\pi r_+^{\,2} \tag{2.17}$$

and

$$A_C = 4\pi r_{++}^{\,2}. \tag{2.18}$$

If one keeps Λ constant and increases M, r_+ will increase and r_{++} will decrease. One can understand this in the following way. When $M = 0$ the gravitational potential $g(\partial/\partial t, \partial/\partial t)$ is $1 - \Lambda r^2 3^{-1}$. The introduction of a mass M at the origin produces an additional potential of $-2Mr^{-1}$. Horizons occur at the two values of r at which $g(\partial/\partial t, \partial/\partial t)$ vanishes. Thus as M increases, the black-hole horizon r_+ increases and the cosmological horizon r_{++} decreases. When $9\Lambda M^2 = 1$ the two horizons coincide. The surface gravity K can be thought of as the gravitational field or gradient of the potential at the horizons. As M increases both κ_H and κ_C decrease.

The Kerr–Newman–de Sitter space can be expressed in Boyer-Lindquist-type coordinates as[20,21]

$$\begin{aligned} ds^2 = {} & \rho^2(\Delta_r^{-1}dr^2 + \Delta_\theta^{-1}d\theta^2) \\ & + \rho^{-2}\Xi^{-2}\Delta_\theta[a\,dt - (r^2 + a^2)d\phi]^2 \\ & - \Delta_r \Xi^{-2}\rho^{-2}(dt - a\sin^2\theta\, d\phi)^2, \end{aligned} \tag{2.19}$$

where

$$\rho^2 = r^2 + a^2\cos^2\theta, \tag{2.20}$$

$$\Delta_r = (r^2 + a^2)(1 - \Lambda r^2 3^{-1}) - 2Mr + Q^2, \tag{2.21}$$

$$\Delta_\theta = 1 + \Lambda a^2 3^{-1}\cos^2\theta, \tag{2.22}$$

$$\Xi = 1 + \Lambda a^2 3^{-1}. \tag{2.23}$$

The electromagnetic vector potential A_a is given by

$$A_a = Qr\rho^{-2}\Xi^{-1}(\delta_a^t - a\sin^2\theta\,\delta_a^\phi). \tag{2.24}$$

Note that our Λ has the opposite sign to that in Ref. 21.

There are apparent singularities in the metric at the values of r for which $\Delta_r = 0$. As before, these correspond to horizons and can be removed by using appropriate coordinate patches. The Penrose-Carter diagram of the symmetry axis ($\theta = 0$) of these spaces is shown in Fig. 5 for the case that Δ_+ has 4 distinct roots: r_{--}, r_-, r_+, and r_{++}. As before, r_{++} and r_+ can be regarded as the cosmological and black-hole event horizons, respectively. In addition, however, there is now an inner black-hole horizon at $r = r_-$. Passing through this, one comes to the ring singularity at $r = 0$, on the other side of which there is another cosmological horizon at $r = r_{--}$ and another infinity. The diagram shown is the simplest one to draw but it is not simply connected; one can take covering spaces. Alternatively one can identify regions in this diagram.

The Killing vector $\tilde{K} = \partial/\partial\phi$ is uniquely defined by the condition that its orbits should be closed curves with parameter length 2π. The other Kill-

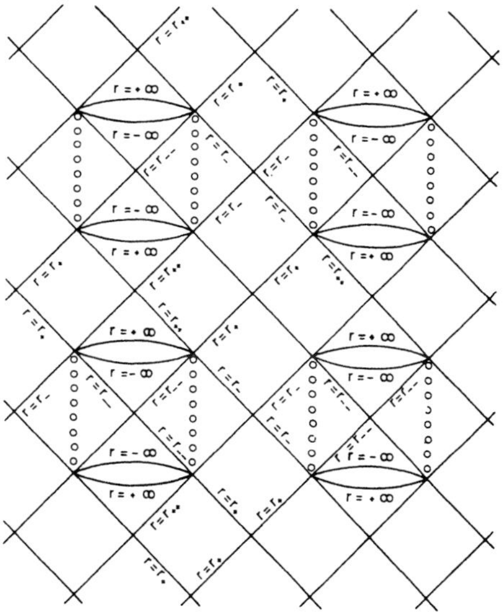

FIG. 5. The Penrose-Carter diagram of the symmetry axis of the Kerr-Newman–de Sitter solution for the case that Δ_r has four distinct real roots. The infinities $r = +\infty$ and $r = -\infty$ are not joined together. The external cosmological horizon occurs at $r = r_{++}$ the exterior black-hole horizon at $r = r_+$, the inner black-hole horizon at $r = r_-$. The open circles mark where the ring singularity occurs, although this is not on the symmetry axis. On the other side of the ring at negative values of r there is another cosmological horizon at $r = r_{--}$ and another infinity.

ing vector $K=\partial/\partial t$ is not so specially picked out. One can add different constants multiples of $\tilde{K}$ to K to obtain Killing vectors which are null on the different horizons and one can then define surface gravities as before. We shall be interested only in those for the r_+, r_{++} horizons. They are

$$\kappa_H=\Lambda 6^{-1}\Xi^{-1}(r_+-r_{--})(r_+-r_-)(r_{++}-r_+)(r_+{}^2+a^2)^{-1}, \tag{2.25}$$

$$\kappa_C=\Lambda 6^{-1}\Xi^{-1}(r_{++}-r_+)(r_{++}-r_-)(r_{++}-r_-)(r_+{}^2+a^2)^{-1}. \tag{2.26}$$

The areas of these horizons are

$$A_H=4\pi(r_+{}^2+a^2), \tag{2.27}$$

$$A_C=4\pi(r_{++}{}^2+a^2). \tag{2.28}$$

III. CLASSICAL PROPERTIES OF EVENT HORIZONS

In this section we shall generalize a number of results about black-hole event horizons in the classical theory to spacetimes which are not asymptotically flat and may have a nonzero cosmological constant, and to event horizons which are not black-hole horizons. The event horizon of a black hole in asymptotically flat spacetimes is normally defined as the boundary of the region from which one can reach future null infinity, $\mathscr{I}^+$, along a future-directed timelike or null curve. In other words it is $\dot{J}^-(\mathscr{I}^+)$ [or equivalently $\dot{I}^-(\mathscr{I}^+)$], where an overdot indicates the boundary and J^- is the causal past (I^- is the chronological past). However, one can also define the black-hole horizon as $\dot{I}^-(\lambda)$, the boundary of the past of a timelike curve λ which has a future end point at future timelike infinity, i^+ in Fig. 3. One can think of λ as the world line of an observer who remains outside the black hole and who does not accelerate away to infinity. The event horizon is the boundary of the region of spacetime that he can see if he waits long enough. It is this definition of event horizon that we shall extend to more general spacetimes which are not asymptotically flat.

Let λ be a future inextensible timelike curve representing an observer's world line. For our considerations of particle creation in the next section we shall require that the observer have an indefinitely long time in which to detect particles. We shall therefore assume that λ has infinite proper length in the future direction. This means that it does not run into a singularity. The past of λ, $I^-(\lambda)$, is a terminal indecomposable past set, or TIP in the language of Geroch, Kronheimer, and Penrose.[22] It represents all the events that the observer can ever see. We shall assume that what the observer sees at late times can be predicted (classically at least) from a spacelike surface $\mathcal{S}$, i.e., $I^-(\lambda)\cap J^+(\mathcal{S})$ is contained in the future Cauchy development $D^+(\mathcal{S})$.[12] We shall also assume that $\dot{I}^-(\lambda)\cap J^+(\mathcal{S})$, the portion of the event horizon to the future of $\mathcal{S}$, is contained in $D^+(\mathcal{S})$. Such an event horizon will be said to be predictable. The event horizon will be generated by null geodesic segments which have no future end points but which have past end points if and where they intersect other generators.[12] In another paper[23] it is shown that the generators of a predictable event horizon cannot be converging if the Einstein equations hold (with or without cosmological constant), provided that the energy-momentum tensor satisfies the strong energy condition $T_{ab}u^a u^b \geq \frac{1}{2}T^a_a u^b u_b$ for any timelike vector u_a, i.e., provided that $\mu+P_i\geq 0$, $\mu+\sum_{i=1}^{i=3}P_i\geq 0$, where μ is the energy density and P_i are the principal pressures. This gives immediately the following result, which, because of the very suggestive analogy with thermodynamics, we call:

The second law of event horizons: The area of any connected two-surface in a predictable event horizon cannot decrease with time. The area may be infinite if the two-dimensional cross section is not compact. However, in the examples in Sec. II, the natural two-sections are compact and have constant area.

In the case of gravitational collapse in asymptotically flat spacetimes one expects the spacetime eventually to settle down to a quasistationary state because all the available energy will either fall through the event horizon of the black hole (thereby increasing its area) or be radiated away to infinity. In a similar way one would expect that where the intersection of $I^-(\lambda)$ with a spacelike surface $\mathcal{S}$ had compact closure (which we shall assume henceforth), there would only be a finite amount of energy available to be radiated through the cosmological event horizon of the observer and that therefore this spacetime would eventually approach a stationary state. One is thus lead to consider solutions in which there is a Killing vector K which is timelike in at least some region of $I^-(\lambda)\cap J^+(\mathcal{S})$. Such solutions would represent the asymptotic future limit of general spacetimes with predictable event horizons.

Several results about stationary empty asymptotically flat black-hole solutions can be generalized to stationary solutions of the Einstein equations, with cosmological constant, which contain predictable event horizons. The first such theorem is that the null geodesic generators of each connected component of the event horizon must coincide with orbits of some Killing vector.[24,21]

These Killing vectors may not coincide with the original Killing vector K and may be different for different components of the horizon. In either of

$$F(W,x,x')=\int \delta x[w]\exp\left[\frac{i}{4}\int_0^W g(\dot{x},\dot{x})dw\right] \quad (4.2)$$

and the integral is taken over all paths $x(w)$ from x to x'.

As in the Hartle and Hawking paper,[10] this path integral can be given a well-defined meaning by analtyically continuing the parameter W to negative imaginary values and analytically continuing the coordinates to a region where the metric is positive-definite. A convenient way of doing this is to embed de Sitter space as the hyperboloid

$$-T^2+S^2+X^2+Y^2+Z^2=3\Lambda^{-1} \quad (4.3)$$

in the five-dimensional space with a Lorentz metric:

$$ds^2=-dT^2+dS^2+dX^2+dY^2+dZ^2. \quad (4.4)$$

Taking T to be $i\tau$ (τ real), we obtain a sphere in five-dimensional Euclidean space. On this sphere the function F satisfies the diffusion equation

$$\frac{\partial F}{\partial \Omega}=\bar{\Box}^2 F, \quad (4.5)$$

where $\Omega=iW$ and $\bar{\Box}^2$ is the Laplacian on the four-sphere. Because the four-sphere is compact there is a unique solution of (4.5) for the initial condition

$$F(0,x,x')=\delta(x,x'), \quad (4.6)$$

where $\delta(x,x')$ is the Dirac δ function on the four-sphere. One can then define the propagator $G(x,x')$ from (4.1) by analytically continuing the solution for F back to real values of the parameter W and real coordinates x and x'. Because the function F is analytic for finite points x and x', any singularities which occur in $G(x,x')$ must come from the end points of the integration in (4.1). As shown in Ref. 10, there will be singularities in $G(x,x')$ when, and only when, x and x' can be joined by a null geodesic. This will be the case if and only if

$$(T-T')^2=(S-S')^2+(X-X')^2+(Y-Y')^2+(Z-Z')^2. \quad (4.7)$$

The coordinates, T,S,X,Y,Z can be related to the static coordinates t,r,θ,ϕ used in Sec. II by

$$T=(\Lambda 3^{-1}-r^2)^{1/2}\sinh\Lambda^{1/2}3^{-1/2}t, \quad (4.8)$$

$$S=(\Lambda 3^{-1}-r^2)^{1/2}\cosh\Lambda^{1/2}3^{-1/2}t, \quad (4.9)$$

$$X=r\sin\theta\cos\phi, \quad (4.10)$$

$$Y=r\sin\theta\sin\phi, \quad (4.11)$$

$$Z=r\cos\theta. \quad (4.12)$$

The horizons $\Lambda r^2=3$ are the intersection of the hyperplanes $T=\pm S$ with the hyperboloid. As in Ref. 10 we define the complexified horizon by $\Lambda r^2=3$, θ, ϕ real. On the complexified horizon X, Y, and Z are real and either $T=S=\Lambda^{-1/2}3^{1/2}V$, $U=0$ or $T=-S=\Lambda^{-1/2}3^{1/2}U$, $V=0$. By Eq. (4.7) a complex null geodesic from a real point (T',S',X',Y',Z') on the hyperboloid can intersect the complex horizon only on the real sections $T=\pm S$ real. If the point (T',S',X',Y',Z') is in region I ($S>|T|$) the propagator $G(x',x)$ will have a singularity on the past horizon at the point where the past-directed null geodesic from x' intersects the horizon. As shown in Ref. 10, the ϵ convergence factor in (4.1) will displace the pole slightly below the real axis in the complex plane on the complexified past horizon. The propagator $G(x',x)$ is therefore analytic in the upper half U plane on the past horizon. Similarly, it will be analytic in the lower V plane on the future horizon.

The propagator $G(x',x)$ satisfies the wave equation

$$(\Box_x^2-m^2)G(x',x)=-\delta(x,x') \quad (4.13)$$

Thus if x' is a fixed point in region I, the value $G(x',x)$ for a point in region II will be determined by the values of $G(x',x)$ on a characteristic Cauchy surface for region II consisting of the section of the $U=0$ horizon for real $V\geq 0$ and the section of the $V=0$ horizon for real $U\geq 0$. The coordinates r and t of the point x are related to U and V by

$$e^{2\kappa_C t}=VU^{-1} \quad (4.14)$$

$$r=(1+UV)(1-UV)^{-1}\kappa_C^{-1} \quad (4.15)$$

If one holds r fixed at a real value but lets $t=\tau+i\sigma$, then

$$U=|U|\exp(-i\sigma\kappa_C), \quad (4.16)$$

$$V=|V|\exp(+i\sigma\kappa_C). \quad (4.17)$$

For a fixed value of σ the metric (2.3) of de Sitter space remains real and unchanged. Thus the value of $G(x',x)$ at a complex coordinate t of the point x but real r,θ,ϕ can be obtained by solving the Klein-Gordon equation with real coefficients and with initial data on the Cauchy surface $V=0$, $U=|U|\exp(-i\kappa_C\sigma)$ and $U=0$, $V=|V|\exp(+i\kappa_C\sigma)$. Because $G(x',x)$ is analytic in the upper half U plane on $V=0$ and the lower half V plane on $U=0$, the data and hence the solution will be regular provided that

$$-\pi\kappa_C^{-1}\leq\sigma\leq 0. \quad (4.18)$$

The operator

$$\left(\frac{\partial}{\partial t}\right)_r=\kappa_C\left(V\frac{\partial}{\partial V}-U\frac{\partial}{\partial U}\right) \quad (4.19)$$

commutes with the Klein-Gordon operator $\Box_x^2-m^2$ and is zero when acting on the initial data for σ

satisfying (4.18). Thus the solution $G(x', x)$ determined by the initial data will be analytic in the coordinates t of the point x for σ satisfying Eq. (4.18).

This is the basic result which enables us to show that an observer moving on a timelike geodesic in de Sitter space will detect thermal radiation.

The propagator we have defined appears to be similar to that constructed by other authors.[28-30] However, our use of the propagator will be different: Instead of trying to obtain some observer-independent measure of particle creation, we shall be concerned with what an observer moving on a timelike geodesic in de Sitter space would measure with a particle detector which is confined to a small tube around his world line. Without loss of generality we can take the observer's world line to be at the origin of polar coordinates in region I. Within the world tube of the particle detector the spacetime can be taken as flat.

The results we shall obtain are independent of the detailed nature of the particle detector. However, for explicitness we shall consider a particle model of a detector similar to that discussed by Unruh[14] for uniformly accelerated observers in flat space. This will consist of some system such as an atom which can be described by a nonrelativistic Schrödinger equation

$$i\frac{\partial\Psi}{\partial t'}=H_0\Psi+g\phi\Psi\ ,$$

where t' is the proper time along the observer's world line, H_0 is the Hamiltonian of the undisturbed particle detector and $g\phi\Psi$ is a coupling term to the scalar field ϕ. The undisturbed particle detector will have energy levels E_i and wave functions $\Psi_i(\vec{R}')e^{-iE_it}$, where $\vec{R}'$ represents the spatial position of a point in the detector.

By first-order perturbation theory the amplitude to excite the detector from energy level E_i to a higher-energy level E_j is proportional to

$$\int dt' \int d^3\vec{R}'\bar{\Psi}_j g\phi\Psi_i \exp[-i(E_i-E_j)t'] .$$

In other words, the detector responds to components of field ϕ which are positive frequency along the observer's world line with respect to his proper time. By superimposing detector levels with different energies one can obtain a detector response function of a form

$$f(t')h(\vec{R}),$$

where $f(t')$ is a purely positive-frequency function of the observer's proper time t' and h is zero outside some value of r' corresponding to the radius of the particle detector. Let $\mathcal{P}$ be a three-surface which completely surrounds the observer's world line. If the observer detects a particle, it must have crossed $\mathcal{P}$ in some mode k_j which is a solution of the Klein-Gordon equation with unit Klein-Gordon norm over the hypersurface $\mathcal{P}$. The amplitude for the observer to detect such a particle will be

$$\int\int fh(x')G(x',x)\overleftrightarrow{\partial}_a \bar{k}_j(x)dV'd\Sigma^a , \qquad (4.20)$$

where the volume integral in x' is taken over the volume of the particle detector and the surface integral in x is taken over $\mathcal{P}$.

The hypersurface $\mathcal{P}$ can be taken to be a spacelike surface of large constant r in the past in region III and a spacelike surface of large constant r in the future in region II. In the limit that r tends to infinity these surfaces tend to past infinity $\mathscr{I}^-$ and future infinity $\mathscr{I}^+$, respectively. We shall assume that there were no particles present on the surface in the distant past. Thus the only contribution to the amplitude (4.20) comes from the surface in the future. One can interpret this as the spontaneous creation of a pair of particles, one with positive and one with negative energy with respect to the Killing vector $K=\partial/\partial t$. The particle with positive energy propagates to the observer and is detected. The particle with negative energy crosses the event horizon into region II where K is spacelike. It can exist there as a real particle with timelike four-momentum. Equivalently, one can regard the world lines of the two particles as being the world line of a single particle which tunnels through the event horizon out of region II and is detected by the observer.

Suppose the detector is sensitive to particles of a certain energy E. In this case the positive-frequency-response function $f(t)$ will be proportional to e^{-iEt}. By the stationarity of the metric, the propagator $G(x',x)$ can depend on the coordinates t' and t only through their difference. This means that the amplitude (4.20) will be zero except for modes k_j of the form $\chi(r,\theta,\varphi)e^{-iEt}$. If one takes out a δ function which arises from the integral over $t-t'$, the amplitude for detection is proportional to

$$\mathcal{E}_E(\vec{R}',\vec{R})=\int_{-\infty}^{+\infty} dt\, e^{-iEt}G(0,\vec{R}';t,\vec{R}) , \qquad (4.21)$$

where $\vec{R}'$ and $\vec{R}$ denote respectively (r',θ',ϕ') and (r,θ,φ) and the radial and angular integrals over the functions h and χ have been factored out. Using the result derived above that $G(x',x)$ is analytic in a strip of width $\pi\kappa_C^{-1}$ below the real t axis, one can displace the contour in (4.21) down $\pi\kappa_C^{-1}$ to obtain

$$\mathcal{E}_E(\vec{R}',\vec{R}) = \exp(-\pi E_C \kappa_C^{-1})\int_{-\infty}^{+\infty} dt\, e^{-iEt} G(0,\vec{R}', t - i\pi\kappa_C^{-1}, \vec{R}). \tag{4.22}$$

By Eqs. (4.16) and (4.17) the point $(t - i\pi\kappa_C^{-1}, r, \theta, \varphi)$ is the point in region III obtained by reflecting in the origin of the U, V plane. Thus

$$\begin{pmatrix}\text{amplitude for particle of energy } E \text{ to propagate}\\ \text{from region II and be absorbed by observer}\end{pmatrix} = \exp(-\pi E\kappa_C^{-1})\begin{pmatrix}\text{amplitude for particle with energy}\\ E \text{ to propagate from region III and}\\ \text{be absorbed by observer}\end{pmatrix}. \tag{4.23}$$

By time-reversal invariance the latter amplitude is equal to the amplitude for the observer's detector in an excited state to emit a particle with energy E which travels to region II. Therefore

$$\begin{pmatrix}\text{probability for detector to absorb}\\ \text{a particle from region II}\end{pmatrix} = \exp(-2\pi E\kappa_C^{-1})\begin{pmatrix}\text{probability for detector to emit}\\ \text{a particle to region II}\end{pmatrix}. \tag{4.24}$$

This is just the condition for the detector to be in thermal equilibrium at a temperature

$$T = (2\pi)^{-1}\kappa_C = (12)^{-1/2}\pi^{-1}\Lambda^{1/2}. \tag{4.25}$$

The observer will therefore measure an isotropic background of thermal radiation with the above temperature. Because all timelike geodesics are equivalent under the de Sitter group, any other observer will also see an isotropic background with the same temperature even though he is moving relative to the first observer. This is yet another illustration of the fact that different observers have different definitions of particles. It would seem that one cannot, as some authors have attempted, construct a unique observer-independent renormalized energy-momentum tensor which can be put on the right-hand side of the classical Einstein equations. This subject will be dealt with in another paper.[16]

Another way in which one can derive the result that a freely moving observer in de Sitter space will see thermal radiation is to note that the propagator $G(x, x')$ is an analytic function of the coordinates T, S, T', S', or alternatively U, V, U', V' except when x and x' can be joined by null geodesics. On the other hand, the static-time coordinate t is a multivalued function of T and S or U and V, being defined only up to an integral multiple of $2\pi i\kappa_C^{-1}$. Thus the propagator $G(x', x)$ is a periodic function of t with period $2\pi i\kappa_C^{-1}$. This behavior is characteristic of what are known as "thermal Green's functions."[33] These may be defined (for interacting fields as well as the noninteracting case considered here) as the expectation value of the time-ordered product of the field operators, where the expectation value is taken not in the vacuum state but over a grand canonical ensemble at some temperature $T = \beta^{-1}$. Thus

$$G_T(x', x) = i\,\mathrm{Tr}[e^{-\beta H}\mathcal{J}\,\phi(\dot{x})\varphi(x)]/\mathrm{Tr}\,e^{-\beta H}, \tag{4.26}$$

where $\mathcal{J}$ denotes Wick time-ordering and H is the Hamiltonian in the observer's static frame. ϕ is the quantum field operator and Tr denotes the trace taken over a complete set of states of the system. Therefore

$$\begin{aligned}-iG_T(\vec{R}', t', \vec{R}, t) &= \mathrm{Tr}[e^{-\beta H}\mathcal{J}\,\phi(\vec{R}, t)\varphi(\vec{R}, t')]/\mathrm{Tr}\,e^{-\beta H}\\ &= \mathrm{Tr}[e^{-\beta H}\mathcal{J}\,\varphi(\vec{R}', t')e^{\beta H}e^{-\beta H}\phi(\vec{R}, t)]/\mathrm{Tr}\,e^{-\beta H}\\ &= \mathrm{Tr}[e^{-\beta H}\mathcal{J}\,\phi(\vec{R}', t' + i\beta)\phi(\vec{R}', t)]/\mathrm{Tr}\,e^{-\beta H}\\ &= -iG_T(\vec{R}', t' + i\beta; \vec{R}, t).\end{aligned} \tag{4.27}$$

Since

$$\phi(\vec{R}, t) = e^{-\beta H}\phi(\vec{R}, t - i\beta)e^{\beta H}. \tag{4.28}$$

Thus the thermal propagator is periodic in $t - t'$ with period iT^{-1}. One would expect $G_T(x', x)$ to have singularities when x and x' can be connected by a null geodesic and these singularities would be repeated periodically in the complex $t' - t$ plane. It therefore seems that the propagator $G(x', x)$ that we have defined by a path integral is the same as the thermal propagator $G_T(x', x)$ for a grand canonical ensemble at temperature T $T = (2\pi)^{-1}\kappa_C$ in the observer's static frame. Thus to the observer it will seem as if he is in a bath of blackbody radiation at the above temperature. It is interesting to note that a similar result was found for two-dimensional de Sitter space by Figari, Hoegh-Krohn, and Nappi[34] although they did not appreciate its significance in terms of particle creation.

The correspondence between $G(x',x)$ and the thermal Green's function is the same as that which has been pointed out in the black-hole case by Gibbons and Perry.[35] As in their paper, one can argue that because the free-field propagator $G(x',x)$ is identical with the free-field thermal propagator $G_T(x',x)$, any n-point interacting Green's function $\hat{G}$ which can be constructed by perturbation theory from G in a renormalizable field theory will be identical to the n-point interacting thermal Green's function constructed from G_T in a similar manner. This means that the result that an observer will think himself to be immersed in blackbody radiation at temperature $T=\kappa_C(2\pi)^{-1}$ will be true not only in the free-field case that we have treated but also for fields with mutual interactions and self-interactions. In particular, one would expect it to be true for the gravitational field, though this is, of course, not renormalizable, at least in the ordinary sense.

It is more difficult to formulate the propagator for higher-spin fields in terms of a path integral. However, it seems reasonable to define the propagators for such fields as solutions of the relevant inhomogeneous wave equation with the boundary conditions that the propagator from a point x' in region I is an analytic function of x in the upper half U plane and lower half V plane on the complexified horizon. With this definition one obtains thermal radiation just as in the scalar case.

V. PARTICLE CREATION IN BLACK-HOLE DE SITTER SPACES

For the reasons given in Sec. III one would expect that a solution of Einstein's equations with positive cosmological constant which contained a black hole would settle down eventually to one of the Kerr–Newman–de Sitter solutions described in Sec. II. We shall therefore consider what would be seen by an observer in such a solution. Consider first the Schwarzschild–de Sitter solution. Suppose the observer moves along a world line λ of constant r, θ, and ϕ in region I of Fig. 4. The world line λ coincides with an orbit of the static Killing vector $K=\partial/\partial t$. Let $\varphi^2=g(K,K)$ on λ. One would expect that the observer would see thermal radiation with a temperature $T_C=(2\pi\psi)^{-1}\kappa_C$ coming from all directions except that of the black hole and thermal radiation of temperature $T_H=(2\pi\varphi)^{-1}\kappa_H$ coming from the black hole. The factor ψ appears in order to normalize the static Killing vector to have unit magnitude at the observer. The variation of ψ with r can be interpreted as the normal red-shifting of temperature.

There are, however, certain problems in showing that this is the case. These difficulties arise from the fact that when one has two or more sets of horizons with different surface gravities one has to introduce separate Kruskal-type coordinate patches to cover each set of horizons. The coordinates of one patch will be real analytic functions of the coordinates of the next patch in some overlap region between the horizons in the real manifold. However, branch cuts arise if one continues the coordinates to complex values. To see this, let U_1, V_1 be Kruskal coordinates in a patch covering a pair of intersecting horizons with a surface gravity κ_1 and let U_2, V_2 be a neighboring coordinate patch covering horizons with surface gravity κ_2. In the overlap region one has

$$V_1U_1^{-1}=-e^{2\kappa_2 t}, \tag{5.1}$$

$$V_2U_2^{-1}=-e^{2\kappa_2 t}. \tag{5.2}$$

Thus

$$-V_2U_2^{-1}=(-V_1)^P U_1^{-P}, \tag{5.3}$$

where $P=\kappa_2\kappa_1^{-1}$. There is thus a branch cut in the relation between the two coordinate patches if $\kappa_2\neq\kappa_1$.

One way of dealing with this problem would be to imagine perfectly reflecting walls between each black-hole horizon and each cosmological horizon. These walls would divide the manifold up into a number of separate regions each of which could be covered by a single Kruskal-coordinate patch. In each region one could construct a propagator as before but with perfectly reflecting boundary conditions at the walls. By arguments similar to those given in the previous section, these propagators will have the appropriate periodic and analytic properties to be thermal Green's functions with temperatures given by the surface gravities of the horizons contained within each region. Thus an observer on the black-hole side of a wall will see thermal radiation with the black-hole temperature, while an observer on the cosmological side of the wall will see radiation with the cosmological temperature. One would expect that, if the walls were removed, an observer would see a mixture of radiation as described above.

Another way of dealing with the problem would be to define the paopagator $G(x',x)$ to be a solution of the inhomogeneous wave equation on the real manifold which was such that if the point were extended to complex values of a Krushal-type-coordinate patch covering one set of intersecting horizons, it would be analytic on the complexified horizon in the upper half or lower half U or V plane depending on whether the point x was respectively to the future or the past of $V=0$ or $U=0$. Then, using a similar argument to that in the previous section about the dependence of the propagator on initial data on the complexified horizon, one can show that the propagator $G(x',x)$ between a point x' in region I and a point x in region II_C is analytic in a strip of width $\pi\kappa_C^{-1}$ below the real axis of the complex t plane. Similarly, the propagator $G(x',x)$ between a point x' in region I and a point x in region II_H will be analytic in a strip of width $\pi\kappa_H^{-1}$. Using these results one can show that

$$\begin{pmatrix}\text{probability of a particle of energy } E,\\ \text{relative to the observer, propagating}\\ \text{from } \mathscr{I}^+ \text{ to observer}\end{pmatrix} = \exp[-(E2\pi\psi\kappa_C^{-1})]\begin{pmatrix}\text{probability of a particle of energy } E,\\ \text{relative to the observer, propagating}\\ \text{from observer to } \mathscr{I}^+\end{pmatrix}, \tag{5.4}$$

and similarly the probability of propagating from the future singularity of the black hole will be related by the appropriate factor to the probability for a similar particle to propagate from the observer into the black hole. These results establish the picture described at the beginning of this section.

One can derive similar results for the Kerr–de Sitter spaces. There is an additional complication in this case because there is a relative angular velocity between the black hole and the cosmological horizon. An observer in region I who is at a constant distance r from the black hole and who is nonrotating with respect to distant stars will move on an orbit of the Killing vector K which is null on the cosmological horizon. For such an observer the probability of a particle of energy E, relative to the observer, propagating to him from beyond the future cosmological horizon will be $\exp[-(2\pi\psi E\kappa_C^{-1})]$ times the probability for a similar particle to propagate from the observer to beyond the cosmological horizon. The probabilities for emission and absorption by the black hole will be similarly related except that in this case the energy E will be replaced by $E-n\Omega_H$, where n is the aximuthal quantum number or angular momentum of the particle about the axis of rotation of the black hole and Ω_H is the angular velocity of the black-hole horizon relative to the cosmological horizon. As in the ordinary black-hole case, the black hole will exhibit superradiance for modes for which $E<n\Omega_H$. In the case that the observer is moving on the orbit of a Killing vector K which is rotating with respect to the cosmological horizon, one again gets similar results for the radiation from the cosmological and black-hole horizons with E replaced by $E-n\Omega_C$ and $E-n\Omega_H$, respectively. Where Ω_C and Ω_H are the angular velocities of the cosmological and black-hole horizons relative to the observers frame and are defined by the requirement that $K+\Omega_C\tilde{K}$ and $K+\Omega_H\tilde{K}$ should be null on the cosmological and black-hole horizons.

VI. IMPLICATIONS AND CONCLUSIONS

We have shown that the close connection between event horizons and thermodynamics has a wider validity than the ordinary black-hole situations in which it was first discovered. As observer in a cosmological model with a positive cosmological constant will have an event horizon whose area can be interpreted as the entropy or lack of information that the observer has about the regions of the universe that he cannot see. When the solution has settled down to a stationary state, the event horizon will have associated with it a surface gravity κ which plays a role similar to temperature in the classical first law of event horizons derived in Sec. III. As was shown in Sec. IV., this similarity is more than an analogy: The observer will detect an isotropic background of thermal radiation with temperature $(2\pi)^{-1}\kappa$ coming, apparently, from the event horizon. This result was obtained by considering what an observer with a particle detector would actually measure rather than by trying to define particles in an observer-independent manner. An illustration of the observer dependence of the concept of particle is the result that the thermal radiation in de Sitter space appears isotropic and at the same temperature to every geodesic observer. If particles had an observer-independent existence and if the radiation appeared isotropic to one geodesic observer, it would not appear isotropic to any other geodesic observer. Indeed, as an observer approached the first observer's future event horizon the radiation would diverge. It seems clear that this observer dependence of particle creation holds in the case of black holes as well: An observer at constant distance from a black hole will observe a steady emission of thermal radiation but an observer falling into a black hole will not observe any divergence in the radiation as he approaches the first-observer's event horizon.

A consequence of the observer dependence of particle creation would seem to be that the back reaction must be observer-dependent also, if one assumes, as seems reasonable, that the mass of the detector increases when it absorbs a particle and therefore the gravitational field changes. This will be discussed further in another paper,[16] but we remark here that it involves the abandoning of the concept of an observer-independent metric for spacetime and the adoption of something like the Everett-Wheeler interpretation of quantum mechanics.[36] The latter viewpoint seems to be required anyway when dealing with the quantum mechanics of the whole universe rather than an isolated system.

If a geodesic observer in de Sitter space chooses not to absorb any of the thermal radiation, his energy and entropy do not change and so one would not expect any change in the solution. However, if he does absorb some of the radiation, his energy and hence his gravitational mass will increase. If the solution now settles down again to a new stationary state, it follows from the first law of event horizons that the area of the cosmological event horizon will be less than it appeared to be before. One can interpret this as a reduction in the entropy of the universe beyond the event horizon caused by the propagation of some radiation from this region to the observer. Unlike the black-hole case, the surface gravity of the cosmological horizon decreases as the horizon shrinks. There is thus no danger of the observer's cosmological event horizon shrinking catastrophically around him because of his absorbing too much thermal radiation. He has, however, to be careful that he does not absorb so much radiation that his particle detector undergoes gravitational collapse to produce a black hole. If this were to happen, the black hole would always have a higher temperature than the surrounding universe and so would radiate energy faster than it absorbs it. It would therefore evaporate, leaving the universe as it was before the observer began to absorb radiation.

---

*Present address: Max-Planck-Institute für Physik and Astrophysik, 8 München 40, Postfach 401212, West Germany. Telephone: 327001.

[1]W. Israel, Phys. Rev. 164, 1776 (1967).
[2]H. Muller zum Hagen *et al.*, Gen. Relativ. Gravit. 4, 53 (1973).
[3]B. Carter, Phys. Rev. Lett. 26, 331 (1970).
[4]S. W. Hawking, Commun. Math. Phys. 25, 152 (1972).
[5]D. C. Robinson, Phys. Rev. Lett. 34, 905 (1975).
[6]D. C. Robinson, Phys. Rev. D 10, 458 (1974).
[7]J. Bekenstein, Phys. Rev. D 7, 2333 (1973).
[8]J. Bekenstein, Phys. Rev. D 9, 3292 (1974).
[9]S. W. Hawking, Phys. Rev. D 13, 191 (1976).
[10]J. Hartle and S. W. Hawking, Phys. Rev. D 13, 2188 (1976).
[11]R. Penrose, in *Relativity, Groups a d Topology*, edited by C. DeWitt and B. DeWitt (Gordon and Breach, New York, 1964).
[12]S. W. Hawking and G. F. R. Ellis, *Large Scale Structure of Spacetime* (Cambridge Univ. Press, New York, 1973).
[13]J. Bardeen, B. Carter, and S. W. Hawking, Commun. Math Phys. 31, 162 (1973).
[14]W. Unruh, Phys. Rev. D 14, 870 (1976).
[15]A. Ashtekar and A. Magnon, Proc. R. Soc. London A346, 375 (1975).
[16]S. W. Hawking, in preparation.
[17]J. D. North, *The Measure of the Universe* (Oxford Univ. Press, New York, 1965).
[18]C. Kahn and F. Kahn, Nature 257, 451 (1975).
[19]E. Schrödinger, *Expanding Universes* (Cambridge Univ. Press, New York, 1956).
[20]B. Carter, Commun. Math. Phys. 17, 233 (1970).
[21]B. Carter, in *Les Astre Occlus* (Gordon and Breach, New York, 1973).
[22]R. Geroch, E. H. Kronheimer, and R. Penrose, Proc. R. Soc. London A327, 545 (1972).
[23]S. W. Hawking, in preparation.
[24]S. W. Hawking, Commun. Math. Phys. 25, 152 (1972).
[25]P. Hajicek, Phys. Rev. D 7, 2311 (1973).
[26]L. Smarr, Phys. Rev. Lett. 30, 71 (1973); 30, 521(E) (1973).
[27]O. Nachtmann, Commun. Math. Phys. 6, 1 (1967).
[28]E. A. Tagirov, Ann. Phys. (N.Y.) 76, 561 (1973).
[29]P. Candelas and D. Raine, Phys. Rev. D 12, 965 (1975).
[30]J. S. Dowker and R. Critchley, Phys. Rev. D 13, 224 (1976).
[31]S. W. awking, Nature 248, 30 (1974).
[32]S. W. Hawking, Commun. Math. Phys. 43, 199 (1975).
[33]A. L. Fetter and J. P. Walecka, *Quantum Theory of Many Particle Systems* (McGraw-Hill, New York, 1971).
[34]R. Figari, R. Hoegh-Krohn, and C. Nappi, Commun. Math. Phys. 44, 265 (1975).
[35]G. W. Gibbons and M. J. Perry, Phys. Rev. Lett. 36, 985 (1976).
[36]*The Many Worlds Interpretation of Quantum Mechanics* edited by B. S. DeWitt and N. Graham (Princeton Univ. Press, Princeton, N. J., 1973).

PHYSICAL REVIEW D VOLUME 21, NUMBER 12 15 JUNE 1980

Gravitational effects on and of vacuum decay

Sidney Coleman*
Stanford Linear Accelerator Center, Stanford University, Stanford, California 94305

Frank De Luccia
Institute for Advanced Study, Princeton, New Jersey 88548
(Received 4 March 1980)

It is possible for a classical field theory to have two stable homogeneous ground states, only one of which is an absolute energy minimum. In the quantum version of the theory, the ground state of higher energy is a false vacuum, rendered unstable by barrier penetration. There exists a well-established semiclassical theory of the decay of such false vacuums. In this paper, we extend this theory to include the effects of gravitation. Contrary to naive expectation, these are not always negligible, and may sometimes be of critical importance, especially in the late stages of the decay process.

I. INTRODUCTION

Consider the theory of a single scalar field defined by the action

$$S=\int d^4x\left[\tfrac{1}{2}(\partial_\mu\phi)^2-U(\phi)\right], \tag{1.1}$$

where U is as shown in Fig. 1. That is to say, U has two local minima, $\phi_\pm$, only one of which, ϕ_-, is an absolute minimum. The classical field theory defined by Eq. (1.1) possesses two stable homogeneous equilibrium states, $\phi=\phi_+$ and $\phi=\phi_-$. In the quantum version of the theory, though, only the second of these corresponds to a truly stable state, a true vacuum. The first decays through barrier penetration; it is a false vacuum. This is a prototypical case; false vacuums occur in many field theories. In particular, they occur in some unified electroweak and grand unified theories, and it is this that gives the theory of vacuum decay possible physical importance. For simplicity, though, we will restrict ourselves here to the theory defined by Eq. (1.1); the extension of our methods to more elaborate field theories is straightforward.

The decay of the false vacuum is very much like the nucleation processes associated with first-order phase transitions in statistical mechanics.[1] The decay is initiated by the materialization of a bubble of true vacuum within the false vacuum. This is a quantum tunneling event, and has a certain probability of occurrence per unit time per unit volume, Γ/V. Once the bubble materializes, it expands with a speed asymptotically approaching that of light, converting false vacuum into true as it grows.

In the semiclassical (small $\hbar$) limit, Γ/V admits an expansion of the form

$$\Gamma/V=Ae^{-B/\hbar}[1+O(\hbar)]. \tag{1.2}$$

There exist algorithms for computing the coefficients A and B; indeed, in the limit of small energy-density difference between the two vacuums, it is possible to compute B in closed form. Also, in this same limit, it is possible to give a closed-form description of the growth of the bubble after its quantum formation. We will recapitulate this analysis later in this paper.

In this paper, we extend the theory of vacuum decay to include the effects of gravitation. At first glance, this seems a pointless exercise. In any conceivable application, vacuum decay takes place on scales at which gravitational effects are utterly negligible. This is a valid point if we are talking about the formation of the bubble, but not if we are talking about its subsequent growth. The energy released by the conversion of false vacuum to true is proportional to the volume of the bubble; thus, so is the Schwarzschild radius associated with this energy. Hence, as the bubble grows, the Schwarzschild radius eventually becomes comparable to the radius of the bubble.

This can easily be made quantitative. A sphere of radius Λ and energy density ϵ has Schwarzschild radius $2G\epsilon(4\pi\Lambda^3/3)$, where G is Newton's constant. This is equal to Λ when

$$\Lambda=(8\pi G\epsilon/3)^{-1/2}. \tag{1.3}$$

For an ϵ of (1 GeV)^4, the associated Λ is 0.8 km. Of course, in a typical unified electroweak or grand unified field theory, the relevant energy densities are larger than this and the associated lengths correspondingly smaller. We are dealing here with phenomena which take place on scales neither subnuclear nor astronomical, but rather civic, or even domestic, scales far too small to neglect if we are interested in the cosmological consequences of vacuum decay. Contrary to naive expectation, the inclusion of gravitation is not point-

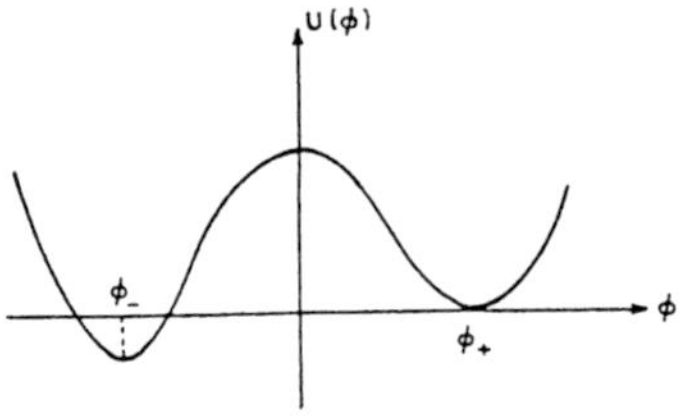

FIG. 1. The potential $U(\phi)$ for a theory with a false vacuum.

less; indeed, any description of vacuum decay that neglects gravitation is seriously incomplete.

Not only does gravitation affect vacuum decay, vacuum decay affects gravitation. In Eq. (1.1), there is no absolute zero of energy density; adding a constant to U has no effect on physics. This is not so when we include gravitation:

$$S=\int d^4x\sqrt{-g}\,[\tfrac{1}{2}g^{\mu\nu}\partial_\mu\phi\partial_\nu\phi-U(\phi)-(16\pi G)^{-1}R], \tag{1.4}$$

where R is the curvature scalar. Here, adding a constant to U is equivalent to adding a term proportional to $\sqrt{-g}$ to the gravitational Lagrangian, that is to say, to introducing a cosmological constant.[2] Thus, once the vacuum decays, gravitational theory changes; the cosmological constant inside the bubble is different from the one outside the bubble. Hence, in our computations, we need an initial condition not needed in the absence of gravitation. We must specify the initial value of the cosmological constant; equivalently, we must specify the absolute zero of energy density.

The experimental observation that the current value of the cosmological constant is zero gives two cases special interest. (1) $U(\phi_+)$ is zero. This would be the appropriate case to study if we were currently living in a false vacuum whose apocalyptic decay is yet to occur. (2) $U(\phi_-)$ is ero. This would be the appropriate case to study if we were living after the apocalypse, in the debris of a false vacuum which decayed at some early time in the history of the universe. Although our methods are applicable to arbitrary initial value of the cosmological constant, we pay special attention to these two cases.

The organization of the remainder of this paper is as follows: In Sec. II we summarize the theory of vacuum decay in the absence of gravitation. We emphasize the thin-wall approximation, the approximation that is valid in the limit of small energy-density difference between true and false vacuum. In Sec. III we begin the extension to gravitation, again emphasizing the thin-wall approximation. In the two special cases described above, we explicitly compute the effects of gravitation on the decay coefficient B of Eq. (1.2). As we have just argued, for any conceivable application, these effects are too small to worry about. Nevertheless, when we discovered it was within our power to compute them, we were unable to resist the temptation to do so. We have made no attempt to study the effects of gravitation on the coefficient A. This computation would involve the evaluation of a functional determinant; even if we had the courage to attempt such an evaluation, we would be frustrated by the nonrenormalizability of our theory. In Sec. IV we study the growth of the bubble. Section V states our conclusions. We have tried to write it in such a way that it will be intelligible to a reader who has skipped the intervening sections.

II. OLD RESULTS SUMMARIZED

In this section, we recapitulate the known results on vacuum decay in the theory defined by Eq. (1.1). The reader who wishes the arguments that lie behind our assertions is referred to the original literature.[1]

The Euclidean action is defined as minus the formal analytic continuation of Eq. (1.1) to imaginary time,

$$S_E=\int d^4x[\tfrac{1}{2}(\partial_\mu\phi)^2+U(\phi)], \tag{2.1}$$

where the metric is the usual positive-definite one of Euclidean four-space. The Euclidean equation of motion is the Euler-Lagrange equation associated with S_E. Let ϕ be a solution of this equation such that (1) ϕ approaches the false vacuum, ϕ_+, at Euclidean infinity, (2) ϕ is not a constant, and (3) ϕ has Euclidean action less than or equal to that of any other solution obeying (1) and (2). Then the coefficient B in the vacuum decay amplitude is given by

$$B=S_E(\phi)-S_E(\phi_+). \tag{2.2}$$

ϕ is called "the bounce." (The name has to do with the corresponding entity in particle mechanics.)

For the theories at hand, it can be shown[3] that the bounce is always O(4) symmetric, that is to say, ϕ is a function only of ρ, the Euclidean distance from an appropriately chosen center of coordinates. The Euclidean action then simplifies,

$$S_E=2\pi^2\int_0^\infty \rho^3 d\rho[\tfrac{1}{2}(\phi')^2+U], \tag{2.3}$$

as does the equation of motion

$$\phi''+\frac{3}{\rho}\phi'=\frac{dU}{d\phi}, \tag{2.4}$$

where the prime denotes $d/d\rho$.

It is possible to obtain an explicit approximation for ϕ in the limit of small energy-density difference between the two vacuums. Let us define ϵ by

$$\epsilon = U(\phi_+) - U(\phi_-)\,, \tag{2.5}$$

and let us write U as

$$U(\phi) = U_0(\phi) + O(\epsilon)\,, \tag{2.6}$$

where U_0 is a function chosen such that $U_0(\phi_-) = U_0(\phi_+)$, and such that $dU_0/d\phi$ vanishes at both ϕ_+ and ϕ_-.

The approximate ϕ obeys the equation

$$\phi'' = \frac{dU_0}{d\phi}\,. \tag{2.7}$$

Note that we have not only discarded the term in Eq. (2.4) proportional to ϵ, we have also discarded the term proportional to ϕ'. We will justify this shortly. Equation (2.7) admits a first integral,

$$[\tfrac{1}{2}(\phi')^2 - U_0]' = 0\,. \tag{2.8}$$

Its value is determined by the condition that $\phi(\infty)$ is ϕ_+:

$$\tfrac{1}{2}(\phi')^2 - U_0 = -U_0(\phi_\pm)\,. \tag{2.9}$$

Thus, as ρ traverses the real line, ϕ goes monotonically from ϕ_- to ϕ_+. Equation (2.9) determines ϕ in terms of a single integration constant. We will choose this to be $\bar{\rho}$, the point at which ϕ is the average of its two extreme values:

$$\int_{(\phi_+ + \phi_-)/2}^{\phi} d\phi[2(U_0 - U_0(\phi_\pm))]^{-1/2} = \rho - \bar{\rho}\,. \tag{2.10}$$

Thus, for example, if

$$U_0 = \tfrac{1}{8}\lambda(\phi^2 - \mu^2/\lambda)^2\,, \tag{2.11}$$

then

$$\phi = \frac{\mu}{\sqrt{\lambda}}\tanh[\tfrac{1}{2}\mu(\rho - \bar{\rho})]\,. \tag{2.12}$$

All that remains is to determine $\bar{\rho}$. We will do this on the assumption that $\bar{\rho}$ is large compared to the length scale on which ϕ varies significantly. For the example of Eq. (2.11), this means that $\bar{\rho}\mu$ is much greater than one. This assumption will be justified (for sufficiently small ϵ) at the end of the computation.

If $\bar{\rho}$ is large, the bounce looks like a ball of true vacuum, $\phi = \phi_-$, embedded in a sea of false vacuum, $\phi = \phi_+$, with a transition region ("the wall") separating the two. The wall is small in thickness compared to the radius of the ball; in our example, its thickness is $O(\mu^{-1})$. It is for this reason that the approximation we are describing is called "the thin-wall approximation." When we justify the thin-wall approximation, we will also justify our earlier neglect of the ϕ' term in Eq. (2.4). Away from the wall, this term is negligible because ϕ' is negligible; at the wall, it is negligible because $\bar{\rho}$ is large.

We will now determine $\bar{\rho}$ by computing B from Eq. (2.2) and demanding that it be stationary under variations of $\bar{\rho}$. The region of integration breaks naturally into three parts: outside the wall, inside the wall, and the wall itself. We divide B accordingly. Outside the wall, $\phi = \phi_+$. Hence,

$$B_{\text{outside}} = 0\,. \tag{2.13}$$

Inside the wall, $\phi = \phi_-$. Hence,

$$B_{\text{inside}} = -\frac{\pi^2}{2}\bar{\rho}^4\epsilon\,. \tag{2.14}$$

Within the wall, in the thin-wall approximation,

$$\begin{aligned} B &= 2\pi^2\bar{\rho}^3\int d\rho\,[\tfrac{1}{2}\phi'^2 + U_0(\phi) - U_0(\phi_+)] \\ &\equiv 2\pi^2\bar{\rho}^3 S_1\,. \end{aligned} \tag{2.15}$$

Equation (2.9) gives us an integral expression for S_1,

$$S_1 = \int_{\phi_-}^{\phi_+} d\phi\{2[U_0(\phi) - U_0(\phi_+)]\}^{1/2}\,. \tag{2.16}$$

For future reference, we note that we can also write S_1 as

$$S_1 = 2\int d\rho\,[U_0(\phi) - U_0(\phi_+)]\,. \tag{2.17}$$

We can now compute

$$B = -\tfrac{1}{2}\pi^2\bar{\rho}^4\epsilon + 2\pi^2\bar{\rho}^3 S_1\,. \tag{2.18}$$

This is stationary at

$$\bar{\rho} = 3S_1/\epsilon\,. \tag{2.19}$$

We have justified our approximation: $\bar{\rho}$ indeed becomes large when ϵ becomes small. We now know

$$B = 27\pi^2 S_1^4/2\epsilon^3\,. \tag{2.20}$$

We have used the bounce to compute a coefficient which enters into the probability for the quantum materialization of a bubble of true vacuum within the false vacuum. We can also use the bounce to describe the classical growth of the bubble after its materialization. The surface $t = 0$ is the intersection of Euclidean space (imaginary time) and Minkowski space (real time). It can be shown that the value of ϕ on this surface can be thought of as the configuration of the field at the moment the bubble materializes. Also, at this moment, the time derivative of the field is zero. These initial-value data, together with the classical field equations in Minkowski space, suffice to determine the growth of the bubble.

Of course, there is no need to go to the bother of explicitly solving the classical field equations. All we need to do is analytically continue the Euclidean solution we already possess; that is to say, all we need to do is make the substitution

$$\rho \to (|\vec{x}|^2 - t^2)^{1/2}.$$

Thus, Euclidean O(4) invariance becomes Minkowskian O(3, 1) invariance. In the thin-wall approximation, the bubble materializes at rest with radius $\bar{\rho}$. As it grows, its surface traces out the hyperboloid $\rho = \bar{\rho}$. Since $\bar{\rho}$ is typ cally a quantity of subnuclear magnitude, this means that from the viewpoint of macrophysics, almost immediately upon its materialization the bubble accelerates to essentially the speed of light and continues to grow indefinitely at that speed.

III. INCLUSION OF GRAVITATION: MATERIALIZATION OF THE BUBBLE

In this section, we begin the extension of the analysis of Sec. II to the theory described by Eq. (1.4), the theory of a scalar field interacting with gravity.

As before, we begin by constructing a bounce, a solution of the Euclidean field equations obeying appropriate bounda y conditions. This is apparently a formidable task; we now have to keep track of not just a single scalar field but also of the ten components of the metric tensor. However, things are not as bad as they seem, for there is no reason for gravitation to break the symmetries of the purely scalar problem. Thus it is reasonable to assume that, in the presence of gravity as in its absence, the bounce is invariant under four-dimensional rotations.

We emphasize that, in contrast to the case of a single scalar field, we have no theorem to back up this assumption. We will shortly construct, in the thin-wall approximation, an invariant bounce, but this will still leave open the possibility that there exist noninvariant bounces of lower Euclidean action. We do not think it likely that such objects exist, but we cannot prove they do not, and the reader should be warned that if they do exist, they dominate vacuum decay, and all our conclusions are wrong.

We begin by constructing the most general rotationally invariant Euclidean metric. The orbits of the rotation group are three-dimensional manifolds with the geometry of three-spheres. On each of these spheres, we introduce angular coordinates in the canonical way. We define a radial curve to be a curve of fixed angular coordinates. By rotational invariance, radial curves must be normal to the three-spheres through which they pass. We choose our radial coordinate ξ to measure distance along these radial curves. Thus, the element of length is of the form

$$(ds)^2 = (d\xi)^2 + \rho(\xi)^2 (d\Omega)^2 \,, \tag{3.1}$$

where $(d\Omega)^2$ is the element of distance on a unit three-sphere and ρ gives the radius of curvature of each three-sphere. Note that rotational invariance has made its usual enormous simplification; ten unknown functions of four variables have been reduced to one unknown function of one variable. Note also that we can redefine ξ by the addition of a constant without changing the form of the metric; equivalently, we can begin measuring ξ from wherever we choose.

Given Eq. (3.1), it is a straightforward exercise in the manipulation of Christoffel symbols to compute the Euclidean equations of motion. We will give only the results here. The scalar field equation is

$$\phi'' + \frac{3\rho'}{\rho}\phi' = \frac{dU}{d\phi}, \tag{3.2}$$

where the prime denotes $d/d\xi$. The Einstein equation

$$G_{\xi\xi} = -\kappa T_{\xi\xi}\,, \tag{3.3}$$

where $\kappa = 8\pi G$, becomes

$$\rho'^2 = 1 + \tfrac{1}{3}\kappa\rho^2(\tfrac{1}{2}\phi'^2 - U). \tag{3.4}$$

The other Einstein equations are either identities or trivial consequences of these equations. Finally,

$$S_E = 2\pi^2 \int d\xi \left(\rho^3(\tfrac{1}{2}\phi'^2 + U) + \frac{3}{\kappa}(\rho^2\rho'' + \rho\rho'^2 - \rho)\right). \tag{3.5}$$

In the thin-wall approximation, the construction of the bo nce from these equations is astonishingly simple. Equation (3.2) differs from its counterpart in the pure scalar case, Eq. (2.4), in only two respects. Firstly, the independent variable is called ξ rather than ρ. This is a trivial change. Secondly, the coefficient of the ϕ' term involves a factor of ρ'/ρ rather than one of $1/\rho$. But this is also a triv al change, since in the thin-wall approximation we neglect this term anyway. (This is a bit facile. Of course, because the term is differe t in form, the eventual self-consistent justification of the approximation must also be different in form. We will deal with this pr blem when we come to it.) Thus, in the thin-wall approximation, we need only to copy Eq. (2.10),

$$\int_{(\phi_+ + \phi_-)/2}^{\phi} d\phi \{2[U_0 - U_0(\phi_\pm)]\}^{-1/2} = \xi - \bar{\xi}. \quad (3.6)$$

Here $\bar{\xi}$ is an integration constant, but, as we have explained, one with no convention-independent meaning.

Once we have ϕ, we can solve Eq. (3.4) to find ρ. This is a first-order differential equation; to specify its solution, we need one integration constant. We will choose this to be

$$\bar{\rho} \equiv \rho(\bar{\xi}). \quad (3.7)$$

This does have a convention-independent meaning; it is the radius of curvature of the wall separating false vacuum from true. We do not need the explicit expression for ρ for our immediate purposes, so we will not pause now to construct it.

Our next task is to find $\bar{\rho}$. The computation is patterned on that in Sec. II: First we compute B, the difference in action between the bounce and the false vacuum. Then we find $\bar{\rho}$ by demanding that B be stationary.

We first eliminate the second-derivative term from Eq. (3.5) by integration by parts; the surface term from the parts integration is harmless because we are only interested in the action difference between two solutions that agree at infinity. We thus obtain

$$S_E = 4\pi^2 \int d\xi \left(\rho^3(\tfrac{1}{2}\phi'^2 + U) - \frac{3}{\kappa}(\rho\rho'^2 + \rho)\right). \quad (3.8)$$

We now use Eq. (3.4) to eliminate ρ'. We find

$$S_E = 4\pi^2 \int d\xi \left(\rho^3 U - \frac{3\rho}{\kappa}\right). \quad (3.9)$$

So far, we have made no approximations. We now evaluate B, from Eq. (3.9), in the thin-wall approximation. As before, we divide the integration region into three parts. Outside the wall, bounce and false vacuum are identical; thus, as before,

$$B_{\text{outside}} = 0. \quad (3.10)$$

In the wall, we can replace ρ by $\bar{\rho}$, and U by U_0. Thus,

$$B_{\text{wall}} = 4\pi^2\bar{\rho}^3 \int d\xi\,[U_0(\phi) - U_0(\phi_+)]$$
$$= 2\pi^2\bar{\rho}^3 S_1 \quad (3.11)$$

by Eq. (2.17). Inside the wall, ϕ is a constant. Hence,

$$d\xi = d\rho(1 - \tfrac{1}{3}\kappa\rho^2 U)^{-1/2} \quad (3.12)$$

and

$$B_{\text{inside}} = -\frac{12\pi^2}{\kappa}\int_0^{\bar{\rho}} \rho\, d\rho \{[1 - \tfrac{1}{3}\kappa\rho^2 U(\phi_-)]^{1/2} - (\phi_- \to \phi_+)\}$$
$$= \frac{12\pi^2}{\kappa^2}\left(U(\phi_-)^{-1}\{[1 - \tfrac{1}{3}\kappa\bar{\rho}^2 U(\phi_-)]^{3/2} - 1\} - (\phi_- \to \phi_+)\right). \quad (3.13)$$

[As a consistency check, it is easy to verify that this reduces to Eq. (2.14) when κ goes to zero.]

This is an ugly expression, and to continue our investigation in full generality would quickly involve us in a monstrous algebraic tangle. Thus we now restrict our attention to the two cases of special interest identified in Sec. I.

The first special case is decay from a space of positive energy density into a space of zero energy density. This is the case that is relevant if we are now in a postapocalyptic age. In this case

$$U(\phi_+) = \epsilon, \quad U(\phi_-) = 0. \quad (3.14)$$

It is then a trivial exercise to show that B is stationary at

$$\bar{\rho} = \frac{12 S_1}{4\epsilon + 3\kappa S_1^2}$$
$$= \frac{\bar{\rho}_0}{1 + (\bar{\rho}_0/2\Lambda)^2}, \quad (3.15)$$

where $\bar{\rho}_0 = 3S_1/\epsilon$, the bubble radius in the absence of gravity, and $\Lambda = (\kappa\epsilon/3)^{-1/2}$, as in Eq. (1.3). At this point,

$$B = \frac{B_0}{[1 + (\bar{\rho}_0/2\Lambda)^2]^2}, \quad (3.16)$$

where $B_0 = 27\pi^2 S_1^4/2\epsilon^3$, the decay coefficient in the absence of gravity.

These equations have some interesting properties, but we will postpone discussing them until we write down the corresponding equations for the second special case. This is decay from a space of zero energy density into a space of negative energy density, the case that is relevant if we are now in a preapocalyptic age. In this case

$$U(\phi_+) = 0, \quad U(\phi_-) = -\epsilon. \quad (3.17)$$

As before, trivial algebra shows that

$$\bar{\rho} = \frac{\bar{\rho}_0}{1 - (\bar{\rho}_0/2\Lambda)^2} \quad (3.18)$$

and

$$B = \frac{B_0}{[1 - (\bar{\rho}_0/2\Lambda)^2]^2}. \quad (3.19)$$

These equations have been derived in the thin-wall approximation. Before we discuss their implications, we should discuss the reliability of the approximation. In the absence of gravitation, the thin-wall approximation was valid if $\bar{\rho}$ was large compared to the characteristic range of variation of ϕ; the significant quantity was $\bar{\rho}$ because $1/\rho$ multiplied the neglected ϕ' term in Eq. (2.4). In the presence of gravitation, $1/\rho$ is replaced by ρ'/ρ [see Eq. (3.2)]; thus it is this quantity that must be small at the wall.

By Eq. (3.4),

$$\frac{\rho'^2}{\rho^2} = \frac{1}{\rho^2} + \frac{\kappa}{3}\left(\tfrac{1}{2}\phi'^2 - U\right). \tag{3.20}$$

The left-hand side of this equation is certainly small if both terms on the right are small. The first term is just $(1/\rho)^2$, as before. As for the second term, the quantity in parentheses is approximately constant over the wall, vanishes on one side of the wall, and has magnitude ϵ on the other. Thus, it is certainly an overestimation to replace it by ϵ everywhere; this turns the second term into $(1/\Lambda)^2$.

Thus, the thin-wall approximation is justified if both $\bar{\rho}$ and Λ are large compared to the characteristic range of variation of ϕ. This condition puts no restraint on $\bar{\rho}_0/\Lambda$, the ratio that measures the importance of gravitation; thus, it is not senseless to discuss our results for arbitrarily large values of this ratio. (Although it is not senseless, it is useless; as we said in Sec. I, in any conceivable application this ratio is negligible. We will proceed anyway.)

In the first special case, decay into the present condition, we see that gravitation makes the materialization of the bubble more likely (diminishes B), and makes the radius of the bubble at its moment of materialization smaller (diminishes $\bar{\rho}$). In the second special case, decay from the present condition, things are just the other way around; gravitation makes the materialization of the bubble less likely and its radius larger. Indeed, gravitation can totally quench vacuum decay; at $\rho_0 = 2\Lambda$ or, equivalently,

$$\epsilon = \tfrac{3}{4}\kappa S_1^2, \tag{3.21}$$

the bubble radius becomes infinite and the decay probability vanishes. For higher values of $\bar{\rho}_0$ or, equivalently, smaller values of ϵ, our equations admit of no sensible solution at all. Gravitation has stabilized the false vacuum.

We believe we understand this surprising phenomenon. Our explanation begins with a computation of the energy of a thin-walled bubble, in the absence of gravitation, at the time of its materialization. For the moment, we will give the bubble an arbitrary radius $\bar{\rho}$, postponing use of our knowledge that $\bar{\rho}$ is $\bar{\rho}_0$. The energy is the sum of a negative volume term and a positive surface term,

$$E = -\frac{4\pi}{3}\epsilon\bar{\rho}^3 + 4\pi S_1\bar{\rho}^2$$
$$= \frac{4\pi}{3}\epsilon\bar{\rho}^2(\bar{\rho}_0 - \bar{\rho}). \tag{3.22}$$

We see that this vanishes for the actual bubble, $\bar{\rho} = \bar{\rho}_0$. This is as it should be. The energy of the world vanishes before the bubble materializes, and, whatever else barrier penetration may do, it does not violate the conservation of energy.

We now see how to compute the effects of gravitation on the bubble radius, in the limit that these effects are small. All we have to do is to compute the effects of gravitation on the total energy of the unperturbed bubble. If the gravitational contribution is positive, the bubble radius will have to grow, so Eq. (3.22) will develop a small negative contribution and total energy will remain zero. On the other hand, if the gravitational contribution is negative, the bubble will have to shrink. Of course, we already know that it is the former alternative that prevails, and not just in the limit of small gravitational effects. However, the point of the computation is to understand why it prevails.

There are two terms in the gravitational contribution to the energy of the unperturbed bubble. The first is the ordinary Newtonian potential energy of the bubble. This is easily computed by integrating over all space the square of the gravitational field, itself easily computed from Gauss's law. The answer is

$$E_{\text{Newton}} = -\epsilon\pi\bar{\rho}_0^5/15\Lambda^2. \tag{3.23}$$

Note that this is negative, as a gravitational potential energy should be. The second term comes from the fact that the nonzero energy density inside the bubble distorts its geometry. Thus, there is a correction to the volume of the bubble and thus to the volume term in the bubble energy. We can determine this correction to the volume from Eq. (3.12); the infinitesimal element of volume is

$$4\pi\rho^2 d\xi = 4\pi\rho^2 d\rho\,(1 - \tfrac{1}{2}\rho^2/\Lambda^2) + O(G^2). \tag{3.24}$$

Note that this is smaller than the Euclidean formula; thus this geometrical correction reduces the magnitude of the negative volume energy and hence makes a positive contribution to the total energy. Integration yields

$$E_{\text{geom}} = 2\pi\epsilon\bar{\rho}_0^5/5\Lambda^2 \tag{3.25}$$

in the small-G limit. The total gravitational correction is the sum of these two terms,

$$E_{\text{grav}} = \pi\epsilon\bar{\rho}_0^5/3\Lambda^2. \tag{3.26}$$

This is positive; hence, the bubble is larger in the presence of gravitation than in its absence.

We now understand what is happening. Vacuum

decay proceeds by the materialization of a bubble. By energy conservation, this bubble always has energy zero, the sum of a negative volume term and a positive surface term. In the absence of gravity, we can always make a zero-energy bubble no matter how small ϵ is; we just have to make the bubble large enough, and the volume/surface ration will do the job. However, in the presence of gravity, the negative energy density inside the bubble distorts the geometry of space in such a way as to diminish the volume/surface ratio. Thus it is possible that, for sufficiently small ϵ, no bubble, no matter how big, will have energy zero. What Eq. (3.18) is telling us is that this is indeed the case.

We cannot exclude the possibility that decay could proceed through nonspherical bubbles, although we think this is unlikely; we guess that such configurations would only worsen the volume/surface ratio. Nor can we exclude the possibility that decay proceeds through some nonsemiclassical process, one that does not involve bubble formation at all, and whose probability vanishes more rapidly than exponentially in the small-$\hbar$ limit. About this possibility we cannot even make guesses.

IV. INCLUSION OF GRAVITATION: GROWTH OF THE BUBBLE

In Sec. II, we explained how to obtain a description of the classical growth of the bubble after its quantum materialization, in the absence of gravitation. All we had to do was analytically continue the scalar field ϕ from Euclidean space to Minkowskian space. Because of the enormous symmetry of the bounce—O(4)-invariant in Euclidean space, O(3,1)-invariant in Minkowski space—for much of Minkowski space this continuation was trivial. To be more precise, if we choose the center of the bubble at its moment of materialization to be the center of coordinates, for all spacelike points the contribution was a mere reinterpretation of ρ as Minkowskian spacelike separation rather than Euclidean radial distance. It was only for timelike points that nontrivial continuation (to imaginary ρ) was needed.

All of this carries over to the case in which gravitation is present. The only difference is that we have to continue the metric as well as the scalar field, turning an O(4)-invariant Euclidean manifold into an O(3,1)-invariant Minkowskian manifold.

Thus, a large part of the manifold is analogous to the spacelike region described above. Here,

$$ds^2 = -d\xi^2 - \rho(\xi)^2(d\Omega_S)^2\,, \tag{4.1}$$

where $d\Omega_S$ is the element of length on a unit hyperboloid with spacelike normal vector in Minkowski space. [The overall minus sign has appeared because we adhere to the convention that a Minkowski metric has signature $(+---)$.] In this region, ϕ is $\phi(\rho)$. In the thin-wall approximation, the bubble wall always lies within this region at $\rho=\bar{\rho}$. Thus, this is all the manifold we need if we are only interested in studying the expansion of the bubble as it appears from the outside.

However, if we wish to go inside the bubble, we may encounter vanishing ρ. This is a pure coordinate singularity analogous to reaching the light cone, the boundary of the spacelike region, in Minkowski space. We get beyond the singularity by continuing to the timelike region. We choose ξ to be zero when ρ is zero, and continue to $\xi = i\tau$, with τ real. We thus obtain

$$ds^2 = d\tau^2 - \rho(i\tau)^2(d\Omega_T)^2\,, \tag{4.2}$$

where $d\Omega_T$ is the element of length for a unit hyperboloid with timelike normal in Minkowski space. One way of describing this equation is to say that the interior of the bubble always contains a Robertson-Walker universe of open type.

Let us now apply this general prescription to the special cases we have analyzed in Sec. III. In the thin-wall approximation, no analytic continuation is needed for ϕ; it is equal to ϕ_+ outside the bubble and ϕ_- inside the bubble. The metric outside the bubble is obtained by solving Eq. (3.4):

$$\rho'^2 = 1 - \frac{\kappa\rho^2}{3}U(\phi_+)\,. \tag{4.3}$$

Inside the bubble, it is obtained by solving the same equation with ϕ_+ replaced by ϕ_-. The two metrics are joined at the bubble wall, not at the same ξ, but at the same ρ, $\rho=\bar{\rho}$.

In our first special case, decay into the present condition, U vanishes inside the bubble. Thus the interior metric is $\rho=\xi$, ordinary Minkowski space. Outside the bubble, Eq. (4.3) becomes

$$\rho'^2 = 1 - \rho^2/\Lambda^2\,. \tag{4.4}$$

The solution is

$$\rho = \Lambda \sin(\xi/\Lambda)\,. \tag{4.5}$$

We shall now show that this is ordinary de Sitter space, written in slightly unconventional coordinates.[4]

We begin by recapitulating the definition of de Sitter space. Consider a five-dimensional Minkowski space with O(4,1)-invariant metric

$$ds^2 = -(dw)^2 + (dt)^2 - (dx)^2 - (dy)^2 - (dz)^2\,. \tag{4.6}$$

In this space, consider the hyperboloid defined by

$$\Lambda^2 = w^2 - t^2 + x^2 + y^2 + z^2, \qquad (4.7)$$

with Λ some positive number. This is a four-dimentional manifold with a Minkowskian metric; it is de Sitter space. Note that de Sitter space is as homogeneous as Minkowski space; any point in the space can be transformed into any other by an O(4, 1) transformation.

To put the metric of de Sitter space into our standard form, we must choose the location of the center of the bubble at its moment of materialization. Since de Sitter space is homogeneous, we can without loss of generality choose this point to be $(\Lambda, 0, 0, 0, 0)$. The O(3, 1) group of the vacuum decay problem is then the Lorentz group acting on the last four coordinates. Thus we replace these by "angular" coordinates, as in Eq. (4.1):

$$ds^2 = -(dw)^2 - (d\rho)^2 - \rho^2(d\Omega_S)^2. \qquad (4.8)$$

Equation (4.7) then becomes

$$\Lambda^2 = w^2 + \rho^2.$$

If we now define ξ by

$$w = \Lambda\cos(\xi/\Lambda), \quad \rho = \Lambda\sin(\xi/\Lambda), \qquad (4.9)$$

the metric falls into the desired form.

In this metric, ρ is bounded above by Λ. The geometrical reason for this is clear from Eq. (4.7). A spacelike slice of de Sitter space (say the hypersurface $t=0$) is a hypersphere of radius Λ; on a hypersphere, no circle has greater circumference than a great circle. This also explains a curious feature of Eq. (3.15),

$$\bar{\rho} = \frac{\bar{\rho}_0}{1+(\bar{\rho}_0/2\Lambda)^2}. \qquad (3.15)$$

No matter how we choose $\bar{\rho}_0$, $\bar{\rho}$ is always less than or equal to Λ. The reason for this is now obvious; the bubble cannot be bigger than this because a bigger bubble could not fit into the false vacuum.

We now go to our second special case, the decay of the present vacuum. Here U vanishes outside the bubble, so it is the exterior metric that is ordinary Minkowski space. Inside the bubble, Eq. (4.3) becomes

$$\rho'^2 = 1 + \rho^2/\Lambda^2. \qquad (4.10)$$

The solution is

$$\rho = \Lambda\sinh(\xi/\Lambda). \qquad (4.11)$$

Since we are now inside the bubble, we will also need the continuation of this to the timelike region, the Robertson-Walker universe inside the bubble. By Eq. (4.2) this is

$$ds^2 = d\tau^2 - \Lambda^2\sin^2(\tau/\Lambda)[d\Omega_T]^2. \qquad (4.12)$$

This is an open expanding-and-contracting universe. We normally think of oscillating universes as necessarily closed, but this rule depends on the positivity of energy, very much violated here.

The metric defined by Eq. (4.12) has singularities when τ is an integral multiple of $\pi\Lambda$. We shall now show that these singularities are spurious, mere coordinate artifacts.

Consider a five-dimensional Minkowski space with O(3, 2)-invariant metric

$$ds^2 = (dw)^2 + (dt)^2 - (dx)^2 - (dy)^2 - (dz)^2. \qquad (4.13)$$

In this space, consider the hyperboloid defined by

$$\Lambda^2 = w^2 + t^2 - x^2 - y^2 - z^2, \qquad (4.14)$$

with Λ some positive number. This is a four-dimensional manifold with a Minkowski metric; indeed, by applying the same coordinate transformations we used in our analysis of de Sitter space, one can easily show that the metric is that defined by Eqs. (4.11) and (4.12).

Although the hyperboloid is free of singularities, it is not totally without pathologies; it contains closed timelike curves, for example, circles in the w-t plane. These can easily be eliminated. The hyperboloid is homeomorphic by $R_3 \times S^1$. (That is to say, once we have freely given x, y, and z, w and t must lie on a circle.) Thus, it is not simply connected, and we may replace it by its simply connected covering space. [In Eq. (4.12), this corresponds to interpreting τ as an ordinary real variable rather than an angular variable.] This covering has neither singularities nor closed timelike curves. It is much like de Sitter space, except that its symmetry group is O(3, 2) rather than O(4, 1); it is called anti-de Sitter space.[5]

Anti-de Sitter space is the universe inside the bubble. We shall now show that this universe is dynamically unstable; even the tiny corrections to the thin-wall approximation are sufficient to convert the coordinate singularity in Eq. (4.12) into a genuine singularity, to cause gravitational collapse.

For our discussion, we need the exact field equations in the timelike region. These are

$$\ddot{\phi} + \frac{3\dot{\rho}}{\rho}\dot{\phi} + \frac{dU}{d\phi} = 0 \qquad (4.15)$$

and

$$\dot{\rho}^2 = 1 + \frac{\kappa\rho^2}{3}(\tfrac{1}{2}\dot{\phi}^2 + U), \qquad (4.16)$$

where the dot indicates differentiation with respect to τ. In general, initial-value data for this system consists of a point in the ϕ-$\dot{\phi}$ plane at some value of ρ. (The associated value of τ has no convention-independent meaning.) In the special case of vanishing ρ, a nonsingular solution must have vanishing $\dot{\phi}$, just as, in Euclidean space, a rotationally invariant function must have vanishing gradient at the origin of coordinates.

For notational simplicity in what follows, we choose the zero of ϕ to be the true vacuum $\phi_- = 0$. Also, we define

$$\mu^2 \equiv \left.\frac{d^2U}{d\phi^2}\right|_0 . \tag{4.17}$$

Until now, we took the initial-value data at $\rho = \tau = 0$ to be $\phi = \dot{\phi} = 0$. This led to the solution

$$\phi = 0 , \tag{4.18}$$

and

$$\rho = \Lambda \sin\tau/\Lambda . \tag{4.19}$$

Now, vanishing $\dot{\phi}$ at $\rho = 0$ is an exact result; the bounce is rotationally invariant. But vanishing ϕ is just an approximation; as can be seen from the explicit formula for the bounce, at $\rho = 0$, ϕ is $O(\exp[-\mu\rho])$, exponentially small but not zero.

As long as ϕ remains exponentially small, we can neglect its effects on ρ and continue to use Eq. (4.19). Also, we can replace Eq. (4.15) by its linear approximation

$$\ddot{\phi} + \frac{3\dot{\rho}}{\rho}\dot{\phi} + \mu^2\phi = 0 . \tag{4.20}$$

Of course, if in the course of time ϕ grows large, we can no longer make these approximations and must return to the exact equations.

We begin by solving Eq. (4.20) for $\rho/\Lambda \ll 1$. In this region, ρ is approximately τ, anf Eq. (4.20) is a Bessel equation. Thus, ϕ is simply an exponentially small multiple of the nonsingular oscillatory solution of this equation and remains exponentially small throughout this region.

We next turn to the region $\rho\mu \gg 1$. Note that under the conditions of the thin-wall approximation this overlaps the preceding region. In this region, Eq. (4.20) is the Newtonian equation of motion for a weakly damped harmonic oscillator, with a slowly varying damping coefficient $3\dot{\rho}/\rho$. Thus,

$$\phi \propto \cos(\mu\tau + \theta)\exp\left(-\int d\tau\, 3\dot{\rho}/2\rho\right) \tag{4.21}$$

or

$$\phi = a\rho^{-3/2}\cos(\mu\tau + \theta) . \tag{4.22}$$

Here a is an exponentially small coefficient and θ is an angle independent of any of the parameters of the theory, whose explicit value is of no interest to us. We see that ϕ remains exponentially small throughout the expansion and subsequent contraction of the universe, all the way down to the region where ρ/Λ is once again much less than one. In this region, we may rewrite Eq. (4.22) as

$$\phi = a\rho^{-3/2}\cos(\pi\Lambda\mu - \rho\mu + \theta) . \tag{4.23}$$

If we attempt to continue all the way to the second zero of ρ predicted by Eq. (4.19), ϕ becomes large and our approximations break down. However, whatever happens to ϕ, a second zero is inevitable. From Eq. (4.16) and the assumed properties of U,

$$\dot{\rho}^2 \geq 1 - \rho^2/\Lambda^2 . \tag{4.24}$$

Thus, once ρ is much less than Λ and diminishing, it must continue to diminish all the way to zero, and it must do so quickly, in a time of order ρ.

Of course, vanishing ρ does not imply singular behavior. On the contrary, we can always obtain a one-parameter family of nonsingular solutions by integrating the equations of motion backward from the second zero of ρ, using as final-value data any point on the line $\dot{\phi} = 0$. If we continue this family of solutions into the region of validity of Eq. (4.23), then, for any fixed ρ, they will define a curve in the ϕ-$\dot{\phi}$ plane. Because we are integrating the equations of motion over a very short time interval, this curve can have only a negligible dependence on the small parameter ϵ, the energy-density difference between the two vacuums. On the other hand, the angle in the ϕ-$\dot{\phi}$ plane obtained from Eq. (4.23) at fixed ρ is a very rapidly varying function of ϵ because Λ is proportional to $\epsilon^{-1/2}$.

Thus for general ϵ, Eq. (4.23) does not continue to a nonsingular solution. This argument does not eliminate the possibility that there might be special values of ϵ for which the singularity may be avoided. However, we believe this possibility is of little interest; an instability that can be removed only by fine tuning the parameters of the theory is going to return once we consider other corrections to our approximation (e.g., the effects of a small initial matter density). This completes the argument for gravitational collapse of the bubble interior.

V. CONCLUSIONS

Although some of the results in the body of this paper hold in more general circumstances, we will restrict ourselves here to the thin-wall approximation, the approximation that is valid in the limit of small energy-density difference between true and false vacuum, and to the two cases of special interest identified in Sec. I.

The first special case is decay from a space of vanishing cosmological constant, the case that applies if we are currently living in a false vacuum. In the absence of gravitation, vacuum decay proceeds through the quantum materialization of a bubble of true vacuum, separated by a thin wall from the surrounding false vacuum. The bubble is at rest at the moment of materialization, but it rapidly grows; its wall traces out a hyperboloid in Minkowski space, asymptotic to the light cone.

If all we are interested in is vacuum decay as seen from the outside, not a word of the preceding description needs to be changed in the presence of gravitation. At least at first glance, this is surprising because one would imagine that gravitation would have some effect on the growth of the bubble. There are two ways of understanding why this does not happen. The first is a mathematical way: The growth of the bubble in the absence of gravitation is O(3, 1) invariant; the inclusion of gravitation does not spoil this invariance; the only O(3, 1)-invariant hypersurfaces are hyperboloids with light cones as their asymptotes. The second is a physical way: Quantum tunneling does not violate the law of conservation of energy; thus the total energy of the expanding bubble is always identically zero, the negative energy of the interior being canceled by the positive energy of the wall. Because the bubble is spherically symmetric, the gravitational field at the outer edge of the wall is determined exclusively by the total energy within. That is to say, it is zero, and neither accelerates nor retards the growth of the edge.

Of course, gravitation affects the quantitative features of vacuum decay. In any conceivable application, these effects are totally negligible, but we have computed them anyway. In general, gravitation makes the probability of vacuum decay smaller; in the extreme case of very small energy-density difference, it can even stabilize the false vacuum, preventing vacuum decay altogether. We believe we understand this. For the vacuum to decay, it must be possible to build a bubble of total energy zero. In the absence of gravitation, this is no problem, no matter how small the energy-density difference; all one has to do is make the bubble big enough, and the volume/surface ratio will do the job. In the presence of gravitation, though, the negative energy density of the true vacuum distorts geometry within the bubble with the result that, for a small enough energy density, there is no bubble with a big enough volume/surface ratio.

Within the bubble, the effects of gravitation are more dramatic. The geometry of space-time within the bubble is that of anti–de Sitter space, a space much like conventional de Sitter space except that its group of symmetries is O(3, 2) rather than O(4, 1). Although this space-time is free of singularities, it is unstable under small perturbations, and inevitably suffers gravitational collapse of the same sort as the end state of a contracting Friedmann universe. The time required for the collapse of the interior universe is on the order of the time Λ discussed in Sec. I, microseconds or less.

This is disheartening. The possibility that we are living in a false vacuum has never been a cheering one to contemplate. Vacuum decay is the ultimate ecological catastrophe; in a new vacuum there are new constants of nature; after vacuum decay, not only is life as we know it impossible, so is chemistry as we know it. However, one could always draw stoic comfort from the possibility that perhaps in the course of time the new vacuum would sustain, if not life as we know it, at least some structures capable of knowing joy. This possibility has now been eliminated.

The second special case is decay into a space of vanishing cosmological constant, the case that applies if we are now living in the debris of a false vacuum which decayed at some early cosmic epoch. This case presents us with less interesting physics and with fewer occasions for rhetorical excess than the preceding one. It is now the interior of the bubble that is ordinary Minkowski space, and the inner edge of the wall that continues to trace out a hyperboloid. The mathematical reason for this is the same as before. The physical reason is even simpler than before: Within a spherically symmetric shell of energy there is no gravitational field.

As before, the effects of gravitation are negligible in any conceivable application, but we have computed them anyway. The sign is opposite to that in the previous case; gravitation makes vacuum decay more likely.

The space-time outside the bubble is now conventional de Sitter space. Of course, neither this space nor the Minkowski space inside is subject to catastrophic gravitational collapse initiated by small perturbations.

Finally, we must comment on the problem of the cosmological constant in the context of spontaneous symmetry breakdown. This problem was raised some years ago[2] and we have little new to say about it, but our work here has brought it home to us with new force.

Normally, when something is strictly zero, there is a reason for it. Vector-meson squared masses vanish because of gauge invariance, those of Dirac fields because of chiral symmetry. There is nothing to keep the squared masses of scalar mesons zero, but it is no disaster if they go negative, merely a sign that we are expanding about the wrong ground state.

But there is no reason for the cosmological constant (equivalently, the absolute energy density of the vacuum) to vanish. Indeed, if it were not for the irrefutable empirical evidence, one would expect it to be a typical microphysical number, the radius of the universe to be less than a kilometer. Even worse, zero energy density is the edge of disaster; even the slightest negativity would be enough to initiate catastrophic gravitational collapse.

There is something we do not understand about gravitation, and this something has nothing to do with loops of virtual gravitons. There has to be change, and change at a length scale much larger than the Planck length.

ACKNOWLEDGMENTS

This work was supported in part by the Department of Energy under Contract No. DE-AC03-76SF00515 and Contract No. DE-AS02-76ER02220 and by the National Science Foundation under Grant No. PHY77-22864.

---

*Permanent address: Department of Physics, Harvard University, Cambridge, Mass. 02138.

[1]We follow here the treatment given in S. Coleman, Phys. Rev. D 15, 2929 (1977); 16, 1248(E) (1977); and C. G. Callan and S. Coleman, *ibid.* 16, 1762 (1977). These contain references to the earlier literature.

[2]A. Linde, Pis'ma Zh. Eksp. Teor. Fiz. 19, 320 (1974) [JETP Lett. 19, 183 (1974)]; M. Veltman, Phys. Rev. Lett. 34, 777 (1975).

[3]S. Coleman, V. Glaser, and A. Martin, Commun. Math. Phys. 58, 211 (1978).

[4]Our treatment of de Sitter space closely follows that of S. Hawking and G. Ellis, *The Large Scale Structure of Space-Time* (Cambridge University Press, New York, 1973).

[5]For more on anti–de Sitter space, see Hawking and Ellis (Ref. 4).

SUPERCOOLED PHASE TRANSITIONS IN THE VERY EARLY UNIVERSE

S.W. HAWKING and I.G. MOSS
Department of Applied Mathematics and Theoretical Physics, Cambridge, UK

Received 12 January 1982

The universe might have had a prolonged exponentially expanding phase caused by its being stuck in a metastable state of the grand unified phase transition. The only way that it could exit from this exponential expansion without introducing too much inhomogeneity or spatial curvature would be through a homogeneous "bubble" solution in which quantum tunnelling occured everywhere at the same time. This would produce more baryons than the conventional scenarios.

There has been considerable interest recently in the possibility that the universe might have undergone an exponentially expanding "inflationary" stage caused by the vacuum energy of a supercooled metastable state of a first-order phase transition associated with the breakdown of the grand unification symmetry [1–4]. Such an exponential expansion might provide the answer to a number of problems in cosmology such as

(1) Why are there so few magnetic monopoles [1,5].

(2) Why is the universe so spatially homogeneous and isotropic?

(3) Why is the universe so nearly flat, i.e. why is it expanding at almost exactly the critical rate to avoid recollapse?

The main difficulty with this inflationary scenario has been to find some way of getting out of the exponentially expanding stage without creating large-scale inhomogeneities that would be incompatible with observation. In his original paper Guth [2] suggested that small bubbles of the broken symmetry phase would form by vacuum tunelling [6] and then expand by a large factor before colliding with other bubbles. However, as he himself admitted, and others have confirmed [7], one would end up with a very inhomogeneous universe dominated by a few very large bubbles.

Most of the papers that have discussed the exit from the inflationary stage have treated it essentially as a phase transition problem in flat spacetime and have neglected the curvature and finite horizon size of the universe. The aim of this letter is to show that these effects play a very important role and can lead to a phase transition which occurs simultaneously at all points of space thus not creating any inhomogeneity or magnetic monopoles. If the adjustable parameters in the effective potential for the Higgs fields lie in a certain range, the spatial flatness of the universe would also be accounted for.

The effective potential of the Higgs fields has the general form

$$V(\phi) = \tfrac{1}{2}(m^2 + \xi R + cT^2)\phi^2 + \tfrac{1}{4}\alpha^2\phi^4(\log \phi^2/\phi_0^2 - \tfrac{1}{2} + m^2/\alpha^2\phi_0^2) + \tfrac{1}{8}\alpha^2\phi_0^4 - \tfrac{1}{4}m^2\phi_0^2 , \quad (1)$$

where ϕ is some measure of the magnitude of the Higgs field and ϕ_0 is its expectation value at zero temperature in flat space with $V(\phi_0) = 0$. The quantities m^2 and ξ are renormalised parameters which can be given any value. $\xi = \frac{1}{6}$ is a natural choice because it corresponds to conformal invariance when $m = 0$ and it is not renormalized at one loop though it is at two loops. The quantity c is a numerical factor of order 1 which depends on the group and the representation. Neglecting small corrections arising from the ξR term, the energy density of the universe will be

$$\rho = (\pi^2/30)g(T)T^4 + V(\phi,T) , \quad (2)$$

where $g(T)$ is the effective number of spin states of relativistic particles. The rate of expansion will be given by

$$H^2 = (8\pi/3m_p^2)\rho - k/S^2 \quad (3)$$

where $H = \dot{S}/S$ and S is the scale factor of the universe.

At very high temperatures the universe will be in the symmetric phase $\phi = 0$. If $m^2 + \xi R \geqslant 0$, the symmetric phase will remain a local minimum of the potential as the temperature falls although it will cease to be a global minimum at some critical temperature T_c of the order of $\alpha\phi_0$. However, in realistic models the probability to tunnel to the broken symmetry phase with $\phi \approx \phi_0$ is very low [8]. The universe will therefore continue to expand in the symmetric phase until the thermal energy, the first term in eq. (2), becomes small compared to the vacuum energy $V(0, T)$. This will lead to an exponential expansion and the universe will rapidly approach a de Sitter state which depends only on the value of $V(0)$ through the relation

$$H = [8\pi V(0)/3m_p^2]^{1/2} \quad (4)$$

but which is otherwise independent of the initial conditions and the initial value of k. This is very similar to the way that a gravitational collapse rapidly approaches a stationary black-hole state which depends only on the mass and angular momentum but which is otherwise independent of the nature of the collapsing body. Following the black-hole analogy, we shall therefore assume a cosmological "no hair" theorem and treat the universe as a de Sitter model.

In the gravitational collapse case, the black hole has an effective temperature [9] $T_B = m_p^2(8\pi M)^{-1}$. Similarly, de Sitter space has an effective temperature [10] $T_s = H/2\pi$. In the case of grand unified phase transitions this will be of the order of 10^{11} GeV. In a de Sitter universe it is impossible to define a temperature smaller than T_s. It is therefore incorrect to use in the effective potential (1) the redshifted temperature of the universe when this falls below T_s. In fact one can regard the ξR term as representing the contribution of T_s to the effective potential.

In accordance with our viewpoint that de Sitter space has no hair, we shall calculate the probability of tunnelling from the local minimum at $\phi = 0$ to the global minimum by looking for solutions of the classical coupled scalar and gravitational field equations that are near the euclidean version of de Sitter space, i.e. a four-sphere of radius H^{-1}.

If $m^2 > 2H^2(1-6\xi)$, there are inhomogeneous "bubble" solutions as there are in the decay of the false vacuum in flat spacetime [6]. Gravity and the curvature of the universe do not have such effect except to make the bubble slightly smaller [11]. However, as mentioned earlier, this case would lead to unacceptably large inhomogeneities in the universe today.

If $2H^2(1 - 6\xi) \geqslant m^2 \geqslant - 12H^2\xi$, the potential (1) would lead to inhomogenous bubble solutions in flat spacetime whose radius was greater than that of the de Sitter space, H^{-1}. In this case the only euclidean solution of the coupled equations, apart from $\phi = 0$, is the homogeneous solution $\phi = \phi_1$ where $V(\phi_1)$ is the local maximum of V on a four-sphere of radius H_1^{-1} where $H_1^2 = 8\pi V(\phi_1)/3m_p^2$. The tunnelling probability per unit four-volume is of the order of $(m^2 + \xi R)^2 e^{-B}$ where

$$B = \tfrac{1}{8} m_p^4 \,[1/V(0) - 1/V(\phi_1)] \quad (5)$$

is the difference between the combined gravitational and scalar field actions of the $\phi = \phi_1$, and the $\phi = 0$ solutions. Even though B may be very large and the tunnelling probability very small, this does not matter because the universe will continue in the essentially stationary de Sitter state until it makes a quantum transition everywhere to the $\phi = \phi_1$ solution.

The $\phi = \phi_1$ solution is unstable as is reflected by the fact that it has a negative perturbative mode $\phi = \phi_1 + \epsilon$ where ϵ is a constant on the four-sphere. One would therefore expect the solution in lorentzian spacetime to evolve according to the classical field equations with ϕ running downhill from the maximum at $\phi = \phi_1$ to the global minimum at $\phi = \phi_0$. The solution will presumably be spatially homogeneous but evolving in time so it can be represented by a Friedmann–Robertson–Walker metric with ϕ a functionof time only. One can introduce Robertson–Walker coordinates with $K = +1,0$ or -1 into de Sitter space though

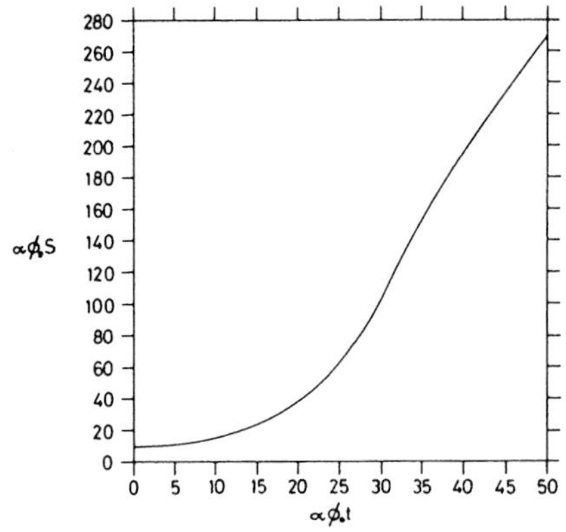

Fig. 1. The scale factor of the universe starting from a k = +1 time symmetric state with $\phi = \phi_1 + \epsilon$.

the coordinates cover the whole space only in the first case. It would seem reasonable to start the classical evolution of the solution on a spacelike surface whose curvature (positive or negative) is of the order of the radius H_1^{-1} of the de Sitter space. The most natural choice would seem to be the surface of time symmetry in the K = +1 coordinates but the conclusions in the K = −1 case are very similar.

While the scalar field remains near $\phi = \phi_1$, the expansion of the universe will continue to be nearly exponential or, in the K = +1 case, $S = H_1^{-1} \cosh H_1 t$ where S is the scale factor. The timescale for ϕ to run downhill from the maximum at $\phi = \phi_1$ is

$$3H_1/(m^2 + 12\xi H_1^2) . \qquad (6)$$

When ϕ approaches the global minimum ϕ_0 it will start to perform damped harmonic osciallations around $\phi = \phi_0$ and the expansion of the universe will change to that of a model filled with non relativistic matter. Figs. 1 and 2 show the numerical solutions for S and ϕ in the case $K = +1, m^2 + 12\xi H_1^2 = 24H_1^2$.

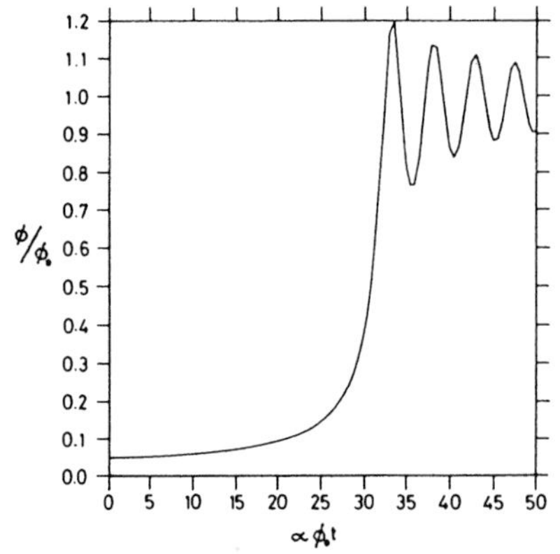

Fig. 2. The value of the scalar field in the same universe.

One can think of the oscillations of ϕ about ϕ_0 as corresponding to a very large number of Higgs particles in a coherent state. Presumably they will decay into light gauge particles and fermions which will have an approximately thermal distribution at a temperature T_2 given by $(\pi^2/30)g(T_2)T_2^4 \approx V(\phi_1)$. The decay of the Higgs particles would probably give a much larger baryon asymmetry than in the normal scenario.

The radius of the curvature of the space sections of the universe at the time of thermalization will be of the order $H_1^{-1} \cosh H_1 t_2$ where t_2 is the time from the $\phi = \phi_1$ state to thermalization. If one assumes that this is a few times the timescale (6) for ϕ to run downhill, one finds that the universe would have been sufficiently spatially flat to expend to its present size with a nearly critical density provided that $m^2 + 12\xi H_1^2 < \frac{1}{12} H_1^2$.

This seems the only situation in which one can have a significant exponential expansion at the GUT era without introducing either too much inhomogeneity or too much spatial curvature.

Finally, if $m^2 + 12\xi H_1^2 < 0$, the symmetric state $\phi = 0$ will cease to be a local minimum of the potential before the temperature drops to T_s so there will not be a prolonged exponential expansion phase.

We are very grateful to A.D. Linde for discussions which gave rise to this paper.

References

[1] A.H. Guth and S.H. Tye, Phys. Rev. Lett. 44 (1980) 631, 963.
[2] A.H. Guth, Phys. Rev. D23 (1981) 347.
[3] A.H. Guth and E.J. Weinberg, Phys. Rev. D23 (1981) 876.
[4] A.D. Linde, Lebedev Institute preprint 229 (1981).
[5] Ya.B. Zeldovich and M.Yu Khlopov, Phys. Lett. 79B (1978) 239.

J. Preskill, Phys. Rev. Lett. 43 (1979) 1365.
[6] S. Coleman, Phys. Rev. D15 (1977) 2929;
C. Callan and S. Coleman, Phys. Rev. D16 (1977) 1762.
[7] S.W. Hawking, I.G. Moss and J.M. Stewart, Bubble collisions in the very early universe, Cambridge preprint (1981).
[8] A.D. Linde, Phys. Lett. 99B (1981) 391.
G.P. Cook and K.T. Mahanthappa, Phys. Rev. D23 (1981) 1321;
A. Billoire and K. Tamvakis, CERN preprint TH.3019 (1981)
[9] S.W. Hawking, Commun. Math. Phys. 43 (1975) 199.
[10] G.W. Gibbons and S.W. Hawking, Phys. Rev. D15 (1977) 2738.
[11] S. Coleman and F. De Luccia, Phys. Rev. D21 (1980) 3305.

PHYSICAL REVIEW D VOLUME 28, NUMBER 12 15 DECEMBER 1983

Wave function of the Universe

J. B. Hartle
Enrico Fermi Institute, University of Chicago, Chicago, Illinois 60637
and Institute for Theoretical Physics, University of California, Santa Barbara, California 93106

S. W. Hawking
Department of Applied Mathematics and Theoretical Physics, Silver Street, Cambridge, England
and Institute for Theoretical Physics, University of California, Santa Barbara, California 93106
(Received 29 July 1983)

The quantum state of a spatially closed universe can be described by a wave function which is a functional on the geometries of compact three-manifolds and on the values of the matter fields on these manifolds. The wave function obeys the Wheeler-DeWitt second-order functional differential equation. We put forward a proposal for the wave function of the "ground state" or state of minimum excitation: the ground-state amplitude for a three-geometry is given by a path integral over all compact positive-definite four-geometries which have the three-geometry as a boundary. The requirement that the Hamiltonian be Hermitian then defines the boundary conditions for the Wheeler-DeWitt equation and the spectrum of possible excited states. To illustrate the above, we calculate the ground and excited states in a simple minisuperspace model in which the scale factor is the only gravitational degree of freedom, a conformally invariant scalar field is the only matter degree of freedom and $\Lambda>0$. The ground state corresponds to de Sitter space in the classical limit. There are excited states which represent universes which expand from zero volume, reach a maximum size, and then recollapse but which have a finite (though very small) probability of tunneling through a potential barrier to a de Sitter-type state of continual expansion. The path-integral approach allows us to handle situations in which the topology of the three-manifold changes. We estimate the probability that the ground state in our minisuperspace model contains more than one connected component of the spacelike surface.

I. INTRODUCTION

In any attempt to apply quantum mechanics to the Universe as a whole the specification of the possible quantum-mechanical states which the Universe can occupy is of central importance. This specification determines the possible dynamical behavior of the Universe. Moreover, if the uniqueness of the present Universe is to find any explanation in quantum gravity it can only come from a restriction on the possible states available.

In quantum mechanics the state of a system is specified by giving its wave function on an appropriate configuration space. The possible wave functions can be constructed from the fundamental quantum-mechanical amplitude for a complete history of the system which may be regarded as the starting point for quantum theory.[1] For example, in the case of a single particle a history is a path $x(t)$ and the amplitude for a particular path is proportional to

$$\exp(iS[x(t)])\,, \tag{1.1}$$

where $S[x(t)]$ is the classical action. From this basic amplitude, the amplitude for more restricted observations can be constructed by superposition. In particular, the amplitude that the particle, having been prepared in a certain way, is located at position x and nowhere else at time t is

$$\psi(x,t)=N\int_C \delta x(t)\exp(iS[x(t)])\,. \tag{1.2}$$

Here, N is a normalizing factor and the sum is over a class of paths which intersect x at time t and which are weighted in a way that reflects the preparation of the system. $\psi(x,t)$ is the wave function for the state determined by this preparation. As an example, if the particle were previously localized at x' at time t' one would sum over all paths which start at x' at t' and end at x at t thereby obtaining the propagator $\langle x,t\,|\,x',t'\rangle$. The oscillatory integral in Eq. (1.2) is not well defined but can be made so by rotating the time to imaginary values.

An alternative way of calculating quantum dynamics is to use the Schrödinger equation,

$$i\partial\psi/\partial t=H\psi\,. \tag{1.3}$$

This follows from Eq. (1.2) by varying the end conditions on the path integral. For a particular state specified by a weighting of paths C, the path integral (1.2) may be looked upon as providing the boundary conditions for the solution of Eq. (1.3).

A state of particular interest in any quantum-mechanical theory is the ground state, or state of minimum excitation. This is naturally defined by the path integral, made definite by a rotation to Euclidean time, over the class of paths which have vanishing action in the far past. Thus, for the ground state at $t=0$ one would write

$$\psi_0(x,0)=N\int \delta x(\tau)\exp(-I[x(\tau)])\,, \tag{1.4}$$

where $I[x(\tau)]$ is the Euclidean action obtained from S by sending $t \to -i\tau$ and adjusting the sign so that it is positive.

In cases where there is a well-defined time and a corresponding time-independent Hamiltonian, this definition of ground state coincides with the lowest eigenfunction of the Hamiltonian. To see this specialize the path-integral expression for the propagator $\langle x,t \mid x',t' \rangle$ to $t=0$ and $x'=0$ and insert a complete set of energy eigenstates between the initial and final state. One has

$$\langle x,0 \mid 0,t' \rangle = \sum_n \psi_n(x)\bar{\psi}_n(0)\exp(iE_n t') = \int \delta x(t)\exp(iS[x(t)]) , \quad (1.5)$$

where $\psi_n(x)$ are the time-independent energy eigenfunctions. Rotate $t' \to -i\tau'$ in (1.5) and take the limit as $\tau' \to -\infty$. In the sum only the lowest eigenfunction (normalized to zero energy) survives. The path integral becomes the path integral on the right of (1.4) so that the equality is demonstrated.

The case of quantum fields is a straightforward generalization of quantum particle mechanics. The wave function is a functional of the field configuration on a spacelike surface of constant time, $\Psi = \Psi[\phi(\vec{x}),t]$. The functional Ψ gives the amplitude that a particular field distribution $\phi(\vec{x})$ occurs on this spacelike surface. The rest of the formalism is similarly generalized. For example, for the ground-state wave functional one has

$$\Psi_0[\phi(\vec{x}),0] = N \int \delta\phi(x)\exp(-I[\phi(x)]) , \quad (1.6)$$

where the integral is over all Euclidean field configurations for $\tau < 0$ which match $\phi(\vec{x})$ on the surface $\tau = 0$ and leave the action finite at Euclidean infinity.

In the case of quantum gravity new features enter. For definiteness and simplicity we shall restrict our attention throughout this paper to spatially closed universes. For these there is no well-defined intrinsic measure of the location of a spacelike surface in the spacetime beyond that contained in the intrinsic or extrinsic geometry of the surface itself. One therefore labels the wave function by the three-metric h_{ij} writing $\Psi = \Psi[h_{ij}]$. Quantum dyanmics is supplied by the functional integral

$$\Psi[h_{ij}] = N \int_C \delta g(x)\exp(iS_E[g]) . \quad (1.7)$$

S_E is the classical action for gravity including a cosmological constant Λ and the functional integral is over all four-geometries with a spacelike boundary on which the induced metric is h_{ij} and which to the past of that surface satisfy some appropriate condition to define the state. In particular for the amplitude to go from a three-geometry h'_{ij} on an initial sp celike surface to a three-geometry h''_{ij} on a final spacelike surface is

$$\langle h''_{ij} \mid h'_{ij} \rangle = \int \delta g \exp(iS_E[g]) , \quad (1.8)$$

where the sum is over all four-geometries which match h'_{ij} on the initial surface and h''_{ij} on the final surface. Here one clearly sees that one cannot specify time in these states. The proper time between the surfaces depends on the four-geometries in the sum.

As in the mechanics of a particle the functional integral (1.7) implies a differential equation on the wave function. This is the Wheeler-DeWitt equation[2] which we shall derive from this point of view in Sec. II. With a simple choice of factor ordering it is

$$\left[-G_{ijkl}\frac{\delta^2}{\delta h_{ij}\delta h_{kl}} - {}^3R(h)h^{1/2} + 2\Lambda h^{1/2} \right] \Psi[h_{ij}] = 0 , \quad (1.9)$$

where G_{ijkl} is the metric on superspace,

$$G_{ijkl} = \tfrac{1}{2}h^{-1/2}(h_{ik}h_{jl} + h_{il}h_{jk} - h_{ij}h_{kl}) \quad (1.10)$$

and ${}^3R$ is the scalar curvature of the intrinsic geometry of the three-surface. The problem of specifying cosmological states is the same as specifying boundary conditions for the solution of the Wheeler-DeWitt equation. A natural first question to ask is what boundary conditions specify the ground state?

In the quantum mechanics of closed universes we do not expect to find a notion of ground state as a state of lowest energy. There is no natural definition of energy for a closed universe just as there is no independent standard of time. Indeed in a certain sense the total energy for a closed universe is always zero—the gravitational energy canceling the matter energy. It is still reasonable, however, to expect to be able to define a state of minimum excitation corresponding to the classical notion of a geometry of high symmetry. This paper contains a proposal for the definition of such a ground-state wave function for closed universes. The proposal is to extend to gravity the Euclidean-functional-integral construction of nonrelativistic quantum mechanics and field theory [Eqs. (1.4) and (1.6)]. Thus, we write for the ground-state wave function

$$\Psi_0[h_{ij}] = N \int \delta g \exp(-I_E[g]) , \quad (1.11)$$

where I_E is the Euclidean action for gravity including a cosmological constant Λ. The Euclidean four-geometries summed over must have a boundary on which the induced metric is h_{ij}. The remaining specification of the class of geometries which are summed over determines the ground state. Our proposal is that the sum should be over compact geometries. This means that the Universe does not have any boundaries in space or time (at least in the Euclidean regime) (cf. Ref. 3). There is thus no problem of boundary conditions. One can interpret the functional integral over all compact four-geometries bounded by a given three-geometry as giving the amplitude for that three-geometry to arise from a zero three-geometry, i.e., a single point. In other words, the ground state is the amplitude for the Universe to appear from nothing.[4] In the following we shall elaborate on this construction and show in simple models that it indeed supplies reasonable wave functions for a state of minimum excitation.

The specification of the ground-state wave function is a constraint on the other states allowed in the theory. They must be such, for example, as to make the Wheeler-DeWitt equation Hermitian in an appropriate norm. In analogy with ordinary quantum mechanics one would expect to be able to use these constraints to extrapolate the boundary conditions which determine the excited states of the theory from those fixed for the ground state by Eq. (1.7). Thus, one can in principle determine all the allowed cosmological states.

The wave functions which result from this specification will not vanish on the singular, zero-volume three-geometries which correspond to the big-bang singularity. This is analogous to the behavior of the wave function of the electron in the hydrogen atom. In a classical treatment, the situation in which the electron is at the proton is singular. However, in a quantum-mechanical treatment the wave function in a state of zero angular momentum is finite and nonzero at the proton. This does not cause any problems in the case of the hydrogen atom. In the case of the Universe we would interpret the fact that the wave function can be finite and nonzero at the zero three-geometry as allowing the possibility of topological fluctuations of the three-geometry. This will be discussed further in Sec. VIII.

After a general discussion of this proposal for the ground-state wave function we shall implement it in a minisuperspace model. The geometrical degrees of freedom in the model are restricted to spatially homogeneous, isotropic, closed universes with S^3 topology, the matter degrees of freedom to a single, homogeneous, conformally invariant scalar field and the cosmological constant is assumed to be positive. A semiclassical evaluation of the functional integral for the ground-state wave function shows that it indeed does possess characteristics appropriate to a "state of minimum excitation."

Extrapolating the boundary conditions which allow the ground state to be extracted from the Wheeler-DeWitt equation, we are able to go further and identify the wave functions in the minisuperspace models corresponding to excited states of the matter field. These wave functions display some interesting features. One has a complete spectrum of excited states which show that a closed universe similar to our own and possessed of a cosmological constant can escape the big crunch and tunnel through to an eternal de Sitter expansion. We are able to calculate the probability for this transition.

In addition to the excited states we make a proposal for the amplitudes that the ground-state three-geometry consists of disconnected three-spheres thus giving a meaning to a gravitational state possessing different topologies.

Our conclusion will be that the Euclidean-functional-integral prescription (1.7) does single out a reasonable candidate for the ground-state wave function for cosmology which when coupled with the Wheeler-DeWitt equation yields a basis for constructing quantum cosmologies.

II. QUANTUM GRAVITY

In this section we shall review the basic principles and machinery of quantum gravity with which we shall explore the wave functions for closed universes. For simplicity we shall represent the matter degrees of freedom by a single scalar field ϕ, more realistic cases being straightforward generalizations. We shall approach this review from the functional-integral point of view although we shall arrive at many canonical results.[5] None of these are new and for different approaches to the same ends the reader is referred to the standard literature.[6]

A. Wave functions

Our starting point is the quantum-mechanical amplitude for the occurrence of a given spacetime and a given field history. This is

$$\exp(iS[g,\phi]) , \tag{2.1}$$

where $S[g,\phi]$ is the total classical action for gravity coupled to a scalar field. We are envisaging here a fixed manifold although there is no real reason that amplitudes for different manifolds may not be considered provided a rule is given for their relative phases. Just as the interesting observations of a particle are not typically its entire history but rather observations of position at different times, so also the interesting quantum-mechanical questions for gravity correspond to observations of spacetime and field on different spacelike surfaces. Following the general rules of quantum mechanics the amplitudes for these more restricted sets of observations are obtained from (2.1) by summing over the unobserved quantities.

It is easy to understand what is meant by fixing the field on a given spacelike surface. What is meant by fixing the four-geometry is less obvious. Consider all four-geometries in which a given spacelike surface occurs but whose form is free to vary off the surface. By an appropriate choice of gauge near the surface (e.g., Gaussian normal coordinates) all these four-geometries can be expressed so that the only freedom in the four-metric is the specification of the three-metric h_{ij} in the surface. Specifying the three-metric is therefore what we mean by fixing the four-geometry on a spacelike surface. The situation is not unlike gauge theories. There a history is specified by a vector potential $A_\mu(x)$ but by an appropriate gauge transformation $A_0(x)$ can be made to vanish so that the field on a surface can be completely specified by the $A_i(x)$.

As an example of the quantum-mechanical superposition principle the amplitude for the three-geometry and field to be fixed on two spacelike surfaces is

$$\langle h_{ij}'',\phi'' \mid h_{ij}',\phi' \rangle = \int \delta g\, \delta\phi \exp(iS[g,\phi]) , \tag{2.2}$$

where the integral is over all four-geometries and field configurations which match the given values on the two spacelike surfaces. This is the natural analog of the propagator $\langle x'',t' \mid x',t' \rangle$ in the quantum mechanics of a single particle. We note again that the proper time between the two surfaces is not specified. Rather it is summed over in the sense that the separation between the surfaces depends on the four-geometry being summed over. It is not that one could not ask for the amplitude to have the three-geometry and field fixed on two surfaces *and* the proper time between them. One could. Such an amplitude, however, would not correspond to fixing observations on just two surfaces but rather would involve a set of intermediate observations to determine the time. It would therefore not be the natural analog of the propagator.

Wave functions Ψ are defined by

$$\Psi[h_{ij},\phi] = \int_C \delta g\, \delta\phi \exp(iS[g,\phi]) . \tag{2.3}$$

The sum is over a class C of spacetimes with a compact boundary on which the induced metric is h_{ij} and field configurations which match ϕ on the boundary. The remaining specification of the class C is the specification of the state.

If the Universe is in a quantum state specified by a wave function Ψ then that wave function describes the correlations between observables to be expected in that state. For example, in the semiclassical wave function describing a universe like our own, one would expect Ψ to be large when ϕ is big and the spatial volume is small, large when ϕ is small and the spatial volume is big, and small when these quantities are oppositely correlated. This is the only interpretative structure we shall propose or need.

B. Wheeler-DeWitt equation

A differential equation for Ψ can be derived by varying the end conditions on the path integral (2.3) which defines it. To carry out this derivation first recall that the gravitational action appropriate to keeping the three-geometry fixed on a boundary is

$$l^2 S_E = 2\int_{\partial M} d^3x\, h^{1/2} K + \int_M d^4x (-g)^{1/2}(R-2\Lambda) . \tag{2.4}$$

The second term is integrated over spacetime and the first over its boundary. K is the trace of the extrinsic curvature K_{ij} of the boundary three-surface. If its unit normal is n^i, $K_{ij}=-\nabla_i n_j$ in the usual Lorentzian convention. l is the Planck length $(16\pi G)^{1/2}$ in the units with $\hbar=c=1$ we use throughout. Introduce coordinates so that the boundary is a constant t surface and write the metric in the standard $3+1$ decomposition:

$$ds^2 = -(N^2 - N_i N^i)dt^2 + 2N_i dx^i dt + h_{ij}dx^i dx^j . \tag{2.5}$$

The action (2.4) becomes

$$l^2 S_E = \int d^4x\, h^{1/2} N[K_{ij}K^{ij} - K^2 + {}^3R(h) - 2\Lambda] , \tag{2.6}$$

where explicitly

$$K_{ij} = \frac{1}{N}\left[-\frac{1}{2}\frac{\partial h_{ij}}{\partial t} + N_{(i|j)} \right] \tag{2.7}$$

and a stroke and ${}^3R$ denote the covariant derivative and scalar curvature constructed from the three-metric h_{ij}. The matter action S_M can similarly be expressed as a function of N, N_i, h_{ij}, and the matter field.

The functional integral defining the wave function contains an integral over N. By varying N at the surface we push it forward or backward in time. Since the wave function does not depend on time we must have

$$0 = \int \delta g\, \delta\phi \left[\frac{\delta S}{\delta N} \right] \exp(iS[g,\phi]) . \tag{2.8}$$

More precisely, the value of the integral (2.3) should be left unchanged by an infinitesimal translation of the integration variable N. If the measure is invariant under translation this leads to (2.8). If it is not, there will be in addition a divergent contribution to the relation which must be suitably regulated to zero or cancel divergences arising from the calculation of the right-hand side of (2.8).

Classically the field equation $H \equiv \delta S/\delta N = 0$ is the Hamiltonian constraint for general relativity. It is

$$H = h^{1/2}(K^2 - K_{ij}K^{ij} + {}^3R - 2\Lambda - l^2 T_{nn}) = 0 , \tag{2.9}$$

where T_{nn} is the stress-energy tensor of the matter field projected in the direction normal to the surface. Equation (2.8) shows how $H=0$ is enforced as an operator identity for the wave function. More explicitly one can note that the K_{ij} involve only first-time derivatives of the h_{ij} and therefore may be completely expressed in terms of the momenta π_{ij} conjugate to the h_{ij} which follow from the Lagrangian in (2.6):

$$\pi_{ij} = -h^{1/2}(K_{ij} - h_{ij}K) . \tag{2.10}$$

In a similar manner the energy of the matter field can be expressed in terms of the momentum conjugate to the field π_ϕ and the field itself. Equation (2.8) thus implies the operator identity $H(\pi_{ij}, h_{ij}, \pi_\phi, \phi)\Psi = 0$ with the replacements

$$\pi^{ij} = -i\frac{\delta}{\delta h_{ij}}, \quad \pi_\phi = -i\frac{\delta}{\delta\phi} . \tag{2.11}$$

These replacements may be viewed as arising directly from the functional integral, e.g., from the observation that when the time derivatives in the exponent are written in differenced form

$$-i\frac{\delta}{\delta h_{ij}} \int \delta g\, \delta\phi\, e^{iS} = \int \delta g\, \delta\phi\, \pi^{ij} e^{iS} . \tag{2.12}$$

Alternatively, they are the standard representation of the canonical commuation relations of h_{ij} and π^{ij}.

In translating a classical equation like $\delta S/\delta N = 0$ into an operator identity there is always the question of factor ordering. This will not be important for us so making a convenient choice we obtain

$$\left\{ -G_{ijkl}\frac{\delta^2}{\delta h_{ij}\delta h_{kl}} + h^{1/2}\left[-{}^3R(h) + 2\Lambda + l^2 T_{nn}\left[-i\frac{\delta}{\delta\phi}, \phi \right] \right] \right\} \times \Psi[h_{ij}, \phi] = 0 . \tag{2.13}$$

This is the Wheeler-DeWitt equation which wave functions for closed universes must satisfy. There are also the other constraints of the classical theory, but the operator versions of these express the gauge invariance of the wave function rather than any dynamical information.[6]

We should emphasize that the ground-state wave function constructed by a Euclidean functional-integral prescription [(Eq. (1.11)] will satisfy the Wheeler-DeWitt equation in the form (2.13). Indeed, this can be demonstrated explicitly by repeating the steps in the above demonstration starting with the Euclidean functional integral.

C. Boundary conditions

The quantity G_{ijkl} can be viewed as a metric on superspace—the space of all three-geometries (no connection with supersymmetry). It has signature

$(-,+,+,+,+,+)$ and the Wheeler-DeWitt equation is therefore a "hyperbolic" equation on superspace. It would be natural, therefore, to expect to impose boundary conditions on two "spacelike surfaces" in superspace. A convenient choice for the timelike direction is $h^{1/2}$ and we therefore expect to impose boundary conditions at the upper and lower limits of the range of $h^{1/2}$. The upper limit is infinity. The lower limit is zero because if h_{ij} is positive definite or degenerate, $h^{1/2}\geq 0$. Positive-definite metrics are everywhere spacelike surfaces; degenerate metrics may signal topology change. Summarizing the remaining functions of h_{ij} by the conformal metric $\widetilde{h}_{ij}=h_{ij}/h^{1/3}$ we may write an important boundary condition on Ψ as

$$\Psi[\widetilde{h}_{ij},h^{1/2},\phi]=0,\quad h^{1/2}<0\,. \tag{2.14}$$

Because $h^{1/2}$ has a semidefinite range it is for many purposes convenient to introduce a representation in which $h^{1/2}$ is replaced by its canonically conjugate variable $-\frac{4}{3}Kl^{-2}$ which has an infinite range. The advantages of this representation have been extensively discussed.[7] In the case of pure gravity since $-\frac{4}{3}Kl^{-2}$ and $h^{1/2}$ are conjugate, we can write for the transformation to the epresentation where $\widetilde{h}_{ij}$ and K are definite

$$\Phi[\widetilde{h}_{ij},K]=\int_0^{\infty}\delta h^{1/2}\exp\left[-i\tfrac{4}{3}l^{-2}\int d^3x\,h^{1/2}K\right]\Psi[h_{ij}] \tag{2.15}$$

and inversely,

$$\Psi[h_{ij}]=\int_{-\infty}^{+\infty}\delta K\exp\left[+i\tfrac{4}{3}l^{-2}\int d^3x\,h^{1/2}K\right]\Phi[\widetilde{h}_{ij},K]\,. \tag{2.16}$$

In each case the functional integrals are over the values of $h^{1/2}$ or K at each point of the spacelike hypersurface and we have indicated limits of integration.

The condition (2.14) implies through (2.15) that $\Phi[\widetilde{h}_{ij},K]$ is analytic in the lower-half K plane. The contour in (2.16) can thus be distorted into the lower-half K plane. Conversely, if we are given $\Phi[\widetilde{h}_{ij},K]$ we can reconstruct the wave function Ψ which satisfies the boundary condition (2.14) by carrying out the integration in (2.16) over a contour which lies below any singularities of $\Phi[\widetilde{h}_{ij},K]$ in K.

In the presence of matter K and $\widetilde{h}_{ij}$ remain convenient labels for the wave functional provided the labels for the matter-field amplitudes $\widetilde{\phi}$ a e chosen so that a multiple of K is canonically conjugate to $h^{1/2}$. In cases where the matter-field action itself involves the scalar curvature this means that the label $\widetilde{\phi}$ will be the field amplitude rescaled by some power of $h^{1/2}$. For example, in the case of a conformally invariant scalar field the appropriate label is $\widetilde{\phi}=\phi h^{1/6}$. With this understanding we can write for the functionals

$$\Psi=\Psi[h_{ij},\widetilde{\phi}],\quad \Phi=\Phi[\widetilde{h}_{ij},K,\widetilde{\phi}] \tag{2.17}$$

and the transformation formulas (2.15) and (2.16) remain unchanged.

D. Hermiticity

The introduction of wave functions as functional integrals [Eq. (2.3)] allows the definition of a scalar product with a simple geometric interpretation in terms of sums over spacetime histories. Consider a wave function Ψ defined by the integral

$$\Psi[h_{ij},\phi]=N\int_C \delta g\,\delta\phi\,\exp(iS[g,\phi])\,, \tag{2.18}$$

over a class of four-geometries and fields C, and a second wave function Ψ' defined by a similar sum over a class C'. The scalar product

$$(\Psi',\Psi)=\int \delta h\,\delta\phi\,\overline{\Psi}'[h_{ij},\phi]\Psi[h_{ij},\phi] \tag{2.19}$$

has the geometric interpretation of a sum over all histories

$$(\Psi',\Psi)=\overline{N}'N\int \delta g\,\delta\phi\,\exp(iS[g,\phi])\,, \tag{2.20}$$

where the sum is over histories which lie in class C to the past of the surface and in the time reversed of class C' to its future.

The scala product (2.19) is not the p oduct that would be required by canonical theory to define the Hilbert space of physical states. That would presumably involve integration over a hypersurface in the space of all three-geometries rather than over the whole space as in (2.19). Rather, Eq. (2.19) is a mathematical construction made natural by the functional-integral formulation of quantum gravity.

In gravity we expect the field equations to be satisfied as identities. An extension of the argument leading to Eq. (2.8) will give

$$\int \delta g\,\delta\phi\,H(x)\exp(iS[g,\phi])=0 \tag{2.21}$$

for any class of geometries summed over and for any intermediate spacelike surface on which $H(x)$ is evaluated. Equation (2.21) can be evaluated for the particular sum which ente s Eq. (2.20). $H(x)$ can be interpreted in the scalar product as an operator acting on either Ψ' or Ψ. Thus,

$$(H\Psi',\Psi)=(\Psi',H\Psi)=0\,. \tag{2.22}$$

The Whee er-DeWitt operator must therefore be Hermitian in the scalar product (2.19).

Since the Wheeler-DeWitt operator is a second-order functional-differential operator, the requirement of Hermiticity will essentially be a requirement that certain surface terms on the boundary of the space of three-metrics vanish and, in particular, at $h^{1/2}=0$ and $h^{1/2}=\infty$. As in ordinary quantum mechanics these co ditions will prove useful in providing boundary conditions for the solution of the equation.

III. GROUND-STATE WAVE FUNCTION

In this section, we shall put forward in detail our proposal for the ground-state wave function for closed cosmologies. The wave function depends on the topology and the three-metric of the spacelike surface and on the values of the matter field on the surface. For simplicity we shall begin by considering only S^3 topo ogy. Other possibilities will be considered in Sec. VIII.

As discussed in the Introduction, the ground-state wave function is to be constructed as a functional integral of the form

$$\Psi_0[h_{ij},\phi]=N\int \delta g\,\delta\phi\,\exp(-I[g,\phi])\,, \quad (3.1)$$

where I is the total Euclidean action and the integral is over an appropriate class of Euclidean four-geometries with compact boundary on which the induced metric is h_{ij} and an appropriate class of Euclidean field configurations which match the value given on the boundary. To complete the definition of the ground-state wave function we need to give the class of geometries and fields to be summed over. Our proposal is that the geometries should be compact and that the fields should be regular on these geometries. In the case of a positive cosmological constant Λ any regular Euclidean solution of the field equations is necessarily compact.[8] In particular, the solution of greatest symmetry is the four-sphere of radius $3/\Lambda$, whose metric we write as

$$ds^2=(\sigma/H)^2(d\theta^2+\sin^2\theta\,d\Omega_3{}^2)\,, \quad (3.2)$$

where $d\Omega_3{}^2$ is the metric on the three-sphere. $H^2=\sigma^2\Lambda/3$ and we have introduced the normalization factor $\sigma^2=l^2/24\pi^2$ for later convenience. Thus, it is clear that compact four-geometries are the only reasonable candidates for the class to be summed over when $\Lambda>0$.

If Λ is zero or negative there are noncompact solutions of the field equations. The solutions of greatest symmetry are Euclidean space ($\Lambda=0$) with

$$ds^2=\sigma^2(d\theta^2+\theta^2 d\Omega_3{}^2) \quad (3.3)$$

and Euclidean anti–de Sitter space ($\Lambda<0$) with

$$ds^2=(\sigma/H)^2(d\theta^2+\sinh^2\theta\,d\Omega_3{}^2)\,. \quad (3.4)$$

One might therefore feel that the ground state for $\Lambda\leq 0$ should be defined by a functional integral over geometries which are asymptotically Euclidean or asymptotically anti–de Sitter. This is indeed appropriate to defining the ground state for scattering problems where one is interested in particles which propagate in from infinity and then out to infinity again.[9] However, in the case of cosmology, one is interested in measurements that are carried out in the interior of the spacetime, whether or not the interior points are connected to some infinite regions does not matter. If one were to use asymptotically Euclidean or anti–de Sitter four-geometries in the functional integral that defines the ground state one could not exclude a contribution from four-geometries that consisted of two disconnected pieces, one of which was compact with the three-geometry as boundary and the other of which was asymptotically Euclidean or anti–de Sitter with no interior boundary. Such disconnected geometries would in fact give the dominant contribution to the ground-state wave function. Thus, one would effectively be back with the prescription given above.

The ground-state wave function obtained by summing over compact four-geometries diverges for large three-geometries in the cases $\Lambda\leq 0$ and the wave function cannot be normalized. This is because the Λ in the action damps large four-geometries when $\Lambda>0$, but it enhances them when $\Lambda<0$. We shall therefore consider only the case $\Lambda>0$ in this paper and shall regard $\Lambda=0$ as a limiting case of $\Lambda>0$.

An equivalent way of describing the ground state is to specify its wave function in the $\tilde{\phi},\tilde{h}_{ij},K$ representation. Here too it can be constructed as a functional integral:

$$\Phi_0[\tilde{h}_{ij},K,\tilde{\phi}]=N\int \delta g\,\delta\phi\,\exp(-I^K[g,\phi])\,. \quad (3.5)$$

The sum is over the same class of fields and geometries as before except that now $\tilde{\phi}$, $\tilde{h}_{ij}$, and K are fixed on the boundary rather than ϕ and h_{ij}. The action I^K is therefore the Euclidean action appropriate to holding $\tilde{\phi}$, $\tilde{h}_{ij}$, and K fixed on a boundary. It is a sum of the appropriate pure gravitational action which up to an additive constant is

$$l^2 I_E^K[g]=-\tfrac{2}{3}\int_{\partial M} d^3x\, h^{1/2}K-\int_M d^4x\, g^{1/2}(R-2\Lambda) \quad (3.6)$$

and a contribution from the matter. The latter is well illustrated by the action of a single conformally invariant scalar field, an example which we shall use exclusively in the rest of this paper. We have

$$I_M^K[g,\phi]=\tfrac{1}{2}\int_M d^4x\, g^{1/2}[(\nabla\phi)^2+\tfrac{1}{6}R\phi^2]\,. \quad (3.7)$$

These actions differ from the more familiar ones in which ϕ and h_{ij} are fixed only in having different surface terms. Indeed, these surface terms are just those required to ensure the equivalence of (3.1) and (3.5) as a consequence of the transformation formulas (2.15) and (2.16). In the case of the matter action of a conformally invariant scalar field with $\tilde{\phi},h_{ij},K$ fixed the additional surface term conveniently cancels that required in the action when ϕ and h_{ij} are fixed.

It is important to recognize that the functional integral (3.5) does not yield the wave function at the Lorentzian value of K but rather at a Euclidean value of K. For the moment denote the Lorentzian value by K_L. If the hypersurfaces of interest were labeled by a time coordinate t in a coordinate system with zero shift [$N_i=0$ in Eq. (2.5)] then the rotation $t\to i\tau$ and the use of the traditional conventions $K_L=-\nabla\cdot n$ and $K=\nabla\cdot n$ will send $K_L\to -iK$. In terms of the Euclidean K the transformation formulas (2.15) and (2.16) can be rewritten to read

$$\Phi[\tilde{h}_{ij},K,\tilde{\phi}]=\int_0^\infty \delta h^{1/2}\exp\left[-\tfrac{4}{3}l^{-2}\int d^3x\, h^{1/2}K\right]\times\Psi[h_{ij},\tilde{\phi}]\,, \quad (3.8)$$

$$\Psi[h_{ij},\phi]=-\frac{1}{2\pi i}\int_C \delta K\,\exp\left[\tfrac{4}{3}l^{-2}\int d^3x\, h^{1/2}K\right]\times\Phi[\tilde{h}_{ij},K,\tilde{\phi}]\,, \quad (3.9)$$

where the contour C runs from $-i\infty$ to $+i\infty$. At the risk of some confusion we shall continue to use K in the remainder of this paper to denote the Euclidean K despite having used the same symbol in Secs. I and II for the Lorentzian quantity.

There is one advantage to constructing the ground-state wave function from the functional integral (3.5) rather than (3.1) and it is the following: the integral in Eq. (3.9) will always yield a wave function $\Psi_0[h_{ij},\phi]$ which vanishes for $h^{1/2}<0$ if the contour C is chosen to the right of any singularities of $\Phi_0[\tilde{h}_{ij},K,\tilde{\phi}]$ in K provided Φ does not diverge too strongly in K. The boundary condition (2.14) is thus automatically enforced. This is a considerable advantage when the wave function is only evaluated approximately.

The Euclidean gravitational action [Eq. (3.6)] is not positive definite. The functional integrals in Eqs. (3.1) and (3.5) therefore require careful definition. One way of doing this is to break the integration up into an integral over conformal factors and over geometries in a given conformal equivalence class. By appropriate choice of the contour of integration of the conformal factor the integral can probably be made convergent. If this is the case a pro rly convergent functional integral can be constructed.

This then is our prescription for the ground state. In the following sections we shall derive some of its properties and demonstrate its reasonableness in a simple minisuperspace model.

IV. SEMICLASSICAL EXPECTATIONS

An important advantage of a functional-integral prescription for the ground-state wave function is that it yields the semiclassical approximation for that wave function directly. In this section, we shall examine the semiclassical approximation to the ground-state wave function defined in Sec. III. For simplicity we shall consider the case of pure gravity. The extension to include matter is straightforward.

The semiclassical approximation is obtained by evaluating the functional integral by the method of steepest descents. If there is only one stationary-phase point the semiclassical approximation is

$$\Psi_0[h_{ij}]=N\Delta^{-1/2}[h_{ij}]\exp(-I_{\rm cl}[h_{ij}]) \ . \tag{4.1}$$

Here, $I_{\rm cl}$ is the Euclidean gravitational action evaluated at the stationary-phase point, that is, at that solution $g^{\rm cl}_{\mu\nu}$ of the Euclidean field equations

$$R_{\mu\nu}=\Lambda g_{\mu\nu} \ , \tag{4.2}$$

which induces the metric h_{ij} on the closed three-surface boundary and satisfies the asymptotic conditions discussed in Sec. III. $\Delta^{-1/2}$ is a combination of determinants of the wave operators defining the fluctuations about $g^{\rm cl}_{\mu\nu}$ including those contributed by the ghosts. We shall focus mainly on the exponent. For further information on Δ in the case without boundary see Ref. 10.

If there is more than one stationary-phase point, it is necessary to consider the contour of integration in the path integal more carefully in order to decide which gives the dominant contribution. In general this will be the stationary-phase point with the lowest value of ReI although it may not be if there are two stationary-phase points which correspond to four-metrics that are conformal to one another. We shall see an example of this in Sec. VI. The ground-state wave function is real. This means that if the stationary-phase points have complex values of the action, there will be equal contributions from stationary-phase points with complex-conjugate values of the action. If there is no four-geometry which is a stationary-phase point, the wave function will be zero in the semiclassical approximation.

The semiclassical approximation for Ψ_0 can also be obtained by first evaluating the semiclassical approximation to Φ_0 from the functional integral (3.5) and then evaluating the transformation integral (3.9) by steepest descents. This will be more convenient to do when the boundary conditions of fixing $\tilde{h}_{ij}$ and K yield a unique dominant stationary-phase solution to (4.2) but fixing h_{ij} does not.

One can fix the normalization constant N in (4.1) by the requirement

$$\int \delta h\, \bar{\Psi}_0[h_{ij}]\Psi_0[h_{ij}]=1 \ . \tag{4.3}$$

As explained in Sec. II, one can interpret (4.3) geometrically as a path integral over all four-geometries which are compact on *both* sides of the three-surface with the metric h_{ij}. The semiclassical approximation to this path integral will thus be given by the action of the compact four-geometry without boundary which is the solution of the Einstein field equation. In the case of $\Lambda>0$ the solution with the most negative action is the four-sphere. Thus,

$$N^2=\exp\left[-\frac{2}{3H^2}\right] . \tag{4.4}$$

The semiclassical approximation for the wave function gives one considerable insight into the boundary conditions for the Wheeler-DeWitt equation, which are implied by the functional-integral prescription for the wave function. As discussed in Sec. II, these are naturally imposed on three-geometries of very large volumes and vanishing volumes.

Consider the limit of small three-volumes first. If the limiting three-geometry is such that it can be embedded in flat space then the classical solution to (4.2) when $\Lambda>0$ is the four-sphere and remains so as the three-geometry shrinks to zero. The action approaches zero. The value of the wave function is therefore controlled by the behavior of the determinants governing the fluctuations away from the classical solution. These fluctuations are to be computed about a vanishingly small region of a space of constant positive curvature. In this limit one can neglect the curvature and treat the fluctuations as about a region of flat space. The determinant can therefore be evaluated by considering its behavior under a constant conformal rescaling of the four-metric and the boundary three-metric. The change in the determinant under a change of scale is given by the value of the associated ζ function at zero argument.[11]

Regular four-geome ries contain many hypersurfaces on which the three-volume vanishes. For example, consider the four-sphere of radius R embedded in a five-dimensional flat space. The three-surfaces which are the intersection of the four-sphere with surfaces of x^5 equals constant have a regular three-metric for $|x^5|<R$. The volume vanishes when $|x^5|=R$ at the north and south poles even though these are perfectly regular points of the four-geometry. One therefore would not expect the wave function to vanish at vanishing three-volume. Indeed, the three-volume will have to vanish somewhere if the topolo-

gy of the four-geometry is not that of a product of a three-surface with the real line or the circle. When the volume does vanish, the topology of the three-geometry will change. One cannot calculate the amplitude for such topology change from the Wheeler-DeWitt equation but one can do so using the Euclidean functional integral. We shall estimate the amplitude in some simple cases in Sec. VIII.

A qualitative discussion of the expected behavior of the wave function at large three-volumes can be given on the basis of the semiclassical approximation when $\Lambda>0$ as follows. The four-sphere has the largest volume of any real solution to (4.2). As the volume of the three-geometry becomes large one will reach three-geometries which no longer fit anywhere in the four-sphere. We then expect that the stationary-phase geometries become complex. The ground-state wave function will be a real combination of two expressions like (4.1) evaluated at the complex-conjugate stationary-phase four-geometries. We thus expect the wave function to oscillate as the volume of the three-geometry becomes large. If it oscillates without being strongly damped this corresponds to a universe which expands without limit.

The above considerations are only qualitative but do suggest how the behavior of the ground-state wave function determines the boundary conditions for the Wheeler-DeWitt equation. In the following we shall make these considerations concrete in a minisuperspace model.

V. MINISUPERSPACE MODEL

It is particularly straightforward to construct minisuperspace models using the functional-integral approach to quantum gravity. One simply restricts the functional integral to the restricted degrees of freedom to be quantized. In this and the following sections, we shall illustrate the general discussion of those preceding with a particularly simple minisuperspace model. In it we restrict the cosmological constant to be positive and the four-geometries to be spatially homogeneous, isotropic, and closed so that they are characterized by a single scale factor. An explicit metric in a useful coordinate system is

$$ds^2=\sigma^2[-N^2(t)dt^2+a^2(t)d\Omega_3{}^2]\,, \tag{5.1}$$

where $N(t)$ is the lapse function and $\sigma^2=l^2/24\pi^2$. For the matter degrees of freedom, we take a single conformally invariant scalar field which, consistent with the geometry, is always spatially homogeneous, $\phi=\phi(t)$. The wave function is then a function of only two variables:

$$\Psi=\Psi(a,\phi),\quad \Phi=\Phi(K,\tilde{\phi})\,. \tag{5.2}$$

Models of this general structure have been considered previously by DeWitt,[12] Isham and Nelson,[13] and Blyth and Isham.[14]

To simplify the subsequent discussion we introduce the following definitions and rescalings of variables:

$$\phi=\frac{\tilde{\phi}}{a}=\frac{\chi}{(2\pi^2\sigma^2)^{1/2}a}\,, \tag{5.3}$$

$$\Lambda=3\lambda/\sigma^2,\quad H^2=|\lambda|\,. \tag{5.4}$$

The Lorentzian action keeping χ and a fixed on the boundaries is

$$S^a=\tfrac{1}{2}\int dt\left[\frac{N}{a}\right]\left[\left[\frac{a}{N}\frac{da}{dt}\right]^2+a^2-\lambda a^4+\left[\frac{a}{N}\frac{d\chi}{dt}\right]^2-\chi^2\right]. \tag{5.5}$$

From this action the momenta π_a and π_χ conjugate to a and χ can be constructed in the usual way. The Hamiltonian constraint then follows by varying the action with respect to the lapse function and expressing the result in terms of a, χ, and their conjugate momenta. One finds

$$\tfrac{1}{2}(-\pi_a{}^2-a^2+\lambda a^4+\pi_\chi{}^2+\chi^2)=0\,. \tag{5.6}$$

The Wheeler-DeWitt equation is the operator expression of this classical constraint. There is the usual operator-ordering problem in passing from classical to quantum relations but its particular resolution will not be central to our subsequent semiclassical considerations. A class wide enough to remind oneself that the issue exists can be encompassed by writing

$$\pi_a{}^2=-\frac{1}{a^p}\frac{\partial}{\partial a}\left[a^p\frac{\partial}{\partial a}\right], \tag{5.7}$$

although this is certainly not the most general form possible. In passing from the classical constraint to its quantum operator form there is also the possibility of a matter-energy renormalization. This will lead to an additive arbitrary constant in the equation. We thus write for the quantum version of Eq. (5.6)

$$\frac{1}{2}\left[\frac{1}{a^p}\frac{\partial}{\partial a}\left[a^p\frac{\partial}{\partial a}\right]-a^2+\lambda a^4-\frac{\partial^2}{\partial\chi^2}+\chi^2-2\epsilon_0\right]\times\Psi(a,\chi)=0\,. \tag{5.8}$$

A useful property stemming from the conformal invariance of the scalar field is that this equation separates. If we assume reasonable behavior for the function Ψ in the amplitude of the scalar field we can expand in harmonic-oscillator eigenstates

$$\Psi(a,\chi)=\sum_n c_n(a)u_n(\chi)\,, \tag{5.9}$$

where

$$\frac{1}{2}\left[-\frac{d^2}{d\chi^2}+\chi^2\right]u_n(\chi)=(n+\tfrac{1}{2})u_n(\chi)\,. \tag{5.10}$$

The consequent equation for the $c_n(a)$ is

$$\frac{1}{2}\left[-\frac{1}{a^p}\frac{d}{da}\left[a^p\frac{dc_n}{da}\right]+(a^2-\lambda a^4)c_n\right]=(n+\tfrac{1}{2}-\epsilon_0)c_n\,. \tag{5.11}$$

For small a this equation has solutions of the form

$$c_n\approx\text{constant},\quad c_n\approx a^{1-p} \tag{5.12}$$

[if p is an integer there may be a $\log(a)$ factor]. For large a the possible behaviors are

$$c_n \sim a^{-(p/2+1)} \exp(\pm \tfrac{1}{3} i H a^3) . \quad (5.13)$$

To construct the solution of Eq. (5.11) which corresponds to the ground state of the minisuperspace model we turn to our Euclidean functional-integral prescription. As applied to this minisuperspace model, the prescription of Sec. III for $\Psi_0(a_0,\chi_0)$ would be to sum $\exp(-I[g,\phi])$ over those Euclidean geometries and field configurations which are represented in the minisuperspace and which satisfy the ground-state boundary conditions. The geometrical sum would be over compact geometries of the form

$$ds^2=\sigma^2[d\tau^2+a^2(\tau)d\Omega_3^2] \quad (5.14)$$

for which $a(\tau)$ matches the prescribed value of a_0 on the hypersurface of interest. The prescription for the matter field would be to sum over homogeneous fields $\chi(\tau)$ which match the prescribed value χ_0 on the surface and which are regular on the compact geometry. Explicitly we could write

$$\Psi_0(a_0,\chi_0)=\int \delta a\, \delta\chi \exp(-I[a,\chi]) , \quad (5.15)$$

where, defining $d\eta=d\tau/a$, the action is

$$I=\tfrac{1}{2}\int d\eta \left[-\left(\frac{da}{d\eta}\right)^2 -a^2+\lambda a^4+\left(\frac{d\chi}{d\eta}\right)^2+\chi^2\right] . \quad (5.16)$$

A conformal rotation [in this case of $a(\eta)$] is necessary to make the functional integral in (5.15) converge.[15]

An alternative way of constructing the ground-state wave function for the minisuperspace model is to work in the K representation. Here, introducing

$$k=\sigma K/9 \quad (5.17)$$

as a simplifying measure of K, one would have

$$\Phi_0(k_0,\chi_0)=\int \delta a\, \delta\chi \exp(-I^k[a,\chi]) . \quad (5.18)$$

The sum is over the same class of geometries and fields as in (5.15) except they must now assume the given value of k on the bounding three-surface. That is, on the boundary they must satisfy

$$k_0=\frac{1}{3a}\frac{da}{d\tau} . \quad (5.19)$$

The action I^k appropriate for holding k fixed on the boundary is

$$I^k=k_0 a_0^3+I \quad (5.20)$$

[cf. Eq. (3.6)]. Once $\Phi_0(k_0,\chi_0)$ has been computed, the ground-state wave function $\Psi_0(k_0,\chi_0)$ may be recovered by carrying out the contour integral

$$\Psi_0(a_0,\chi_0)=-\frac{1}{2\pi i}\int_C dk\, e^{ka_0^3}\Phi_0(k,\chi_0) , \quad (5.21)$$

where the contour runs from $-i\infty$ to $+i\infty$ to the right of any singularities of $\Phi_0(k_0,\chi_0)$.

From the general point of view there is no difference between computing $\Psi_0(a_0,\chi_0)$ directly from (5.15) or via the K representation from (5.21). In Sec. VI we shall calculate the semiclassical approximat on to $\Psi_0(a_0,\chi_0)$ both ways with the aim of advancing arguments that the rules of Sec. III define a wave function which may reasonably be considered as the state of minimal excitation and of displaying the boundary conditions under which Eq. (5.11) is to be solved.

VI. GROUND-STATE COSMOLOGICAL WAVE FUNCTION

In this section, we shall evaluate the ground-state wave function for our minisuperspace model and show that it possesses properties appropriate to a state of minimum excitation. We shall first evaluate the wave function in the semiclassical approximation from the steepest-descents approximation to the defining functional integral as described in Sec. IV. We shall then solve the Wheeler-DeWitt equation with the boundary conditions implied by the semiclassical approximation to obtain the precise wave function.

It is the exponent of the semiclassical approximation which will be most important in its interpretation. We shall calculate only this exponent from the extrema of the action and leave the determination of the prefactor [cf. Eq. (4.1)] to the solution of the differential equation. Thus, for example, if there were a single real Euclidean extremum of least action we would write for the semiclassical approximation to the functional integral in Eq. (5.15)

$$\Psi_0(a_0,\chi_0)\approx N e^{-I(a_0,\chi_0)} . \quad (6.1)$$

Here, $I(a_0,\chi_0)$ is the action (5.16) evaluated at the extremum configurations $a(\tau)$ and $\chi(\tau)$ which satisfy the ground-state boundary conditions spelled out in Sec. III and which match the arguments of the wave function on a fixed-τ hypersurface.

A. The matter wave function

A considerable simplification in evaluating the ground-state wave function arises from the fact that the energy-momentum tensor of an extremizing conformally invariant field vanishes in the compact geometries summed over as a consequence of the ground-state boundary conditions. One can see this because the compact four-geometries of the class we are considering are conformal to the interior of three-spheres in flat Euclidean space. A constant scalar field is the only solution of the conformally invariant wave equation on flat space which is a constant on the boundary three-sphere. The energy-momentum tensor of this field is zero. This implies that it is zero in any geometry of the class (5.14) because the energy-momentum tensor of a conformally invariant field scales by a power of the conformal factor under a conformal transformation.

More explicitly in the minisuperspace model we can show that the matter and gravitational functional integrals in (5.15) may be evaluated separately. The ground-state boundary conditions imply that geometries in the sum are conformal to half of a Euclidean Einstein-static universe, i.e., that the range of η is $(-\infty,0)$. The boundary conditions at infinite η are that $\chi(\eta)$ and $a(\eta)$ vanish. The boundary conditions at $\eta=0$ are that $a(0)$ and $\chi(0)$ match the arguments of the wave function a_0 and χ_0. Thus, not only does the action (5.16) separate into a sum of a gravitational part and a matter part, but the boundary conditions on the $a(\eta)$ and $\chi(\eta)$ summed over do not depend on one another. The matter and gravitational integrals can thus be evaluated separately.

Let us consider the matter integral first. In Eq. (5.16) the matter action is

$$I_M=\frac{1}{2}\int d\eta\left[\left(\frac{d\chi}{d\eta}\right)^2+\chi^2\right]. \tag{6.2}$$

This is the Euclidean action for the harmonic oscillator. Evaluation of the matter field integral in (5.15) therefore gives

$$\Psi_0(a_0,\chi_0)=e^{-\chi_0^2/2}\psi_0(a_0). \tag{6.3}$$

Here, $\psi_0(a)$ is the wave function for gravity alone given by

$$\psi_0(a_0)=\int \delta a \exp(-I_E[a]), \tag{6.4}$$

I_E being the gravitational part of (5.16). Equivalently we can write in the K representation

$$\Phi_0(k_0,\chi_0)=e^{-\chi_0^2/2}\phi_0(k_0), \tag{6.5}$$

where

$$\phi_0(k_0)=\int \delta a \exp(-I_E^k[a]). \tag{6.6}$$

$I_E{}^k[a]$ is related to I_E as in (5.20) and the sum is over $a(\tau)$ which satisfy (5.19) on the boundary. Equation (6.3) shows that as far as the matter field is concerned, $\Psi_0[a_0,\chi_0]$ is reasonably interpreted as the ground-state wave function. The field oscillators are in their state of minimum excitation—the ground state of the harmonic oscillator. We now turn to a semiclassical calculation of the gravitational wave function $\psi_0(a_0)$.

B. The semiclassical ground-state gravitational wave function

The integral in (6.4) is over $a(\tau)$ which represent [through (5.14)] compact geometries with three-sphere boundaries of radius a. The integral in (6.6) is over the same class of geometries except that the three-sphere boundary must possess the given value of k. The compact geometry which extremizes the gravitational action in these cases is a part of the Euclidean four-sphere of radius $1/H$ with an appropriate three-sphere boundary. In the case where the three-sphere radius is fixed on the boundary there are two extremizing geometries. For one the part of the four-sphere bounded by the three-sphere is greater than a hemisphere and for the other it is less. A careful analysis must therefore be made of the functional integral to see which of these extrema contributes to the semiclassical approximation. We shall give such an analysis below but first we show that the correct answer is achieved more directly in the K representation from (6.6) because there is a single extremizing geometry with a prescribed value of k on a three-sphere boundary and thus no ambiguity in constructing the semiclassical approximation to (6.6).

For three-sphere hypersurfaces of the four-sphere with an outward pointing normal, k ranges from approaching $+\infty$ for a surface encompassing a small region about a pole to approaching $-\infty$ for the whole four-sphere (see Fig. 1). More exactly, in the notation of Eq. (3.7)

$$k=\frac{H}{3}\cot\theta. \tag{6.7}$$

The extremum action is constructed through (5.20) with the integral in (5.16) being taken over that part of the four-sphere bounded by the three-sphere of given k. It is

$$I_E^k(k)=-\frac{1}{3H^2}\left[1-\frac{\kappa}{(\kappa^2+1)^{1/2}}\right], \tag{6.8}$$

where

$$k=\tfrac{1}{3}\kappa H. \tag{6.9}$$

The semiclassical approximation to (6.6) is now

$$\phi_0(k_0)\approx N\exp[-I_E^k(k_0)]. \tag{6.10}$$

The wave function $\psi_0(a_0)$ in the same approximation can be constructed by carrying out the contour integral

$$\psi_0(a_0)=-\frac{N}{2\pi i}\int_C dk\exp[ka_0^3-I_E{}^k(k)] \tag{6.11}$$

by the method of steepest descents. The exponent in the integrand of Eq. (6.11) is minus the Euclidean action for pure gravity with a kept fixed instead of k:

$$I_E(a)=-ka^2+I_E{}^k(k). \tag{6.12}$$

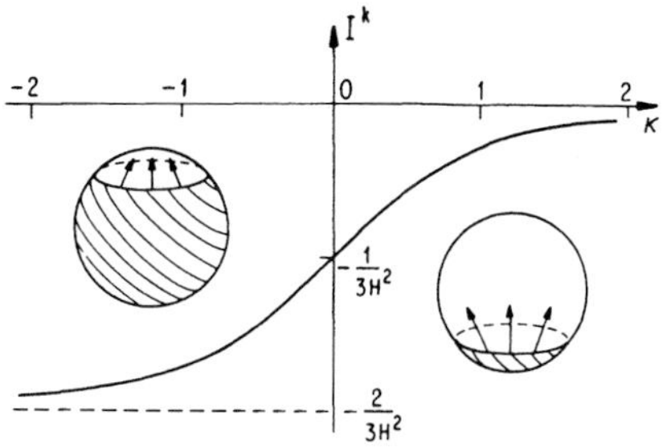

FIG. 1. The action I^k for the Euclidean four-sphere of radius $1/H$. The Euclidean gravitational action for the part of a four-sphere bounded by a three-sphere of definite K is plotted here as a function of κ (a dimensionless measure of K [Eq. (6.9)]). The action is that appropriate for holding K fixed on the boundary. The shaded regions of the inset figures show schematically the part of the four-sphere which fills in the three-sphere of given K used in computing the action. A three-sphere of given K fits in a four-sphere at only one place. Three-spheres with positive K (diverging normals) bound less than a hemisphere of four-sphere while those with negative K (converging normals) bound more than a hemisphere. The action tends to its flat-space value (zero) as K tends to positive infinity. It tends to the Euclidean action for all of de Sitter space as K tends to negative infinity.

To evaluate (6.11) by steepest descents we must find the extrema of Eq. (6.12). There are two cases depending on whether Ha_0 is greater or less than unity.

For $Ha_0 < 1$ the extrema of $I_E(k)$ occur at real values of k which are equal in magnitude and opposite in sign. They are the values of k at which a three-sphere of radius a_0 would fit into the four-sphere of radius $1/H$. That is, they are those values of k for which Eq. (6.7) is satisfied with $a_0{}^2=(\sin\theta/H)^2$. This is not an accident; it is a consequence of the Hamilton-Jacobi theory. The value of I_E at these extrema is

$$I_{\pm}=-\frac{1}{3H^2}[1\pm(1-H^2a_0{}^2)^{3/2}]\ , \qquad (6.13)$$

where the upper sign corresponds to $k<0$ and the lower to $k>0$, i.e., to filling in the three-sphere with greater than a hemisphere of the four-sphere or less than a hemisphere, respectively.

There are complex extrema of I_E but all have actions whose real part is greater than the real extrema described above. The steepest-descents approximation to the integral (6.11) is therefore obtained by distorting the contour into a steepest-descents path (or sequence of them) passing through one or the other of the real extrema. The two real extrema and the corresponding steepest-descents directions are shown in Fig. 2. One can distort the contour into a steepest-descents path passing through only one of them—the one with positive k as shown. The functional integral thus singles out a unique semiclassical approximation to $\psi_0(a_0)$ which is

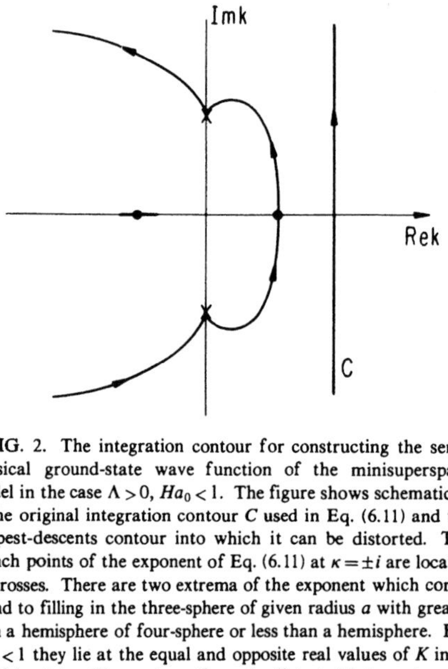

FIG. 2. The integration contour for constructing the semiclassical ground-state wave function of the minisuperspace model in the case $\Lambda>0$, $Ha_0<1$. The figure shows schematically the original integration contour C used in Eq. (6.11) and the steepest-descents contour into which it can be distorted. The branch points of the exponent of Eq. (6.11) at $\kappa=\pm i$ are located by crosses. There are two extrema of the exponent which correspond to filling in the three-sphere of given radius a with greater than a hemisphere of four-sphere or less than a hemisphere. For $Ha_0<1$ they lie at the equal and opposite real values of K indicated by dots. The contour C can be distorted into a steepest-descents contour through the extremum with positive K as shown. It cannot be distorted to pass through the extremum with negative K in the steepest-descents direction indicated. The contour integral thus picks out the extremum corresponding to less than a hemisphere of four-sphere (cf. Fig. 1) as the leading term in the semiclassical approximation.

$$\psi_0(a_0)\approx N\exp[-I_-(a_0)],\quad Ha_0<1\ , \qquad (6.14)$$

corresponding to filling in the three-sphere with less than a hemisphere's worth of four-sphere.

From Eq. (4.4) we recover the normalization factor N:

$$N=\exp(-\tfrac{1}{3}H^{-2})\ . \qquad (6.15)$$

Thus, for $Ha_0 \ll 1$

$$\psi_0(a_0)=\exp(\tfrac{1}{2}a_0{}^2-\tfrac{1}{3}H^{-2})\ . \qquad (6.16)$$

One might have thought that the extremum I_+, which corresponds to filling in the three-geometry with more than a hemisphere, would provide the dominant contribution to the ground-state wave function as $\exp(-I_+)$ is greater than $\exp(-I_-)$. However, the steepest-descents contour in the integral (6.7) does not pass through the extremum coresponding to I_+. This is related to the fact that the contour of integration of the conformal factor has to be rotated in the complex plane in order to make the path integral converge as we shall show below.

For $Ha_0>1$ there are no real extrema because we cannot fit a three-sphere of radius $a_0>1/H$ into a four-sphere of radius $1/H$. There are, however, complex extrema of smallest real action located at

$$k=\pm\frac{i}{3}H\left[1-\frac{1}{H^2a_0{}^2}\right]^{1/2}\ . \qquad (6.17)$$

It is possible to distort the contour in Eq. (6.11) into a steepest-descents contour passing through both of them as shown in Fig. 3. The resulting wave function has the form

$$\psi_0(a_0)=2\cos\left[\frac{(H^2a_0{}^2-1)^{3/2}}{3H^2}-\frac{\pi}{4}\right]\ ,\quad Ha_0>1 \qquad (6.18)$$

or for $Ha_0 \gg 1$

$$\psi_0(a_0)\approx e^{+iHa_0{}^3/3}+e^{-iHa_0{}^3/3}\ . \qquad (6.19)$$

The semiclassical approximation to the ground-state gravitational wave function $\psi_0(a)$ contained in Eqs. (6.16) and (6.19) may also be obtained directly from the functional integral (6.4) without passing through the k representation. We shall now sketch this derivation. We must consider explicitly the conformal rotation which makes the gravitational part of the action in (5.16) positive definite. The gravitational action is

$$I_E[a]=\tfrac{1}{2}\int d\eta\left[-\left[\frac{da}{d\eta}\right]^2-a^2+H^2a^4\right]\ . \qquad (6.20)$$

If one performed the functional integration

$$\psi_0(a_0)=\int \delta a(\eta)\exp(-I_E[a]) \qquad (6.21)$$

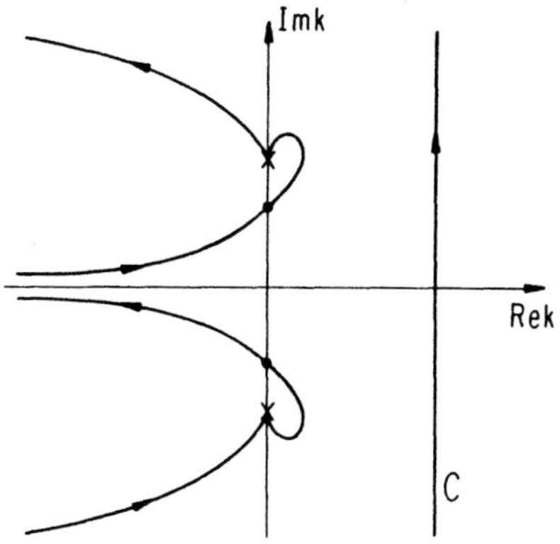

FIG. 3. The integration contour for constructing the semiclassical ground-state wave function of the minisuperspace model in the case $\Lambda > 0$, $Ha_0 > 1$. The figure shows schematically the original contour C used in Eq. (6.11) and the steepest-descents contour into which it can be distorted. The branch points of the exponent of Eq. (6.11) at $\kappa = \pm i$ are located by crosses. There are two complex-conjugate extrema of the exponent as indicated by dots and the contour C can be distorted to pass through both along the steepest-descents directions at 45° to the real axis as shown.

over real values of a, one would obtain a divergent result because the first term in (6.20) is negative definite. One could make the action infinitely negative by choosing a rapidly varying a. The solution to this problem seems to be to integrate the variable a in Eq. (6.21) along a contour that is parallel to the imaginary axis.[15] For each value of η, the contour of integration of a will cross the real axis at some value. Suppose there is some real function $\bar{a}(\eta)$ which maximizes the action. Then if one distorts the contour of integration of a at each value of η so that it crosses the real axis at $\bar{a}(\eta)$, the value of the action at the solution $\bar{a}(\eta)$ will give the saddle-point approximation to the functional integral (6.21), i.e.,

$$\psi_0(a_0) \approx N \exp(-I_E[\bar{a}(\eta)]) \ . \tag{6.22}$$

If there were another real function $\hat{a}(\eta)$ which extremized the action but which did not give its maximum value there would be a nearby real function $\hat{a}(\eta)+\delta a(\eta)$ which has a greater action. By choosing the contour of integration in (6.21) to cross the real a axis at $\hat{a}(\eta)+\delta a(\eta)$, one would get a smaller contribution to the ground-state wave function. Thus, the dominant contribution comes from the real function $\bar{a}(\eta)$ with the greatest value of the action.

It may be that there is no real $a(\eta)$ which maximizes the action. In this case the dominant contribution to the ground-state wave function will come from complex functions $a(\eta)$ which extremize the action. These will occur in complex-conjugate pairs because the wave function is real.

In the case of $Ha_0 < 1$, we have already seen that there are two real functions $a(\eta)$ which extremize the action and which correspond to less than or more than a hemisphere of the four-sphere. Their actions are I_- and I_+, respectively, given by (6.13). In fact, I_- is the maximum value of the action for real $a(\eta)$ and therefore gives the dominant contribution to the ground-state wave function. Thus, we again recover Eqs. (6.14) and (6.16). In the case of $Ha_0 > 1$, there is no maximum of the action for real $a(\eta)$. In this case the dominant contribution to the ground-state wave function comes from a pair of complex-conjugate $a(\eta)$ which extremize the action. Thus, we would expect an oscillatory wave function like that given by Eq. (6.19).

C. Ground-state solution of the Wheeler-DeWitt equation

The ground-state wave function must be a solution of the Wheeler-DeWitt equation for the minisuperspace model [Eqs. (5.8) or (5.11)]. The $\exp(-\chi^2/2)$ dependence of the wave function on the matter field deduced in Sec. VI A shows that in fact $\psi_0(a)$ must solve Eq. (5.11) with $n=0$. There are certainly solutions of this equation which have the large-a combination of exponentials required of the semiclassical approximation by Eq. (6.19) as a glance at Eq. (5.13) shows. In fact the prefactor in these asymptotic behaviors shows that the ground-state wave function will be *normalizable* in the norm

$$(\psi_0', \psi_0) = \int_0^\infty da\, a^p \bar{\psi}_0(a)\psi_0(a) \tag{6.23}$$

in which the Wheeler-DeWitt operator is Hermitian.

The Wheeler-DeWitt equation enables us to determine the prefactor in the semiclassical approximation from the standard WKB-approximation formulas. With $p=0$, for example, this would give when $Ha_0 > 1$

$$\psi_0(a_0) = 2(H^2 a_0^4 - a_0^2 + \epsilon_0 + \tfrac{1}{2})^{-1/4} \times \cos\left[\frac{(H^2 a_0^2 - 1)^{3/2}}{3H^2} - \frac{\pi}{4}\right] . \tag{6.24}$$

We could also solve the equation numerically. Figure 4 gives an example when $p=0$ and $\epsilon_0 = -\frac{1}{2}$. There we have assumed that the wave function vanishes at $a=0$. The dotted lines represent graphs of the prefactor in Eq. (6.24) and show that the semiclassical approximation becomes rapidly more accurate as Ha increases beyond 1. We shall return to an interpretation of these facts below.

D. Correspondence with de Sitter space

Having obtained $\psi_0(a)$, we are now in a position to assess its suitability as the ground-state wave function. Classically the vacuum geometry with the highest symmetry, hence minimum excitation, is de Sitter space—the surface of a Lorentz hyperboloid in a five-dimensional Lorentz-signatured flat spacetime. The properties of the wave function contained in Eqs. (6.16) and (6.19) are those one would expect to be semiclassically associated with this geometry. Sliced into three-spheres de Sitter space contains spheres only with a radius greater than $1/H$. Equation (6.16) shows that the wave function is an exponential-

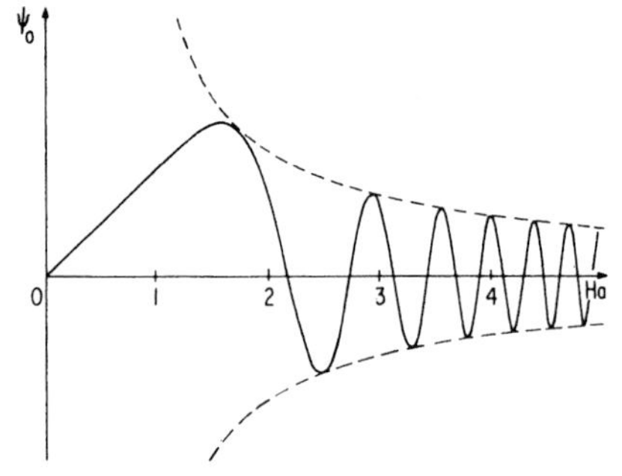

FIG. 4. A numerical solution of the Wheeler-DeWitt equation for the ground-state wave function $\psi_0(a)$. A solution of Eq. (5.11) is shown for $H=1$ in Planck units. We have assumed for definiteness $p=0$, $\epsilon_0=-\frac{1}{2}$, and a vanishing wave function at the origin. The wave function is damped for $Ha<1$ corresponding to the absence of spheres of radii smaller than H^{-1} in Lorentzian de Sitter space. It oscillates for $Ha>1$ decaying only slowly for large a. This reflects the fact that de Sitter space expands without limit. In fact, the envelope represented by the dotted lines is the distribution of three-spheres in Lorentzian de Sitter space: $[Ha(H^2a^2-1)^{1/2}]^{-1}$.

ly decreasing function with decreasing a for radii below that radius. Equation (6.24) shows the spheres of radius larger than $1/H$ are found with an amplitude which varies only slowly with the radius. This is a property expected of de Sitter space which expands both to the past and the future without limit. Indeed, tracing the origin of the two terms in (6.19) back to extrema with different signs of k one sees that one of these terms corresponds to the contracting phase of de Sitter space while the other corresponds to the expanding phase. The slow variation in the amplitude of the ground-state wave function reflects precisely the distribution of three-spheres in classical de Sitter space. Lorentzian de Sitter space is conformal to a finite region of the Einstein static universe

$$ds^2=\sigma^2a^2(\eta)(-d\eta^2+d\Omega_3^2)\,, \tag{6.25}$$

where $a(t)=(\cosh Ht)/H$ and $dt=ad\eta$. Three-spheres are evidently distributed uniformly in η in the Einstein static universe. The distribution of spheres in a in Lorentzian de Sitter space is therefore proportional to

$$[a(H^2a^2-1)^{1/2}]^{-1}\,. \tag{6.26}$$

This is the envelope of the probability distribution $a^p|\psi(a)|^2$ for spheres of radius a deduced from the semiclassical wave function and shown in Fig. 4. The wave function constructed from the Euclidean prescription of Sec. III appropriately reflects the properties of the classical vacuum solution of highest symmetry and is therefore reasonably called the ground-state wave function.

VII. EXCITED STATES

Our Universe does not correspond to the ground state of the simple minisuperspace model. It might be that the inclusion of more degrees of freedom in the model would produce a ground state which resembles our Universe more closely or it might be that we do not live in the ground state but in an excited state. Such excited states are not to be calculated by a simple path-integral prescription, but rather by solving the Wheeler-DeWitt equation with the boundary conditions that are required to maintain Hemiticity of the Hamiltonian operator between these states and the ground state. In this section, we shall construct the excited states for the minisuperspace model discussed in Sec. VI.

In the minisuperspace model where the spacelike sections are metric three-spheres all excitations in the gravitational degrees of freedom have been frozen out. We can study, however, excitations in the matter degrees of freedom. These are labeled by the harmonic-oscillator quantum number n as we have already seen [cf. Eq. (5.10)]. The issue then is what solution of Eq. (5.11) for $c_n(a)$ corresponds to this excited state. The equation can be written in the form of a one-dimensional Schrödinger equation

$$-\frac{1}{2a^p}\frac{d}{da}\left[a^p\frac{dc_n}{da}\right]+V(a)c_n=(n+\tfrac{1}{2}+\epsilon_0)c_n\,, \tag{7.1}$$

where

$$V(a)=\tfrac{1}{2}(a^2-\lambda a^4)\,. \tag{7.2}$$

At $a=0$ Eq. (7.1) will in general have two types of solutions one of which is more convergent than the other [cf. Eq. (5.12)]. The behavior for the ground state which corresponds to the functional-integral prescription could be deduced from an evaluation of the determinant in the semiclassical approximation as discussed in Sec. IV. Whatever the result of such an evalution, the solution must be purely of one type or the other in order to ensure the Hermiticity of the Hamiltonian constraint. The same requirement ensures a similar behavior for the excited-state solutions. In the following by "regular" solutions we shall mean those conforming to the boundary conditions arising from the functional-integral prescription. The exact type will be unimportant to us.

The potential $V(a)$ is a barrier of height $1/(4\lambda)$. At large a, the cosmological-constant part of the potential dominates and one has solutions which are linear combinations of the oscillating functions in (5.13). As we have already seen in the analysis of the ground state, the two possibilities correspond to a de Sitter contraction and a de Sitter expa sion. With either of these asymptotic behaviors, a wave p cket constructed by superimposing states of different n to produce a wave function with narrow support about some mean value of the scalar field would show this mean value increasing as one moved from large to small a.

Since each of the asymptotic behaviors in (5.13) is physically acceptable there will be solutions of (7.1) for all n. If, however, λ is small and n not too large, there are some values of n which are more important than others. These are the values which make the left-hand side of (7.1) at or close to those values of the energy associated with the metastable states (resonances) of the Schrödinger Hamiltonian on the right-hand side. To make this precise write

$$-\frac{1}{2}\frac{1}{a^p}\frac{d}{da}\left[a^p\frac{dc}{da}\right]+V(a)c=\epsilon c \ . \tag{7.3}$$

This is the zero angular momentum Schrödinger equation in $d=p+1$ dimensions for single-particle motion in the potential $V(a)$. Classically, for $\epsilon<1/(4\lambda)$ there are two classes of orbits: bound orbits with a maximum value of a and unbound orbits with a minimum value of a. Quantum mechanically there are no bound states. For discrete values of $\epsilon \ll 1/(4\lambda)$, however, there are metastable states. They lie near those values of ϵ which would be bound states if $\lambda=0$ and the barrier had infinite height. Since when $\lambda=0$ (7.3) is the zero angular momentum Schrödinger equation for a particle in a "radial" harmonic-oscillator potential in $d=p+1$ dimensions, these values are

$$\epsilon_N=2N+d/2, \quad N=0,1,2,\ldots \ . \tag{7.4}$$

For nonzero λ, if the particle has an energy near one of these values and much less than $1/(4\lambda)$ it can execute many oscillations inside the well but eventually it will tunnel out.

For the cosmological problem the classical Hamiltonian corresponding to (7.3) describes the evolution of homogeneous, isotropic, spatially closed cosmologies with radiation and a cosmological constant. The bound orbits correspond to those solutions for which the radiation density is sufficiently high that its attractive effect causes an expanding universe to recollapse before the repulsive effect of the cosmological constant becomes important. By contrast the unbound orbits correspond to de Sitter evolutions in which a collapsing universe never reaches a small enough volume for the increasing density of radiation to reverse the effect of the cosmological constant. There are thus two possible types of classical solutions. Quantum mechanically the Universe can tunnel between the two.

We can calculate the tunneling probability for small λ by using the usual barrier-penetration formulas from ordinary quantum mechanics. Let P be the probability for tunneling from inside the barrier to outside per transversal of the potential inside from minimum to maximum a. Then

$$P\approx e^{-B}, \tag{7.5}$$

where

$$B=2\int_{a_0}^{a_1} da[V(a)-\epsilon]^{1/2} \tag{7.6}$$

and a_0 and a_1 are the two turning points where $V(a)=\epsilon$. In the limit of $\epsilon \ll 1/(4\lambda)$ the barrier-penetration factor becomes

$$B=\frac{2}{3\lambda}=\frac{2}{3H^2} \ . \tag{7.7}$$

In magnitude this is just the total gravitational action for the Euclidean four-sphere of radius $1/H$ which is the analytic continuation of de Sitter space. This is familiar from general semiclassical results.[16]

Our own Universe corresponds to a highly excited state of the minisuperspace model. We know that the age of the Universe is about 10^{60} Planck times. The maximum expansion, assuming a radiation dominated model, is therefore at least of order $a_{\max}{}^2\approx 10^{120}$. A wave packet describing our Universe would therefore have to be superpositions of states of definite n, with n at least $\approx a_{\max}{}^2\approx 10^{120}$. As large as this number is, the dimensionless limit on the inverse cosmological constant is even larger. In order to have such a large radiation dominated Universe λ must be less than 10^{-120}. The probability for our Universe to tunnel quantum mechanically at the moment of its maximum expansion to a de Sitter-type phase rather than recollapse is $P\approx\exp(-10^{120})$. This is a very small number but of interest if only because it is nonzero.

VIII. TOPOLOGY

In the preceding sections we have considered the amplitudes for three-geometries with S^3 topology to occur in the ground state. The functional-integral construction of the ground-state wave function, however, permits a natural extension to calculate the amplitudes for other topologies. We shall illustrate this extension in this section with some simple examples in the semiclassical approximation.

There is no compelling reason for restricting the topologies of the Euclidean four-geometries which enter in the sum defining the ground-state wave function. Whatever one's view on this question, however, there must be a ground-state wave function for every topology of a three-geometry which can be embedded in a four-geometry which enters the sum. In the general case this will mean all possible three-topologies—disconnected as well as connected, multiply connected as well as simply connected. The general ground-state wave function will therefore have N arguments representing the possibility of N compact disconnected three-geometries. The functional-integral prescription for the ground-state wave function in the case of pure gravity would then read

$$\Psi_0[\partial M^{(1)},h_{ij}^{(1)},\ldots,\partial M^{(N)},h_{ij}^{(N)}]=\int \delta g \exp(-I_E[g]) \ , \tag{8.1}$$

where the sum is over all compact Euclidean four-geometries which have N disconnected compact boundaries $\partial M^{(i)}$ on which the induced three-metrics are $h_{ij}^{(i)}$. Since there is nothing in the sum which distinguishes one three-boundary from another the wave function must be symmetric in its arguments.

The wave function defined by (8.1) obeys a type of Wheeler-DeWitt equation in each argument but this is no longer sufficient to determine its form—in particular the correlations between the three-geometries. The functional integral is here the primary computational tool.

It is particularly simple to construct the semiclassical approximations to ground-state wave functions for those three-geometries with topologies which can be embedded in a compact Euclidean solution of the field equations. Consider for example the four-sphere. If the three-geometry has a single connected component and can be embedded in the four-sphere, then the extremal geometry at which the action is evaluated to give the semiclassical approximation is the smaller part of the four-sphere bounded by this three-geometry. The semiclassical ground-state wave function is

$$\Psi_0[h_{ij}] \approx N\Delta^{-1/2}[h_{ij}] \times \exp\left[2l^{-2}\left\{\int_{\partial M} dx^3 h^{1/2}K + \int_M d^4x\, g^{1/2}\Lambda\right\}\right] \tag{8.2}$$

where M is the smaller part of the four-sphere and K is the trace of the extrinsic curvature of the three-surface computed with outward-pointing normals. Since there is a large variety of topologies of three-surfaces which *can* be embedded in the four-sphere—spheres, toruses, etc.,— we can easily compute their associated wave functions. Of course, these are many interesting three-surfaces which cannot be so embedded and for which the extremal solution defining the semiclassical approximation is not part of the four-sphere. In general one would expect to find wave functions for arbitrary topologies since any three-geometry is cobordant to zero and therefore there is some compact four-manifold which has it as its boundary. The problem of finding solutions of the field equations on these four-manifolds which match the given three-geometry and are compact thus becomes an interesting one.

Similarly, the semiclassical approximation for wave functions representing N disconnected three-geometries are equally easily computed when the geometries can be embedded in the four-sphere. The extremal geometry defining the semiclassical approximation is then simply the four-sphere with the N three-geometries cut out of it. The symmetries of the solution guarantee that as far as the exponent of the semiclassical approximation is concerned, it does not matter where the three-geometries are cut out provided that they do not overlap. To give a specific example, we calculate the amplitude for two disconnected three-spheres of radius $a_{(1)}$ and $a_{(2)}$ assuming $a_{(1)} < a_{(2)} < H^{-1}$. One possible extremal geometry is two disconnected portions of a four-sphere attached to the two three-spheres. This gives a product wave function with no correlation. Another extremal geometry is the smaller half of the four-sphere bounded by the spheres of radius $a_{(2)}$ with the portion interior to a sphere of radius $a_{(1)}$ removed. This gives an additional contribution to the wave function which expresses the correlation between the spheres. The correlated part in the semiclassical approximation is

$$\Psi_0^c(a_{(1)},a_{(2)}) = N\Delta^{-1/2}(a_{(1)},a_{(2)}) \times \exp\left\{\frac{1}{3H^2}[-(1-H^2a_{(2)}{}^2)^{3/2} + (1-H^2a_{(1)}{}^2)^{3/2}]\right\}. \tag{8.3}$$

While the exponent is simple, the calculation of the determinant is now more complicated—it does not factor.

Equation (8.3) shows that the amplitude to have two correlated three-spheres of radius $a_{(1)} < a_{(2)} < H^{-1}$ is smaller than the amplitude to have a single three-sphere of radius $a_{(2)}$. In this crude sense topological complexity is suppressed. The amplitude for the Universe to bifurcate is of the order $\exp[-1/(3H^2)]$—a very large factor.

IX. CONCLUSIONS

The ground-state wave function for closed universes constructed by the Euclidean functional-integral prescription put forward in this paper can be said to represent a state of minimal excitation for these universes for two reasons. First, it is the natural generalization to gravity of the Euclidean functional integral for the ground-state wave function of flat-spacetime field theories. Second, when the prescription is applied to simple minisuperspace models, it yields a semiclassical wave function which corresponds to the classical solution of Einstein's equations of highest spacetime symmetry and lowest matter excitation.

The advantages of the Euclidean function-integral prescription are many but perhaps three may be singled out. First it is a complete prescription for the wave function. It implies not only the Wheeler-DeWitt equation but also the boundary conditions which determine the ground-state solution. The requirement of Hermiticity of the Wheeler-DeWitt operator extends these boundary conditions to the excited states as well.

A second advantage of this prescription for the ground-state wave function is common to all functional-integral formulations of quantum amplitudes. They permit the direct and explicit calculation of the semiclassical approximation. At the current stage of the development of quantum gravity where qualitative understanding is more important than precise numerical results, this is an important advantage. It is well illustrated by our minisuperspace model in which we were able to calculate semiclassically the probability of tunneling between a universe doomed to end in a big crunch and an eternal de Sitter expansion.

A final advantage of the Euclidean functional-integral prescription for the ground-state wave function is that it naturally generalizes to permit the calculation of amplitudes not usually considered in the canonical theory. In particular, we have been able to provide a functional-integral prescription for amplitudes for the occurrence of three-geometries with multiply connected and disconnected topologies in the ground state. In the semiclassical approximation we have been able to evaluate simple examples of such amplitudes.

The Euclidean functional-integral prescription sheds light on one of the fundamental problems of cosmology: the singularity. In the classical theory the singularity is a place where the field equations, and hence predictability, break down. The situation is improved in the quantum theory. An analogous improvement occurs in the problem of an electron orbiting a proton. In the classical theory there is a singularity and a breakdown of predictability when the electron is at the same position as the proton. However, in the quantum theory there is no singularity or breakdown. In an s-wave state, the amplitude for the electron to coincide with the proton is finite and nonzero, but the electron just carries on to the other side. Similarly, the amplitude for a zero-volume three-sphere in our minisuperspace model is finite and nonzero. One might interpret this as implying that the universe could continue through the singularity to another expansion period, although the classical concept of time would break down so that one could not say that the expansion happened after the contraction.

The ground-state wave function in the simple minisuperspace model that we have considered with a conformally invariant field does not correspond to the quantum state of the Universe that we live in because the matter wave function does not oscillate. However, it seems that this may be a consequence of using only zero rest mass fields and that the ground-state wave function for a universe with a massive scalar field would be much more complicated and might provide a model of quantum state of the observed Universe. If this were the case, one would have solved the problem of the initial boundary conditions of the Universe: the boundary conditions are that it has no boundary.[3]

ACKNOWLEDGMENTS

The authors are grateful for the hospitality of the Institute of Theoretical Physics, Santa Barbara, California, where part of this work was carried out. The research of one of us (J.H.) was supported in part by NSF Grants Nos. PHY81-07384 and PHY80-26043.

---

[1]See, e.g., R. P. Feynman, Rev. Mod. Phys. 20, 367 (1948); R. P. Feynman and A. R. Hibbs, *Quantum Mechanics and Path Integrals* (McGraw-Hill, New York, 1965) for discussions of quantum mechanics from this point of view.

[2]B. S. DeWitt, Phys. Rev. 160, 1113 (1967); J. A. Wheeler, in *Battelle Rencontres*, edited by C. DeWitt and J. A. Wheeler (Benjamin, New York, 1968).

[3]S. W. Hawking, in *Astrophysical Cosmology*, Pontificia Academiae Scientarium Scripta Varia, 48 (Pontificia Academiae Scientarium, Vatican City, 1982).

[4]For related ideas, see A. Vilenkin, Phys. Lett. 117B, 25 (1982); Phys. Rev. D 27, 2848 (1983).

[5]The connection between the canonical and functional-integral approaches to quantum gravity has been extensively discussed. See, in particular, H. Leutwyler, Phys. Rev. 134, B1155 (1964); L. Faddeev and V. Popov, Usp. Fiz. Nauk. 111, 427 (1973) [Sov. Phys. Usp. 16, 777 (1974)]; E. S. Fradkin and G. A. Vilkovisky, CERN Report No. TH-2332, 1977 (unpublished).

[6]For reviews of the canonical theory, see K. Kuchar, in *Quantum Gravity 2*, edited by C. Isham, R. Penrose, and D. W. Sciama (Clarendon, Oxford, 1981); A Hanson, T. Regge, and C. Teitelboim, *Constrained Hamiltonian Systems* (Academia Nazionale dei Lincei, Rome, 1976); K. Kuchar, in *Relativity, Astrophysics and Cosmology*, edited by W. Israel (Reidel, Dordrecht, 1973).

[7]J. M. York, Phys. Rev. Lett. 28, 1082 (1972).

[8]J. Milnor, *Morse Theory* (Princeton University Press, Princeton, New Jersey, 1962).

[9]For example, J. B. Hartle (unpublished).

[10]See, e.g., G. W. Gibbons and M. J. Perry, Nucl. Phys. B146, 90 (1978).

[11]S. W. Hawking, Commun. Math. Phys. 55, 133 (1977).

[12]B. DeWitt, Phys. Rev. 160, 1113 (1967).

[13]C. Isham and J. E. Nelson, Phys. Rev. D 10, 3226 (1974).

[14]W. E. Blyth and C. Isham, Phys. Rev. D 11, 768 (1975).

[15]G. W. Gibbons, S. W. Hawking, and M. J. Perry, Nucl. Phys. B138, 141 (1978).

[16]See, e.g., S. Coleman, Phys. Rev. D 15, 2929 (1977).

Nuclear Physics B239 (1984) 257–276

THE QUANTUM STATE OF THE UNIVERSE

S.W. HAWKING

Department of Applied Mathematics and Theoretical Physics, Silver Street, Cambridge CB3 9EW, England

Received 28 December 1983

The quantum state of the universe is determined by the specification of the class of metrics and matter field configurations that are summed over in the path integral. The only natural choice of this class seems to be compact euclidean (i.e. positive definite) metrics and matter fields that are regular on them. This choice incorporates the idea that the universe is completely self-contained and has no boundary or asymptotic region. I show that in a simple "minisuperspace" model this boundary condition leads to a wave function which can be interpreted as a superposition of quantum states which are peaked around a family of classical solutions of the field equations. These solutions are non-singular and represent oscillating universes with a long inflationary period. They could be a good description of the observed universe. I also show that the features of the minisuperspace model that give rise to such a wave function are also present in models that contain all the degrees of freedom of the gravitational and matter fields.

1. Introduction

The task of theoretical physics is to construct a mathematical model of the universe which agrees with all the observations made so far and which predicts the results of future observations. This model usually consists of two parts:

(1) Local laws which govern the physical fields in the model. In classical physics, these laws are normally expressed as differential equations which can be derived from an action I. In quantum physics the laws can be obtained from a path integral over all field configurations weighted with $\exp(iI)$.

(2) Boundary conditions which pick out one particular state from among the set of those allowed by the local laws. The classical state can be specified by boundary conditions for the differential equations and the quantum state can be determined by asymptotic conditions on the class C of field configurations that are summed over in the path integral.

A great deal of work has been done on the first part, particularly in the last twenty years. Although we do not yet have a complete, consistent field theory, we probably know most of the features that it should have. The outstanding remaining problem is to construct a quantum theory of gravity. The standard approach to quantizing general relativity by a perturbation expansion produces infinities which are "unrenormalizable", that is, they cannot be removed by redefining the quantities in the action. There are three possible solutions to this problem:

(i) It may be possible to find a theory, such as $N=8$ supergravity, in which all the infinities cancel to all orders in the perturbation expansion.

(ii) The infinities may simply be a result of the perturbation expansion which is known to break down at high energies. Gravitational effects may cut off the high energies and give rise to a finite theory.

(iii) It may be necessary to add higher derivative terms to the gravitational action. This is known to give a renormalizable theory but there are problems with negative energies, unitarity and runaway solutions.

It seems likely that the solution to the problems of quantum gravity lies with one of the three alternatives listed above. My personal opinion is that it is probably the second possibility: the path integral over the conformal factor seems to damp out the contributions of metrics with a lot of conformal curvature. However, whichever of the three solutions is the correct one, it is likely that we will know before too long.

By contrast with all the work on the local laws, very little has been done on the boundary conditions. Indeed, many people would claim that the boundary conditions are not part of physics but belong to metaphysics or religion. They would claim that nature had complete freedom to start the universe off any way it wanted. That may be so, but it could also have made it evolve in a completely arbitrary and random manner. Yet all the evidence is that it evolves in a regular way according to certain laws. It would therefore seem reasonable to suppose that there are also laws governing the boundary conditions. One might also argue that the observed universe is so complicated that it could not possibly have arisen from some simple boundary condition. I do not think that this is a valid objection: the laws of quantum electrodynamics are simple but they give rise to all the complexities of chemistry and biology. I shall show in a simplified model that a simple boundary condition for the universe can still give rise to very complicated behaviour. Indeed, a number of classically allowed behaviours occur with relative probabilities determined by the boundary conditions.

It might seem premature to speculate about the boundary conditions when we are not yet sure of the exact form of the local laws. However, we know that the large-scale structure of the universe is determined by gravity and we are fairly sure that gravity is described by general relativity or some variant, such as supergravity or higher derivative theories. We also believe that quantum theory can be formulated in terms of path integrals. The problem of boundary conditions then becomes one of specifying the class of spacetime metrics and matter field configurations that are summed over in the path integral. I shall adopt the point of view that the path integral should be evaluated in the euclidean regime, that is, over positive definite metrics and matter fields in these metrics. There are then only two natural choices for the class of metrics:

(a) compact metrics and matter fields that are regular on them;

(b) non-compact metrics that are asymptotic to metrics of maximal symmetry, i.e. to flat euclidean space or to euclidean anti-de Sitter space, and matter fields that are asymptotically zero.

Boundary conditions of type (b) give rise to the usual vacuum state that one uses in scattering problems in which one sends particles in from infinity and measures what comes back out to infinity. However, I shall show in sect. 3 that they are not suitable as boundary conditions for the whole universe. This leaves only boundary conditions of type (a). They incorporate the idea that the universe is completely self-contained and has no boundaries, either at singularities at finite distance or at infinity. One could paraphrase them as: "the boundary conditions of the universe are that it has no boundary".

Boundary conditions of type (a) were proposed in [1] but at that time I did not know how to calculate their consequences. The necessary formalism was developed in collaboration with Hartle in [2] and is summarised in sect. 2. The formalism is applied in sect. 4 to a simple "minisuperspace" toy model of a homogeneous isotropic universe containing a massive scalar field. It is shown that the boundary conditions (a) give rise to a wave function that can be interpreted as a superposition of quantum states peaked around different classical solutions representing oscillating universes. One might object that the quantum state should be peaked around only a single classical solution: after all, we observe only one universe. The answer is that if one had a quantum state that was peaked around a particular classical solution, that quantum state would describe not only the universe but also any observers who measured its properties. Suppose one now had a different quantum state that was peaked around a different classical solution of the field equations. That quantum state would also contain observers who would measure that properties of the second solution. Then consider a quantum state that was the sum of the two previous states. Because of the linearity of quantum mechanics, there would be no interference; observers in the first state would measure the properties of the first solution and observers in the second state would measure those of the second. The classical solutions that correspond to the wave function of the model can have a long period of exponential expansion and could be a reasonable description of the universe that we live in. They would have most of the desirable features of the inflationary model [3–7] and the advantage of having been derived from a definite proposal for the boundary conditions of the universe.

One could easily object to the minisuperspace toy model on the grounds that it truncated the infinite number of degrees of freedom of the gravitational and matter fields down to a finite number and that it ignored the problems of the divergences of quantum gravity. However, I show in sect. 5 that the features of the minisuperspace model that lead to a wave function of the right form would also be present in a model that did not restrict the degrees of freedom provided that it contains fields of non-zero rest-mass. The quantum corrections to the gravitational action will produce terms of the form R^2 and $C_{\mu\nu\rho\lambda}\, C^{\mu\nu\rho\lambda}$. These will be finite in the case of possibilities (i) and (ii) and infinite but renormalized in the case of possibility (iii). Such terms can act like massive scalar or spin-2 fields. It will be shown in a future paper [8] that the R^2 term at least leads to a minisuperspace model which is very similar to that with a massive scalar field.

These results and the fact that there does not seem to be any other natural candidate leads me, at least, to believe that the quantum state of the universe is determined by a path integral over compact positive definite metrics. If this is correct, the second part of the problem of constructing a mathematical model of the universe will have been solved. It would then only remain to determine the exact form of the local laws and we should, in principle, be able to predict the probability of anything in the universe. In practice of course, the equations will be much too complicated to solve in any but very simple situations.

2. Quantum gravity

In certain special spacetime metrics, such as those of the Schwarzschild or de Sitter solutions, it is possible to define a new time-coordinate $\tau = it$ and to change the metric from a lorentzian signature $(-+++)$ to a euclidean or positive definite signature $(++++)$. However, this is not possible in general: a general lorentzian metric will not have a section in the complexified spacetime manifold on which the metric is real and positive definite. Similarly, a general positive definite metric will not have a section on which the metric is real and lorentzian. In this respect, the spacetime metric behaves like other fields in ordinary quantum field theory in flat space: a Yang–Mills field which is real in Minkowski space will not, in general, be real when analytically continued to euclidean space and a field which is real in euclidean space will not be real when analytically continued to Minkowski space. The point about the euclidean approach to quantum field theory in flat space is not that individual configurations in euclidean space correspond with configurations in Minkowski space, but that the path integral over all real field configurations in euclidean space is equivalent, in the sense of contour integration, to the path integral over all real field configurations in Minkowski space. If this works for other quantum fields, it seems reasonable to suppose that it works for gravity also. I shall therefore assume that the quantum theory of the universe can be derived from a path integral over euclidean, i.e. positive definite, metrics and matter fields which are regular on the manifolds defined by these metrics. I shall show that the Wheeler–DeWitt equation, the analogue of the Schrödinger equation, is the same whether it is derived from a path integral over euclidean or lorentzian metrics but that the wave function of the universe itself determines whether it corresponds to a euclidean or a lorentzian geometry in the classical limit.

As in refs. [2, 9], my starting point is the assumption that the probability for a 4-metric $g_{\mu\nu}$ and a matter field configuration ϕ is proportional to

$$\exp(-\tilde{I}[g_{\mu\nu}, \phi]) , \tag{2.1}$$

where $\tilde{I}$ is the euclidean action

$$\tilde{I}[g_{\mu\nu}, \phi] = \frac{m_{\mathrm{p}}^2}{16\pi}\left(-\int_{\partial \mathrm{M}} 2Kh^{1/2}\,\mathrm{d}^3x - \int_{\mathrm{M}}\left(R - 2\Lambda + \frac{16\pi}{m_{\mathrm{p}}^2} L(g_{\mu\nu}, \phi)\right) g^{1/2}\,\mathrm{d}^4x\right), \tag{2.2}$$

where h_{ij} is the 3-metric on the boundary ∂M and K is the trace of the second fundamental form of the boundary. The surface term in the action is necessary because the curvature scalar R contains second derivatives of the metric [10–12]. The physics of the universe is governed by probabilities of the form (2.1) for all 4-metrics $g_{\mu\nu}$ and matter field configurations belonging to a certain class C. The specification of the class C determines the quantum state of the universe. This will be considered further in sect. 3.

In practice, one is normally interested in the probability, not of the entire 4-metric, but of a more restricted set of observables. Such a probability can be derived from the basic probability (2.1) by integrating over the unobserved quantities. In cosmology, one is concerned with observables, not at infinity, but in some finite region in the interior of the 4-geometry. A particularly important case is the probability $P[h_{ij}, \phi]$ of finding a closed compact 3-submanifold S which divides the 4-manifold M into two parts $\mathrm{M}_\pm$ and on which the induced 3-metric is h_{ij} and the matter field configuration is ϕ:

$$P[h_{ij}, \phi] = \int \mathrm{d}[g_{\mu\nu}]\,\mathrm{d}[\phi] \exp(-\tilde{I}[g_{\mu\nu}, \phi]), \tag{2.3}$$

where the integral is over all 4-metrics and matter field configurations belonging to the class C which contain a 3-submanifold S on which the induced 3-metric is h_{ij} and the matter field configuration is ϕ. This probability can be factorized into the product of two amplitudes or wave functions $\Psi_\pm[h_{ij}, \phi]$, $P[h_{ij}, \phi] = \Psi_+[h_{ij}, \phi]\Psi_-[h_{ij}, \phi]$, where

$$\Psi_\pm[h_{ij}, \phi] = \int_{\mathrm{C}_\pm} \mathrm{d}[g_{\mu\nu}]\,\mathrm{d}[\phi] \exp(-\tilde{I}[g_{\mu\nu}, \phi]). \tag{2.4}$$

The path integral is over the classes $\mathrm{C}_\pm$ of 4-metrics and matter field configurations on $\mathrm{M}_\pm$ which agree with the given 3-metric h_{ij} and matter field configuration ϕ on S. For example, if the class C consisted of asymptotically euclidean metrics, then C_+ would consist of 4-metrics and matter field configurations on compact manifolds M_+ with boundary S and C_- would consist of metrics and fields on asymptotically euclidean manifolds M with inner boundary S. In this case, the two wave functions $\Psi_\pm$ will be different from each other but, if the class C consists of compact metrics, they will be the same.

One can regard $\Psi_\pm(h_{ij}, \phi]$ as the "wave functions of the universe" [2]. Note that they do not depend on time or, more generally, on the position of the 3-manifold S in the 4-manifold M. In general, there is no invariant way to specify this. In fact, the 3-metric h_{ij} usually determines the position of S in M or determines it up to a finite ambiguity. In what follows I shall drop the subscripts $\pm$ on the wave function Ψ.

The wave function $\Psi[h_{ij}, \phi]$ obeys a functional differential equation, the Wheeler–DeWitt equation, which is the analogue of the Schrödinger equation. This can be derived from the path integral definition (2.4) as follows. In the neighbourhood of the surface S, one can introduce a time-coordinate τ which is constant on S so that the metric takes the standard 3+1 form:

$$ds^2 = (N^2 - N_i N^i)\, d\tau^2 + 2N_i\, dx^i\, d\tau + h_{ij}\, dx^i\, dx^j . \qquad (2.5)$$

The action then becomes

$$\tilde{I}[g_{\mu\nu}, \phi] = \frac{m_P^2}{16\pi} \int d^3x\, d\tau\, h^{1/2} N[K_{ij}K^{ij} - K^2 - {}^3R(h_{ij}) + 2\Lambda - 16\pi m_P^{-2} L(\phi)] , \qquad (2.6)$$

where K_{ij} is the second fundamental form

$$K_{ij} = \frac{1}{N}\left[\frac{1}{2}\frac{\partial h_{ij}}{\partial \tau} + N_{(i|j)}\right] \qquad (2.7)$$

and a stroke and ${}^3R$ denote the covariant derivative and scalar curvature constructed from the 3-metric h_{ij}.

The functional integral defining the wave function contains an integral over N. By varying N at the surface one pushes it forward or backward in time. Since the wave function does not depend on time one must have

$$0 = \int d[g_{\mu\nu}]\, d[\phi] \left[\frac{\delta \tilde{I}}{\delta N}\right] \exp(-\tilde{I}[g_{\mu\nu}, \phi]) . \qquad (2.8)$$

Classically the field equation $H \equiv \delta \tilde{I}/\delta N = 0$ is the hamiltonian constraint for general relativity. It is

$$H = -\frac{m_P^2 h^{1/2}}{8\pi}(K^{ij}K_{ij} - K^2 + {}^3R - 2\Lambda + 8\pi m_P^{-2} T_{nn}) = 0 , \qquad (2.9)$$

where T_{nn} is the euclidean stress energy tensor of the matter field projected in the direction normal to the surface. One can substitute the expression (2.9) for $\delta\tilde{I}/\delta N$ in (2.8). The second fundamental form K_{ij} can be expressed in terms of the functional operator $\delta/\delta h_{ij}$:

$$\frac{\delta \tilde{I}}{\delta h_{ij}} = \frac{m_P^2}{16\pi} h^{1/2}(K_{ij} - h_{ij}K) . \qquad (2.10)$$

This identity can be derived by varying the metric in the definition of the action (2.2). The time derivative of the variation of the metric can be removed by integrating by parts in the normal manner. This produces a surface term which is δh_{ij} times the right-hand side of (2.10). Similarly, the time derivatives of the matter fields ϕ can be expressed in terms of the operator $\delta/\delta\phi$. Using these substitutions in (2.8), one obtains the Wheeler-DeWitt equation

$$\left[-G_{ijkl}\frac{\delta^2}{\delta h_{ij}\delta h_{kl}}-h^{1/2}\left({}^3R(h)-2\Lambda+8\pi m_{\rm p}^{-2}T_{\rm nn}\left(\frac{\delta}{\delta\phi},\phi\right)\right)\right]\Psi[h_{ij},\phi]=0\,, \tag{2.11}$$

where G_{ijkl} is the metric on superspace, the space of all 3-metrics h_{ij}:

$$G_{ijkl}=\tfrac{1}{2}h^{1/2}(h_{ik}h_{jl}+h_{il}h_{jk}-h_{ij}h_{kl})\,. \tag{2.12}$$

There is a question of the factor ordering in the functional differential equation or, equivalently, of the measure in the path integral. This will not be important in the situation that I shall consider. The Wheeler–DeWitt equation is the same as the equation one would have obtained had one started out with a path integral over lorentzian metrics. One can also consider variations within the 3-surface S generated by the "Shift" vector N_i:

$$0=\int\frac{\delta\tilde{I}}{\delta N_i}\exp\left(-\tilde{I}[g_{\mu\nu},\phi]\right)\,. \tag{2.13}$$

This equation implies that the wave function is invariant under diffeomorphisms, i.e. that Ψ is a functional of the 3-geometry and not of the particular 3-metric h_{ij}.

The metric $-G_{ijkl}$ on superspace, the space of 3-geometries, has a signature $(+-----)$. The Wheeler–DeWitt equation can therefore be thought of as a hyperbolic equation on superspace with $h^{1/2}$ as the time coordinate. I shall show that the semiclassical approximation to the path integral gives the boundary conditions for the Wheeler–DeWitt equation at small $h^{1/2}$. The Wheeler–DeWitt equation can then be solved to give the values of the wave function at larger values of $h^{1/2}$. In a lorentzian metric, $h^{1/2}=0$ would be a singularity. However, this is not necessarily the case in a euclidean metric as can be seen by considering the example of a 4-sphere of radius R embedded in flat 5-dimensional space: a surface $|x^5|<R$ intersects the 4-sphere in a 3-sphere of non-zero radius. However, when $|x^5|=R$, the 3-sphere shrinks to zero radius but there is no singularity of the 4-geometry. Indeed, $h^{1/2}$ has to go to zero if the topology of the 3-surfaces is to change.

By its construction, the wave function $\Psi[h_{ij},\phi]$ vanishes for 3-metrics h_{ij} which are not positive definite, i.e. for which the determinant h of the metric is negative. This means that h_{ij} is not a quantum observable with an unrestricted range and that in many cases it is more convenient to replace it by $\tilde{h}_{ij}$, the 3-metric up to a conformal factor, and $Km_{\rm p}^2/12\pi$, the momentum conjugate to $h^{1/2}$. One can then define a wave function Φ in this representation by

$$\Phi[\tilde{h}_{ij},K,\phi]=\int d[g_{\mu\nu}]\,d[\phi]\exp\left(-I^k[g_{\mu\nu},\phi]\right)\,, \tag{2.14}$$

where $I^k[g_{\mu\nu},\phi]$ is the action appropriate to the situation in which K and $\tilde{h}_{ij}$ are kept fixed on the boundary rather than h_{ij}. It differs from the action $\tilde{I}[g_{\mu\nu}, \phi]$ in that the surface term has a different coefficient, $\frac{2}{3}K$ rather than $2K$. The wave function $\Phi[\tilde{h}_{ij}, K, \phi]$ is the Laplace transform of $\Phi[h_{ij}, \phi]$:

$$\Phi[\tilde{h}_{ij}, K, \phi] = \int_0^\infty \mathrm{d}[h^{1/2}] \exp\left[-\frac{m_\mathrm{P}^2}{12\pi}\int \mathrm{d}^3x\, h^{1/2}K\right]\Psi[h_{ij}, \phi]. \qquad (2.15)$$

Similarly, $\Psi[h_{ij}, \phi]$ is the inverse Laplace transform of $\Phi[\tilde{h}_{ij}, K, \phi]$.

$$\Psi[h_{ij}, \phi] = \int_\Gamma \mathrm{d}\left[\frac{m_\mathrm{P}^2}{24\pi i}K\right] \exp\left[\frac{m_\mathrm{P}^2}{12\pi}\int \mathrm{d}^3x\, h^{1/2}K\right]\Phi[\tilde{h}_{ij}, K, \phi], \qquad (2.16)$$

where for each point of S the contour Γ runs from $-i\infty$ to $+i\infty$ to the right of any singularities of $\Phi[\tilde{h}_{ij}, K, \phi]$ in the complex K plane. Providing that $\Phi[\tilde{h}_{ij}, K, \phi]$ does not diverge exponentially for large Re K, this choice of contour will ensure that $\Psi[h_{ij}, \phi] = 0$ for $h^{1/2} < 0$ because one can close the contour in the right half K plane.

The square of the trace of the second fundamental form K^2 can be expressed as an operator on the wave function Ψ:

$$K^2 = \frac{64\pi^2}{m_\mathrm{P}^4}\left[h^{-1/2}h_{ij}\frac{\delta}{\delta h_{ij}}\right]h^{-1/2}h_{kl}\frac{\delta}{\delta h_{kl}}. \qquad (2.17)$$

If $K^2\Psi/\Psi$ is positive, one can interpret the wave function Ψ in terms of a euclidean 4-geometry in the classical limit. On the other hand, if $K^2\Psi/\Psi$ is negative, then the observable K is imaginary which corresponds to a lorentzian 4-geometry in the classical limit. The quantity $K^2\Psi/\Psi$ is positive or negative according to whether the wave function Ψ depends in an exponential or oscillatory manner on the scale $h^{1/2}$. Thus the form of the wave function determines whether it corresponds to a euclidean or lorentzian 4-geometry in the classical limit.

3. Boundary conditions

As was stated in the introduction, the boundary conditions on the class C of metrics that are summed over in the path integral determine the quantum state of the universe. There seem to be only two natural boundary conditions for positive definite metrics:

(a) compact metrics;

(b) non-compact metrics which are asymptotic to metrics of maximal symmetry, i.e. flat euclidean space or euclidean anti-de Sitter space.

Boundary conditions of type (b) define the usual vacuum state. In this state, the expectation values of most quantities are defined to be zero so the vacuum state is not of much interest as the quantum state of the universe. In particle scattering calculations, one starts with the vacuum state and one changes the state by creating particles by the action of field operators at infinity in the infinite past. One lets the particles interact and then annihilates the resultant particles by the action of other field operators at future infinity. This gets one back to the vacuum state. If one supposed that the quantum state of the universe was some such particle scattering state, one would lose all ability to predict the state of the universe because one would have no idea of what was coming in. One would also expect that the matter in the universe would be concentrated in a certain region and would decrease to zero at large distances instead of the roughly homogeneous universe that we observe.

In particle scattering problems, one is interested in observables at infinity. One is therefore concerned only with metrics which are connected to infinity: any disconnected compact parts of the metric would not contribute to the scattering of particles from infinity. In cosmology, on the other hand, one is concerned with observables in a finite region in the middle of the space and it does not matter whether this region is connected to an infinite asymptotic region. For example, one might ask for the amplitudes $\Psi_{\pm}[h_{ij}, \phi]$ for a closed compact 3-manifold S with a 3-metric h_{ij} and a matter field configuration ϕ. As explained in sect. 2, $\Psi_{\pm}$ is given by a path integral over all metrics belonging to classes $C_{\pm}$ which are bounded by S with the given 3-metric and matter field configuration. If the class C that determines the quantum state consists of metrics of type (b), then C_+ would consist of compact metrics and C_- would contain two different kinds of metrics:

(i) connected asymptotically euclidean or anti-de Sitter metrics which had an inner boundary at S;

(ii) disconnected metrics which consisted of a compact part with boundary at S and an asymptotically euclidean or anti-de Sitter part without any inner boundary.

One cannot exclude disconnected metrics from the path integral because they can be approximated by connected metrics in which the different parts were joined by thin tubes. The tubes could be chosen to have negligible action. Similarly, topologically non-trivial metrics cannot be excluded because they can be approximated by topologically trivial metrics.

One would expect the dominant contribution to the path integral for the wave function to come from metrics which were near to solutions of the field equations with the appropriate boundary conditions. In the case of metrics of kind (i), it is shown in the appendix however that such solutions cannot give the dominant contribution to the wave function. This is borne out by calculations in specific examples. Thus, the dominant contribution to the wave function would come from metrics of kind (ii). These in general have a smaller action and hence give a bigger contribution than metrics of kind (i). One sees therefore that even if one adopted boundary conditions of type (b), the result would be almost the same as if one had adopted type (a), as far as the amplitude for finding a compact 3-submanifold S was concerned. Similar results would apply to other observations in a finite region. It would therefore seem more natural to suppose that the boundary conditions of the universe were of type (a), i.e. that the quantum state of the universe was defined by a path integral over compact positive definite metrics. In this case, the two amplitudes $\Psi_{\pm}$ are equal. I shall therefore drop the subscript $\pm$.

It should be emphasized that this is a *proposal* for the quantum state of the universe. One cannot derive it from some other principle but merely show that it is a natural choice. The ultimate test, however, is not whether it is aesthetically appealing but whether it enables one to make predictions that agree with observations. I shall endeavour to do this in the next section for a simple model.

4. Minisuperspace

In order to investigate whether the boundary conditions proposed in the last section correspond to the quantum state of the universe that we live in, one would like to calculate the wave function $\Psi[h_{ij}, \phi]$ and see if it can be interpreted as predicting what we observe. This is equivalent to finding a solution of the Wheeler–DeWitt equation obeying certain boundary conditions. The Wheeler–DeWitt equation is a second order functional differential equation on an infinite dimensional manifold, superspace, the space of all three metrics and matter field configurations. We do not know how to solve such an equation but we can hope to get some idea of the nature of the solution by considering the equation restricted to a finite dimensional submanifold called, minisuperspace. In other words, one restricts the infinite number of degrees of freedom of the gravitational and matter fields down to a finite number. The Wheeler–DeWitt equation then becomes a wave equation on a finite dimensional manifold and can be solved by standard techniques. The boundary conditions for the Wheeler–DeWitt equation can be obtained from the semiclassical approximation to the path integral, i.e. by expressing the wave function in the form

$$\Psi[h_{ij}, \phi] = N_0 \sum_i A_i \exp(-B_i)\,, \tag{4.1}$$

where N_0 is a normalization constant, and B_i are the actions of classical solutions of the field equations which are compact and which have the given 3-metric h_{ij} and matter field configuration ϕ on the boundary. The prefactors A_i are given by the determinants of small fluctuations about the classical solutions. I shall not bother with them but only with the more important quantities B_i which appear in the exponential.

The simplest example of a minisuperspace model is that for a spatially homogeneous and isotropic universe. Such models have been considered by [10–13]. The euclidean metric can be written in the form

$$\mathrm{d}s^2 = \sigma^2[N^2(\tau)\,\mathrm{d}\tau^2 + a^2(\tau)\,\mathrm{d}\Omega_3^2]\,, \tag{4.2}$$

where $N(\tau)$ is the lapse function, $\sigma^2 = 2/(3\pi m_\mathrm{p}^2)$ and $\mathrm{d}\Omega_3^2$ is the metric on a 3-sphere of unit radius. The gravitational action is

$$\tilde{I} = \tfrac{1}{2}\int \mathrm{d}\tau \left\{\frac{N}{a}\right\}\left[-\left\{\frac{a}{N}\frac{\mathrm{d}a}{\mathrm{d}\tau}\right\}^2 - a^2\right]. \tag{4.3}$$

There is no solution of the classical lorentzian field equations for an empty closed homogeneous and isotropic universe. One therefore has to include some form of matter. The simplest equations are obtained for a conformally invariant scalar field ϕ which is constant on the 3-sphere sections. The euclidean action of such a field is

$$\tilde{I}=\tfrac{1}{2}\int \mathrm{d}\tau\left(\frac{N}{a}\right)\left[\left(\frac{a}{N}\frac{\mathrm{d}\chi}{\mathrm{d}\tau}\right)^2-\chi^2\right], \tag{4.4}$$

where $\chi=\sqrt{2}/\pi a\sigma\phi$. The Wheeler–DeWitt equation is then

$$\frac{1}{2}\left(\frac{1}{a^p}\frac{\partial}{\partial a}\left[a^p\frac{\partial}{\partial a}\right]-a^2-\frac{\partial^2}{\partial\chi^2}+\chi^2\right)\Psi[a,\chi]=0\,. \tag{4.5}$$

The advantage of using a conformally invariant scalar field ϕ is that the Wheeler–DeWitt equations separates, i.e. there are solutions of the form

$$\Psi(a,\chi)=C(a)f(\chi)\,. \tag{4.6}$$

The function f obeys the harmonic oscillator equation

$$\frac{1}{2}\left(-\frac{\mathrm{d}^2}{\mathrm{d}\chi^2}+\chi^2\right)f=Ef\,. \tag{4.7}$$

It is therefore natural to take f to be the harmonic oscillator wave function with eigenvalues $E=n+\frac{1}{2}$. The disadvantage of using a conformally invariant scalar field is that the boundary condition of integrating over compact metrics uniquely picks out the ground state $n=0$ of the harmonic oscillator [2]. The gravitational part of the wave function $C(a)$ is then almost the same as if there were no matter present. Such a wave function could not represent the universe we live in. For this reason, Hartle and I proposed in [2] that the quantum state defined by integrating over compact metrics should be regarded as the "ground state" of the universe but that the universe that we lived in did not obey these boundary conditions. Rather it was a linear superposition of other "excited state" solutions with higher values of n. This meant that one lost any ability to predict the quantum state of the universe because one had no means of determining the complex coefficients of the excited states in the superposition.

With hindsight it is clear that Hartle and I were misled by the mathematical simplicity of the conformally invariant scalar field. If one fills in a 3-sphere of radius a with a topologically trivial euclidean 4-geometry of the form (4.2), the action of the scalar field is independent of the form of $a(\tau)$. This means that there is no coupling between the scalar field and the gravitational degree of freedom which is a purely conformal one. Such a model cannot describe the observed universe in which the expansion is strongly coupled to the matter content. In order to obtain a model which corresponds to what we observe, it is necessary to introduce non-conformally invariant fields. In this section I shall describe the simplest such model, a homogeneous isotropic universe with a spatially constant massive scalar field. A future paper [8] will deal with a model in which the scalar field is replaced by an effective R^2 term generated by quantum corrections.

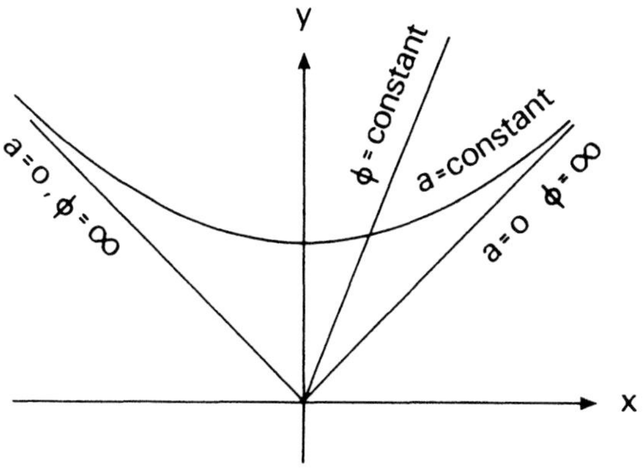

Fig. 1. The causal structure of the Wheeler–DeWitt equation can be seen in coordinates (x, y) in which the second derivatives take the form $\partial^2/\partial y^2 - \partial^2/\partial x^2$. The region $a > 0$ is mapped into the interior of the future light-cone of the origin.

The Wheeler–DeWitt equation for a homogeneous isotropic universe with a scalar field $\tilde{\phi}$ with mass $\tilde{m}$ is

$$\frac{1}{2}\left(\frac{1}{a^p}\frac{\partial}{\partial a}a^p\frac{\partial}{\partial a} - a^2 - \frac{1}{a^2}\frac{\partial^2}{\partial \phi^2} + a^4 m^2 \phi^2\right)\Psi(a, \phi) = 0\,, \tag{4.8}$$

where $\phi = \sigma\tilde{\phi}$ and $m = \sigma\tilde{m}$. The exponent p represents some, but not all of the uncertainty in the factor ordering. Its precise value will not be important for what follows. One can regard (4.8) as a wave equation in the (a, ϕ) plane. The causal structure of the Wheeler–DeWitt equation is seen more easily by introducing new coordinates

$$x = a \sinh \phi, \qquad y = a \cosh \phi\,. \tag{4.9}$$

In these coordinates, the second derivatives take the form $\partial^2/\partial y^2 - \partial^2/\partial x^2$. Thus, the characteristics are lines at 45° in the (x, y) plane. The region $a > 0$ is mapped into the interior of the future light cone of the origin (see fig. 1). The surfaces of constant ϕ are straight lines through the origin and the surfaces of constant a are spacelike hyperbolae within the light cone.

If one knew the solution on the light cone of the origin, one could solve the wave equation by standard techniques. The light cone is the surface $a = 0$, $\phi = \pm\infty$. It is therefore rather hard to apply the boundary conditions of the previous section there, but what I shall do is use the semiclassical approximation to the path integral to estimate the form of the solution on lines of constant large positive or negative ϕ.

In order to apply the semiclassical approximation, one needs to know the solutions to the classical euclidean field equations. One can express the metric in the form

$$ds^2 = \sigma^2(d\tau^2 + a^2(\tau)\, d\Omega_3^2)\,. \tag{4.10}$$

One is interested in metrics which are compact. Thus a will be zero at some value of τ which can be chosen to be zero. At $\tau = 0$, $d\phi/d\tau$ must be zero and $da/d\tau = 1$. One can then integrate the equations for $a(\tau)$ and $\phi(\tau)$ with these initial conditions

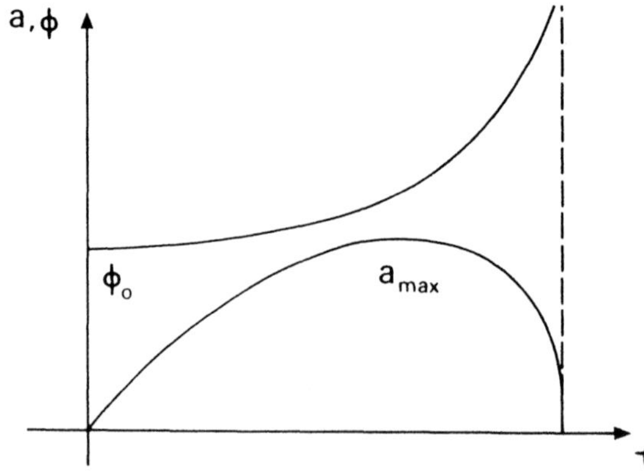

Fig. 2. Graphs of the radius a and the scalar field ϕ for a solution of the euclidean field equations.

and $\phi = \phi_0$ at $\tau = 0$. If $\phi_0 > 0$, $\phi(\tau)$ and $d\phi/d\tau$ will monotonically increase with τ and $da/d\tau$ will monotonically decrease (see fig. 2). What happens is that the positive energy density of the $m^2\phi^2(\tau)$ term in the action behaves like a positive cosmological constant and causes the 4-geometry to have positive curvature. The 3-sphere radius $a(\tau)$ rises to a maximum a_{max} which is a monotonically decreasing function of ϕ_0. The radius $a(\tau)$ then decreases and becomes zero at a singularity of the 4-geometry. If $\phi_0 \gg 1$, i.e. $\ddot{\phi} \gg m_p$, then $\phi(\tau)$ does not increase much by the time $a(\tau)$ reaches a_{max} and $a_{max} \simeq 1/m\phi_0$ (see fig. 3).

In order to estimate the wave function $\Psi(a, \phi)$, one looks for a solution $(a(\tau), \phi(\tau))$ which matches the given values of a, and ϕ at some value $\tau = \tau_0$. For sufficiently small values of a and ϕ there will be a unique trajectory $a(\tau)$, $\phi(\tau)$ which passes through the given values of a and ϕ and which does not cross any other trajectory (see fig. 3). For such a, ϕ the path integral for the wave function $\Psi(a, \phi)$ will be dominated by the contribution from the euclidean solution represented by the trajectory $a(\tau)$, $\phi(\tau)$, i.e.

$$\Psi(a, \phi) \approx N_0 \exp(-\tilde{I}[a(\tau), \phi(\tau)]) . \tag{4.11}$$

If two neighbouring trajectories $a(\tau)$, $\phi(\tau)$ in the a, ϕ plane intersect at a point a_1, ϕ_1, the path integral for the wave function $\Psi(a_1, \phi_1)$ will have a zero mode at the euclidean solution $a(\tau)$, $\phi(\tau)$. If a_2, ϕ_2 is a point on the same trajectory at a larger value of τ, the path integral for $\Psi(a_2, \phi_2)$ will have a negative mode at the euclidean solution $a(\tau)$, $\phi(\tau)$. This means that the dominant contribution to the path integral for Ψ will not come from that euclidean solution. Another way of obtaining this result is to notice that the neighbouring trajectories intersect at about the maximum value of $a(\tau)$. Beyond that value of τ, the radius a will be decreasing and the trace of the second fundamental form K will be negative. It is shown in the appendix that a solution with negative K cannot give the dominant contribution to the wave function Ψ.

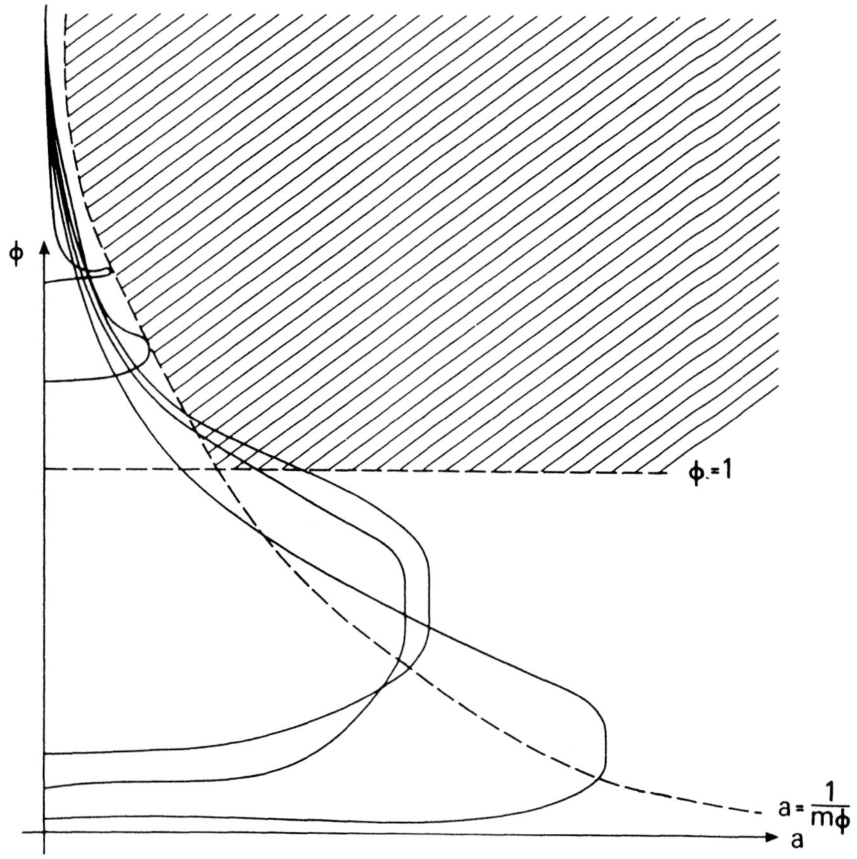

Fig. 3. Solutions of the euclidean field equations in the (a, ϕ) plane. There are no solutions which reach the region $\phi > 1$, $a < (m\phi)^{-1}$ without crossing another solution.

There is a region of the (a, ϕ) plane with $\phi > 1$, $a > (m\phi)^{-1}$ through which there are no euclidean trajectories which have not intersected another trajectory (see fig. 3).* The path integral (2.4) over compact euclidean metrics for the wave function $\Psi(a, \phi)$ for a point (a, ϕ) in that region will not receive its dominant contribution from any real euclidean metric. Instead, one can distort the contour of integration in the path integral so that the dominant contribution comes from a complex solution of the field equations. The action of such a solution will be complex and so the wave function will be oscillatory though it will be real because there will be an equal contribution from the complex conjugate solution.

One can estimate the form of the wave function in the region $\phi > 1$, $a > (m\phi)^{-1}$ by using the K representation $\Phi(K, \phi)$. The semiclassical approximation to the

* Note added in proof: D.N. Page has pointed out that there are trajectories through this region which have not intersected another trajectory. However, they correspond to solutions with large positive actions. They therefore do not give much contribution to the wave function.

wave function $\Phi(K, \phi)$ is

$$\Phi(K, \phi) \approx N_k \sum_i A_i \exp(-B_i)\,, \tag{4.12}$$

where N_k is a normalization factor. B_i are the actions I^k of classical solutions which are bounded by a 3-sphere with the trace of the second fundamental form equal to K and the prefactors A_i are given by the determinants of small fluctuations about the classical solutions. If $\phi \gg 1$ and $K > 0$, the action $I^k(K, \phi)$ is almost the same as that for de Sitter space with $\Lambda = 3m^2\phi^2/\sigma^2$, i.e.

$$I^k(k, \phi) \approx -\frac{1}{3m^2\phi^2}\left(1 - \frac{3k}{(9k^2 + m^2\phi^2)^{1/2}}\right), \tag{4.13}$$

where $k = \frac{1}{9}K\sigma$. The wave function $\Psi(a, \phi)$ can then be obtained by an inverse Laplace transform of $\Phi(k, \phi)$

$$\Psi(a, \phi) \approx \frac{N_k}{2\pi i}\int_\Gamma \mathrm{d}k \exp(ka^3 - I^k)\,, \tag{4.14}$$

where the contour Γ is parallel to the imaginary k axis. The integral (4.14) can be evaluated by the method of steepest descent.

If $a < (m\phi)^{-1}$, the saddle point occurs at a real positive value of k

$$k = \tfrac{1}{3}m\phi\left(\frac{1}{m^2\phi^2a^2} - 1\right)^{1/2}. \tag{4.15}$$

This value of k corresponds to a real euclidean solution $a(\tau)$, $\phi(\tau)$ bounded by the given values of a, ϕ. The wave function is therefore

$$\begin{aligned}\Psi(a, \phi) &\approx N_k \exp(ka^3 - I^k) \\ &\approx N_k \exp(-\tilde{I}(a, \phi))\,,\end{aligned} \tag{4.16}$$

where $\tilde{I}(a, \phi)$ is the action of the solution bounded by the given values of a and ϕ:

$$\tilde{I}(a, \phi) = -\frac{1}{3m^2\phi^2}(1 - (1 - m^2\phi^2a^2)^{3/2})\,. \tag{4.17}$$

If the radius a is greater than $(m\phi)^{-1}$ there will not be a stationary phase point in (4.14) for real values of k. This corresponds to the fact that there is no real euclidean solution. However, if a is only slightly greater than $(m\phi)^{-1}$, one can analytically extend the approximate expression (4.13) for I^k and find stationary phase points at imaginary values of k:

$$k = \pm\tfrac{1}{3}im\phi\left(1 - \frac{1}{m^2\phi^2a^2}\right)^{1/2}. \tag{4.18}$$

It is possible to distort the contour Γ to pass through both of these stationary phase points. The wave function therefore has the form:

$$\Psi(a, \phi) \approx N_k \exp\left(\frac{1}{3m^2\phi^2}\right)\cos\left((m^2\phi^2a^2 - 1)^{3/2}/(3m^2\phi^2) - \tfrac{1}{4}\pi\right)\,. \tag{4.19}$$

One can use the behaviour (4.16) and (4.19) of the wave function as the boundary conditions for the Wheeler–DeWitt equation near the light cone in the (x, y) plane. One can then solve the Wheeler–DeWitt equation as a wave equation with these boundary conditions to determine the behaviour of the wave function at other values of a and ϕ. Because the wave function is rapidly oscillating, one can use the WKB method. One writes the wave function in the form

$$\Psi(a, \phi) = C(a, \phi) \cos[S(a, \phi)], \tag{4.20}$$

where $S(a, \phi)$ is a rapidly varying phase and $C(a, \phi)$ is a slowly varying amplitude. The trajectories of ∇S in the (a, ϕ) plane correspond to classical lorentzian solutions $(a(t), \phi(t))$ of the field equations. The semi-classical approximation for large ϕ corresponds to a classical solution which starts with $\phi(t) = \phi_1$, $\mathrm{d}\phi/\mathrm{d}t = 0$, at a minimum radius $a(t) = (m\phi_1)^{-1}$ at $t = 0$ and then expands in a de Sitter-like manner. The mass term tends to make $\phi(t)$ decrease with a timescale $3H/m^2$. However, the expansion rate $H = m\phi_1$. Thus if $\phi_1 > 1$ the universe will expand by a factor of order $\mathrm{e}^{3\phi_1^2}$ before ϕ decreases significantly. For large ϕ_1 this gives a long inflationary period.

Eventually $\phi(t)$ decreases to zero and starts oscillating in time. The universe goes over to a matter-dominated phase with $a(t) \propto t^{2/3}$. The universe will expand to a maximum radius of order $\exp(6\phi_1^2)/m\phi_1$ and then recollapse. A classical solution that starts with a general value of ϕ_1 will recollapse to a singularity. It will be time symmetric about $t = 0$ but not in general time symmetric about $t = t_1$, the time of maximum expansion. However, for discrete values of the scalar field ϕ_1 the field $\phi(t)$ is zero at $t = t_1$. In these cases, the solution will be time symmetric about $t = t_1$ but with $\phi(t)$ replaced by $-\phi(t_1 - t)$. These solutions will therefore oscillate indefinitely. There are also periodic solutions which are time symmetric about either the minimum or the maximum radius, but not both and there are oscillating solutions which are not time symmetric at all and which are aperiodic. The orbits of ∇S in the (a, ϕ) plane correspond to these non-singular oscillating solutions. One thus obtains a wave function $\Psi(a, \phi)$ which can be interpreted as a superposition of wave functions corresponding to oscillating universes in the classical limit. Note that the wave function does not represent a single classical universe but a whole ensemble of classical universes which expand out to different maximum radii.

The wave function $\Psi(a, \phi)$ will grow exponentially in a region near the y-axis in which $a^2 m^2 \phi^2 < 1$. This part of the wave function can be interpreted as corresponding to a flat euclidean 4-geometry. If one wishes the wave function to be normalizable, one could introduce a tiny Λ term. This would give a λa^4 term in the Wheeler–DeWitt equation (4.8) where $\lambda = \frac{1}{3}\sigma^2 \Lambda$. It would cut off the exponential growth of Ψ at very large values of a. Superimposed upon this exponentially growing wave function will be an oscillating component which corresponds to the periodic solutions described above. The oscillating component in the wave function should be interpreted as corresponding to a lorentzian geometry and the exponentially growing component in the wave function should be interpreted as corresponding to a euclidean geometry. We live in a lorentzian geometry and therefore we are interested really only in the oscillatory part of the wave function.

5. Beyond minisuperspace

In the previous section it was shown that the quantum state defined by the path integral over all compact euclidean metrics was a reasonable representation of the observed universe in the case of a simple minisuperspace model in which nearly all the degrees of freedom of the gravitational and matter fields were frozen out. The question naturally arises as to whether the quantum state would still be a good description of the universe if one included all those extra degrees of freedom or whether it was a misleading result produced by the truncation.

The important feature of the wave function in the model of the previous section was that it was an oscillatory rather than an exponential function of the coordinates (a, ϕ) in a certain region of minisuperspace. This allowed it to be interpreted as corresponding to lorentzian 4-geometries in the classical limit. The reasons that the wave function oscillated were:

(1) the Wheeler–DeWitt equation was a hyperbolic wave equation on minisuperspace;

(2) there was a region of minisuperspace which could not be reached by real euclidean solutions of the field equations without negative modes. This meant that the dominant contribution to the path integral for the wave function came from complex solutions of the field equations.

The "metric" $-G_{ijkl}$ in the full Wheeler–DeWitt equation that multiplies the second functional derivatives of the wave function $\partial^2\Psi/\partial h_{ij}\partial h_{kl}$ has a signature $(+-----)$ at each point of the 3-manifold S. It might therefore appear that the gravitational part of the Wheeler–DeWitt equation was ultra-hyperbolic, having a signature with one plus and five minuses at each point of S. However, all but one of these infinite number of pluses can be regarded as arising from the gauge invariance of the theory. Gauge invariance manifests itself in two ways: there is the obvious freedom to make coordinate transformations in the 3-surface S. This is reflected in the equation (2.13) which states that the wave function is unchanged by coordinate transformations in S. The other gauge invariance, which is less apparent, is the freedom to choose S at different positions in the 4-geometry $g_{\mu\nu}$. Given a 4-geometry, one could, locally at least, restrict the positions of S in it to a one parameter family of non-intersecting 3-surfaces by imposing one gauge condition on the 3-metric h_{ij} such as ${}^3R = \text{constant}$. With this restriction on h_{ij}, the gravitational part of the Wheeler–DeWitt equation is hyperbolic: the one plus sign corresponds to the overall size of the 3-surface S. The matter field variables in the Wheeler–DeWitt equation all have negative signature provided that they obey reasonable positive energy conditions. Thus the full Wheeler–DeWitt equation can be regarded as a hyperbolic wave equation, in a certain sense.

The other important property of the model of the previous section, that there should be a region of superspace or minisuperspace which cannot be reached by real euclidean solutions, seems to depend on the presence of fields with non-zero rest-mass such as a massive scalar field. Investigations of minisuperspace models containing a conformally invariant scalar field [2] or empty homogeneous anisotropic models [14] indicate that there is always a real euclidean solution of the field equations which gives the dominant contribution to the path integral for the wave function. The wave function is therefore an exponential rather than an oscillatory function of the superspace variables and so corresponds to a euclidean rather than a lorentzian 4-geometry in the classical limit.

In the massive scalar field minisuperspace model the potential $-a^2+a^4m^2\phi^2$, the non-derivative term in the Wheeler–DeWitt equation (4.8), changes sign from negative to positive on a surface in the (a, ϕ) plane which is "spacelike" for large $|\phi|$, i.e. which has a negative signature with respect to the second derivative terms in the Wheeler–DeWitt equation. Points below this surface can be reached by euclidean solutions, so the wave function is exponential in this region. The wave function on the surface $-a^2+a^4m^2\phi^2=0$ is of the form $\Psi \approx N_0 \exp(1/3m^2\phi^2)$. A solution of the wave equation with this initial data must oscillate in a region of positive potential. This is why one obtains an oscillatory wave function which can be interpreted in terms of lorentzian 4-geometries in the classical limit. This feature of the massive scalar field model does not depend on the truncation to a finite number of degrees of freedom by the imposition of symmetries: it would also be present in a model in which the 3-metric h_{ij} and the scalar field ϕ were fully general and did not have any symmetries. Thus one would expect that the solution of the Wheeler–DeWitt equation on the full superspace of the massive scalar field model would be oscillatory if it obeyed the boundary condition that it was given by a path integral over compact metrics.

Under a scale transformation $g_{\mu\nu} \to \Omega^2 g_{\mu\nu}$, $\Omega=$constant, the gravitational action transforms as $\tilde{I} \to \Omega^2 \tilde{I}$. This means that the potential terms in the gravitational part of the Wheeler–DeWitt equation cannot change sign under a "timelike" motion in superspace that corresponds simply to a scale transformation of the 3-metric h_{ij}. In other words, the gravitational part of the potential can change sign only on a "timelike" surface in superspace. The action of a conformal invariant field, such as the electromagnetic field or a conformally invariant scalar field, is unchanged under a scale transformation. Thus the potential terms in the Wheeler–DeWitt equation arising from such fields can also change sign only on "timelike" surfaces in superspace. In fact, under reasonable positive energy assumptions, the potential terms produced by such fields will always be positive. Thus the potential cannot change from negative to positive in a model that contains only zero, rest-mass fields, either conformally invariant fields or the gravitational field. One would therefore not expect minisuperspace models that contained only zero rest-mass fields to produce oscillatory wave functions.

On the other hand, if one includes all the infinite number of gravitational and matter degrees of freedom, one will generate R^2 and C^2 terms in the effective action. These terms will be finite in the case of a supergravity theory in which all the divergences cancel or in the case that gravity provides its own cutoff (possibilities (i) and (ii) of sect. 1) and will be infinite but renormalized in the case of a higher derivative theory (possibility (iii)). Such terms in the effective action behave like massive scalar or spin-2 particles. One might therefore hope that they would give rise to oscillatory wave functions. This has been verified, at least in the case of a minisuperspace model with an R^2 term. This will be described in a forthcoming paper [8]. It may well be therefore that the observed universe owes its existence to quantum gravitational effects.

I am grateful for discussions with J.B. Hartle, A.D. Linde, J. Luttrell and I.G. Moss.

Appendix

THE DOMINANT CONTRIBUTION TO THE PATH INTEGRAL

The wave function Ψ is given by the inverse Laplace transform

$$\Psi[h_{ij}, \phi] = \int_{\Gamma} \mathrm{d}\left(\frac{m_{\mathrm{P}}^2}{24\pi i} K\right) \exp\left(\frac{m_{\mathrm{P}}^2}{12\pi} \int \mathrm{d}^3x\, h^{1/2} K\right) \Phi[\tilde{h}_{ij}, K, \phi] , \tag{A.1}$$

where the wave function Φ in the K representation is

$$\Phi[\tilde{h}_{ij}, K] = \int \mathrm{d}[g_{\mu\nu}] \exp(-I^k[g_{\mu\nu}, \phi]) . \tag{A.2}$$

One expects the dominant contribution to Φ to come from metrics near a solution $\tilde{g}_{\mu\nu}$ of the field equations with the given values of $\tilde{h}_{ij}$ and K on S. The dominant contribution to Ψ will then come from the saddle point in the contour integral (A.1). At the saddle point the exponent of e in (A.1)

$$\frac{m_{\mathrm{P}}^2}{12\pi} \int_{\mathrm{S}} \mathrm{d}^3x\, h^{1/2} K - I^k[\tilde{g}_{\mu\nu}, \phi] \tag{A.3}$$

will be an extremum under small variations of K:

$$\frac{m_{\mathrm{P}}^2}{12\pi} h^{1/2} = \frac{\delta I^k}{\delta K} . \tag{A.4}$$

Eq. (A.4) implies that $\tilde{g}_{\mu\nu}$ is a solution of the field equations with the boundary conditions h_{ij}, ϕ on S. However, if $\tilde{g}_{\mu\nu}$ is a real euclidean metric, then in order for it to be a saddle point in (A.1), the exponent (A.3) must be, not only an extremum, but a minimum on the real K-axis at each point of S, i.e.

$$\frac{\delta h^{1/2}}{\delta K} < 0 , \tag{A.5}$$

where $h^{1/2}$ is regarded as a function of the boundary data $(\tilde{h}_{ij}, K, \phi)$ on S that determine the solution $\tilde{g}_{\mu\nu}$.

Consider the case $\Lambda = 0$ and $\phi = 0$ on S. Under a scale transformation

$$\delta K = -\varepsilon K, \qquad \delta h^{1/2} = 3\varepsilon h^{1/2}. \tag{A.6}$$

If the 3-metric $\tilde{h}_{ij}$ is conformally flat and the 4-metric $g_{\mu\nu}$ is asymptotically euclidean, the solution $\tilde{g}_{\mu\nu}$ will be flat euclidean space outside a round 3-sphere S. For this solution K will be constant and negative. Thus under a scale transformation

$$\frac{\delta h^{1/2}}{\delta K} > 0. \tag{A.7}$$

This shows that the metric $\tilde{g}_{\mu\nu}$ cannot be a saddle point in the contour integral (A.1) and so cannot give the dominant contribution to the wave function Ψ. One can now consider a continuous variation of the values of Λ and the matter field configuration ϕ on S from zero. Under such a variation the left-hand side of (A.7) will change but it cannot become negative everywhere on S unless there is some Λ and ϕ for which the left-hand side of (A.7) is either infinite or zero everywhere on S. In the first case, the solution $\tilde{g}_{\mu\nu}$ will have a zero mode for fixed $\tilde{h}_{ij}$, K and ϕ on S, and in the second case it will have a zero mode for fixed h_{ij} and ϕ. If one continues the variation until the left side of (A.7) is negative, the solution $\tilde{g}_{\mu\nu}$ will have a negative mode for the given boundary conditions. In the first case, this indicates that $\tilde{g}_{\mu\nu}$ does not provide the dominant contribution to the wave function Φ and hence does not provide the dominant contribution to Ψ. In the second case, the direct representation of Ψ in terms of a path integral shows that $\tilde{g}_{\mu\nu}$ does not provide the dominant contribution to Ψ.

References

[1] S.W. Hawking, in Astrophysical cosmology, Pontificiae Academiae Scientarium Scripta Varia, 48 (1982) 563
[2] J.B. Hartle and S.W. Hawking, Phys. Rev. D28 (1983) 2960
[3] A.H. Guth, Phys. Rev. D23 (1981) 347
[4] A.D. Linde, Phys. Lett. 108B (1982) 389
[5] S.W. Hawking and I.G. Moss, Phys. Lett. B110 (1982) 35
[6] A. Albrecht and P.J. Steinhardt, Phys. Rev. Lett. 48 (1982) 120
[7] A.H. Guth and S.Y. Pi, Phys. Rev. Lett. 49 (1982) 1110
[8] S.W. Hawking and J. Luttrell, paper in preparation
[9] S.W. Hawking, Quantum cosmology, Les Houches lectures (1983), to be published
[10] B.S. DeWitt, Phys. Rev. 160 (1967) 113
[11] C.J. Isham and J.E. Nelson, Phys. Rev. D10 (1974) 3226
[12] W.E. Blyth and C.J. Isham, Phys. Rev. D11 (1975) 768
[13] I.G. Moss and W.A. Wright, The wave function of the inflationary universe, Newcastle Univ. preprint (1983)
[14] J. Luttrell, private communication

PHYSICAL REVIEW D
PARTICLES AND FIELDS

THIRD SERIES, VOLUME 31, NUMBER 8 — 15 APRIL 1985

Origin of structure in the Universe

J. J. Halliwell and S. W. Hawking
Department of Applied Mathematics and Theoretical Physics, Silver Street, Cambridge CB3 9EW, United Kingdom and Max Planck Institut for Physics and Astrophysics, Foehringer Ring 6, Munich, Federal Republic of Germany
(Received 17 December 1984)

It is assumed that the Universe is in the quantum state defined by a path integral over compact four-metrics. This can be regarded as a boundary condition for the wave function of the Universe on superspace, the space of all three-metrics and matter field configurations on a three-surface. We extend previous work on finite-dimensional approximations to superspace to the full infinite-dimensional space. We treat the two homogeneous and isotropic degrees of freedom exactly and the others to second order. We justify this approximation by showing that the inhomogeneous or anisotropic modes start off in their ground state. We derive time-dependent Schrödinger equations for each mode. The modes remain in their ground state until their wavelength exceeds the horizon size in the period of exponential expansion. The ground-state fluctuations are then amplified by the subsequent expansion and the modes reenter the horizon in the matter- or radiation-dominated era in a highly excited state. We obtain a scale-free spectrum of density perturbations which could account for the origin of galaxies and all other structure in the Universe. The fluctuations would be compatible with observations of the microwave background if the mass of the scalar field that drives the inflation is 10^{14} GeV or less.

I. INTRODUCTION

Observations of the microwave background indicate that the Universe is very close to homogeneity and isotropy on a large scale. Yet we know that the early Universe cannot have been completely homogeneous and isotropic because in that case galaxies and stars would not have formed. In the standard hot big-bang mode the density perturbations required to produce these structures have to be assumed as initial conditions. However, in the inflationary model of the Universe[1-4] it was possible to show that the ground-state fluctuations of the scalar field that causes the exponential expansion would lead to a spectrum of density per urbations that was almost scale free.[5-7] In the simplest grand-unified-theory (GUT) inflationary model the amplitude of the density perturbations was too large but an amplitude that was consistent with observation could be obtained in other models with a different potential for the scalar field.[8] Similarly, ground-state fluctuations of the gravitational-wave modes would lead to a spectrum of long-wavelength gravitational waves that would be consistent with observation provided that the Hubble constant H in the inflationary period was not more than about 10^{-4} of the Planck mass.[9]

One cannot regard these results as a completely satisfactory explanation of the origin of structure in the Universe because the inflationary model does not make any assumption about the initial or boundary conditions of the Universe. In particular, it does not guarantee that there should be a period of exponential expansion in which the scalar field and the gravitational-wave modes would be in the ground state. In the absence of some assumption about the boundary conditions of the Universe, any present state would be possible: one could pick an arbitrary state for the Universe at the present time and evolve it backward in time to see what initial conditions it arose from. It has recently been proposed[10-13] that the boundary conditions of the Universe are that it has no boundary. In other words, the quantum state of the Universe is defined by a path integral over compact four-metrics without boundary. The quantum state can be described by a wave function Ψ which is a function on the infinite-dimensional space W called superspace which consists of all three-metrics h_{ij} and matter field configurations Φ_0 on a three-surface S. Because the wave function does not depend on time explicitly, it obeys a system of zero-energy Schrödinger equations, one for each choice of the shift N_i and the lapse N on S. The Schrödinger equations can be decomposed into the momentum constraints, which imply that the wave function is the same at all points of W that are related by coordinate t ansformations, and the Wheeler-DeWitt equations, which can be regarded as a system of second-order differential equations for Ψ on W. The requirement that the wave function be given by a path integral over compact four-metrics then becomes a set of boundary conditions for the Wheeler-DeWitt equations which determines a unique solution for Ψ.

It is difficult to solve differential equations on an infinite-dimensional manifold. Attention has therefore been concentrated on finite-dimensional approximations to W, called "minisuperspaces." In other words, one restricts the number of gravitational and matter degrees of freedom to a finite number and then solves the Wheeler-DeWitt equations on a finite-dimensional manifold with boundary conditions that reflect the fact that the wave function is given by a path integral over compact four-metrics. In particular,[12–15] it has been shown that in the case of a homogeneous isotropic closed universe of radius a with a massive scalar field ϕ the wave function corresponds in the classical limit to a family of classical solutions which have a long period of exponential or "inflationary" expansion and then go over to a matter-dominated expansion, reach a maximum radius, and then collapse in a time-symmetric manner. This model would be in agreement with observation but, because it is so restricted, the only prediction it can make is that the observed value of the density parameter Ω should be exactly one.[15] The aim of this paper is to extend this minisuperspace model to the full number of degrees of freedom of the gravitational and scalar fields. We treat the 2 degrees of freedom of the minisuperspace model exactly and we expand the other inhomogeneous and anisotropic degrees of freedom to second order in the Hamiltonian. In the region of W in which Ψ oscillates rapidly, one can use the WKB approximation to relate the wave function to a family of classical solutions and so introduce a concept of time. As in the minisuperspace case, the family includes solutions with a long period of exponential expansion. We show that the gravitational-wave and density-perturbation modes obey decoupled time-dependent Schrödinger equations with respect to the time parameter of the classical solution. The boundary conditions imply that these modes start off in the ground state. While they remain within the horizon of the exponentially expanding phase, they can relax adiabatically and so they remain in the ground state. However, when they expand outside the horizon of the inflationary period, they become "frozen" until they reenter the horizon in the matter-dominated era. They then give rise to gravitational waves and a scale-free spectrum of density perturbations. These would be consistent with the observations of the microwave background and could be large enough to explain the origins of galaxies if the mass of the scalar field were about 10^{-5} of the Planck mass. Thus the proposal that the quantum state of the Universe is defined by a path integral over compact four-metrics seems to be able to account for the origin of structure in the Universe: it arises, not from arbitrary initial conditions, but from the ground-state fluctuations that have to be present by the Heisenberg uncertainty principle.

In Sec. II we review the Hamiltonian formalism of classical general relativity, and in Sec. III we show how this leads to the canonical treatment of the quantum theory. In Sec. IV we summarize earlier work[13] on a homogeneous isotropic minisuperspace model with a massive scalar field. We extend this to all the matter and gravitational degrees of freedom in Sec. V, treating the inhomogeneous modes to second order in the Hamiltonian. In Sec. VI we decompose the wave function into a background term which obeys an equation similar to that of the unperturbed minisuperspace model, and perturbation terms which obey time-dependent Schrödinger equations. We use the path-integral expression for the wave function in Sec. VII to show that the perturbation wave functions start out in their ground states. Their subsequent evolution is described in Sec. VIII. In Sec. IX we calculate the anisotropy that these perturbations would produce in the microwave background and compare with observation. In Sec. X we summarize the paper and conclude that the proposed quantum state could account not only for the large-scale homogeneity and isotropy but also for the structure on smaller scales.

II. CANONICAL FORMULATION OF GENERAL RELATIVITY

We consider a compact three-surface S which divides the four-manifold M into two parts. In a neighborhood of S one can introduce a coordinate t such that S is the surface $t=0$ and coordinates x^i ($i=1,2,3$). The metric takes the form

$$ds^2=-(N^2-N_iN^i)dt^2+2N_idx^idt+h_{ij}dx^idx^j\,. \tag{2.1}$$

N is called the lapse function. It measure the proper-time separation of surfaces of constant t. N_i is called the shift vector. It measures the deviation of the lines of constant x^i from the normal to the surface S. The action is

$$I=\int (L_g+L_m)d^3x\,dt\ , \tag{2.2}$$

where

$$L_g=\frac{m_P{}^2}{16\pi}N(G^{ijkl}K_{ij}K_{kl}+h^{1/2}\,{}^3R)\ , \tag{2.3}$$

$$K_{ij}=\frac{1}{2N}\left[-\frac{\partial h_{ij}}{\partial t}+2N_{(i\,|\,j)}\right], \tag{2.4}$$

is the second fundamental form of S, and

$$G^{ijkl}=\tfrac{1}{2}h^{1/2}(h^{ik}h^{jl}+h^{il}h^{jk}-2h^{ij}h^{kl})\ . \tag{2.5}$$

In the case of a massive scalar field Φ

$$L_m=\tfrac{1}{2}Nh^{1/2}\left[N^{-2}\left[\frac{\partial\Phi}{\partial t}\right]^2-2\frac{N^i}{N^2}\frac{\partial\Phi}{\partial t}\frac{\partial\Phi}{\partial x^i}-\left[h^{ij}-\frac{N^iN^j}{N^2}\right]\frac{\partial\Phi\partial\Phi}{\partial x^i\partial x^j}-m^2\Phi^2\right]. \tag{2.6}$$

In the Hamiltonian treatment of general relativity one regards the components h_{ij} of the three-metric and the field Φ as the canonical coordinates. The canonically conjugate momenta are

$$\pi^{ij}=\frac{\partial L_g}{\partial \dot{h}_{ij}}=-\frac{h^{1/2}m_P{}^2}{16\pi}(K^{ij}-h^{ij}K)\ , \qquad (2.7)$$

$$\pi_\Phi=\frac{\partial L_m}{\partial \dot{\Phi}}=N^{-1}h^{1/2}\left[\dot{\Phi}-N^i\frac{\partial\Phi}{\partial x^i}\right]. \qquad (2.8)$$

The Hamiltonian is

$$H=\int(\pi^{ij}\dot{h}_{ij}+\pi_\Phi\dot{\Phi}-L_g-L_m)d^3x$$
$$=\int(NH_0+N_iH^i)d^3x\ , \qquad (2.9)$$

where

$$H_0=16\pi m_P{}^{-2}G_{ijkl}\pi^{ij}\pi^{kl}-\frac{m_P{}^2}{16\pi}h^{1/2}\,{}^3R$$
$$+\tfrac{1}{2}h^{1/2}\left[\frac{\pi_\Phi{}^2}{h}+h^{ij}\frac{\partial\Phi\partial\Phi}{\partial x^i\partial x^j}+m^2\Phi^2\right], \qquad (2.10)$$

$$H^i=-2\pi^{ij}{}_{|j}+h^{ij}\frac{\partial\Phi}{\partial x^j}\pi_\Phi\ , \qquad (2.11)$$

and

$$G_{ijkl}=\tfrac{1}{2}h^{-1/2}(h_{ik}h_{jl}+h_{il}h_{jk}-h_{ij}h_{kl})\ . \qquad (2.12)$$

The quantities N and N_i are regarded as Lagrange multipliers. Thus the solution obeys the momentum constraint

$$H^i=0 \qquad (2.13)$$

and the Hamiltonian constraint

$$H_0=0\ . \qquad (2.14)$$

For given fields N and N^i on S the equations of motion are

$$\dot{h}_{ij}=\frac{\partial H}{\partial\pi^{ij}},\quad \dot{\pi}^{ij}=-\frac{\partial H}{\partial h_{ij}}\ ,$$
$$\dot{\Phi}=\frac{\partial H}{\partial\pi_\Phi},\quad \dot{\pi}_\Phi=-\frac{\partial H}{\partial\Phi}\ . \qquad (2.15)$$

III. QUANTIZATION

The quantum state of the Universe can be described by a wave function Ψ which is a function on the infinite-dimensional m nifold W of all three-metrics h_{ij} and matter fields Φ on S. A tangent vector to W is a pair of fields (γ_{ij},μ) on S where γ_{ij} can be regarded as a small ch nge of the metric h_{ij} and μ can be regarded as a small change of Φ. For each choice of $N>0$ on S there is a natural metric $\Gamma(N)$ on W:[15]

$$ds^2=\int N^{-1}\left[\frac{m_P{}^2}{32\pi}G^{ijkl}\gamma_{ij}\gamma_{kl}+\tfrac{1}{2}h^{1/2}\mu^2\right]d^3x\ . \qquad (3.1)$$

The wave function Ψ does not depend explicitly on the time t because t is just a coordinate which can be given arbitrary values by different choices of the undetermined multipliers N and N_i. This means that Ψ obeys the zero-energy Schrödinger equation:

$$H\Psi=0\ . \qquad (3.2)$$

The Hamiltonian operator H is the classical Hamiltonian with the usual substitutions:

$$\pi^{ij}(x)\to -i\frac{\delta}{\delta h_{ij}(x)},\quad \pi_\phi(x)\to -i\frac{\delta}{\delta\phi(x)}\ . \qquad (3.3)$$

Because N and N_i are regarded as independent Lagrange multipliers, the Schrödinger equation can be decomposed into two parts. There is the momentum constraint

$$H_-\Psi\equiv\int N_iH^id^3x\ \Psi$$
$$=\int h^{1/2}N_i\left[2\left[\frac{\delta}{\delta h_{ij}(x)}\right]_{|j}-h^{ij}\frac{\partial\Phi}{\partial x^j}\frac{\delta}{\delta\Phi(x)}\right]d^3x\ \Psi$$
$$=0\ . \qquad (3.4)$$

This implies that Ψ is the same on three-metrics and matter field configurations that are related by coordinate transformations in S. The other part of the Schrödinger equation, corresponding to $H_|\Psi=0$, where $H_|=\int NH_0d^3x$ is called the Wheeler-DeWitt equation. There is one Wheeler-DeWitt equation for each choice of N on S. One can regard them as a system of second-order partial differential equations for Ψ on W. There is some ambiguity in the choice of operator ordering in these equations but this will not affect the results of this paper. We shall assume that $H_|$ has the form[15]

$$(-\tfrac{1}{2}\nabla^2+\xi\mathbf{R}+V)\Psi=0\ , \qquad (3.5)$$

where ∇^2 is the Laplacian in the metric $\Gamma(N)$. R is the curvature scalar of this metric and the potential V is

$$V=\int h^{1/2}N\left[-\frac{m_P{}^2}{16\pi}{}^3R+\epsilon+U\right]d^3x\ , \qquad (3.6)$$

where $U=T^{00}-\frac{1}{2}\pi_\Phi{}^2$. The constant ϵ can be regarded as a renormalization of the cosmologic l constant Λ. We shall assume that the renormalized Λ is zero. We shall also assume that the coefficient ξ of the scalar curvature $\mathbf{R}$ of W is zero.

Any wave function Ψ which satisfies the momentum constraint and the Wheeler-DeWitt equation for each choice of N and N_i on S describes a possible quantum state of the Universe. We sh ll be concerned with the particular solution which represents the quantum state defined by a path integral over compact four-metrics without boundary. In this case[11-13]

$$\Psi=\int d[g_{\mu\nu}]d[\Phi]\exp[-\hat{I}(g_{\mu\nu},\Phi)]\ , \qquad (3.7)$$

where $\hat{I}$ is the Euclidean action obtained by setting N negative imaginary and the path integral is taken over ll compact four-metrics $g_{\mu\nu}$ and matter fields Φ which are bounded by S on which the three-metric is h_{ij} and the matter field is Φ. One can regard (3.7) as a boundary condition on the Wheeler-DeWitt equations. It implies that Ψ tends to a constant, which can be normalized to one, as h_{ij} goes to zero.

IV. UNPERTURBED FRIEDMANN MODEL

References 12–14 considered the minisuperspace model which consisted of a Friedmann model with metric

$$ds^2=\sigma^2(-N^2dt^2+a^2d\Omega_3^2)\,, \tag{4.1}$$

where $d\Omega_3^2$ is the metric of the unit three-sphere. The normalization factor $\sigma^2=2/3\pi m_P^2$ has been included for convenience. The model contains a scalar field $(2^{1/2}\pi\sigma)^{-1}\phi$ with mass $\sigma^{-1}m$ which is constant on surfaces of constant t. One can easily generalize this to the case of a scalar field with a potential $V(\phi)$. Such generalizations include models with higher-derivative quantum corrections.[16] The action is

$$I=-\tfrac{1}{2}\int dt\,Na^3\left[\frac{1}{N^2a^2}\left(\frac{da}{dt}\right)^2-\frac{1}{a^2}-\frac{1}{N^2}\left(\frac{d\phi}{dt}\right)^2+m^2\phi^2\right]. \tag{4.2}$$

The classical Hamiltonian is

$$H=\tfrac{1}{2}N(-a^{-1}\pi_a{}^2+a^{-3}\pi_\phi{}^2-a+a^3m^2\phi^2)\,, \tag{4.3}$$

where

$$\pi_a=-\frac{a\,da}{N\,dt},\quad \pi_\phi=\frac{a^3d\phi}{N\,dt}\,. \tag{4.4}$$

The classical Hamiltonian constraint is $H=0$. The classical field equations are

$$N\frac{d}{dt}\left(\frac{1}{N}\frac{d\phi}{dt}\right)+\frac{3}{a}\frac{da}{dt}\frac{d\phi}{dt}+N^2m^2\phi=0\,, \tag{4.5}$$

$$N\frac{d}{dt}\left(\frac{1}{N}\frac{da}{dt}\right)=N^2am^2\phi^2-2a\left(\frac{d\phi}{dt}\right)^2. \tag{4.6}$$

The Wheeler-DeWitt equation is

$$\tfrac{1}{2}Ne^{-3\alpha}\left(\frac{\partial^2}{\partial\alpha^2}-\frac{\partial^2}{\partial\phi^2}+2V\right)\Psi(\alpha,\phi)=0\,, \tag{4.7}$$

where

$$V=\tfrac{1}{2}(e^{6\alpha}m^2\phi^2-e^{4\alpha}) \tag{4.8}$$

and $\alpha=\ln a$. One can regard Eq. (4.7) as a hyperbolic equation for Ψ in the flat space with coordinates (α,ϕ) with α as the time coordinate. The boundary condition that gives the quantum state defined by a path integral over compact four-metrics is $\Psi\to 1$ as $\alpha\to-\infty$. If one integrates Eq. (4.7) with this boundary condition, one finds that the wave function starts oscillating in the region $V>0$, $|\phi|>1$ (this has been confirmed numerically[14]). One can interpret the oscillatory component of the wave function by the WKB approximation:

$$\Psi=\mathrm{Re}(Ce^{iS})\,, \tag{4.9}$$

where C is a slowly varying amplitude and S is a rapidly varying phase. One chooses S to satisfy the classical Hamilton-Jacobi equation:

$$H(\pi_\alpha,\pi_\phi,\alpha,\phi)=0\,, \tag{4.10}$$

where

$$\pi_\alpha=\frac{\partial S}{\partial\alpha},\quad \pi_\phi=\frac{\partial S}{\partial\phi}\,. \tag{4.11}$$

One can write (4.10) in the form

$$\tfrac{1}{2}f^{ab}\frac{\partial S\partial S}{\partial q^a\partial q^b}+e^{-3\alpha}V=0\,, \tag{4.12}$$

where f^{ab} is the inverse to the metric $\Gamma(1)$:

$$f^{ab}=e^{-3\alpha}\mathrm{diag}(-1,1)\,. \tag{4.13}$$

The wave function (4.9) will then satisfy the Wheeler-DeWitt equation if

$$\nabla^2C+2if^{ab}\frac{\partial C\partial S}{\partial q^a\partial q^b}+iC\nabla^2S=0\,, \tag{4.14}$$

where ∇^2 is the Laplacian in the metric f_{ab}. One can ignore the first term in Eq. (4.14) and can integrate the equation along the trajectories of the vector field $X^a=dq^a/dt=f^{ab}\partial S/\partial q^b$ and so determine the amplitude C. These trajectories correspond to classical solutions of the field equations. They are parametrized by the coordinate time t of the classical solutions.

The solutions that correspond to the oscillating part of the wave function of the minisuperspace model start out at $V=0$, $|\phi|>1$ with $da/dt=d\phi/dt=0$. They expand exponentially with

$$S=-\tfrac{1}{3}e^{3\alpha}m\,|\phi|(1-m^{-2}e^{-2\alpha}\phi^{-2})$$

$$\approx-\tfrac{1}{3}e^{3\alpha}m\,|\phi|\,, \tag{4.15}$$

$$\frac{d\alpha}{dt}=m\,|\phi|,\quad \frac{d\,|\phi|}{dt}=-\tfrac{1}{3}m\,. \tag{4.16}$$

After a time of order $3m^{-1}(|\phi_1|-1)$, where ϕ_1 is the initial value of ϕ, the field ϕ starts to oscillate with frequency m. The solution then becomes matter dominated and expands with e^α proportional to $t^{2/3}$. If there were other fields present, the massive scalar particles would decay into light particles and then the solution would expand with e^α proportional to $t^{1/2}$. Eventually the solution would reach a maximum radius of order $\exp(9\phi_1{}^2/2)$ or $\exp(9\phi_1{}^2)$ depending on whether it is radiation or matter dominated for most of the expansion. The solution would then recollapse in a similar manner.

V. THE PERTURBED FRIEDMANN MODEL

We assume that the metric is of the form (2.1) except the right hand side has been multiplied by a normalization factor σ^2. The three-metric h_{ij} has the form

$$h_{ij}=a^2(\Omega_{ij}+\epsilon_{ij})\,, \tag{5.1}$$

where Ω_{ij} is the metric on the unit three-sphere and ϵ_{ij} is a perturbation on this metric and may be expanded in harmonics:

$$\epsilon_{ij}=\sum_{n,l,m}[6^{1/2}a_{nlm}\tfrac{1}{3}\Omega_{ij}Q^n_{lm}+6^{1/2}b_{nlm}(P_{ij})^n_{lm}+2^{1/2}c^0_{nlm}(S^0_{ij})^n_{lm}+2^{1/2}c^e_{nlm}(S^e_{ij})^n_{lm}+2d^0_{nlm}(G^0_{ij})^n_{lm}+2d^e_{nlm}(G^e_{ij})^n_{lm}]\ . \quad (5.2)$$

The coefficients $a_{nlm}, b_{nlm}, c^0_{nlm}, c^e_{nlm}, d^0_{nlm}, d^e_{nlm}$ are functions of the time coordinate t but not the three spatial coordinates x^i.

The $Q(x^i)$ are the standard scalar harmonics on the three-sphere. The $P_{ij}(x^i)$ are given by (suppressing all but the i,j indices)

$$P_{ij}=\frac{1}{(n^2-1)}Q_{|ij}+\tfrac{1}{3}\Omega_{ij}Q\ . \quad (5.3)$$

They are traceless, $P_i{}^i=0$. The S_{ij} are defined by

$$S_{ij}=S_{i|j}+S_{j|i}\ , \quad (5.4)$$

where S_i are the transverse vector harmonics, $S_i{}^{|i}=0$. The G_{ij} are the transverse traceless tensor harmonics $G_i{}^i=G_{ij}{}^{|j}=0$. Further details about the harmonics and their normalization can be found in Appendix A.

The lapse, shift, and the scalar field $\Phi(x^i,t)$ can be expanded in terms of harmonics:

$$N=N_0\left[1+6^{-1/2}\sum_{n,l,m}g_{nlm}Q^n_{lm}\right]\ , \quad (5.5)$$

$$N_i=e^{\alpha}\sum_{n,l,m}[6^{-1/2}k_{nlm}(P_i)^n_{lm}+2^{1/2}j_{nlm}(S_i)^n_{lm}]\ , \quad (5.6)$$

$$\Phi=\sigma^{-1}\left[\frac{1}{2^{1/2}\pi}\phi(t)+\sum_{n,l,m}f_{nlm}Q^n_{lm}\right]\ , \quad (5.7)$$

where $P_i=[1/(n^2-1)]Q_{|i}$. Hereafter, the labels n, l, m, o, and e will be denoted simply by n. One can then expand the action to all orders in terms of the "background" quantities a,ϕ,N_0 but only to second order in the "perturbations" $a_n,b_n,c_n,d_n,f_n,g_n,k_n,j_n$:

$$I=I_0(a,\phi,N_0)+\sum_n I_n\ , \quad (5.8)$$

where I_0 is the action of the unperturbed model (4.2) and I_n is quadratic in the perturbations and is given in Appendix B.

One can define conjugate momenta in the usual manner. They are

$$\pi_a=-N_0{}^{-1}e^{3\alpha}\dot{\alpha}+\text{quadratic terms}\ , \quad (5.9)$$

$$\pi_\phi=N_0{}^{-1}e^{3\alpha}\dot{\phi}+\text{quadratic terms}\ , \quad (5.10)$$

$$\pi_{a_n}=-N_0{}^{-1}e^{3\alpha}[\dot{a}_n+\dot{\alpha}(a_n-g_n)+\tfrac{1}{3}e^{-\alpha}k_n]\ , \quad (5.11)$$

$$\pi_{b_n}=N_0{}^{-1}e^{3\alpha}\frac{(n^2-4)}{(n^2-1)}(\dot{b}_n+4\dot{\alpha}b_n-\tfrac{1}{3}e^{-\alpha}k_n)\ , \quad (5.12)$$

$$\pi_{c_n}=N_0{}^{-1}e^{3\alpha}(n^2-4)(\dot{c}_n+4\dot{\alpha}c_n-e^{-\alpha}j_n)\ , \quad (5.13)$$

$$\pi_{d_n}=N_0{}^{-1}e^{3\alpha}(\dot{d}_n+4\dot{\alpha}d_n)\ , \quad (5.14)$$

$$\pi_{f_n}=N_0{}^{-1}e^{3\alpha}[\dot{f}_n+\dot{\phi}(3a_n-g_n)]\ . \quad (5.15)$$

The quadratic terms in Eqs. (5.9) and (5.10) are given in Appendix B. The Hamiltonian can then be expressed in terms of these momenta and the other quantities:

$$H=N_0\left[H_{|0}+\sum_n H^n_{|2}+\sum_n g_nH^n_{|1}\right]+\sum_n(k_n{}^SH^n_{\_1}+j_n{}^VH^n_{\_1})\ . \quad (5.16)$$

The subscripts 0,1,2 on the $H_|$ and $H_\_$ denote the orders of the quantities in the perturbations and S and V denote the scalar and vector parts of the shift part of the Hamiltonian. $H_{|0}$ is the Hamiltonian of the unperturbed model with $N=1$:

$$H_{|0}=\tfrac{1}{2}e^{-3\alpha}(-\pi_a{}^2+\pi_\phi{}^2+e^{6\alpha}m^2\phi^2-e^{4\alpha})\ . \quad (5.17)$$

The second-order Hamiltonian is given by

$$H_{|2}=\sum_n H^n_{|2}=\sum_n({}^SH^n_{|2}+{}^VH^n_{|2}+{}^TH^n_{|2})\ ,$$

where

$$\begin{aligned}{}^SH^n_{|2}=\tfrac{1}{2}e^{-3\alpha}\Bigg[&\left[\tfrac{1}{2}a_n{}^2+\frac{10(n^2-4)}{(n^2-1)}b_n{}^2\right]\pi_a{}^2+\left[\tfrac{15}{2}a_n{}^2+\frac{6(n^2-4)}{(n^2-1)}b_n{}^2\right]\pi_\phi{}^2\\&-\pi_{a_n}{}^2+\frac{(n^2-1)}{(n^2-4)}\pi_{b_n}{}^2+\pi_{f_n}{}^2+2a_n\pi_{a_n}\pi_a+8b_n\pi_{b_n}\pi_a-6a_n\pi_{f_n}\pi_\phi\\&-e^{4\alpha}\left[\tfrac{1}{3}(n^2-\tfrac{5}{2})a_n{}^2+\frac{(n^2-7)}{3}\frac{(n^2-4)}{(n^2-1)}b_n{}^2+\tfrac{2}{3}(n^2-4)a_nb_n-(n^2-1)f_n{}^2\right]\\&+e^{6\alpha}m^2(f_n{}^2+6a_nf_n\phi)+e^{6\alpha}m^2\phi^2\left[\tfrac{3}{2}a_n{}^2-\frac{6(n^2-4)}{(n^2-1)}b_n{}^2\right]\Bigg]\ ,\end{aligned} \quad (5.18)$$

$${}^VH^n_{|2}=\tfrac{1}{2}e^{-3\alpha}\left[(n^2-4)c_n{}^2(10\pi_a{}^2+6\pi_\phi{}^2)+\frac{1}{(n^2-4)}\pi_{c_n}{}^2+8c_n\pi_{c_n}\pi_a+(n^2-4)c_n{}^2(2e^{4\alpha}-6e^{6\alpha}m^2\phi^2)\right]\ , \quad (5.19)$$

$${}^TH^n_{|2}=\tfrac{1}{2}e^{-3\alpha}\{d_n{}^2(10\pi_a{}^2+6\pi_\phi{}^2)+\pi_{d_n}{}^2+8d_n\pi_{d_n}\pi_a+d_n{}^2[(n^2+1)e^{4\alpha}-6e^{6\alpha}m^2\phi^2]\}\ . \quad (5.20)$$

The first-order Hamiltonians are

$$H^n_{|1}=\tfrac{1}{2}e^{-3\alpha}\{-a_n(\pi_\alpha{}^2+3\pi_\phi{}^2)+2(\pi_\phi\pi_{f_n}-\pi_\alpha\pi_{a_n})+m^2e^{6\alpha}(2f_n\phi+3a_n\phi^2)-\tfrac{2}{3}e^{4\alpha}[(n^2-4)b_n+(n^2+\tfrac{1}{2})a_n]\}\ . \tag{5.21}$$

The shift parts of the Hamiltonian are

$$^SH^n_{\_1}=\tfrac{1}{3}e^{-3\alpha}\left[-\pi_{a_n}+\pi_{b_n}+\left[a_n+\frac{4(n^2-4)}{(n^2-1)}b_n\right]\pi_\alpha+3f_n\pi_\phi\right], \tag{5.22}$$

$$^VH^n_{\_1}=e^{-\alpha}[\pi_{c_n}+4(n^2-4)c_n\pi_\alpha]\ . \tag{5.23}$$

The classical field equations are given in Appendix B.

Because the Lagrange multipliers N_0, g_n, k_n, j_n are independent, the zero energy Schrödinger equation

$$H\Psi=0 \tag{5.24}$$

can be decomposed as before into momentum constraints and Wheeler-DeWitt equations. As the momentum constraints are linear in the momenta, there is no ambiguity in the operator ordering. One therefore has

$$^SH^n_{\_1}\Psi=-\tfrac{1}{3}e^{-3\alpha}\left[\frac{\partial}{\partial a_n}-\left[a_n+\frac{4(n^2-4)}{(n^2-1)}b_n\right]\frac{\partial}{\partial\alpha}-\frac{\partial}{\partial b_n}-3f_n\frac{\partial}{\partial\phi}\right]\Psi=0\ , \tag{5.25}$$

$$^VH^n_{\_1}\Psi=e^{-\alpha}\left[\frac{\partial}{\partial c_n}+4(n^2-4)c_n\frac{\partial}{\partial\alpha}\right]\Psi=0\ . \tag{5.26}$$

The first-order Hamiltonians $H^n_{|1}$ give a series of finite dimensional second-order differential equations, one for each n. In the order of approximation that we are using, the ambiguity in the operator ordering will consist of the possible addition of terms linear in $\partial/\partial\alpha$. The effect of such terms can be compensated for by multiplying the wave function by powers of e^α. This will not affect the relative probabilities of different observations at a given value of α. We shall therefore ignore such ambiguities and terms:

$$\tfrac{1}{2}e^{-3\alpha}\left[a_n\left[\frac{\partial^2}{\partial\alpha^2}+3\frac{\partial^2}{\partial\phi^2}\right]-2\left[\frac{\partial^2}{\partial f_n\partial\phi}-\frac{\partial^2}{\partial a_n\partial\alpha}\right]+m^2e^{6\alpha}[2\phi f_n+3a_n\phi^2]-\tfrac{2}{3}e^{4\alpha}[(n^2-4)b_n+(n^2+\tfrac{1}{2})a_n]\right]\Psi=0\ . \tag{5.27}$$

Finally, one has an infinite-dimensional second-order differential equation

$$\left[H_{|0}+\sum_n({}^SH^n_{|2}+{}^VH^n_{|2}+{}^TH^n_{|2})\right]\Psi=0\ , \tag{5.28}$$

where $H_{|0}$ is the operator in the Wheeler-DeWitt equation of the unperturbed Friedmann minisuperspace model:

$$H_{|0}=\tfrac{1}{2}e^{-3\alpha}\left[\frac{\partial^2}{\partial\alpha^2}-\frac{\partial^2}{\partial\phi^2}+e^{6\alpha}m^2\phi^2-e^{4\alpha}\right] \tag{5.29}$$

and

$$\begin{aligned}^SH^n_{|2}=\tfrac{1}{2}e^{-3\alpha}\Bigg[&-\left[\tfrac{1}{2}a_n{}^2+\frac{10(n^2-4)}{(n^2-1)}b_n{}^2\right]\frac{\partial^2}{\partial\alpha^2}-\left[\tfrac{15}{2}a_n{}^2+\frac{6(n^2-4)}{(n^2-1)}b_n{}^2\right]\frac{\partial^2}{\partial\phi^2}\\&+\frac{\partial^2}{\partial a_n{}^2}-\frac{(n^2-1)}{(n^2-4)}\frac{\partial^2}{\partial b_n{}^2}-\frac{\partial^2}{\partial f_n{}^2}-2a_n\frac{\partial^2}{\partial a_n\partial\alpha}-8b_n\frac{\partial^2}{\partial b_n\partial\alpha}+6a_n\frac{\partial^2}{\partial f_n\partial\phi}\\&-e^{4\alpha}\left[\tfrac{1}{3}(n^2-\tfrac{5}{2})a_n{}^2+\frac{(n^2-7)}{3}\frac{(n^2-4)}{(n^2-1)}b_n{}^2+\tfrac{2}{3}(n^2-4)a_nb_n-(n^2-1)f_n{}^2\right]\\&+e^{6\alpha}m^2(f_n{}^2+6a_nf_n\phi)+e^{6\alpha}m^2\phi^2\left[\tfrac{3}{2}a_n{}^2-\frac{6(n^2-4)}{(n^2-1)}b_n{}^2\right]\Bigg]\ ,\end{aligned} \tag{5.30}$$

$$^VH^n_{|2}=\tfrac{1}{2}e^{-3\alpha}\left[-(n^2-4)c_n{}^2\left[10\frac{\partial^2}{\partial\alpha^2}+6\frac{\partial^2}{\partial\phi^2}\right]-\frac{1}{(n^2-4)}\frac{\partial^2}{\partial c_n{}^2}-8c_n\frac{\partial^2}{\partial c_n\partial\alpha}+(n^2-4)c_n{}^2(2e^{4\alpha}-6e^{6\alpha}m^2\phi^2)\right], \tag{5.31}$$

$$^TH^n_{|2}=\tfrac{1}{2}e^{-3\alpha}\left[-d_n{}^2\left[10\frac{\partial^2}{\partial\alpha^2}+6\frac{\partial^2}{\partial\phi^2}\right]-\frac{\partial^2}{\partial d_n{}^2}-8d_n\frac{\partial^2}{\partial d_n\partial\alpha}+d_n{}^2[(n^2+1)e^{4\alpha}-6e^{6\alpha}m^2\phi^2]\right]\ . \tag{5.32}$$

We shall call Eq. (5.28) the master equation. It is not hyperbolic because, as well as the positive second derivatives $\partial^2/\partial\alpha^2$ in $H_{|0}$, there are the positive second derivatives $\partial^2/\partial a_n{}^2$ in each ${}^SH^n_{|2}$. However, one can use the momentum constraint (5.25) to substitute for the partial derivatives with respect to a_n and then solve the resultant differential equation on $a_n=0$. Similarly, one can use the momentum constraint (5.26) to substitute for the partial derivatives with respect to c_n and then solve on $c_n=0$. One thus obtains a modified equation which is hyperbolic for small f_n. If one knows the wave function on $a_n=0=c_n$, one can use the momentum constraints to calculate the wave function at other values of a_n and c_n.

VI. THE WAVE FUNCTION

Because the perturbation modes are not coupled to each other, the wave function can be expressed as a sum of terms of the form

$$\Psi=\mathrm{Re}\left[\Psi_0(\alpha,\phi)\prod_n \Psi^{(n)}(\alpha,\phi,a_n,b_n,c_n,d_n,f_n)\right]$$
$$=\mathrm{Re}(C\,e^{iS}) , \tag{6.1}$$

where S is a rapidly varying function of α and ϕ and C is a slowly varying function of all the variables. If one substitutes (6.1) into the master equation and divides by Ψ, one obtains

$$-\frac{\nabla_2{}^2\Psi_0}{2\Psi_0}-\sum_n\frac{\nabla_2{}^2\Psi^{(n)}}{2\Psi^{(n)}}-\sum_{n\lessgtr m}\frac{(\nabla_2\Psi^{(n)})\cdot(\nabla_2\Psi^{(m)})}{2\Psi^{(n)}\Psi^{(m)}}$$
$$-\frac{(\nabla_2\Psi_0)}{\Psi_0}\cdot\left[\sum_n\frac{\nabla_2\Psi^{(n)}}{\Psi^{(n)}}\right]$$
$$+\sum_n\frac{H^n_{|2}\Psi}{\Psi}+e^{-3\alpha}V(\alpha,\phi)=0 , \tag{6.2}$$

where $\nabla_2{}^2$ is the Laplacian in the minisuperspace metric $f_{ab}=e^{3\alpha}\mathrm{diag}(-1,1)$ and the dot product is with respect to this metric.

An individual perturbation mode does not contribute a significant fraction of the sums in the third and fourth terms in Eq. (6.2). Thus these terms can be replaced by

$$-\frac{(\nabla_2\Psi)}{\Psi}\cdot\sum_n\frac{(\nabla_2\Psi^{(n)})}{\Psi^{(n)}}+\frac{1}{2}\left[\sum_n\frac{\nabla_2\Psi^{(n)}}{\Psi^{(n)}}\right]^2$$
$$\approx -i(\nabla_2 S)\cdot\sum_n\frac{(\nabla_2\Psi^{(n)})}{\Psi^{(n)}}+\frac{1}{2}\left[\sum_n\frac{\nabla_2\Psi^{(n)}}{\Psi^{(n)}}\right]^2 . \tag{6.3}$$

In order that the ansatz (6.1) be valid, the terms in (6.2) that depend on a_n,b_n,c_n,d_n,f_n have to cancel out. This implies

$$\frac{(\nabla_2\Psi)}{\Psi}\cdot(\nabla_2\Psi^{(n)})+\tfrac{1}{2}\nabla_2{}^2\Psi^{(n)}=\frac{H^n_{|2}\Psi}{\Psi}\Psi^{(n)} , \tag{6.4}$$

$$(-\tfrac{1}{2}\nabla_2{}^2+e^{-3\alpha}V+\tfrac{1}{2}J\cdot J)\Psi_0=0 , \tag{6.5}$$

where

$$J=\sum_n\frac{\nabla_2\Psi^{(n)}}{\Psi^{(n)}} .$$

In regions in which the phase S is a rapidly varying function of α and ϕ, one can neglect the second term in (6.4) in comparison with the first term. One can also replace the π_α and π_ϕ which appear in $H^n_{|2}$ by $\partial S/\partial\alpha$ and $\partial S/\partial\phi$, respectively. The vector $X^a=f^{ab}\partial S/\partial q^b$ obtained by raising the covector $\nabla_2 S$ by the inverse minisuperspace metric f^{ab} can be regarded as $\partial/\partial t$ where t is the time parameter of the classical Friedmann metric that corresponds to Ψ by the WKB approximation. One then obtains a time dependent Schrödinger equation for each mode along a trajectory of the vector field X^a:

$$i\frac{\partial\Psi^{(n)}}{\partial t}=H^n_{|2}\Psi^{(n)} . \tag{6.6}$$

Equation (6.5) can be interpreted as the Wheeler-DeWitt equation for a two-dimensional minisuperspace model with an extra term $\frac{1}{2}J\cdot J$ arising from the perturbations. In order to make J finite, one will have to make subtractions. Subtracting out the ground-state energies of the $H^n_{|2}$ corresponds to a renormalization of the cosmological constant Λ. There is a second subtraction which corresponds to a renormalization of the Planck mass m_P and a third one which corresponds to a curvature-squared counterterm. The effect of such higher-derivative terms in the action has been considered elsewhere.[16]

One can write $\Psi^{(n)}$ as

$$\Psi^{(n)}={}^S\Psi^{(n)}(\alpha,\phi,a_n,b_n,f_n)\,{}^V\Psi^{(n)}(\alpha,\phi,c_n)\,{}^T\Psi^{(n)}(\alpha,\phi,d_n) , \tag{6.7}$$

where ${}^S\Psi^{(n)}$, ${}^V\Psi^{(n)}$, and ${}^T\Psi^{(n)}$ obey independent Schrödinger equations with ${}^SH^n_{|2}$, ${}^VH^n_{|2}$, and ${}^TH^n_{|2}$, respectively.

VII. THE BOUNDARY CONDITIONS

We want to find the solution of the master equation that corresponds to

$$\Psi[h_{ij},\Phi]=\int d[g_{\mu\nu}]d[\Phi]\exp(-\hat{I}) , \tag{7.1}$$

where the integral is taken over all compact four-metrics and matter fields which are bounded by the three-surface S. If one takes the scale parameter α to be very negative but keeps the other parameters fixed, the Euclidean action $\hat{I}$ tends to zero like $e^{2\alpha}$. Thus one would expect Ψ to tend to one as α tends to minus infinity.

One can estimate the form of the scalar, vector, and tensor parts ${}^S\Psi^{(n)}$, ${}^V\Psi^{(n)}$, ${}^T\Psi^{(n)}$ of the perturbation $\Psi^{(n)}$ from the path integral (7.1) One takes the four-metric $g_{\mu\nu}$ and the scalar field Φ to be of the background form

$$ds^2=\sigma^2(-N^2dt^2+e^{2\alpha(t)}d\Omega_3{}^2) \tag{7.2}$$

and $\phi(t)$, respectively, plus a small perturbation described by the variables (a_n,b_n,f_n), c_n, and d_n as functions of t. In order for the background four-metric to be compact, it has to be Euclidean when $\alpha=-\infty$, i.e., N has to be purely negative imaginary at $\alpha=-\infty$, which we shall take to be $t=0$. In regions in which the metric is Lorentzian, N

will be real and positive. In order to allow a smooth transition from Euclidean to Lorentzian, we shall take N to be of the form $-ie^{i\mu}$ where $\mu=0$ at $t=0$. In order that the four-metric and the scalar field be regular at $t=0, a_n, b_n, c_n, d_n, f_n$ have to vanish there.

The tensor perturbations d_n have the Euclidean action

$$ {}^T\hat{I}_n=\tfrac{1}{2}\int dt\, d_n {}^T D d_n + \text{boundary term}\ , \tag{7.3} $$

where

$$ {}^T D=\left[-\frac{d}{dt}\left[\frac{e^{3\alpha}d}{iN_0 dt}\right]+iN_0 e^{\alpha}(n^2-1)\right]+4iN_0 e^{3\alpha}\left[+\tfrac{1}{2}e^{-2\alpha}-\tfrac{3}{2}m^2\phi^2-\frac{3\dot\phi^2}{2(iN_0)^2}-\frac{3\dot\alpha^2}{2(iN_0)^2}-\frac{1d}{iN_0 dt}\left[\frac{\dot\alpha}{iN_0}\right]\right]\ . \tag{7.4} $$

The last term in (7.4) vanishes if the background metric satisfies the background field equations. The action is extremized when d_n satisfies the equation

$$ {}^T D d_n=0\ . \tag{7.5} $$

For a d_n that satisfies (7.5), the action is just the boundary term

$$ {}^T\hat{I}_n^{\,\mathrm{cl}}=\frac{1}{2iN_0}e^{3\alpha}(d_n\dot d_n+4\dot\alpha d_n^2)\ . \tag{7.6} $$

The path integral over d_n will be

$$ \int d[d_n]\exp(-{}^T\hat{I}_n)=(\det{}^T D)^{-1/2}\exp(-{}^T\hat{I}_n^{\,\mathrm{cl}})\ . \tag{7.7} $$

One now has to integrate (7.7) over different background metrics to obtain the wave function ${}^T\Psi^{(n)}$. One expects the dominant contribution to come from background metrics that are near a solution of the classical background field equations. For such metrics one can employ the adiabatic approximation in which one regards α to be a slowly varying function of t. Then the solution of (7.5) which obeys the boundary condition $d_n=0$ at $t=0$ is

$$ d_n=A(e^{\nu\tau}-e^{-\nu\tau})\ , \tag{7.8} $$

where $\nu=e^{-\alpha}(n^2-1)^{1/2}$ and $\tau=\int iN_0 dt$. This approximation will be valid for background fields which are near a solution of the background field equations and for which

$$ \left|\frac{\dot\alpha}{N_0}\right| \ll n\,e^{-\alpha}\ . \tag{7.9} $$

For a regular Euclidean metric, $|\dot\alpha/N_0|=e^{-\alpha}$ near $t=0$. If the metric is a Euclidean solution of the background field equations, then $|\dot\alpha/N_0|<e^{-\alpha}$. Thus the adiabatic approximation should hold for large values of n into the region in which the solution of the background field equations becomes Lorentzian and the WKB approximation can be used. The wave function ${}^T\Psi^{(n)}$ will then be

$$ {}^T\Psi^{(n)}=B\exp\left[-\left[\tfrac{1}{2}n\,e^{2\alpha}\coth(\nu\tau)+\frac{2}{iN_0}\dot\alpha\, e^{3\alpha}\right]d_n^2\right]\ . \tag{7.10} $$

In the Euclidean region, τ will be real and positive. For large values of n, $\coth(\nu\tau)\approx 1$. In the Lorentzian region where the WKB approximation applies, τ will be complex but it will still have a positive real part and $\coth(\nu\tau)$ will still be approximately 1 for large n. Thus

$$ {}^T\Psi^{(n)}=B\exp\left[-2i\frac{\partial S}{\partial\alpha}d_n^2-\tfrac{1}{2}n\,e^{2\alpha}d_n^2\right]\ . \tag{7.11} $$

The normalization constant B can be chosen to be 1. Thus, apart from a phase factor, the gravitational-wave modes enter the WKB region in their ground state.

We now consider the vector part ${}^V\Psi^{(n)}$ of the wave function. This is pure gauge as the quantities c_n can be given any value by gauge transformations parametrized by the j_n. The freedom to make gauge transformations is reflected quantum mechanically in the constraint

$$ e^{-\alpha}\left[\frac{\partial}{\partial c_n}+4(n^2-4)c_n\frac{\partial}{\partial\alpha}\right]\Psi=0\ . \tag{7.12} $$

One can integrate (7.12) to give

$$ \Psi(\alpha,\{c_n\})=\Psi\left[\alpha-2\sum_n(n^2-4)c_n^2,0\right]\ , \tag{7.13} $$

where the dependence on the other variables has been suppressed. One can also replace $\partial\Psi/\partial\alpha$ by $i(\partial S/\partial\alpha)\Psi$. One can then solve for ${}^V\Psi^{(n)}$:

$$ {}^V\Psi^{(n)}=\exp\left[2i(n^2-4)c_n^2\frac{\partial S}{\partial\alpha}\right]\ . \tag{7.14} $$

The scalar perturbation modes a_n, b_n, and f_n involve a combination of the behavior of the tensor and vector perturbations. The scalar part of the action is given in Appendix B. The action is extremized by solutions of the classical equations

$$ N_0\frac{d}{dt}\left[e^{3\alpha}\frac{\dot a_n}{N_0}\right]+\tfrac{1}{3}(n^2-4)N_0^2e^{\alpha}(a_n+b_n)+3e^{3\alpha}(\dot\phi\dot f_n-N_0^2m^2\phi f_n) $$
$$ =N_0^2[3e^{3\alpha}m^2\phi^2-\tfrac{1}{3}(n^2+2)e^{\alpha}]g_n+e^{3\alpha}\dot\alpha\dot g_n-\tfrac{1}{3}N_0\frac{d}{dt}\left[e^{2\alpha}\frac{k_n}{N_0}\right] \tag{7.15} $$

$$N_0\frac{d}{dt}\left[e^{3\alpha}\frac{\dot b_n}{N_0}\right]-\tfrac{1}{3}(n^2-1)N_0{}^2e^{\alpha}(a_n+b_n)=\tfrac{1}{3}(n^2-1)N_0{}^2e^{\alpha}g_n+\tfrac{1}{3}N_0\frac{d}{dt}\left[e^{2\alpha}\frac{k_n}{N_0}\right], \tag{7.16}$$

$$N_0\frac{d}{dt}\left[e^{3\alpha}\frac{\dot f_n}{N_0}\right]+3e^{3\alpha}\dot\phi\,\dot a_n+N_0{}^2[m^2e^{3\alpha}+(n^2-1)e^{\alpha}]f_n=e^{3\alpha}(-2N_0{}^2m^2\phi g_n+\dot\phi\,\dot g_n-e^{-\alpha}\dot\phi k_n)\,. \tag{7.17}$$

There is a three-parameter family of solutions to (7.15)–(7.17) which obey the boundary condition $a_n=b_n=f_n=0$ at $t=0$. There are however, two constraint equations:

$$\dot a_n+\frac{(n^2-4)}{(n^2-1)}\dot b_n+3f_n\dot\phi=\dot\alpha g_n-\frac{e^{-\alpha}}{(n^2-1)}k_n\,, \tag{7.18}$$

$$3a_n(-\dot\alpha^2+\dot\phi^2)+2(\dot\phi\dot f_n-\dot\alpha\dot a_n)+N_0{}^2m^2(2f_n\phi+3a_n\phi^2)-\tfrac{2}{3}N_0{}^2e^{-2\alpha}[(n^2-4)b_n+(n^2+\tfrac{1}{2})a_n]$$
$$=\tfrac{2}{3}\dot\alpha e^{-\alpha}k_n+2g_n(-\dot\alpha^2+\dot\phi^2)\,. \tag{7.19}$$

These correspond to the two gauge degrees of freedom parametrized by k_n and g_n, respectively. The Euclidean action for a solution to Eqs. (7.15)–(7.19) is

$$S\!\!\!\!\!\!\;I_n^{\rm cl}=\frac{1}{2iN_0}e^{3\alpha}\left[-a_n\dot a_n+\frac{(n^2-4)}{(n^2-1)}b_n\dot b_n+f_n\dot f_n+\dot\alpha\left[-a_n{}^2+\frac{4(n^2-4)}{(n^2-1)}b_n{}^2\right]+3\dot\phi a_nf_n+g_n(\dot\alpha a_n-\dot\phi f_n)\right.$$
$$\left.-\tfrac{1}{3}e^{-\alpha}k_n\left[a_n+\frac{(n^2-4)}{(n^2-1)}b_n\right]\right], \tag{7.20}$$

where the background field equations have been used.

In many ways the simplest gauge to work in is that with $g_n=k_n=0$. However, this gauge does not allow one to find a compact four-metric which is bounded by a three-surface with arbitrary values of a_n, b_n, and f_n and which is a solution of the Eqs. (7.15)–(7.17) and the constraint equations. Instead, we shall use the gauge $a_n=b_n=0$ and shall solve the constraint Eqs. (7.18) and (7.19) to find g_n and k_n:

$$g_n=3\frac{(n^2-1)\dot\alpha\dot\phi f_n+\dot\phi\dot f_n+N_0{}^2m^2\phi f_n}{(n^2-4)\dot\alpha^2+3\dot\phi^2}\,, \tag{7.21}$$

$$k_n=3(n^2-1)e^{\alpha}\frac{\dot\alpha\dot\phi\dot f_n+N_0{}^2m^2\phi f_n\dot\alpha-3f_n\dot\phi(-\dot\alpha^2+\dot\phi^2)}{(n^2-4)\dot\alpha^2+3\dot\phi^2}\,. \tag{7.22}$$

With these substituted, (7.17) becomes a second-order equation for f_n,

$$N_0\frac{d}{dt}\left[e^{3\alpha}\frac{\dot f_n}{N_0}\right]+N_0{}^2[m^2e^{3\alpha}+(n^2-1)e^{\alpha}]f_n=e^{3\alpha}(-2N_0{}^2m^2\phi g_n+\dot\phi\,\dot g_n-e^{-\alpha}\dot\phi k_n)\,. \tag{7.23}$$

For large n we can again use the adiabatic approximation to estimate the solution of (7.23) when $|\phi|>1$:

$$f_n=A\sinh(\nu\tau)\,, \tag{7.24}$$

where $\nu^2=e^{-2\alpha}(n^2-1)$. Thus for these modes

$${}^{S}\Psi^{(n)}(\alpha,\phi,0,0,f_n)\approx\exp\left[-\tfrac{1}{2}n\,e^{2\alpha}f_n{}^2-\tfrac{1}{2}i\frac{\partial S}{\partial\phi}g_nf_n\right]. \tag{7.25}$$

This is of the ground-state form apart from a small phase factor. The value of ${}^{S}\Psi^{(n)}$ at nonzero values of a_n and b_n can be found by integrating the constraint equations (5.25) and (5.27).

The tensor and scalar modes start off in their ground states, apart possibly from the modes at low n. The vector modes are pure gauge and can be neglected. Thus the total energy

$$E=\sum_n\frac{H_{|2}^{(n)}\Psi^{(n)}}{\Psi^{(n)}}$$

of the perturbations will be small when the ground-state energies are subtracted. But $E=i(\nabla_2S)\cdot J$ where $J=\sum_n\nabla_2\Psi^{(n)}/\Psi^{(n)}$. Thus J is small. This means that the wave function Ψ_0 will obey the Wheeler-DeWitt equation of the unperturbed minisuperspace model and the phase factor S will be approximately $-i\ln\Psi_0$. However the homogeneous scalar field mode ϕ will not start out in its ground state. There are two reasons for this: first, regularity at $t=0$ requires $a_n=b_n=c_n=d_n=f_n=0$, but does not require $\phi=0$. Second, the classical field equation for ϕ is of the form of a damped harmonic oscillator with a constant frequency m rather than a decreasing frequency $e^{-\alpha}n$. This means that the adiabatic approximation is not valid at small t and that the solution of the classical field equation is ϕ approximately constant. The action of such solutions is small, so large values of $|\phi|$ are not damped as they are for the other variables. Thus the WKB trajectories which start out from large values of $|\phi|$ have high probability. They will correspond to classical solutions which have a long inflationary period and then go over to a matter-dominated expansion. In a realistic model which included other fields of low rest mass, the matter energy in the oscillations of the massive scalar field would decay into light particles with a thermal spectrum. The model would then expand as a radiation-dominated universe.

VIII. GROWTH OF PERTURBATIONS

The tensor modes will obey the Schrödinger equation

$$i\frac{\partial\,{}^{T}\Psi^{(n)}}{\partial t}={}^{T}H^{n}_{|2}\,{}^{T}\Psi^{(n)} \tag{8.1}$$

$$=\tfrac{1}{2}e^{-3\alpha}\left\{+d_n{}^2\left[10\left[\frac{\partial S}{\partial\alpha}\right]^2+6\left[\frac{\partial S}{\partial\phi}\right]^2\right]-\frac{\partial^2}{\partial d_n{}^2}-8d_n i\frac{\partial S}{\partial\alpha}\frac{\partial}{\partial d_n}+d_n{}^2[(n^2+1)e^{4\alpha}-6e^{6\alpha}m^2\phi^2]\right\}. \tag{8.2}$$

One can write

$${}^{T}\Psi^{(n)}=\exp(-2\alpha)\exp\left[-2i\frac{\partial S}{\partial\alpha}d_n{}^2\right]{}^{T}\Psi_0^{(n)}, \tag{8.3}$$

then

$$i\frac{\partial\,{}^{T}\Psi_0^{(n)}}{\partial t}=\tfrac{1}{2}e^{-3\alpha}\left[-\frac{\partial^2}{\partial d_n{}^2}+d_n{}^2(n^2-1)e^{4\alpha}\right]{}^{T}\Psi_0^{(n)}. \tag{8.4}$$

The WKB approximation to the background Wheeler-DeWitt equation has been used in deriving (8.4). Then (8.4) has the form of the Schrödinger equation for an oscillator with a time-dependent frequency $\nu=(n^2-1)^{1/2}e^{-\alpha}$. Initially the wave function ${}^{T}\Psi_0^{(n)}$ will be in the ground state (apart from a normalization factor) and the frequency ν will be large compared to $\dot\alpha$. In this case one can use the adiabatic approximation to show that ${}^{T}\Psi_0^{(n)}$ remains in the ground state

$${}^{T}\Psi_0^{(n)}\approx\exp(-\tfrac{1}{2}n\,e^{2\alpha}d_n{}^2). \tag{8.5}$$

The adiabatic approximation will break down when $\nu\approx\dot\alpha$, i.e., the wave length of the gravitational mode becomes equal to the horizon scale in the inflationary period. The wave function ${}^{T}\Psi_0^{(n)}$ will then freeze

$${}^{T}\Psi_0^{(n)}\approx\exp(-\tfrac{1}{2}n\,e^{2\alpha_*}d_n{}^2), \tag{8.6}$$

where α_* is the value of α at which the mode goes outside the horizon. The wave function ${}^{T}\Psi_0^{(n)}$ will remain of the form (8.6) until the mode reenters the horizon in the matter- or radiation-dominated era at the much greater value α_e of α. One can then apply the adiabatic approximation again to (8.4) but ${}^{T}\Psi_0^{(n)}$ will no longer be in the ground state; it will be a superposition of a number of highly excited states. This is the phenomenon of the amplification of the ground-state fluctuations in the gravitational-wave modes that was discussed in Refs. 9, 17, and 18.

The behavior of the scalar modes is rather similar but their description is more complicated because of the gauge degrees of freedom. In the previous section we evaluated the wave function ${}^{S}\Psi^{(n)}$ on $a_n=b_n=0$ by the path-integral prescription. The ground-state form (in f_n) that we found will be valid until the adiabatic approximation breaks down, i.e., until the wavelength of the mode exceeds the horizon distance during the inflationary period. In order to discuss the subsequent behavior of the wave function. It is convenient to use the first-order Hamiltonian constraint (5.27) to evaluate ${}^{S}\Psi^{(n)}$ on $a_n\neq0, b_n=f_n=0$. One finds that

$${}^{S}\Psi^{(n)}(\alpha,\phi,a_n,0,0)=B\exp[iCa_n{}^2]\,{}^{S}\Psi_0^{(n)}(\alpha,\phi,a_n). \tag{8.7}$$

The normalization and phase factors B and C depend on α and ϕ but not a_n:

$$C=\frac{1}{2}\left[\frac{\partial S}{\partial\alpha}\right]^{-1}\left[\left[\frac{\partial S}{\partial\alpha}\right]^2-\tfrac{1}{3}(n^2-4)e^{4\alpha}\right]. \tag{8.8}$$

At the time the wavelength of the mode equals the horizon distance during the inflationary period, the wave function ${}^{S}\Psi_0^{(n)}$ has the form

$${}^{S}\Psi_0^{(n)}=\exp(-\tfrac{1}{2}ny_*{}^{-2}e^{2\alpha_*}a_n{}^2), \tag{8.9}$$

where y_* is the value of $y=(\partial S/\partial\alpha)[\partial S/\partial\phi]^{-1}$ when the mode leaves the horizon, $y_*=3\phi_*$. More generally, in the case of a scalar field with a potential $V(\phi)$, $y=6V(\partial V/\partial\phi)^{-1}$.

One can obtain a Schrödinger equation for ${}^{S}\Psi_0^{(n)}$ by putting $b_n=f_n=0$ in the scalar Hamiltonian ${}^{S}H^{n}_{|2}$ and substituting for $\partial/\partial b_n$ and $\partial/\partial f_n$ from the momentum constraint (5.25) and the first-order Hamiltonian constraint (5.27), respectively. This gives

$$i\frac{\partial\,{}^{S}\Psi_0^{(n)}}{\partial t}=\tfrac{1}{2}e^{-3\alpha}\left\{-y^2\frac{\partial^2}{\partial a_n{}^2}+e^{4\alpha}(n^2-4)\times\left[\frac{1}{y^2}-\tfrac{1}{3}e^{4\alpha}\left[\frac{\partial S}{\partial\alpha}\right]^{-2}\right]a_n{}^2\right\}{}^{S}\Psi_0^{(n)}, \tag{8.10}$$

where terms of order $1/n^2$ have been neglected. The term $e^{4\alpha}[\partial S/\partial\alpha]^{-2}$ will be small compared to $1/y^2$ except near the time of maximum radius of the background solution. The Schrödinger equation for ${}^{S}\Psi_0^{(n)}(a_n)$ is very similar to the equation for ${}^{T}\Psi_0^{(n)}(d_n)$, (8.4), except that the kinetic term is multiplied by a factor y^2 and the potential term is divided by a factor y^2. One would therefore expect that for wavelengths within the horizon. ${}^{S}\Psi_0^{(n)}$ would have the ground-state form $\exp(-\frac{1}{2}ny^{-2}e^{2\alpha}a_n{}^2)$ and this is borne out by (8.9). On the other hand, when the wavelength becomes larger than the horizon, the Schrödinger equation (8.10) indicates that ${}^{T}\Psi_0^{(n)}$ will freeze in the form (8.9) until the mode reenters the horizon in the matter-dominated era. Even if the equation of state of the Universe changes to radiation dominated during the period that the wavelength of the mode is greater than the horizon size, it will still be true that ${}^{S}\Psi_0^{(n)}$ is frozen in the form (8.9). The ground-state fluctuations in the scalar modes will therefore be amplified in a similar manner to the tensor modes. At the time of reentry of the horizon the rms fluctuation in the scalar modes, in the gauge in which $b_n=f_n=0$, will be greater by the factor y_* than the rms fluctuation in the tensor modes of the same wavelength.

IX. COMPARISON WITH OBSERVATION

From a knowledge of ${}^{T}\Psi_0^{(n)}$ and ${}^{S}\Psi_0^{(n)}$ one can calculate the relative probabilities of observing different values of d_n and a_n at a given point on a trajectory of the vector field X^i, i.e., at a given value of α and ϕ in a background metric which is a solution of the classical field equations. In fact, the dependence on ϕ will be unimportant and we shall neglect it. One can then calculate the probabilities of observing different amounts of anisotropy in the microwave background and can compare these predictions with the upper limits set by observation.

The tensor and scalar perturbation modes will be in highly excited states at large values of α. This means that we can treat their development as an ensemble evolving according to the classical equations of motion with initial distributions in d_n and a_n proportional to $|{}^{T}\Psi_0^{(n)}|^2$ and $|{}^{S}\Psi_0^{(n)}|^2$, respectively. The initial distributions in $\dot{d}_n$ and $\dot{a}_n$ will be proportional to $|{}^{T}\Psi_0^{(n)}\pi_{d_n}{}^{T}\Psi_0^{(n)}|$ and $|{}^{S}\Psi_0^{(n)}\pi_{a_n}{}^{S}\Psi_0^{(n)}|$, respectively. In fact, at the time that the modes reenter the horizon, the distributions will be concentrated at $\dot{d}_n=\dot{a}_n=0$.

The surfaces with $b_n=f_n=0$ will be surfaces of constant energy density in the classical solution during the inflationary period. By local conservation of energy, they will remain surfaces of constant energy density in the era after the inflationary period when the energy is dominated by the coherent oscillations of the homogeneous background scalar field ϕ. If the scalar particles decay into light particles and heat up the Universe, the surfaces with $b_n=f_n=0$ will be surfaces of constant temperature. The surface of last scattering of the microwave background will be such a surface with temperature T_s. The microwave radiation can be considered to have propagated freely to us from this surface. Thus the observed temperature will be

$$T_0=\frac{T_s}{1+z}\,, \tag{9.1}$$

where z is the red-shift of the surface of last scattering. Variations in the observed temperature will arise from variations in z in different directions of observation. These are given by

$$1+z=l^{\mu}n_{\mu} \tag{9.2}$$

evaluated at the surface of last scattering where n_μ is the unit normal to the surfaces of constant t in the gauge $g_n=k_n=j_n=0$ and $b_n=f_n=0$ on the surface of last scattering and l^μ is the parallel propagated tangent vector to the null geodesic from the observer normalized by $l^\mu n_\mu=1$ at the present time. One can calculate the evolution of $l^\mu n_\mu$ down the past light cone of the observer:

$$\frac{d}{d\lambda}[l^\mu n_\mu]=n_{\mu;\nu}l^\mu l^\nu\,, \tag{9.3}$$

where λ is the affine parameter on the null geodesic. The only nonzero components of $n_{\mu;\nu}$ are

$$n_{i;j}=e^{2\alpha}\left[\dot{\alpha}\Omega_{ij}+\sum_n(\dot{a}_n+\dot{\alpha}a_n)\tfrac{1}{3}\Omega_{ij}Q + \sum_n(\dot{b}_n+\dot{\alpha}b_n)P_{ij}+\sum_n(\dot{d}_n+\dot{\alpha}d_n)G_{ij}\right]. \tag{9.4}$$

In the gauge that we are using, the dominant anisotropic terms in (9.4) on the scale of the horizon, will be those involving $\dot{\alpha}a_n$ and $\dot{\alpha}d_n$. These will give temperature anisotropies of the form

$$\langle(\Delta T/T)^2\rangle\approx\langle a_n{}^2\rangle \text{ or } \approx\langle d_n{}^2\rangle\,. \tag{9.5}$$

The number of modes that contribute to anisotropies on the scale of the horizon is of the order of n^3. From the results of the last section

$$\langle a_n{}^2\rangle=y_*{}^2n^{-1}e^{-2\alpha_*}\,, \tag{9.6}$$

$$\langle d_n{}^2\rangle=n^{-1}e^{-2\alpha_*}\,. \tag{9.7}$$

The dominant contribution comes from the scalar modes which give

$$\langle(\Delta T/T)^2\rangle\approx y_*{}^2n^2e^{-2\alpha_*}\,. \tag{9.8}$$

But $ne^{-\alpha_*}\approx\dot{\alpha}_*$, the value of the Hubble constant at the time that the present horizon size left the horizon during the inflationary period. The observational upper limit of about 10^{-8} on $\langle(\Delta T/T)^2\rangle$ restricts this Hubble constant to be less than about $5\times10^{-5}m_P$ (Ref. 8) which in turn restricts the mass of the scalar field to be less than 10^{14} GeV.

X. CONCLUSION AND SUMMARY

We started from the proposal that the quantum state of the Universe is defined by a path integral over compact four-metrics. This can be regarded as a boundary condition for the Wheeler-DeWitt equation for the wave function of the Universe on the infinite-dimensional manifold, superspace, the space of all three-metrics and matter field configurations on a three-surface S. Previous papers had considered finite-dimensional approximations to superspace and had shown that the boundary condition led to a wave function which could be interpreted as corresponding to a family of classical solutions which were homogeneous and isotropic and which had a period of exponential or inflationary expansion. In the present paper we extended this work to the full superspace without restrictions. We treated the two basic homogeneous and isotropic degrees of freedom exactly and the other degrees of freedom to second order. We justified this approximation by showing that the inhomogeneous or anisotropic modes started out in their ground states.

We derived time-dependent Schrödinger equations for each mode. We showed that they remained in the ground state until their wavelength exceeded the horizon size during the inflationary period. In the subsequent expansion the ground-state fluctuations got frozen until the wavelength reentered the horizon during the radiation- or matter-dominated era. This part of the calculation is similar to earlier work on the development of gravitational waves[9] and density perturbations[5,6] in the inflationary Universe but it has the advantage that the assumptions of a period of exponential expansion and of an initial ground state for the perturbations are justified. The perturbations would be compatible with the upper limits set by observations of the microwave background if the scalar field that drives the inflation has a mass of 10^{14} GeV or less.

In Sec. VIII we calculated the scalar perturbations in a gauge in which the surfaces of constant time are surfaces of constant density. There are thus no density fluctuations in this gauge. However, one can make a transformation to a gauge in which $a_n = b_n = 0$. In this gauge the density fluctuation at the time that the wavelength comes within the horizon is

$$\langle (\Delta\rho/\rho)^2 \rangle \approx y^2 \frac{\dot{\rho}_e^{\,2}}{\dot{\alpha}_e^{\,2}\rho_e^{\,2}} \dot{\alpha}_*^{\,2} . \qquad (10.1)$$

Because y and $\dot{\alpha}_*$ depend only logarithmically on the wavelength of the perturbations, this gives an almost scale-free spectrum of density fluctuations. These fluctuations can evolve according to the classical field equations to give rise to the formation of galaxies and all the other structure that we observe in the Universe. Thus all the complexities of the present state of the Universe have their origin in the ground-state fluctuations in the inhomogeneous modes and so arise from the Heisenberg uncertainty principle.

APPENDIX A: HARMONICS ON THE THREE-SPHERE

In this appendix we describe the properties of the scalar, vector, and tensor harmonics on the three-sphere S^3. The metric on S^3 is Ω_{ij} and so the line element is

$$dl^2 = \Omega_{ij} dx^i dx^j$$
$$= d\chi^2 + \sin^2\chi (d\theta^2 + \sin^2\theta \, d\phi^2) . \qquad \text{(A1)}$$

A vertical bar will denote covariant differentiation with respect to the metric Ω_{ij}. Indices i,j,k are raised and lowered using Ω_{ij}.

Scalar harmonics

The scalar spherical harmonics $Q^n_{lm}(\chi,\theta,\phi)$ are scalar eigenfunctions of the Laplacian operator on S^3. Thus, they satisfy the eigenvalue equation

$$Q^{(n)}{}_{|k}{}^{|k} = -(n^2-1) Q^{(n)}, \quad n = 1,2,3,\ldots . \qquad \text{(A2)}$$

The most general solution to (A2), for given n, is a sum of solutions

$$Q^{(n)}(\chi,\theta,\phi) = \sum_{l=0}^{n-1} \sum_{m=-l}^{l} A^n_{lm} Q^n_{lm}(\chi,\theta,\phi) , \qquad \text{(A3)}$$

where A^n_{lm} are a set of arbitrary constants. The Q^n_{lm} are given explicitly by

$$Q^n_{lm}(\chi,\theta,\phi) = \Pi^n_l(\chi) Y_{lm}(\theta,\phi) , \qquad \text{(A4)}$$

where $Y_{lm}(\theta,\phi)$ are the usual harmonics on the two-sphere, S^2, and $\Pi^n_l(\chi)$ are the Fock harmonics.[19,20] The spherical harmonics Q^n_{lm} constitute a complete orthogonal set for the expansion of any scalar field on S^3.

Vector harmonics

The transverse vector harmonics $(S_i)^n_{lm}(\chi,\theta,\phi)$ are vector eigenfunctions of the Laplacian operator on S^3 which are transverse. That is, they satisfy the eigenvalue equation

$$S_i^{(n)}{}_{|k}{}^{|k} = -(n^2-2) S_i^{(n)}, \quad n = 2,3,4,\ldots \qquad \text{(A5)}$$

and the transverse condition

$$S_i^{(n)|i} = 0 . \qquad \text{(A6)}$$

The most general solution to (A5) and (A6) is a sum of solutions

$$S_i^{(n)}(\chi,\theta,\phi) = \sum_{l=1}^{n-1} \sum_{m=-l}^{l} B^n_{lm} (S_i)^n_{lm}(\chi,\theta,\phi) , \qquad \text{(A7)}$$

where B^n_{lm} are a set of arbitrary constants. Explicit expressions for the $(S_i)^n_{lm}$ are given in Ref. 20 where it is also explained how they are classified as odd (o) or even (e) using a parity transformation. We thus have two linearly independent transverse vector harmonics S_i^o and S_i^e $(n,l,m$ suppressed).

Using the scalar harmonics Q^n_{lm} we may construct a third vector harmonics $(P_i)^n_{lm}$. defined by $(n,l,m$ suppressed)

$$P_i = \frac{1}{(n^2-1)} Q_{|i}, \quad n = 2,3,4,\ldots . \qquad \text{(A8)}$$

It may be shown to satisfy

$$P_{i|k}{}^{|k} = -(n^2-3) P_i \quad \text{and} \quad P_i{}^{|i} = -Q . \qquad \text{(A9)}$$

The three vector harmonics S_i^o, S_i^e, and P_i constitute a complete orthogonal set for the expansion of any vector field on S^3.

Tensor harmonics

The transverse traceless tensor harmonics $(G_{ij})^n_{lm}(\chi,\theta,\phi)$ are tensor eigenfunctions of the Laplacian operator on S^3 which are transverse and traceless. That is, they satisfy the eigenvalue equation

$$G_{ij}^{(n)}{}_{|k}{}^{|k} = -(n^2-3) G_{ij}^{(n)}, \quad n = 3,4,5,\ldots \qquad \text{(A10)}$$

and the transverse and traceless conditions

$$G_{ij}^{(n)|i}=0,\quad G_i^{(n)i}=0\ . \tag{A11}$$

The most general solution to (A11) and (A12) is a sum of solutions

$$G_{ij}^{(n)}(\chi,\theta,\phi)=\sum_{l=2}^{n-1}\sum_{m=-l}^{l}C_{lm}^{n}(G_{ij})_{lm}^{n}(\chi,\theta,\phi)\ , \tag{A12}$$

where C_{lm}^n are a set of arbitrary constants. As in the vector case they may be classified as odd or even. Explicit expressions for $(G_{ij}^o)_{lm}^n$ and $(G_{ij}^e)_{lm}^n$ are given in Ref. 20.

Using the transverse vector harmonics $(S_i^o)_{lm}^n$ and $(S_i^e)_{lm}^n$, we may construct traceless tensor harmonics $(S_{ij}^o)_{lm}^n$ and $(S_{ij}^e)_{lm}^n$ defined, both for odd and even, by $(n,l,m$ suppressed)

$$S_{ij}=S_{i|j}+S_{j|i} \tag{A13}$$

and thus $S_i{}^i=0$ since S_i is transverse. In addition, the S_{ij} may be shown to satisfy

$$S_{ij}{}^{|j}=-(n^2-4)S_i\ , \tag{A14}$$

$$S_{ij}{}^{|ij}=0\ , \tag{A15}$$

$$S_{ij|k}{}^{|k}=-(n^2-6)S_{ij}\ . \tag{A16}$$

Using the scalar harmonics Q_{lm}^n, we may construct two tensors $(Q_{ij})_{lm}^n$ and $(P_{ij})_{lm}^n$ defined by $(n,l,m$ suppressed)

$$Q_{ij}=\tfrac{1}{3}\Omega_{ij}Q,\quad n=1,2,3 \tag{A17}$$

and

$$P_{ij}=\frac{1}{(n^2-1)}Q_{|ij}+\tfrac{1}{3}\Omega_{ij}Q,\quad n=2,3,4\ . \tag{A18}$$

The P_{ij} are traceless, $P_i{}^i=0$, and in addition, may be shown to satisfy

$$P_{ij}{}^{|j}=-\tfrac{2}{3}(n^2-4)P_i\ , \tag{A19}$$

$$P_{ij|k}{}^{|k}=-(n^2-7)P_{ij}\ , \tag{A20}$$

$$P_{ij}{}^{|ij}=\tfrac{2}{3}(n^2-4)Q\ . \tag{A21}$$

The six tensor harmonics Q_{ij}, P_{ij}, S_{ij}^o, S_{ij}^e, G_{ij}^o, and G_{ij}^e constitute a complete orthogonal set for the expansion of any symmetric second-rank tensor field on S^3.

Orthogonality and normalization

The normalization of the scalar, vector, and tensor harmonics is fixed by the orthogonality relations. We denote the integration measure on S^3 by $d\mu$. Thus

$$d\mu=d^3x(\det\Omega_{ij})^{1/2}=\sin^2\chi\,\sin\theta\,d\chi\,d\theta\,d\phi\ . \tag{A22}$$

The Q_{lm}^n are normalized so that

$$\int d\mu Q_{lm}^{n}Q_{l'm'}^{n'}=\delta^{nn'}\delta_{ll'}\delta_{mm'}\ . \tag{A23}$$

This implies

$$\int d\mu(P_i)_{lm}^n(P^i)_{l'm'}^{n'}=\frac{1}{(n^2-1)}\delta^{nn'}\delta_{ll'}\delta_{mm'} \tag{A24}$$

and

$$\int d\mu(P_{ij})_{lm}^n(P^{ij})_{l'm'}^{n'}=\frac{2(n^2-4)}{3(n^2-1)}\delta^{nn'}\delta_{ll'}\delta_{mm'}\ . \tag{A25}$$

The $(S_i)_{lm}^n$, both odd and even, are normalized so that

$$\int d\mu(S_i)_{lm}^n(S^i)_{l'm'}^{n'}=\delta^{nn'}\delta_{ll'}\delta_{mm'}\ . \tag{A26}$$

This implies

$$\int d\mu(S_{ij})_{lm}^n(S^{ij})_{l'm'}^{n'}=2(n^2-4)\delta^{nn'}\delta_{ll'}\delta_{mm'}\ . \tag{A27}$$

Finally, the $(G_{ij})_{lm}^n$, both odd and even, are normalized so that

$$\int d\mu(G_{ij})_{lm}^n(G^{ij})_{l'm'}^{n'}=\delta^{nn'}\delta_{ll'}\delta_{mm'}\ . \tag{A28}$$

The information given in this appendix about the spherical harmonics is all that is needed to perform the derivations presented in the main text. Further details may be found in Refs. 19 and 20.

APPENDIX B: ACTION AND FIELD EQUATIONS

The action (5.8) is

$$I=I_0(\alpha,\phi,N_0)+\sum_n I_n\ , \tag{B1}$$

where I_0 is the action of the unperturbed model (4.2):

$$I_0=-\tfrac{1}{2}\int dt\,N_0e^{3\alpha}\left[\frac{\dot\alpha^2}{N_0{}^2}-e^{-2\alpha}-\frac{\dot\phi^2}{N_0{}^2}+m^2\phi^2\right]\ . \tag{B2}$$

I_n is quadratic in the perturbations and may be written

$$I_n=\int dt(L_g^n+L_m^n)\ , \tag{B3}$$

where

$$L_g^n=\tfrac{1}{2}e^{\alpha}N_0\left[\tfrac{1}{3}(n^2-\tfrac{5}{2})a_n{}^2+\frac{(n^2-7)}{3}\frac{(n^2-4)}{(n^2-1)}b_n{}^2-2(n^2-4)c_n{}^2-(n^2+1)d_n{}^2+\tfrac{2}{3}(n^2-4)a_nb_n\right.$$

$$\left.+g_n[\tfrac{2}{3}(n^2-4)b_n+\tfrac{2}{3}(n^2+\tfrac{1}{2})a_n]+\frac{1}{N_0{}^2}\left[-\frac{1}{3(n^2-1)}k_n{}^2+(n^2-4)j_n{}^2\right]\right]$$

$$+\frac{1}{2}\frac{e^{3\alpha}}{N_0}\left\{-\dot a_n{}^2+\frac{(n^2-4)}{(n^2-1)}\dot b_n{}^2+(n^2-4)\dot c_n{}^2+\dot d_n{}^2\right.$$

$$+\dot{\alpha}\left[-2a_n\dot{a}_n+8\frac{(n^2-4)}{(n^2-1)}b_n\dot{b}_n+8(n^2-4)c_n\dot{c}_n+8d_n\dot{d}_n\right]$$

$$+\dot{\alpha}^2\left[-\tfrac{3}{2}a_n{}^2+6\frac{(n^2-4)}{(n^2-1)}b_n{}^2+6(n^2-4)c_n{}^2+6d_n{}^2\right]+g_n[2\dot{\alpha}\dot{a}_n+\dot{\alpha}^2(3a_n-g_n)]$$

$$+e^{-\alpha}\left[k_n\left[-\tfrac{2}{3}\dot{a}_n-\tfrac{2}{3}\frac{(n^2-4)}{(n^2-1)}\dot{b}_n+\tfrac{2}{3}\dot{\alpha}g_n\right]-2(n^2-4)\dot{c}_nj_n\right]\Bigg\} \quad \text{(B4)}$$

and

$$L_m^n=\tfrac{1}{2}N_0e^{3\alpha}\left[\frac{1}{N_0{}^2}(\dot{f}_n{}^2+6a_n\dot{f}_n\dot{\phi})-m^2(f_n{}^2+6a_nf_n\phi)-e^{-2\alpha}(n^2-1)f_n{}^2\right.$$

$$+\frac{3}{2}\left[\frac{\dot{\phi}^2}{N_0{}^2}-m^2\phi^2\right]\left[a_n{}^2-\frac{4(n^2-4)}{(n^2-1)}b_n{}^2-4(n^2-4)c_n{}^2-4d_n{}^2\right]+\frac{\dot{\phi}^2}{N_0{}^2}g_n{}^2$$

$$\left.-g_n\left[2m^2f_n\phi+3m^2a_n\phi^2+2\frac{\dot{f}_n\dot{\phi}}{N_0{}^2}+3\frac{a_n\dot{\phi}^2}{N_0{}^2}\right]-2\frac{e^{-\alpha}}{N_0{}^2}k_nf_n\dot{\phi}\right]. \quad \text{(B5)}$$

The full expressions for π_α and π_ϕ are

$$\pi_\alpha=\frac{e^{3\alpha}}{N_0}\left[-\dot{\alpha}+\sum_n\left[-a_n\dot{a}_n+\frac{4(n^2-4)}{(n^2-1)}b_n\dot{b}_n+4(n^2-4)c_n\dot{c}_n+4d_n\dot{d}_n\right]\right.$$

$$\left.+\dot{\alpha}\sum_n\left[-\tfrac{3}{2}a_n{}^2+\frac{6(n^2-4)}{(n^2-1)}b_n{}^2+6(n^2-4)c_n{}^2+6d_n{}^2\right]+\sum_n g_n[\dot{a}_n+\dot{\alpha}(3a_n-g_n)+\tfrac{1}{3}e^{-\alpha}k_n]\right], \quad \text{(B6)}$$

$$\pi_\phi=\frac{e^{3\alpha}}{N_0}\left\{\dot{\phi}+\sum_n\left[3a_n\dot{f}_n+\dot{\phi}\left[\tfrac{3}{2}a_n{}^2-\frac{4(n^2-4)}{(n^2-1)}b_n{}^2-4(n^2-4)c_n{}^2-4d_n{}^2\right]\right]\right.$$

$$\left.+\sum_n[\dot{\phi}g_n{}^2-g_n(\dot{f}_n+3a_n\dot{\phi})-e^{-\alpha}k_nf_n]\right\}. \quad \text{(B7)}$$

The classical field equations may be obtained from the action (B1) by varying with respect to each of the fields in turn. Variation with respect to α and ϕ gives two field equations, similar to those obtained in Sec. IV, but modified by terms quadratic in the perturbations:

$$N_0\frac{d}{dt}\left[\frac{1}{N_0}\frac{d\phi}{dt}\right]+3\frac{d\alpha}{dt}\frac{d\phi}{dt}+N_0{}^2m^2\phi=\text{quadratic terms}, \quad \text{(B8)}$$

$$N_0\frac{d}{dt}\left[\frac{\dot{\alpha}}{N_0}\right]+3\dot{\phi}^2-N_0{}^2e^{-2\alpha}-\tfrac{3}{2}(-\dot{\alpha}^2+\dot{\phi}^2-N_0{}^2e^{-2\alpha}+N_0{}^2m^2\phi^2)=\text{quadratic terms}. \quad \text{(B9)}$$

Variation with respect to the perturbations a_n, b_n, c_n, d_n, and f_n leads to five field equations:

$$N_0\frac{d}{dt}\left[e^{3\alpha}\frac{\dot{a}_n}{N_0}\right]+\tfrac{1}{3}(n^2-4)N_0{}^2e^{\alpha}(a_n+b_n)+3e^{3\alpha}(\dot{\phi}\dot{f}_n-N_0{}^2m^2\phi f_n)=N_0{}^2[3e^{3\alpha}m^2\phi^2-\tfrac{1}{3}(n^2+2)e^{\alpha}]g_n$$

$$+e^{3\alpha}\dot{\alpha}\dot{g}_n-\tfrac{1}{3}N_0\frac{d}{dt}\left[e^{2\alpha}\frac{k_n}{N_0}\right], \quad \text{(B10)}$$

$$N_0\frac{d}{dt}\left[e^{3\alpha}\frac{\dot{b}_n}{N_0}\right]-\tfrac{1}{3}(n^2-1)N_0{}^2e^{\alpha}(a_n+b_n)=\tfrac{1}{3}(n^2-1)N_0{}^2e^{\alpha}g_n+\tfrac{1}{3}N_0\frac{d}{dt}\left[e^{2\alpha}\frac{k_n}{N_0}\right], \quad \text{(B11)}$$

$$\frac{d}{dt}\left[e^{3\alpha}\frac{\dot{c}_n}{N_0}\right]=\frac{d}{dt}\left[e^{2\alpha}\frac{j_n}{N_0}\right], \quad \text{(B12)}$$

$$N_0\frac{d}{dt}\left[e^{3\alpha}\frac{\dot{d}_n}{N_0}\right]+(n^2-1)N_0{}^2e^{\alpha}d_n=0\ , \tag{B13}$$

$$N_0\frac{d}{dt}\left[e^{3\alpha}\frac{\dot{f}_n}{N_0}\right]+3e^{3\alpha}\dot{\phi}\,\dot{a}_n+N_0{}^2[m^2e^{3\alpha}+(n^2-1)e^{\alpha}]f_n=e^{3\alpha}(-2N_0{}^2m^2\phi g_n+\dot{\phi}\,\dot{g}_n-e^{-\alpha}\phi k_n)\ . \tag{B14}$$

In obtaining (B10)–(B14), the field equations (B8) and (B9) have been used and terms cubic in the perturbations have been dropped.

Variation with respect to the Lagrange multipliers k_n, j_n, g_n, and N_0 leads to a set of constraints. Variation with respect to k_n and j_n leads to the momentum constraints:

$$\dot{a}_n+\frac{(n^2-4)}{(n^2-1)}\dot{b}_n+3f_n\dot{\phi}=\dot{\alpha}g_n-\frac{e^{-\alpha}}{(n^2-1)}k_n\ , \tag{B15}$$

$$\dot{c}_n=e^{-\alpha}j_n\ . \tag{B16}$$

Variation with respect to g_n gives the linear Hamiltonian constraint:

$$3a_n(-\dot{\alpha}^2+\dot{\phi}^2)+2(\dot{\phi}\dot{f}_n-\dot{\alpha}\dot{a}_n)+N_0{}^2m^2(2f_n\phi+3a_n\phi^2)-\tfrac{2}{3}N_0{}^2e^{-2\alpha}[(n^2-4)b_n+(n^2+\tfrac{1}{2})a_n]$$
$$=\tfrac{2}{3}\dot{\alpha}\,e^{-\alpha}k_n+2g_n(-\dot{\alpha}^2+\dot{\phi}^2)\ . \tag{B17}$$

Finally, variation with respect to N_0 yields the Hamiltonian constraint, which we write as

$$\tfrac{1}{2}e^{3\alpha}\left[-\frac{\dot{\alpha}^2}{N_0{}^2}+\frac{\dot{\phi}^2}{N_0{}^2}-e^{-2\alpha}+m^2\phi^2\right]=\text{quadratic terms}\ . \tag{B18}$$

[1]A. H. Guth, Phys. Rev. D 23, 347 (1981).
[2]A. D. Linde, Phys. Lett. **108B**, 389 (1982).
[3]S. W. Hawking and I. G. Moss, Phys. Lett. **110B**, 35 (1982).
[4]A. Albrecht and P. J. Steinhardt, Phys. Rev. Lett. **48**, 120 (1982).
[5]S. W. Hawking, Phys. Lett. **115B**, 295 (1982).
[6]A. H. Guth and S. Y. Pi, Phys. Rev. Lett. **49**, 1110 (1982).
[7]J. M. Bardeen, P. J. Steinhardt, and M. S. Turner, Phys. Rev. D **28**, 679 (1983).
[8]S. W. Hawking, Phys. Lett. B **150B**, 339 (1985).
[9]V. A. Rubakov, M. V. Sazhin, and A. V. Veryaskin, Phys. Lett. **115B**, 189 (1982).
[10]S. W. Hawking, Pontif. Accad. Sci. Varia **48**, 563 (1982).
[11]J. B. Hartle and S. W. Hawking, Phys. Rev. D **28**, 2960 (1983).
[12]S. W. Hawking, in *Relativity, Groups and Topology II*, Les Houches 1983, Session XL, edited by B. S. DeWitt and R. Stora (North-Holland, Amsterdam, 1984).
[13]S. W. Hawking, Nucl. Phys. **B239**, 257 (1984).
[14]S. W. Hawking and Z. C. Wu, Phys. Lett. **151B**, 15 (1985).
[15]S. W. Hawking and D. N. Page, DAMTP report, 1984 (unpublished).
[16]S. W. Hawking and J. C. Luttrell, Nucl. Phys. **B247**, 250 (1984).
[17]L. P. Grishchuk, Zh. Eksp. Teor. Fiz. **67**, 825 (1974) [Sov. Phys. JETP **40**, 409 (1975)]; Ann. N.Y. Acad. Sci. **302**, 439 (1977).
[18]A. A. Starobinsky, Pis'ma Zh. Eksp. Teor. Fiz. **30**, 719 (1979) [JETP Lett. **30**, 682 (1979)].
[19]E. M. Lifschitz and I. M. Khalatnikov, Adv. Phys. **12**, 185 (1963).
[20]U. H. Gerlach and U. K. Sengupta, Phys. Rev. D **18**, 1773 (1978).

IV. WORMHOLES

PHYSICAL REVIEW D VOLUME 37, NUMBER 4 15 FEBRUARY 1988

Wormholes in spacetime

S. W. Hawking
Department of Applied Mathematics and Theoretical Physics, University of Cambridge, Silver Street, Cambridge CB3 9EW, England

(Received 28 October 1987)

Any reasonable theory of quantum gravity will allow closed universes to branch off from our nearly flat region of spacetime. I describe the possible quantum states of these closed universes. They correspond to wormholes which connect two asymptotically Euclidean regions, or two parts of the same asymptotically Euclidean region. I calculate the influence of these wormholes on ordinary quantum fields at low energies in the asymptotic region. This can be represented by adding effective interactions in flat spacetime which create or annihilate closed universes containing certain numbers of particles. The effective interactions are small except for closed universes containing scalar particles in the spatially homogeneous mode. If these scalar interactions are not reduced by sypersymmetry, it may be that any scalar particles we observe would have to be bound states of particles of higher spin, such as the pion. An observer in the asymptotically flat region would not be able to measure the quantum state of closed universes that branched off. He would therefore have to sum over all possibilities for the closed universes. This would mean that the final state would appear to be a mixed quantum state, rather than a pure quantum state.

I. INTRODUCTION

In a reasonable theory of quantum gravity the topology of spacetime must be able to be different from that of flat space. Otherwise, the theory would not be able to describe closed universes or black holes. Presumably, the theory should allow all possible spacetime topologies. In particular, it should allow closed universes to branch off, or join onto, our asymptotic flat region of spacetime. Of course, such behavior is not possible with a real, nonsingular, Lorentzian metric. However, we now all know that quantum gravity has to be formulated in the Euclidean domain. There, it is no problem: it is just a question of plumbing. Indeed, it is probably necessary to include all possible topologies for spacetime to get unitarity.

Topology change is not something that we normally experience, at least, on a macroscopic scale. However, one can interpret the formation and subsequent evaporation of a black hole as an example: the particles that fell into the hole can be thought of as going off into a little closed universe of their own. An observer in the asymptotically flat region could not measure the state of the closed universe. He would therefore have to sum over all possible quantum states for the closed universe. This would mean that the part of the quantum state that was in the asymptotically flat region would appear to be in a mixed state, rather than a pure quantum state. Thus, one would lose quantum coherence.[1,2]

If it is possible for a closed universe the size of a black hole to branch off, it is also presumably possible for little Planck-size closed universes to branch off and join on. The purpose of this paper is to show how one can describe this process in terms of an effective field theory in flat spacetime. I introduce effective interactions which create, or destroy, closed universes containing certain numbers of particles. I shall show that these effective interactions are small, except for scalar particles. There is a serious problem with the very large effective interactions of scalar fields with closed universes. It may be that these interactions can be reduced by supersymmetry. If not, I think we will have to conclude that any scalar particles that we observe are bound states of fermions, like the pion. Maybe this is why we have not observed Higgs particles.

I base my treatment on general relativity, even though general relativity is probably only a low-energy approximation to some more fundamental quantum theory of gravity, such as superstrings. For closed universes of the Planck size, any higher-order corrections induced from string theory will change the action by a factor ~ 1. So the effective field theory based on general relativity should give answers of the right order of magnitude.

In Sec. II, I describe how closed universes or wormholes can join one asymptotically Euclidean region to another, or to another part of the same region. Solutions of the Wheeler-DeWitt equation that correspond to such wormholes are obtained in Sec. III. These solutions can also be interpreted as corresponding to Friedmann universes. It is an amusing thought that our Universe could be just a rather large wormhole in an asymptotically flat space.

In Sec. IV, I calculate the vertex for the creation or annihilation of a wormhole containing a certain number of particles. Section V contains a discussion of the initial quantum state in the closed-universe Fock space. There are two main possibilities: either there are no closed universes present initially, or there is a coherent state which is an eigenstate of the creation plus annihilation operators for each species of closed universe. There will be loss of quantum coherence in the first case, but not the second. This is described in Sec. VI. The interactions between wormholes and particles of different spin in asymptotically flat space are discussed in Sec.

VII. Finally, in Sec. VIII, I conclude that wormholes will have to be taken into account in any quantum theory of gravity, including superstrings.

This paper supercedes earlier work of mine[3-5] on the loss of quantum coherence. These papers were incorrect in associating loss of coherence with simply connected spaces with nontrivial topology, rather than with wormholes.

II. WORMHOLES

What I am aiming to do is to calculate the effect of closed universes that branch off on the behavior of ordinary, nongravitational particles in asymptotically flat space at energies low compared to the Planck mass. The effect will come from Euclidean metrics which represent a closed universe branching off from asymptotically flat space. One would expect that the effect would be greater, the larger the closed universe. Thus one might expect the dominant contribution would come from metrics with the least Euclidean action for a given size of closed universe. In the $R=0$ conformal gauge, these are conformally flat metrics:

$$ds^2=\Omega^2 dx^2 ,$$

$$\Omega=1+\frac{b^2}{(x-x_0)^2} .$$

At first sight, this looks like a metric with a singularity at the point x_0. However, the blowing up of the conformal factor near x_0 means that the space opens out into another asymptotically flat region, joined to the first asymptotically flat region by a wormhole of coordinate radius b and proper radius $2b$. The other asymptotic region can be a separate asymptotically flat region of the Universe, or it can be another part of the first asymptotic region. In the latter case, the conformal factor will be modified slightly by the interaction between the two ends of the wormhole, or handle to spacetime.[6] However, the change will be small when the separation of the two ends is large compared to $2b$, the size of the wormhole. Typically, b will be of the order of the Planck length, so it will be a good approximation to neglect the interactions between wormholes. This conformally flat metric is just one example of a wormhole. There are, of course, nonconformally flat closed universes that can join onto asymptotically flat space. Their effects will be similar, but will involve gravitons in the asymptotically flat space. Since it is difficult to observe gravitons, I shall concentrate on conformally flat closed universes.

I shall consider a set of matter fields ϕ in the closed universe. Spin-1 gauge fields are conformally invariant. In the case of matter fields of spin $\frac{1}{2}$ and 0, the effect of any mass will be small for wormholes of the Planck size. I shall therefore take the matter fields ϕ to be conformally invariant. The effect of mass could be included as a perturbation.

In order to find the effect of the closed universe or wormhole on the matter fields ϕ in the asymptotically flat spaces, one should calculate the Green's functions

$$\langle \phi(y_1)\phi(y_2)\cdots\phi(y_r)\phi(z_1)\phi(z_2)\cdots\phi(z_s)\rangle ,$$

where $y_1,\ldots,y_r$ and $z_1,\ldots,z_s$ are points in the two asymptotic regions (which may be the same region). This can be done by performing a path integration over all matter fields ϕ and all metrics $g_{\mu\nu}$ that have one or two asymptotically flat regions and a handle or wormhole connecting them. Let S be a three-sphere, which is a cross section of the closed universe or wormhole. One can then factorize the path integral into a part

$$\langle 0|\phi(y_1)\cdots\phi(y_r)|\psi\rangle ,$$

which depends on the fields on one side of S, and a part

$$\langle \psi|\phi(z_1)\cdots\phi(z_s)|0\rangle ,$$

which depends on the fields on the other side of S. Strictly speaking, one can factorize in this way only when the regions at the two ends of the wormhole are separate asymptotic regions. However, even when they are the same region, one can neglect the interaction between the ends and factorize the path integral if the ends are widely separated.

In the above $|0\rangle$ represented the usual particle scattering vacuum state defined by a path integral over asymptotically Euclidean metrics and matter fields that vanish at infinity. $|\psi\rangle$ represented the quantum state of the closed universe or wormhole on the surface S. This can be described by a wave function Ψ which depends on the induced metric h_{ij} and the values ϕ_0 of the matter fields on S. The wave function obeys the Wheeler-DeWitt equation

$$\left[-m_P^{-2}G_{ijkl}\frac{\delta^2}{\delta h_{ij}\delta h_{kl}}-m_P^2 h^{1/2}\,{}^3R+\tfrac{1}{2}h^{1/2}T^{nn}\left[\phi_0,-i\frac{\delta}{\delta\phi_0}\right]\right]\Psi[h_{ij},\phi_0]=0 ,$$

where

$$G_{ijkl}=\tfrac{1}{2}h^{1/2}(h_{ik}h_{jl}+h_{il}h_{jk}-h_{ij}h_{kl}) .$$

The wave function also obeys the momentum constraint

$$\left[-2im_P^2\left[\frac{\delta}{\delta h_{ij}}\right]_{|j}+T^{ni}\left[\phi_0,-i\frac{\delta}{\delta\phi_0}\right]\right]\Psi[h_{ij},\phi_0]=0 .$$

III. WORMHOLE EXCITED STATES

The solutions of the Wheeler-DeWitt equation that correspond to wormholes, that is, closed universes connecting two asymptotically Euclidean regions, form a Hilbert space $\mathcal{H}_w$ with the inner product

$$\langle \psi_1 | \psi_2 \rangle = \int d[h_{ij}] d[\phi_0] \Psi_1^* \Psi_2 .$$

Let $| \psi_i \rangle$ be a basis for $\mathcal{H}_w$. Then one can write the Green's function in the factorized form

$$\langle \phi(y_1) \cdots \phi(y_r) \phi(z_1) \cdots \phi(z_s) \rangle = \sum \langle 0 | \phi(y_1) \cdots \phi(y_r) | \psi_i \rangle \langle \psi_i | \phi(z_1) \cdots \phi(z_s) | 0 \rangle .$$

What are these wormhole excited states $| \psi_i \rangle$? To find them one would have to solve the full Wheeler-DeWitt and momentum constraint equations. This is too difficult, but one can get an idea of their nature from mode expansions.[7] One can write the three-metric h_{ij} on the surface S as

$$h_{ij} = \sigma^2 a^2 (\Omega_{ij} + \epsilon_{ij}) .$$

Here $\sigma^2 = 2/3\pi m_P^2$ is a normalization factor, Ω_{ij} is the metric on the unit three-sphere, and ϵ_{ij} is a perturbation, which can be expanded in harmonics on the three-sphere:

$$\epsilon_{ij} = \sum_{n,l,m} [6^{1/2} a_{nlm} \tfrac{1}{3} \Omega_{ij} Q_{lm}^n + 6^{1/2} b_{nlm} (P_{ij})_{lm}^n + 2^{1/2} c_{nlm}^0 (S_{ij}^0)_{lm}^n + 2^{1/2} c_{nlm}^e (S_{ij}^e)_{lm}^n + 2 d_{nlm}^0 (G_{ij}^0)_{lm}^n + 2 d_{nlm}^e (G_{ij}^e)_{lm}^n] .$$

The $Q(x^i)$ are the standard scalar harmonics on the three-sphere. The $P_{ij}(x^i)$ are given by (suppressing all but i,j indices)

$$P_{ij} = \frac{1}{n^2 - 1} Q_{|ij} + \tfrac{1}{3} \Omega_{ij} Q .$$

They are traceless, $P_i^i = 0$. The S^{ij} are defined by

$$S_{ij} = S_{i|j} + S_{j|i} ,$$

where S_i are the transverse vector harmonics, $S_i^{|i} = 0$. The G_{ij} are the transverse traceless tensor harmonics $G_i^i = G_{ij}^{|j} = 0$. Further details about harmonics and their normalization can be found in Ref. 7.

Consider a conformally invariant scalar field ϕ. One can describe it in terms of hyperspherical harmonics on the surface S:

$$\phi_0 = \sigma^{-1} a^{-1} \sum f_n Q_n .$$

The wave function Ψ is then a function of coefficients a_n, b_n, c_n, d_n, and f_n and the scale factor a.

One can expand the Wheeler-DeWitt operator to all orders in a and to second order in the other coefficients. In this approximation, the different modes do not interact with each other, but only with the scale factor a. However, the conformal scalar coefficients f_n do not even interact with a. One can therefore write the wave function as a sum of products of the form

$$\Psi = \Psi_0(a, a_i, b_i, c_i, d_i) \prod \psi_n(f_n) .$$

The part of the Wheeler-DeWitt operator that acts on ψ_n is

$$-\frac{d^2}{df_n^2} + (n^2+1) f_n^2 .$$

It is therefore natural to take them to be harmonic-oscillator wave functions

$$\psi_{nm} = \left[\frac{\beta^2}{\pi 2^{2m} (m!)^2} \right]^{1/4} e^{-\beta^2 f_n^2/2} H_m(\beta f_n) ,$$

where $\beta^4 = (n^2+1)$ and H_m are Hermite polynomials. The wave functions ψ_{nm} can then be interpreted as corresponding to the closed universe containing m scalar particles in the nth harmonic mode.

The treatment for spin-$\frac{1}{2}$ and -1 fields is similar. The appropriate data for the fields on S can be expanded in harmonics on the three-sphere. The main difference is that the lowest harmonic is not the $n=0$ homogeneous mode, as in the scalar case, but has $n = \frac{1}{2}$ or 1. Again, the coefficients of the harmonics appear in the Wheeler-DeWitt equation to second order only as fermionic[8] or bosonic harmonic oscillators, with a frequency independent of a. One can therefore take the wave functions to be fermion or boson harmonic-oscillator wave functions in the coefficients of the harmonics. They can then be interpreted as corresponding to definite numbers of particles in each mode.

In the gravitational part of the wave function, Ψ_0, the coefficients a_n, b_n, and c_n reflect gauge degrees of freedom. They can be made zero by a diffeomorphism of S and suitable lapse and shift functions. The coefficients d_n correspond to gravitational wave excitations of the closed universe. However, gravitons are very difficult to observe. I shall therefore take these modes to be in their ground state.

The scale factor a appears in the Wheeler-DeWitt equation as the operator

$$\frac{\partial^2}{\partial a^2} - a^2 .$$

I shall assume that the zero-point energies of each mode are either subtracted or canceled by fermions in a supersymmetric theory. The total wave function Ψ will then satisfy the Wheeler-DeWitt equation if the gravitational part Ψ_0 is a harmonic-oscillator wave function in a with unit frequency and level equ l to the sum E of the energies of the matter-field harmonic oscillators.

The wave function Ψ_0 will oscillate for $a<r_0 =(2E)^{1/2}$. In this region one can use the WKB approximation[7,9,10] to relate it to a Lorentzian solution of the classical field equations. This solution will be a $k=+1$ Friedmann model filled with conformally invariant matter. The maximum radius of the Friedmann model will be $a=r_0$. For $a>r_0$, the wave function will be exponential. Thus, in this region it will correspond to a Euclidean metric. This will be the wormhole metric described in Sec. II, with $b=1/2\sigma r_0$. These excited state solutions were first found in Ref. 11, but their significance as wormholes was not re lized. Notice that the wave function is exponentially damped at large a, whereas the cosmological wave functions described in Refs. 7, 9, and 10 tend to grow exponentially at large a. The difference here is that one is looking at the closed universe from an asymptotically Euclidean region, instead of from a compact Euclidean space, as in the cosmological case. This changes the sign of the trace K surface term in the gravitational action.

IV. THE WORMHOLE VERTEX

One now wants to calculate the matrix element of the product of the values of ϕ at the points $y_1, y_2, \ldots, y_r$ between the ordinary, flat-space vacuum $\langle 0|$ and the closed-universe state $|\psi\rangle$. This is given by the path integral

$$\langle 0|\phi(y_1)\cdots\phi(y_r)|\psi\rangle=\int d[h_{ij}]d[\phi_0]\Psi[h_{ij},\phi_0]\int d[g_{\mu\nu}]d[\phi]\phi(y_1)\cdots\phi(y_r)e^{-I[g,\phi]} .$$

The gravitational field is required to be asymptotic lly flat at infinity, and to have a three-sphere S with induced metric h_{ij} as its inner boundary. The scalar field ϕ is required to be zero at infinity, and to have the value ϕ_0 on S.

In general, the positions of the points y_i cannot be specified in a gauge-invariant manner. However, I shall be concerned only with the effects of the wormholes on low-energy particle physics. In this case the separation of the points y_i can be taken to be large compared to the Planck length, and they can be taken to lie in flat Euclidean space. Their positions can then be specified up to an overall translation and rotation of Euclidean sp ce.

Consider first a wormhole state $|\psi\rangle$ in which only the $n=0$ homogeneous scalar mode is excited above its ground st te. The integral over the wave function Ψ of the wormhole can then be replaced in the above by

$$\int da\, df_0\, \psi_E(a)\psi_{0m}(f_0) .$$

T e path integral will then be over asymptotically Euclidean metrics whose inner boundary is a three-sphere S of radius a and scalar fields with the constant value f_0 on S. The saddle point for the path integral will be flat Euclidean space outside a three-sphere of radius a centered on a point x_0 and the scalar field

$$\phi=\frac{a\sigma f_0}{(x-x_0)^2}$$

(the energy-momentum tensor of this scalar field is zero). The action of this saddle point will be $(a^2+f_0^2)/2$. The determinant Δ of the small fluctuations about the saddle point will be independent of f_0. Its precise form will not be important.

The integral over the coefficient f_0 of the $n=0$ scalar harmonic will contain a factor of

$$\int df_0 f_0^r e^{-f_0^2}H_m(f_0) .$$

This will be zero when m, the number of particles in the mode $n=0$, is greater than r, the number of points y_i in the correlation function. This is what one would expect, because e ch particle in the closed universe must be created or annihilated at a point y_i in the asymptotically flat region. If $r>m$, particles may be created at one point y_i and annihilated at another point y_j without going into the closed universe. However, such matrix elements are just products of flat-space propagators with matrix elements with $r=m$. It is sufficient herefore to consider only the case with $r=m$.

The integral over the radius a will contain a factor

$$\int da\, a^m e^{-a^2}H_E(a)\Delta(a) ,$$

where $E=m$ is the level number of the radial harmonic oscillator. For small m, the dominant contribution will come from $a\sim 1$, that is, wormholes of the Planck size. The value $C(m)$ of this integral will be ~ 1.

The matrix element will then be

$$D(m)\prod\frac{\sigma}{(y_i-x_0)^2} ,$$

where $D(m)$ is another factor ~ 1. One now has to integrate over the position x_0 of the wor hole, with a measure of the form $m_P^4 dx_0^4$, and over an orthogonal matrix O which specifies its orientation with respect to the points y_i. The $n=0$ mode is invariant under O, so this second integral will have no effect, but the integral over x_0 will ensure the energy and momentum are conserved in the asymptotically flat region. This is what one would expect, because the Wheeler-DeWitt and momentum constraint equations imply that a closed universe has no energy or momentum.

The matrix element will be the same as if one was in

flat space with an effective interaction of the form

$$F(m) m_P^{4-m} \phi^m (c_{0m} + c_{0m}^\dagger) ,$$

where $F(m)$ is another coefficient ~ 1 and c_{0m} and $c_{0m}^\dagger$ are the annihilation and creation operators for a closed universe containing m scalar particles in the $n=0$ homogeneous mode.

In a similar way, one can calculate the matrix elements of products of ϕ between the vacuum and a closed-universe state containing m_0 particles in the $n=0$ mode, m_1 particles in the $n=1$ mode, and so on. The energy-momentum tensor of scalar fields with higher harmonic angular dependence will not be zero. This will mean that the saddle-point metric in the path integral for the matrix element will not be flat space, but will be curved near the surface S. In fact, for large particle numbers, the saddle-point metric will be the conformally flat wormhole metrics described in Sec. II. However, the saddle-point scalar fields will have a Q_n angular dependence and a $\sigma^{n+1}/(x-x_0)^{n+2}$ radial dependence in the asymptotic flat region. This radial decrease is so fast that the closed universes with higher excited harmonics will not give significant matrix elements, except for that containing two particles in the $n=1$ modes. By the constraint equations, or, equivalently, by averaging over the orientation O of the wormhole, the matrix element will be zero unless the two particles are in a state that is invariant under O. The matrix element for such a universe will be the same as that produced by an effective interaction of the form

$$\nabla\phi\nabla\phi(c_{12}+c_{12}^\dagger)$$

with a coefficient ~ 1.

In a similar way one can calculate the matrix elements for universes containing particles of spin $\frac{1}{2}$ or higher. Again, the constraint equations or averaging over O mean that the matrix element is nonzero only for closed-universe states that are invariant under O. This means that the corresponding effective interactions will be Lorentz invariant. In particular, they will contain even numbers of spinor fields. Thus, fermion number will be conserved mod 2: the closed universes are bosons.

The matrix elements for universes containing spin-$\frac{1}{2}$ particles will be equivalent to effective interactions of the form

$$m_P^{4-3m/2}\psi^m d_m + \text{c.c.} ,$$

where ψ^m denotes some Lorentz-invariant combination of m spinor fields ψ or their adjoints $\bar{\psi}$, and d_m is the annihilation operator for a closed universe containing m spin-$\frac{1}{2}$ particles in $n=\frac{1}{2}$ modes. One can neglect the effect of closed universes with spin-$\frac{1}{2}$ particles in higher modes.

In the case of spin-1 gauge particles, the effective interaction would be of the form

$$m_P^{4-2m}[(F_{\mu\nu})^m(g_m+g_m^\dagger)] ,$$

where g_m is the annihilation operator for a closed universe containing m spin-1 particles in $n=1$ modes. As before, the higher modes can be neglected.

V. THE WORMHOLE INITIAL STATE

What I have done is introduce a new Fock space $\mathcal{F}_w$ for closed universes, which is based on the one wormhole Hilbert space $\mathcal{H}_w$. The creation and annihilation operators $c_{nm}^\dagger$, c_{nm}, etc., act on $\mathcal{F}_w$ and obey the commutation relations for bosons. The full Hilbert space of the theory, as far as asymptotically flat space is concerned, is isomorphic to $\mathcal{F}_p \otimes \mathcal{F}_w$, where $\mathcal{F}_p$ is the usual flat-space particle Fock space.

The distinction between annihilation and creation operators is a subtle one because the closed universe does not live in the same time as the asymptotically flat region. If both ends of the wormhole are in the same asymptotic region, one can say that a closed universe is created at one point and is annihilated at another. However, if a closed universe branches off from our asymptotically flat region, and does not join back on, one would be free to say either (1) it was present in the initial state and was annihilated at the junction point x_0, (2) it was not present initially, but was created at x_0 and is present in the final state, or (3) as Sidney Coleman (private communication) has suggested, one might have a coherent state of closed universes in both the initial and final states, in such a way that they were both eigenstates of the annihilation plus creation operators $c_{nm}+c_{nm}^\dagger$, etc., with some eigenvalue q.

In this last case, the closed-universe sector of the state would remain unchanged and there would be no loss of quantum coherence. However, the initial state would contain an infinite number of closed universes. Such eigenstates would not form a basis for the Fock space of closed universes.

Instead, I shall argue that one should adopt the second possibility: there are no closed universes in the initial state, but closed universes can be created and appear in the final state. If one takes a path-integral approach, the most natural quantum state for the Universe is the so-called "ground" state, or, "no boundary" state.[8] This is the state defined by a path integral over all compact metrics without boundary. Calculations based on minisuperspace models[7–11] indicate that this choice of state leads to a universe like we observe, with large regions that appear nearly flat. One can then formulate particle scattering questions in the following way: one asks for the conditional probability that one observes certain particles on a nearly flat surface S_2 given that the region is nearly asymptotically Euclidean and is in the quantum state defined by conditions on the surfaces S_1 and S_3 to either side of S_2, and at great distance from it in the positive and negative Euclidean-time directions, respectively. One then analytically continues the position of S_2 to late real time. It then measures the final state in the nearly flat region. One continues the positions of both S_1 and S_3 to early real time. One gives the time coordinate of S_1 a small positive imaginary part, and the time coordinate of S_3 a small negative imaginary part. The initial state is then defined by data on the surfaces S_1 and S_3.

If one adopts the formulation of particle scattering in terms of conditional probabilities, one would impose the conditions on the surfaces S_1 and S_3 in the nearly flat region. However, one would not impose conditions on any closed universes that branched off or joined on between S_1 and S_3, because one could not observe them. Thus, the initial or conditional state would not contain any closed universes. A closed universe that branched off between S_1 and S_2 (or between S_2 and S_3) would be regarded as having been created. If it joined up again between S_1 and S_2 (S_2 and S_3, respectively), it would be regarded as having been annihilated again. Otherwise, it would be regarded as part of the final state. An observer in the nearly flat region would be able to measure only the part of the final state on S_2 and not the state of the closed universe. He would therefore have to sum over all possibilities for the closed universes. This summation would mean that the part of the final state that he could observe would appear to be in a mixed state rather than in a pure quantum state.

VI. THE LOSS OF QUANTUM COHERENCE

Let $|\alpha_i\rangle$ be a basis for the flat-space Fock space $\mathcal{F}_p$ and $|\beta\rangle_j$ be a basis for the wormhole Fock space $\mathcal{F}_w$. In case (2) above, in which there are no wormholes initially, the initial, or conditional, state can be written as the state

$$\lambda^i|\alpha_i\rangle|O\rangle_w \ ,$$

where $|O\rangle_w$ is the zero closed-universe state in $\mathcal{F}_w$. The final state can be written as

$$\mu^{ij}|\alpha_i\rangle|\beta_j\rangle \ .$$

However, an observer in the nearly flat region can measure only the states $|\alpha_i\rangle$ on S_2, and not the closed-universe states $|\beta_j\rangle$. He would therefore have to sum over all possible states for the closed universes. This would give a mixed state in the $\mathcal{F}_p$ Fock space with density matrix

$$\rho^i_k=\mu^{ij}\bar{\mu}_{kj} \ .$$

The matrix ρ^{ik} will be Hermitian and positive semidefinite, if the final state is normalized in $\mathcal{H}$:

$$\mathrm{tr}\rho=\mu^{ij}\bar{\mu}_{ij}=1 \ .$$

These are the properties required for it to be interpreted as the density matrix of a mixed quantum state. A measure of the loss quantum coherence is

$$1-\mathrm{tr}(\rho^2)=1-\mu^{ij}\mu^{kl}\bar{\mu}_{il}\bar{\mu}_{kj} \ .$$

This will be zero if the final state is a pure quantum state. Another measure is the entropy which can be defined as

$$-\mathrm{tr}(\rho\ln\rho) \ .$$

This again will be zero for a pure quantum state.

If case (3) above is realized, the initial closed-universe state is not the no-wormhole state $|O\rangle_w$, but a coherent state $|q\rangle_w$ such that

$$(c_{nm}+c^\dagger_{nm})|q\rangle_w=q_{nm}|q\rangle_w \ .$$

The effective interactions would leave the closed-universe sector in the same coherent state. Thus the final state would be the product of some state in $\mathcal{F}_p$ with the coherent state $|q\rangle_w$. There would be no loss of quantum coherence, but one would have effective ϕ^m and other interactions whose coefficients would depend on the eigenvalues q_{nm}, etc. It would seem that these could have any value.

VII. WORMHOLE EFFECTIVE INTERACTIONS

There will be no significant interaction between wormholes, unless they are within a Planck length of each other. Thus, the creation and annihilation operators for wormholes are practically independent of the positions in the asymptotically flat region. This means that the effective propagator of a wormhole excited state is $\delta^4(p)$. Using the propagator one can calculate Feynman diagrams that include wormholes, in the usual manner.

The interactions of wormholes with m scalar particles in the $n=0$ mode are alarmingly large. The $m=1$ case would be a disaster; it would give the scalar field a propagator that was independent of position because a scalar particle could go into a wormhole whose other end was at a great distance in the asymptotically flat region. Suppose, however, that the scalar field were coupled to a Yang-Mills field. One would have to average over all orientations of the gauge group for the closed universe. This would make the matrix element zero, except for closed-universe states that were Yang-Mills singlets. In particular, the matrix element would be zero for $m=1$. A special case is the gauge group Z_2. Such fields are known as twisted scalars. They can reverse sign on going round a closed loop. They will have zero matrix elements for m odd because one will have to sum over both signs.

Consider now the matrix element for the scalar field, and its complex conjugate, between the vacuum and a closed universe containing a scalar particle and antiparticle in the $n=0$ mode. This will be nonzero, because a particle-antiparticle state contains a Yang-Mills singlet. It would give an effective interaction of the form

$$m_P^2\,\mathrm{tr}(\phi\bar{\phi})(c_{011}+c^\dagger_{011}) \ ,$$

where c_{011} is the annihilation operator for a closed universe with one scalar particle and one antiparticle in the $n=0$ mode. This again would be a disaster; with two of these vertices one could make a closed loop consisting of a closed universe [propagator, $\delta^4(p)$] and a scalar particle (propagator, $1/p^2$). This closed loop would be infrared divergent. One could cut off the divergence by giving the scalar particle a mass, but the effective mass would be the Planck mass. One might be able to remove this mass by renormalization, but the creation of closed universes would mean that a scalar particle would lose quantum coherence within a Planck length. The

$m = 4$ matrix element will give a large ϕ^4 effective vertex.

There seems to be four possibilities in connection with wormholes containing only scalar particles in the $n = 0$ mode.

(1) They may be reduced or canceled in a supersymmetric theory.

(2) The scalar field may be absorbed as a conformal factor in the metric. This could happen, however, only for one scalar field that was a Yang-Mills singlet.

(3) It may be that any scalar particle that we observe is a bound state of particles of higher spin, such as the pion.

(4) The universe may be in a coherent state $|q\rangle_w$ as described above. However, one would then have the problem of why the eigenvalues q should be small or zero. This is similar to the problem of why the θ angle should be so small, but there are now an infinite number of eigenvalues.

In the case of particles of spin $\frac{1}{2}$, the exclusion principle limits the occupation numbers of each mode to zero or 1. Averaging over the orientation O of the wormhole will mean that the lowest-order interaction will be for a wormhole containing one fermion and one antifermion. This would give an effective interaction of the form

$$m_P \psi\bar{\psi}(d_{11} + d_{11}^{\dagger}) ,$$

where d_{11} is the annihilation operator for a closed universe containing a fermion and an antifermion in $n = \frac{1}{2}$ modes. This would give the fermion a mass of the order of the Planck mass. However, if the fermion were chiral, this interaction would cancel out under averaging over orientation and gauge groups. This is because there is no two-chiral-fermion state that is a singlet under both groups. This suggests that supersymmetry might ensure the cancellation of the dangerous interactions with wormholes containing scalar particle in the $n = 0$ mode. Conformally flat wormholes, such as those considered in this paper, should not break supersymmetry.

For chiral fermions, the lowest-order effective interaction will be of the four-Fermi form

$$m_P^{-2} \operatorname{tr}(\psi_1 \gamma^{\mu} \bar{\psi}_1 \psi_2 \gamma_{\mu} \bar{\psi}_2)(d_{1111} + d_{1111}^{\dagger}) ,$$

where d_{1111} is the annihilation operator for a wormhole containing a fermion and an antifermion each of species 1 and 2. This would lead to baryon decay, but with a lifetime $\sim 10^{50}$ yr. There will also be Yukawa-type effective interactions produced by closed universes containing one scalar particle, one fermion, and one antifermion.

VIII. CONCLUSION

It would be tempting to dismiss the idea of wormholes by saying that they are based on general relativity, and we now all know that string theory is the ultimate theory of quantum gravity. However, string theory, or any other theory of quantum gravity, must reduce to general relativity on scales large compared to the Planck length. Even at the Planck length, the differences from general relativity should be only ~ 1. In particular, the ultimate theory of quantum gravity should reproduce classical black holes and black-hole evaporation. It is difficult to see how one could describe the formation and evaporation of a black hole except as the branching off of a closed universe. I would therefore claim that any reasonable theory of quantum gravity, whether it is supergravity, or superstrings, should allow little closed universes to branch off from our nearly flat region of spacetime.

The effect of these closed universes on ordinary particle physics can be described by effective interactions which create or destroy closed universes. The effective interactions are small, apart from those involving scalar fields. The scalar field interactions may cancel because of supersymmetry. Or, any scalar particles that we observe may be bound states of particles of higher spin. Near a wormhole of the Planck size, such a bound state would behave like the higher-spin particles of which it was made. A third possibility is that the universe is in a coherent $|q\rangle_w$ state. I do not like this possibility because it does not seem to agree with the "no boundary" proposal for the quantum state of the Universe. There also would not seem to be any way to specify the eigenvalues q. Yet the values of the eigenvalues for large particle numbers cannot be zero if these interactions are to reproduce the results of semiclassical calculations on the formation and evaporation of macroscopic black holes.

The effects of little closed universes on ordinary particle physics may be small, apart, possibly, for scalar particles. Nevertheless, it raises an important matter of principle. Because there is no way in which we could measure the quantum state of closed universes that branch off from our nearly flat region, one has to sum over all possible states for such universes. This means that the part of the final state that we can measure will appear to be in a mixed quantum state, rather than a pure state. I think even Gross[12] will agree with that.

[1] R. M. Wald, Commun. Math. Phys. **45**, 9 (1975).

[2] S. W. Hawking, Phys. Rev. D **14**, 2460 (1976).

[3] S. W. Hawking, D. N. Page, and C. N. Pope, Nucl. Phys. **B170**, 283 (1980).

[4] S. W. Hawking, in *Quantum Gravity 2: A Second Oxford Symposium*, edited by C. J. Isham, R. Penrose, and D. W. Sciama (Clarendon, Oxford, 1981).

[5] S. W. Hawking, Commun. Math. Phys. **87**, 395 (1982).

[6] C. W. Misner, Ann. Phys. (N.Y.) **24**, 102 (1963).

[7] J. J. Halliwell and S. W. Hawking, Phys. Rev. D **31**, 1777 (1985).

[8] P. D. D'Eath and J. J. Halliwell, Phys. Rev. D **35**, 1100 (1987).

[9] S. W. Hawking, Nucl. Phys. **B239**, 257 (1984).

[10] S. W. Hawking and D. N. Page, Nucl. Phys. **B264**, 185 (1986).

[11] J. B. Hartle and S. W. Hawking, Phys. Rev. D **28**, 2960 (1983).

[12] D. J. Gross, Nucl. Phys **B236**, 349 (1984).

Nuclear Physics B306 (1988) 890–907
North-Holland, Amsterdam

AXION-INDUCED TOPOLOGY CHANGE IN QUANTUM GRAVITY AND STRING THEORY

Steven B. GIDDINGS

*Joseph Henry Laboratories, Princeton University, Princeton, NJ 08544, USA, and Lyman Laboratory of Physics, Harvard University, Cambridge, MA 02138, USA**

Andrew STROMINGER

Department of Physics, University of California, Santa Barbara, CA 93106, USA

Received 23 October 1987
(Revised 23 November 1987)

We consider a system compr sed of an axion (described by a rank-three antisymmetric tensor field strength) coupled to gravity. Instantons are found which describe the nucleation of a Planck-sized baby Robertson–Walker universe. Information loss to the baby universes can lead to an effective loss of quantum coherence. An estimate of the magnitude of this effect on particle propagation is made in the semi-classical approximation. This magnitude depends on the parameters of the theory (which includes a cutoff since the theory is non-renormalizable) and on the quantum state of the many-universe system. In contrast to the naive expectation that Planck-scale dynamics should lead to very small effects at low energies, the effects of these instantons can be large. The case of string theory is considered in some detail, and it is found that a massless dilaton can suppress the tunneling.

1. Introduction

Current experiments cannot reach energies high enough to directly probe the laws of physics at the grand unification or Planck scales. It might neve theless be possible to learn something about these laws if they violate symmetries of low-energy physics. A classic example of this is proton decay. On rather general grounds one expects baryon number not to be an exact symmetry of a grand unified model, and proton decay should occur at some rate [1]. While it has not actually been measured, the lower bounds on the rate do provide strong constraints on grand unified models.

Another example of this is the possible effective loss of quantum coherence when quantum mechanics is combined with general relativity. As first emphasized by Wheeler [2], when the space-time metric is treated as a quantum field, the topology of space-time is expected to fluctuate on scales of the order of the Planck length.

* Present address.

0550-3213/88/$03.50 © Elsevier Science Publishers B.V.
(North-Holland Physics Publishing Division)

Baby universes might be pinched off and carry away information. This effective information loss can lead to an effective loss of quantum coherence as viewed by the macroscopic observer who cannot measure the quantum state of the baby universes [3]. (The possible loss of quantum coherence in quantum gravity was first discussed from a different point of view by Hawking [4], following his discovery of incoherence in semi-classical black-hole evaporation.) One thus expects, on rather general grounds, that Planck-scale dynamics could lead to quantum incoherence.

While at first glance it may seem that effective quantum incoherence is an inevitable consequence of baby universes, this does not appear to be the case. The rate of coherence loss depends on the state of the many-universe system. Coleman* has in fact argued that if the many universe system is in an equilibrium state quantum incoherence will *not* be observed. Whether or not the system is in such a state depends both on the dynamics and on the boundary conditions. This fascinating issue will be only briefly addressed in this paper – although we do hope that the results presented herein will be useful in further investigations along these lines.

When quantum coherence is lost, it can be shown that a current associated to a symmetry of the laws of physics is not necessarily exactly conserved, although it is conserved on the average [4–7]. Thus, if quantum gravity violates quantum coherence in the most general fashion there would be no exact conservation laws. It is very hard to see how this could be consistent with experimental observation. For example, Banks et al. [5] consider the effects of a small amount of energy non-conservation on β decay. An enormous amount of phase space is available for decays which violate energy conservation; so even an energy non-conserving interaction, which is suppressed by many powers of the weak scale divided by the Planck scale, would lead to observable effects. This is a beautiful example of how Planck-scale physics might lead to observable effects at low energies. In fact, it rules out the most general form of quantum incoherence, unless it is suppressed by a fantastically small dimensionless parameter.

Thus, if quantum incoherence does occur, it probably takes a more subtle form. One possibility is that currents associated with local symmetries (such as the energy–momentum tensor) are exactly conserved while those associated only with global conservation laws are not. Ellis et al. [6] have considered some of the experimental consequences of this type of incoherence. The type of incoherence, if any, that actually occurs in nature should, in principle, be deduced from the microscopic laws of physics.

One of the difficulties in considering this issue is that there have been no models in which the loss of quantum coherence has been calculable as an expansion in a small parameter, so much of the discussion has been rather vague. In this paper, we shall present just such a model. The model is simply gravity coupled to an axion described by a rank-three antisymmetric tensor field strength. Previously unnoticed

* S. Coleman, private communication.

peculiar properties of such axions in euclidian space lead to instantons which describe the nucleation of baby universes. The loss of quantum coherence can then, in principle, be computed in a semi-classical expansion around these instantons. (To actually compute beyond leading order requires a real quantum theory of gravity, such as string theory.) We do find that conservation laws associated with a local symmetry are conserved, but those associated with a global conservation law are not always conserved. Phase information of a quantum state can also be lost. This leads, for example, to effects on kaon oscillation (at least for some states of the many-universe system) which can be estimated in terms of the parameters of our model and the kaon–axion couplings; we sketch an estimate of one such effect. Despite the fact that such effects arise from Planck-scale physics, they are *not* a priori small and experimentally inaccessible. In fact, they may be *enhanced* by powers of the ratio of the Planck mass to the other masses in the problem. Thus, the opposite problem occurs; in order to avoid conflict with experiment there must be a suppression factor. In our model, this suppression can be obtained by taking the topological coupling constant governing the strength of the effect to be extremely small or by cutting off the integration over instanton sizes above the Planck length. Alternatively there may also be states of the many-universe system for which quantum incoherence does not occur. Indeed, it may be that there are no physically reasonable states in which loss of quantum coherence can be observed. Since our results may be rather surprising, it is worth emphasizing at the outset several important assumptions underlying our calculation. These are as follows.

(1) Non-trivial topologies should be included in the euclidean functional integral for quantum gravity. This is certainly debatable since there is no obvious inconsistency if one simply excludes them from the functional integral, and there is no known set of first principles from which the necessity of inclusion can be derived.

(2) The effects of these topologies can be reliably computed in a euclidean semi-classical expansion using the action (2.1). Since (2.1) is not positive definite and the semi-classical approximation may not be reliable, this is again debatable.

2. Axionic instantons

Consider the action

$$S=\int d^4x\sqrt{g}\left(-\frac{M_P^2}{16\pi}R+f^2H_{\mu\nu\lambda}H^{\mu\nu\lambda}\right)-\gamma(\chi-1)-\frac{M_P^2}{8\pi}\int_{\partial V}d^3S^a(\kappa_a-\kappa_{0a}),\tag{2.1}$$

where M_p is the Planck mass and f is the Peccei–Quinn scale*. The three-form

* In the absence of matter couplings, f can be eliminated from the action by rescaling H. However, it is useful to retain the f dependence in this form for our later consideration of matter couplings.

$H = dB$ so that $dH = 0$. χ is the Euler number of the manifold, and γ is a complex topological coupling constant whose imaginary part is an integral multiple of $i\pi$. $\int(\kappa - \kappa_0)$ is the integral of the trace of the extrinsic curvature of the boundary, minus that of the boundary embedded in flat space. This term is required by unitarity [8], but turns out to be zero in the following. The equations of motion are

$$M_p^2 G_{\mu\nu} = 16\pi f^2\left(3H_{\mu\alpha\beta}H_\nu{}^{\alpha\beta} - \tfrac{1}{2}g_{\mu\nu}H_{\alpha\beta\gamma}H^{\alpha\beta\gamma}\right), \tag{2.2}$$

$$d^*H = 0 \tag{2.3}$$

where * is the Hodge dual. The axion field is sometimes rewritten as follows. First, define the conserved axion current $j = {}^*H$. Since dj vanishes, locally $j = da$. (Our normalization of H has been chosen so that when gauge fields are coupled, the non-anomalous symmetry is $a \to a + 2\pi$.) However, caution must be exercised; if the manifold is not simply connected, the pseudoscalar a may not be globally defined*. We will in fact encounter just such a situation. Locally, however, Einstein's equation may be written

$$M_p^2 G_{\mu\nu} = 16\pi f^2 T_{\mu\nu}(a), \tag{2.4}$$

$$T_{\mu\nu}(a) = -\nabla_\mu a\, \nabla_\nu a + \tfrac{1}{2}g_{\mu\nu}\, \nabla_\lambda a\, \nabla^\gamma a\,. \tag{2.5}$$

This is to be compared with the euclidean stress energy of a massless minimally coupled scalar field

$$T_{\mu\nu}(\phi) = \nabla_\mu\phi\, \nabla_\nu\phi - \tfrac{1}{2}g_{\mu\nu}\, \nabla_\lambda\phi\, \nabla^\lambda\phi\,, \tag{2.6}$$

which is identical to (2.5) except for a relative minus sign**. Thus, the euclidean behavior of an axion field described by a three-form field strength coupled to gravity is radically different from that of a scalar field. This is ultimately responsible for the fact that interesting instantons do exist for gravity coupled to axions.

The connection between negative euclidean stress energy and the existence of instantons can be made more precise with the aid of the following

* Demanding that coupling of a to $F \wedge F$ be well-defined then gives a quantization condition. We will not discuss this further, except to note that for the wormhole solution mentioned later in sect. 2, the quantization condition becomes one for M_p/f

** Note that had we done the same calculation in Minkowski space, there would be no relative minus sign. Put another way, rewriting $H = {}^*da$ and continuing to euclidean space do not commute. The reason for this is that the tensor $\varepsilon_{\mu\nu\rho\sigma}$ obeys $\varepsilon^2 = 24$ in euclidean space and $\varepsilon^2 = -24$ in Minkowski space. It is thus essential, in everything that follows, that the euclidean functional integral is defined by analytic continuation of (2.1) rather than the action $\int \nabla_\mu a\, \nabla^\mu a$. Of course, the two different axion formalisms are only equivalent on shell, so it is not too surprising that they could have different behavior in euclidean space.

Theorem. Consider a four manifold M with $n \geqslant 1$ boundaries of arbitrary topology and one boundary which is topologically S^3, and a euclidean signature metric g on M which is asymptotically euclidean near the S^3 boundary and has vanishing extrinsic curvature on the other n boundaries. Then the Ricci tensor of g has negative eigenvalues somewhere on M.

This theorem follows almost trivially from theorems proved in ref. [9]. As we shall see in sect. 3, the conditions described above on M and g are exactly the conditions which a geometry must satisfy in order to have an interpretation as a gravitational instanton describing topology change in the asymptotically flat context. (Similar theorems exist which pertain to other types of gravitational instantons.) On the other hand, for an ordinary scalar field the Ricci tensor obeys $R_{\mu\nu} = \nabla_\mu \phi \nabla_\nu \phi$ and therefore cannot have any negative eigenvalues. Contrariwise, for an axion the eigenvalues are non-positive and instantons are indeed possible.

To construct an explicit example, consider the following spherically symmetric ansatz for the metric and H:

$$ds^2 = dt^2 + a^2(t)\, d^2\Omega_3 , \tag{2.7}$$

$$H = h(t)\varepsilon , \tag{2.8}$$

where $d^2\Omega_3$ is the line element on the three-sphere and ε is the volume, three-form normalized so that

$$\int_{S^3} \varepsilon = 2\pi^2 a^3(t) . \tag{2.9}$$

The H equation $dH = d^*H = 0$ is obeyed by setting

$$h(t) = n/f^2 a^3(t) , \tag{2.10}$$

so that

$$\int_{S^3} H = 2\pi^2 n/f^2 , \tag{2.11}$$

for any S^3 surrounding the origin. The time–time component of Einstein's equation is then

$$\frac{3M_P^2}{16\pi}\left(\frac{\dot{a}^2}{a^2} - \frac{1}{a^2}\right) = -\frac{3n^2}{a^6 f^2} . \tag{2.12}$$

The solution of this equation is

$$\frac{t}{r} = \sqrt{\tfrac{1}{2}}\, F\left[\cos^{-1}\left(\frac{r}{a}\right), \sqrt{\tfrac{1}{2}}\right] - \sqrt{2}\, E\left[\cos^{-1}\left(\frac{r}{a}\right), \sqrt{\tfrac{1}{2}}\right] + \frac{1}{ra}\sqrt{(a^4 - r^4)} , \tag{2.13}$$

where F and E are the elliptic integrals of the first and second kinds, respectively, and $r^2 = 4\sqrt{\pi}\, n/fM_P$ is the $t = 0$ value of a^2. The remaining equations follow from

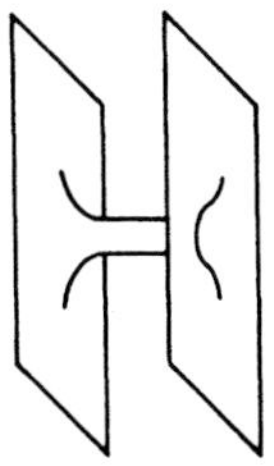

Fig. 1. A gravitational instanton with two asymptotically euclidean regions. Two dimensions are suppressed; each circle around the throat represents a three-sphere.

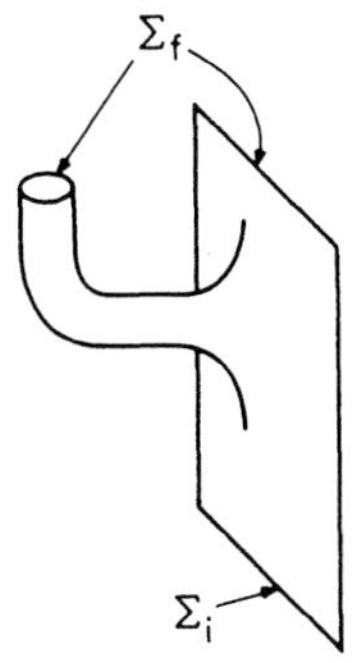

Fig. 2. A gravitational instanton representing topology change from R^3 to $R^3 \oplus S^3$.

the Bianchi identities. This solution is characterised by one free parameter, n (which in string theory is quantized [10]).

This instanton is drawn in fig. 1. Since $a^2(t) \to t^2$ for $t \to \pm\infty$ there are two asymptotically euclidean regions. They are joined by a throat whose cross sections are S^3's. The axion current j has total integrated flux n/f^2 through the throat. As it stands, it is difficult to ascribe a physical interpretation to this instanton because of the two asymptotic regions. (However, it might represent communication between two different universes.)

The situation can be improved by slicing the instanton in half through the minimal surface of the throat. The resulting half-throat solution is suggestively redrawn in fig. 2*. It represents tunnelling between an initial surface Σ_i which is topologically R^3, to a final surface Σ_f which is topologically $R^3 \oplus S^3$**. It describes nucleation of a baby Robertson–Walker universe*** created at its moment of time

* A similar instanton has been previously considered by E. Witten, who identified opposite points of the minimal S^3. A non-orientable asymptotically euclidean instanton with one boundary is thereby obtained (private communication, 1984).

** Or the reverse process $R^3 \oplus S^3 \to R^3$.

*** A similar process of nucleation of a small universe was described using instantons in ref. [3].

symmetry and maximal radius. (In fact, our instanton is itself an analytic continuation of a Robertson–Walker cosmology.) It is critical that the fields on Σ_i and Σ_f take values appropriate for the description of a tunnelling process. Namely, it must be true that the fields and their first times derivatives on Σ_i and Σ_f are all real when analytically continued back to a Lorentzian space–time. This is obvious for the R^3 parts of Σ_i and Σ_f. On the S^3 portion, the time derivative of the metric vanishes because it is a minimal surface. The time components of H vanish because it is a three-form tangent to the three-sphere. Thus, the instanton does obey exactly the right boundary conditions for description of the tunnelling $R^3 \to R^3 \oplus S^3$*.

An additional important feature which characterizes this instanton is the axion charge on the non-contractible three-spheres. The (dimensionless) axion charge on a three-surface Σ is defined by

$$q = f^2 \int_\Sigma H . \tag{2.14}$$

If Σ is any three-sphere surrounding $t = 0$, the axion charge q equals n. In string theory, global anomalies in the string sigma model lead to quantization of this charge [10].

The observer on R^3 will measure a change of $-n$ in the axion charge, since the baby universe pinches off n units of axion charge. This is something of a puzzle to the R^3 observer, who cannot observe the charge on the baby universe, since he may believe that axion charge is conserved due to the (unbroken at our level of analysis) Peccei–Quinn symmetry of the action (2.1). As we shall see in sect. 3 this effective charge non-conservation can be understood as a result of the breakdown of quantum coherence.

Using the trace of Einstein's equation

$$M_p^2 R = -16\pi f^2 H^2 \tag{2.15}$$

and the fact that $\chi - 1 = -1$ the action of this instanton is easily seen to be

$$S = 2f^2 \int d^4x \sqrt{g}\, H^2 + \gamma , \tag{2.16}$$

$$= \frac{24\pi^2 n^2}{f^2} \int_0^\infty \frac{dt}{a^3} + \gamma , \tag{2.17}$$

$$= \frac{6\pi^3 n^2}{r^2 f^2} + \gamma , \tag{2.18}$$

* An approximate wormhole solution can be obtained by taking two widely separated instantons with opposite charge and identifying the minimal S^3's. This is a configuration for which a is not globally defined. It represents the contribution of a virtual Robertson–Walker universe to the vacuum–vacuum amplitude.

where $r = a(0)$ is the radius of the baby universe. Using the constraint equation (2.12) at $t = 0$

$$n = r^2 f \frac{M_p}{4\sqrt{\pi}}. \tag{2.19}$$

The action may alternatively be written as

$$S = \frac{3\pi^2}{8} r^2 M_p^2 + \gamma, \tag{2.20}$$

$$= \frac{3\pi^{5/2} n M_p}{2f} + \gamma. \tag{2.21}$$

We see that the nucleation of the baby universes large relative to the Planck length is highly suppressed – a comforting thought! A typical baby universe will have a radius of order $\sqrt{(8/3\pi^2)}\,M_p$. The nucleation rate of baby universes per Planck three-volume per Planck time is of order e^{-S}. Equivalently, there is, on the average, one tunneling event in every region of volume e^S/M_p^4. Another quantity which is important in determining the low-energy effects of these instantons is the axion charge carried away by the baby universe. This has a typical value of $n \sim 2f/3\pi^{5/2}M_p$.

If γ is large, the action S is large for any size instanton. The instantons are then widely separated and the dilute gas approximation is justified*. This allows one to ignore interactions between instantons.

Justification of the semi-classical expansion requires an additional small dimensionless parameter. Since the action (2.1) is not renormalizable, the semi-classical loop expansion is not very well defined. If we define the theory with a momentum space cutoff Λ, then the loop expansion is an expansion in (Λ/M_p). Since the presence of such a cutoff is presumably implied by new physics at the scale Λ (not described by the action (2.1)), it is perhaps appropriate to also cutoff the integration over instanton sizes at $r = 1/\Lambda$. We emphasize that this integration is ultraviolet finite and we are not forced to cut it off by divergences. If we do cut off the instanton size at $1/\Lambda \gg 1/M_p$, the maximum nucleation rate occurs for instantons of size $r \sim 1/\Lambda$ and charge $q \sim fM_p/4\sqrt{\pi}\Lambda^2$. This implies a suppression factor $\exp(-3\pi^2 M_p^2/8\Lambda^2)$. This factor is highly Λ-dependent, and is extremely small for $\Lambda/M_p \ll 1$. For example, changing Λ from M_p to $M_p/10$ changes the rate by a hundred orders of magnitude! Because the rate is so highly cutoff dependent, we cannot obtain a reliable estimate without a full quantum theory of gravity, although we can argue that the process should occur at some rate. In string theory, the loop

* However, large positive γ may lead to other difficulties since configurations with positive Euler number are enhanced rather than suppressed.

expansion parameter Λ/M_p is the dilaton vacuum expectation value, and can be made arbitrarily small.

3. Quantum incoherence

There are two approaches to describing the effects of these instantons on time evolution. One approach is to work on an enlarged Hilbert space which includes all the baby universes. The hamiltonian* $\mathscr{H}$ on this space contains an interaction term which describes the splitting off and rejoining of the baby universes. $\mathscr{H}$ describes unitary evolution (assuming no gravitational singularities) and the vacuum state of $\mathscr{H}$ is a pure state.

A second approach is to construct an effective hamiltonian $\mathscr{H}_{\text{eff}}$ which acts on a smaller Hilbert space whose states depend only on observables on R^3, and not on the baby universes. $\mathscr{H}_{\text{eff}}$ is appropriate for computing the results of measurements not involving measuring devices on the baby universes. $\mathscr{H}_{\text{eff}}$ does not, in general, describe unitary evolution since there can be information loss to the baby universes. Pure states can evolve into mixed states, but probability is conserved. Quantum incoherence due to information loss to baby universes was first discussed in ref. [3] and is similar in spirit to information loss in black-hole evaporation and from "virtual black holes" discussed earlier by Hawking [4]. Recent discussions of this type of incoherence can be found in refs. [11,12].

The vacuum state of $\mathscr{H}_{\text{eff}}$ is a density matrix ρ representing a mixed state; i.e. $\operatorname{tr}\rho^2/|\operatorname{tr}\rho|^2 < 1$. To compute ρ, consider first the simplified model in which the enlarged Hilbert space consists of states denoted $|m\rangle|n\rangle$, where m is the axion charge on R^3, and n is the axion charge on the baby universes. Let the Hamiltonian $\mathscr{H}$ act on the state $|m\rangle|n\rangle$ as

$$\mathscr{H}|m\rangle|n\rangle = e^{-S}M_p(|m+1\rangle|n-1\rangle + |m-1\rangle|n+1\rangle). \tag{3.1}$$

$\mathscr{H}$ transfers one unit of axion charge between R^3 and the baby universes in a characteristic time e^S/M_p. The eigenstates of $\mathscr{H}$ are

$$|\theta, p\rangle = \sum_n e^{i\theta n}|n, p-n\rangle, \tag{3.2}$$

with eigenvalues

$$\mathscr{H}|\theta, p\rangle = 2\cos\theta\, e^{-S}|\theta, p\rangle. \tag{3.3}$$

The vacuum is described by the density matrix (no sum on p)

$$\rho = |\theta, p\rangle\langle\theta, p|, \tag{3.4}$$

* The construction of a hamiltonian requires a definition of time. This is only well defined on scales which are large compared to the Planck scale.

which obeys

$$\frac{\operatorname{tr}\rho^2}{|\operatorname{tr}\rho|^2}=1, \tag{3.5}$$

and represents a pure state.

To obtain the density matrix ρ_{eff} which involves only the axion charge on R^3, we must trace ρ over all values of the unobservable axion charge $p-n$ on the baby universe, or equivalently over all values of the unobservable total charge p

$$\begin{aligned}\rho_{\text{eff}}&=\operatorname{tr}_{\mathrm{p}}|\theta,p\rangle\langle\theta,p|\\&=\sum_n \mathrm{e}^{i\theta n}|n\rangle\langle n|\mathrm{e}^{-i\theta n}\\&=\sum_n |n\rangle\langle n|,\end{aligned} \tag{3.6}$$

where the state $|n\rangle$ describes only the axion charge on R^3. This result can be derived by a more general argument. The observer on R^3 views the instanton as a random stochastic process which changes the axion charge. Asymptotically, all information about the dependence of the quantum state on the axion charge will be lost, and the state of maximum entropy will be reached. ρ_{eff} is indeed the state of maximum entropy since there is equal probability for any value of the axion charge.

While the details of the loss of axion charge incurred by our instanton differ slightly from this simplified example, the observer on R^3 will still view the instantons as stochastic sources and sinks of axion charge. The equilibrium state of the system will, therefore, be described by the same density matrix $\rho_{\text{eff}}=\Sigma_n|n\rangle\langle n|$. The rate of approach to equilibrium is governed by the instanton action. The characteristic time is e^S in Planck units (10^{-43} s).

Of course, incoherence of the axion vacuum is not the only type of incoherence produced by these instantons. Any field which interacts with axions can lose phase information, as we shall see momentarily for the case of kaons. The baby universes can also carry off other types of particles as well as axions. A systematic method for calculating these processes has recently been given by Lavrelashvili et al. [11] and by Hawking [12]; the former, in particular, discuss the case where a quark–antiquark pair goes down the hole and use this process to put a bound of $\mathscr{A}\lesssim 10^{-17}M_{\mathrm{p}}^4$ on the amplitude for such a topology change to occur.

One might worry that other low-energy conservation laws, such as energy conservation, would be violated in this process [5]. However, our tunnelling process conserves energy exactly, since energy conservation is enforced by integrating over the instanton location. Indeed, since energy is given by a surface integral at spatial infinity, its conservation is insensitive to localized processes in space–time. It similarly appears that any charge associated with a local conservation law will be exactly conserved.

We now turn to the issue of how to observe the effects of axionic instantons. There are two types of effects one might measure. In our model, space–time is filled with a bath of incoherent axions created by the gas of instantons. One might look for particle oscillations or depolarizations due to interactions with this bath. Such effects can be detected by measuring only a few components of the density matrix and can arise from either coherent or incoherent sources. Of perhaps more interest is observation of the quantum incoherence which is the characteristic signal of our process. This requires measuring all the components of the density matrix in order to determine whether or not $\mathrm{tr}\,\rho^2 < (\mathrm{tr}\,\rho)^2$. The types of processes in which loss of quantum coherence might be observed, even if charges associated with local conservation laws are exactly conserved, were discussed by Ellis et al. [6], as well as in refs. [11, 12]. One possibility is to look for incoherence in neutron interferometry. It is natural to look for measurable effects due to the basic couplings $a\bar{n}\gamma_5 n$, $j^\mu \bar{n}\gamma_5\gamma_\mu n$. Another possibility is the K^0–$\overline{K}^0$ system. The density matrix is a two-by-two hermitian matrix

$$\rho = \begin{pmatrix} \alpha|K^0\rangle\langle K^0| & \gamma^*|\overline{K}^0\rangle\langle K^0| \\ \gamma|K^0\rangle\langle\overline{K}^0| & \beta|\overline{K}^0\rangle\langle\overline{K}^0| \end{pmatrix}, \tag{3.7}$$

where α and β are real. Measurement of $K^0(\overline{K}^0)$ decays determines $\alpha(\beta)$, while decays of the CP eigenstates $K_{1,2} = \sqrt{\tfrac{1}{2}}\,(K^0 \pm \overline{K}^0)$ measure $\alpha + \beta \pm (\gamma + \gamma^*)$. Ellis et al. use this to obtain bounds on the impurification (loss of coherence) rate of ρ.

The kaon impurification rate can not be computed using kaons directly since they are not relevant physical variables at the length scale characterizing the instantons. Instead, one should use quarks or whatever else is the relevant field at the scale Λ or M_p. However, working with spinors introduces technical complications which are inessential for understanding the qualitative features of the process. We, therefore, for illustrative purposes momentarily pretend that kaons are fundamental particles. The corrections to our final answer deduced from a proper treatment using quarks will be discussed later. In a hypothetical world in which kaons are fundamental, the kaon impurification rate can be calculated in terms of the axion–kaon couplings. Presumably, virtually all couplings lead to some form of impurification. We consider only the CP invariant, P non-invariant coupling of the axion current to a pseudoscalar current

$$\tfrac{1}{2} i g j_\mu \nabla^\mu\left[(K^0)^2 - (\overline{K}^0)^2\right] \tag{3.8}$$

(recall that j_μ is the axion current). We do not know if this is the most important term – we consider it because the effect is particularly easy to calculate.

We have performed a semi-classical calculation of the impurification rate – in a particular state of the many-universe system – using the methods of, e.g. ref. [11]. This involves solving the Euclidean kaon wave equation in the instanton background. Here, we shall content ourselves with presenting a simple order of magnitude estimate of the rate. As will be evident, this is sufficient to establish that it is very high unless the instantons themselves are suppressed by a small dimensionless parameter.

The kaon equation of motion is (neglecting the standard a mixing terms)

$$(\Box - M_K^2)K^0 = -ig\overline{K}^0 \nabla_\mu j^\mu ,$$

$$(\Box - M_K^2)\overline{K}^0 = igK^0 \nabla_\mu j^\mu . \tag{3.9}$$

where M_K is the kaon mass. A K^0 propagating in an axion background will rotate into a $\overline{K}^0$ and vice versa. We therefore expect that the nucleation of a baby universe (which requires a non-zero axion field), in the presence of a kaon, to be accompanied by some K^0–$\overline{K}^0$ rotation. The rotation rate due to instantons with axion charge q can be estimated by simply replacing $g\nabla_\mu j^\mu$ by its expectation value in a dilute gas of charge q instantons

$$M_q^2 \equiv g\langle \nabla_\mu j^\mu \rangle_q = \frac{gq}{f^2} e^{-S_q} M_p^4 , \tag{3.10}$$

where S_q is the action of an instanton with charge q. The expectation value acts something like an off-diagonal mass term (we shall see momentarily that its effect is incoherent) which shifts the kaon mass eigenvalues by $\pm M_q^2/2M_K$. The rotation rate is, therefore, of order M_q^2/M_K.

This can be discussed using a generalization of the interaction hamiltonian (3.1) describing transfer of axion charge from the large to the baby universe. Consider a simplified Hilbert space spanned by states of the form

$$|K^0, m\rangle |n\rangle , \tag{3.11}$$

where m is the axion charge on R^3 and n the unobservable charge on the baby universes. The portion $\mathscr{H}_q$ of the interaction hamilton $\mathscr{H}$, which transfers q units of axion charge, then acts as

$$\mathscr{H}_q |K^0, m\rangle |n\rangle = e^{-S_q} M_p |K^0, m+q\rangle |n-q\rangle$$
$$+ i\frac{M_q^2}{M_K} |\overline{K}^0, m+q\rangle |n-q\rangle . \tag{3.12}$$

We now approximate the interaction hamiltonian by $\mathscr{H}_Q + \mathscr{H}_{-Q}$ where $q = \pm Q$ are the axion charges of the dominant instantons (as discussed in sect. 2 and to be summarized shortly). From this we find that after a time of order $1/M_p$, the initial state $|K^0, m\rangle|n\rangle$ evolves into

$$|K^0, m\rangle|n\rangle + e^{-S_Q}\big(|K^0, m+Q\rangle|n-Q\rangle + |K^0, m-Q\rangle|n+Q\rangle\big) + i\frac{M_Q^2}{M_K M_p}\big(|K^0, m+Q\rangle|n-Q\rangle - |\overline{K}^0, m-Q\rangle|n+Q\rangle\big). \tag{3.13}$$

Tracing over the unobservable charge on the baby universe, we obtain the effective density matri (to lowest order in M_Q^2)

$$\begin{aligned}\rho_{\text{eff}} = {} & |K^0, m\rangle\langle K^0, m| + \left(e^{-S_Q}|K^0, m+Q\rangle + i\frac{M_Q^2}{M_K M_p}|\overline{K}^0, m+Q\rangle\right) \\ & \times \left(e^{-S_Q}\langle K^0, m+Q| - i\frac{M_Q^2}{M_K M_p}\langle\overline{K}^0, m+Q|\right) \\ & + \left(e^{-S_Q}|K^0, m-Q\rangle - i\frac{M_Q^2}{M_K M_p}|\overline{K}^0, m-Q\rangle\right) \\ & \times \left(e^{-S_Q}\langle K^0, m-Q| + i\frac{M_Q^2}{M_K M_p}\langle\overline{K}^0, m-Q|\right),\end{aligned} \tag{3.14}$$

which obeys $\operatorname{tr}\rho^2/(\operatorname{tr}\rho)^2 < 1$. Notice that there is no destructive interference between instantons with charge $\pm Q$, even though they rotate kaons in the opposite direction. Interference can not occur because the final states of the baby universes differ in the two cases.

The outcome of an experiment which does not measure the axion charge is described by a density matrix involving only K^0 and $\overline{K}^0$. In the K^0–$\overline{K}^0$ basis (3.14) becomes

$$\tilde{\rho} = \begin{pmatrix} 1 + e^{-2S_Q} & 0 \\ 0 & 2\dfrac{M_Q^4}{M_K^2 M_p^2} \end{pmatrix}. \tag{3.15}$$

In summary, interactions of a kaon with an incoherent axion bath cause an initially pure state $|K^0\rangle\langle K^0|$ to approach the mixed state $|K^0\rangle\langle K^0| + |\overline{K}^0\rangle\langle\overline{K}^0|$. The characteristic impurification time scale is of order $(M_K M_p/M_Q^2) \times 10^{-43}$ s. A typical value of Q is $Q = (f/M_p)$, so that M_Q^2/M_K is of order $g(M_p^3/fM_K)e^{-\gamma}$ for large γ. Approximating the charge q by its typical values, we thus find a time scale of $fM_{\rm p}e^{\gamma}/M_{\rm p}^2 g \times 10^{-43}$ s. This time scale decreases, and becomes more observable, if $M_{\rm p}$ is increased. This result is modified by a factor $(\Lambda^2/M_{\rm p}^2)\exp(3\pi^2 M_{\rm p}/8\Lambda^2)$ if we put a cutoff $\Lambda \ll M_{\rm p}$ on the instanton size. Incoherence arises because the magnitude and phase of K^0–$\bar{K}^0$ rotation is correlated with the axion charge on the baby universe, and the latter quantity is experimentally inaccessible.

It is interesting to note that, even though the interaction term (3.8) and the entire action are CP invariant, CP is not conserved by our incoherent process. This is because the kaons interact with axions which carry CP down the hole onto the baby universe. This is an example of the more general phenomenon that symmetries do not lead to conservation laws in the absence of quantum coherence.

It is natural to ask if this type of CP violation could in fact account for the observed CP violation in nature. We do not know the answer to this question. It would be interesting to determine, by analysis of the density matrix that is measured in the experimental tests of CP violation, whether CP violation is intrinsically incoherent.

We now consider and discuss how this result is altered if the calculation is done properly in terms of quarks. This of course depends on the quark–axion couplings. The largest effects come from terms which respect the Peccei–Quinn symmetry such as

$$\frac{\lambda}{M_{\rm p}} \nabla_\mu j^\mu \bar{q} T q \,, \tag{3.16}$$

where $\bar{q}Tq$ is a general quark bilinear and λ is a dimensionless coupling constant. Peccei–Quinn violating couplings such as $a\bar{q}Tq$ will be further suppressed by powers of m_q/f. Replacing $\nabla_\mu j^\mu$ by its typical value and setting $f = M_{\rm p}$, we see that this is roughly equivalent to altering the quark mass matrix with terms of order $\lambda M_{\rm p} e^{-S_Q}$. This, in turn, alters the kaon mass eigenvalues by amounts of the same order (up to ratios of nonplanckian masses). Thus, we see that treating kaons as fundamental led us to overestimate impurification rates by a factor of $M_{\rm p}/M_{\rm K}$, but it still grows with $M_{\rm p}$. The fact that fermions couple less strongly than scalars in this context has been noted previously [4, 11, 12]*. In ref. [11], rates similar to those we obtain for fermions and scalars were derived, in a slightly different context, from both a detailed calculation and dimensional arguments.

Correcting our answer by $M_{\rm K}/M_{\rm p}$, the rate we obtain can be compared with e.g. the inverse kaon mass. Since this is of order 10^{-24} s, a minimum requirement is an upper bound of order 10^{-19} on the dimensionless factor

$$g \frac{M_{\rm p}^2}{\Lambda^2} e^{-\gamma} e^{-3\pi^2 M_{\rm p}^2/8\Lambda^2} \,, \tag{3.17}$$

* The coupling (3.16) violates chiral symmetry. The rate could be suppressed if chiral symmetry is imposed. We have not checked this.

where we have taken the Peccei–Quinn scale f to equal M_p. Possibly stronger bounds could be obtained by considering the effects on kaon oscillations or CP violation [6, 11]. There are at least two possible ways that the factor (3.17) could be small but non-zero

(1) The topological coupling γ is large.

(2) The effective cutoff on instanton size is large. The expression (3.17) is clearly very sensitive to Λ/M_p. A Λ an order of magnitude smaller than M_p would be sufficient.

Finally, there is the important issue of how these results depend on the choice of vacuum for the many-universe system. Arguments have been given that there are certain vacua for which the incoherence rate vanishes [13] and others for which it is infinite*! For example, if we use the $|\theta, p\rangle$ vacua of (3.2) rather than the $|m, n\rangle$ vacua in our kaon computation, no incoherence occurs in the process we consider. The reason is that the state of the baby universes is an eigenstate with respect to the addition of another baby universe. The kaons which are rotated in opposite directions from the nucleation of baby universes with opposite charge then end up in the same state (up to a phase) as far as the baby universes are concerned. It is, therefore, possible for them to interfere, unlike in the case we consider. (In this case, there may still be other effects of the instantons, such as CP violation or mass shifts.) Clearly, in order to assess the relevance of our results it is essential to determine the physically correct vacuum state. This depends on the details of the dynamics and the initial conditions for the universe(s). It is not at all clear that the state we employ in (3.12) is physically reasonable – it changes under time evolution. We will not attempt to answer these interesting questions in this paper.

4. Instantons in low-energy effective string field theory

String theory provides the ideal setting in which to calculate this type of effect since it is apparently a consistent quantum theory of gravity. Unfortunately, as we shall see, in the simplest case of one axion and one dilaton in the bosonic or heterotic string these instantons are suppressed in a rather strange fashion by the non-linear coupling of the dilaton. This suppression depends crucially on the exact coefficients of the couplings, so it may be that instantons do exist in other string theories or with more than one axion. They may also be relevant in the case that supersymmetry breaking gives the dilaton a large mass but the axion remains light. Instantons could contribute on scales above the axion mass and below the dilaton mass. This would lead to a large suppression factor of $\exp(M_p^2/M_D^2)$ (where M_D is

* M. Peskin, private communication. Peskin considers the case where the baby universes are forever separated from their mothers and the system never reaches equilibrium. The number of babies then diverges with time and stimulated emission might then occur as for an atom in a bath of many photons.

the dilaton mass) and might provide a natural explanation of the large factor required in sect. 3 from phenomenological considerations.

Ideally, a string field theory instanton should be described as an extrema of a string field theory action. However, extrema of the classical low-energy effective string field theory correspond, to leading order in α', to conformally invariant sigma models. In ref. [14] it was shown that, given a conformal sigma model, one can construct a corresponding solution to the string field equations of the string field theory of ref. [14]. We, therefore, henceforth consider only the low-energy effective action.

The four-dimensional low-energy effective action is

$$S=\int d^4x\sqrt{g}\,e^{\phi}\left[-R-(\nabla\phi)^2+H^2+\frac{\alpha'}{8}(R^2-F^2)+\cdots\right]$$
$$-2\int_{\partial V}d^3S^a e^{\phi}(\kappa_a-\kappa_{0a})+\gamma(\chi-1)+\text{boundary terms}, \quad (4.1)$$

where

$$dH=-\frac{\sqrt{3}\,\alpha'}{8}(R\wedge R-F\wedge F). \quad (4.2)$$

There may be additional massless particles arising from compactification, but we suppress these. The boundary terms are fixed by the requirement that no second derivatives appear in the total action. (Otherwise, the sum of the actions of two field configurations, glued along the boundaries along which the fields agree, will not equal the action of the sum [8].) These boundary terms are quite important – in fact the dilaton equation of motion implies that

$$S=-2\int_{\partial V}d^3S^a\left[e^{\phi}(\kappa_a-\kappa_{0a})+\nabla_a e^{\phi}\right]. \quad (4.3)$$

Thus, the action is entirely a boundary term. (A similar result holds in 10 or 26 dimensions.)

Note that we have retained the parameter γ in (4.1) as a free topological coupling in the action. This can obtain a contribution from dimensional reduction of a coupling to the ten-dimensional Euler character as well as from the second-order corrections to S. As far as we know, γ does not correspond to the vacuum expectation value of any field, nor is it fixed by any principle of string theory.

The action S can be rewritten in the form

$$S=\int d^4x\sqrt{\hat{g}}\left(-R(\hat{g})+\tfrac{1}{2}(\nabla\phi)^2+e^{\beta\phi}H^2+\cdots\right), \quad (4.4)$$

where

$$\hat{g} = e^{\phi} g , \tag{4.5}$$

and all indices are contracted with $\hat{g}$. For the bosonic, heterotic, or type I string, $\beta = 2$. However, since the result depends dramatically on β, we shall present results for arbitrary β.

We consider the ansatze:

$$d\hat{s}^2 = dt^2 + a^2(t)\, d\Omega_3^2 ,$$

$$H = h(t)\varepsilon ,$$

$$\phi = \phi(t), \tag{4.6}$$

as in eqs. (2.7) and (2.8). The tt component of Einstein's equation is now

$$3\left(\frac{\dot{a}^2}{a^2} - \frac{1}{a^2}\right) = \tfrac{1}{4}\dot{\phi}^2 - \frac{3e^{\beta\phi}n^2}{a^6} . \tag{4.7}$$

Conservation of stress energy still implies that, as in (2.12), the right-hand-side behaves like $1/a^6$. The solutions for a and H are completely unchanged; $a(t)$ is once again given by eq. (2.13), and $h(t)$ by eq. (2.10). However, the dilaton field obeys, as a function of a,

$$e^{-\beta\phi(a)} = e^{-\beta\phi(r)} \cos\left[\frac{\sqrt{3}\,\beta}{2} \arccos\left(\frac{r}{a}\right)^2\right] . \tag{4.8}$$

For $\sqrt{3}\,\beta/2 > 1$, $e^{-\beta\phi}$ vanishes* when $a = r[\cos(\pi/\sqrt{3}\,\beta)]^{-1/2}$. This means that the action, as given by the volume integral (4.1), is infinite. (The surface-integral form is invalid due to the divergence in ϕ.) Furthermore, this singularity corresponds to weak (in fact zero) coupling, so the semi-classical approximation is valid. Thus, t nnelling does not appear to occur for the action (4.1) with $\beta = 2$.

A similar divergence occurs for Yang–Mills instantons in string theory. To leading order in α', the usual Yang–Mills instanton is an extrema of the action with $\phi = 0$. ϕ becomes non-zero at the next order in α', and $e^{-2\phi}$ diverges at the origin. The action remains finite in this case, but the coupling is strong at the origin.

It seems rather surprising that the apparently minor modification of adding a scalar dilaton to the graviton–axion system should have such a dramatic effect. We do not know whether or not axionic instantons do exist when more than one axion or scalar is considered, or in other string theories.

* We are grateful to V. Rubakov for pointing this out.

It is a pleasure to thank V. Rubakov for valuable discussions. We also benefited from discussions with S. Coleman, H. Georgi, M. Peskin, M. Srednicki and D. Witt. We would like to thank Liz Keate-Giddings for TEXing this paper. We also appreciate the hospitality of the Aspen Center for Physics, where part of this work was carried out. S.B.G. was supported in part by NSF grant PHY82-15249 and A.S. supported in part by DOE Outstanding Junior Investigator Grant 8-484062-25000-3, and an A.P. Sloan Foundation Fellowship.

Note added in proof

Recent work by S. Coleman and the authors (Nucl. Phys. B, to appear) has shown that within the context of simple models for baby universe dynamics like those described herein, the coherent states discussed at the end of sect. 3 are indeed the only physically reasonable states. Thus, in this context, quantum incoherence is not experimentally observable.

References

[1] H. Georgi and S.L. Glashow, Phys. Rev. Lett. 32 (1974) 438
[2] J.A. Wheeler, Geometrodynamics (Academic, New York, 1962)
[3] A. Strominger, Phys. Rev. Lett. 52 (1984) 1733
[4] S.W. Hawking, Commun. Math. Phys. 87 (1982) 395; *in* Quantum Gravity 2: a Second Oxford Symposium, ed. C.J. Isham, R. Penrose and D.W. Sciama (Clarendon, Oxford, 1981);
S.W. Hawking, D.N. Page and C.N. Pope, Nucl. Phys. B170 (1980) 283
[5] T. Banks, M. Peskin and L. Susskind, Nucl. Phys. B244 (1984) 125
[6] J. Ellis, J. Hagelin, D.V. Nanopoulos and M. Srednicki, Nucl. Phys. B241 (1984) 381
[7] D.J. Gross, Nucl. Phys. B236 (1984) 349
[8] S.W. Hawking, *in* General Relativity, an Einstein Centenary Survey, eds. S.W. Hawking and W. Israel (Cambridge University Press, Cambridge, 1979)
[9] J. Cheeger and D. Grommol, Ann. Math 96(3) (1972) 413
[10] R. Rohm and E. Witten, Ann. Phys. 170 (1986) 454
[11] GV. Lavrelashvili, V.A. Rubakov and P.G. Tinyakov, Particle creation and destruction of quantum coherence by topological change, Institute for Nuclear Research Preprint, Moscow (1984), Nucl. Phys. B, to be published
[12] S.W. Hawking, Phys. Lett. B195 (1987) 337; Wormholes in spacetime, DAMTP preprint (1987)
[13] S. Coleman, private communication
[14] A. Strominger, Nucl. Phys. B294 (1987) 93

Nuclear Physics B310 (1988) 643–668
North-Holland, Amsterdam

WHY THERE IS NOTHING RATHER THAN SOMETHING:
A theory of the cosmological constant*

Sidney COLEMAN

Lyman Laboratory of Physics, Harvard University, Cambridge, MA 02138, USA

Received 9 May 1988

Wormholes are topology-changing configurations in euclidean quantum gravity whose importance has recently been advocated by several authors. I argue here that if wormholes exist, they have the effect of making the cosmological constant vanish. The argument involves approximations in dealing with physics at the wormhole energy scale (assumed to be somewhat below the Planck mass) but is exact in all interactions at all lower energies.

1. Introduction and conclusions

Fig. 1 is a sketch of a wormhole, a field configuration that arises in the euclidean path-integral formulation of quantum gravity. Two asymptotically flat spaces are connected by a narrow tube. Fig. 2 shows a closely related configuration; half a wormhole has been attached to a manifold that is approximately flat on the wormhole scale, although it may have curvature on larger scales. Fig. 2 could be described as a picture of a small piece of space breaking off from the main body of the universe to form a disconnected "baby universe". However, since the manifold is euclidean, not minkowskian, a better description would be that fig. 2 depicts the mixing (through quantum tunneling) of a state with no baby universes and a state with one baby universe.

Last year, Hawking [1], Giddings and Strominger [2], and Lavrelashvili et al. [3] proposed that wormholes could lead to extraordinary phenomena. The original belief was that the production of baby universes would implement the longstanding idea that geometrical fluctuations on a small scale lead to an apparent loss of quantum coherence on a large scale [4]. This turned out not to be the case [5]. (Although opinion on this point is not unanimous [6].)

Despite this disappointment, wormholes are still interesting. Indeed, in this paper I shall argue that wormholes lead to a phenomenon even more extraordinary that the loss of quantum coherence, and one that, unlike coherence loss, has been observed: the vanishing of the cosmological constant.

* Research supported in part by the National Science Foundation under grant no. PHY-87-14654

0550-3213/88/$03.50 © Elsevier Science Publishers B.V.
(North-Holland Physics Publishing Division)

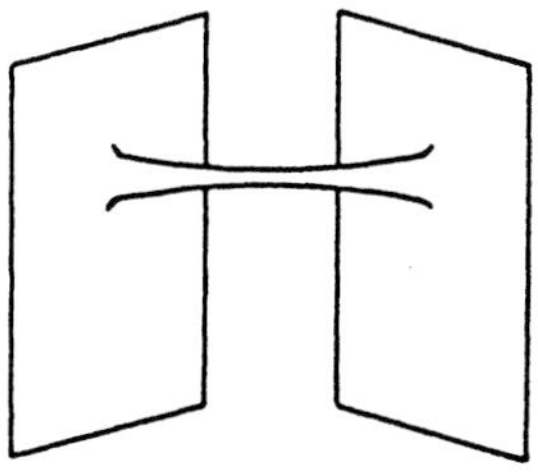

Fig. 1. A wormhole.

Fig. 2. A semiwormhole effecting a transition between a state with no baby universes and one with one baby universe.

1.1. OUTLINE OF THE PAPER

In the computation, I shall treat the wormholes semiclassically and sum over wormholes in the dilute-gas approximation. However, there will be no further approximations; all physics below the wormhole energy scale will be treated exactly. Thus, what vanishes is not some tree-approximation cosmological constant, nor some effective cosmological constant associated with a metastable inflationary phase, but the real thing, the bottom-line cosmological constant, the absolute ground-state energy density, including all interactions to all orders.

Giddings and Strominger [2] found wormholes as actual instantons, solutions of the euclidean equations of motion, in a theory of gravity coupled to a massless axion. Hawking [1] found them as constrained instantons in pure gravity, stationary points of the action subject to a constraint, which later had to be integrated over. For both analyses, the size of the dominant wormholes is on the order of the Planck length, and their action is of order $\hbar$. These are not good parameters for justifying semiclassical approximations and dilute-gas sums. Therefore, in this paper, I shall pretend that I am working in some (as yet unconstructed) theory in which there is a clear separation between the Planck scale and the wormhole scale, in which the size of the dominant wormholes is an order of magnitude or so greater than the Planck length, and their action is a large multiple of $\hbar$. Possibly such a theory can be constructed and does indeed describe the world. Possibly the qualitative conclusions

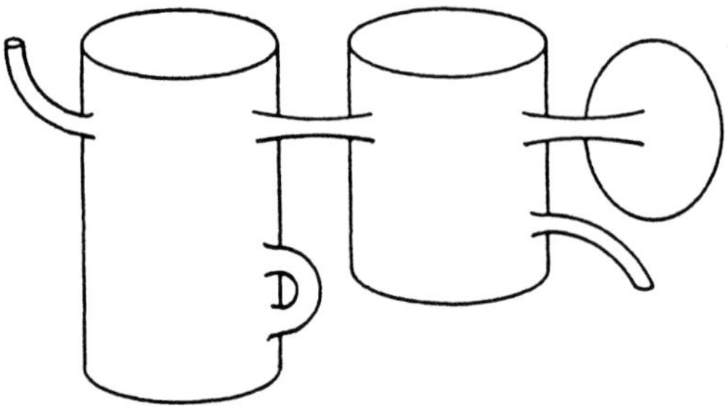

Fig. 3. A manifold with wormholes and baby universes.

Fig. 4. The manifold of fig. 3, with wormholes and baby universes stripped away.

of the dilute-gas approximation remain valid even when semiclassical methods cannot be guaranteed. (This last is an ancient if not honorable QCD argument.) In any case, making this pretense helps me to keep my various assumptions clearly separated.

The dilute-gas sum enables us to write the functional integral over manifolds full of baby universes and wormholes (fig. 3) in terms of an integral over manifolds st ipped of these (fig. 4). In sect. 2 I explain how this works, following the papers of ref. [5]. Because the two ends of a wormhole can be very far apart, one might expect the wor hole sum to induce nonlocal interactions, but this is not what happens. The only effect of the wormholes is to add local interactions to the Lagrange density, one for each independent kind of wormhole. However, the coefficients of these terms are not numbers, but operators, A_i, where i labels the kind of wormhole. The A's commute with each other and with all local fields on the stripped manifolds; they act only on the variables describing the baby universes. An observer who can only do experiments at distance scales larger than the wormhole scale interprets the A eigenvalues as constants of nature, constants whose values depends on something over which he has no control, what A-eigenspace he happens to be in. The A eigenvalues are like the θ angle in gauge field theory, arbitrary and in principle undeterminable parameters. Furthermore, depending on how many of them there are, they can mask some or all of the information about the constants of nature given by Planck-scale physics. The predictive power of the theory of everything has gone down the wormhole.

In sect. 3 I pick up another strand. Following the work of Hartle and Hawking [7], I review the path-integral construction of solutions of the Wheeler–DeWitt equation, possible wave functions of the universe. In this method, different wave functions arise from different choices of boundary conditions on the path integral. Hartle and Hawking advocated an especially simple choice of boundary conditions, so-called "no boundary" boundary conditions. This section also contains a discussion of normalization of the wave function which may be original. (It may also be wrong.)

Sect. 4 is the heart of the argument. The material in the previous two sections is put together to show that the Hartle–Hawking solution has a delta-function peak on the submanifold of A-eigenvalue space for which the cosmological constant vanishes.

Even without knowing the details of the argument, we can see how wormhole dynamics has the possibility of resolving one of the questions associated with the cosmological constant, what might be called the question of prearrangement. We believe that ordinary quantum field theory describes physics from shortly after the Planck time to the current epoch. Thus, in principle, all the constants of nature could have been measured very early on, say during an inflationary phase. How could they have known to adjust themselves so that when everything settled down the cosmological constant would be zero? Wormholes answer this by saying that, on an extremely small scale, our universe is in contact with other universes, otherwise disconnected, but governed by the same physics as ours (see fig. 3). Even if our universe is small and hot, these can be large and cool, and see the cosmological constant. Prearrangement is replaced by precognition.

The detailed argument of sect. 4 shows that it is an infrared divergence, a blow-up when the size of the otherwise disconnected universe becomes infinite, that drives the extinction of the cosmological constant. This is reminiscent of the old low-energy theorems of quantum electrodynamics, which are also driven by infrared divergences, in this case as the photon wavelength becomes infinite. In this sense the vanishing of the cosmological constant is the ultimate low-energy theorem.

All of this has been for the Hartle–Hawking solution. But although the "no boundary" boundary condition may be pretty, it is not divinely ordained, and thus in sect. 5 I investigate alternative boundary conditions. I find that it is possible to find boundary conditions such that the wave function does not have the delta-function peak, but that these are unnatural, in the technical sense. To be more precise, if I keep the boundary conditions fixed, but alter physics on the Planck scale, the delta-function peak reappears. Of course, it is always possible to make it go away again by readjusting the boundary conditions. This is what is usually called fine tuning. Thus, without wormholes, we have to fine tune to keep the cosmological constant zero; with wormholes, we have to fine tune to keep it nonzero. I believe this is progress.

Sect. 6 is a discussion of possible extensions of the methods of this paper to establish further relations among the constants of nature. There are obstacles to doing this, but if they could be overcome, this would offer the most direct method of confirming or falsifying the ideas presented here.

Although I find this theory in many ways very attractive, I must in honesty stress its speculative character. It rests on wormhole dynamics and the euclidean formulation of quantum gravity. Thus it is doubly a house built on sand. Wormholes may not exist, or, if they do exist, their effects may be overwhelmed by those of some more exotic configurations. Likewise, the euclidean formulation of gravity is not a subject with firm foundations and clear rules of procedure; indeed, it is more like a trackless swamp. I think I have threaded my way through it safely, but it is always possible that unknown to myself I am up to my neck in quicksand and sinking fast.

1.2. CONNECTION WITH OTHER WORK

There are two papers that directly influenced this work. One is Linde's recent theory of the cosmological constant [8]. Linde begins with the theory of a scalar field ϕ, with dynamics governed by a Lagrange density of the usual type, $\mathscr{L}(\phi, \partial_\mu\phi)$, interacting minimally with the gravitational field, $g_{\mu\nu}$. All these are functions of the usual space-time variable, x. To these he adds a second set of fields, $\tilde{\phi}$ and $\tilde{g}_{\mu\nu}$, which are functions of a second space-time variable, $\tilde{x}$. The (real-time) action for the theory is

$$S = \int \mathrm{d}^4x\, \mathrm{d}^4\tilde{x} \sqrt{-g}\sqrt{-\tilde{g}}\left[-R + \mathscr{L}(\phi, \partial_\mu\phi) + \tilde{R} - \mathscr{L}(\tilde{\phi}, \partial_\mu\tilde{\phi})\right], \qquad (1.1)$$

in units where $16\pi G = 1$. Thus, for each of the two sets of fields he obtains the usual equations of motion, with an additional term of cosmological-constant form, whose coefficient is a space-time average over the other set of fields. Linde shows that if he assumes things eventually settle down to a state of constant ϕ, this additional term cancels out whatever cosmological constant was originally present. This is an extraordinarily clever scheme; to my knowledge, it is the first appearance of the idea I described earlier as the replacement of prearrangement by precognition. Nevertheless, it has some serious difficulties. (1) The form of the theory is completely ad hoc, devised just to make the cosmological constant vanish. (2) The theory is completely classical. It is not obvious how to consistently quantize it, nor is it clear whether the cancellation of the cosmological constant survives quantum corrections. (3) The canc llation depends on ϕ eventually settling down; it is not clear how to generalize the mechanism to chaotic inflation, for example. I think the theory advanced here is free of these difficulties; it can be thought of as an attempt to improve on Linde's original idea.

The other is Hawking's 1984 paper, "The cosmological constant is probably zero" [9]. Hawking considers theories in which the cosmological constant is connected to a dynamical variable, but one that is independent of space-time position, like the θ angle of gauge field theory. He then attempts to compute the dependence of the wave function of the universe on θ. He finds an infrared divergence (very closely related to the one I find) that causes the wave function to have a peak at that value of θ that makes the cosmological constant vanish. This also is impressively clever: to my knowledge, it is the first appearance of the idea that the cosmological constant could be fixed by the shape of the wave function of the universe. Unfortunately, the original proposal has serious difficulties; Hawking's wave-function multiplied by any function of θ is a solution of the Wheeler–DeWitt equation obeying Hartle–Hawking boundary conditions. Thus one could argue that the peaking is just a reflection of normalization conventions, having no physical meaning. I think the theory presented here avoids this objection; it can be thought of as an attempt to improve on Hawking's original idea.

Finally, Banks has independently constructed a theory in which topology-changing processes are responsible for the suppression of the cosmological constant [10]. It is not easy to compare Banks's theory to mine, because his is phrased in terms of minkowskian rather than euclidean concepts. However, unless one of us has made a mistake, it is sure that the two theories are not equivalent. I find a total extinction of the cosmological constant, while Banks finds merely a large suppression (by a factor connected with the existence of an inflationary epoch).

2. The effects of wormholes

The material in this section is discussed in detail in the papers of ref. [5]. Thus I will be brief here, saying just enough to define the terms that appear in the formulas we will need later.

In any given theory, baby universes (and the associated wormholes) may come in many types, distinguished by their size, intrinsic geometry, state of internal excitation, etc. For notational simplicity, I will assume the types are labeled by a discrete index, i. The generalization to a continuous index is trivial. For bookkeeping purposes, it is convenient to introduce annihilation and creation operators for each type of baby universe, a_i and $a_i^\dagger$, normalized to obey the usual commutation relations

$$\left[a_i, a_j^\dagger\right] = \delta_{ij}. \tag{2.1}$$

Because baby universes are closed universes, they always carry zero energy, linear momentum, and angular momentum. They also carry zero value of any conserved charge coupled to a gauge field, like electric charge. However, if the theory admits ungauged continuous symmetries, baby universes can carry nonzero values of the associated charges.

Let us denote the *TCP* conjugate of a baby universe of type i by i^*. Because baby universes carry zero angular momentum, $i^{**} = i$. Thus, we can always choose a basis for the one-baby-universe subspace such that *TCP* is complex conjugation. In this basis, $i^* = i$. Of course, in such a basis, the basic baby universes are not in general charge eigenstates.

Let the action of a wormhole of type i be $2S_i$. Then the action of a semiwormhole (one that terminates on a baby universe, as in fig. 2) is S_i. If we have a manifold on which all the fields (including the metric) are slowly varying on a wormhole length scale, then the amplitude for inserting the end of a wormhole of type i in some point on the manifold is

$$K_i \sqrt{g}\, \mathrm{d}^4 x, \tag{2.2}$$

where K_i is some function of the fields on the manifold. If we expand K_i in a Taylor series in the fields and their derivatives, the higher terms will be suppressed ompared with the leading terms by factors (given by dimensional analysis) of the wormhole length scale. If the wormhole leads to a baby universe that carries zero charge, we expect the leading term to be a constant. If the baby universe carries nonzero charge, we expect it to be the operator of lowest dimension carrying the appropriate charge.

We are now in a position to discuss the wormhole summation formula. We assume that we are working in some theory that at the Planck length is described by a local field theory. At a somewhat larger length, the wormholes appear. We wish to describe physics at some even larger length scale, L. If it were not for the wormholes, we would proceed in the familiar manner of renormalization-group analysis. We would write our fields as the sum of long-wavelength (larger than L) backgrounds and short-wavelength (between L and the Planck length) fluctuations. We would then integrate over the fluctuations to obtain a new local field theory. However, because of the wormholes, qualitatively new processes appear, the annihilation and creation of baby universes, and the connection of distantly separated (or even otherwise disconnected) parts of the space-time manifold. (I stress that I am explicitly assuming that the wormholes are the only configurations that have unconventional effects in this energy range.)

Thus, we begin with some (possibly disconnected) manifold, M, with long-wavelength background fields, like the manifold of fig. 4. We also specify the initial number of baby universes of each t pe, n_i, and the final number, n'_i. We then inte rate over the fluctuations and sum over all possible locations of the wormholes, as shown in fig. 3. In the papers of ref. [5] it is shown that the result of this process is

$$\sum_{\substack{\text{fluctuations}\\ \text{and wormholes,}\\ n_i,\, n'_i \text{ fixed}}} \mathrm{e}^{-S} = \langle n'_1, n'_2 \ldots | \mathrm{e}^{-S_{\mathrm{eff}}} | n_1, n_2 \ldots \rangle \,. \tag{2.3}$$

Here

$$S_{\text{eff}} = \int_M d^4x \sqrt{g}\, \mathscr{L}_{\text{eff}} , \tag{2.4}$$

where

$$\mathscr{L}_{\text{eff}} = \mathscr{L}_0 + \sum_i \left(a_i^\dagger + a_{i*}\right) \mathscr{L}_i , \tag{2.5}$$

$\mathscr{L}_0$ is the result of integration over the fluctuations, and

$$\mathscr{L}_i = e^{-S_i} K_i . \tag{2.6}$$

I do not wish to give the proof of this formula here, but it is not hard to convince yourself that it is correct. If we use Wick's theorem to normal-order S_{eff}, the annihilation terms take care of the initial baby universes, the creation terms take care of the final baby universes, and the contraction takes care of the wormholes that begin and end on M.

(As I have said, this formula is obtained by summing over all possible locations of the wormholes. However, if some of the wormholes carry nonzero values of conserved charges, this is not the right thing to do. For example, if M contains a disconnected closed component, as in fig. 4, the sum of all the charges carried off this component by wormholes must be zero. However, the error we make by ignoring this constraint is harmless. If we expand in annihilation and creation operators, the undesired terms are multiplied by strings of $\mathscr{L}_i$'s which carry nonzero charge. Thus, they disappear when we integrate over the charged fields on M.)

If we adopt a basis in which $i^* = i$, then the effective lagrangian only involves the hermitian operators

$$A_i = a_i + a_i^\dagger . \tag{2.7}$$

In this basis, the $\mathscr{L}_i$'s are also all hermitian. The A's are a set of commuting hermitian operators; thus they can be simultaneously diagonalized. We denote their eigenvalues by α_i.

For later purposes, it will be useful to have an explicit formula [6] relating A eigenstates to baby-universe-number eigenstates

$$\langle \alpha_1, \alpha_2 \ldots | n_1, n_2 \ldots \rangle = \prod_i \psi_{n_i}\left(\alpha_i/\sqrt{2}\right), \tag{2.8}$$

where $\psi_n(q)$ is the nth energy eigenfunction of the harmonic oscillator, $H = \frac{1}{2}(p^2 + q^2)$.

If we rewrite eq. (2.3) in terms of A-eigenstates, we find

$$\langle \alpha_1' \dots | e^{-S_{\rm eff}} | \alpha_1 \dots \rangle = e^{-S_{\rm eff}(\alpha)} \prod_i \delta(\alpha_i' - \alpha_i), \tag{2.9}$$

where

$$S_{\rm eff}(\alpha) = \int_M d^4x \sqrt{g} \left(\mathscr{L}_0 + \sum_i \mathscr{L}_i \alpha_i \right). \tag{2.10}$$

No local operator connects any A-eigenspace with any other A-eigenspace. Thus, from the viewpoint of an observer who cannot detect baby universes (e.g., from a human viewpoint), the α's are simply arbitrary parameters that appear in the lagrangian of the world, like the θ angle in gauge field theory. The predictive properties of the theory of everything on the Planck scale (encoded in $\mathscr{L}_0$), may be partially or totally obscured by the effects of the α's. As we shall see, though, a new predictive principle will come into play, one that will fix the combined effect of $\mathscr{L}_0$ and the α's.

3. The wave function of the universe*

The wave function of the universe is a common concept in quantum field theory, although in theories without gravitation it is not usually given such a grand name. As an example, let me review the theory of scalar fields on some fixed three-space quantized in temporal gauge.

I will define boundary values, B, to be the values of the scalar fields and the space components of the gauge fields at fixed time. (I give them this name because they are the boundary values in Hamilton's principle.) Then the wave function of the universe is a function (more properly a functional), $\Psi(B, t)$, where t is (real or imaginary) time. It obeys the Gauss-law constraints, which state that it is unchanged by (three-dimensional) gauge transformations, and an equation of motion, Schrödinger's equation.

Let us define an amplitude

$$A(B_1, B_2, t) \equiv \sum e^{-S}, \tag{3.1}$$

where the S is the (gauge-invariant) euclidean action and the sum is over all motions that go from B_2 to B_1 in euclidean time t. This amplitude obeys a composition law

$$\int A(B_1, B_2, t) \, d\mu(B_2) A(B_2, B_3, t') = A(B_1, B_3, t + t'), \tag{3.2}$$

* Most of the material in the first part of this section comes from Hartle and Hawking [7] and from Hawking's review of quantum cosmology [11]. These contain references to the earlier literature.

for appropriate measure $\mu(B)$. Further, given any function $\Phi(B)$

$$\Psi(B,t) = \int A(B,B',t)\Phi(B')\,\mathrm{d}\mu(B'), \tag{3.3}$$

is a possible wave function of the universe for $t > 0$. That is to say, it obeys the Gauss-law constraints and the (imaginary time) Schrödinger equation. (The constraints follow immediately from the gauge invariance of S, the Schrödinger equation takes a bit of work.)

(This treatment of the path integral is very slapdash; in particular, I have totally ignored the notorious divergence of this integral over gauge orbits. For the case at hand, we know how to take care of this, either by Faddeev–Popov gauge fixing or by replacing the continuum theory by a lattice theory. For gravity it is not so clear that things can be made right; I will say more about this shortly.)

When gravity becomes part of the dynamics, this picture changes in several ways. (1) The boundary values B, include the three-geometry, g_{ij}. There are three new constraint equations, which state that Ψ is unchanged by (three-dimensional) general coordinate transformations. (2) The boundary three-geometry may be disconnected, a union of several connected components. Indeed, we had better allow for this possibility if we are to have any hope of describing the distribution of baby universes. For the same reason that it is convenient in ordinary field theory to include the vacuum, the zero-particle state, in Fock space, it is convenient here to include, in the set of possible boundary values, no boundary, the zero-component case, which we denote by $B = 0$. (3) There is no time in the formalism; Ψ depends on B only. The Schrödinger equation is replaced by a new equation, the Wheeler–DeWitt equation, but this is not an equation of time evolution but another equation of constraint.

The disappearance of time is a novelty, but not as radical a one as might be thought. All the questions we ask in quantum theory that involve time can be rephrased without it, although usually at some cost in awkwardness. For example, instead of asking, "At four o'clock, what is the probability distribution of electron position?", we can ask "If we project the wave function of the universe on to the subspace in which the hands of the clock are pointing to four, what is the probability distribution of electron position?". Even questions about sequences of measurements in time, like a string of Stern–Gerlach experiments, can be rephrased in time-free language; as Augustine observed, the past is present memory. Thus, instead of asking about the probabilities of sequences of observations in time, we ask about the probability distribution of experimental records, bubble-chamber photographs and counter positions.

Hartle and Hawking proposed constructing the wave function of the universe from a euclidean path integral, rather as in eq. (3.3) [7]. For gravity, the definition of the euclidean path integral is plagued by difficulties. The theory is nonrenormalizable and must be cut off, but I know of no nonperturbative cutoff that preserves general covariance. Also, the euclidean action is not bounded below. (Special rules for rotating the integration over conformal factors into the complex plane are usually invoked to take care of this [11].) Finally, there is the divergence of the integral over gauge orbits, mentioned earlier. Despite these difficulties, I believe that there should be a consistent way of defining the euclidean path integral. My reason is that superstring theory is finite and consistent, and should reduce to the theory of gravity (plus other fields) at the Planck mass and below. Thus whether superstring theory is true or false, its euclidean formulation should automatically generate a consistent formulation of euclidean gravity. Just to be safe, though, I will organize my arguments so they depend as little as possible on the details of this hoped-for formulation.

Let us define $A(B_1, B_2)$ as in eq. (3.1), but with no time; that is to say, we sum over all four-manifolds that go from B_1 to B_2. (Just to keep things simple and to avoid having to worry about the behavior of fields at infinity, I will restrict myself to compact three-geometries in the boundary values and compact four-manifolds in the path integral.) Which boundary values we think of as initial and which we think of as final depends only on whether the normal vector to the boundary points into or out of the four-manifold, that is to say, upon the orientation of the three-geometry on the boundary. If we denote the operation or reversing the orientation by an overbar, we thus have

$$A(B_1, B_2) = A(\bar{B}_2, \bar{B}_1). \tag{3.4}$$

We can also change the orientation of the boundary values in another way, not by reidentifying initial and final boundaries, but by changing the orientation of the four-manifold. Operators that change sign under a change of orientation, parity-odd operators like $\epsilon^{\mu\nu\lambda\sigma}F_{\mu\nu}F_{\lambda\sigma}$, are imaginary in euclidean space if they are real in minkowskian space. Thus

$$A(\bar{B}_2, \bar{B}_1) = A(B_2, B_1)^*. \tag{3.5}$$

Note that these last two equations imply that $A(B_1, B_2)$ defines a hermitian bilinear form.

Hartle and Hawking showed that

$$\Psi(B) = \int A(B, B')\Phi(B')\,\mathrm{d}\mu(B'), \tag{3.6}$$

is a possible wave function of the universe, for any Φ and any measure μ; that is to say, it obeys all the equations of constraint, including the Wheeler–DeWitt equation. For brevity, I shall call such functions "allowable wave functions" from now on.

Hartle and Hawking paid special attention to the wave function defined by the simplest boundary condition of all, no boundaries. In our notation, $\Phi(B)$ is proportional to δ_{B_0}, and

$$\Psi^{\mathrm{HH}}(B)=A(B,0)=A(0,B)^{*}. \tag{3.7}$$

They suggested that this was in a sense the ground-state wave function of the universe.

A naive generalization of the composition law, eq. (3.2) would lead us to believe that

$$A(B_1,B_3)=\int A(B_1,B_2)\,\mathrm{d}\mu(B_2)A(B_2,B_3), \tag{3.8}$$

for appropriate measure μ. Indeed, one can give a hand-waving argument for this equation. One takes all manifolds that go from B_3 to B_1, slices them in two along some arbitrarily chosen hypersurface in parameter space, and integrates over all boundary values on the hypersurface*.

If we define an inner product between allowable wave functions by

$$(\Psi_1,\Psi_2)=\int \Psi_1^{*}(B)\Psi_2(B)\,\mathrm{d}\mu(B), \tag{3.9}$$

then it follows from eqs. (3.6) and (3.8) that

$$(\Psi_1,\Psi_2)=\int \Phi_1(B_1)^{*}A(B_1,B_2)\Phi_2(B_2)\,\mathrm{d}\mu(B_1)\,\mathrm{d}\mu(B_2). \tag{3.10}$$

We can derive a similar path-integral formula for the matrix element of a local observable. Let $\phi(x)$ be a gauge-invariant local scalar field. (ϕ could be a quite complicated object, some function of high-order derivatives of fundamental fields (including the Riemann tensor).) In any generally covariant theory, the expectation value of ϕ must be independent of x, because any x may be turned into any other by a general coordinate transformation. We define

$$(\Psi_1,\phi\Psi_2)=\int \Psi_1^{*}(B)\phi(x)\Psi_2(B)\,\mathrm{d}\mu(B), \tag{3.11}$$

where x is anywhere on the boundary. (The invariance of the wave functions under three-dimensional coordinate transformations implies the desired independence of x.)

Let us also define

$$A^{\phi}(B_1,B_2)=\sum \mathrm{e}^{-S}\phi(x), \tag{3.12}$$

* If you are expert in this field, please be patient; I will redo these arguments in a less outrageous way (although not that much less outrageous) in a few paragraphs.

where the sum is over all the four-manifolds that go from B_2 to B_1. For each topological class of four-manifolds, we parameterize the manifolds by some coordinate system, and place x at an arbitrary chosen interior location. When we sum over all manifolds in a class, the dependence on x disappears, because of the general covariance of S. The same reasoning that led to eq. (3.10) now leads to

$$(\Psi_1, \phi\Psi_2) = \int \Phi_1(B_1)^* A^{\phi}(B_1, B_2)\Phi_2(B_2)\, d\mu(B_1)\, d\mu(B_2). \qquad (3.13)$$

If we apply these formulas to the Hartle–Hawking wave function, eq. (3.9), we find

$$\langle\phi\rangle^{HH} = \frac{(\Psi^{HH}, \phi\Psi^{HH})}{(\Psi^{HH}, \Psi^{HH})} = \frac{\Sigma e^{-S}\phi}{\Sigma e^{-S}}, \qquad (3.14)$$

where the sum in both numerator and denominator is over compact manifolds without boundaries. Hearteningly, this looks very much like the expression for the ground-state expectation value of a local field in a field theory without dynamical gravity.

There is a serious problem with all this reasoning. Eq (3.8) is probably false; there is no good argument that such a measure μ exists, and plenty of evidence that it does not [12]. I will first explain why the hand-waving argument that led to eq. (3.8) is misleading. I will then argue that our final formulas for inner products and matrix elements, eqs. (3.10) and (3.13), make perfect sense even in the absence of eq. (3.8) and may be adopted as the primary definitions of these entities. Note that both these equations, and in eq. (3.6), the equation linking Ψ and Φ, the choice of the measure μ is unimportant. Any nonsingular change of μ can be compensated for by an appropriate redefinition of Φ. The argument is rather dry and technical, and the reader uninterested in nuts and bolts may well skip ahead at this point to sect. 4.

Why is the hand-waving argument for eq. (3.8) misleading? In eq. (3.2), there was a unique way of slicing a path in two, at a point of fixed time. In the case at hand, there is no time, and there is no unique way of slicing a four-geometry in two. Of course, no matter how you slice it, you always get the same action, so one might think this is just a problem of overcounting, to be absorbed in the normalization of the measure. However, there is no guarantee that one overcounts by the same factor for different four-geometries, and, if one does not, there is no way to absorb things in the measure.

Without actually doing the calculations in full detail, there is no way to eliminate the possibility that eq. (3.8) is saved by miraculous cancellations. However, it is possible to construct a trivial model that clearly shows what could go wrong. Let us consider a square lattice of points on the plane, and let us define

$$A(x_1, x_2) = \sum_{\text{paths}} e^{-S}, \qquad (3.15)$$

where the sum is over all paths that go from x_1 to x_2 (two points on the lattice), and S is some function of path length only. (The analogy to quantum gravity should be clear.) The only translation-invariant measure of a point set on a lattice is the number of points in the set, times a normalization constant. Thus, the same hand-waving argument that led to eq. (3.8) would lead us to believe that

$$A(x_1, x_3) = N \sum_{x_2} A(x_1, x_2) A(x_2, x_3), \tag{3.16}$$

for some choice of N. But this is clear nonsense; the right-hand side overcounts each path from x_1 to x_3 by a factor proportional to its length, and there is no way to make things right just by adjusting N.

I shall now show that eqs. (3.10) and (3.13) are consistent and sensible definition of inner products and matrix elements, even if eq. (3.8) is false. It will be convenient to write things in an operator notation. Thus, eq. (3.6) becomes

$$\Psi = A\Phi . \tag{3.17}$$

If we define an inner product by

$$\langle \Phi_1, \Phi_2 \rangle = \int \Phi_1(B) \Phi_2^*(B)\, \mathrm{d}\mu(B), \tag{3.18}$$

then eqs. (3.10) and (3.13) become

$$(\Psi_1, \Psi_2) = \langle \Phi_1, A\Phi_2 \rangle, \qquad (\Psi_1, \phi\Psi_2) = \langle \Phi_1, A^{\phi}\Phi_2 \rangle. \quad (3.19), (3.20)$$

To show that these are sensible definitions, I will need two assumptions. The first is that any allowable wave function Ψ is of the form $A\Phi$ for some Φ. This is clearly necessary if we are to be able to give meaning at all to these equations. The second is that A is positive semidefinite. (We already know that it is hermitian.) Both of these are true in a theory without dynamical gravity, eq. (3.3), because there the place of A is taken by $Pe^{-Ht} = e^{-Ht}P$, where P is the projection on states obeying the Gauss-law constraints.

Eq. (3.17) does not uniquely determine Φ in terms of Ψ. We can always make the transformation

$$\Phi \to \Phi + \Delta , \tag{3.21}$$

where Δ is in the null set of A,

$$A\Delta = 0, \tag{3.22}$$

without changing Ψ. We must make sure our definitions, eqs. (3.19) and (3.20), are invariant under such transformations.

The invariance is obvious for eq. (3.19). Also, the positive semidefiniteness of A implies that eq. (3.19) defines a positive-definite inner product among allowable states.

Proof. Semidefiniteness is obvious. Let Ψ be a state of zero norm. Thus, $(\Psi, \Psi) = 0 = \langle \Phi, A\Phi \rangle$. By the semidefiniteness of A, this implies that $A\Phi = \Psi = 0$.

Rather than analyzing eq. (3.20) directly, I shall analyze a stronger statement that defines the action of a local field on an allowable state

$$\phi\Psi = A^{\phi}\Phi . \tag{3.23}$$

It follows from eq. (3.19) that

$$(\Psi_1, \Psi_2) = \langle \Phi_1, \Psi_2 \rangle . \tag{3.24}$$

Thus eq. (3.23) implies eq. (3.20).

The arguments of Hartle and Hawking apply without alteration to $A^{\phi}\Phi$, and show that this is an allowable wave function for any Φ. The presence of an additional ϕ in the interior of the four-manifolds over which we sum has no effect, no more than an operator insertion in the interior of the region $(0, t)$ would keep eq. (3.3) from defining a solution of the Schrödinger equation. Thus the range of A^{ϕ} is contained in the range of A. This implies that the null set of A^{ϕ} contains the null set of A, that is to say, that eq. (3.23) is invariant under the transformation (3.21). This completes the argument.

The preceding paragraphs have been an attempt to define inner products and matrix elements in a way that is independent of the suspect eq. (3.8). However, even without eq. (3.8), there is still a lot of conjecture and hand-waving in the argument, and I am by no means sure I have got everything right even now. Thus I should identify those pieces of this machinery that are essential for my subsequent reasoning. Only the formula of the expectation value of a scalar field, eq. (3.14), is needed for sect. 4 (the vanishing of the cosmological constant) and sect. 6 (further relations among the constants of nature). The generalization of eq. (3.14) for states defined by an arbitrary Φ is needed for sect. 5 (naturalness). Of all our equations, these look most like the familiar formulas of ordinary field theory; it is possible they will survive even if the rest of the structure crumbles.

4. The effects of wormholes on the wavefunction of the universe

In this section I shall attempt to compute the Hartle–Hawking wave function of the universe in the presence of wormholes. To keep my equations compact, I shall compress my notation, and denote by α the string $\alpha_1, \alpha_2, \ldots$. Likewise, $d\alpha$, $\delta(\alpha)$, and α^2 will denote the corresponding products and sum.

To begin, let us consider a theory in which the action is $S_{\text{eff}}(\alpha)$, for some fixed value of the α's, and in which we integrate only over configurations that are slowly varying on the wormhole scale. In this case, the Hartle–Hawking wave function is

$$\Psi_\alpha^{\mathrm{HH}}(B) = \sum \mathrm{e}^{-S_{\mathrm{eff}}(\alpha)}, \tag{4.1}$$

where the sum is over all manifolds that go from no boundary to B. (B is, of course, slowly varying on the wormhole scale.) A general manifold will have several components. Some of these will be connected to B (only one, if B is itself connected). However, there may be other components which are closed, that is to say, which have no boundary at all. The action is a sum over the various components. Thus the sum over four-manifolds factorizes

$$\Psi_\alpha^{\mathrm{HH}}(B) = \psi_\alpha^{\mathrm{HH}}(B) Z(\alpha), \tag{4.2}$$

where $\psi_\alpha^{\mathrm{HH}}$ is given by the sum over manifolds connected to B, and

$$Z(\alpha) = \sum_{\mathrm{CM}} \mathrm{e}^{-S_{\mathrm{eff}}(\alpha)}, \tag{4.3}$$

where CM denotes closed manifolds.

Let us now compute the expectation value of some scalar field, ϕ. By eq. (3.15)

$$\langle\phi\rangle_\alpha^{\mathrm{HH}} = \frac{\sum_{\mathrm{CM}} \mathrm{e}^{-S_{\mathrm{eff}}(\alpha)} \phi(x)}{\sum_{\mathrm{CM}} \mathrm{e}^{-S_{\mathrm{eff}}(\alpha)}}. \tag{4.4}$$

The denominator is just $Z(\alpha)$ again. Hence

$$\sum_{\mathrm{CM}} \mathrm{e}^{-S_{\mathrm{eff}}(\alpha)} \phi(x) = \langle\phi\rangle_\alpha^{\mathrm{HH}} Z(\alpha). \tag{4.5}$$

This identity will be useful to us shortly.

It is possible to simplify the sum in eq. (4.4). I will spare the reader the combinatorics, which are straightforward, and simply state the result

$$\langle\phi\rangle_\alpha^{\mathrm{HH}} = \frac{\sum_{\mathrm{CCM}} \mathrm{e}^{-S_{\mathrm{eff}}(\alpha)} \phi(x)}{\sum_{\mathrm{CCM}} \mathrm{e}^{-S_{\mathrm{eff}}(\alpha)}}, \tag{4.6}$$

where CCM denotes closed connected manifolds. This equation tells us that, if the α's are constants, we make no error if we simply ignore disconnected closed components in the path integral. As we shall see immediately, though, the situation is quite different for wormhole dynamics, where the α's are variables. The reason is not deep: the disconnected components are not really disconnected; they are connected by wormholes.

Now let us turn to the real thing, the theory with wormholes. The argument of Ψ is now not just B, but also the number of baby universes, or equivalently, the α's.

Likewise, the no-boundary condition states not only that there is no slowly-varying boundary, but also that there are no baby universes. Eq. (2.8) tells us how to write the no-baby-universe state in terms of the α's

$$\langle \alpha | 0 \rangle = \mathrm{e}^{-\alpha^2/4}, \tag{4.7}$$

times an irrelevant normalization constant. Thus we can directly apply the worm-hole summation formula, eq. (2.9), to find

$$\Psi^{\mathrm{HH}}(B, \alpha) = \mathrm{e}^{-\alpha^2/4} \psi_\alpha^{\mathrm{HH}}(B) Z(\alpha). \tag{4.8}$$

This equation strongly suggests that $Z(\alpha)$ governs the probability of finding given values of α in the Hartle–Hawking state. To get a more precise idea of what is going on, let us compute $\langle \phi \rangle^{\mathrm{HH}}$. For this computation, we must sum over closed manifolds. Thus both the initial and final state contain no baby universes. It then follows from the wormhole summation formula and eq. (4.5) that

$$\langle \phi \rangle^{\mathrm{HH}} = \frac{\int \mathrm{d}\alpha\, \mathrm{e}^{-\alpha^2/2} \langle \phi \rangle_\alpha^{\mathrm{HH}} Z(\alpha)}{\int \mathrm{d}\alpha\, \mathrm{e}^{-\alpha^2/2} Z(\alpha)}. \tag{4.9}$$

We see that the probability distribution in α is

$$\mathrm{d}P = \mathrm{e}^{-\alpha^2/2} Z(\alpha)\, \mathrm{d}\alpha, \tag{4.10}$$

up to a normalization. (Note that the factor is Z, not Z^2, as one might naively expect from eq. (4.8).)

I shall now show that $Z(\alpha)$ displays the announced peak. It is a well-known result in diagrammatic perturbation theory that the sum of all vacuum-to-vacuum graphs is the exponential of the sum of connected graphs. Exactly the same combinatorics apply here; thus

$$Z(\alpha) = \exp\left[\sum_{\mathrm{CCM}} \mathrm{e}^{-S_{\mathrm{eff}}(\alpha)} \right]. \tag{4.11}$$

The sum over closed connected manifolds can be expressed in terms of a background-gravitational-field effective action, Γ. (Please do not confuse this with S_{eff}, which is an effective action in quite another sense.)

Let me remind you how the corresponding object is defined in the theory of a single scalar field, ϕ. The field is written as the sum of a background field and a remainder, $\phi = \phi_{\mathrm{b}} + \phi_{\mathrm{r}}$. The action is expanded in powers of ϕ_{r} with ϕ_{b} fixed. $\Gamma(\phi_{\mathrm{b}})$ is def ned as minus the sum of all one-particle-irreducible connected vacuum-to-vacuum graphs. (I define Γ this way only for brevity; it can also be defined nonperturbatively.) The path integral of e^{-S} is then equal to $e^{-\Gamma}$, evaluated at the stationary point of Γ. (If there are several stationary points, the one of minimum action is selected.) If there are other fields in the theory, we may either introduce background fields for them also, or just integrate over them. Also, if we have background gauge fields, proper gauge-fixing conditions must be imposed to preserve the gauge-invariance of Γ, but I will not go into detail on this here.

In addition to all the usual horrible technical problems, there is a real problem of principle in extending this to gravity. The sum in eq. (4.11) runs over manifolds of all possible topologies. No matter how careful I am with my gauge-fixing conditions, there is no way I can write a metric defined on a four-torus (for example) as a metric defined on a four-sphere plus a remainder. Thus I will define an independent effective action for each topology, and write

$$\sum_{\mathrm{CCM}} e^{-S_{\mathrm{eff}}(\alpha)} = \sum_{\mathrm{topologies}} e^{-\Gamma_\alpha(g)}, \tag{4.12}$$

where g denotes the background metric on each topology, and each term on the right is to be evaluated at its stationary point. Note that I have chosen to introduce only a background for the metric; all other fields have just been integrated over.

Eq. (4.12) may strike you as an egregious example of empty formalism; I have succeeded in writing a path integral which I can not do as a sum over the stationary points of an infinite set of functions which I do not know. But this is an overstatement; there is one class of manifolds for which we do know Γ, large smooth manifolds. In particular, the leading term in Γ for large volume is given by

$$\Gamma = \lambda \int d^4x \sqrt{g} + \dots, \tag{4.13}$$

where λ is the cosmological constant*. (Although I have not indicated it explicitly, λ is, of course, a function of the α's.)

I stress that λ is the fully renormalized cosmological constant, including all effects of all interactions. To convince yourself of this, imagine doing the computation with a fixed gravitational field, but still integrating over all the other interactions. Then λ would clearly be the exact ground-state energy density. All higher energy states would be negligible for large volumes, for the same reason that, in nonrelativistic quantum mechanics, $\mathrm{Tr}\, e^{-HT}$ is dominated by the ground-state energy for large T.

* I have defined λ here as an energy density. The usual cosmological constant of general relativity is $8\pi G\lambda$.

The first correction to eq. (4.12) is also known

$$\Gamma = \int d^4x \sqrt{g} \left[\lambda - \frac{1}{16\pi G} R \right] + \dots, \qquad (4.14)$$

where G is Newton's constant, again including all renormalization effects of all interactions. I shall neglect for the moment the higher terms in the large-volume expansion of Γ; as we shall see, they are in fact negligible for our immediate interests.

The stationary points of eq. (4.14) are Einstein spaces*

$$R_{\mu\nu} = 8\pi G \lambda g_{\mu\nu}. \qquad (4.15)$$

For these

$$\Gamma = -\lambda \int d^4x \sqrt{g}. \qquad (4.16)$$

Thus for positive λ we want the Einstein space of maximum volume, for negative λ that of minimum volume. For positive λ, the space of maximum volume is known; it is the four-sphere of radius $\sqrt{3/8\pi G\lambda}$, for which

$$\Gamma = -3/8G^2\lambda. \qquad (4.17)$$

As promised, for sufficiently small λ, the neglected terms in eq. (4.14) are negligible compared to this. For negative λ, the minimum volume space is not known. Nevertheless, whatever it is, it makes a positive contribution to Γ proportional to $1/\lambda$. Thus

$$\ln Z \begin{cases} \propto e^{3/(8G^2\lambda)}, & \lambda \to 0^+, \\ \to 0, & \lambda \to 0^-. \end{cases} \qquad (4.18)$$

This is the infrared divergence referred to in sect. 1. It comes from very large manifolds. If an infrared cutoff is introduced, say by restricting the path integral to manifolds with diameters less than some maximum value, D, Γ approaches a finite limit, proportional to D^2/G, as λ goes to zero. If, in the presence of such a cutoff, we normalize the probability distribution in α, eq. (4.10), and then let D go to infinity, the probability distribution becomes concentrated on that submanifold of α space on which λ vanishes. This is the delta-function peaking announced in sect. 1.

Comments.

(1) The appearance of delta-function in the infrared limit is nothing new. For

* In this paragraph my argument is tangent to Hawking's in his 1984 paper [9]. I shall say more about this in sect. 5.

example, consider a free massive scalar field in a periodic box. The ground-state probability distribution for the spatial average of the field is a gaussian with variance inversely proportional to the volume of the box; this becomes a delta-function when the volume becomes infinite.

(2) This argument assumes that there are some values of α for which λ vanishes. At first glance, this does not seem to be a stringent constraint. For example, if there exist baby universes which carry zero values of all conserved charges, the corresponding α's multiply Lagrange densities for which the term of lowest dimension is a constant, i.e. an addition to λ. Unfortunately, this argument rests on neglecting effects of higher order in α; there is no reason to believe it for large α. (Indeed, if α is too large, S_{eff} might describe a silly theory. For example, a kinetic term might go negative, producing ghost states*.) This is a potential problem only if the cosmological constant produced by $\mathscr{L}_0$ is of wormhole scale or larger; if it is smaller than this (as occurs, for example, in theories with supersymmetry breaking below the GUT scale), there is no difficulty.

(3) I have found stationary points of Γ among large smooth geometries, but there might be others, hiding among tiny wrinkled geometries to which eq. (4.14) does not apply. If these have lower effective action than the four-sphere, they could undercut my conclusions, by producing a peak at some location that has nothing to do with vanishing λ. I stress that I have no reason to believe that this disgusting possibility occurs, but neither can I exclude it.

(4) There is a temptation to think of the vanishing of the cosmological constant as something that takes place in time; we might think that the constants of nature somehow adjust themselves to make λ vanish. This is wrong. To see this, we must formulate questions involving time in time-free language, as explained in sect. 3. Then the factorized structure of the wave function, eq. (4.8), tells us that even at a very early epoch (i.e., for boundary values B corresponding to very high-energy densities) the constants of nature were the same as the ones that currently (i.e., for B's corresponding to low-energy densities) make λ vanish.

This statement needs some modification. Eq. (4.8) is valid only at distance scales larger than the wormhole scale, or, roughly speaking, for times later than the "wormhole epoch". Time should be a fairly useful concept from the Planck epoch on. Thus, there might be a description of the vanishing of λ in terms of a process taking place in time between the Planck epoch and the wormhole epoch. This might give us useful insights; for example, it might tell us how to get beyond the dilute-gas approximation. Unfortunately, I've not been able to construct such a description.

5. Naturalness

Let me return to Hawking's 1984 paper [9], briefly discussed in sect. 1. Hawking considers theories in which the cosmological constant is a dynamical variable, but

* I thank T. Banks for this observation.

one which has a vanishing gradient. (For example, this could come about through the dependence of λ on the θ angle of axionless QCD.) He computes, in leading semi-classical approximation, the wave function of the universe, and finds that it contains a factor of $\exp(3/8G^2\lambda)$*. From this he concludes that "the most probable configurations will be those with very small values of λ".

The problem with Hawking's exponential factor is that there is no way of telling whether it is a real enhancement or just a reflection of an arbitrary normalization convention. An example may clarify this point. Consider a particle moving in three dimensions under the influence of the potential

$$V(r)=\begin{cases}0, & r<r_1,\\ V_0, & r_1<r<r_2,\\ 0, & r_2<r,\end{cases} \tag{5.1}$$

where r_1, r_2, and V_0 are positive numbers. Let us normalize the s-wave energy eigenfunctions to be equal to $\sin(kr)$ for $r<r_1$. Then for $r>r_2$, an eigenfunction with energy less than V_0 is exponentially large, while one with energy greater than V_0 is of order one. Anyone who attempted to use this to argue that the most probable energies are less than V_0 would be making a serious error.

For the particle problem, the general solution to the Schrödinger equation is a superposition of energy eigenfunctions with arbitrary coefficients; we can always choose the coefficients to cancel the exponential factor. Likewise, for Hawking's problem, the general allowable wave function obeying Hartle–Hawking boundary conditions is a superposition of the solutions for different λ's with arbitrary coefficients; we can always choose the coefficients to cancel the exponential factor. This is not the case for wormhole theory. We are not free to superpose wave functions for different α's with arbitrary coefficients, if we wish to preserve the Hartle–Hawking boundary conditions. "No boundary" implies no baby universes, and this fixes the coefficient to be $e^{-\alpha^2/4}$.

This is a real difference, but is it a significant one? The Hartle–Hawking wave function may be elegant, but it is not divinely ordained. What would happen if we chose other boundary conditions, if we gave $\Phi(B)$ support other than at $B=0$?

Clearly, giving Φ support on boundary values that are slowly varying on the wormhole scale does no damage. This only changes $\psi_\alpha(B)$ in eq. (4.8); the crucial factors are untouched. However, giving Φ support on boundary values correspond-

* In sect. 4 I have found something much more singular, the exponential of Hawking's factor. Both computations are correct. With wormholes, we have to sum not only over manifolds connected to B, but also over disconnected components, because they are really connected (by wormholes). It is this sum that turns Hawking's single exponential into my double exponential. Even if we were to sum over disconnected components in Hawking's theory, it would have no effect. Without wormholes, disconnected components are truly disconnected, and can have independent values of cosmological constant. Thus, the sum merely produces a numerical factor, independent of the value of the cosmological constant on the component connected to B.

ing to baby universes can destroy the whole argument. By these means we can replace the factor of $e^{-\alpha^2/4}$ by an arbitrary function of α, $F(\alpha)$, and change eq. (4.10) to

$$dP = |F(\alpha)|^2 Z(\alpha)\, d\alpha. \tag{5.2}$$

If we choose $|F|^2$ to have an essential zero, we can cancel the essential singularity in Z. We can even choose $|F|^2$ to be a delta-function at some randomly selected value of α. I shall now show that all such choices of F are unnatural in the technical sense.

When we describe a phenomenon as unnatural we may mean either that it requires fine tuning of short-distance physics or that it requires fine tuning of initial conditions. The original cosmological-constant problem was that vanishing cosmological constant was unnatural in the first sense; the slightest alteration in the parameters of microphysics would produce an enormous cosmological constant. The flatness problem of pre-inflationary cosmology was that a flat universe was unnatural in the second sense; the slightest departure from the critical mass density in the early universe would produce an enormous departure later.

Of course, if one believes wholeheartedly in a unique theory of everything, naturalness in the first sense is an irrelevant criterion. At every energy scale, the parameters of microphysics are what they are because they can not be other; to ask what would happen if they were slightly different is to ask what would happen if two times two were slightly different from four. But even a fanatic on this point would admit the relevance of naturalness in the second sense. Even if the theory of everything predicted a unique quantum state at the Planck scale, to us, able to measure only a limited set of observables, it might well appear to be a density matrix, a statistical superposition of states. For this reason I shall attempt to argue that the undesired F's are unnatural in both senses.

It will clarify the argument to imagine that we are dealing with three widely-separated length scales. The smallest is the Planck scale, where we imagine the parameters of our theory being slightly altered. Next, there is an intermediate scale, where we obtain an effective field theory by integrating over fluctuations down to the Planck scale. Finally, there is the scale of wormholes, instantons in the intermediate-scale effective field theory.

Naturalness in the first sense is the easiest to investigate. If we make a small change in the parameters of Planck-scale physics, this induces a small change in the parameters of the intermediate-scale theory. Naturalness in the second sense is harder to investigate, because it is not clear what plays the role of initial conditions. If we compare eqs. (3.3) and (3.6), Φ seems the natural candidate; thus I will consider changes of the form

$$\delta\Psi = A\delta\Phi. \tag{5.3}$$

The time-free way to say that we change initial conditions at the Planck epoch is to say that $\delta\Phi$ has support only on Planck-sized three-geometries. (I stress that this identification is based on analogy and guesswork, so this part of the analysis is less firmly based than the rest.)

For simplicity, let me consider the case where Φ is concentrated on a single three-geometry; more complicated cases can be treated by superposition. Given any four-manifold that enters into the construction of an allowable wave function in the intermediate-scale theory, we can patch in the new boundary at any point. All we have to do is remove a Planck-size ball around the point, and replace the ball with a four-manifold that goes from the new boundary to a three-sphere. (Such a four-manifold always exists because all three-geometries are in the same cobordism class.) This is very much like the operation of inserting a wormhole end in a manifold, described in sect. 2. As in that case, the result is the same as that of inserting a local operator. (This should be familiar from string theory, where insertion of a boundary can be replaced by insertion of an operator.)

Thus, although eq. (5.3) on the Planck scale looks very different from a change in the lagrangian, once we integrate over Planck-size fluctuations, it has the same kind of effect. This is very much like the situation in condensed-matter physics, where any small change of whatever kind on the microscopic scale can always be absorbed in a change of the coupling constants in the continuum-limit field theory. But in condensed-matter physics, this is not quite the whole story; there is also wave-function renormalization. As we shall see, there is also wave-function renormalization here.

For simplicity, I will deal with a theory in which there is only one kind of baby universe. In sect. 2 I explained how to compute the amplitude for the creation or annihilation of a baby universe by patching a semiwormhole onto a larger amplitude. Why semi? Any slice through a wormhole (and anything near such a slice) is a possible configuration of a baby universe. The midpoint slice is special because it is the maximum of the baby-universe wave function, in leading semiclassical approximation. (I do not want to take the time to show this in detail here; the computation is essentially identical to the minisuperspace studies of expanding universes in refs. [7,11].) Any other slice would have led to a smaller amplitude, and an erroneous one; the wormhole amplitude would not have been the product of baby universe creation and annihilation amplitudes. This is exactly like the situation in field theory, where improperly normalized fields lead to erroneous amplitudes which violate unitarity. For either field theory or wormhole theory, the error is corrected by introducing an explicit wave-function renormalization factor for each asymptotic particle or baby universe.

When we change the theory at the Planck scale, we change the field equations the baby universes must satisfy at the intermediate scale. Thus, boundary conditions that were formerly just right for producing n baby universes are now a little bit mismatched, because the baby universe has changed a little bit. In the operator language of sect. 2, this is equivalent to renormalizing the n-baby-universe state,

$$|n\rangle \to e^{-\epsilon n}|n\rangle, \tag{5.4}$$

where ϵ is a small number. In the case I have been considering, ϵ is always positive, because we start with optimal boundary conditions. However, we could imagine starting with boundary conditions that are less than optimal and explicit wave-function renormalization factors to compensate for this. Then, changing the coupling constants in some directions would improve the boundary conditions (negative ϵ); of course, changing them in the opposite direction would make them even worse (positive ϵ). Thus, there are always many directions of change for which eq. (5.4) holds with positive ϵ, and I will restrict myself to these from now on.

The transformation (5.4) leaves the Hartle–Hawking boundary condition, $n = 0$, unchanged. For more general $F(\alpha)$

$$\begin{aligned} F(\alpha) &\to e^{-\epsilon a^\dagger a}F(\alpha) \\ &= (2\pi \sinh\epsilon)^{-1/2}\int d\alpha'\, F(\alpha')\exp\left[-\frac{(\alpha^2+\alpha'^2)\cosh\epsilon - 2\alpha\alpha'}{2\sinh\epsilon}\right], \end{aligned} \tag{5.5}$$

by a standard harmonic-oscillator formula. This is a remarkable transformation. For any nonzero ϵ, it transforms any tempered distribution (not even necessarily square-integrable) into an analytic function. In particular, if F has an essential zero that cancels the essential singularity in Z, this destroys it. This is the result announced in sect. 1. Even if matters are carefully arranged so the cosmological constant is nonzero, the slightest change in Planck-scale physics drives it to zero; a nonzero cosmological constant requires fine tuning.

This is in striking contrast to the situation for the theories without wormholes discussed at the beginning of this section. The θ angle is not a geometrical variable like the α's, so we can not arrange from the wave function of the universe to be an eigenstate of θ by adjusting the boundary conditions. However, we can do the trick by the usual gauge-field-theory technology: we add $in\theta$ to the action, where n is the winding number, and sum over gauge-field configurations with no constraint on n. If we then make a small change in Planck-scale physics, the θ eigenstate remains a θ eigenstate; the only effect is to make a small change in the cosmological constant.

6. The big fix

Eq. (4.14) displays the leading terms in the expansion of the effective action, the cosmological term and the curvature term. For a four-sphere of radius r, the first of these grows like r^4 and the second like r^2. The higher terms in the expansion grow less rapidly with r; for small λ they make only small corrections.

Let us consider the effects of these small corrections. Because the four-sphere is a maximally symmetric space, the stationary point of Γ remains a four-sphere, but the radius of the four-sphere and the value of Γ may change by a small amount.

It will be convenient to make a nonlinear change of variables in α-space, and define

$$\alpha_0 = \tfrac{8}{3} G^2 \lambda . \tag{6.1}$$

I will denote all the other α's by $\hat{\alpha}$. We would expect the stationary value of Γ to have an expansion something like

$$\Gamma = -\frac{1}{\alpha_0} + \Gamma_0(\hat{\alpha}) + \alpha_0 \Gamma_1(\hat{\alpha}) + \dots . \tag{6.2}$$

I have written an expansion in integral powers of α_0 just for simplicity. There might well be fractional powers and/or logarithms; my reasoning (with slight modifications) will hold for these as well.

Now let us compute

$$\ln\left[\frac{Z(\alpha_0, \hat{\alpha})}{Z(\alpha_0, \hat{\alpha}')}\right] = e^{1/\alpha_0}\left[e^{-\Gamma_0(\hat{\alpha}) + \dots} - e^{-\Gamma_0(\hat{\alpha}') + \dots}\right] . \tag{6.3}$$

This expression shows that the argument of sect. 4 was incomplete. As the infrared cutoff is removed, not only is the probability distribution in α concentrated on the manifold on which α_0 vanishes, within that manifold, it is concentrated on the submanifold on which Γ_0 is minimum. (This assumes that Γ_0 is bounded below, and that its minimum is attained for finite $\hat{\alpha}$; if this is not so, the whole scheme collapses.) And this is not the end; within this submanifold, the distribution is further concentrated on the submanifold on which Γ_1 is minimum, etc. This goes on forever, or until we run out of α's to adjust. (Note that we would have had none of these further developments if Z had been a simple exponential rather than the exponential of an exponential.)

This is the big fix. If there are only a finite number of α's, they are all fixed by these conditions. (If there are an infinite number of α's, there may or may not be enough conditions to fix them.) The apparent arbitrariness introduced by the wormholes has completely disappeared, at least in the case where there are only a finite number of wormhole types. Wormholes still do some dirty work, though. Some or all of the information in the theory of everything at the Planck scale is still lost down the wormholes, because we can still make some or all of the terms in $\mathcal{L}_0$ vanish by adding constants to the α's.

The obvious next step is to compute Γ_0. If Γ_0 could be expressed completely in terms of physics at accessible energies, its minimization would lead to new relations among the observable constants of nature. Unfortunately, this does not seem likely*. There is no reason why S_{eff} could not contain an α-dependent term proportional to the Euler density, the contraction of the curvature tensor with its double dual. We have been assuming all along that only topologically trivial manifolds are important below the wormhole energy scale; thus this term has absolutely no effect on physics at accessible energies and its value can not be expressed in terms of such physics. But it makes a nontrivial contribution to Γ_0.

Thus it would seem that Γ_0 depends sensitively on physics all the way up to the wormhole scale, and to test the consequences of the minimization of Γ_0 requires detailed knowledge of wormhole physics. Possibly this is too gloomy; it may be that only a small amount of information about wormhole physics will be enough to fix Γ_0. I have no grounds for optimism here, but clearly there is more to be done on this point, and, indeed, more to be done on the whole complex of ideas explored in this paper.

I am grateful to Tom Banks for keeping me informed of his work in progress, to Stanley Deser, Jim Hartle, and Andy Strominger for conversations on the lore of quantum gravity, to S.T. Yau for information on Einstein spaces, and to Andy Cohen, Steve Giddings, Paul Ginsparg, and Frank Wilczek for vigorous if not always successful attempts to keep me honest.

References

[1] S.W. Hawking, Phys. Lett. B195 (1987) 337; Phys. Rev. D37 (1988) 904
[2] S. Giddings and A. Strominger, Nucl. Phys. B306 (1988) 890
[3] G.V. Lavrelashvili, V.A. Rubakov, and P.-G. Tinyakov, JETP Lett. 46 (1987) 167
[4] S.W. Hawking, D.N. Page, and C.N. Pope, Nucl. Phys. B170 (1980) 283;
S.W. Hawking, Comm. Math. Phys. 87 (1982) 395;
A. Strominger, Phys. Rev. Lett. 52, (1984) 1733;
D. Gross, Nucl. Phys. B236 (1984) 349
[5] S. Coleman, Nucl. Phys. B307 (1988) 867;
S. Giddings and A. Strominger, Nucl. Phys. B307 (1988) 854
[6] S.W. Hawking and R. Laflamme, Baby universes and the non-renormalizability of gravity, DAMTP preprint
[7] J. Hartle and S.W. Hawking, Phys. Rev. D28 (1983) 2960
[8] A. Linde, Phys. Lett. B200 (1988) 272
[9] S.W. Hawking, Phys. Lett. B134 (1984) 403
[10] T. Banks, Nucl. Phys. B309 (1988) 493
[11] S.W. Hawking, Quantum cosmology *in* Relativity, groups, and topology II, ed. B. DeWitt, R. Stora (North-Holland, Amsterdam, 1983)
[12] S. Giddings, private communication;
J. Hartle, private communication;
A. Strominger, private communication

* I am very grateful to Steve Weinberg for closing my mind on this point.

Nuclear Physics B317 (1989) 665–692
North-Holland, Amsterdam

WORMHOLES AND THE COSMOLOGICAL CONSTANT*

Igor KLEBANOV

Stanford Linear Accelerator Center, Stanford University, Stanford, CA 94309, USA

Leonard SUSSKIND**

Physics Department, Stanford University, Stanford, CA 94309, USA

Tom BANKS

Santa Cruz Institute for Particle Physics, University of California, Santa Cruz, CA 95064, USA

Received 14 September 1988

We review Coleman's wormhole mechanism for the vanishing of the cosmological constant. We show that in a minisuperspace model wormhole-connected universes dominate the path integral. We also provide evidence that the euclidean path integral over geometries with spherical topology is unstable with respect to formation of infinitely many wormhole-connected 4-spheres. Consistency is restored by summing over all topologies, which leads to Coleman's result. Coleman's argument for determination of other parameters is reviewed and applied to the mass of the pion. A discouraging result is found that the pion mass is driven to zero. We also consider qualitatively the implications of the wormhole theory for cosmology. We argue that a small number of universes containing matter and energy may exist in contact with infinitely many cold and empty universes. Contact with the cold universes insures that the cosmological constant in the warm ones is zero.

1. Introduction

The cosmological constant plays two roles in physics. The first role is that of a coupling constant, similar to other mass and coupling parameters in microscopic physics. Its origin is likely to include short distance physics including wavelengths down to the Planck scale. The other role, as its name suggests, is that of a macroscopic parameter controlling the large scale behaviour of the universe. From the microscopic point of view we have no explanation of why the cosmological constant vanishes. From the cosmic viewpoint it vanishes so that the universe can be big and flat, as observed. Thus, it seems a miracle that microscopic physics should

* Work supported by the Department of Energy, contract DE-AC03-76SF00515.
** Work supported by NSF PHY 812280.

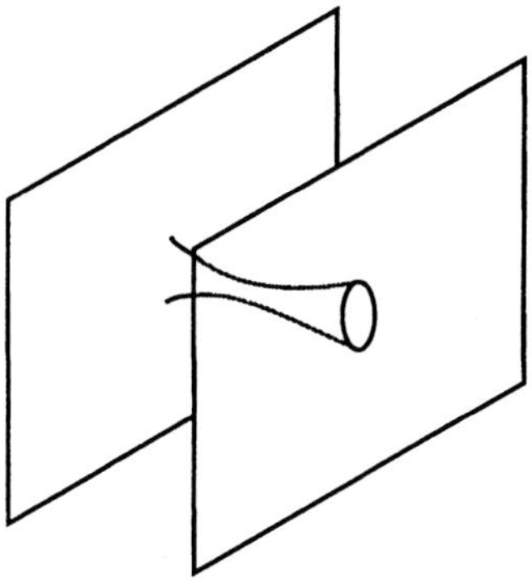

Fig. 1. A wormhole is a microscopic configuration connecting two asymptotically flat regions of space-time.

be fine tuned with practically infinite precision just so that the large scale structure of space-time can look as it does. What seems to be needed, as emphasized by Linde [1], is a direct connection between the cosmic scale physics and the microscopic machinery which creates coupling constants. Wormholes provide just such a large distance–small distance connection. As far as we know, early speculations about wormholes date back to Wheeler. Hawking and others [2] have emphasized that unusual and surprising effects can be associated with them [3]. In particular, Hawking speculated that quantum fluctuations in spacetime topology at small scales may play an important role in shifting the cosmological constant to zero [4, 5].

A wormhole is a microscopic connection between two otherwise smooth and large regions of space-time. For example, in fig. 1 a wormhole is shown connecting two flat two-dimensional sheets. The two sheets may actually be portions of the same sheet, as in fig. 2, or may be parts of otherwise disconnected universes, as in fig. 3. The important thing about wormholes is that they are small and cost little action but can connect arbitrarily distant regions of space-time. Evidently, there is a potential connection between the very large and the very small. Recently, Coleman [6], Giddings and Strominger [7] have considered the effects of wormholes in the euclidean path integral of quantum gravity. Similar ideas have been explored in refs. [9, 10]. Remarkably, it was shown that the entire effect of wormholes is to modify coupling constants and to provide a probability distribution for them. Even more remarkable is Coleman's claim [11] that the probability for a given value of the cosmological constant is overwhelmingly concentrated at zero*. One purpose of this paper is to review Coleman's arguments and discuss some subtle points about the euclidean path integral and the wave function of the universe. A second purpose is

* The idea that the probability for the cosmological constant is peaked at zero appears in earlier papers by E. Baum, Phys. Lett. B133 (1983) 185 and S. Hawking (ref. [5]).

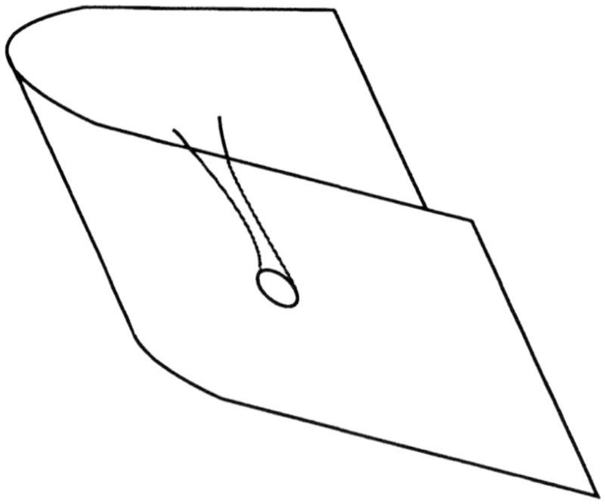

Fig. 2. The two asymptotically flat regions can actually be connected even in the absence of a wormhole.

to clarify the implications of Coleman's theory for other parameters as well as for physics of the early universe. In sect. 2 we review Coleman's arguments and rederive his results using a somewhat different method. In sect. 3 we point to some of the subtleties in defining the euclidean path integral for gravity using a minisuperspace model as an example. In sect. 4 we argue that some of these subtleties are actually clarified once the wormholes are taken into account. Sect. 5 is an attempt to use Coleman's approach to fix other fundamental parameters. The results of a naive treatment turn out to be quite discouraging: wormholes shift the pion mass to zero and the neutrino mass away from zero. Large wormholes may be generated to implement these shifts. We speculate on how one might avoid these unphysical conclusions. Finally, in sect. 6 we address the issue of whether generation of heat in the early universe is consistent with the mechanism that shifts the cosmological constant to zero.

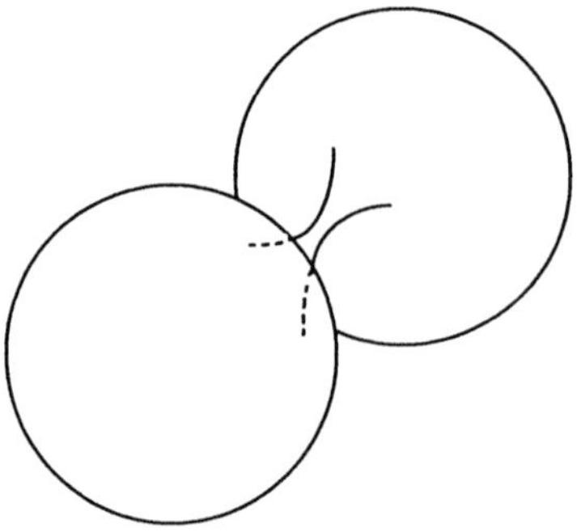

Fig. 3. A wormhole between two otherwise disconnected universes.

2. Coleman's mechanism

In this section we will derive Coleman's results in a way that some people have found more transparent than Coleman's original arguments. Consider the euclidean path integral version of quantum gravity. We integrate over all compact topologies of space-time; in particular, we focus on geometries which consist of some number of large universes connected by tiny wormholes. To begin with, we will assume that the wormholes can be treated as dilute so that their emissions are independent. This means that their average space-time separation is much greater than their size, which we take to be of the order of the Planck scale. For definiteness, we take the large universes to have spherical topology. Let us first focus on a single universe with no wormholes. The euclidean path integral for the expectation value of some observable M is

$$\langle M\rangle_\lambda = \frac{\int \mathrm{d}g\, \mathrm{e}^{-I(g,\lambda)} M}{\int \mathrm{d}g\, \mathrm{e}^{-I(g,\lambda)}}, \tag{2.1}$$

where the symbols have the following meaning. The parameters, such as couplings, masses and the cosmological constant, are collectively indicated by λ. The expression $\langle M\rangle_\lambda$ denotes the expectation value of M in a theory with only a single large universe without wormholes and with parameters λ. The integration $\int \mathrm{d}g$ indicates a sum over metrics and other local fields and $I(g, \lambda)$ is the action functional.

Now consider the effects of wormholes connecting distant regions of a single large universe. In particular, suppose that the two points connected by the wormhole are x and x'. Let $\phi_i(x)$ be a basis for the local operators at x. We assume that the effect of a wormhole is to insert the expression

$$\sum_{ij} C_{ij}\phi_i(x)\phi_j(x') \tag{2.2}$$

into the integrand of the path integral, where $C_{ij} \sim \exp(-S_\mathrm{w})$ and S_w is the wormhole action. Thus, for example, the numerator of eq. (2.1) would be replaced by

$$\int \mathrm{d}g M \mathrm{e}^{-I(g,\lambda)} \int \mathrm{d}x\, \mathrm{d}x' \sum_{ij} \tfrac{1}{2} C_{ij}\phi_i(x)\phi_j(x'). \tag{2.3}$$

The process in eq. (2.3) can be represented by a figure in which a line connects x and x' (see fig. 4). It is important to distinguish such processes from ordinary propagators connecting x and x'. The ordinary processes propagate through a large region of space-time and depend on space-time separation between x and x'. Instead, wormholes "short circuit" space-time. Therefore, the coefficients C_{ij} do not depend on x and x', at least when the two points are distant. This lack of dependence on space-time separation makes the wormhole amplitudes very different

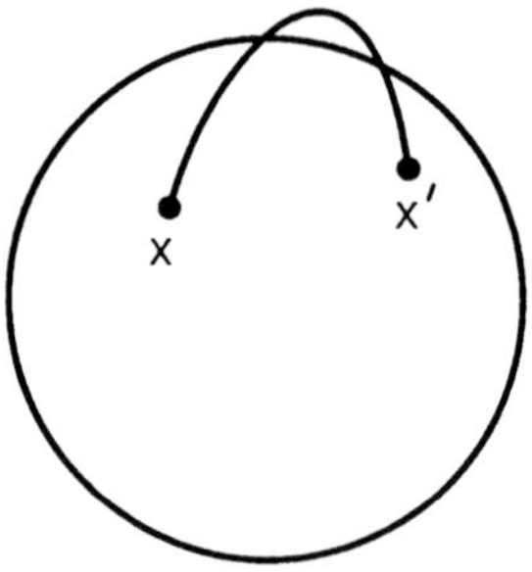

Fig. 4. A large universe with one wormhole.

from amplitudes for the ordinary processes. In fact, if the wormholes are sufficiently dilute, the amplitude for each wormhole will typically scale like the square of the space-time volume instead of the volume.

Next consider the sum over any number of wormholes, as in fig. 5. It is easy to see that the sum exponentiates to yield

$$\int \mathrm{d}g M \mathrm{e}^{-I(g,\lambda)} \exp\left[\tfrac{1}{2}\int \mathrm{d}x\,\mathrm{d}x' \sum_{ij} C_{ij}\phi_i(x)\phi_j(x')\right]. \tag{2.4}$$

Let us write this in a different form through the use of the identity

$$\exp\left(\tfrac{1}{2}C_{ij}V_iV_j\right) \sim \int \prod_k \mathrm{d}\alpha_k \exp\left(-\tfrac{1}{2}D_{ij}\alpha_i\alpha_j\right)\exp\left(-\alpha_l V_l\right), \tag{2.5}$$

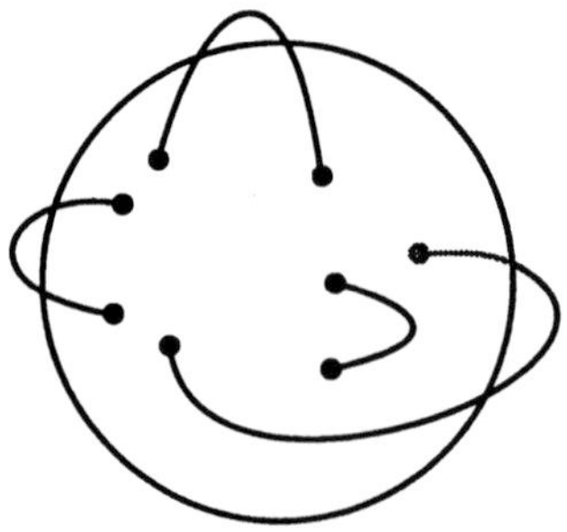

Fig. 5. A large universe with multiple wormholes.

where D_{ij} is the inverse of C_{ij}. The matrix element in question becomes

$$\int \prod_k \mathrm{d}\alpha_k \exp\left(-\tfrac{1}{2}D_{ij}\alpha_i\alpha_j\right) \int \mathrm{d}g\, M \exp\left[-I(g,\lambda) - \alpha_l \int \mathrm{d}x\, \phi_l(x)\right]. \qquad (2.6)$$

If λ_i are the coefficients of $\int \mathrm{d}x\, \phi_i$ in the lagrangian, then eq. (2.6) takes the form

$$\int \prod_k \mathrm{d}\alpha_k \exp\left(-\tfrac{1}{2}D_{ij}\alpha_i\alpha_j\right) \int \mathrm{d}g\, M \exp[-I(g,\lambda+\alpha)]\,. \qquad (2.7)$$

In a similar manner we can take into account processes involving additional closed universes. Each additional universe gives a factor $\int \mathrm{d}g \exp[-I(g,\lambda+\alpha)]$ in the α-integrand. The combinatorics again exponentiate giving

$$\langle M\rangle = (1/N)\int \mathrm{d}\alpha \exp\left(-\tfrac{1}{2}D_{ij}\alpha_i\alpha_j\right) \mathrm{d}g\, M\, \mathrm{e}^{-I(g,\lambda+\alpha)} \exp\left[\int \mathrm{d}g'\, \mathrm{e}^{-I(g',\lambda+\alpha)}\right], \qquad (2.8)$$

where N is a normalization factor. Let us compare eq. (2.8) and eq. (2.1). We see that $\langle M\rangle$ can be written in the form

$$\langle M\rangle = \int \mathrm{d}\alpha\, \rho(\alpha) \langle M\rangle_{\lambda+\alpha}, \qquad (2.9)$$

where

$$\rho(\alpha) = (1/N)\exp\left(-\tfrac{1}{2}D_{ij}\alpha_i\alpha_j\right) \int \mathrm{d}g\, \mathrm{e}^{-I(g,\lambda+\alpha)} \exp\left[\int \mathrm{d}g'\, \mathrm{e}^{-I(g',\lambda+\alpha)}\right]. \qquad (2.10)$$

Eq. (2.9) has a remarkable form. It says that any expectation value computed in our universe is a weighted average over expectation values in universes without wormholes but with couplings $\lambda + \alpha$. This is precisely the formula for an ensemble of worlds with a statistical distribution of coupling constants. In other words, if God created a large number of big smooth worlds, each with couplings $\lambda + \alpha$, drawn from a statistical distribution with weight $\rho(\alpha)$, exactly the same formula would result. Needless to say, an observer in one of the members of the ensemble would have no way to deduce the existence of the others.

One might wonder if two experiments at different locations of space-time would agree on the values of the couplings. The answer is yes because the integration variables α_i are not functions of position. There is a single overall integral over α_i. Thus, one of the effects of the wormholes is to equalize the couplings in all regions of space-time, even in large universes which would otherwise be disconnected. Furthermore, there is no restriction that the operator M must only involve a single region of space-time. In fact, M could be a product of observables in our region and some vastly different place and time. The formula states that, in that case too, an integration over a single set of α_i defines the expectation value. Therefore, there can be no disparity between the values of the coupling constants in distant regions of space-time. This completes the first part of Coleman's argument. The second part involves the computation of the probability function ρ. Let us define

$$X(\alpha)=\int \mathrm{d}g\,\mathrm{e}^{-I(g,\lambda+\alpha)}, \tag{2.11}$$

where the integration is over geometries with spherical topology and no wormholes. There are geometries of spherical topology which can be described as several large spheres connected by wormholes (see fig. 6). We do not wish to include these in the definition of X. We will return to this point later. For now, let us assume that this separation can be made. Then the probability function is

$$\rho(\alpha)=(1/N)X\mathrm{e}^{X}\mathrm{e}^{-(1/2)D_{ij}\alpha_i\alpha_j}. \tag{2.12}$$

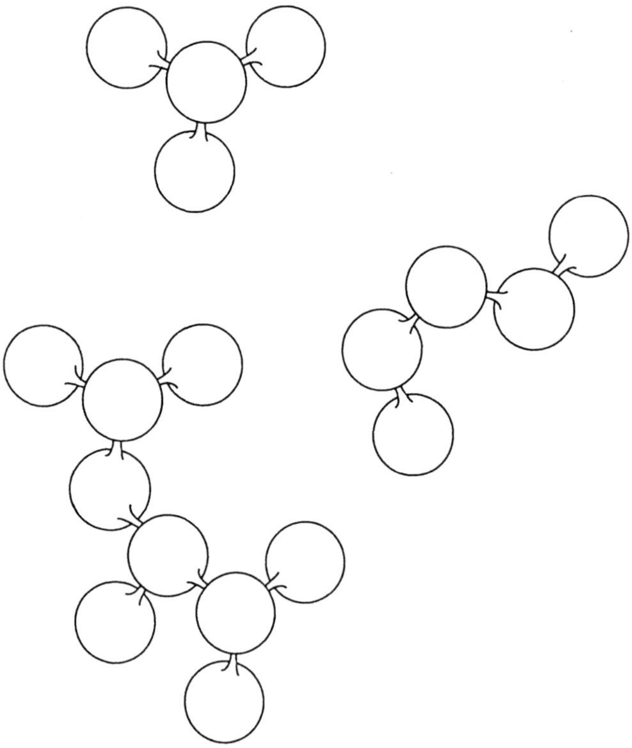

Fig. 6. Geometries of spherical topology which consist of many wormhole-connected spheres.

Coleman suggests that the leading approximation to X is the contribution from the classical stationary point associated with euclidean de Sitter space. The euclidean de Sitter space is a 4-sphere whose radius is controlled by the physical cosmological constant Λ. Consider a large smooth universe of spherical topology with metric g_{ij}. Let us compute the effective action for gravity by integrating over all fluctuations including matter and gauge fields. The result can be expanded in powers of the curvature tensor and its derivatives*

$$S_{\mathrm{eff}} = \int \mathrm{d}^4x\sqrt{g}\left[\Lambda - (1/16\pi G)R + aR_{abcd}R^{abcd} + bR_{ab}R^{ab} + cR^2 + \dots\right], \quad (2.13)$$

where Λ, G, a, etc., are functions of the wormhole-shifted fundamental parameters $\lambda + \alpha$. If we approximate S_{eff} by Einstein gravity, then the variational equation is

$$R_{ij} = 8\pi G\Lambda g_{ij}. \quad (2.14)$$

The maximum volume solution of this equation is the 4-sphere whose radius becomes large as $\Lambda \to 0$. Therefore, let us in general restrict our attention to large 4-spheres of radius r. Then

$$R_{abcd} = (1/r^2)(g_{ac}g_{bd} - g_{ad}g_{bc}). \quad (2.15)$$

Substituting this into eq. (2.13), we find

$$S_{\mathrm{eff}}(r) = \tfrac{8}{3}\pi^2\left(\Lambda r^4 - \frac{3}{4\pi G}r^2 + A_1 + \frac{A_2}{r^2} + \dots\right). \quad (2.16)$$

A dominant contribution to the euclidean path integral comes from the stationary point of $S_{\mathrm{eff}}(r)$. For large r (small Λ), this occurs at $r^2 \approx (3/8\,\pi G\Lambda)$. Plugging this into eq. (2.16) gives

$$S_{\mathrm{eff}} = -3/8G^2\Lambda. \quad (2.17)$$

Approximating X by such a saddle point, we find

$$\rho \sim \exp\left(-\tfrac{1}{2}D_{ij}\alpha_i\alpha_j\right)\exp\left(\frac{3}{8G^2\Lambda}\right)\exp\left[\exp\left(\frac{3}{8G^2\Lambda}\right)\right], \quad (2.18)$$

where Λ is the physical value of the cosmological constant which, in general, depends in a complicated way on many wormhole parameters α_i. However, Λ has a simple linear dependence on the α which shifts the "bare value" of the cosmological constant. Obviously, the function ρ is infinitely peaked at $G^2(\alpha)\Lambda(\alpha) = 0$. This

* We thank Steve Weinberg for emphasizing this to us.

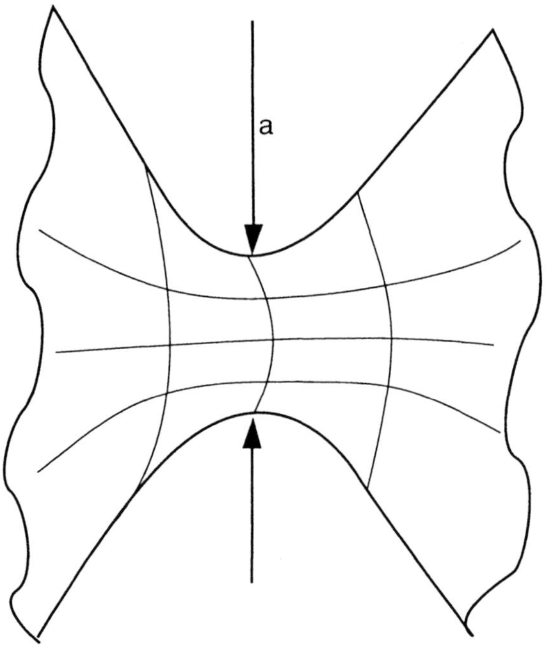

Fig. 7. A more detailed picture of a wormhole of size a.

is the basis for Coleman's claim that wormholes provide a mechanism for setting the cosmological constant to zero. How much does the above argument depend on the details of physics at and below Planck scale? It seems to us that the answer is: very little. In particular, it does not depend on the existence of classical wormhole solutions of Planck size. To see this, consider a geometry consisting of two smooth asymptotically flat regions connected by a wormhole of size a much larger than the Planck scale (see fig. 7). Certainly, including such configurations does not depend on peculiar small distance effects. How does the action depend on a? As a approaches zero, it varies like a^2, as long as a is large enough to ignore the effects of terms of order R^3 in the effective lagrangian (2.13). When a becomes of order Planck scale, we do not know how the action varies. In fact, we do not know whether the notion of a smooth metric or even a 4-dimensional manifold describes the wormhole adequately. The only important assumption is that the integral over a makes sense and gives rise to an effective description in terms of bilocal operators, as in eq. (2.2). For example, in Einstein gravity, the contribution of a wormhole with positive action is maximized at the endpoint $a = 0$. There, in place of a saddle point associated with a classical solution, the path integral is dominated by the endpoint contribution.

Let us now consider what happens if the dilute gas approximation breaks down. This is likely to occur if the shift of the cosmological constant is of the order M_p^4. Under these circumstances we must introduce interactions among wormholes. For example, a process in which two wormholes are absorbed close to one another may have to be taken into account. We can always represent this by the two wormholes coalescing to form a third one, which is then absorbed. This can be accounted for by adding a cubic term in the α's to $D_{ij}\alpha_i\alpha_j$ in eq. (2.8). More generally, this term should be replaced by some unknown function of α_i. It may be that the natural variables are not α_i but some θ_i which are non-linear functions of the α_i. In fact, in this regime, the space of α_i may be compact so that the shifts of couplings are bounded.

3. Euclidean path integrals in minisuperspace

In this section we will review the Hartle–Hawking definition of the euclidean path integral. There are two main components involved in this definition. One is the idea of a sum over compact euclidean geometries. This idea is basic to our entire discussion. The other component is a method of definition of divergent integrals by continuation to complex values of the conformal factor due to Gibbons, Hawking and Perry [12]. We will find that this second component is not consistent with Coleman's treatment*. Rather than discuss the full path integral for quantum gravity, we will restrict our attention to a minisuperspace model. The main points can be easily understood in this context. Let us consider the euclidean geometries with metric

$$ds^2 = (2G/3\pi)\left[dt^2 + a^2(t)\, d\Omega_3^2\right], \tag{3.1}$$

where $d\Omega_3^2$ is the metric of a unit three-sphere. The dynamical variable is the scale factor $a(t)$. We would like to integrate over the compact euclidean geometries, i.e. the trajectories which begin with $a = 0$ at $t = 0$ and end with $a = 0$ at some final time $t = T$. The euclidean action of such a geometry is

$$I = -\tfrac{1}{2}\int_0^T dt\,(a\dot{a}^2 + a - h^2a^3), \tag{3.2}$$

where $h = 4G\sqrt{\Lambda}/3$. This action defines a hamiltonian

$$H = \tfrac{1}{2}\left[-p^2/a - a + h^2a^3\right]. \tag{3.3}$$

The total euclidean path integral contains a functional integral over $a(t)$ and an

* This was also mentioned in a footnote in ref. [10].

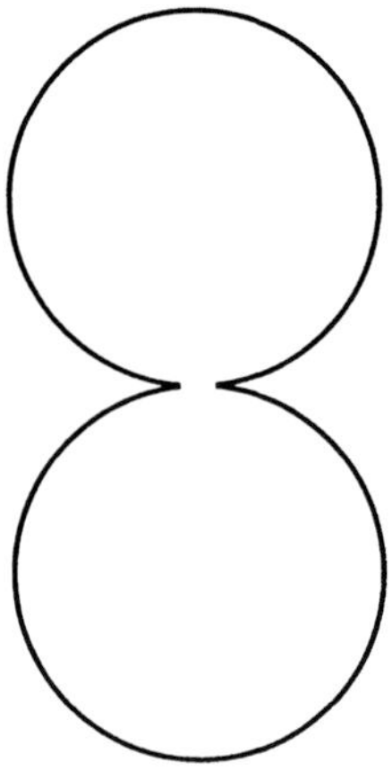

Fig. 8. Two 4-spheres connected by a wormhole.

integration over T [18]*:

$$X=\int_0^{\infty} \mathrm{d}T \int \mathrm{d}a(t)\, \mathrm{e}^{-I(a(t))}. \tag{3.4}$$

The integral is only a formal expression because the action in eq. (3.2) is not bounded from below. First of all, it can become large and negative because of rapid oscillations of a which cause $\dot{a}$ to be large. This ultraviolet instability can be cured by adding, for example, an R^2 term to the gravitational action. Perhaps of greater interest are configurations of very long duration. First, consider the classical de Sitter solution which is a stationary point of eq. (3.2):

$$a(t)=(1/h)\sin(ht) \tag{3.5}$$

with duration $T=\pi/h$. This is the configuration Coleman uses to compute the probability function $\rho(\alpha)$. It describes a 4-sphere. The action of this solution is $I_1=-3/8G^2\Lambda$. Next consider the configuration with duration slightly less than $2\pi/h$. The new configuration is made by joining two 4-spheres as in fig. 8. The neck where the two spheres join has the minimum value of the scale factor a_{min}. This is the simplest wormhole configuration included in the minisuperspace model. The action of such a configuration is

$$I_2\approx -6/(8G^2\Lambda)+\tfrac{1}{2}a_{\mathrm{min}}^2. \tag{3.6}$$

* The range of integration over T is not a priori fixed. We have made the choice which is natural in the euclidean formalism.

Similarly, N spheres can be joined by narrow necks with the euclidean action

$$I_N \approx -3N/8G^2\Lambda + \tfrac{1}{2}(N-1)a_{\min}^2. \tag{3.7}$$

These configurations are what remains in the minisuperspace of the wormhole-connected universes of Coleman. If Coleman is right, we expect them to dominate the euclidean path integral in the minisuperspace.

Hartle and Hawking have proposed a definition of the euclidean path integral which somehow regulates the divergences associated with the unbounded action [13]. When applied to the minisuperspace model considered in this paper, their prescription amounts to the following. Following ref. [12], the contour of integration over a is rotated to ia and $t \to it$. The resulting path integral is given by eq. (3.4), with I replaced by

$$\mathscr{I} = \tfrac{1}{2}\int_0^T \mathrm{d}t\left(a\dot{a}^2 + a + h^2a^3\right). \tag{3.8}$$

Eq. (3.8) is the action for a quantum mechanical problem with the hamiltonian

$$\mathscr{H} = \tfrac{1}{2}\left(p^2/a + a + h^2a^3\right). \tag{3.9}$$

Note the difference between H and $\mathscr{H}$. All the terms in $\mathscr{H}$ are positive and in the limit $\Lambda \to 0$ nothing special happens. The quantum mechanics problem with $\Lambda = 0$ is completely stable. Therefore, it is not possible that the resulting path integral is of order $\exp(3/8G^2\Lambda)$. This property of the Hartle–Hawking definition of the euclidean path integral continues to be true if we integrate over geometries more general than in the minisuperspace model. Consider, for example, the path integral over conformally flat geometries of spherical topology: $g_{ij} = \phi^2\delta_{ij}$. As explained in sect. 4, this set includes networks of wormhole-connected spherical universes which constitute the tree approximation to the path integral considered by Coleman. The euclidean path integral for Einstein gravity reduces to

$$X = \int[\mathrm{d}\phi]\exp\left(\int \mathrm{d}^4x\left[(3/8\pi G)(\partial\phi)^2 - \Lambda\phi^4\right]\right). \tag{3.10}$$

Clearly, this expression is formal due to the unconventional sign of the kinetic term for ϕ. With the conformal rotation $\phi \to i\phi$, it is defined to be

$$X = \int[\mathrm{d}\phi]\exp\left(-\int \mathrm{d}^4x\left[(3/8\pi G)(\partial\phi)^2 + \Lambda\phi^4\right]\right). \tag{3.11}$$

This is just the euclidean path integral for the stable ϕ^4-theory*. Therefore, it

* This is not a conventional theory since it must be regulated in a conformally invariant way.

cannot develop an exponential singularity as $\Lambda \to 0$. The non-appearance of singular terms in the minisuperspace euclidean path integral as $\Lambda \to 0$ is related to properties of the Hartle–Hawking wave function of the universe. This wave function can be defined in the following way. Let $\tilde{\psi}(a)$ be the euclidean path integral for the analytically continued problem (3.8) over all trajectories which begin at $a = 0$ and end at a.

$$\tilde{\psi}(a) = \int_0^\infty \mathrm{d}T \int \mathrm{d}a(t) \exp[-\mathscr{I}(a(t))]\, \delta(a(T) - a). \tag{3.12}$$

Another representation of this formula is

$$\tilde{\psi}(a) = \langle a| \int_0^\infty \mathrm{d}T \mathrm{e}^{-\mathscr{H}T} |a = 0\rangle = \langle a| \frac{1}{\mathscr{H}} |a = 0\rangle \tag{3.13}$$

A subsequent continuation $a \to -ia$ in the argument of $\tilde{\psi}$ gives a Wheeler–De Witt wave function for the original problem defined by eq. (3.2). It can be written formally as

$$\psi(a) = \langle a| \int_0^\infty \mathrm{d}T \mathrm{e}^{-HT} |a = 0\rangle = \langle a| \frac{1}{H} |a = 0\rangle. \tag{3.14}$$

In terms of the original problem, specified by the hamiltonian H, there are two stationary paths that can contribute to the wave function significantly. They correspond to the two classical trajectories which begin at $a = 0$ and end at a. The first trajectory terminates at a while $\dot{a} > 0$, as in fig. 9. Its euclidean action is $S_1(a) \approx -\frac{1}{4}a^2$ for $a \ll h^{-1}$. In the same limit the second trajectory, shown in fig. 10, has action

$$S_2(a) \approx -3/(8G^2\Lambda) + \tfrac{1}{4}a^2. \tag{3.15}$$

Thus, in the semiclassical approximation, the two wave functions are

$$\psi_1 \sim \mathrm{e}^{-S_1} = \exp\left(\tfrac{1}{4}a^2\right), \qquad \psi_2 \sim \mathrm{e}^{-S_2} = \exp\left[3/(8G^2\Lambda) - \tfrac{1}{4}a^2\right]. \tag{3.16}$$

From the definition of the wave function in eq. (3.12) it is clear that the total path integral over all closed compact geometries is just $\psi(0)$. Coleman's saddle point

Fig. 9. The minisuperspace trajectory which corresponds to the smaller part of the euclidean sphere.

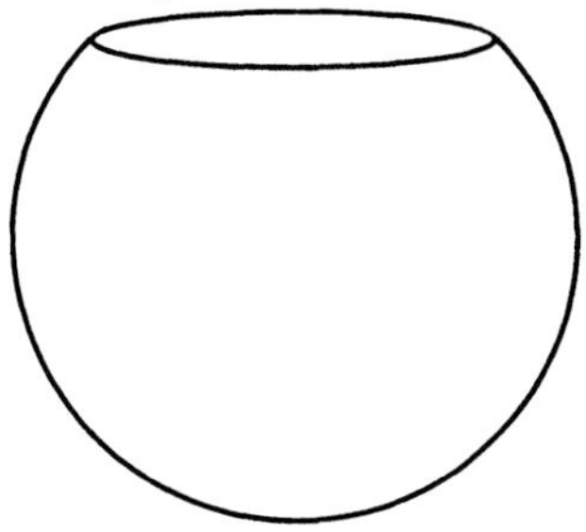

Fig. 10. The minisuperspace trajectory which corresponds to the bigger part of the euclidean sphere.

estimate obviously comes from $\psi_2(0)$. However, ref. [13] claims that rotating the contours of a and t integrations entirely eliminates ψ_2 and picks out ψ_1 as the Hartle–Hawking wave function! Can this prescription be consistent with Coleman's theory? Hartle and Hawking claim a different connection between $\psi(a)$ and the total euclidean path integral. They argue that the euclidean path integral is the norm of the Hartle–Hawking wave function

$$X=\int_0^{\infty}\psi_1^*(a)\,\psi_1(a)\,\mathrm{d}a\,. \tag{3.17}$$

Indeed, this quantity is of the order $\exp(3/8G^2\Lambda)$. To see this, let us study the behaviour of the Hartle–Hawking wave function ψ_1. For small a, ψ_1 grows exponentially. It is easy to see that for $a>h^{-1}$ we enter the classically allowed region where ψ_1 oscillates with the amplitude $\sim\exp(3/16G^2\Lambda)$. Therefore, the norm in eq. (3.17) seemingly gives the answer required for Coleman's theory. Unfortunately, we see no basis for the claim that the euclidean path integral in eq. (3.4) is the conventional norm of the Wheeler–De Witt wave function*. To show that the two are different we make use of eq. (3.14):

$$\int_0^{\infty}\psi^*(a)\,\psi(a)\,\mathrm{d}a=\int_0^{\infty}\mathrm{d}a\langle a=0|\frac{1}{H}|a\rangle\langle a|\frac{1}{H}a=0\rangle=\langle a=0|\frac{1}{H^2}|a=0\rangle\,. \tag{3.18}$$

Obviously, this is not the same as

$$X=\psi(0)=\langle a=0|\frac{1}{H}|a=0\rangle\,. \tag{3.19}$$

* This is related to Coleman's difficulty in defining the composition law for wave functions.

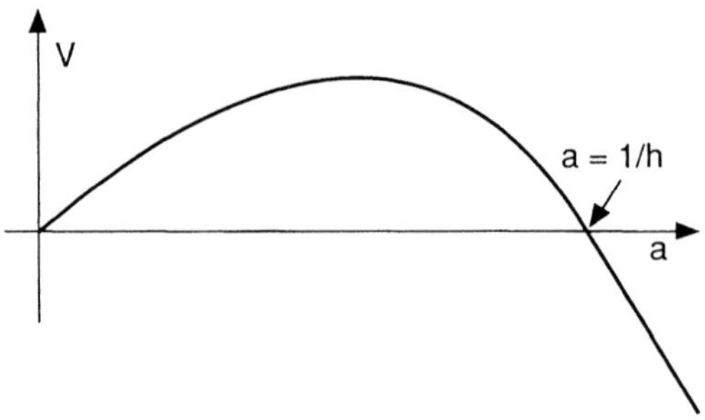

Fig. 11. The potential energy in the hamiltonian $-H$. Note that the mass of the "particle" is also position dependent: $m = a$.

Perhaps, one should simply conclude that Coleman is just wrong and the path integral does not exhibit the $\exp(3/8G^2\Lambda)$ required for his argument. On the other hand, maybe the Hartle–Hawking prescription is not the only possible way to make sense of the euclidean path integral. Below we propose a different regularization of the euclidean path integral in the minisuperspace, which leads to consistency with Coleman's sum over wormhole-connected universes. If the Wheeler-De Witt hamiltonian was bounded from below, the wave function could be obtained in two steps. First, we must solve the imaginary time Schrödinger equation

$$H\psi(a,T) = -\partial\psi(a,T)/\partial T \tag{3.20}$$

with initial condition

$$\psi(a,0) = \delta(a). \tag{3.21}$$

Integrating $\psi(a, T)$ over T gives the wave function $\psi(a)$

$$\psi(a) = \int_0^\infty \psi(a,T)\,\mathrm{d}T. \tag{3.22}$$

These steps can be formally expressed by eq. (3.14). However, the above procedure for defining $\psi(a)$ is not meaningful due to the unconventional sign of the kinetic term in the hamiltonian of eq. (3.20). We propose to define the wave function formally by real time (minkowskian) path integration for the quantum mechanics problem associated with the hamiltonian $-H$:

$$\psi(a) = -i\langle a|\frac{1}{-iH}|a=0\rangle = -i\langle a|\int_0^\infty \mathrm{d}T\mathrm{e}^{iHT}|a=0\rangle. \tag{3.23}$$

The potential energy contained in $-H$ is plotted in fig. 11. The wave function defined by eq. (3.23) satisfies [18]

$$H\psi(a)=\delta(a), \tag{3.24}$$

i.e. it satisfies the Wheeler–De Witt equation everywhere but the origin. In fact, the precise meaning of the Wheeler–De Witt equation at $a=0$ is obscure for several reasons. First of all, since no meaning is attached to $a<0$, the derivatives in H are ambiguous. Secondly, the boundary conditions at $a=0$ involve sub-Planckian physics. However, the path integral in eq. (3.23) does imply eq. (3.24). Calculation of this path integral reveals two interesting results. The first conclusion is that, as with the Hartle–Hawking prescription, $\psi(0)$ does not become large or small in the limit $\Lambda \to 0$. To see this we note that, as $\Lambda \to 0$, the hamiltonian becomes

$$-H=\tfrac{1}{2}(p^2/a+a), \tag{3.25}$$

which is once again the hamiltonian of a stable quantum mechanics problem. The second conclusion is that, unlike the Hartle–Hawking wave function, $\psi(a)$ is complex and corresponds to an outgoing wave at $a \gg 1/h$. This should not come as a surprise since the physical problem we have set up corresponds to the quantum mechanics of a particle starting at $a=0$ at $t=0$. Therefore, our wave function describes a particle tunneling from under the barrier, which extends from $a=0$ to $a=1/h$, to large values of a [14]*. One might object to our definition, which is essentially substituting a minkowskian path integral for a euclidean in the minisuperspace context**. Then, a "manifestly real" euclidean path integral gives rise to a complex wave function. Actually, this is a standard phenomenon in problems which involve an instability, such as the quantum mechanics of minisuperspace. With our alternate regularization we once again find no evidence for $\exp(3/8G^2\Lambda)$ in the normalization of the wave function. This is despite the fact that both ψ_1 and ψ_2 are present in the wave function

$$\psi(a)\approx -\exp(-3/8G^2\Lambda)[\psi_2(a)+i\psi_1(a)]. \tag{3.26}$$

The naive saddle point approximation to $\psi(a)$ would have given

$$\psi(a)\sim\psi_2+i\psi_1. \tag{3.27}$$

What is the origin of the normalization factor in eq. (3.26) and how does it affect Coleman's arguments? The answer is given in sect. 4.

* This interpretation has long been advocated by A. Vilenkin.

** In fact, if the euclidean continuation makes sense, the two path integrals must give the same answer.

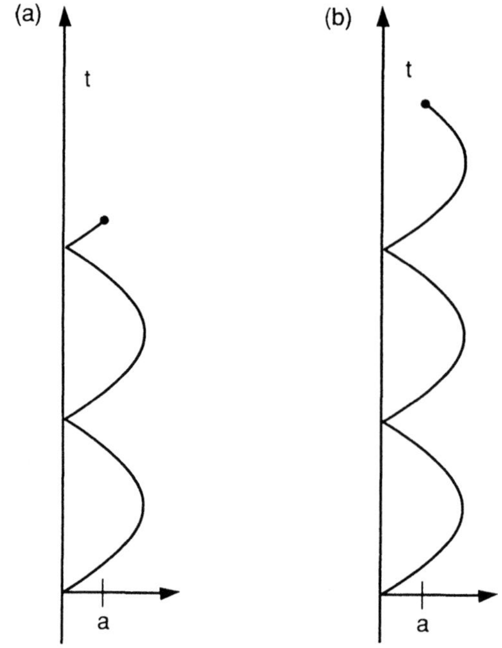

Fig. 12. Minisuperspace trajectories which involve several reflections off the endpoint at $a = 0$.

4. Wormholes in minisuperspace and beyond

In this section we will show that a consistent treatment of wormholes in minisuperspace, analogous to Coleman's, explains the normalization factor of the wave function given by the path integral. As we have emphasized, the euclidean path integral is divergent due to unboundedness of the action. A 4-sphere of radius $r = \sqrt{3/8\pi G\Lambda}$ is a stationary point where the action has the value $-3/8G^2\Lambda$. We can, however, easily generate configurations with much larger negative action. As described in sect. 3, N almost complete 4-spheres glued together in a sequence have the action $\approx -3N/8G^2\Lambda$. Each of these configurations is also a stationary point of the action, and there is no consistent reason to ignore them. But do we even want to discard contributions such as those in fig. 12? Obviously not, since they are just the wormhole-connected universes of Coleman, or what is left of them in minisuperspace. What we want to do is to apply Coleman's reasoning and sum them up. In the saddle point approximation the euclidean path integral becomes

$$\psi(0+) = i\sum_{N=0}^{\infty} \exp\left(\frac{3N}{8G^2\Lambda}\right) + \sum_{N=1}^{\infty} \exp\left(\frac{3N}{8G^2\Lambda}\right). \tag{4.1}$$

The first term sums up the processes in which any number of bounces precede a termination of a trajectory at a point where $\dot{a} > 0$ and $a = 0+$. The second series involves trajectories terminating with $\dot{a} < 0$. The reason for the factor of i in the first term is the presence of an extra negative mode in the fluctuations about the trajectories which end with $\dot{a} > 0$. Although divergent, the series in eq. (4.1) can be formally summed up to

$$\psi(0+) \approx \frac{\exp(3/8G^2\Lambda) + i}{1 - \exp(3/8G^2\Lambda)}. \tag{4.2}$$

This manipulation can be made more convincing by introducing an R^2 stabilizer term into the action. Then, for Λ greater than some critical value, the geometric series which sums up the multiple bounces converges. Analytic continuation to small Λ essentially reproduces eq. (4.2), in agreement with eq. (3.26). Note that eq. (4.2) approaches -1 as $\Lambda \to 0$. Thus, in this limit, only if we sum over the wormholes do we get the correct saddle point approximation to the wave function. This is not to say that tunneling amplitudes cannot be understood without multiple bounces. In particular, if the usual euclidean path integral is applied to a hamiltonian with the conventional sign of the kinetic term, such as $-H$, where H is given in eq. (3.3), the multiple bounce has the amplitude $\sim \exp(-3N/8G^2\Lambda)$. Thus, the successive bounces are strongly suppressed. The reader can verify that the path integral carried out this way leads to the same wave function. Our point is that, if we insist on using the standard sign for the gravitational kinetic energy and the usual saddle point definition of the euclidean path integral, the sum over the geometric series generated by the multiple bounces is necessary. This supports the view that the wormholes provide important contributions to the euclidean path integral. We also see that, if Λ is small, no serious error is made by ignoring wormhole-connected universes since their only effect is to change normalization of the wave function. In Coleman's case, the analogue of summing up the geometric series is just the calculation of sect. 2. The effects of the wormhole summation and the exponentially large contribution from each 4-sphere result in a finite prescription: quantum gravity with zero cosmological constant. As far as we can tell, similar arguments do not apply to the Hartle–Hawking prescription, where it appears that trajectories with multiple bounces do not dominate the euclidean path integral. The sum over multiple bounces in fig. 12a results in normalization $\sim \exp(-3/8G^2\Lambda)$ for $\psi_1(0)$. Since the Hartle–Hawking method of evaluating the euclidean path integral gives $\psi_1(0)$ of order 1, it appears that it is not consistent with the idea that the wormhole-connected universes dominate the euclidean path integral.

Another interesting question is what is the role of higher 4-topologies in Coleman's argument. Let us consider the euclidean path integral over all geometries with a spherical topology. An important subset of these is formed by the conformally flat geometries discussed in sect. 3. Among these geometries there are approximate saddle points, which are generalizations of the wormhole-connected series of

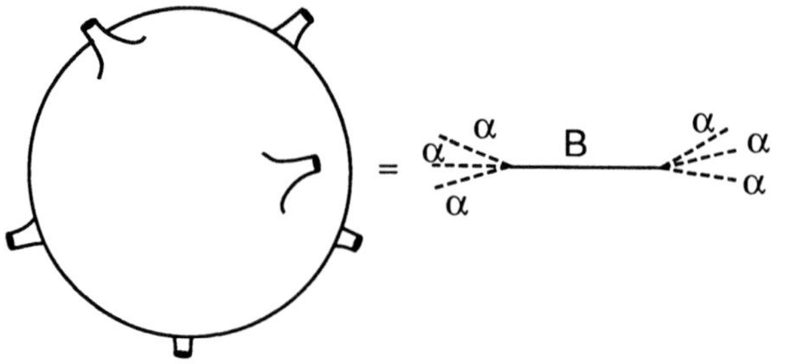

Fig. 13. A representation of emission of baby universes in terms of a Feynman graph.

4-spheres encountered in the minisuperspace. These are the tree diagrams of the "universal field theory". Typical examples are shown in fig. 6. In ordinary field theory the sum over such diagrams is found by a saddle point approximation to the euclidean path integral. Let us therefore construct a simple path integral which sums up the effects of wormhole-connected universes. Define a "field" $B(V)$ which describes large spherical universes of volume V and a field α for wormholes. The emission of wormholes by a universe will be represented by a graph like that shown in fig. 13. The path integral which reproduces the wormhole sum is*

$$\int \mathrm{d}\alpha\, \mathrm{d}B(V) \exp\left[-\tfrac{1}{2}D\alpha^2 + \int_0^\infty \frac{\mathrm{d}V}{V}\left[B(V)\,\mathrm{e}^{1/2[(1/G)\sqrt{3V/2}-\Lambda V+\alpha V]} - \tfrac{1}{2}B^2(V)\right]\right]. \tag{4.3}$$

The reader can check that the expansion in powers of $1/D$ generates the wormhole summation including loops. Integrating out $B(V)$ with some particular integration measure** reduces eq. (4.3) to

$$\int \mathrm{d}\alpha \exp\left[-\tfrac{1}{2}D\alpha^2 + (1/G)\int_0^\infty V^{-1/2}\mathrm{d}V\,\mathrm{e}^{(1/G)\sqrt{3V/2}-\Lambda V+\alpha V}\right]. \tag{4.4}$$

For $\alpha < \Lambda$ the volume integration in the exponent converges to

$$\sqrt{\frac{\pi}{G^2(\Lambda-\alpha)}}\exp\left(\frac{3}{8G^2(\Lambda-\alpha)}\right). \tag{4.5}$$

* A similar path integral appears in the work of Giddings and Strominger (ref. [10]).
** Our conclusions do not depend sensitively on the choice of measure.

Ordinarily, the sum of connected tree graphs would be given by the extremum of

$$-\tfrac{1}{2}D\alpha^2+\sqrt{\frac{\pi}{G^2(\Lambda-\alpha)}}\,\exp\left(\frac{3}{8G^2(\Lambda-\alpha)}\right). \tag{4.6}$$

In our case this sum diverges for any D due to the singularity of (4.6) at $\Lambda=\alpha$ [10]. We observe, however, that, for a sufficiently large Λ, there exists a local maximum of (4.6) with

$$\alpha \sim 1/GD\Lambda^{3/2}. \tag{4.7}$$

We can regard the value of (4.6) at this extremum as the definition of the sum of connected tree graphs. With this definition, the cosmological constant is only weakly shifted by wormholes. As Λ decreases, the local maximum in eq. (4.7) disappears. This happens well before the wormhole-shifted cosmological constant vanishes. Thus, the sum of tree diagrams ceases to make sense below some value of the physical cosmological constant. It should be noted that all the tree diagrams are spheres topologically, while metrically they are very far from spherical. This raises an interesting possibility. It suggests that the euclidean path integral over geometries of spherical topology is unstable with respect to break up into infinitely many wormhole-connected 4-spheres. For Λ less than some critical value, this integral is uncontrollably divergent. On the other hand, the result of integration over α, which sums up the loop diagrams in eq. (4.4), is well-defined in the sense that all the physical amplitudes are dominated by the singularity at $\Lambda-\alpha=0$. In the region where $\Lambda-\alpha<0$ the volume integral in eq. (4.4) is problematic. However, we can formally continue from the region where $\Lambda-\alpha>0$ to obtain the same answer as in eq. (4.5). This procedure seems to indicate that there is no preference for a negative physical cosmological constant. Let us consider what the Hartle–Hawking prescription says about the euclidean path integral over geometries of spherical topology. An important class of these geometries, which includes the wormhole-connected spheres, is the conformally flat geometries $g_{ij}=\phi^2\delta_{ij}$ considered in sect. 3. The theory specified by eq. (3.10) has an instanton

$$\phi(r)=\left(1+\tfrac{2}{3}\pi G\Lambda r^2\right)^{-1}. \tag{4.8}$$

This is simply the 4-sphere of radius $\sqrt{8/8\pi G\Lambda}$ which constitutes the euclidean de Sitter space. Multi-instanton configurations are the wormhole-connected 4-spheres. Coleman's treatment assumes that the euclidean path integral is saturated by the instanton sum. As we have seen above, this sum diverges. Even if defined by the extremum of the action in eq. (4.6), it becomes singular at some critical value of Λ, below which it does not make sense. On the other hand, the integration contour rotation to imaginary values of the conformal factor leads to the euclidean path integral in eq. (3.11) which defines a stable ϕ^4-theory and is presumably well-behaved as $\Lambda \to 0$. This once again suggests that this prescription eliminates the instability which leads to Coleman's wonderful effect. This is not meant to imply that the Hartle–Hawking prescription, which is based on the integration contour rotation for the conformal factor, is incorrect. We believe, however, that it cannot coexist with Coleman's results. Some other prescription must be found to justify them.

5. Fixing other coupling constants

Once the cosmological constant has been set to zero, it is natural to ask if Coleman's theory predicts some or all of the additional parameters, such as the mass and coupling constants. Coleman's answer is yes. Let us review his argument. We imagine carrying out the path integral over all fields in a background euclidean de Sitter space of radius r. Explicit examples indicate that the result has the form $\exp[-S_{\text{eff}}(r)]$ with $S_{\text{eff}}(r)$ given by the eq. (2.16). As in sect. 2, we eliminate the radius by solving

$$\partial S_{\text{eff}}/\partial r = 0\,. \tag{5.1}$$

This gives $r^2 = 3/8\pi G\Lambda + O(G^3\Lambda)$. S_{eff} can then be written in terms of G, Λ, and a set of parameters A_i which depend on the wormhole-shifted fundamental constants

$$S_{\text{eff}} = -3/8G^2\Lambda + \tfrac{8}{3}\pi^2 A_1 + O(G^2\Lambda)A_2 + \dots\,. \tag{5.2}$$

The probability for the wormhole-shifted couplings and masses is proportional to $\exp[\exp[-S_{\text{eff}}]]$, which for small $G^2\Lambda$ reduces to

$$\rho(\alpha) \sim \exp\left[\exp\left[3/8G^2\Lambda - \tfrac{8}{3}\pi^2 A_1\right]\right]. \tag{5.3}$$

The absolute maximum of this function will generally occur at $G^2\Lambda(\alpha) = 0$. This defines some subspace in the space of wormhole-shifted constants. On this surface the probability varies, being infinitely sharply peaked at the place where $A_1(\alpha)$ achieves its minimum. If this occurs at a point in the space α_i, then all the parameters are determined. If it occurs on some higher-dimensional surface, then only some relations between couplings are fixed. In this case the process can be continued by minimizing A_2 and so on. Whether the process determines all the interesting couplings is not known*. It is important to know whether symmetries

* This procedure requires that we introduce a lower cut-off on $G^2\Lambda$, and take it to zero at the end of the calculation. Grinstein and Wise [17] have suggested that the proper quantity to be cut off is $G_0^2\Lambda$, where G_0 is the "bare" Newton's constant, which does not depend on the α's. With this regulator the probability is maximized at $G/G_0 = 0$. Unfortunately, this seems to imply that wormhole effects make gravity a free theory. However, the full consequences of this approach have not yet been worked out.

restrict the allowable couplings which can be generated by wormhole effects. For example, suppose that the fundamental lagrangian is invariant under some global symmetry, such as the chiral symmetry or baryon number conservation. Wormholes can then break this symmetry. The mechanism involves a wormhole through which the conserved current flows. For example, a unit of baryon number can pass through a wormhole. This induces a bilocal operator

$$\int \mathrm{d}x\,\mathrm{d}x'\,O^\dagger(x)O(x'), \tag{5.4}$$

where x and x' are the ends of the wormhole and O is a baryon number violating operator. The arguments in sect. 2 indicate that the phenomenology of a single large universe will require the operator $\alpha_0(O+O^\dagger)$ in the lagrangian. Of course, it may happen that the probability is maximized at $\alpha_0=0$. However, this is not a priori implied by the symmetry of the theory without the wormholes. The obvious question is to what extent the A_i can be computed from a knowledge of low-energy physics alone. Power counting indicates that A_1 depends on the short-distance physics. Since it is dimensionless, it will generally be logarithmically divergent in the ultraviolet when expressed in terms of integrals over wave numbers. Therefore, it is sensitive to physics at arbitrarily short distances. Nevertheless, it appears that, when applied to masses of spin-0 and spin-$\frac{1}{2}$ particles, Coleman's procedure leads to some discouraging conclusions. Consider a light pseudoscalar particle, such as the pion. We wish to study the euclidan path integral on a 4-sphere of radius r as a function of the pion mass. Let us assume that m_π is much smaller than the QCD scale f_π. The low-energy interactions of pions are well described by the SU(2) × SU(2) non-linear sigma model with the cut-off set around f_π:

$$\mathscr{L}=\tfrac{1}{4}f_\pi^2\,\mathrm{tr}\left(\nabla_i U\,\nabla^i U^\dagger\right)+\mathrm{tr}\left(MU+U^\dagger M^\dagger\right), \tag{5.5}$$

where the SU(2) variable U is related to $\boldsymbol{\pi}$ by

$$U=\exp\left(i\boldsymbol{\sigma}\cdot\boldsymbol{\pi}/f_\pi\right). \tag{5.6}$$

The portion of the path integral with momenta $\le f_\pi$ will be dealt with using this model. The rest of the euclidean path integral should make use of a more fundamental theory involving quarks, gluons, Z's, W's, etc. Now consider the limit of large f_π in which the pion becomes a free minimally coupled point particle with lagrangian

$$\mathscr{L}=\tfrac{1}{2}\left[\left(\nabla\boldsymbol{\pi}\right)^2+m_\pi^2\boldsymbol{\pi}^2\right]. \tag{5.7}$$

In this case the path integral reduces to evaluation of the determinant of $\nabla^2+m_\pi^2$. Methods and results of such a calculation in arbitrary curved backgrounds can be found in ref. [16]. Recently, the calculation was performed directly on a 4-sphere by

Grinstein and Wise [15] with the purpose of application to Coleman's theory. They isolated the coefficient A_1 in eq. (5.3)

$$A_1 = \frac{1}{32\pi^2}\left(2 - \tfrac{1}{15}\right)\log\frac{m_\pi^2}{f_\pi^2} + O\left(\frac{m_\pi^2}{f_\pi^2}\right). \tag{5.8}$$

The logarithmic dependence on m_π originates in the infrared modes and is totally independent of the ultraviolet cut-off. Thus, corrections to eq. (5.8), coming from the interaction terms in eq. (5.5), depend on the strength of these perturbations among the infrared modes. We have checked by an explicit calculation that, to lowest order, the interactions do not generate additional infrared logarithms. The reason is that Goldstone bosons decouple at low momenta. Although we have no general proof of this, we expect the coefficient of $\log m_\pi$ in eq. (5.8) to be universal. Thus, it appears that the probability for the pion mass is infinitely peaked at $m_\pi = 0$. Unless a way around this conclusion is found, the theory of wormholes is in trouble. It is also interesting to do a similar calculation for a free massive fermion. Using the knowledge of eigenvalues and degeneracies of the Dirac operator on a 4-sphere [19], we find

$$\log\det\left(i\gamma^\mu D_\mu + m\right) \sim \sum_{l=0}^{\infty} \frac{(l+3)!}{l!} \exp\left(-\frac{l+2}{rM}\right) \log\left(\frac{(l+2)^2}{r^2M^2} + \frac{m^2}{M^2}\right), \tag{5.9}$$

where M is the cut-off mass. We would like to calculate the sign of the coefficient of $\log(m/M)$ in A_1. Since this term originates in the infrared modes, the coefficient is cut-off independent. In contrast with the result for a free scalar, we find that the sign is negative. Therefore, wormholes drive the free fermion mass toward the cut-off scale. Let us apply this to neutrino physics. Since wormholes break chiral symmetry, ν acquires Majorana mass through operators like

$$\phi^\dagger\psi_L\gamma_2\phi^\dagger\psi_L + \text{h.c.}, \tag{5.10}$$

where ϕ is the Higgs field. At low momenta, neutrinos are almost free and the above calculation should be applicable. Our calculation indicates that wormholes drive the neutrino mass away from zero. A possible way out of this predicament involves assuming that the dilute wormhole approximation is bad. If the shift of the cosmological constant produced by wormholes is of order M_p^4, then wormholes must develop a significant density. As we have mentioned, under these circumstances the dynamics of α's becomes non-linear. In fact, the α's may be poor global coordinates for the space of wormhole fields. This space may be better described by some coordinates $\theta(\alpha)$. The manifold may even be compact. In general, the parameters α_i which multiply the specific operators are highly non-linear functions of θ_i. Now

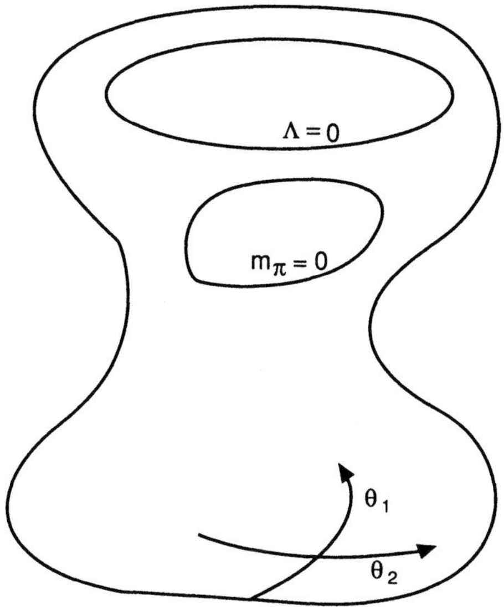

Fig. 14. Schematic drawing of $\Lambda = 0$ and $m_\pi = 0$ surfaces in the space of wormhole variables $\theta(\alpha)$.

consider the surface $\Lambda(\theta) = 0$. This will be some curved surface in θ-space, as shown in fig. 14. Similar remarks apply to the surface $m_\pi(\theta) = 0$. Since we know very little about the θ-space, we see no general reason for the two surfaces to intersect. This might happen if the required shift of the cosmological constant is fairly large in Planck units and the wormholes required for shifting Λ become dense. Furthermore, the wormholes required for shifting m_π carry different quantum numbers from those that shift Λ, since the former must carry off chiral charge while the latter should be chiral singlets. A high density of neutral wormholes could leave very little space for the ones that shift m_π. On the other hand, the probability distribution near $m_\pi = 0$ is

$$\rho \sim \exp\left[(f_\pi/m_\pi)^{0.32}\exp(3/8G^2\Lambda)\right]. \tag{5.11}$$

Evidently, as $\Lambda \to 0$, the driving force on the chiral wormholes diverges and it is not clear that they can resist increasing their density. Kaplunovsky suggested an even more disturbing possibility that larger scale wormholes may be forced to occur if the Planck scale wormholes are not sufficient to shift m_π to zero. Once we have integrated out the effects of wormholes and other fluctuations above a given scale, there is no reason why a new round of wormholes cannot be important at larger scale. One might object that the factor $\exp(-S_w)$ in the amplitude for each wormhole decreases rapidly with the wormhole size. This causes the coefficient D in eq. (2.8) to be of order $\exp(S_w)$. Unfortunately, the factor in eq. (5.3) is so strong as

$\Lambda \to 0$ that it easily overwhelms any finite value of D, no matter how large. This suggests that, if m_π is not driven to zero by microscopically small wormholes, then wormholes should occur with maximum possible density at every scale up to f_π. This seems unphysical since it would almost certainly adversely affect predictions of the standard model. At the moment, we have no answer to this puzzle.

6. Wormholes and cosmology

An important question about Coleman's theory is whether it is consistent with a reasonable cosmology. It will be disappointing if the theory truly predicts nothing rather than something: namely, a cold universe devoid of matter and energy. We must hope that there is at least a finite number of universes which have undergone an interesting cosmological development. Obviously, the relevant issue is the absolute number of such universes and not the fractional number. We will argue that a possible outcome of the wormhole theory is that the number of warm universes is finite while the number of cold ones diverges. As a result, the expectation values of all observables will be dominated by cold empty universes. Under these circumstances the quantities of physical interest are conditional probabilities given that one is in a warm universe. Let us begin with extending the minisuperspace model by including a scalar variable $\phi(t)$ which represents the state of all matter fields in the universe. The euclidean action becomes

$$I = \int_0^T \mathrm{d}t \left\{ -\tfrac{1}{2} a \dot{a}^2 - \tfrac{1}{2} a + a^3 \left[\tfrac{1}{2} \dot{\phi}^2 + V(\phi) \right] \right\}. \tag{6.1}$$

For our purposes we will restrict the shape of $V(\phi)$ to be the one in fig. 15. A warm universe corresponds to a euclidean trajectory which emerges in the classically allowed region of a with ϕ in the vicinity of ϕ_b. Subsequently, it can tunnel to the lower well around ϕ_a dumping the energy difference into heat. We are making no attempt here to model a realistic cosmology. We prefer $V(\phi)$ because this potential makes it easy to separate the important saddle point trajectories. The euclidean equations of motion are

$$\ddot{\phi} + 3\frac{\dot{a}}{a}\dot{\phi} = \frac{\mathrm{d}V}{\mathrm{d}\phi}, \qquad \left(\frac{\dot{a}}{a}\right)^2 = \frac{1}{a^2} + \dot{\phi}^2 - 2V(\phi). \tag{6.2}$$

The two important saddle points correspond to classical trajectories which start at $a = 0$ with $\dot{\phi} = 0$ and $\phi = \phi_a$ or ϕ_b. We observe that, if a trajectory starts at the bottom of the well and has $\dot{\phi} = 0$ initially, it never acquires any non-zero $\dot{\phi}$.

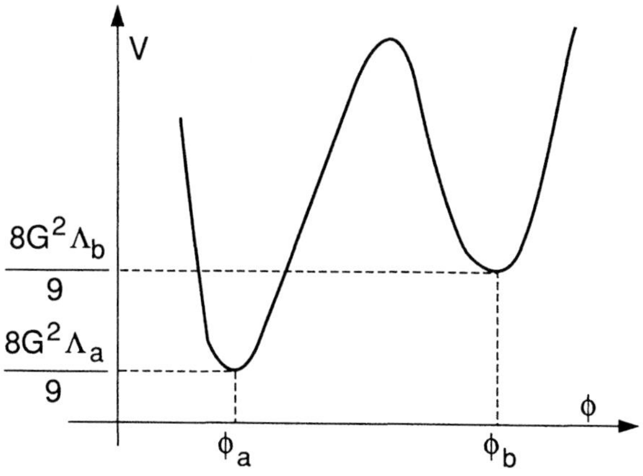

Fig. 15. A convenient potential for the scalar variable ϕ.

Therefore, the solutions are

$$\phi = \phi_a, \qquad a(t) = \frac{1}{h_a}\sin(h_a t), \tag{6.3}$$

and

$$\phi = \phi_b, \qquad a(t) = \frac{1}{h_b}\sin(h_b t), \tag{6.4}$$

where $h_a^2 = 2V(\phi_a)$ and similarly for b. Ignoring wormholes, these two trajectories approximately saturate the euclidean path integral for the Wheeler–De Witt wave function in the classically forbidden region $a^2 < \frac{1}{2}V(\phi)$ (neglecting $\dot{\phi}$). To extrapolate to large values of a, the euclidean trajectories must be matched on to the solutions of the Minkowski equations of motion for $a^2 > \frac{1}{2}V(\phi)$. Eventually, within the classically allowed region tunneling to the lower well, accompanied by heat generation, takes place. We assume that the tunneling rate between the two wells is much slower than the rate for a to tunnel from $a = 0$ into the classically allowed region*. Under these circumstances it is clear that a warm universe can be recognized in the euclidean path integral as the extremum in eq. (6.4), and a cold universe – as the extremum in eq. (6.3).

Let us now consider the effect of wormholes in the euclidean path integral. In analogy with the discussion of sect. 4, a wormhole is just a reflection off $a = 0$. There are three types of wormholes. The two diagonal processes take place entirely at ϕ_a or ϕ_b. In addition, transitions between ϕ_a and ϕ_b are possible. The reason

* This is true if the wells are separated by a high barrier.

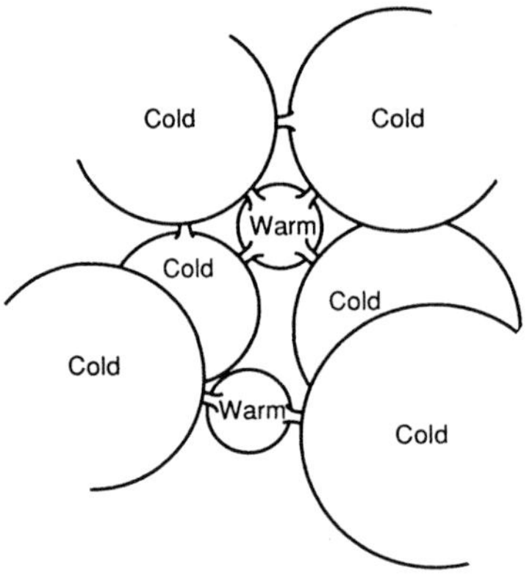

Fig. 16. Important contributions to the euclidean path integral in a theory with two types of large universes.

appreciable transitions exist, even if the barrier is high, is that the ϕ-motion at $a=0$ costs no action. If the wormholes are taken into account, the euclidean path integral is saturated by series of large and small bubbles which correspond to cold and warm universes. When we depart from minisuperspace, the path integral will contain arbitrarily connected bubbles of type a (cold) and of type b (warm), as in fig. 16. The probability in eq. (5.3) generalizes to a function of $\Lambda_a = 9V(\phi_a)/8G^2$ and $\Lambda_b = 9V(\phi_b)/8G^2$, as well as other constants. Typically, it will have the form

$$\rho \sim \exp\left(F_a \exp\left(\frac{3}{8G^2\Lambda_a}\right) + F_b \exp\left(\frac{3}{8G^2\Lambda_b}\right)\right), \tag{6.5}$$

where F_a and F_b are the prefactors which depend on the wormhole-shifted fundamental constants of the theory.

An important question is whether both Λ_a and Λ_b are shifted to zero by wormholes. If this is the case, we will be forced to conclude that both a and b become cold and uninteresting universes. However, it is quite possible that the prefactors in eq. (6.5) are such that this does not occur. Recall that both F and Λ depend on the fundamental constants $\lambda + \alpha$. Therefore, the F's have an implicit dependence on the Λ's. Let us imagine a simplified model where the prefactors are functions of just Λ_a and Λ_b.

$$F_a = F_a(\Lambda_a, \Lambda_b), \qquad F_b = F_b(\Lambda_a, \Lambda_b). \tag{6.6}$$

Suppose that the prefactors satisfy

$$F_a(0,0) + F_b(0,0) < F_a(0,x), \tag{6.7}$$

where $F_a(0, x)$ is the maximum of $F_a(0, \Lambda_b)$. Then the probability is infinitely larger at $\Lambda_a = 0$ and $\Lambda_b = x$ than at $\Lambda_a = 0$ and $\Lambda_b = 0$. The property (6.7) holds for a broad variety of smooth functions F. Therefore, we do not consider this to be a "fine-tuned" possibility.

It is interesting to determine the mean numbers of universes of type a and b in a connected diagram. The result is

$$N_a \sim \exp(3/8G^2\Lambda_a), \qquad N_b \sim \exp(3/8G^2\Lambda_b). \tag{6.8}$$

Thus, if (6.7) holds, the number of cold universes is driven to infinity, while the warm ones stay finite in number. However, the cosmological constant in these warm universes is driven to zero by contact with the infinity of cold universes. This follows from the fact that $\Lambda_a = 0$ and that eventually the universe tunnels to the well at ϕ_a. The details of this scenario may vary depending on the specific mechanism for inflation. However, the idea that the cosmological constant in our warm universe is driven to zero by contact with an infinity of cold universes can be quite general. It suggests that the reason why the cosmological constant is zero lies outside our own universe.

We are extremely grateful to A. Linde for discussions on the importance of the long distance–short distance connection. Conversations with S. Weinberg, V. Kaplunovsky and M. Wise are also gratefully acknowledged.

References

[1] A. Linde, Phys. Lett. B200 (1988) 272
[2] S. Hawking, Phys. Lett. B195 (1987) 337;
G. Horowitz, M. Perry and A. Strominger, Nucl. Phys. B238 (1984) 653;
A. Strominger, Phys. Rev. Lett. 52 (1984) 1733
[3] See also more recent work by S. Giddings and A. Strominger, Harvard preprint HUTP-87/A067;
G. Lavrelashvili, V. Rubakov and P. Tinyakov, JETP Lett. 46 (1987) 167
[4] S. Hawking, Nucl. Phys. B144 (1978) 349
[5] S. Hawking, Phys. Lett. B134 (1984) 403
[6] S. Coleman, Harvard preprint HUTP-88/A008
[7] S. Giddings and A. Strominger, Harvard preprint HUTP-88/A006
[8] T. Banks, I. Klebanov, L. Susskind, talk delivered by T. Banks at the Yerevan Conference on High Energy Physics, in preparation
[9] T. Banks, Santa Cruz preprint SCIPP-88/09, to be published in Nucl. Phys. B. The relation between this work and the present paper will be discussed in ref. [8]
[10] S. Giddings and A. Strominger, Harvard preprint HUTP-88/A036
[11] S. Coleman, Harvard preprint HUTP-88/A022
[12] G. Gibbons, S. Hawking and M. Perry, Nucl. Phys. B138 (1978) 141
[13] J. Hartle and S. Hawking, Phys. Rev. D28 (1983) 2960
[14] A. Vilenkin, Phys. Rev. D37 (1988) 888
[15] B. Grinstein and M. Wise, Caltech preprint CALT-68-1505
[16] N. Birrel and P. Davies, Quantum Fields in Curved Space, Cambridge University Press, 1982
[17] P. Candelas and S. Weinberg, Nucl. Phys. B237 (1983), 397
[18] J. Halliwell, ITP Santa Barbara preprint NSF-ITP-88-25

Nuclear Physics B323 (1989) 141–186
North-Holland, Amsterdam

WORMHOLES IN SPACETIME AND THE CONSTANTS OF NATURE*

JOHN PRESKILL**, ***

California Institute of Technology, Pasadena, CA 91125, USA

Received 24 November 1988

The proposal that wormholes in spacetime cause the cosmological constant Λ to vanish is reviewed, and its implications are studied. Wormholes also drive Newton's constant G to the lowest possible value. The requirement that G is at its minimum determines, in principle, all other constants of Nature. In practice, the values of fundamental constants other than Λ cannot be predicted without a detailed knowledge of Planck-scale physics.

1. Introduction

A wormhole in spacetime is a gravitational quantum fluctuation that links two distantly separated spacetime points [1–3]. Although it is not yet clearly understood whether such fluctuations must be included in a sensible quantum theory of gravity, it is at least a plausible hypothesis that wormhole fluctuations occur, and it is interesting to contemplate the consequences of this hypothesis.

The euclidean path integral approach to quantum gravity [4–6] provides a formalism in which wormhole effects can be systematically discussed; among the four-dimensional geometries that contribute to the path integral are wormhole configurations in which two points on a smooth background spacetime are connected by a narrow tube, as in fig. 1. But even if the euclidean path integral as currently formulated does not provide an adequate description of quantum gravity, it is certainly possible that wormholes exist and have remarkable effects.

It is natural to wonder whether wormholes can induce an apparent failure of locality in the physics on the background spacetime, or, as Hawking [7] has advocated, an apparent loss of quantum coherence. Coleman, however, has argued persuasively that the physical effects of wormholes are quite different than we might

* This work supported in part by the U.S. Department of Energy under Contract No. DE-AC0381-ER40050.

** NSF Presidential Young Investigator.
*** Bitnet address: Preskill@CALTECH.

0550-3213/89/$03.50
(North-Holland Physics Publishing Division)

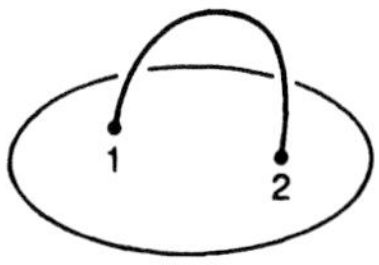

Fig. 1. A wormhole connects points 1 and 2 that are distantly separated on the background spacetime.

naively expect [8, 9]. He concludes that, because of wormhole effects, the fundamental constants of Nature are afflicted by an intrinsic quantum indeterminacy; we must regard our universe as having been chosen at random from an ensemble of possible universes, each with different values of the fundamental constants. Quantum gravity, then, may threaten our ability, even in principle, to make precise predictions about how Nature behaves.

Having created this predicament, Coleman also, in a subsequent paper [10], suggested a means of escaping it*. He argued that the probability distribution of possible universes is in fact a very sharply peaked function, so that it is overwhelmingly likely that a randomly selected universe will have values of the constants of Nature that are precisely determined. If correct, this claim restores in principle our hopes of predicting how Nature behaves. In practice, predictive power is reclaimed only if we can *compute* the values of the fundamental constant at which the sharp peak occurs. Coleman has computed one such constant; he predicts that the cosmological constant Λ is exactly zero. This prediction is in accord with observation.

In this paper, I will examine whether Coleman's reasoning can be applied to determine other constants of Nature, aside from the cosmological constant**. I will also address some challenges to the consistency of Coleman's arguments that one might raise. For example, Coleman assumed that wormholes have a characteristic "thickness" that is not much greater than the Planck length. One may wonder whether wormholes of arbitrarily large thickness can in fact contribute significantly to low-energy processes. Large wormholes, if their effects are unsuppressed, would be hard to reconcile with the well-tested successes of local field theory in describing low-energy physics [20, 16, 18]. One may also worry that assumptions similar to those that lead to the successful prediction $\Lambda = 0$ will lead to unsuccessful predictions concerning the values of other fundamental constants [15, 16]. Any such wrong prediction would cast doubt upon the foundation that underlies Coleman's solution to the cosmological constant problem.

The main conclusions of this paper are as follows: First of all, I argue that all of the constants of Nature do indeed have precisely determined values. That is, for each constant there is a "standard" value that is in principle calculable, and in a

* Related ideas appeared in ref. [11].

** This issue and other related issues have also been addressed in refs. [12–19].

randomly selected universe the probability is one that each constant assumes its standard value. Thus, the quantum indeterminacy that afflicts the fundamental constants turns out to be very mild. The standard value of the cosmological constant Λ can be calculated; it is zero, as Coleman claimed. This result holds quite generally, and is not sensitive to the detailed properties of short-distance physics. But I argue that the standard values of all other fundamental constants are quite sensitive to the details of physics at the Planck scale; hence, no constants other than the cosmological constant can be explicitly calculated. This conclusion is rather disappointing, if not entirely surprising. To put it in a positive light, one might be encouraged that, because of our inability to extract other precise predictions, Coleman's beautiful explanation for $\Lambda = 0$ has not been found to be inconsistent with anything that we know about Nature. Indeed, I will also propose a mechanism that can account naturally for the suppression of large wormholes.

Most of the analysis reported in this paper is premised on the claim that all of the constants of Nature other than the cosmological constant can be determined by requiring that Newton's gravitational constant G must assume the smallest possible value*. The potential quantum indeterminacy of fundamental physics is evaded, because a randomly selected universe is overwhelmingly likely to have G at its minimum. This criterion is sufficient to determine all of the fundamental constants because all contribute to G through renormalization effects. The analysis consists of two main parts. In the first part, I study the effects of wormhole interactions, which were neglected in Coleman's original analysis. These interactions are essential in ensuring that G actually has a nonzero minimum. They are also the key to understanding why the effects of large wormholes are suppressed. In the second part, the renormalization group method is applied to study the dependence of G on the other fundamental constants. It is because the renormalization of G is dominated by "nonuniversal" short-distance effects that we are unable in the end to make precise predictions about the values of the other constants.

The remainder of this paper is organized as follows: In sect. 2, the connection is derived between wormholes in spacetime and the potential quantum indeterminacy of the constants of Nature. As Coleman showed, one can "integrate out" the wormhole fluctuations and obtain an effective field theory with an explicit short-distance cutoff in which wormhole fluctuations no longer occur; the cutoff is the characteristic wormhole "thickness." The effects of the wormholes can then be incorporated into renormalization of the parameters of the effective theory. But there is another, more surprising, remnant of the wormhole fluctuations in low-energy physics. The effective field theory is actually a superposition of many "superselection" sectors that do not communicate with each other through any local physics.

* This criterion was previously suggested by Grinstein and Wise [15]. A different criterion was envisioned in refs. [16, 18]; this accounts for some of the differing conclusions of refs. [16, 18] and the present paper.

These sectors are labeled by a set of parameters, denoted by α, that are presumably infinite in number. In each sector, the "bare" couplings that specify the effective field theory at the wormhole scale are functions of α; this is the origin of the quantum indeterminacy of the fundamental constants.

The derivation presented here of this α-dependence of the effective theory emphasizes the generality of the result. In particular, the argument does not rely at all on the semiclassical approximation invoked by Coleman. We will also see that the interpretation of α as the label of a coherent superposition of "baby universe" states, which played a central role in Coleman's discussion, is really peripheral to the argument, and may not even make sense beyond the semiclassical approximation.

In sect. 3, Coleman's explanation for the vanishing of the cosmological constant is reviewed. The superposition of α-dependent effective theories is shown to be described by a probability distribution that is very sharply peaked at $\Lambda = 0$. This argument requires some additional assumptions beyond those needed in sect. 2. In particular, the argument seems to rely quite heavily on the validity of the euclidean path integral approach to quantum gravity*.

In sect. 4, I argue that the probability distribution of the "α-universes" strongly favors not only that Λ vanish, but also that Newton's constant G assume the smallest possible (α-dependent) value**. Again, this argument seems to make essential use of the euclidean path integral formalism. I claim that the requirement that $G(\alpha)$ is at its minimum suffices to determine all of the α's, and hence to determine all of the constants of Nature. The rest of the paper further explores the implications of this claim.

If we hope to actually determine the α's by minimizing G, we must compute the function $G(\alpha)$. We can imagine carrying out such a calculation in two stages. In the first stage, we integrate out the wormhole fluctuations to obtain an effective theory with α-dependent bare couplings; this effective theory is cut off at the wormhole mass scale M_w. In the second stage, we allow the cutoff to "float" down from M_w into the far infrared; we thus obtain an expression for the renormalized G that describes gravity at long distances in terms of the bare parameters of the effective theory. The first stage of this calculation is the subject of sect. 5, and the second stage is the subject of sect. 6.

If the mechanism proposed by Coleman for ensuring that $\Lambda = 0$ is to apply to Nature, and if this mechanism requires $G(\alpha)$ to be at its minimum, then it is evidently necessary for $G(\alpha)$ to have a *nonzero* minimum. Otherwise G will want to vanish, in conflict with observation. But in the approximation assumed by Coleman, in which wormholes are treated as dilute and the interactions among wormholes are

* The observation that $\Lambda = 0$ is highly "probable" in euclidean quantum gravity had been made earlier by Hawking [21] and by Baum [22]. They, however, had not appreciated the crucial role of wormholes in making Λ an *adjustable* quantity.

** More precisely, $G(\alpha)$ is minimized as a function of α on that surface in α-space where $\Lambda(\alpha) = 0$.

neglected, G is *not* bounded away from zero. Thus, it is necessary to improve on this dilute approximation. The systematic corrections to the dilute approximation are described in sect. 5. I indicate how wormhole interactions can generate a lower bound on G, and illustrate this possibility with a simple toy model.

The study of wormhole interactions will also lead me to propose a mechanism that can account for the suppression of the low-energy effects of large wormholes. This mechanism relies on the observation that the α's are determined by the requirement that $G(\alpha)$ is at its minimum, and it arises due to the interactions of large wormholes with small wormholes. The mechanism has the unusual feature that the suppression of large wormholes cannot be understood in terms of long-distance physics alone. In this sense, it may be regarded as a violation of the principle that short-distance physics is effectively "decoupled" from long-distance physics.

In sect. 6, I describe how an effective theory of gravity is renormalized as the ultraviolet cutoff floats down from the wormhole scale toward the far infrared. I note that some nonperturbative properties of this renormalization can be extracted if I make the rather mild assumption that Einstein gravity has no nontrivial continuum limit. From these properties, I then infer the conditions that the α-dependent bare theory at the wormhole scale must satisfy, in order that Coleman's derivation of $\Lambda = 0$ be applicable. These conditions are found to be quite weak.

I then examine whether calculations of fundamental constants other than Λ can be carried out in practice. I argue that the values of other constants cannot be precisely predicted without a detailed knowledge of physics at the wormhole scale (which is comparable to the Planck scale). This argument is based on straightforward power-counting of Feynman diagrams, which shows that the renormalization of G is dominated by short-distance quantum fluctuations that are sensitive to the details of Planck-scale physics.

The analysis of renormalization effects leads to one rather surprising conclusion. I find that if an elementary scalar or fermion has an α-dependent bare mass that can be adjusted at the wormhole scale, then the physical mass of this particle cannot be far below the wormhole scale, *unless* the physical mass vanishes *exactly*. Among other consequences, this observation implies that if Coleman's mechanism is to be compatible with the known features of particle physics, then we must insist that the light mass scales of particle physics (like the weak interaction scale) are determined *dynamically*. The weak interaction scale must *not* be determined by adjusting a bare mass parameter at the wormhole scale.

2. The effects of wormholes

A wormhole in spacetime is a gravitational quantum fluctuation that links two distantly separated spacetime points. Hence, it is natural to wonder whether wormholes can induce an apparent violation of locality. But Coleman [8] has argued persuasively that the physical effects of wormholes are quite different than we might naively expect. He concludes that, because of wormhole effects, the fundamental constants of Nature are afflicted with an intrinsic quantum indeterminacy. Quantum gravity, then, may threaten our ability, even in principle, to make precise predictions about how Nature behaves.

In this section, I will briefly review how this remarkable conclusion is reached, following closely a reformulation of Coleman's original argument due to Klebanov, Susskind, and Banks [16]. I include this section to establish notation, and also to make two important points. First, although Coleman originally presented his argument in the context of a semiclassical approximation, the validity of this approximation is not at all a necessary ingredient in the argument. This point is important because we will see when we apply Coleman's ideas to the cosmological constant problem that the semiclassical approximation used by Coleman cannot be justified. Second, while the notion of a "baby universe" associated with a wormhole played a central role in Coleman's discussion, this notion is actually peripheral to the argument, and may not make sense beyond the semiclassical approximation.

The basic assumptions underlying Coleman's analysis are that wormholes exist, and that the wormholes have two essential properties: (i) Wormholes have a characteristic "thickness" R_{w}. This thickness is presumably of order the Planck length M_{P}^{-1}, the characteristic length scale of gravitational quantum fluctuations. We assume that wormholes much thicker than R_{w} are rare and can be neglected*. (ii) Wormholes have no characteristic length. We assume that the wormhole is completely indifferent to the separation between its two ends in the background spacetime; it takes a short cut.

It follows from assumption (i) that, for the purpose of discussing physics at energies well below R_{w}^{-1}, wormholes may be integrated out; their effects may be incorporated into an effective field theory in which wormhole fluctuations no longer occur. In this low-energy effective theory, the contribution to a quantum mechanical expectation value due to a wormhole that connects the spacetime points x_1 and x_2 can be represented by an expansion in local operators at x_1 and x_2, of the form

$$\sum_{ab} \Delta_{ab}\mathcal{O}_a(x_1)\mathcal{O}_b(x_2). \tag{2.1}$$

Here, the $\mathcal{O}_a$'s are a suitable basis for the gauge-invariant local operators, and Δ is a real symmetric matrix. By the assumption (ii) above, Δ is independent of x_1 and x_2. The bilocal expansion eq. (2.1) is understood to be valid when the separation between x_1 and x_2 is large compared to R_{w}.

* We will examine more closely in sect. 5 the justification of this assumption.

After an integration over all possible positions of the wormhole ends on the background spacetime, the contribution eq. (2.1) becomes*

$$C \equiv \tfrac{1}{2}\int d^4x_1\sqrt{g_1}\int d^4x_2\sqrt{g_2}\sum_{ab}\Delta_{ab}\mathcal{O}_a(x_1)\mathcal{O}_b(x_2). \tag{2.2}$$

Now, in treating the effect of a configuration of N wormholes, let us imagine that the wormholes are sufficiently "dilute" that they may be considered to be independent of one another. The N-wormhole contribution then factorizes and can be expressed as $C^N/N!$, where the $1/N!$ compensates for overcounting of identical wormholes. (In sect. 5 we will see how this dilute approximation can be systematically improved.)

The contribution due to an arbitrary wormhole configuration in the dilute limit can now be obtained by summing over N. The sum exponentiates, and can be expressed as a gaussian integral,

$$e^C = \int(d\alpha)\exp\Big(-\tfrac{1}{2}\sum_{a,b}\alpha_a(\Delta^{-1})_{ab}\,\alpha_b\Big)\exp\Big(\sum_a\alpha_a\int d^4x\sqrt{g}\,\mathcal{O}_a(x)\Big). \tag{2.3}$$

We see that, in the dilute approximation, the process of integrating out wormholes induces an α-dependent shift in the local (euclidean) effective action,

$$\delta S(\alpha) = -\sum_a\alpha_a\int d^4x\sqrt{g}\,\mathcal{O}_a(x). \tag{2.4}$$

Quantum mechanical expectation values in the effective theory are expressed as integrals over α of expectation values computed in universes with α-dependent couplings**. The various α-universes are weighted by a probability distribution

$$P(\alpha) = \exp\Big(-\tfrac{1}{2}\sum_{a,b}\alpha_a(\Delta^{-1})_{ab}\,\alpha_b\Big). \tag{2.5}$$

However, our physical observations are actually carried out in a universe with a fixed value of the α_a's, for we in principle can measure the constants of Nature to arbitrary accuracy.

The loss of quantum coherence declaimed by Hawking [7] may be interpreted as a loss of information about the other α-universes that occur in the superposition eq. (2.3). Because physics is done with all α_a's fixed, this loss of coherence is not

* We neglect a small error due to the breakdown of the bilocal expansion for $x_1 \sim x_2$.

** The number of independent α's is the same as the number of nonzero eigenvalues of the matrix Δ in eq. (2.1). "Generically", this is the same as the number of gauge-invariant local operators; that is, it is infinite.

accessible to any observer. Instead of a failure of quantum mechanics or of locality, wormhole effects introduce into physics a fundamental quantum indeterminacy of the constants of Nature. Wormholes threaten to inhibit our ability to predict with certainty the constants of Nature on the basis of fundamental theory.

That the α-distribution $P(\alpha)$ is gaussian, and that the shift of the effective action $\delta S(\alpha)$ is linear in α, are artifacts of the dilute approximation that we have made. Corrections to this approximation will be discussed in sect. 5. But the conclusion that the effects of wormholes can be incorporated into an α-dependent renormalization of the effective action seems to be much more general. I believe that this conclusion follows from just the two assumptions stated above (subject to some caveats that will be mentioned later).

The above reformulation of Coleman's argument was previously presented in ref. [16]. It differs from Coleman's original presentation [8] in two significant respects. First, Coleman's argument was carried out within the context of the semiclassical approximation, in which the only wormhole configurations that contribute to the euclidean path integral are solutions to the euclidean equations of motion (stationary points of the euclidean action) and small fluctuations about such solutions. The above argument is more satisfying because it makes no reference to the semiclassical approximation. This point is important, because we will see in sect. 3 that, if we are to invoke Coleman's arguments to solve the cosmological constant problem, then the semiclassical approximation cannot be justified. (We *have*, so far, regarded the wormholes as dilute, but our conclusions survive even when the dilute approximation is abandoned.)

The second respect in which our formulation differs from Coleman's original argument is that Coleman attaches to α an interpretation that was eschewed above. Coleman includes in his analysis, in addition to wormholes attached to the background spacetime, "semiwormholes" that connect the background spacetime to a tiny closed three-geometry, or "baby universe." He is thus able to interpret α as the label of a coherent superposition of baby universe states. As we have seen, this interpretation is not essential in the derivation of the main result, that wormhole effects can be absorbed into an α-dependent effective action. This is a good thing, because it seems that the notion of a baby universe, or at least the connection between baby universes and the α's associated with wormhole physics, makes sense only within the semiclassical approximation.

Although this is somewhat off the main line of the argument in this paper, I will describe in the remainder of this section how the connection between α's and baby universes can be established. This discussion will clarify why the semiclassical approximation appears to be essential.

Let the index i label the states of a basis for the baby universe states. We may define S_{ij} as the amplitude for the baby universe state i to evolve to the baby universe state j determined, for example, by a euclidean path integral (fig. 2). We may also consider the amplitude for a semiwormhole to connect baby universe i to

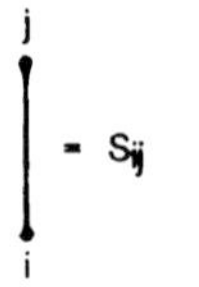

Fig. 2. The amplitude for baby universe state i to evolve to baby universe state j.

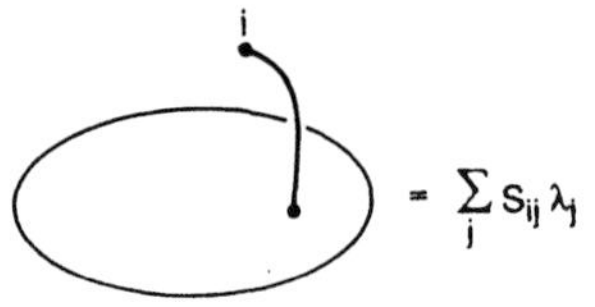

Fig. 3. The amplitude for baby universe state i to connect via a semiwormhole to the background spacetime.

a background spacetime (fig. 3); quantities λ_j can be defined (if the matrix S is invertible) so that this amplitude is denoted

$$\sum_j S_{ij}\lambda_j .$$

In this expression, we may think of S as a baby universe "propagator," and λ_j as the strength of the "coupling" of baby universe j to the background spacetime.

So far this is just definitions. But now we make the nontrivial assumption that the amplitude for a wormhole to attach to the background spacetime is *determined* by the baby universe propagator and the semiwormhole amplitude to be

$$\sum_{ij} \lambda_i S_{ij}\lambda_j .$$

Heuristically, then, the wormhole amplitude (fig. 4) obeys a "cutting rule" or "completeness relation," denoted schematically in fig. 5; it can be reproduced by a sum over baby universe "intermediate states."

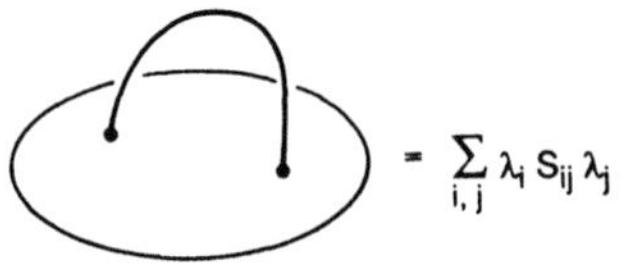

Fig. 4. The amplitude for a wormhole to attach to the background spacetime.

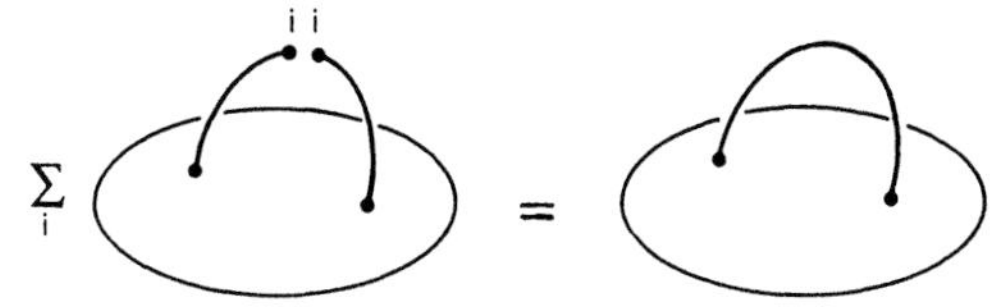

Fig. 5. The baby universe completeness relation.

In the semiclassical approximation, such a completeness relation *is* satisfied, because there is a natural way to associate a baby universe with a wormhole. The baby universe is the three-geometry obtained when we slice the classical wormhole solution through its midpoint [2, 8, 9]. But there is no unambiguous way of associating a general wormhole configuration (that is not a classical solution) with a baby universe, and I see no reason to expect that there exists a basis for the baby universe states such that the identity indicated in fig. 5 is satisfied, except in the semiclassical approximation.

This completeness relation provides the logical connection that is needed to relate wormhole physics to baby universes. To establish this relationship, assume the validity of the completeness relation and sum over wormhole and semiwormhole configurations. (A typical configuration is shown in fig. 6.) In the dilute approximation the sum exponentiates, and we have

$$Z[J]=\exp\left(\sum_{i,j}\left(\tfrac{1}{2}J_iS_{ij}J_j+J_iS_{ij}\lambda_j+\tfrac{1}{2}\lambda_iS_{ij}\lambda_j\right)\right), \tag{2.6}$$

where J_i is a source that couples to the baby universe i. This generating function can be expressed as a gaussian integral

$$Z[J]=\int(\mathrm{d}\alpha)\exp\left(-\tfrac{1}{2}\sum_{i,j}\alpha_i(S^{-1})_{ij}\,\alpha_j+\sum_i\alpha_i(J_i+\lambda_i)\right). \tag{2.7}$$

Now it is convenient to redefine our basis for the baby universe states so that S_{ij} becomes δ_{ij}. (This is a conventional "renormalization" of the baby universe wave

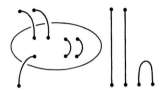

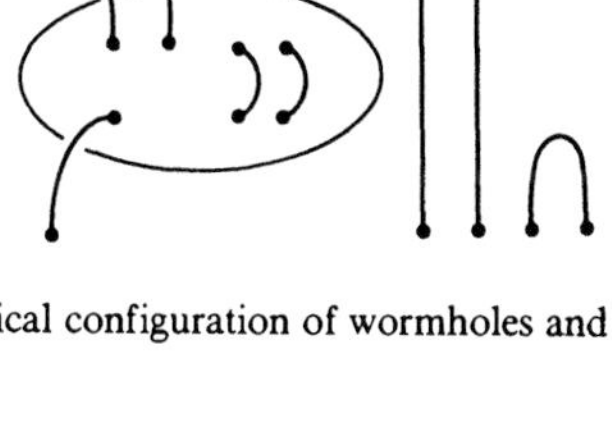

Fig. 6. A typical configuration of wormholes and semiwormholes.

functions.) In this new basis, the sources and couplings become

$$J_i' = (S^{1/2})_{ij} J_j,$$

$$\lambda_i' = (S^{1/2})_{ij} \lambda_j, \tag{2.8}$$

and after a change of variable in the integral, our expression for the generating function becomes

$$Z[J'] = \prod_i \int (\mathrm{d}\alpha') \exp\left[-\tfrac{1}{2}\alpha_i'\alpha_i' + \alpha_i'(J_i' + \lambda_i')\right] \tag{2.9}$$

up to a normalization factor, independent of J and λ, that can be absorbed into the integration measure.

Wick's theorem tells us that

$$\int (\mathrm{d}\alpha)\,\mathrm{e}^{-(1/2)\alpha^2}\alpha^n = \left\langle 0 \middle| (a + a^\dagger)^n \middle| 0 \right\rangle, \tag{2.10}$$

where $a, a^\dagger$ are conventionally normalized annihilation and creation operators satisfying $[a, a^\dagger] = 1$. We therefore have

$$Z[J'] = \prod_i \left\langle 0 \middle| \exp\left[(\lambda_i' + J_i')(a_i + a_i^\dagger)\right] \middle| 0 \right\rangle$$

$$= \prod_i \left[\mathrm{e}^{(1/2)J_{\mathrm{in},i}^2}\,\mathrm{e}^{(1/2)J_{\mathrm{out},i}^2} \left\langle 0 \middle| \mathrm{e}^{J_{\mathrm{out},i} a_i}\,\mathrm{e}^{\lambda_i'(a_i + a_i^\dagger)}\,\mathrm{e}^{J_{\mathrm{in},i} a_i^\dagger} \middle| 0 \right\rangle \right]. \tag{2.11}$$

In the second equality in eq. (2.11) we have written $J_i' = J_{\mathrm{in},i} + J_{\mathrm{out},i}$, in order to distinguish "incoming" and "outgoing" baby universe states, and have invoked the identity $\mathrm{e}^{A+B} = \mathrm{e}^A \mathrm{e}^B \mathrm{e}^{-(1/2)[A,B]}$ that is satisfied by operators A and B that commute with $[A, B]$. The prefactor $\mathrm{e}^{(1/2)J_{\mathrm{in},i}^2}\mathrm{e}^{(1/2)J_{\mathrm{out},i}^2}$ in this identity is associated with incoming (outgoing) states that propagate to other incoming (outgoing) states (fig. 7). We did not intend to include such processes. (It is inappropriate to include them in the semiclassical approximation.) So we suppress the prefactor and obtain

$$\langle m_1, m_2, \ldots, \mathrm{out} | n_1, n_2, \ldots, \mathrm{in} \rangle = \prod_i \left\langle m_i \middle| \mathrm{e}^{\lambda_i'(a_i + a_i^\dagger)} \middle| n_i \right\rangle. \tag{2.12}$$

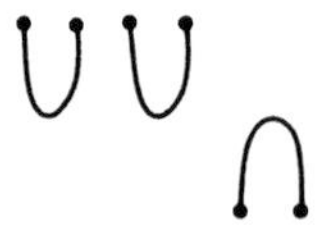

Fig. 7. Processes that are excluded from the wormhole sum.

This result was derived by Coleman [8] and by Strominger and Giddings [9]. It shows that α_i' can be interpreted as the operator $a_i + a_i^\dagger$ acting on baby universe states of type i, and that physics at fixed α_i' is physics in a "condensate" of baby universes, an eigenstate of the "position" operator of the baby universe "oscillator." But to justify it, we needed to assume not only the dilute approximation, but also the baby universe completeness relation (fig. 5).

Thus, the identification of α as the label of a baby universe state may make sense only within the semiclassical approximation. That is why I emphasize that Coleman's main conclusion, that wormhole effects induce an α-dependent renormalization of the local effective action, does not rely on this interpretation of the α's.

3. The cosmological constant

The discussion of the previous section indicates that, because of wormhole effects, the constants of Nature are afflicted with an inescapable indeterminacy. Our universe has been chosen at random from an ensemble of possible universes, all with different values of the α's, and hence of the fundamental constants. Wormholes in spacetime threaten to render unachievable the greatest ambition of physics – to predict the constants of Nature as consequences of a complete fundamental theory.

Having created this predicament, Coleman [10] also suggested a means of escaping it. He argued that the distribution of possible universes is a very sharply peaked function of the α's. Thus, it is overwhelmingly likely that a randomly selected universe will have values of the constants of Nature that are precisely known.

This claim, if correct, restores in principle our hopes of predicting how Nature behaves. In practice, predictive power is reclaimed only if we can *compute* the values of the fundamental constants at which the sharp peak occurs. Coleman has computed one such constant; he predicts that the cosmological constant is exactly zero. This prediction is confirmed by observation.

Our main objective in this paper is to examine how Coleman's reasoning can be applied to determine the other constants of Nature, aside from the cosmological constant. But as a prelude to that main topic, I will review in this section how the cosmological constant is calculated.

This calculation relies more heavily than the results of the previous section of the specific features of the euclidean quantum gravity formalism. To begin, recall that in this formalism [4–6, 10], the expectation value of a gauge-invariant local observable $\mathcal{O}$ is expressed as

$$\langle \mathcal{O} \rangle = N \int (\mathrm{d}g)\, \mathrm{e}^{-S[g]} \mathcal{O}; \qquad (3.1)$$

here $S[g]$ is the euclidean action, N is a normalization factor, and the path integral

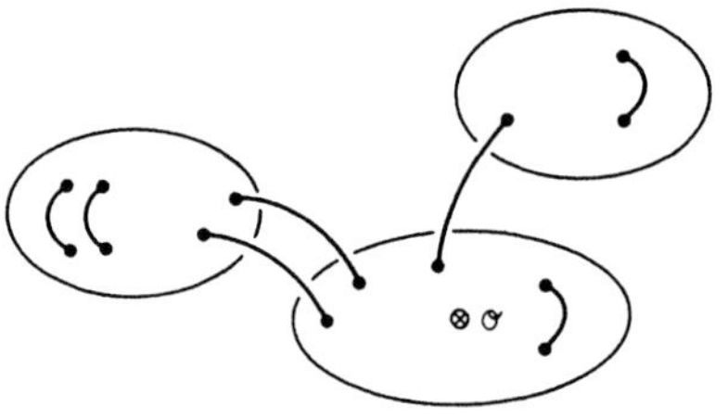

Fig. 8. A connected geometry that becomes disconnected after wormholes are integrated out. The operator $\mathcal{O}$ whose expectation value is to be evaluated acts on one of the connected components.

is over four-dimensional geometries that are closed and connected*. (The integration over fields other than the metric has not been explicitly indicated, but is understood.) We assume as in sect. 2 that this path integral includes a sum over configurations in which narrow wormholes link distantly separated points on a smooth background spacetime, and we imagine "integrating" out the wormholes and incorporating their effects into a local effective action for the background spacetime. But now we note that the connected closed manifolds include geometries in which many distinct large smooth manifolds are linked together by narrow wormholes. Thus, when we integrate out wormholes, a typical connected closed manifold may yield many disconnected components (fig. 8). Integrating out wormholes, then, affects the euclidean path integral in two ways. As we already saw in sect. 2, expectation values become expressed in the effective theory as integrals over the α parameters of expectation values computed in various universes with α-dependent couplings. But furthermore, the sum over connected closed manifolds in the underlying theory becomes a sum over all closed manifolds, including disconnected ones, in the effective theory.

Each disconnected manifold can be factored into a connected component on which the operator $\mathcal{O}$ acts, and a remainder on which no local operator acts. Since the sum over all closed four-dimensional geometries can be written as the exponential of a sum over all *connected* closed four-geometries, the expression for the expectation value of $\mathcal{O}$ may be written in the effective theory as

$$\langle\mathcal{O}\rangle = N\int(\mathrm{d}\alpha)P(\alpha)\exp\left(\int(\mathrm{d}g')\,\mathrm{e}^{-S[g',\alpha]}\right)\int(\mathrm{d}g)\,\mathrm{e}^{-S[g,\alpha]}\mathcal{O}, \tag{3.2}$$

where again the integrals over g and g' denote sums over *connected* closed four-geometries, but with wormholes now excluded. The distribution $P(\alpha)$ is that considered in sect. 2, which would be gaussian in the dilute approximation. But we

* In this discussion, the Hartle–Hawking boundary condition [6] has been implicitly adopted. The results are not sensitively dependent on this choice.

now see that the measure for the integration over α contains the additional factor

$$Q(\alpha) \equiv \exp\left(\int (\mathrm{d}g)\, \mathrm{e}^{-S[g,\alpha]} \right). \tag{3.3}$$

It is Coleman's proposal that this factor is a sharply peaked function of the α's; since $P(\alpha)$ is presumably a smooth function, the factor $Q(\alpha)$ determines where the support of the α-distribution lies.

To compute the factor $Q(\alpha)$ in eq. (3.3), Coleman makes a further hypothesis – that the sum over connected closed four-geometries is dominated by large smooth geometries with small curvature (in Planck units). The idea is that by integrating out wormholes we have satiated the desire of a typical four-geometry to crinkle up on the Planck scale. This assumption is in the same spirit as, though logically independent of, the assumption that it makes sense to integrate out wormholes at all. Surely, given the current status of our understanding of quantum gravity, both assumptions must be regarded as unproved working hypotheses.

Anyway, given that the integral $\int (\mathrm{d}g) \mathrm{e}^{-S[g,\alpha]}$ is dominated by large smooth geometries, we may compute it by the following procedure: Recall that $S[g,\alpha]$ is the action of an effective theory with a cutoff of order $M_{\mathrm{w}} \sim R_{\mathrm{w}}^{-1}$, the characteristic mass scale of the wormhole fluctuations. Imagine allowing this cutoff to float down to a mass scale $M \ll M_{\mathrm{w}}$; we integrate out all quantum fluctuations with wave number $k > M$, and incorporate the effects of these fluctuations into a renormalized effective action $S_M[g,\alpha]$. This renormalized effective action can be expanded in operators that are local on the distance scale M^{-1} [23].

For $M \ll M_{\mathrm{w}}$, the loop corrections in the effective theory with cutoff M, and the effects of higher dimension operators in S_M, are suppressed by powers of M/M_{w}. Because loop corrections are small, the path integral of the effective theory can be evaluated to good accuracy semiclassically, with the result

$$\int (\mathrm{d}g)_M \mathrm{e}^{-S_M[g,\alpha]} = \mathrm{e}^{-S_M[\bar{g},\alpha]}, \tag{3.4}$$

where $\bar{g}$ is the four-geometry at which S_M is stationary. If this stationary point is a large smooth geometry with volume of order R^4, we improve on the semiclassical approximation by allowing the cutoff to float down to $M \sim R^{-1}$. Then all fluctuations at wavelengths less than R have been absorbed into the renormalization of the effective action, and fluctuations on larger wavelengths are absent because the volume acts as an infrared cutoff*.

* I have described the procedure in terms of a floating cutoff in order to emphasize that the action may always be regarded as effectively local and also that infrared divergent loop integrals need not be encountered; they arise only as $R \to \infty$ or $M \to 0$.

Of course, it was a bit disingenuous to say above that $\bar{g}$ is the stationary point of S_M; we really want $\bar{g}$ to *minimize* S_M. It is well known that the stationary points of the euclidean action of gravity are not even local minima; we can always perform a conformal transformation on the metric that lowers the action. In order that the above discussion apply I must assume that there is *some* correct procedure for dealing with the conformal fluctuations, and that when this procedure is invoked, conformal fluctuations can be integrated out and absorbed into a renormalized effective action.

The effective action S_M can be expanded in terms of local operators, and, for $M \ll M_w$, operators with n derivatives acting on the metric have effects that are suppressed by the factor $(M/M_w)^n$. This expansion has the form*

$$S_M[g] = \frac{1}{16\pi G}\int d^4x\sqrt{g}\,(2\Lambda - R + \dots), \tag{3.5}$$

where terms with more than two derivatives have been neglected; the parameters G and Λ have implicit dependence on M_w, M, and α. At the stationary points of S_M, the metric obeys the euclidean Einstein equation

$$R_{\mu\nu} = \Lambda g_{\mu\nu}, \tag{3.6}$$

and the action is

$$S_M[\bar{g}] = -\frac{\Lambda}{8\pi G}\int d^4x\sqrt{\bar{g}} = -\frac{\Lambda}{8\pi G}V \tag{3.7}$$

where V is the volume of the euclidean spacetime. For $\Lambda > 0$, the solution of eq. (3.6) with maximal volume is known to be a four-sphere with

$$V = 24\pi^2/\Lambda^2. \tag{3.8}$$

Hence the factor eq. (3.3) is found to be

$$Q(\alpha) = \exp(e^{-S_M[\bar{g},\alpha]}) = \exp[\exp(3\pi/G\Lambda)], \tag{3.9}$$

for $\Lambda > 0$. Here Λ is the cosmological constant renormalized at the scale $M \sim V^{-1/4}$, with V given by eq. (3.8).

Eq. (3.9) is the sharp peak at $\Lambda = 0$ found by Coleman. Λ is a function of α, and eq. (3.9) together with eq. (3.2) tells us that it is overwhelmingly likely that a universe selected at random will have a value of α such that $\Lambda = 0$, assuming that Λ vanishes for *some* choice of α. We note that the quantity that has been found to

* Massive fields (heavier than M) have been integrated out, and light fields have been set equal to values that minimize S_M, so S_M is expressed as a functional of the metric only.

vanish is the *renormalized* Λ that one would actually measure in a large smooth universe; it includes the contributions from the fluctuations of all quantum fields. We also note that it was safe to neglect the higher dimension operators in eq. (3.5); these give contributions to $S_M[\tilde{g}]$ that are independent of Λ for Λ small, and do not modify the $\Lambda = 0$ singularity in eq. (3.9).

(The procedure that we have used to calculate $Q(\alpha)$ may make sense for $\Lambda > 0$, but it is ill-defined for $\Lambda < 0$. For $\Lambda < 0$, the best approach may be to analytically continue eq. (3.9) from positive values of Λ to negative values. Our implicit assumption that G is non-negative will be defended in sect. 6.)

It may be worthwhile to comment on the use of "semiclassical" reasoning in this derivation. We treated gravity semiclassically in the sense that we argued that loop effects are small when the floating ultraviolet cutoff M is much less than the cutoff M_w of the bare theory (and in fact comparable to the infrared cutoff $V^{-1/4}$). But we certainly did not require that the renormalization of S as the cutoff floats from M_w to M be susceptible to a perturbative treatment. Indeed, it may be a misstatement to say that the path integral is dominated by large smooth geometries, for we do not mean to exclude the possibility that the effective action is strongly renormalized by quantum fluctuations at scales just below M_w. Nor, of course, have we required that the process of integrating out wormholes, which generated the theory with cutoff M_w, be accurately described by the semiclassical approximation.

The observation that we made in sect. 2 – that the notion of a baby universe makes sense only within the context of an untrustworthy approximation – has a consequence that should be mentioned now. We have noted that a "superselection rule" applies to wormhole physics; universes with different values of α cannot communicate with each other through any local physical process. But Coleman's explanation for $\Lambda = 0$ can be invoked only if we consider a superposition of these α-universes as in eq. (3.2). One might object to such a superposition, citing the analogy with the θ-vacua of QCD. It would be perverse to say that the ground state of QCD is a superposition of θ-vacua. Why shouldn't we take the same attitude toward a superposition of α-universes?

The key distinction between θ-vacua and α-universes, I suspect, is that the α superselection rule is an approximate rule that holds only in the low-energy limit. It would have been easier to imagine that the α superselection rule were exact if the interpretation of α as the label of a baby universe were also exact. If this interpretation fails, it is hard to see how α-sectors can be identified in sub-Planck scale physics. This, perhaps, is the proper way to interpret Coleman's claim [10] that an a priori distribution in α with support at a single point in α-space would be "unnatural."

The argument in this section evidently required stronger assumptions than the discussion in sect. 2. Here we apparently needed to take more seriously than in sect. 2 the proposal that expectation values can be evaluated as euclidean path integrals (and that a sensible prescription exists for dealing with conformal fluctuations).

And we further assumed that, after wormholes are integrated out, the path integral is dominated by large smooth geometries. That these assumptions lead to the exciting conclusion that the cosmological constant vanishes already provides sufficient motivation for exploring their further consequences. But it is obviously an urgent matter to understand better the justification for these assumptions, if indeed any can be found.

4. Newton's constant

We have emphasized that the factor $Q(\alpha)$ in the measure for the integration over α-universes overwhelmingly favors $\Lambda = 0$, if the renormalized cosmological constant can vanish for some value of α. This measure factor has a further consequence; Newton's constant G wants to assume the smallest possible value, preferably zero*. In the remainder of this paper, we will explore the consequences of this tendency of G to seek its minimum allowed value.

One consequence is immediate: the renormalized G as a function of α must be bounded away from zero

$$G(\alpha) \geqslant G_{\min} > 0 . \tag{4.1}$$

Otherwise, the same assumptions that lead to the successful prediction $\Lambda = 0$ will also lead to the unsuccessful prediction $G = 0$. (A gravitational interaction is observed in our universe.)

One might object, at first, to the assertion that the probability distribution

$$Q(\alpha) \sim \exp[\exp(3\pi/G\Lambda)] \tag{4.2}$$

is capable of determining both G and Λ. This claim sounds strange, because $Q(\alpha)$ is a function of only the product $G\Lambda$. Thus, one might argue, the very sharp peak in $Q(\alpha)$ occurs at $G\Lambda = 0$, and $Q(\alpha)$ is completely indifferent to the value of G once Λ has assumed the value $\Lambda = 0$.

Indeed, to extract from eq. (4.2) a probability distribution on the surface $G\Lambda = 0$ (where $Q(\alpha)$ is singular), we must specify a preferred way of approaching this surface. I will argue below that the correct procedure is to evaluate $Q(\alpha)$ on surfaces of constant Λ, and then take the limit $\Lambda \to 0$. If Λ assumes a very small positive value, $Q(\alpha)$ strongly favors that G^{-1} increase. This tendency of G^{-1} to increase is not suddenly lost when Λ is exactly zero.

We claim, then, that the support of the distribution $Q(\alpha)$ will lie at that point (or those points) in α-space where $G^{-1}(\alpha)$ assumes its maximal value on the surface defined by $\Lambda(\alpha) = 0$. There is another objection that one could raise against this claim. One might question whether the requirement that G^{-1} assumes its maximum

* This point was stressed previously by Grinstein and Wise [15].

has any physical content. Since G^{-1} is a dimensional quantity, the argument might go, its value of course depends on the units in which we choose to express it. Nothing prevents us from choosing our unit of mass to be $G^{-1/2}$. Then G^{-1} is one, and it makes no sense to "maximize" it.

This argument is also misleading, for several reasons. First of all, the distribution $Q(\alpha)$ quantifies the relative probabilities of *different* "theories" with different values of the constants of Nature. When two theories are compared, we should express the physical constants of both in the *same* units; it is valid, for example, to say that G in theory A is smaller than G in theory B. In fact, all of the theories that we are comparing share the same α-independent ultraviolet cutoff, the wormhole scale M_w. Thus, when we compare different theories, we may express all mass scales (like $G^{-1/2}$) as dimensionless quantities, in units of M_w. Most important, the criterion that $G(\alpha)M_w^2$ is at its minimum has consequences that (in principle) relate to experimentally accessible low-energy physics. Indeed, we will argue that this criterion actually fixes all of the dimensionless quantities that specify the properties of low-energy physics.

To support the claim that $Q(\alpha)$ should be evaluated on the surface $G\Lambda = 0$ as a $\Lambda \to 0$ limit, we will regulate the very singular dependence of $Q(\alpha)$ on $\Lambda(\alpha)$ by introducing an infrared cutoff into the calculation of $Q(\alpha)$. The point is that Λ controls the volume of the four-geometry that dominates the path integral*.

Imagine that we repeat the calculation of $Q(\alpha)$ in sect. 3, but that the closed four-geometries that contribute to the path integral eq. (3.3) are now restricted to geometries with volume V less than a maximum volume V_{max}; V_{max}, then, is our infrared cutoff. We assume again that the path integral is dominated by the four-sphere geometry. Then we have

$$Q(\alpha) \sim \exp[\exp(-S_0)], \tag{4.3}$$

where S_0 is the minimal action of a four-sphere that satisfies the volume constraint $V \leqslant V_{max}$.

For a four-sphere of radius R, the action eq. (3.5) becomes**

$$S(R) = \frac{1}{16\pi G}\Omega(2\Lambda R^4 - 12R^2 + \dots), \tag{4.4}$$

where $\Omega = 8\pi^2/3$ is the volume of the unit four-sphere. The minimum*** of $S(R)$

* We choose to regulate $Q(\alpha)$ by means of an infrared cutoff because such a regulator can be formulated in the bare theory, without any reference to α. Since it is our desire to extract an unambiguous function of α on the surface $G\Lambda = 0$, it seems sensible to describe the approach to this surface in a manner that does not introduce implicit α-dependence.

** For sufficiently large R, Λ and G may be taken to be R-independent constants, because they "stop running" in the infrared limit. This will be explained at greater length in sect. 6.

*** As in sect. 3, this minimum exists only if $\Lambda > 0$. For $\Lambda < 0$, $Q(\alpha)$ may be defined by analytic continuation in Λ.

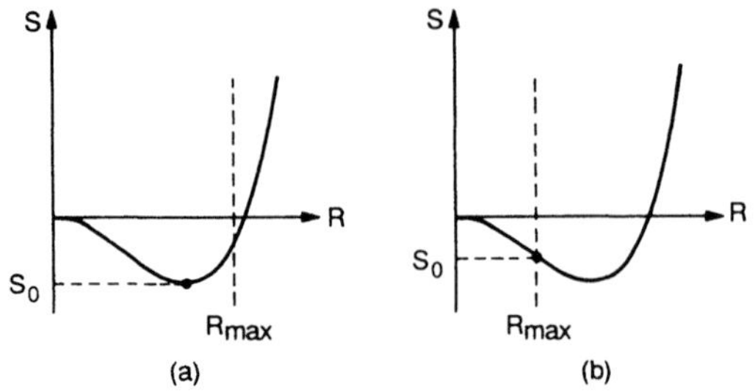

Fig. 9. The action S for a four-sphere of radius R, with $\Lambda > 0$. The minimum S_0 of the action is indicated (a) for $\Lambda \geqslant 3R_{\max}^{-2}$ and (b) for $\Lambda \leqslant 3R_{\max}^{-2}$.

occurs at $R^2 = 3\Lambda^{-1}$ for $\Lambda \geqslant 3R_{\max}^{-2}$ and at $R^2 = R_{\max}^2$ for $\Lambda \leqslant 3R_{\max}^{-2}$, where $V_{\max} = \Omega R_{\max}^4$ (see fig. 9). Therefore, the minimum of the action is

$$S_0 = -3\pi/G\Lambda, \qquad \Lambda \geqslant 3R_{\max}^{-2},$$

$$S_0 = -(3\pi/G\Lambda)\left[2\left(\tfrac{1}{3}\Lambda R_{\max}^2\right) - \left(\tfrac{1}{3}\Lambda R_{\max}^2\right)^2\right], \quad 0 < \Lambda \leqslant 3R_{\max}^{-2}. \tag{4.5}$$

The typical value of Λ, then, in the distribution $Q(\alpha)$ is

$$\Lambda \sim R_{\max}^{-2}, \tag{4.6}$$

which is independent of G. As the infrared cutoff is removed, $Q(\alpha)$ becomes very sharply peaked at $\Lambda = 0$.

Now suppose that both Λ and G are functions of the α's. We wish to minimize S_0 in eq. (4.5) as a function of the α's. To accomplish this, we may find the minimum of $G(\alpha)$ on each $\Lambda(\alpha) = \text{constant}$ surface, and then subsequently minimize with respect to $\Lambda(\alpha)$. Obviously if $G(\alpha) = 0$ for any value of α, then $S_0 = -\infty$ for that value of α, and $Q(\alpha)$ has a sharp peak at $G = 0$. A randomly selected universe, then, is overwhelmingly likely to have $G = 0$, in conflict with observation.

We must assume, therefore, that the bound eq. (4.1) applies. Then $Q(\alpha)$ favors that G^{-1} assume a value that is close to its maximum in the region of α-space where $\Lambda \sim R_{\max}^{-2}$; the typical deviation of G^{-1} from its maximum is of order $R_{\max}^{-2}$. As the infrared cutoff is removed, $Q(\alpha)$ becomes a very sharply peaked function of *both* Λ and G. The peak occurs precisely at the maximal value of $G^{-1}(\alpha)$ on the $\Lambda(\alpha) = 0$ surface.

Presumably there are many α's. The number of α_a's is the number of nonzero eigenvalues of the matrix Δ_{ab} in eq. (2.1), and this number is likely to be infinite.

The condition $\Lambda(\alpha)=0$ determines a surface of codimension one in the α-space. But we should expect that the condition that $G(\alpha)$ assumes its minimum on the $\Lambda(\alpha)=0$ surface is sufficient to determine all of the remaining α's. A function, even a function of many variables, generically attains its global minimum at an isolated point, assuming that a global minimum exists.

Therefore, we have in principle recovered the ability to calculate the constants of Nature. All of the arbitrariness that is potentially introduced by wormhole physics can be overcome. We need only compute all of the constants of Nature as functions of the α's; then we find the α that minimizes G on the $\Lambda=0$ surface. This determines the α's, and hence all of the other constants. In practice, this calculation will be challenging.

The α-dependence of G, or of any coupling constant, comes from two sources, and we can imagine carrying out the calculation of $G(\alpha)$ in two stages. First, we integrate out wormholes at the wormhole scale M_w, and thus obtain a "bare" theory with cutoff M_w and α-dependent "bare" parameters. Second, we allow the cutoff to float from M_w to $M \ll M_w$; the effects of quantum fluctuations with wave number k satisfying $M_w > k > M$ are incorporated into a renormalization of the effective action. The fluctuations that are integrated out in the second stage are not wormholes, but conventional "loop" fluctuations. The $G(\alpha)$ we want to calculate is extracted from the effective action in the infrared or $M \to 0$ limit.

It is clear that if this $G(\alpha)$ is to have a nonvanishing minimum, then so must the bare Newton's constant $G_0(\alpha)$, renormalized at the scale M_w. The point is that $G=0$ is stable under renormalization; Newton's constant is unrenormalized when gravity is free. (Renormalization in quantum gravity is discussed in more detail in sect. 6.) Therefore, if the bare Newton's constant $G_0(\alpha)$, obtained by integrating out wormholes, were to vanish for some α, then the physical Newton's constant $G(\alpha)$ would also vanish for that value of α. Since $G(\alpha)$ must satisfy the bound eq. (4.1), we require the bare coupling to satisfy a similar bound,

$$G_0(\alpha) \geqslant G_{0,\,\mathrm{min}} > 0\,. \tag{4.7}$$

In the dilute approximation described in sect. 2, we saw that the effective action is linear in α. Thus $G_0^{-1}(\alpha)$ is linear in α, and G_0 vanishes when a particular combination of the α's is infinite. In view of eq. (4.7), we conclude that the dilute approximation is invalid. To understand why the gravitational interaction has a nonvanishing strength we must go beyond the dilute approximation. The corrections to the dilute approximation are discussed in sect. 5, with particular attention devoted to the origin of the lower bound on $G_0(\alpha)$.

Is it possible that the requirement that $G(\alpha)$ attains its minimum on the $\Lambda(\alpha)=0$ surface fails to determine all the α's? If there are undetermined α's, then there is presumably a symmetry that prevents G from depending on those α's. But everything couples to gravity, and all other interactions contribute to the renormalization of G, so it is hard to think of such a symmetry.

We have emphasized in this section that Newton's constant $G(\alpha)$ must obey a positive lower bound; otherwise, on the basis of the same assumptions that underlie Coleman's explanation for $\Lambda = 0$, we predict $G = 0$, in conflict with observation. In fact, it is also necessary for the coefficients of higher derivative terms in the gravitational effective action to obey *upper* bounds. Just as the newtonian gravitational interaction must not be permitted to be arbitrarily *weak* for any value of α, so the higher derivative gravitational interactions must not be permitted to be arbitrarily *strong* for any value of α.

In the derivation of $\Lambda = 0$ in sect. 3 we required that the contribution from higher derivative terms to the value of the effective action at the stationary point be negligible. A higher derivative term in the action with arbitrarily large coefficient can upset Coleman's mechanism by destabilizing the stationary point. Or the α-distribution may favor that this coefficient be infinite; then the theory of gravity in a typical α-universe will not agree with Newton's theory in the low-energy limit.

The parameters that must obey such upper bounds are coupling constants of the renormalized effective action that describes gravity at low energy. Unlike the lower bound on G, these upper bounds probably place no restrictions on the "bare" parameters of the effective theory defined at the wormhole scale M_w. Rather, the upper bounds are an automatic consequence of the renormalization procedure as the cutoff floats from M_w to $M \ll M_w$. All of the renormalized couplings of gravity obey upper bounds provided only that, as we expect, gravity has no nontrivial continuum limit*. This will be explained further in sect. 6.

5. Beyond the dilute approximation

Up to now in this paper we have for the most part treated wormholes as dilute, and have ignored the interactions among wormholes. But we have also noted that this dilute approximation cannot really be justified. In this section we will discuss the corrections to the dilute approximation due to interactions.

For the purpose of this discussion, we will find it useful to make a distinction between wormholes and "instantons." I will use the term instanton for the region of the background spacetime where a wormhole hooks on; the term instanton is appropriate because this is a small localized region in the background (euclidean) spacetime. A wormhole connects together two instantons.

We must distinguish two types of interactions that arise when corrections to the dilute approximation are considered; these may be called "wormhole interactions" and "instanton interactions." Wormhole interactions are described by vertices at

* A similar suggestion was made by Weinberg [14].

Fig. 10. A *wormhole* interaction.

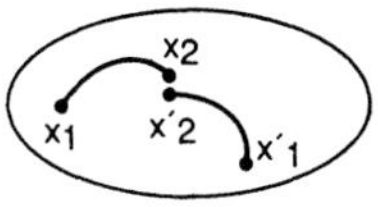

Fig. 11. A short-range interaction between *instantons* at spacetime positions x_2 and x_2'.

which a branching of wormholes occurs (fig. 10). These interactions induce non-gaussian corrections to the probability distribution $P(\alpha)$ in eq. (2.5) and eq. (3.2).

Of far greater interest are instanton interactions. These occur when two instantons closely approach one another on the background spacetime (fig. 11). The instanton interactions induce corrections to the α-dependent effective action $S[g, \alpha]$ that are nonlinear in α. Hence, the instanton interactions also correct the factor $Q(\alpha)$ in eq. (3.3). The instanton interactions are of greater interest than the wormhole interactions (at least for our present purposes) because, by modifying $Q(\alpha)$, they can change where the support of the α-distribution lies.

If we assume that the instanton interactions are of short range, then the effects of these interactions can be systematically expanded in powers of a "density" of instantons. This expansion is precisely analogous to the cluster expansion of statistical mechanics [24]. The cluster expansion, when truncated at a finite order, gives accurate results if the range of the instanton interactions is small compared to the mean separation between instantons*.

Actually, in a theory of gravity, we should not be surprised to find that instanton interactions are *long*-range. The instanton is likely to be surrounded by a gravitational field that decays like an inverse power of the distance from the instanton, and the gravitational interaction between instantons, then, will also decay like a power. (Indeed, there might be other light fields that are excited in the vicinity of the instanton, and these would also induce a long-range interaction.) The long-range interactions cause one to worry about the validity of the cluster expansion.

However, as Gupta and Wise [17] have recently emphasized, the long-range instanton interactions are not the effects that we are really interested in, because they do not generate corrections to the effective action $S[g, \alpha]$ that are nonlinear in α. Instead, the physical effects of these long-range instanton interactions can be

* The cluster expansion for QCD instantons was described in, for example, ref. [25].

reproduced exactly in an effective field theory with an action linear in α; in this effective theory, they are represented by exchanges of light field quanta between interaction vertices that are linear in α. The terms in $S[g, \alpha]$ that are nonlinear in α are generated only by the genuinely short-range instanton interactions that occur when the separation between instantons is comparable to the instanton size. It *is* sensible to carry out an expansion of the effects of *these* interactions in powers of the instanton density. It will be understood, then, that the instanton interactions considered below are actually the short-range portion of interactions that may also include a long-range tail.

In order to formulate the cluster expansion, we must first understand how the notion of a density of instantons arises. For this purpose, we reconsider the dilute approximation that was described in sect. 2. In sect. 2 we made the hypothesis that the contribution C of a single wormhole to a quantum mechanical expectation value can be expressed as a bilocal expansion in local operators, as in eq. (2.2). If the curvature of the background spacetime is small, and all fields on spacetime are weak (in units of the wormhole scale R_w^{-1}), then the sum over local operators is dominated by the operator **1**, and the contribution becomes

$$C = \tfrac{1}{2}\Delta V^2. \tag{5.1}$$

Here V is the volume of the background spacetime, and Δ is the coefficient of the leading term in eq. (2.2), presumably of order R_w^{-8}. In the semiclassical approximation discussed by Coleman [8] and Giddings and Strominger [9], Δ would be further suppressed by the small factor e^{-2S}, where S is the (large) action of a semiwormhole.

In the dilute approximation, the contribution due to N_w noninteracting wormholes is $C^{N_w}/N_w!$ The sum over N_w is thus dominated by N_w of order C. In other words, the number N_w of wormholes in a typical configuration that gives a significant contribution to the euclidean path integral is

$$N_w \sim \tfrac{1}{2}\Delta V^2. \tag{5.2}$$

It is therefore possible to regard Δ as a "density of wormholes" per *volume squared*. On the other hand, a configuration with N wormholes has $2N$ instantons; the notion of a "density of instantons" per *volume* does not appear to make sense in the $V \to \infty$ limit.

However, the above discussion applies to the combinatorics of wormholes only *after* the α integration in eq. (2.3) has been performed. (The effect of the integration is to match up $2N_w$ instantons with N_w wormholes in all possible ways.) Instead, we are interested in studying physics at a fixed value of α. The leading factor in the integrand of eq. (2.3), arising from the operator **1** in the expansion, has the form

$$e^{(1/2)\alpha^2} e^{\alpha\Delta^{1/2}V}. \tag{5.3}$$

(Here we have rescaled α so that the wormhole "propagator" is one; α is now dimensionless.) The term in eq. (5.3) of order V^N may be interpreted as the contribution at fixed α due to an N-instanton configuration, and the sum over N is dominated by

$$N/V \sim \alpha \Delta^{1/2}. \tag{5.4}$$

Thus, $\alpha \Delta^{1/2}$ may be regarded as a density of instantons.

There is a peculiar complementarity at work here. We began this discussion by describing the physical effects of bilocal objects, the wormholes. But then, for sound physics reasons, we focused attention on a sector with a fixed value of α. For α fixed, it is far more appropriate to describe the physics in terms of local objects, the instantons, instead of in terms of wormholes. Furthermore, while the density of wormholes is, of course, α-independent, the density of instantons depends on α as in eq. (5.4).

The instanton density $\alpha \Delta^{1/2}$ will be the expansion parameter of our cluster expansion, and the dilute approximation is a good approximation of $\alpha \Delta^{1/2}$ sufficiently small. Thus, whatever the value of $\Delta^{1/2}$, the dilute approximation can be justified for small enough α. (And this in no way requires the validity of a *semiclassical* approximation; that is, wormhole effects need not be dominated by the contribution of a classical wormhole solution.) On the other hand, whatever the value of $\Delta^{1/2}$, the dilute approximation will always break down for α sufficiently *large*. In particular, the dilute approximation *cannot* be used to study the behavior of the distribution $Q(\alpha)$ for asymptotically large α.

In the dilute approximation, it is assumed that instantons can be freely superposed, and that a contribution due to two instantons factorizes into a product of one instanton contributions. But this assumption is expected to fail when the instantons closely approach one another. More generally, the two-wormhole configuration depicted in fig. 11, in which instantons centered at x_2 and x_2' have a small separation, gives a contribution

$$\frac{1}{8}\int \mathrm{d}^4x_1\sqrt{g_1}\,\mathrm{d}^4x_2\sqrt{g_2}\,\mathrm{d}^4x_1'\sqrt{g_1'}\,\mathrm{d}^4x_2'\sqrt{g_2'}$$

$$\times \sum_{a,b,a',b'} \mathcal{O}_a(x_1)\Delta_{ab}\left[\mathcal{O}_b(x_2)\mathcal{O}_{b'}(x_2') + R^{(2)}_{bb'}(x_2,x_2')\right]\Delta_{b'a'}\mathcal{O}_{a'}(x_1') + \dots \tag{5.5}$$

Here, the remainder $R^{(2)}_{bb'}(x,x')$ is a small correction to the factorized operator $\mathcal{O}_b(x)\mathcal{O}_b(x')$ when $(x-x')$ is large compared to the characteristic instanton size. (Of course, there are also corrections that occur when other pairs of instantons closely approach one another; these have not been indicated in eq. (5.5).)

If we include the effects of all two-instanton interactions in the wormhole sum, the integrand of the α-integration in eq. (2.3) becomes modified. The modification is a shift in the effective action of order α^2,

$$\delta S^{(2)}(\alpha) = -\tfrac{1}{2}\sum_{a,b} \alpha_a \lambda^{(2)}_{ab} \alpha_b ,$$

$$\lambda^{(2)}_{ab} = \int \mathrm{d}^4x \sqrt{g}\, \mathrm{d}^4x' \sqrt{g'}\, R^{(2)}_{ab}(x, x') . \tag{5.6}$$

It is clear that the "Feynman rule" for evaluating the α-integral tells us that the effect of $\delta S^{(2)}(\alpha)$ is to include the two-instanton interaction for all pairs of instantons.

Similarly, there are (connected) n-body instanton interactions that occur when n instantons closely approach one another. Their effect is to induce an order-α^n shift in the effective action

$$\delta S^{(n)}(\alpha) = -\frac{1}{n!} \sum_{a_1 \dots a_n} \alpha_{a_1} \dots \alpha_{a_n} \lambda^{(n)}_{a_1 \dots a_n} ,$$

$$\lambda^{(n)}_{a_1 \dots a_n} = \int \mathrm{d}^4x_1 \sqrt{g_1} \dots \mathrm{d}^4x_n \sqrt{g_n}\, R^{(n)}_{a_1 \dots a_n}(x_1, \dots, x_n) , \tag{5.7}$$

where $R^{(n)}_{a_1 \dots a_n}$ is a translation-invariant operator-valued function that drops off rapidly when any separation $(x_i - x_j)$ is large.

If the instanton interactions are of short range, then the interaction vertex $R^{(n)}$ in eq. (5.7) is smeared out in spacetime over a region with a size comparable to the size of an instanton. The effects of this interaction at low energy can be well accounted for by a local effective action. Formally, we may perform $n-1$ of the integrals over spacetime in eq. (5.7); we then obtain one remaining integral over spacetime of a function that can be expanded in terms of local operators.

By thus expanding eq. (5.7) in terms of local operators, we may express

$$S(\alpha) = S^{(0)} + \sum_{n=1}^{\infty} S^{(n)}(\alpha) \tag{5.8}$$

as a sum of local operators, with the coefficient of each operator expanded as a power series in α. For example, the coefficient $\tilde{\Lambda}_0$ of $\int \mathrm{d}^4x \sqrt{g}$ may be written*

$$\tilde{\Lambda}_0(\alpha) = \tilde{\Lambda}^{(0)} + \sum_{n=1}^{\infty} \sum_{a_1 \dots a_n} \frac{1}{n!} \tilde{\Lambda}^{(n)}_{a_1, \dots, a_n} \alpha_{a_1} \dots \alpha_{a_n} . \tag{5.9}$$

* $\tilde{\Lambda}_0$ is $\Lambda_0/8\pi G_0$ in the notation of eq. (3.5). It is a "bare" parameter that will be further renormalized by fluctuations below the wormhole scale M_{w}.

Expansions like eq. (5.9) are the basis of our earlier claim that corrections due to instanton interactions may be expanded in powers of the instanton density. We now see that this statement is true only rather schematically. The term $\Sigma_a \Lambda_a^{(1)} \alpha_a$ linear in α in eq. (5.9) is the density of instantons that we identified in eq. (5.4). The terms higher order in α are not really higher powers of the density; each term involves a different combination of the α's*. But, generically, when the mean instanton separation is comparable to the instanton size, all terms in eq. (5.9) are of roughly the same order.

Borrowing metaphorically from the semiclassical picture, we might regard each α_a as being associated with a distinct instanton "type." Then each α_a can be interpreted as the density of instantons of type a, and eq. (5.9) is a power series expansion in the various densities.

There is an expansion similar to eq. (5.9) for G_0^{-1}, where G_0 is the bare Newton's constant at the wormhole scale M_w. We noted in sect. 4 that, if we are to accept Coleman's explanation for the vanishing of the cosmological constant, we must insist that G_0^{-1} is bounded above. Obviously, we cannot understand the origin of this bound by considering the expansion in α to any finite order. We must somehow sum up the series.

Before discussing further the upper bound on G_0^{-1}, let us notice that there is another logically independent reason why we must go beyond the dilute approximation if we are to invoke Coleman's mechanism. If we compute quantum corrections in the effective theory cut off at the wormhole scale M_w, $\tilde{\Lambda}$ will be renormalized by an amount of order M_w^4. It would presumably require an unnatural fine tuning for the α-independent bare parameter $\tilde{\Lambda}^{(0)}$ to cancel this renormalization to high accuracy. Therefore, if the renormalized cosmological constant is to vanish, the α-dependent shift of $\tilde{\Lambda}_0(\alpha)$ in eq. (5.9) must be at least of order M_w^4. But then, the density of instantons is of order M_w^4, or the mean instanton separation is $M_w^{-1} = R_w$, which is just the condition for corrections to the dilute approximation to be important**.

We cannot escape this conclusion by appealing to supersymmetry. If supersymmetry remains unbroken below the wormhole scale, then $\tilde{\Lambda}_0(\alpha) = 0$. But a bare cosmological constant is eventually generated at a lower scale, where supersymmetry is spontaneously broken. Again, unless the α-independent bare parameters are carefully tuned, we expect that a large α-dependent shift of bare parameters is required for the renormalized cosmological constant to vanish. The α-dependent shifts are small when the instantons are dilute.

Let us now consider further the origin of the bound on Newton's constant. Here we wish to argue that it is plausible that $G_0^{-1}(\alpha)$, the bare coupling at the wormhole

* Indeed, for the other coefficients in $S(\alpha)$, even the linear term involves a different linear combination of the α's than appears in the "instanton density."

** This argument emerged from discussions with A. Cohen and M. Wise.

scale M_w, is bounded above as a function of α. The renormalization of G^{-1} by quantum fluctuations below the wormhole scale will be discussed in sect. 6.

Heuristically, one expects instanton effects to "saturate" when the mean instanton separation becomes comparable to the instanton size R_w. When the instantons become dense, the picture on which our analysis has been based – of wormholes attaching to a smooth background spacetime – may no longer apply. Indeed, the notion of an instanton gas, and the interpretation of our expansion parameter as a density of instantons, probably ceases to make sense when the dilute approximation is no longer valid.

It is reasonable to guess that, if the α-dependent shift of a parameter in the effective action does have a maximum as a function of α, then the maximum is attained at a value of α that is comparable to the value of α for which the cluster expansion breaks down. That is, the maximum occurs when the instanton density is of order one. We would estimate then, that the maximal shift is a factor of order one times a power of the wormhole scale M_w determined by dimensional analysis. For example, for Newton's constant we might expect

$$\delta\left[(16\pi G_0)^{-1}\right] \lesssim M_w^2 \tag{5.10}$$

to be satisfied for any α.

To make this discussion a bit more concrete, I will describe a simple toy model in which the cluster expansion can be summed to all orders in α. In this toy model, we can see explicitly that G_0^{-1} is a bounded function of α. For the sake of simplicity, this model will have just one "type" of instanton and just one α-parameter; it is obviously possible to generalize it to a model with many α's and with qualitatively similar behavior. The point of the model is that excluded volume effects can be expected to cause repulsive interactions of instantons at small separations, and that short-range repulsive interactions can significantly alter the α-dependence of the effective action at asymptotically large α.

For the purpose of characterizing the short-range instanton interactions in this toy model, imagine that euclidean spacetime has been divided into many identical tiny cells, each with a physical volume of R_w^4, the volume of an instanton. A configuration of the instanton gas can be described by assigning to each cell a non-negative integer, the number of instantons that occupy the cell (see fig. 12). We will label the cells with an index i, and denote the corresponding integer by n_i.

Now the assumption of our model is that instantons in distinct cells do not interact, but instantons in the same cell *do* interact. An instanton occupying cell i may either encourage more instantons to occupy that cell (attractive interaction), or discourage additional instantons (repulsive interaction).

Suppose, at first, that there are no instanton interactions at all. Then the instantons are noninteracting identical particles, and summing over all configura-

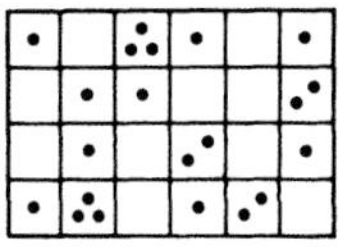

Fig. 12. Euclidean spacetime has been divided into cells; each cell contains a number of instantons that is assumed to be a (non-negative) integer.

tions of the instanton gas generates an effective action $\delta S(\alpha)$ given by

$$e^{-\delta S(\alpha)} = \prod_i \left(\sum_{n_i=0}^{\infty} \frac{1}{n_i!} (\alpha \mathcal{O}_i)^{n_i} \right) = \prod_i e^{\alpha \mathcal{O}_i}$$

$$= \exp\left(\alpha \sum_i \mathcal{O}_i \right). \tag{5.11}$$

Here $\mathcal{O}_i$ quantifies the dependence of the instanton contribution on the background geometry; it can be expanded in terms of local operators in the cell i. In eq. (5.11) we have recovered the result eq. (2.4) that we derived previously in the dilute approximation.

To simulate interactions, we replace the sum over n in eq. (5.11) by

$$\sum_{n=0}^{\infty} \frac{a_n}{n!} (\alpha \mathcal{O}_i)^n. \tag{5.12}$$

Then, for $a_n > 1$, we have introduced an attractive short-range n-body interaction, and for $a_n < 1$, a repulsive n-body interaction. With this modification, eq. (5.11) becomes

$$\delta S(\alpha) = -\sum_i \ln f(\alpha \mathcal{O}_i),$$

$$f(z) = \sum_n \frac{a_n}{n!} z^n. \tag{5.13}$$

To extract the value of δG_0^{-1}, we expand δS in powers of derivatives of the background metric. If the operator $\mathcal{O}$ can be expanded as

$$\mathcal{O} = c_0 + c_1 R + \dots$$

(where R is the Ricci scalar), then we have

$$\delta S(\alpha) = \int d^4x \sqrt{g}\, R_w^{-4} \left(-\ln f(\tilde{\alpha}) - \frac{\tilde{\alpha} f'(\tilde{\alpha})}{f(\tilde{\alpha})} \frac{c_1}{c_0} R(x) + \dots \right), \tag{5.14}$$

where

$$\tilde{\alpha} = c_0 \alpha . \tag{5.15}$$

In eq. (5.14), the sum over cells has been replaced by a volume integral, and the volume R_w^4 of the cell has been inserted.

From eq. (5.14) we see that the shift in Newton's constant is given by

$$\delta\left[(16\pi G_0)^{-1}\right] = \frac{\tilde{\alpha} f'(\tilde{\alpha})}{f(\tilde{\alpha})} \frac{c_1}{c_0 R_w^4} . \tag{5.16}$$

There are many reasonable functions $f(\tilde{\alpha})$ for which this shift is bounded; for example, f may be an (even) real polynomial with no real zeros. A simple function that works is

$$f(z) = 1 + z + \tfrac{1}{2} z^2 . \tag{5.17}$$

In this case, one or two instantons are allowed to occupy a cell without interacting, but a repulsive interaction forbids three or more instantons in a single cell.

Clearly this toy model has many unrealistic features, but it serves to illustrate that instanton interactions can cause α-dependent shifts in the effective action to be bounded as functions of α.

Before concluding this section, I would like to reexamine one of the assumptions underlying our whole analysis, in the light of our observations about instanton interactions. This assumption was explicitly stated in sect. 2 – that wormholes have a characteristic thickness R_w. Then it makes sense to integrate out wormholes, and obtain an effective theory with cutoff R_w^{-1}, in which wormhole fluctuations no longer occur. The heuristic picture underlying this assumption is that thick wormholes have large euclidean action, and so give a contribution to the path integral that is highly suppressed. We therefore make only a very small error by neglecting wormholes that are much thicker than the characteristic size R_w.

This reasoning, however, must be regarded with caution. We may imagine that the α's can be divided into two sets – parameters α_S associated with small instantons and parameters α_L associated with large instantons. On the basis of the heuristic picture described above, we expect that the "propagator" Δ_L for large wormholes is extremely small; the probability distribution $P(\alpha)$ in eq. (2.5) strongly favors small values of α_L. It is in this sense that large α_L-dependent shifts in the effective action are highly improbable.

But as we emphasized in sect. 3, it is not $P(\alpha)$ but the factor

$$Q(\alpha) \sim \exp\{\exp[3\pi / G(\alpha)\Lambda(\alpha)]\} \tag{5.18}$$

that determines where the support of the α-distribution lies. The very singular

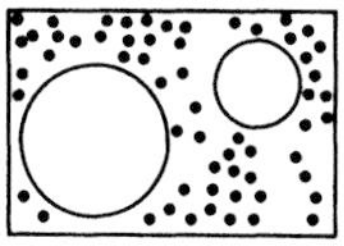

Fig. 13. Large instantons exclude small instantons from spacetime.

dependence of eq. (5.18) on α_L for $\Lambda(\alpha) \sim 0$ overcomes the $\exp(-\alpha_L^2 \Delta_L)$ suppression in $P(\alpha)$. Hence, in spite of their large euclidean action, the large instantons try their best to increase G^{-1} on the $\Lambda = 0$ surface. The contribution due to large instantons to G^{-1} is presumably maximized when the large instantons are dense. And so, in spite of the strong tendency of the factor $P(\alpha)$ to suppress the effects of the large instantons, the large instantons cannot be prevented from becoming dense, and inducing effects that are of order one.

This conclusion is distressing. It appears that wormholes of arbitrarily large thickness will contribute significantly to physical processes at low energy. How are we to reconcile this phenomenon with the well-documented success of local field theory in describing low-energy physics*?

I believe that interactions between large and small instantons are responsible for suppressing the effects of large instantons. I have in mind a picture in which small instantons crowd out the large ones (fig. 13). If our goal is to maximize G^{-1}, then configurations with many large instantons are inefficient, because the large instantons exclude from spacetime a region that the small instantons would like to occupy. A dense gas of small instantons gives a contribution of order R_S^{-2} to G_0^{-1}, where R_S is the size of a small instanton, while a dense gas of large instantons contributes of order R_L^{-2}, where R_L is the size of a large instanton. Therefore, if there is a trade-off between large and small instantons, the instanton gas will favor increasing the abundance of small instantons at the expense of the large instantons. This effect might provide the justification for ignoring the large wormholes.

A very crude model will illustrate how this mechanism works. Imagine that there are only two instanton types, the small ones with size R_S and the large ones with size R_L. Then the α-dependence of the bare Newton's constant has the approximate form

$$G_0^{-1} \cong C_S R_S^{-2} n_S(\alpha_S) + C_L R_L^{-2} n_L(\alpha_L) . \tag{5.19}$$

Here n_S is the dimensionless density of small instantons, the fraction of the volume of spacetime that is occupied by the instantons; n_L is the corresponding density of large instantons, and C_S, C_L are numerical constants of order one. We wish to determine α_S and α_L by maximizing G_0^{-1}. But if small instantons cannot sit on top

* This issue was also raised by Kaplanovsky [20], and has been discussed in refs. [16,18].

Fig. 14. If excluded from attaching to spacetime by a large instanton, a small wormhole may attach to a large wormhole instead.

of large instantons, then the densities are constrained by

$$n_S(\alpha_S) + n_L(\alpha_L) \leqslant 1. \tag{5.20}$$

Evidently, the maximum of G_0^{-1} subject to the constraint eq. (5.20) is $n_S = 1$, $n_L = 0$, assuming that $C_S R_S^{-2} > C_L R_L^{-2}$. The large instantons are completely eliminated in favor of the small instantons.

The distinction between instanton interactions and wormhole interactions is important for understanding why large and small instantons mutually exclude each other. When excluded from spacetime by a large instanton, a small instanton will slide up onto a large wormhole (fig. 14); that is, the small wormhole will attach to the large wormhole instead of attaching to the background spacetime. One might have argued from this perspective that the physics of small wormholes is unaffected by large wormholes; the small wormhole finds the large wormhole indistinguishable from the background spacetime. But we have already learned that the physical consequences of wormhole interactions are quite different than the consequences of instanton interactions. The interactions between large and small wormholes alter the distribution $P(\alpha)$, but have no effect on the crucial factor $Q(\alpha)$ that determines what values of the α's are favored. The instanton interactions, not the wormhole interactions, affect $Q(\alpha)$, and it is therefore the instanton interactions, not the wormhole interactions, that must account for the suppression of the effects of large wormholes.

The suggestion that small instantons are responsible for suppressing the effects of large instantons sounds surprising at first. It is a cherished principle of physics, the decoupling principle, that long-distance physics is relatively insensitive to the details of short-distance physics. Ordinarily, we expect that it is possible to "integrate out" short-wavelength quantum fluctuations and incorporate the effects of these fluctuations into the renormalization of the parameters of an effective field theory that describes long-distance physics. (Indeed, we invoked this strategy in sect. 3, and will do so again in sect. 6.) Hence, we might expect that the effects of large instantons can be understood within the context of a low-energy effective field theory in which small instantons have already been integrated out. But how can the mechanism described above for suppressing large instantons, which arises from interactions with small instantons, be understood in such an effective field theory?

I believe, in fact, that the α-dependence of G^{-1} is an *exception* to the decoupling principle. The dominant contribution to the dependence of G^{-1} on α_L cannot be accounted for in terms of long-distance physics alone. It arises because of the unusual way in which short-distance physics and long-distance physics are intertwined here. Usually, we may think of short-wavelength fluctuations as being superposed on a long-wavelength background field configuration. Because the short-wavelength fluctuations are indifferent to the long-wavelength physics, they induce renormalizations that are independent of the background field; that is why decoupling usually works. From this perspective, the key feature of the instanton interactions is that they prevent small instantons from sitting on top of large instantons. Therefore, the effects of the small instantons are not independent of the configuration of large instantons. The dependence of G^{-1} on α_L is thus dominated by the effect of α_L on the small instantons, and it cannot be understood in terms of long-distance physics alone. That is why decoupling fails.

One should also notice that this mechanism for suppressing large instantons relies crucially on the claim that the α's are determined by maximizing G^{-1}. The point is simply that G^{-1} has the dimensions of mass to a *positive* power. Therefore, merely on dimensional grounds, small instantons are much more effective than large instantons at causing G^{-1} to increase.

6. Renormalization in quantum gravity

In spite of the intrinsic indeterminacy introduced by wormhole effects, it is reasonable to hope that the constants of Nature can, in principle, be computed. We must calculate the dependence on the α's of the various parameters that characterize low-energy physics, and then find the value of α that is overwhelmingly favored by the probability distribution $Q(\alpha)$ in eq. (3.3).

To find the favored value of α, it is enough to know the α-dependence of two quantities, the (renormalized) cosmological constant Λ and the (renormalized) Newton's constant G. As we have emphasized, the α-dependence of these quantities arises from two sources (as does the α-dependence of all the renormalized parameters), and we can imagine calculating the α-dependence in two stages. In the first stage, we integrate out wormhole fluctuations, and thus obtain an effective theory with α-dependent bare couplings that is cut off at the wormhole scale M_w. In the second stage, we integrate out quantum fluctuations with wavelengths greater than $R_w = M_w^{-1}$, to obtain an effective theory that is appropriate for describing the far-infrared behavior of gravity. The first stage was the subject of sect. 5, and the second stage is the subject of this section.

In sect. 5, we argued that it will be difficult to analyze the first stage with precision, because the dilute approximation does not apply. This conclusion already discourages us about the prospects for calculating the α-dependence of Λ and G. But even if it is hopeless to calculate the α-dependence of the *bare* couplings induced at the first stage, one might still nurture the hope that the preferred values of at least some of the *renormalized* couplings can be calculated to reasonable accuracy. There are, presumably, many α's, and so it may not be seriously wrong to assume that, as the α's vary, the bare couplings at the wormhole scale can vary without restriction*. Then we can forget all about the origin of the α-dependence. Our task is simply to find "the best of possible worlds" – that choice of the bare parameters at scale M_w for which Coleman's distribution has its sharp peak.

Unfortunately, this program in its most extreme form suggested above is doomed to fail. As we have seen, in the best of worlds G vanishes as well as Λ, and there is no gravitational interaction. We must accept that our universe is not the most perfect one, and abandon the hypothesis that wormhole effects allow the bare parameters to vary arbitrarily. A milder version of this hypothesis might be reasonable, however; we can assume that while G_0^{-1} is bounded above as a function of the α's, all other bare couplings can be varied without restriction. Surely, this is at best a caricature of the actual α-dependence of the bare couplings, but it may serve as a first approximation.

This point of view suggests some questions that we will address below. For example, will an upper bound on G_0^{-1} suffice to ensure that the renormalized G^{-1} is also bounded above, as we require for Coleman's mechanism to be consistent with observation, even if *no* restrictions are placed on other bare couplings? And also, do the coefficients in the renormalized gravitational effective action of terms that involve *higher* derivatives of the metric obey suitable bounds even as the bare couplings vary without restriction? Such bounds are also required for the consistency of Coleman's mechanism.

I will argue that Coleman's explanation for $\Lambda = 0$ passes both of the consistency tests posed above. That is, only one condition on the α-dependence of the base couplings – the upper bound on G_0^{-1} – suffices to justify Coleman's picture of the infrared behavior of gravity. These arguments are rather heuristic, but, I hope, cogent.

What can be said about the calculability, in practice, of the const nts of Nature (other than the cosmological const nt)? With respect to this issue, there is not much cause for encouragement. We hope to calculate the various renormalized parameters that describe low-energy physics. For this purpose, we must find the dependence of the renormalized G on these renormalized quantities. This dependence might have been calculable if the renormalization of G were dominated by long-wavelength effects that could be naturally expressed in terms of couplings renormalized at low energy. But unfortunately, the renormalization of G is actually dominated by

* This means, of course, that the bare action varies "without restriction" within a space of reasonable quasilocal theories.

short-wavelength quantum fluctuations. Therefore, the preferred values of the constants of Nature are actually sensitive to physics at the wormhole scale M_w and cannot be computed based on low-energy physics alone.

In a more optimistic light, by failing to make predictions beyond $\Lambda = 0$ we have avoided making any *wrong* predictions that would call Coleman's mechanism into question. Indeed, based on a different point of view than advocated here, Grinstein and Wise [15] and Klebanov, Susskind, and Banks [16] reached some unfortunate conclusions – for example, that the pion has vanishing mass. This embarrassment is evaded if, as I claim, the preferred values of the renormalized quantities are actually determined by short-distance physics. Of course, it remains to be seen whether "the best of possible worlds" in the revised sense proposed here will resemble the universe that we observe.

Let us now consider in more detail the relation between the bare theory at the wormhole scale M_w and the renormalized theory that describes physics at very low energy. It is convenient to imagine that the renormalized theory is obtained from the bare theory by means of a renormalization group transformation [23]. The bare theory is cut off at the mass scale M_w and is quasilocal on the distance scale M_w^{-1}. This bare theory might be quite complicated, involving many degrees of freedom and interactions. Now we allow the cutoff to float down from M_w to $M \ll M_w$; we integrate out all quantum fluctuations with wave number between M_w and M, and we incorporate the effects of these fluctuations into renormalized parameters of an effective theory that has cutoff M and is quasilocal on the distance scale M^{-1}. After M has descended below the mass of the lightest massive particle in the original theory, the only remaining degrees of freedom in the effective theory are massless fields, such as the graviton and photon. It is the parameters of this effective theory, in the limit $M \to 0$, that enter into Coleman's calculation of the distribution $Q(\alpha)$.

The $M \to 0$ limit of the effective theory is the "continuum limit" of the bare theory that we started with. It describes the physics of the theory on length scales that are arbitrarily large compared to the original short-distance cutoff. Furthermore, the only surviving interaction in the $M \to 0$ limit is gravity.

Other massless particles (such as the photon) may remain coupled to the graviton, but the self-interactions of matter have been integrated out. (For example, there is no electrodynamic interaction because there are no massless charged particles.) Thus, the $M \to 0$ limit is the continuum limit of a cutoff theory of gravity. In fact, as we allow M to float toward zero, the renormalized cosmological constant Λ must also approach zero, as we discussed in sect. 3. So we are considering the continuum limit of gravity with a vanishing cosmological constant.

This is a useful insight, because Einstein gravity is nonrenormalizable in perturbation theory, and it would be a stunning surprise if it turned out to be possible to take a continuum limit of quantum gravity other than free field theory. In other words, if the strength of gravitational effects is held fixed at the mass scale M, we do not expect to be able to push the cutoff mass M_0 up arbitrarily high, while maintaining a quasilocal theory of gravity at the scale M_0. If this were possible, then we could devise an ordinary field theory of gravity that makes sense at arbitrarily short distances, just as quantum chromodynamics (presumably) can describe the strong interaction at arbitrarily short distances. Then we would have a consistent field theory of quantum gravity; there would be no need for strings.

The conjecture that quantum gravity has no nontrivial continuum limit can be stated from two different points of view. We may consider (as above) physics at scale M to be fixed, and then ask whether a bare theory can be constructed that reproduces the physics at scale M as the cutoff M_0 of the bare theory approaches infinity. Alternatively, we may consider the cutoff M_0 of the bare theory to be fixed, and then ask whether a gravitational interaction at scale M can remain nontrivial as $M \to 0$, for *any* choice of the quasilocal bare theory with cutoff M_0.

The effective theory at scale M has an action S_M that can be expanded in terms of (quasi-)local operators. Each operator has a coefficient that can be made dimensionless by multiplying by an appropriate power of M. Thus, S_M can be specified by an infinite set of dimensionless couplings*, λ^a_M, $a = 1, 2, 3, \ldots$. The conjecture that gravity has no continuum limit, then, in the second form mentioned above, can be stated: For *any* choice of the quasilocal bare theory at the cutoff M_0, *all* dimensionless renormalized couplings λ^a_M approach zero as $M \to 0$**. In fact, one expects that for small nonvanishing M, the λ^a_M's obey a somewhat stronger inequality constraint,

$$\lambda^a_M \leqslant \lambda^a_{\max}\left(\frac{M_0}{M}\right), \quad M \ll M_0, \tag{6.1}$$

where the upper bound $\lambda^a_{\max}$, a dimensionless function of M_0/M, approaches zero for $M_0/M \to \infty$.

The inequality eq. (6.1) holds for *any* choice of the quasilocal bare theory at cutoff scale M_0, even as the bare couplings range without bound; there is no renormalized theory at scale M that is descended from a quasilocal bare theory at scale M_0 and that has λ^a_M greater than $\lambda^a_{\max}(M_0/M)$. One might have supposed that we can specify any renorm lized theory we please at scale M and then obtain the corresponding bare theory at scale M_0 by running the renormalization group transformation in reverse. But this typically fails, because the renormalization group flow reaches a boundary of the space of quasilocal theories before the floating cutoff M_0 is attained.

The claim that the renormalized couplings are bounded even though the bare couplings are not, may sound surprising at first, but this behavior is just the

* Strictly speaking, these are the "essential" couplings that are not changed by local redefinitions of the fields [23, 26].

** Recall that we are considering the infrared behavior of gravity with vanishing renormalized cosmological constant.

(nonperturbative) "Landau ghost" phenomenon. The Landau ghost is familiar in, for example, QED and ϕ^4 theory in four dimensions, two field theories that, though perturbatively renormalizable, are not expected to have nontrivial continuum limits [23,27]. It is characteristic of this phenomenon that a renormalized coupling λ_M, when evolved by means of the renormalization group to a larger renormalization scale, attains an infinite value at a finite scale $M' > M$. Equivalently, Landau ghost behavior means that renormalized couplings at scale M obey the inequality eq. (6.1), even as bare couplings at the cutoff M_0 vary without restriction. Physically, very strong charge screening effects keep the renormalized coupling finite even if the bare coupling is infinite.

We conclude, then, that if no quantum field theory of gravity has a continuum limit, then all gravitational effects must become weak in the $M \to 0$ limit, irrespective of the bare theory defined at scale $M_0 \gg M$; all dimensionless couplings λ^a_M approach zero. Because all dimensionless couplings are small, the renormalization group flow can be reliably computed for $M \ll M_0$ in perturbation theory. This perturbative analysis allows us to justify neglecting terms in the effective action that involve higher derivatives of the metric in the derivation in sect. 3. (The perturbative flow equations will be discussed further below.) Thus, Coleman's explanation for $\Lambda = 0$ has passed a nontrivial consistency test; we need not worry about higher-derivative interactions upsetting his mechanism, regardless of the α-dependence of the bare higher-derivative couplings.

Like the other dimensionless couplings, $G_M M^2$ obeys the inequality eq. (6.1). Perturbative power-counting as described below then shows that G_M stops running as $M \to 0$; it approaches the limit G, the renormalized Newton's constant. Therefore G obeys an inequality

$$G < C M_0^{-2}, \tag{6.2}$$

where C is a constant, presumably of order one, independent of the bare theory at the cutoff M_0. Correspondingly, the coefficient G^{-1} of R in the renormalized effective action eq. (3.5) is bounded from *below*. In particular, then, G^{-1} is *positive*, as we assumed in sect. 3.

But there is no *lower* bound on G that we can infer from a renormalization group argument alone. Indeed, all gravitational interactions are proportional to a positive power of G, and gravity becomes a free theory for $G = 0$. Evidently, then, there is no renormalization for $G = 0$, and $G = 0$ is a fixed point of the renormalization group. If the bare Newton's constant G_0 vanishes, then so does the renormalized coupling G. We arrive again at the conclusion annunciated earlier; Coleman's mechanism favors the lowest possible value of G, and hence is incompatible with observation unless the bare parameter G_0 is bounded away from zero.

May we also say that a lower bound on the bare Newton's constant $G_0 \geq G_{0,\min} > 0$ suffices to ensure that the renormalized G does not vanish? The issue is whether

G_M^{-1} can be *infinitely* renormalized as M floats from M_0 down to zero, or, in other words, whether the metric can undergo an *infinite* field renormalization. Since the renormalization for $M \ll M_0$ can be studied perturbatively, and we know that G_M stops running, the only question is whether *infinite* higher-derivative bare couplings can induce an infinite field renormalization at $M \sim M_0$. This question is highly nonperturbative and is hard to answer definitively, but I believe that the same screening effects that underlie the inequalities eq. (6.1) will also prevent infinite field renormalization, even if some bare couplings are infinite. The point is that infinite bare couplings run infinitely quickly as the cutoff floats, and so always induce finite effects. Therefore, if the bare Newton's constant is $G_0 \ll M_0^{-2}$, then the renormalized coupling is $G \sim G_0$, regardless of the values of the other bare couplings. (For $G_0 \gtrsim M_0^{-2}$, the renormalized coupling is $G \sim M_0^{-2}$, in accord with the bound eq. (6.2).)

If the bare coupling G_0 has a *minimum* as a function of α, one expects on dimensional grounds that the minimum occurs for $G_0 \sim M_0^{-2}$; then the minimal renormalized coupling is also $G \sim M_0^{-2}$. This minimal renormalized coupling should coincide with the observed Newton's constant. It is therefore reasonable to identify the wormhole scale $M_{\mathrm{w}} = M_0$ with the Planck scale $M_{\mathrm{P}} \sim 10^{19}$ GeV.

While the above discussion suggests that the other bare couplings do not renormalize G_0 by an enormous factor, each coupling does give a contribution, however small, to the renormalization of G_0, because all degrees of freedom in the bare theory couple to gravity. In other words, the renormalized coupling G is a function of all the bare parameters. Therefore, the requirement that G attain its minimum value is expected to completely determine the bare theory, and hence to completely determine all low-energy physics. Quantum indeterminacy of the constants of Nature is thus, in principle, avoided.

Of course, this analysis is premised on the assumption that quantum gravity has no nontrivial continuum limit. It is at least a possibility that the renormalization group equations of quantum gravity have a nontrivial fixed point. Then a nontrivial continuum theory could be formulated*. Various dimensionless couplings would be nonvanishing at the fixed pont, and higher-derivative interactions would have important effects in the continuum theory. There would then be a serious danger that these interactions would destabilize the peak at $\Lambda = 0$ that Coleman found in the distr bution $Q(\alpha)$. But if, as one expects, there is no nontrivial continuum limit of quantum gravity, then our arguments buttress Coleman's prediction that $\Lambda = 0$.

To further support these claims, we should now discuss somewhat more explicitly the renormalization group equations of quantum gravity [26]. These equations describe how the effective action S_M flows as the floating cutoff M is lowered. The

* The possible existence of a nontrivial fixed point has been discussed (from quite different points of view) by Weinberg [26] and by Antoniadis and Tomboulis [28]. The conclusions of ref. [28] have been criticized by Johnston [29].

great utility of these equations derives from the observation that, in the infrared limit $M \to 0$, S_M exhibits "universal" behavior that depends only rather weakly on the original bare action S_{M_0}.

If the bare theory S_{M_0} depends upon an infinite number of adjustable bare parameters, one might worry that the theory has no predictive power. But as Wilson emphasized [23], physics at the energy scale $M \ll M_0$ can be predicted, because, up to corrections suppressed by powers of M/M_0, S_M can be expressed in terms of a small number of renormalized parameters. Indeed, even the power corrections can be systematically taken into account; to any finite order in M/M_0, S_M can be expressed in terms of a finite number of renormalized quantities.

In discussing Coleman's mechanism, we consider the value of the renormalized euclidean action at its stationary point. In this context, the expansion of S_M in powers of M becomes an expansion of the stationary value of the action in powers of R^{-1}, where R is the "radius" of the classical solution. For example, the action $S_M[g]$ of eq. (3.5), when evaluated for a four-sphere of radius R, becomes

$$S(R) = \frac{\Omega}{16\pi G}\left(2\Lambda R^4 - 12R^2 + A_4 + A_6 R^{-2} + \dots\right), \tag{6.3}$$

where $\Omega = 8\pi^2/3$ is the volume of a unit four-sphere. Here A_4, $A_6, \dots$ arise from terms in the action involving $4, 6, \dots$ derivatives of the metric. Because of renormalization effects, the parameters Λ, G, A_4, $A_6, \dots$, are all implicitly dependent on the radius R.

Naively, eq. (6.3) shows "universal" behavior in that $S(R)$ can be expressed to order R^{-2m} in terms of $m+3$ renormalized parameters. This statement is naive because we have ignored the dependence of the renormalized parameters on $R \sim M^{-1}$. For example, we may wish to claim that $S(R)$ up to order R^0 can be expressed in terms of only the renormalized Λ and G. We must argue, then, that as the bare parameters of S_{M_0} vary on the surface where Λ_M and G_M have specified values, $A_{4,M}$ varies by an amount of order one (independent of M). We can analyze this problem in perturbation theory using elementary arguments based on dimensional analysis and power-counting of Feynman diagrams.

(There is a technical point that we should comment on here*. All terms in S_M that involve four derivatives of the metric are actually "inessential" [26, 14]; these terms are either topological and can be ignored in perturbation theory or can be eliminated from the action by a local redefinition of the metric $g_{\mu\nu}$. Ordinarily, it is advisable to eliminate inessential couplings when renormalization group equations are derived, but we will find it convenient to consider the renormalization of A_4. In part, this is because we are ultimately interested in minimizing $S(R)$ with respect to R *and* the α-parameters, subject to an infrared cutoff. The redefinition of the metric

* This point was discussed by Grinstein and Wise [15].

that eliminates A_4 preserves the value of the action at its minimum, but modifies the (invariant) radius of the four-sphere solution. Furthermore, this redefinition depends on α; if we change the bare parameters, a change in A_4 is typically induced, and we must redefine the fields again to eliminate it. Thus, if we were to eliminate A_4 and *then* minimize the action with respect to α, we would be minimizing subject to an *α-dependent* infrared cutoff. It seems less confusing to keep the infrared cutoff fixed and α-independent; we then lose the freedom to eliminate A_4.)

Roughly speaking, S_M can be obtained from S_{M_0} in perturbation theory by evaluating Feynman diagrams that are cut off in the ultraviolet at M_0 and in the infrared at M. These diagrams can be divided, according to their sensitivity to the ultraviolet cutoff, into three classes – power-divergent, log-divergent, and convergent.

Power-divergent diagrams are of order M_0^D where $D > 0$ is the superficial degree of divergence of the diagram. These diagrams are "nonuniversal" contributions that are dominated by quantum fluctuations with wave number of order M_0 and that are sensitive to the detailed form of the bare action S_{M_0}. Because of the power divergences, the renormalization group flow is very complex for $M \sim M_0$; many higher dimension operators in S_M contribute to the renormalization of lower dimension operators. But, in perturbation theory, the power-divergent diagrams give a decreasingly important contribution to the flow as M decreases.

Of much greater interest are the log-divergent diagrams that are proportional to a power of $\ln(M_0/M)$. These diagrams receive a significant contribution from quantum fluctuations over a broad range of length scales. Unlike power divergences, the log divergences are universal; the coefficients of leading and nonleading powers of $\ln(M_0/M)$ are calculable in terms of low-energy physics alone and are insensitive to the details of physics at the cutoff.

Finally, there are convergent diagrams. These too are calculable in terms of low-energy physics alone. Their sensitivity to the cutoff is of order M_0^D, where $D < 0$ is the superficial degree of divergence of the diagram. If we are interested in the flow of the renormalized effective action S_M for $M \ll M_0$, we need consider only the log-divergent and convergent diagrams, for the renormalization effects induced by the power-divergent diagrams are not sensitive to M.

Now consider the implications of this classification for the renormalization of quantum gravity. First we consider pure gravity, uncoupled to matter, which we have argued is adequate for studying the asymptotic $M \to 0$ behavior of the effective action S_M. Then we will discuss how our conclusions are modified when gravity is coupled to matter with renormalizable or superrenormalizable interactions.

We have argued that the flow of S_M for M small is determined by log-divergent and convergent diagrams. But in a theory like pure gravity with $\Lambda = 0$, in which all couplings are of nonrenormalizable type, all renormalizations induced by such diagrams are renormalizations of higher dimensional couplings by lower dimen-

Fig. 15. A one-loop diagram that contributes to the renormalization of the gravitational effective action. All lines represent gravitons.

sional couplings. This conclusion follows directly from simple dimensional analysis. Consider, for example, the logarithmic renormalizations. If we scale G^{-1} out of the action as in eq. (6.3), then G is a loop-counting parameter; an L-loop diagram is of order G^{L-1}. Since all other couplings have dimensions of mass to a negative power, it follows immediately on dimensional grounds (for $\Lambda = 0$) that there is no logarithmic renormalization of G^{-1}. Similarly, a logarithmic renormalization of A_4 occurs only in one-loop order, and has the form

$$\frac{\delta A_4}{16\pi G} = b_4 \ln(M_0/M), \tag{6.4}$$

where b_4 is a numerical constant. (This renormalization is generated by the diagram in fig. 15, from which the value of b_4 can be extracted.) Each coupling A_{2n} receives a logarithmic renormalization only up to $(n-1)$-loop order in perturbation theory [26]. For example, A_6 receives two-loop renormalization of order G^2 and a one-loop renormalization of order GA_4.

We see, therefore, that the infrared-sensitive renormalization of each coupling in the effective action is determined by renormalized couplings of lower dimension. Once we have fixed the value of Λ, G, and A_{2m} for $m < n$, the sensitivity of the renormalized A_{2n} to the bare parameters involves only the nonuniversal power-divergent renormalizations. In perturbation theory, GM_0^2 and $A_{2m}M_0^{2m-2}$ are regarded as small, and the power-divergent renormalization of A_{2n} is bounded by a power of M_0 determined by dimensional analysis

$$(\delta A_{2n})_{\text{nonuniversal}} \lesssim CM_0^{-2n+2}. \tag{6.5}$$

Our earlier discussion of nonperturbative effects then suggests that an inequality of this form holds even beyond the domain of validity of perturbation theory*.

Specifically, the renormalization of the cosmological constant Λ is due to power-divergent diagrams only. These diagrams renormalize $\Lambda/8\pi G$ by an amount of order M_0^4, and $\Lambda/8\pi G$ approaches a finite limit as the floating cutoff M approaches zero. Similarly, in pure gravity with vanishing renormalized Λ, G^{-1} is renormalized by power-divergent diagrams only, and it too approaches a finite limit

* A related suggestion was made by Weinberg [14].

for $M \to 0$. We say that Λ and G stop running in the far infrared. On the other hand, A_4 does not stop running; it receives the logarithmic renormalization eq. (6.4). This logarithmic renormalization diverges in the infrared limit $M \to 0$, but the coefficient of the logarithm is determined by the renormalized G. When Λ and G are held fixed, then, $S(R)$ in eq. (6.3) varies as a function of the α's by an amount of order one; it is a bounded function of both α and R. Therefore, the α-dependence that arises from higher-derivative operators can be neglected in the $R \to \infty$ limit, just as we assumed in sect. 3.

The power-counting described above applies to pure gravity, and hence to the renormalization group flow of the gravitational effective action S_M in the far infrared, after all matter has been integrated out. It is also of interest to consider the flow of S_M at intermediate length scales, where the coupling of gravity to matter cannot be neglected. Because the matter action can contain terms of renormalizable or superrenormalizable type, we must modify the earlier analysis, which applied to a theory in which all interactions are of nonrenormalizable type.

Of particular interest is the dependence of the renormalization of G on the masses of matter particles, and this is the only effect of the matter that we will discuss here in any detail. Insofar as the mass-dependent renormalization of G is calculable, we might hope to reach nontrivial conclusions about the masses of elementary particles form the requirement that G^{-1} attains its largest possible value.

Dimensionally, a calculable order-m^2 logarithmic renormalization of G^{-1} is possible in one-loop order, and indeed, one finds that this renormalization *is* generated by the diagram in fig. 16. Calculation of this diagram [30] shows that the logarithmic renormalization is

$$\delta\left(\frac{1}{16\pi G}\right) = -\frac{1}{192\pi^2} m^2 \ln\left(M_0^2/m^2\right) + \mathrm{O}(m^2) + \mathrm{O}(Gm^4), \tag{6.6}$$

where m is the mass of *either* a free (Dirac) fermion or of a free minimally coupled (real) scalar. The order-m^2 correction in eq. (6.6) is a nonuniversal effect (which can be absorbed by shifting M_0^2 in the logarithm), and the order-Gm^4 correction is due to diagrams with two or more loops. Of course, if we include the contributions due to renormalizable interactions, the logarithmic renormalization of G^{-1} is modified. The interactions generate additional logarithms in each order of perturbation theory. These logarithms can be summed, and they have the effect of causing m^2 to

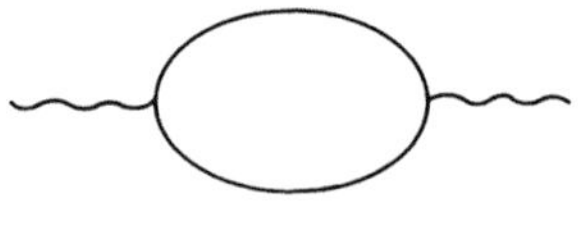

Fig. 16. A one-loop diagram that induces a logarithmic renormalization of G. The solid line represents a *massive* matter particle; it may be either a scalar or a spin $-\frac{1}{2}$ fermion.

become a scale-dependent running parameter. If the matter particle that induces this renormalization is a composite object (like the pion), then the cutoff M_0 in the logarithm is not the wormhole scale, but rather the mass scale (like $4\pi f_\pi$) below which the particle behaves like an effectively elementary object.

The physical mass m of the matter particle appears in the argument of the logarithm in eq. (6.6). This is because the matter loop continues to contribute to the renormalization of G^{-1} as the cutoff M floats down until $M \sim m$, at which point we integrate the matter particle out of the effective action, and it ceases to contribute to loop diagrams. In spite of the logarithm, though, the renormalization eq. (6.6) actually vanishes in the limit $m^2 \to 0$. This observation corroborates our earlier conclusion that G^{-1} stops running in the infrared. Long-wavelength fluctuations of light matter particles, like long-wavelength graviton fluctuations, contribute very little to the renormalization of G.

The *sign* of the calculable logarithmic renormalization in eq. (6.6) is such as to decrease G^{-1}. The criterion of maximizing G^{-1}, then, favors *lighter* matter particles. We ought not to accept this conclusion too readily, however. Because the calculable infrared renormalization is small for small m^2, one must worry about whether this effect is swamped by nonuniversal contributions.

The form of eq. (6.6) should be contrasted with the logarithmic mass-dependent one-loop renormalization of A_4; for a minimally coupled real scalar this is

$$\frac{\delta A_4}{16\pi G} = -\frac{29}{30}\left(\frac{1}{16\pi^2}\right)\ln\left(M_0^2/m^2\right). \tag{6.7}$$

As noted by Grinstein and Wise [15] and Klebanov, Susskind, and Banks [16], then, *if* the correct criterion to determine the α's were to minimize A_4 rather than to maximize G^{-1}, then eq. (6.7) would favor that m^2 approach zero. But furthermore, because the mass-dependent renormalization of A_4 becomes *infinite* for $m^2 \to 0$, we can be confident that the calculable dependence of A_4 on the physical mass m^2 dominates nonuniversal effects for $m^2 \to 0$. We would conclude, therefore, that the pion mass m_π is exactly zero, if $m_\pi = 0$ can be achieved for any value of the α's. (It does not even matter that, for the pion, M_0 in eq. (6.7) is of order $4\pi f_\pi$, rather than the wormhole scale.)

This unpleasant conclusion can be avoided if, as we have proposed, the correct criterion to determine the α's is actually that G^{-1} is at is maximum. Then bare parameters at the wormhole scale must be adjusted so as to maximize the sum of the calculable infrared renormalization of G^{-1} in eq. (6.5) and of the uncalculable (but non-negligible) nonuniversal contributions to the renormalization of G^{-1}. The nature of the nonuniversal contributions depends on whether the light particles have bare masses or acquire their masses from renormalizable interactions.

Consider, for example, an elementary scalar with an adjustable bare mass. In this case we *can* argue that if the bare mass m_0 is chosen to maximize G^{-1}, then the renormalized scalar mass must either vanish or be of the order of the cutoff M_0. If G^{-1} is stationary as a function of m_0^2, we must have

$$0 = \frac{\partial}{\partial m_0^2}(16\pi G)^{-1}$$

$$= \frac{\partial}{\partial m_0^2}(16\pi G_0)^{-1} - \frac{1}{192\pi^2}\ln\left(M_0^2/m^2\right)\left(\frac{\partial}{\partial m_0^2}m^2\right) + \dots, \tag{6.8}$$

where G_0 includes all nonuniversal mass-dependent contributions to the renormalization of G. But because we expect that

$$\frac{\partial}{\partial m_0^2}m^2 \sim 1,$$

$$\frac{\partial}{\partial m_0^2}(16\pi G_0)^{-1} \lesssim \frac{1}{192\pi^2}, \tag{6.9}$$

eq. (6.8) can be satisfied only for

$$\ln\left(M_0^2/m^2\right) \sim 1. \tag{6.10}$$

Thus, while $16\pi G_0^{-1}$ might be a complicated function of m_0^2 that is maximized at a nontrivial value of m_0^2, this maximum will be destabilized by the calculable infrared-sensitive renormalization of G, if $m^2 \ll M_0^2$.

The situation is quite different, however, in the more realistic case of a particle that has no bare mass and acquires its physical mass from a renormalizable interaction, like a Yukawa coupling. In this case, we require G^{-1} to be stationary as a function of the bare Yukawa coupling λ_0, or

$$0 = \frac{\partial}{\partial \lambda_0^2}(16\pi G_0)^{-1} - \frac{1}{192\pi^2}\ln\left(M_0^2/m^2\right)\left(\frac{\partial}{\partial \lambda_0^2}m^2\right) + \dots. \tag{6.11}$$

But now we expect

$$\frac{\partial}{\partial \lambda_0^2}(16\pi G_0)^{-1} \lesssim \frac{1}{(16\pi^2)^2}M_0^2 \tag{6.12}$$

due to diagrams like fig. 17, and

$$\frac{\partial}{\partial \lambda_0^2}m^2 \sim v^2, \tag{6.13}$$

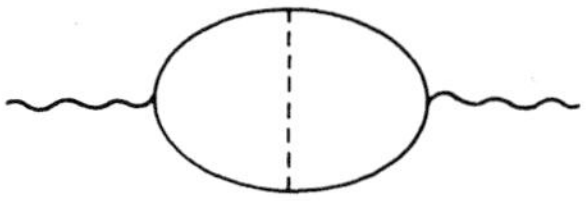

Fig. 17. A two-loop diagram that contributes to the renormalization G. The solid line represents a spin $-\frac{1}{2}$ fermion, and the dotted line represents a scalar.

where v is an expectation value that is determined by other, independent bare parameters*. The maximum of G_0^{-1} will not be destabilized by the infrared renormalization of G, then, provided only that $v^2 \ll M_0^2$.

We see, then, that if one accepts Coleman's explanation for the vanishing of the cosmological constant, then one must also find *unacceptable* any elementary particle that is light compared to the wormhole scale and that has a bare mass that can be adjusted at the wormhole scale, unless the physical mass of the particle vanishes identically**. This argument has nothing to do with the quadratically divergent renormalization of the mass of an elementary scalar. In fact, the same conclusion applies to a light elementary fermion with an adjustable bare mass, even though the fermion mass is protected from any power-divergent renormalization by approximate chiral symmetry.

If Coleman's mechanism is to apply to Nature, we must require therefore that all mass scales of low-energy physics are determined dynamically, rather than by adjusting bare mass parameters in the action S_{M_0} at the wormhole scale. This conclusion is actually unsurprising. If the bare parameters are to be determined by the *infrared-insensitive* renormalization of G, it would evidently require a miracle for the bare Higgs mass, for example, to be fine-tuned just so as to ensure that the weak-interaction mass scale is far below the Planck scale. But if all low-energy mass scales are determined dynamically, then it seems at least conceivable that the low-energy physics that we observe can be accounted for (to excellent accuracy) by maximizing the infrared-insensitive nonuniversal renormalization of G^{-1}.

The curious and rather surprising thing that we have found is that a light elementary particle with an adjustable bare mass *is* compatible with Coleman's mechanism if its physical mass is exactly zero. This observation, unfortunately, does not help us to understand why the weak-interaction scale is so small compared to the Planck scale, but it might have other interesting implications for low-energy physics.

In the end, we have been unable to extract from wormhole physics any quantitative conclusion other than $\Lambda = 0$. Indeed, it seems unlikely that any other precise

* The logarithmic renormalization of λ has been ignored here.

** I am assuming that the α-dependence of the bare mass is such that the physical mass vanishes for some value of α. Note also that, in order to simplify the discussion, I have assumed that elementary scalars are minimally coupled.

predictions can be made without a detailed understanding of Planck-scale physics. This is disappointing. On the other hand, we have probed carefully for fatal flaws in Coleman's beautiful explanation for the vanishing of the cosmological constant, and have found none. This is encouraging.

This research began in collaboration with Andy Cohen and Mark Wise; I am very grateful to both of them for many enlightening discussions. I have also benefited from conversations about wormholes with Jim Cline, Sidney Coleman, Gerry Gilbert, Stephen Hawking, Igor Klebanov, and Frank Wilczek.

Notes Added

There is an interesting exception to the general feature that the dependence of the renormalized Newton's constant G on other renormalized parameters cannot be computed. The dependence of G on the vacuum angle θ of quantum chromodynamics *is* determined by low-energy physics [31]. This exception arises because the θ-dependence of G is entirely due to nonperturbative strong-interaction effects that are extremely weak at very short distances.

More careful consideration indicates that the excluded volume interaction between large and small instantons, discussed in sect. 5, does not by itself suffice to suppress the effects of large wormholes. Rather, these interactions induce a very strong renormalization of the relation between the α-parameters of the large wormholes and the density of large instantons. I am indebted to Joe Polchinski for an enlightening discussion of this point. (See ref. [32].) If the large wormhole problem is to be evaded within the framework described in this paper, it will presumably be necessary to invoke a more subtle interaction between large and small instantons. Other recent discussions of the large wormhole problem may be found in ref. [33].

References

[1] S.W. Hawking, Phys. Lett. B195 (1987) 337; Phys. Rev. D37 (1988) 904

[2] G.V. Lavrelashvili, V. Rubakov and P.G. Tinyakov, JETP Lett. 46 (1987) 167

[3] S.B. Giddings and A. Strominger, Nucl. Phys. B306 (1988) 890;
K. Lee, Phys. Rev. Lett. 61 (1988) 263

[4] S.W. Hawking, *in* Relativity, groups, and topology II, ed. B. DeWitt and R. Stora (North-Holland, Amsterdam, 1983)

[5] J. Hartle, *in* High energy physics 1985, ed. M. Bowick and F. Gürsey (World Scientific, Singapore, 1986)

[6] J. Hartle and S.W. Hawking, Phys. Rev. D28 (1983) 2960

[7] S.W. Hawking, D.N. Page and C.N. Pope, Nucl. Phys. B170 (1980) 283;
S.W. Hawking, Commun. Math. Phys. 87 (1982) 395; Nucl. Phys. B244 (1984) 135

[8] S. Coleman, Nucl. Phys. B307 (1988) 867

[9] S.B. Giddings and A. Strominger, Nucl. Phys. B307 (1988) 854

[10] S. Coleman, Nucl. Phys. B310 (1988) 643

[11] T. Banks, Nucl. Phys. B309 (1988) 493
[12] S.W. Hawking and R. Laflamme, Phys. Lett. B209 (1988) 39
[13] S.B. Giddings and A. Strominger, Baby universes, third quantization, and the cosmological constant, Harvard University preprint HUTP-88/A036 (1988)
[14] S. Weinberg, Rev. Mod. Phys. 61 (1989) 1
[15] B. Grinstein and M.B. Wise, Phys. Lett. B212 (1988) 407
[16] I. Klebanov, L. Susskind and T. Banks, Wormholes and the cosmological constant, SLAC preprint SLAC-PUB-4705 (1988);
I. Klebanov and L. Susskind, Wormholes and cosmology, SLAC preprint SLAC-PUB-4734 (1988)
[17] A. Gupta and M.B. Wise, Comment on wormhole correlations (Revised Version), Caltech preprint CALT-68-1520 (1988)
[18] W. Fischler and L. Susskind, Phys. Lett. B217 (1989) 48
[19] M. Srednicki, Infinite quantization, UC Santa Barbara preprint UCSBTH-88-07 (1988);
S. Adler, On the Banks–Coleman–Hawking argument for the vanishing of the cosmological constant, Institute for Advanced Study preprint IASSNS-HEP 88/35 (1988);
J. Kim and K. Lee, The Scale problem in wormhole physics, FNAL preprint FERMILAB-PUB-88/95-T (1988);
M.B. Mijić, On the probability of having a universe with small cosmological constant, Santa Barbara Institute for Theoretical Physics preprint NSF-ITP-88-128 (1988);
G. Gilbert, Nucl. Phys. B, to be published;
R.C. Myers and V. Periwal, Constraints and correlations in the Coleman calculus, Santa Barbara Institute for Theoretical Physics preprint NSF-ITP-88-151 (1988)
[20] V. Kaplanovsky, unpublished
[21] S.W. Hawking, Phys. Lett. B134 (1984) 403
[22] E. Baum, Phys. Lett. B133 (1983) 185
[23] K. Wilson and J. Kogut, Phys. Rep. 12 (1974) 75
[24] R.P. Feynman, Statistical mechanics (W.A. Benjamin, New York, 1972)
[25] C.G. Callan, R. Dashen and D.J. Gross, Phys. Rev. D17 (1978) 2717
[26] S. Weinberg, *in* General relativity: an Einstein centenary survey, ed. S.W. Hawking and W. Israel (Cambridge University Press, Cambridge, 1979)
[27] M. Aizenman, Phys. Rev. Lett. 47 (1981); Commun. Math. Phys. 86 (1982) 1;
J. Fröhlich, Nucl. Phys. B200 (1982) 281
[28] I. Antoniadis and E.T. Tomboulis, Phys. Rev. D33 (1986) 2756
[29] D. Johnston, Nucl. Phys. B297 (1988) 721
[30] P. Candelas and D.J. Raine, Phys. Rev. D12 (1975) 965
[31] H. B. Nielsen and M. Ninomiya, A solution to the strong CP problem in baby universe theory, INS-721 (1988);
J. Preskill, S. T. Trivedi and M. B. Wise, Wormholes in spacetime and θ_{QCD}, Caltech preprint CALT-68-1539 (1989);
K. Choi and R. Holman, A wormhole solution to the strong CP problem, Carnegie-Mellon preprint CMU-HEP89-04 (1989)
[32] J. Polchinski, Decoupling versus excluded volume or return of the giant wormholes, University of Texas preprint UTTG-06-89 (1989)
[33] B. Grinstein, Charge quantization of wormholes and the finiteness of Newton's constant, FERMILAB-PUB-88/210-T (1988);
S. Coleman and K. Lee, Escape from the menace of the giant wormholes, Harvard preprint HUTP-89/A002 (1989);
S.-J. Rey, The collective dynamics and the correlations of wormholes in quantum gravity, UCSB print-89-0096 (1989);
R. Brustein and S. P. deAlwis, The cluster expansion for wormholes, UTTG-04-89 (1989)

WORMHOLES IN SPACETIME AND θ_{QCD} ☆

John PRESKILL [1], Sandip P. TRIVEDI and Mark B. WISE
California Institute of Technology, Pasadena, CA 91125, USA

Received 7 February 1989

We calculate in chiral perturbation theory the dependence of Newton's gravitational constant G on the θ parameter of quantum chromodynamics, and we find that G, as a function of θ, is minimized at $\theta \simeq \pi$. This calculation suggests that quantum fluctuations in the topology of spacetime would cause θ to assume a value very near π, contrary to the phenomenological evidence indicating that θ is actually near 0.

Wormholes are quantum fluctuations in the topology of spacetime. If such fluctuations occur in quantum gravity, their consequences may be profound. It has been proposed that, because of wormhole effects, the fundamental constants of nature are afflicted by an intrinsic quantum indeterminacy; we must regard our universe as having been chosen at random from an ensemble of possible universes, all with different values of the fundamental constants [1–3]. Coleman has proposed a form for the probability distribution of possible universes [4]. The different universes may be labeled by a set of parameters, denoted α, that are presumably infinite in number; all fundamental constants are functions of α. According to Coleman's proposal, the probability dP that a randomly selected universe has a label in the interval between α and $\alpha+d\alpha$ is

$$dP = d\alpha\, f(\alpha) \exp\{\exp[3/8G^2(\alpha)\Lambda(\alpha)]\} . \quad (1)$$

Here Λ is the cosmological constant, or vacuum energy, and G is Newton's gravitational constant; both are functions of α. The function f is a smooth function of α whose detailed properties need not concern us.

As Coleman observed, the distribution eq. (1) is very sharply peaked on the surface in α-space where $\Lambda(\alpha)=0$. This observation may explain why, in nature, the cosmological constant is indeed found to be very small. In fact, by regulating this distribution in a plausible way as Λ approaches zero, one finds that it strongly favors not only that $\Lambda(\alpha)=0$, but also that $G(\alpha)$ assume the smallest possible value on the surface in α-space where $\Lambda(\alpha)=0$ [5,6]. The requirement that $G(\alpha)$ is at its minimum on the $\Lambda=0$ surface is likely to determine all of the remaining α's, and hence to fix all other constants of nature. If the possible universes are subject to the probability distribution eq. (1), then, the quantum indeterminacy of fundamental physics turns out to be very mild. For each fundamental constant there is a "standard" value that is in principle calculable, and the probability is unity that in a randomly selected universe each constant assumes its standard value.

(Actually, the distribution eq. (1) is not universally accepted [7,8]. Hawking [7], for example, suggests that the correct behavior of the probability distribution for Λ near zero is $dP \sim d\alpha \exp(3/8G^2\Lambda)$, rather than the double exponential in eq. (1). But the conclusion that G assumes, with probability equal to unity, its minimum value on the surface $\Lambda=0$ would apply also in this case.)

While the constants of nature may be determined in principle by the distribution eq. (1), the values of constants other than the cosmological constant cannot be easily computed. To determine other parameters that characterize low-energy physics, we must minimize G as a function of these parameters. But the functional dependence of G on other couplings of

☆ This work is supported in part by the US Department of Energy under Contract No. DE-AC0381-ER40050.
[1] NSF Presidential Young Investigator.

interest is typically very sensitive to the details of physics at the Planck scale; it cannot be computed in terms of low-energy physics alone [6].

Our main purpose in this letter is to point out an exception to the general rule annunciated above. The dependence of G on the θ-parameter of quantum chromodynamics [9] *can* be computed in terms of low-energy physics alone. The point is that θ-dependence arises only through nonperturbative strong interaction effects, and these are presumably exponentially small at the Planck scale, because of asymptotic freedom. We have calculated the θ-dependence of G in an approximation that is valid if the masses of the light up and down quarks are sufficiently small. We find, assuming the validity of this approximation, that the minimum of G occurs for θ very near π. (The minimum would be at exactly $\theta = \pi$ were it not for small CP-violating effects due to the weak interactions.)

We therefore expect that $\theta \simeq \pi$ is overwhelmingly favored by the probability distribution eq. (1). Since the phenomenological evidence suggests that instead $\theta \simeq 0$, our calculation indicates a potentially serious conflict between current ideas about wormholes and observed low-energy physics. We will comment further below about how this conflict might be resolved.

Because the θ parameter is CP odd, and the strong interactions conserve CP to remarkable accuracy, it has long been recognized that θ must be extremely close to either 0 or π. ($\theta = \pi$ is a CP-conserving value because θ is a periodic variable defined modulo 2π.) The experimental limit on the electric dipole moment of the neutron indicates that θ deviates from 0 or π by an amount at most of order 10^{-9} [10]. Given that large CP-violating phases do infect the weak interactions, this inclination of the strong interaction to conserve CP poses a serious puzzle. The most satisfying explanation for the CP conservation by the strong interaction is that originally suggested by Peccei and Quinn [11]. They proposed that θ is actually a dynamical variable, and therefore assumes that value that minimizes the energy density of the vacuum. A powerful nonperturbative argument shows that the vacuum energy density of QCD, as a function of θ, is minimized at $\theta = 0$ [12]. Thus, the Peccei–Quinn mechanism naturally explains why θ is very close to the CP-conserving value. (The minimum of the energy density is perturbed slightly away from $\theta = 0$ by CP-violating effects due to the weak interaction. The amount of the perturbation depends on the detailed nature of CP violation; in the Kobayashi–Maskawa model, one can estimate that the minimum occurs for $\theta \sim 10^{-14}$, which is well within the experimental limit.) Another interesting consequence of this mechanism is that there exists a very light, very weakly interacting particle, the axion, associated with the oscillations of θ about the minimum [13,14]. It has even been proposed that these axions comprise the dark matter of the universe [15].

With the context of wormhole physics, the Peccei–Quinn explanation for $\theta \simeq 0$ is problematic. Their mechanism relies on the existence of an approximate global symmetry, the Peccei–Quinn symmetry, that is intrinsically broken only by a color anomaly. But wormholes have no respect for global symmetries (whether exact or approximate). Rather, wormhole effects are expected to generate α-dependent couplings of all types consistent with the *local* symmetries of fundamental physics [1,2]. It will not do, then, to invoke a Peccei–Quinn symmetry by fiat; the symmetry itself requires an explanation.

(A similar remark applies to another explanation that is sometimes proposed for the small value of the electric dipole moment of the neutron – that the mass of the up quark is zero, or very close to zero. This is no explanation unless one understands *why* the up quark is massless. Indeed, this proposal is closely related to the Peccei–Quinn mechanism, for if the up quark is massless, then there is an approximate global symmetry that is intrinsically broken only by a color anomaly.)

In spite of the above comments, wormholes and the Peccei–Quinn mechanism might be reconcilable. Two possibilities come to mind. Perhaps an approximate Peccei–Quinn mechanism arises in low-energy physics as an accidental consequence of local symmetries, which are not disturbed by wormhole effects. (This would be like the approximate conservation of baryon number in the standard model that is an automatic consequence of local $SU(3) \times SU(2) \times U(1)$ invariance.) It is not so easy to make this idea workable, however. The problem is that it does not suffice for the accidental Peccei–Quinn symmetry to apply to the operators in the effective action that are of renormalizable type (dimension four or less). If the Peccei–Quinn mechanism is to ensure that θ is very small, then nonperturbative strong interaction effects must swamp all other effects that break the Peccei–Quinn symmetry; this constraint typically requires that the symmetry be satisfied by operators of quite high dimension [16]. There is another possible way to rescue the Peccei-Quinn mechanism, in spite of the tendency of wormhole effects to break global symmetries. Although the Peccei-Quinn symmetry is badly broken for generic values of α, it may become a good approximate symmetry for that particular "standard" value of α that minimizes G.

At any rate, there appears to be ample motivation to consider whether, within the context of wormhole physics, the CP conservation of the strong interactions can be explained without appealing to the Peccei–Quinn mechanism, and without requiring the existence of a light axion. Indeed, the crucial feature of the Peccei–Quinn mechanism is that it makes θ an adjustable quantity, a dynamical variable that seeks the minimum of the energy density at $\theta=0$. And wormhole effects also make θ an adjustable quantity, not a dynamical variable, but an α-dependent coupling constant that seeks the sharp peak in the probability distribution eq. (1). Furthermore, as Nielsen and Ninomiya [17] recently stressed, G is CP even while θ is CP odd; therefore strong interaction effects generate a dependence of G on θ that is an even function of θ. This function is stationary at both $\theta=0$ and $\theta=\pi$, and so it is reasonable to expect that its minimum occurs either at $\theta=0$ or at $\theta=\pi$. Since the peak in the probability distribution occurs where G is minimized, the CP conservation by the strong interactions is naturally explained. (As for the Peccei–Quinn mechanism, CP-violating effects due to the weak interventions perturb the minimum, but only slightly.)

The dependence of θ on α arises as follows: The Yukawa couplings of the quarks to the Higgs doublet are modified by wormhole effects, and hence are α-dependent in both modulus and phase. When the electroweak gauge symmetry is spontaneously broken, this α-dependence enters the quark mass matrix. Some of the phases in the mass matrix are unobservable, because they can be removed by a redefinition of the phases of the quark fields. But there remain, as observable parameters, the values of the quark masses and the Kobayashi–Maskawa angles and phases that infect the charged weak current. Finally, there is one phase that can be removed from the quark mass matrix only by a field redefinition that has a color anomaly. This phase is θ. It is irrelevant in all orders of perturbation theory, but nonperturbative strong interactions do depend on θ.

We will assume in the ensuing discussion that it is possible to adjust the α-parameters so that θ changes, while all other couplings in the effective lagrangian remain fixed. It is easy to construct toy models that behave this way, and we expect that this behavior is reasonably generic. When the α-dependence of Newtons's constant G is considered, one finds that perturbative renormalization effects induce large contributions to G that depend on the quark masses and the KM angles. These contributions are of order $M_{\rm Pl}^2$, where $M_{\rm pl}$ is the Planck mass scale, and are sensitive to the details of Planck-scale physics. The criterion that $G(\alpha)$ is at its minimum on the surface $\Lambda(\alpha)=0$, then, determines these quantities, but only in a manner that cannot be computed based on a knowledge of low-energy physics alone [6]. But since the dependence of G on θ arises only from nonperturbative strong interaction effects, θ is calculable based on low-energy physics alone, at least in principle.

Before we proceed with our calculation of $G(\theta)$, one more point needs emphasis. We asserted above that θ can be determined by finding the minimum of $G(\theta)$, but the actual criterion that determines the constants of nature is that $G(\alpha)$ is minimized on the surface where $\Lambda(\alpha)=0$. We must explain why it is an excellent approximation to disregard the requirement that $\Lambda(\alpha)=0$. The crucial point is that the dependence of Λ and G on θ is characterized by the strong interaction scale, rather than the Planck scale. If we perturb θ by a small amount $\delta\theta$, Λ and G change according to

$$\delta\Lambda=a(\theta)\delta\theta,\quad \delta(1/16\pi G)=b(\theta)\delta\theta\,; \tag{2}$$

we will calculate $a(\theta)$ and $b(\theta)$ below, in chiral pertur ation theory. But when a generic α-parameter is perturbed by $\delta\alpha$, we have instead, schematically,

$$\delta\Lambda\sim M_{\rm Pl}^4\,\delta\alpha,\quad \delta(1/16\pi G)\sim M_{\rm Pl}^2\,\delta\alpha\,. \tag{3}$$

Thus, if we perturb θ and adjust α slightly to remain on the $\Lambda=0$ surface, the change of G is given by

$$\delta(1/16\pi G)|_{\delta\Lambda=0}\sim[b(\theta)-a(\theta)/M_{\rm Pl}^2]\delta\theta\,. \tag{4}$$

Because a and b are very small in Planck units, the second term in eq. (4) is negligible. We may just as well minimize $G(\theta)$ without regard for the $\Lambda=0$ constraint.

Now we are finally prepared to describe the calculation of $G(\theta)$, in chiral perturbation theory. The main idea that underlies the calculation is quite simple. If the pion mass were very small, as would be true if the up and down quarks were sufficiently light, then the strong interaction contribution to G would be dominated by a one-pion-loop diagram that has a calculable logarithmic sensitivity to m_π^2. Then, when the pion is light enough, the dependence of G on θ can be calculated from the dependence of m_π^2 on θ. One finds that $G(\theta)$ is minimized when $m_\pi^2(\theta)$ is minimized. And it is easy to see, again in the limit where the pion is sufficiently light, that $m_\pi^2(\theta)$ is minimized at $\theta=\pi$.

To perform the calculation, we make use of a chiral lagrangian that describes the self-interactions at low momenta of the pseudo-Goldstone bosons π^+, π^-, π^0. This chiral lagrangian respects a nonlinearly realized $SU(2)_L\times SU(2)_R$ chiral symmetry. It can be expressed in terms of a field $\Sigma(x)$ that is a 2×2 unitary matrix with determinant one, and that transforms under chiral symmetry as

$$\Sigma\to V_L\Sigma(x)V_R^\dagger , \tag{5}$$

where $V_L\in SU(2)_L$ and $V_R\in SU(2)_R$. In terms of the pion fields, Σ can be expressed as

$$\Sigma=\exp(2i\Pi/f) ,$$
$$\Pi=\begin{pmatrix} \pi^0/\sqrt{2} & \pi^+ \\ \pi^- & -\pi^0/\sqrt{2}\end{pmatrix}, \tag{6}$$

where f is the pion decay constant. The chiral lagrangian can be expanded in powers of the derivatives acting on the Σ field; terms with more derivatives are suppressed at low energy by additional powers of the pion momentum.

The effects of the explicit breaking of chiral symmetry by quark masses can also be systematically incorporated in the chiral lagrangian. If m is the 2×2 mass matrix of the light quarks, then QCD respects a formal symmetry in which eq. (5) is accompanied by

$$m\to V_R m V_L^\dagger . \tag{7}$$

By demanding invariance under this formal symmetry, we find that the leading mass-dependent terms in the chiral lagrangian, in a curved spacetime background, are

$$\mathscr{L}_{mass}=v\,\mathrm{tr}(m\Sigma+\Sigma^\dagger m^\dagger)(1+cR)+\dots . \tag{8}$$

Here v is a quantity with the dimensions of $(\text{mass})^3$, and c is a quantity with the dimensions of $(\text{mass})^{-2}$; both are determined by nonperturbative strong interaction effects. R is the curvative scalar of the background spacetime. In eq. (8), we have neglected terms that contain derivatives of the Σ field, more powers of the light quark mass matrix m, or more powers of the curvature R.

The θ-parameter enters the chiral lagrangian in the light quark mass matrix m, through the relation

$$\theta=\arg(\det m) . \tag{9}$$

We find the precise form of m by performing a chiral rotation of m that ensures that $\mathscr{L}_{mass}$ contains no "tadpole" terms linear in the pion fields. The result is

$$m=\begin{pmatrix} m_u\,e^{i\phi} & 0 \\ 0 & m_d\,e^{i(\theta-\phi)}\end{pmatrix}, \tag{10}$$

where

$$\sin\phi=\frac{m_d\sin\theta}{(m_u^2+m_d^2+2m_u m_d\cos\theta)^{1/2}},$$
$$\sin(\theta-\phi)=\frac{m_u}{m_d}\sin\phi . \tag{11}$$

By expanding Σ in powers of the pion field, we find, in tree approximation in the chiral lagrangian and to lowest order in light quark masses,

$$m_\pi^2(\theta)=(4v/f^2)(m_u^2+m_d^2+2m_u m_d\cos\theta)^{1/2} ,$$
$$\Lambda(\theta)=\Lambda_0-\tfrac{1}{2}f^2 m_\pi^2(\theta) ,$$
$$1/16\pi G=1/16\pi G_0-\tfrac{1}{2}cf^2 m_\pi^2(\theta) \tag{12}$$

here Λ_0 and G_0 are constants independent of θ.

Arguments based on QCD inequalities show that v is nonnegative [12]. Hence, the vacuum energy is evidently minimized at $\theta=0$, as is required for the Peccei–Quinn mechanism to work. But the expression for G in eq. (12) could be minimized at either $\theta=0$ or $\theta=\pi$, depending on the sign of c. Though the sign of c is determined in principle by the nonperturbative strong interactions, we do not know how to compute it reliably. Nonetheless, what we have found is consistent with the expectation of Nielsen and Ninomiya, that the minimum of G occurs at a *CP*-conserving value of θ.

In fact, it is possible to go further, because the tree approximation contribution to the θ-dependence of $(16\pi G)^{-1}$ in eq. (12) is not actually the leading contribution when m_π^2 is very small. There is a contribution from one pion loop that is enhanced by a logarithm of the pion mass. This logarithmically enhanced contribution is [18,5,6]

$$\delta(1/16\pi G)$$

$$= -(1/64\pi^2) m_\pi^2(\theta) \ln[M_{\rm CSB}^2/m_\pi^2(\theta)] \,, \qquad (13)$$

where $M_{\rm CSB}$ is the "chiral symmetry breaking scale" of QCD; a naive estimate of it is $M_{\rm CSB} \sim 4\pi f \sim 1$ GeV. (Eq. (13) is the one-loop contribution to $(16\pi G)^{-1}$ that arises from the minimal coupling of the pion to gravity. There is also a one-loop contribution that involves the nonminimal coupling of the pion to R in eq. (8), but this contribution is of order $(m_\pi^4 \ln m_\pi^2)$ and hence higher order in chiral perturbation theory.)

If the pion mass is sufficiently small, then the one-loop contribution to $(16\pi G)^{-1}$ in eq. (13) dominates the tree contribution in eq. (12). The calculated sign of the one-loop contribution shows that $m_\pi^2(\theta)$ seeks the smallest possible value in order to minimize $G(\theta)$. In view of the expression for $m_\pi^2(\theta)$ in eq. (12), this means that $\theta=\pi$ is the preferred value. We have shown, then, that at least in a world in which the light quark masses are sufficiently small, the criterion that $G(\alpha)$ is at its minimal value on the surface in α-space where $\Lambda(\alpha)=0$ requires θ to be very close to π. (As in the Peccei–Quinn model, weak interactions perturb θ slightly away from the value chosen by QCD, by an amount of order 10^{-14} in the KM model of *CP* violation.)

The approximation of neglecting the contribution to $(16\pi G)^{-1}$ in eq. (12) compared to the contribution in eq. (13) is justified provided that

$$(1/32\pi^2 cf^2) \ln(M_{\rm CSB}^2/m_\pi^2) \gg 1 \,, \qquad (14)$$

it is not clear whether it is justified for realistic values of the light quark masses. To get some insight about whether the conclusion that G is minimized at $\theta=\pi$ survives beyond the approximation of very light quark masses, we have considered the opposite limit of infinite quark masses, or pure Yang–Mills theory. In pure Yang–Mills theory, we have computed $G(\theta)$ in the dilute instanton gas approximation. Unlike chiral perturbation theory, which can be justified when the quark masses are small enough, the dilute instanton gas approximation cannot really be justified. Nonetheless, it is known to give the right answer for the vacuum energy; namely, that the minimum occurs at $\theta=0$, in agreement with the QCD inequality argument.

To calculate $G(\theta)$ we compute the connected two-point function of the energy–momentum tensor and extract its leading behavior at low momentum. The calculation turns out to involve a subtlety concerning the trace anomaly in the presence of instantons; we will not report on the details here. The result is that the minimum of $G(\theta)$ occurs at $\theta=\pi$. Thus, the dilute instanton gas calculation lends support to the view that $G(\theta)$ is minimized at $\theta=\pi$ generically, irrespective of the value of quark masses. Perhaps it will eventually be possible to resolve this issue by doing numerical calculations in lattice QCD.

Finally, let us consider whether our conclusion that $\theta \simeq \pi$ is in conflict with experiment. There is suggestive evidence that θ is actually close to zero in nature [10]. But one should recall that this evidence is based on chiral perturbation theory calculations of the pseudoscalar meson masses that treat the *strange* quark mass as a small parameter, a somewhat dubious procedure [19]. If the corrections to leading order perturbation theory in the strange quark mass turn out to be surprisingly large, then it may be that θ is really close to π in nature after all, as wormhole considerations indicate. Again, this issue may ultimately be resolved by lattice QCD calculations.

To summarize, we have argued that, at least in an approximation in which the masses of the up and down quarks are taken to be very small, wormhole fluctuations in the topology of spacetime drive the θ parameter of QCD to $\theta \simeq \pi$. Since $\theta \simeq 0$ appears to be satisfied in nature, this prediction poses a possible conflict between wormhole physics and experiment. We have noted several ways in which this conflict might be resolved. Perhaps a Peccei–Quinn symmetry can survive in spite of wormhole effects, allowing θ to relax dynamically to the value $\theta=0$. Perhaps chiral perturbation theory is misleading, and wormholes actually prefer $\theta \simeq 0$ for realistic values of the light quark masses. And finally, it is at least conceivable that $\theta \simeq \pi$ really is satisfied in nature, in accord with our prediction.

After completion of this work we found that Choi and Holman have also concluded (using different methods) that wormholes favor $\theta = \pi$ [20].

References

[1] S.W. Hawking, Phys. Rev. D 37 (1988) 904.
[2] S. Coleman, Nucl. Phys. B 307 (1988) 867.
[3] S. Giddings, and A. Strominger, Nucl. Phys. B 307 (1988) 854.
[4] S. Coleman, Nucl. Phys. B 310 (1988) 643.
[5] B. Grinstein and M.B. Wise, Phys. Lett. B 212 (1988) 407.
[6] J. Preskill, preprint CALT-68-31-88 (1988), unpublished.
[7] S.W. Hawking, Phys. Lett. B 134 (1984) 403;
E. Baum, Phys. Lett. B 133 (1983) 185.
[8] J. Polchinski, preprint UTT G-31-88 (1988), unpublished.
[9] G. 't Hooft, Phys. Rev. Lett. 37 (1976) 8; Phys. Rev. D 14 (1976) 3432;
C.G. Callan, R.F. Dashen and D.J. Gross, Phys. Lett. B 63 (1976) 334;
R. Jackiw and C. Rebbi, Phys. Rev. Lett. 37 (1976) 177.
[10] V. Baluni, Phys. Rev. D 19 (1979) 2227;
R.J. Crewther, P. DiVecchia, G. Veneziano and E. Witten, Phys. Lett. B 88 (1979) 123.
[11] R.D. Peccei and H.R. Quinn, Phys. Rev. Lett. 38 (1977) 1440.
[12] C. Vafa and E. Witten, Phys. Rev. Lett. 53 (1984) 535.
[13] S. Weinberg, Phys. Rev. Lett. 46 (1978) 223;
F. Wilczek, Phys. Rev. Lett. 46 (1978) 279.
[14] J.E. Kim, Phys. Rev. Lett. 43 (1979) 103;
M. Dine, W. Fishler and M. Srednicki, Phys. Lett. B 104 (1981) 199.
[15] J. Preskill, M.B. Wise and F. Wilczek, Phys. Lett B 120 (1983) 127;
L.F. Abbott and P. Sikivie, Phys. Lett. B 120 (1983) 133;
M. Dine and W. Fishler, Phys. Lett. B 120 (1983) 137.
[16] H. Georgi, L.M. Hall and M.B. Wise, Nucl. Phys. B 192 (1981) 409.
[17] H.B. Nielsen and M. Ninomiya, preprint INS-Rep-721 (1988), unpublished.
[18] P. Candelas and D.J. Raine, Phys. Rev. D 12 (1975) 965.
[19] D.B. Kaplan and A.V. Manohar, Phys. Rev. Lett. 56 (1986) 2004.
[20] K. Choi and R. Holman, preprint CMU-HEP 89-04 (1989), unpublished.

V. GRAVITATIONAL INSTANTONS

ASYMPTOTICALLY FLAT SELF-DUAL SOLUTIONS TO EUCLIDEAN GRAVITY

Tohru EGUCHI
Stanford Linear Accelerator Center, Stanford University, Stanford, CA 94305, USA

and

Andrew J. HANSON
Lawrence Berkeley Laboratory, University of California, Berkeley, CA 94720, USA

Received 6 February 1978

In an attempt to find gravitational analogs of Yang–Mills pseudoparticles, we obtain two classes of self-dual solutions to the euclidean Einstein equations. These metrics are free from singularities and approach a flat metric at infinity.

The discovery of pseudoparticle solutions to the euclidean SU(2) Yang–Mills theory [1] has suggested the possibility that analogous solutions might occur in Einstein's theory of gravitation. The existence of such solutions would have a profound effect on the quantum theory of gravitation [2,3]. Since the Yang–Mills pseudoparticles possess self-dual field strengths, one likely possibility is that gravitational pseudoparticles are characterized by self-dual curvature.

In fact it has been pointed out by Hawking [3] that the Taub-NUT metric [4], when appropriately continued to euclidean space–time, produces a self-dual curvature and hence is a possible candidate for a gravitational pseudoparticle. He has also given a generalized multi-Taub-NUT metric. However, these metrics do not approach a flat metric at infinity [5]. To see this, let us write the euclidean Taub-NUT solution as

$$(\mathrm{d}s)^2 = [(R+m)/(R-m)]\,\mathrm{d}R^2 + 4(R^2-m^2)\{\sigma_x^2+\sigma_y^2+(2m/(R+m))^2\sigma_z^2\}, \tag{1}$$

where $\sigma_x, \sigma_y, \sigma_z$ form a standard Cartan basis,

$$\begin{aligned}\sigma_x &= \tfrac{1}{2}(-\cos\psi\,\mathrm{d}\theta - \sin\theta\sin\psi\,\mathrm{d}\phi),\\ \sigma_y &= \tfrac{1}{2}(\sin\psi\,\mathrm{d}\theta - \sin\theta\cos\psi\,\mathrm{d}\phi),\\ \sigma_z &= \tfrac{1}{2}(-\mathrm{d}\psi - \cos\theta\,\mathrm{d}\phi),\end{aligned} \tag{2}$$

obeying the structure equations of the exterior algebra [6],

$$\mathrm{d}\sigma_x = 2\sigma_y \wedge \sigma_z, \tag{3}$$

etc. Here θ, ψ and ϕ are Euler angles on S^3 with ranges $0 \leqslant \theta \leqslant \pi$, $0 \leqslant \phi \leqslant 2\pi$, $0 \leqslant \psi \leqslant 4\pi$. Then it is easy to see that the above metric describes a distorted 3-dimensional hypersphere S^3 for any fixed value of $R > m$.

Since a Yang–Mills pseudoparticle approaches a pure gauge at infinity and is interpreted as inducing transitions between topologically inequivalent vacua, one might require that gravitational analogs have a similar asymptotic behavior. In this letter we explore the possibility of gravitational pseudoparticles which possess a self-dual curvature and approach a flat metric at infinity. In the following we present two classes of such solutions. They are both singularity-free in the entire spacetime and their manifolds have a simple topological structure.

In deriving these solutions we exploit a particularly useful choice of gauge (local Lorentz frame). First we define a local orthonormal frame using the vierbeins $e^a{}_\mu$, and take

$$e^a = e^a{}_\mu\,\mathrm{d}x^\mu. \tag{4}$$

In terms of the e^a, the metric is expressed as $\mathrm{d}s^2 = (e^0)^2 + (e^1)^2 + (e^2)^2 + (e^3)^2$. Then the connection one-form $\omega^a{}_b$ is defined by

$$de^a = -\omega^a{}_b \wedge e^b, \quad \omega^a{}_b = -\omega^b{}_a. \tag{5}$$

Latin indices are raised and lowered by a flat metric. Then we define the curvature two-form by

$$R^a{}_b = d\omega^a{}_b + \omega^a{}_c \wedge \omega^c{}_b. \tag{6}$$

Now we note that if $\omega^a{}_b$ is self-dual,

$$\omega^0{}_1 = -\omega^2{}_3, \tag{7}$$

etc., then $R^a{}_b$ is self-dual. This follows directly from the definition (6) of $R^a{}_b$. Since any self-dual curvature gives a vanishing Ricci tensor, any metric yielding a self-dual connection is a solution to the Einstein equation. On the other hand, it is easy to show that any self-dual curvature can be obtained, by a suitable change of gauge, from a metric yielding a self-dual connection ‡1. In this "self-dual gauge", the problem of finding a self-dual solution to the Einstein equation [7] is therefore reduced to one of finding self-dual connections and hence solving first-order differential equations generated by eq. (5). This is quite analogous to the Yang–Mills case [1].

In the following we consider two types of metrics having axial symmetry as in the Taub-NUT case ‡2:

$$\text{I: } (ds)^2 = f^2(r)\,dr^2 + r^2g^2(r)(\sigma_x^2 + \sigma_y^2) + r^2\sigma_z^2, \tag{8}$$

$$\text{II: } (ds)^2 = f^2(r)\,dr^2 + r^2(\sigma_x^2 + \sigma_y^2) + r^2g^2(r)\sigma_z^2. \tag{9}$$

Here we consider these metrics directly in the euclidean space and do not regard them as a result of some continuation from the Minkowski regime. Asymptotic flatness requires that

$$\lim_{r\to\infty} f(r) = \lim_{r\to\infty} g(r) = 1. \tag{10}$$

Taking as our orthonormal frames

$$\text{I: } e^a = (f(r)\,dr, rg(r)\sigma_x, rg(r)\sigma_y, r\sigma_z), \tag{11}$$

$$\text{II: } e^a = (f(r)\,dr, r\sigma_x, r\sigma_y, rg(r)\sigma_z), \tag{12}$$

we find after some simple algebra that the self-duality of the connection implies

$$\text{I: } g^2 = f(2g^2 - 1), \quad f = g(g + rg'), \tag{13}$$

$$\text{II: } fg = 1, \quad f(2 - g^2) = g + rg'. \tag{14}$$

Asymptotically flat solutions are given, respectively, by

$$\text{I: } f(r) = \tfrac{1}{2}(1 + [1 - (a/r)^4]^{-1/2}), \tag{15}$$

$$g(r) = \{\tfrac{1}{2}(1 + [1 - (a/r)^4]^{1/2})\}^{1/2}, \tag{16}$$

$$\text{II: } g(r) = f^{-1}(r) = [1 - (a/r)^4]^{1/2}, \tag{17}$$

where a is an integration constant. The curvature components of case II are given by

$$\begin{aligned} R^0{}_1 &= -R^2{}_3 = -(2a^4/r^6)(e^0 \wedge e^1 - e^2 \wedge e^3),\\ R^0{}_2 &= -R^3{}_1 = -(2a^4/r^6)(e^0 \wedge e^2 - e^3 \wedge e^1),\\ R^0{}_3 &= -R^1{}_2 = +(4a^4/r^6)(e^0 \wedge e^3 - e^1 \wedge e^2). \end{aligned} \tag{18}$$

The curvatures for case I have the same algebraic form with the replacement

$$2a^4/r^6 \to -a^4/2r^6g^6. \tag{19}$$

Hence in both cases the curvatures are regular everywhere for $r \geqslant a$ and fall off like $1/r^6$ at infinity. For comparison, we note that the Taub-NUT curvature produced by eq. (1) is obtained by the replacement

$$2a^4/r^6 \to m/(R + m)^3, \tag{20}$$

and thus goes like $1/R^3$ at infinity.

The manifolds described by the above metrics have the topology $R \times S^3$. Although the metrics have an apparent singularity at $r = a$, it can be eliminated by a change of variable,

$$u^2 = r^2(1 - (a/r)^4). \tag{21}$$

For instance the solution II now takes the form

$$(ds)^2 = du^2/(1 + (a/r)^4)^2 + u^2\sigma_z^2 + r^2(\sigma_x^2 + \sigma_y^2). \tag{22}$$

Our next task is to compute topological invariants of the manifold. Here, as in the Taub-NUT case [8], we have to be careful about possible contributions from the boundary of the manifold.

$\hat{A}$-genus (axial anomaly). The Atiyah–Patodi–Singer theorem [9] gives the $\hat{A}$-genus of the manifold $[r_1, r_2] \times S^3$ as

‡1 The proof involves decomposing any given spin connection $\omega^a{}_b$ into self-dual and anti-self-dual parts. If $R^a{}_b$ is self-dual, the anti-self-dual part of $\omega^a{}_b$ is a pure O(4) gauge transformation, $\Lambda^a{}_c(d\Lambda^{-1})^c{}_b$, and can be gauged away.

‡2 The spherically symmetric ansatz, $ds^2 = f^2dr^2 + r^2g^2(\sigma_x^2 + \sigma_y^2 + \sigma_z^2)$, leads to a trivially flat metric when we impose self-duality.

$$\hat{A}(r_1, r_2) = \hat{A}_{\text{vol}} - (\hat{A}_{\text{surf}} + \tfrac{1}{2}(h_{\text{D}} + \eta_{\text{D}}))|_{r_1}^{r_2}. \tag{23}$$

$\hat{A}_{\text{vol}}$ is the volume integral of the Riemann curvature tensor contracted with its dual and $\hat{A}_{\text{surf}}$ gives the contribution due to the deviation of the metric from a product metric on the boundary [10]. h_{D} is the number of harmonic spinors of the Dirac operator restricted to the boundary and η_{D} gives its spectral asymmetry [9,11]. Using the formulas in refs. [8] and [11] we obtain

$$\hat{A}(r_1 = a, r_2 = \infty) = \tfrac{1}{4} - 0 + (-\tfrac{1}{6} - \tfrac{1}{12}) = 0, \tag{24}$$

for both solutions I and II. Thus these solutions by themselves will not induce chiral symmetry breakdown, just as in the Taub-NUT case [8].

Euler–Poincaré characteristic (trace anomaly). The Euler–Poincaré characteristic χ is related to the thermal effects of gravitational pseudoparticles [3,12]. To calculate χ, we apply the Chern–Gauss–Bonnet theorem [13],

$$\chi = \chi_{\text{vol}} - \chi_{\text{surf}}|_{r_1}^{r_2}, \tag{25}$$

where χ_{vol} and χ_{surf} are the analogs of $\hat{A}_{\text{vol}}$ and $\hat{A}_{\text{surf}}$ in eq. (23). Using the known formulas, we find for both solutions I and II the Euler characteristic ‡3

$$\chi(r_1 = a, r_2 = \infty) = 3 - (-1) + (-4) = 0. \tag{26}$$

This of course agrees with the combinatorial calculation for R X S^3.

We observe that at large r, our curvatures fall like $1/r^6$; in contrast, the euclidean Taub-NUT and Schwarzschild solutions fall like $1/r^3$. This suggests that our metrics describe gravitational "dipoles" while Taub-NUT and Schwarzschild describe monopoles. This is probably a sign that our euclidean solutions will not have a meaningful continuation to Minkowski space, as is the case for the Yang–Mills pseudoparticle.

We are deeply indebted to I. Singer for a number of informative discussions. This research was performed under the auspices of the Division of Physical Research of the U.S. Department of Energy.

‡3 It appears that the manifold of solution II can be compactified by adding an S^2 at $r = a$. In this case (see eq. (22)) the manifold acquires the local topology of $D^2 \times S^2$; since as $r \to a$, the D^2 shrinks to a point, the manifold is homotopic to S^2. If we then omit the $r = a$ boundary term in eq. (26), we obtain $\chi = 4$. However, we know $\chi = 2$ for a manifold homotopic to S^2. Hence the Chern–Gauss–Bonnet theorem requires a "corner" correction in this case. A similar situation occurs if one puts a metric on a cone and tries to compute the Euler characteristic using the Gauss–Bonnet theorem without correcting for the apex. For solution I, analogous arguments indicate that the manifold compactified at $r = a$ is homotopic to the manifold of SO(3). Then the apparent Euler characteristic is 4, while the true value is $\chi = 0$. The compactified manifolds admit a spin structure because the second Stiefel–Whitney classes vanish [14]. However, in practice the "corners" may make it difficult to treat the Dirac operator on the whole manifold. If such an operator can be defined, the $\hat{A}$-genus (axial anomaly) would also require "corner" corrections. This problem is under study.

References

[1] A.A. Belavin, A.M. Polyakov, A.S. Schwarz and Yu.S. Tyupkin, Phys. Lett. 59B (1975) 85.

[2] T. Eguchi and P.G.O. Freund, Phys. Rev. Lett. 37 (1976) 1251;
A.A. Belavin and D.E. Burlankov, Phys. Lett. 58A (1976) 7.

[3] S.W. Hawking, Phys. Lett. 60A (1977) 81.

[4] A. Taub, Ann. Math. 53 (1951) 472;
E. Newman, L. Tamburino and T. Unti, J. Math. Phys. 4 (1963) 915.

[5] C.W. Misner, J. Math. Phys. 4 (1963) 924.

[6] S. Helgason, Differential geometry and symmetric spaces (Academic, 1962);
H. Flanders, Differential forms (Academic, 1963).

[7] For other treatments of self-dual gravitational fields, see e.g.: E.T. Newman, Gen. Rel. Grav. 7 (1976) 107;
R. Penrose, Gen. Rel. Grav. 7 (1976) 31;
J. Plebanski, J. Math. Phys. 16 (1975) 2395;
C.W. Fette, A.I. Janis and E.T. Newman, J. Math. Phys. 17 (1976) 660.

[8] T. Eguchi, P.B. Gilkey and A.J. Hanson, Phys. Rev. D17 (1978) 423;
H. Römer and B. Schroer, Phys. Lett. 71B (1977) 182.

[9] M.F. Atiyah, V.K. Patodi and I. Singer, Bull. London Math. Soc. 5 (1973) 229; Proc. Camb. Phil. Soc. 77 (1975) 43; 78 (1976) 405; 79 (1976) 71.

[10] P.B. Gilkey, Adv. Math. 15 (1975) 334.

[11] N. Hitchin, Adv. Math. 14 (1974) 1.

[12] See, e.g.: M.J. Duff, Nucl. Phys. B125 (1977) 334.

[13] S.S. Chern, Ann. Math. 46 (1945) 674.

[14] J. Milnor, Lectures on characteristic classes (Princeton Univ. Lecture Notes, 1957).

GRAVITATIONAL MULTI-INSTANTONS

G.W. GIBBONS and S.W. HAWKING
Department of Applied Mathematics and Theoretical Physics, University of Cambridge, Cambridge, UK

Received 7 July 1978

We present a new family of self-dual positive definite metrics which are asymptotic to Euclidean space modulo identifications under discrete subgroups of O(4). these solutions contain $3\tau - 3$ parameters where τ is the signature. We show that a fully general self-dual solution with these boundary conditions should have this number of parameters.

There has recently been considerable interest in gravitational instantons. These may be defined as complete nonsingular positive definite metrics which are solutions of the vacuum Einstein equations or the Einstein equations with a Λ term. Solutions which are compact and have a Λ term contribute to the volume canonical ensemble [1]. In the non compact case one is interested in solutions with $\Lambda = 0$ which are asymptotically flat in either the three or four dimensional sense. Examples of the former kind are the Schwarzschild, Kerr and Taub–NUT metrics [2–4]. All of these can be regarded as being asymptotically flat in spatial directions and periodic in the imaginary time direction. They therefore contribute to the thermal canonical ensemble [2]. Metrics which are asymptotically flat in the four-dimensional sense can be regarded as the analogue of the Yang–Mills instantons. They can be interpreted as providing a tunnelling amplitude between distinct gravitational vacua corresponding to topologically inequivalent tetrad fields on a three-dimensional space-like surface in flat space time. The homotopy classes of such tetrad fields correspond to maps from S^3 into SO(4) or SU(2) $\times$ SU(2) and are therefore labelled by pairs of integers (n, m). The transition between an initial vacuum $|n_1, m_1\rangle$ and a final vacuum $|n_2, m_2\rangle$ is effected by a metric which is asymptotically flat in the four dimensional sense and which has Euler number χ and signature τ related to the differences $(n_2 - n_1)$ and $(m_2 - m_1)$.

The *Positive Action Conjecture* [5,6,7] states that the action of any asymptotically Euclidean metric with $R^a_a = 0$ everywhere is positive or zero and is zero if and only if it is flat. Asymptotically Euclidean means that the metric approaches the standard Euclidean metric on R^4 outside some compact region in which the topology may differ from that of R^4. The action is not scale invariant and because it is an extremum at a solution, any asymptotically Euclidean solution would have zero action. Thus there can be no asymptotically Euclidean instantons if the conjecture holds. In fact one can prove [8] that there are no asymptotically Euclidean instantons in which the curvature is self-dual or anti self-dual. However the Positive Action Conjecture does not exclude the possibility of gravitational instantons that are asymptotically locally Euclidean (we shall abbreviate this by "ALE"). This means that outside some compact region the metric approaches the flat metric on R^4 modulo identifications under a discrete subgroup of SO(4). The simplest non trivial example of such an instanton has been found by Eguchi and Hanson [9]. It is self-dual with $\chi = 2$ and $\tau = 1$ and tends asymptotically to Euclidean space in which the point X^a has been identified with $-X^a$. This means it can be regarded as giving the transition from an initial state to its TP image.

In this letter we shall present a family of self-dual gravitational instantons which are asymptotic to Euclidean space modulo the cyclic group Z_s whose action on $(X^0 + iX^1, X^2 + iX^3)$ is $\exp(2\pi i/s)I$ where I is the identity matrix. These solutions have $\chi = s$, $\tau = s - 1$. $s = 1$ gives the flat Euclidean space and $s = 2$ gives the Eguchi Hanson solution but the higher values of s give new solutions. They are a modification of the multi Taub–NUT metrics [3] and can be expressed in the form

$$U^{-1}(d\tau + \boldsymbol{\omega}\cdot d\boldsymbol{x})^2 + U\, d\boldsymbol{x}\cdot d\boldsymbol{x}\,, \qquad (1)$$

where

$$U = \sum_{n=1}^{s} \frac{1}{|x_0 - x|}\,, \qquad (2)$$

$$\mathrm{curl}\, \boldsymbol{\omega} = \mathrm{grad}\, U\,, \qquad (3)$$

and the X_n are s distinct points in 3-dimensional Euclidean space with metric $d\boldsymbol{x}\cdot d\boldsymbol{x}$ in which the curl and grad operations are defined. This differs from the multi Taub–NUT metrics in the absence of a constant term in the potential U. It is thus the analogue of the Jackiw–Nohl–Rebbi [10] generalization of the 't Hooft multi SU(2) Yang–Mills instanton solutions. Omitting the constant term changes the sense in which the solution is asymptotically flat from the three to the four dimensional one.

The vector potential $\boldsymbol{\omega}$ has Dirac string type singularities running from each of the points X_n, however these can be removed by taking two coordinate patches $\{\tau_n, X^i\}$ and $\{\tau_{n+1}, X^i\}$ $(i = 1,2,3)$ around the point X_n where

$$\tau_{n+1} = \tau_n + 2\phi_n\,, \qquad (4)$$

where ϕ_n is the azimuthal angle of the point X^i relative to the line joining X_n to X_{n+1}. This identification also removes the apparent singularity at X_n and makes τ periodic with period 4π. Joining all the s coordinate patches together at large values of X^i gives an asymptotically flat metric on Euclidean space modulo the identification Z_s.

These solutions contain $3s - 6 = 3\tau - 3$ parameters (there are $3s$ parameters for the positions of the s points in 3-dimensional space but one has to subtract 3 for an overall translation and another 3 for an overall rotation except in the case $s = 2$ when only one subtracts 2 because it is necessarily axisymmetric). In fact a fully general ALE self-dual instanton has 3τ self-dual perturbations or zero modes. With these boundary conditions 3 of these correspond to global rotations of the solution (2 in the case $\tau = 1$) so the number of parameters cannot be greater than $3\tau - 3$.

This is again reminiscent of the SU(2) Yang–Mills case where a self-dual instanton with Pontryagin number R has $8k - 3$ [11,12] perturbations whereas the Jackiw–Nohl–Rebbi–'t Hooft solutions have only $5k + 4$ parameters [10].

The method of obtaining the self-dual gravitational modes will be described in more detail elsewhere [13]. From the index theorem for the Maxwell field on a self-dual background there will be τ self-dual Maxwell fields F_n^{ab} that are normalizable – i.e. they die away at large Euclidean distances. In 2-component spinor notation these correspond to τ symmetric fields $\Phi_n{}^{AB}$ which satisfy the spin 1 equation

$$\nabla_{AA'}\Phi_n^{AB} = 0\,. \qquad (5)$$

The self-duality also implies that there are 2 independent covariantly constant spinors $\iota^{A'}$ and $o^{A'}$. One can then form 3τ spin 2 objects of the form:

$$h^{AA'BB'}{}_{nm} = \Phi_n{}^{AB}\,\alpha_m^{A'B'}\,, \qquad (6)$$

where $\alpha_m^{A'B'}$ are the 3 symmetric combinations of $\iota^{A'}$ and $o^{A'}$. The perturbation $h^{ab}{}_{nm}$ is transverse and traceless and satisfies the perturbed self-dual Einstein equations. Some combinations of the 3τ $h^{\alpha\iota}_{mn}$ may be equivalent to a pure gauge transformation, i.e. is equal to

$$V_{a;b} + V_{b;a}\,.$$

By the transverse traceless condition such a vector field would have to satisfy

$$V^a{}_{;a} = 0\,, \qquad (7)$$

and

$$V^{a;b}{}_{;b} = 0\,. \qquad (8)$$

Further, since its symmetrized derivative is square integrable, it must be asymptotic to a Killing vector of Euclidean Space which commutes with the discrete subgroup Z_s.

In the Eguchi-Hanson case $s = 2$ and there are six Killing vectors which commute with Z_s. Four of these correspond to Killing vectors of the Eguchi-Hanson metric and so have zero symmetrized covariant derivative. The other two broken Killing vectors generate 2 of the three transverse traceless zero modes in this space. The remaining zero modes is gauge equivalent to a dilation. We are grateful to D.N. Page for these results.

When $s > 2$, there are four Killing vectors of

Euclidean Space which commute with Z_s. One of these corresponds to the $\partial/\partial\tau$ Killing vector of our metrics but the other three will not in general correspond to Killing vectors. They will thus generate transverse traceless perturbations and so the number of gauge inequivalent modes is $3r - 3$, the same as the number of parameters in our solution.

While this paper was in preparation we heard of the work of Nigel Hitchin who has used Penrose's Non Linear Graviton technique [14] to construct ALE self-dual gravitational instantons. This approach however like that of Ward [15] and Atiyah et al. [16] gives the solutions only implicitly.

We should like to thank Nigel Hitchin, Don Page and Chris Pope for help and discussions.

References

[1] S.W. Hawking, Space time foam, DAMTP preprint.
[2] G.W. Gibbons and S.W. Hawking, Phys. Rev. D15 (1977) 2752.
[3] S.W. Hawking, Phys. Lett. 60A (1977) 81.
[4] G.W. Gibbons and S.W. Hawking, Classification of gravitational instanton symmetries, DAMTP preprint.
[5] G.W. Gibbons, S.W. Hawking and M.J. Perry, Path integrals and the indefiniteness of the gravitational action, DAMTP preprint, Nucl. Phys. B, to be published.
[6] D. Page, The positive action conjecture, Phys. Rev. D, to be published.
[7] S.W. Hawking, Quantum gravity and path integrals, DAMTP and Caltech preprint, Phys. Rev. D, to be published.
[8] G.W. Gibbons and C.N. Pope, The positive action conjecture and asymptotically Euclidean metrics in quantum gravity, DAMTP preprint.
[9] T. Eguchi and A. Hanson, Phys. Lett. 74B (1978) 249.
[10] R. Jackiw, C. Nohl and C. Rebbi, Phys. Rev. D15 (1977) 1642.
[11] M.F. Atiyah, N. Hitchin and I.M. Singer, Proc. Nat. Acad. Sci. 74 (1977) 2662.
[12] R. Jackiw and C. Rebbi, Phys. Lett. 67B (1977) 189.
[13] S.W. Hawking and C.N. Pope, Symmetry breaking by instantons in supergravity, DAMTP preprint.
[14] R. Penrose, G.R.G. 7 (1976) 31.
[15] R. Ward, Phys. Lett. 61A (1977) 81.
[16] M.F. Atyah et al., Phys. Lett. 65A (1978) 185.

Commun. Math. Phys. 66, 267–290 (1979)

Communications in Mathematical Physics

The Positive Action Conjecture and Asymptotically Euclidean Metrics in Quantum Gravity

G. W. Gibbons and C. N. Pope

Department of Applied Mathematics and Theoretical Physics, University of Cambridge, Cambridge CB3 9EW, England

Abstract. The Positive Action conjecture requires that the action of any asymptotically Euclidean 4-dimensional Riemannian metric be positive, vanishing if and only if the space is flat. Because any Ricci flat, asymptotically Euclidean metric has zero action and is local extremum of the action which is a local minimum at flat space, the conjecture requires that there are no Ricci flat asymptotically Euclidean metrics other than flat space, which would establish that flat space is the only local minimum. We prove this for metrics on R^4 and a large class of more complicated topologies and for self-dual metrics. We show that if $R^\mu_\mu \geqq 0$ there are no bound states of the Dirac equation and discuss the relevance to possible baryon non-conserving processes mediated by gravitational instantons. We conclude that these are forbidden in the lowest stationary phase approximation. We give a detailed discussion of instantons invariant under an $SU(2)$ or $SO(3)$ isometry group. We find all regular solutions, none of which is asymptotically Euclidean and all of which possess a further Killing vector. In an appendix we construct an approximate self-dual metric on $K3$ – the only simply connected compact manifold which admits a self-dual metric.

1. Introduction

It has been expected for some time [1, 2, 3] that matter should be unstable when quantum gravity is taken into account. That is one expects gravity at some non-perturbative level to give rise to baryon and lepton number non-conservation. This is most clearly indicated in the external field theory computations of black hole evaporation [2]. One would like to compute processes of this sort using a fully quantized theory of gravity. The version of Quantum Gravity which seems most appropriate to us is the functional approach.

In the functional integral formulation of flat space quantum field theory physical quantities are expressed formally as functional integrals of the form

$$Z = \int_C d[\varphi] O[\varphi] \exp iI[\varphi]. \tag{1.1}$$

0010-3616/79/0066/0267/$04.80

$d[\varphi]$ is some measure on the space of field configurations. $I[\varphi]$ is the classical action functional and $O[\varphi]$ is some classical functional of the field φ whose quantum mechanical expectation one wishes to calculate. C denotes the class of field configurations that enter the sum and is specified by giving suitable boundary conditions for φ. The freedom to choose C corresponds to choosing the states that enter the matrix element. As a preliminary step to defining (1.1) one "Wick rotates" to Euclidean 4-space, $\{\mathbb{R}^4, \delta_{\alpha\beta}\}$ thus making the argument of the exponential real and negative. All fields are then supposed to die away at large spatial distances. The vacuum persistence amplitude corresponds to fields which in addition die away at large positive or negative imaginary times τ. If instead one wishes to compute the grand canonical partition function for a boson system at some temperature T one includes in the sum only those fields which are periodic in imaginary time with period $\beta = T^{-1}$. This case is equivalent to working on a flat space with the time coordinate identified – i.e. on $\{\mathbb{R}^3 \times S^1, \delta_{\alpha\beta}\}$. In both cases one can regard the manifold as the limit of a compact manifold with a boundary which is moved to infinity.

To evaluate the functional integral one first looks for non-singular stationary points of the action functional (classical solutions) and expands about them. Such critical points are called "Instantons".

In Quantum Gravity one might try to imitate this procedure by summing over all 4-dimensional Riemannian spaces $\{M, g_{\alpha\beta}\}$ with arbitrary topology for the manifold M and arbitrary metric $g_{\alpha\beta}$ except that M has a prescribed boundary ∂M and $g_{\alpha\beta}$ induces on ∂M some prescribed geometry with metric $k_{\alpha\beta}$. At finite temperature the relevant instanton is the Schwarzschild solution [4]. Presumably the appropriate boundary condition for the vacuum persistence amplitude corresponds to metrics which are asymptotically Euclidean and whose boundary at infinity can be regarded as a 3-sphere with its standard metric. That is, an asymptotically Euclidean metric is one such that outside a compact set the manifold is diffeomorphic to R^4 with a closed ball removed and the metric tends to the standard flat Euclidean metric at least as fast as r^{-2}, where r is the asymptotic proper radial distance. The condition that the metric tend to flatness as r^{-2} guarantees that the action is finite.

In [5] it was argued that to evaluate the functional integral one should pick in each conformal equivalence class a metric satisfying $R^{\mu}_{\mu} = 0$. One can then explicitly integrate over conformal deformations of that metric. One is then left with the task of summing over all metrics satisfying $R^{\mu}_{\mu} = 0$. The action of these metrics is given by a boundary term (the volume term vanishes). The path integral would be better behaved if the action were positive, vanishing if and only if the metric were flat. This is called the "Positive Action Conjecture" [5, 6]. It is rather plausible since it is the natural generalization to one higher dimension of the familiar Positive Mass Conjecture – a proof of which has recently been announced [7]. If the Positive Action Conjecture is true there can be no non-trivial asymptotically Euclidean solution of the Einstein equations. This is because the action of such a solution would be zero, which may be seen as follows. Under a constant rescaling of the metric: $g_{\alpha\beta} \to \Omega^2 g_{\alpha\beta}$, Ω a constant, $I \to \Omega^2 I$. For a solution I must be stationary and hence must vanish. Furthermore since flat space is a local minimum of the action the non-existence of non-flat Ricci-flat asymptotically Euclidean metrics would establish that the action has a unique local minimum at flat space which is strong evidence (but of course not conclusive proof) that it is always positive. Thus the non-existence of non-flat asymptotically Euclidean solutions is necessary and almost sufficient for the validity of the full Positive Action Conjecture.

This situation is in marked contrast with that in Yang-Mills theory where there exist finite action topologically non-trivial classical solutions which it is believed can be responsible for tunnelling between topologically inequivalent vacua $|O_{YM}, n\rangle$ labelled by an integer n [8]. These classical solutions (instantons) have self-dual field strengths. This implies the existence of bound states of the Dirac equation. Because of these bound states the amplitude for a transition between the state $|O_{YM}, n\rangle$ and $|O_{YM}, n+1\rangle$ whilst the Dirac field remains in the no particle state $|O_D\rangle$ is zero [8].

't Hooft has interpreted this as meaning that tunnelling between topologically inequivalent vacua must be accompanied by a change in chirality of the fermions. In fact he finds that the amplitude

$$\langle O_D|\otimes\langle O_{YM}, n|O_D\rangle\otimes|O_{YM}, n+1\rangle$$

is non-zero only if a suitable external source is provided to alter the chirality. This can in turn lead to baryon and lepton non-conservation.

For gravity things are different. Firstly because of the scaling behaviour of the gravitational action there is no barrier to topology change [9]. Secondly for non-compact metrics with $R^{\mu}_{\mu} \geqq 0$ there will be no bound states of the Dirac equation. This is because such solutions must be covariantly constant and hence not normalizable [10]. Thirdly because of the Positive Action Conjecture we expect no asymptotically Euclidean solutions. Certainly there are no self-dual solutions as will be shown in Sect. 3. One way out of this situation is to consider spin $\frac{3}{2}$ fields on non-asymptotically Euclidean self-dual metrics [11].

A different way in which baryon non-conservation might arise is via metrics with more than one asymptotically Euclidean region. These are the 4-dimensional analogues of the 3-dimensional "wormholes" and Einstein-Rosen throats of black hole physics [12, 13, 14, 15, 16]. Indeed if the Ricci scalar, R^{μ}_{μ}, vanishes they provide initial data for time-symmetric spacetimes in 5-dimensional relativity. The asymptotically Euclidean regions are connected by minimal 3-surfaces which are the intersection of the initial 4-surface with the 4-dimensional apparent horizons in the 5-dimensional Lorentzian manifold which we call, following traditional literary usage, Hyperspace. One may generalize the usual Cosmic Censorship Hypothesis [15, 16, 17, 18] to Hyperspace. If it is true it requires that an asymptotically Euclidean region containing an outermost minimal 3-surface with 3 volume V_3 should have an action I satisfying

$$I \geqq \left(\frac{27}{2^{14}\pi}\right)^{1/3} V_3^{2/3}. \tag{1.2}$$

Thus these wormholes should be far from stationary points and should require large excursions in the action. They would therefore be damped in the functional integral. Nevertheless they may be responsible for baryon and lepton non-conservation. This might occur if for instance a baryon fell in imaginary time from one asymptotic region to another through a hole while a lepton travelled in the opposite direction to replace it [9]. The fact that such configurations are far from stationary points is presumably related to the observed great stability of matter.

In this paper we shall investigate asymptotically Euclidean classical solutions with N asymptotic regions and show that if any do exist their Euler number χ and signature τ must satisfy

$$2\chi-3|\tau|>2N. \tag{1.3}$$

This inequality is sufficient for example to rule out asymptotically Euclidean solutions on R^4 with $N-1$ points removed or on $S^1\times S^3$ with N points removed. Thus manifolds with wormholes and bridges cannot be solutions.

We also give as an illustration of these ideas a detailed discussion of 4-metrics invariant under the action of $SO(3)$ or $SU(2)$ acting on 3 surfaces (i.e. Bianchi IX metrics). The equations for self-dual metrics can be reduced to first order ordinary differential equations for which we find all non-singular solutions. We give a qualitative discussion of the non-self-dual case and argue that the only non-singular solutions must admit a further Killing vector and we find all solutions in this class, relating them to various special cases in the literature.

The paper is in 6 sections. Section 2 contains definitions and examples of asymptotically Euclidean metrics. In 3 we prove our main theorems and apply them. In 4 we set up the Bianchi IX formalism and derive the equations for self-dual solutions. In 5 we discuss the boundary conditions and relate them to the nuts and bolts classification of gravitational instanton symmetries [19] and in Sect. 6 we give the qualitative discussion of the solutions. As an application of the solutions we construct in an appendix an approximate metric on $K3$ – the only simple connected compact manifold to admit a self-dual metric [20].

2. Examples of Asymptotically Euclidean Metrics

We define any asymptotically Euclidean region to be one admitting a chart $\{x^\mu\}$ such that for $(x_\mu x^\mu)^{1/2}=r>r_0$ the metric can be written as

$$g_{\mu\nu}=\left(1+\frac{a^2}{4r^2}\right)^2\delta_{\mu\nu}+h_{\mu\nu} \tag{2.1}$$

where

$$h_{\mu\nu}=O\left(\frac{1}{r^3}\right) \tag{2.2}$$

$$\partial_r^p\partial_T^q h_{\mu\nu}=O\left(\frac{1}{r^{3+p}}\right) \tag{2.3}$$

and ∂_r and ∂_T denote radial and transverse derivatives respectively.

The Gravitational Action is

$$I=-\frac{1}{16\pi}\int_M R\sqrt{g}\,d^4x-\frac{1}{8\pi}\int_{\partial M}[K]\sqrt{k}\,d^3x \tag{2.4}$$

where

$$k_{\mu\nu}=g_{\mu\nu}-n_\mu n_\nu \tag{2.5}$$

is the metric induced on the boundary ∂M whose unit normal is n_μ. $[K]=K-K_0$ is the difference between the trace of the second fundamental form of the boundary ∂M in the metric $g_{\mu\nu}$ and its value in the flat metric $\delta_{\mu\nu}$. In fact

$$\int_{\partial M} K\sqrt{k}d^3x=\frac{\partial V_3}{\partial \xi} \tag{2.6}$$

where V_3 is the 3-volume of ∂M and $\frac{\partial}{\partial \xi}$ denotes derivation with respect to the proper distance along the outward normal to ∂M. For flat space $V_3=2\pi^2\xi^3$. Therefore we take $\int_{\partial M} K_0\sqrt{k}d^3x$ to be $6\pi^2\left(\frac{V_3}{2\pi^2}\right)^{2/3}$.

For metrices of the form (2.1)

$$R_{\mu\nu}=O\left(\frac{1}{r^4}\right) \tag{2.7}$$

but

$$R^\mu_\mu=O\left(\frac{1}{r^5}\right). \tag{2.8}$$

The volume term in (2.4) will therefore converge. The boundary term in (2.4) for a surface of the form $r=R$ is

$$\frac{3\pi a^2}{8}+O\left(\frac{1}{R}\right). \tag{2.9}$$

This evidently converges as $R\to\infty$. The expression

$$I=-\frac{1}{16\pi}\int_M R\sqrt{g}d^4x-\frac{1}{8\pi}\int_{\partial M} K\sqrt{k}d^3x+6\pi^2\left(\frac{1}{2\pi^2}\int_{\partial M}\sqrt{k}d^3x\right)^{2/3} \tag{2.10}$$

is manifestly coordinate independent but its limiting value could in principle depend on the limiting sequence of boundaries chosen. If the surface is sufficiently spherical this will not happen. The shear, $\Sigma_{\mu\nu}$, of the surface ∂M is defined by

$$\Sigma_{\mu\nu}=n_{\alpha;\beta}k^\alpha_\mu k^\beta_\nu-\tfrac{1}{3}k_{\mu\nu}K. \tag{2.11}$$

If

$$\Sigma_{\mu\nu}=O\left(\frac{1}{r^2}\right)$$

then the action will be independent of the limiting sequence.

An example of an asymptotically flat metric with $R^\mu_\mu=0$ is the Tolman wormhole:

$$ds^2=d\xi^2+(a^2+\xi^2)d\Omega_3^2. \tag{2.12}$$

This may be obtained as an analytic continuation of the Friedmann-Robertson-Walker cosmological model filled with a perfect fluid whose pressure is $\frac{1}{3}$ its energy density [21]. It contains two asymptotically Euclidean regions, located at large positive and negative values of ξ, connected by a minimal 3-sphere at $\xi=0$ whose volume is $2\pi^2a^3$. The metric is therefore defined on a manifold of topology $S^3\times R \approx R^4-\{0\}$ with zero Euler number χ and signature τ. The action evaluated at each infinity is

$$I=\frac{3\pi}{8}a^2. \tag{2.13}$$

The metric is conformally flat which is made manifest by introducing a radial coordinate r by

$$\xi=r-\frac{a^2}{4r}. \tag{2.14}$$

This takes the metric (2.12) to

$$ds^2=\left(1+\frac{a^2}{4r^2}\right)^2(dr^2+r^2d\Omega_3^2). \tag{2.15}$$

The asymptotically Euclidean regions are now given by $\infty>r>\frac{a}{2}$ and $\frac{a}{2}>r>0$. The minimal 3-surface is located at $r=\frac{a}{2}$ and the singularity at $r=0$ corresponds to the infinitely distant boundary. The metric is the 4-dimensional analogue of the Einstein-Rosen throat of black hole physics. It may be regarded as a constant time slice of a 5-dimensional hyperspherically symmetric black hole. Such a black hole with vanishing Ricci tensor has the metric

$$ds^2=-\left(1-\frac{a^2}{\varrho^2}\right)(dx^5)^2+\left(1-\frac{a^2}{\varrho^2}\right)^{-1}d\varrho^2+\varrho^2d\Omega_3^2 \tag{2.16}$$

where x^5 is the 5-dimensional time coordinate and ϱ is related to r by

$$\varrho=r+\frac{a^2}{4r}. \tag{2.17}$$

This example may be readily generalized to the case of a conformally flat 4-metric with vanishing Ricci scalar and N asymptotically Euclidean regions. The metric is

$$ds^2=\left(1+\sum_{i=0}^{N-1}a_i^2[(x^\mu-x_i^\mu)(x^\nu-x_i^\nu)\delta_{\mu\nu}]^{-1}\right)^2\delta_{\alpha\beta}dx^\alpha dx^\beta. \tag{2.18}$$

The $N-1$ points $x^\mu=x_i^\mu$ correspond to $N-1$ asymptotic regions. The metric is well defined on $R^4-\{x_i^\nu\}$. It has Euler number $\chi=1-N$ and vanishing signature τ.

One may also readily generalize the Misner Wormhole [13]. This has the metric

$$ds^2=\Omega^2(d\eta^2+d\zeta^2+\sin^2\zeta d\Omega_2^2) \tag{2.19}$$

$$\Omega=a\sum_{n=-\infty}^{\infty}(\cosh(\eta+2n\mu_0)-\cos\zeta)^{-1}. \tag{2.20}$$

If we let $0 \leqq \zeta \leqq \pi$ and $0 \leqq \eta \leqq \eta_0$ we obtain a metric on $S^1 \times S^3$. However the metric is singular at $\eta=0$, $\zeta=0$ which turns out to be an asymptotically Euclidean infinity. The metric is thus well defined on $S^3 \times S^1 - \{0\}$ which has Euler number -1 and signature zero.

3. The Topology of Asymptotically Euclidean Spaces

It is convenient to think of an asymptotically Euclidean space M as being topologically equivalent to a compact manifold M_c with N points removed. A triangulation of M_c will serve as a triangulation of M provided one removes N 4-simplices. Thus the Euler number of an asymptotically Euclidean manifold with N "infinities" is given by

$$\chi[M] = \chi[M_c] - N. \tag{3.1}$$

The signature will be unchanged, i.e.

$$\tau[M] = \tau[M_c]. \tag{3.2}$$

Both χ and τ may be related to integrals over M of curvature invariants. The Gauss-Bonnet Theorem states [22, 23]

$$\chi[M] = \frac{1}{32\pi^2} \int_M \varepsilon_{\alpha\beta\gamma\delta} R^{\alpha\beta} \wedge R^{\gamma\delta} - \frac{1}{32\pi^2} \int_{\partial M} \varepsilon_{\alpha\beta\gamma\delta} (2\vartheta^{\alpha\beta} \wedge R^{\gamma\delta} - \tfrac{4}{3}\vartheta^{\alpha\beta} \wedge \vartheta^{\gamma}_{\ \varepsilon} \wedge \vartheta^{\varepsilon\delta}). \tag{3.3}$$

$R^{\alpha}_{\ \beta}$ are the curvature 2-forms in an orthonormal basis and $\vartheta^{\alpha}_{\ \beta}$ is the second fundamental form of the boundary ∂M. That is if $\omega^{\alpha}_{\ \beta}$ are the actual connection 1-forms and $\omega^{\alpha}_{0\beta}$ are the connection 1-forms if the metric were locally a product near the boundary

$$\vartheta^{\alpha}_{\ \beta} = \omega^{\alpha}_{\ \beta} - \omega^{\alpha}_{0\beta}. \tag{3.4}$$

The second fundamental form may also be defined by constructing a Gaussian coordinate system about each connected component. If n^{α} is the unit normal then

$$k_{\alpha\beta} = g_{\alpha\beta} - n_{\alpha} n_{\beta}. \tag{3.5}$$

One now defines a symmetric tensor $K_{\alpha\beta}$ by

$$K_{\alpha\beta} = \underset{n}{\mathcal{L}} k_{\alpha\beta}. \tag{3.6}$$

The boundary term in the action, K, is

$$K = K_{\alpha\beta} k^{\alpha\beta} \tag{3.7}$$

where

$$k^{\alpha\sigma} k_{\sigma\beta} = \delta^{\alpha}_{\beta}. \tag{3.8}$$

In a basis of 1-forms (ω^0, ω^i) such that

$$\omega^0 = n_{\alpha} dx^{\alpha} \tag{3.9}$$

and

$$k_{\alpha\beta} dx^{\alpha} dx^{\beta} = \sum_{i=1}^{3} (\omega^i)^2 \tag{3.10}$$

we have

$$\frac{\partial\omega^i}{\partial\xi}=\underset{n}{\mathscr{L}}\omega^i=K^i_j\omega^j\,. \tag{3.11}$$

The ϑ^α_β are given by

$$\vartheta^i_0=\frac{\partial\omega^i}{\partial\xi}=K^i_j\omega^j \tag{3.12}$$

$$\vartheta^i_k=0\,. \tag{3.13}$$

For asymptotically Euclidean metrics the second boundary term in (3.3) vanishes and we have

$$\chi[M]=\frac{1}{32\pi^2}\int_M(C_{\mu\nu\alpha\beta}C^{\mu\nu\alpha\beta}-2R_{\mu\nu}R^{\mu\nu}+\tfrac{2}{3}R^2)\sqrt{g}d^4x+\frac{1}{2\pi^2}\int_{\partial M}\det(K^i_j)\sqrt{k}d^3x \tag{3.14}$$

$$=\frac{1}{32\pi^2}\int_M(C_{\mu\nu\alpha\beta}C^{\mu\nu\alpha\beta}-2R_{\mu\nu}R^{\mu\nu}+\tfrac{2}{3}R^2)\sqrt{g}\ {}^4x+N\,. \tag{3.15}$$

If $R_{\mu\nu}=0$,

$$\chi[M]\geqq N \tag{3.16}$$

equality being attained if and only if the metric is flat.

The signature may also be expressed in terms of integrals but one must include a non-local boundary contribution [24]:

$$\tau[M]=\frac{1}{48\pi^2}\int_M R^\beta_\alpha\wedge R^\alpha_\beta-\frac{1}{48\pi^2}\int_{\partial M}\vartheta^\alpha_\beta\wedge R^\beta_\alpha-\eta(0)\,. \tag{3.17}$$

$\eta(s)$ is the η-function of a certain differential operator defined over ∂M. $\eta(0)$ vanishes if ∂M has an orientation reversing isometry, which S^3 does. The other boundary term vanishes for asymptotically Euclidean metrics and we obtain

$$\tau[M]=\frac{1}{48\pi^2}\int_M C_{\alpha\beta\gamma\delta}{}^*C^{\alpha\beta\gamma\delta}\sqrt{g}d^4x\,. \tag{3.18}$$

${}^*C_{\alpha\beta\gamma\delta}$ is the dual of the Weyl tensor $C_{\alpha\beta\gamma\delta}$. Using the fact that

$$C_{\alpha\beta\gamma\delta}C^{\alpha\beta\gamma\delta}\geqq|C_{\alpha\beta\gamma\delta}{}^*C^{\alpha\beta\gamma\delta}|\,, \tag{3.19}$$

equality being attained if and only if the Weyl tensor is self-dual or anti-self-dual we obtain the following inequality for vacuum Einstein solutions

$$2\chi[M]-3|\tau[M]|\geqq 2N \tag{3.20}$$

$$2\chi[M_c]-3|\tau[M_c]|\geqq 4N\,. \tag{3.21}$$

Similar inequalities have previously been obtained by Hitchin in the compact case [20].

One might think that equality could be obtained in the half-flat case. However complete self-dual asymptotically Euclidean metrics (at least with finite fundamental group) do not exist. This can be seen as follows. The index of the Dirac operator $\gamma^\mu \nabla_\mu$ on a manifold with boundary is given by [24]

$$\text{Index}\,[\gamma^\alpha \nabla_\alpha] = \frac{1}{192\pi^2}\int_M R^\beta_\alpha \wedge R^\alpha_\beta - \frac{1}{192\pi^2}\int_{\partial M} \vartheta^\alpha_\beta \wedge R^\beta_\alpha - [\eta_D(0)+h]\tfrac{1}{2}\,. \qquad (3.22)$$

$\eta_D(s)$ is the η function of the Dirac operator restricted to the boundary ∂M and h is the dimension of its kernel. For S^3 these vanish. Now any simply connected half-flat metric posseses spinor structure [25]. Thus on half flat asymptotically Euclidean metric the index of the Dirac operator is

$$\pm\frac{1}{48\pi^2}\int_M R_{\alpha\beta\mu\nu}R^{\alpha\beta\mu\nu}\sqrt{g}\,d^4x\,.$$

This is zero only if the metric is flat.

It follows that any simply connected asymptotically Euclidean half flat metric would admit at least one normalizable solution of the Dirac equation. However by Lichnerowicz's Theorem [10] such a solution would have to be covariantly constant and hence could not be normalizable. The only way out of this contradiction is that there are no simply connected half-flat asymptotically Euclidean solutions. If the manifold is not simply connected we can apply the argument to the universal covering space provided this consists of only finitely many copies of the original manifold. Thus the inequalities (3.20) and (3.21) are saturated only in the trivial flat case. Using them it is easy to rule out (amongst others) solutions with the following topologies:

1) $R^4 - \{N\,\text{points}\} \equiv S^4 - \{N+1\,\text{points}\}$

   $\chi = 1 - N, \quad \tau = 0$

2) $S^1 \times S^3 - \{N\,\text{points}\}$

   $\chi = -N, \quad \tau = 0$

3) $CP^2 - \{N\,\text{points}\}$

   $\chi = 3 - N, \quad \tau = 1\,.$

Thus we see that there are no asymptotically Euclidean solutions (other than flat) on R^4 nor with the topology of the many bridge (2.18) or Misner Wormhole (2.19).

4. Local Bianchi IX Solutions

In the next 3 sections we shall illustrate the ideas we have discussed above by reference to a simple class of solutions of the Einstein Equations. This class we call Bianchi IX metrics. They are defined to be metrics with an $SU(2)$ or $SO(3)$ isometry group acting transitively on 3-surfaces. These are the Euclidean equivalent of a well known class of cosmological models [26]. If $R_{\mu\nu}=0$ the metric may locally be cast in the form

$$ds^2 = (abc)^2 d\eta^2 + a^2\sigma_1^2 + b^2\sigma_2^2 + c^2\sigma_3^2 \qquad (4.1)$$

where a, b and c are functions solely of η and $\{\sigma_i\}$ are 3 basis 1-forms with exterior algebra

$$d\sigma_i = -\tfrac{1}{2}\varepsilon_{ijk}\sigma_j \wedge \sigma_k \tag{4.2}$$

and such that

$$\mathop{\mathscr{L}}_{\partial/\partial n}\sigma_i = 0. \tag{4.3}$$

The vacuum Einstein equations reduce to the form:

$$2\alpha_{\eta\eta} = a^4 - (b^2 - c^2)^2 \tag{4.4}$$

plus the two equations obtained by cyclic permutation of (a, b, c) and

$$4(\alpha_\eta\beta_\eta + \beta_\eta\gamma_\eta + \gamma_\eta\alpha_\eta) = 2a^2b^2 + 2b^2c^2 + 2c^2a^2 - a^4 - b^4 - c^4. \tag{4.5}$$

$$\alpha = \log a \tag{4.6}$$

and cyclically.

(4.5) is a first integral of (4.4) and may be regarded as a constraint on the initial values of (a, b, c) and (a_η, b_η, c_η) which is preserved by the evolution Eq. (4.4). In fact (4.4) are derivable as the Euler Lagrange equations of the action I given by

$$-\frac{8}{\pi}I = \int 2(\alpha_\eta\beta_\eta + \beta_\eta\gamma_\eta + \gamma_\eta\alpha_\eta)\alpha\eta$$
$$+\tfrac{1}{2}\int (2a^2b^2 + 2b^2c^2 + 2c^2a^2 - a^4 - b^4 - c^4)d\eta. \tag{4.7}$$

In fact I coincides with the gravitational action if the 3-volume of the surfaces of constant η is taken to be $16\pi^2(abc)$. The constraint (4.5) may also be derived from the action (4.7) if one imposes the condition that the integral is stationary under the replacement $d\eta \to N(\eta)d\eta$ where $N(\eta)$ is an arbitrary function of η. The constraint says that the "Hamiltonian" corresponding to the Lagrangian (4.7) vanishes. Note that because the time is imaginary the roles of the physical Lagrangian and minus the Hamiltonian are interchanged.

Equations (4.4) and (4.5) may be integrated completely if we impose the condition that two of the invariant directions have equal magnitude. E.g. $a = b$. This leads to the general Taub-NUT family which is invariant under a 4-parameter group with the Lie algebra of $U(2)$:

$$a^2 = b^2 = \tfrac{1}{4}q \sinh q(\eta - \eta_2)\operatorname{cosech}^2 \tfrac{1}{2}q(\eta - \eta_1) \tag{4.8}$$

$$c^2 = q \operatorname{cosech} q(\eta - \eta_2). \tag{4.9}$$

(q, η_1, η_2) are constants of integration. The more familiar form

$$ds^2 = (r^2 - n^2)(r^2 - 2mr + n^2)^{-1}dr^2$$
$$+ 4n^2(r^2 - 2mr + n^2)(r^2 - n^2)^{-1}\sigma_3^2 + (r^2 - n^2)(\sigma_1^2 + \sigma_2^2) \tag{4.10}$$

is obtained using the transformation:

$$n^2 = -\tfrac{1}{4}q \operatorname{cosech} q(\eta_2 - \eta_1) \tag{4.11}$$

$$m = n \cosh q(\eta_2 - \eta_1) \tag{4.12}$$

$$r = \frac{q}{4n}(\coth \tfrac{1}{2}q(\eta - \eta_1) - \coth q(\eta_2 - \eta_1)). \tag{4.13}$$

Two special cases are of note

A) $\eta_1 = \eta_2$. (4.14)

These are the Eguchi-Hanson metrics [27, 28]. Real q corresponds to their type II. Imaginary q corresponds to their type I. One can obtain their metric from (4.10) by the following transformation

$$m = n + \frac{a^4}{128n^3} \tag{4.15}$$

$$r = m + \frac{\varrho^2}{8n} \tag{4.16}$$

where a is Eguchi and Hanson's parameter. One now lets $n \to \infty$. This gives the metric

$$ds^2 = \left(1 - \frac{a^4}{\varrho^4}\right)^{-1} d\varrho^2 + \tfrac{1}{4}\varrho^2\left(1 - \frac{a^4}{\varrho^4}\right)\sigma_3^2 + \tfrac{1}{4}\varrho^2(\sigma_1^2 + \sigma_2^2) \tag{4.17}$$

B) $q = 0$.

This corresponds to the self-dual solution discussed by Hawking [29].

Both A) and B) have self-dual curvature.

All Bianchi IX solutions with self-dual curvature may be obtained systematically. The self-dual conditions leads, after a single integration to the following equations

$$2\frac{d\alpha}{d\eta} = b^2 + c^2 - a^2 - 2\lambda_1 bc \tag{4.18}$$

together with the two others obtained by cyclically permuting (a, b, c) and $(\lambda_1, \lambda_2, \lambda_3)$. The $\{\lambda_i\}$ are constants obeying

$$\lambda_1 = \lambda_2\lambda_3 \quad \text{and cyclically}. \tag{4.19}$$

The possible solutions of (4.19) are

C) $(\lambda_1, \lambda_2, \lambda_3) = (0, 0, 0)$ (4.20)

D) $(\lambda_1, \lambda_2, \lambda_3) = (1, 1, 1)$ (4.21)

E) $(\lambda_1, \lambda_2, \lambda_3) = (-1, -1, +1)$ (4.22)

and cyclic permutations. In fact case E is not distinct from case D since it may be obtained by the substitution $c \to -c$. Both C and D yield in (4.18) first integrals of the evolution Eq. (4.4) which are consistent with the constraint (4.5).

Case C may be obtained directly without integration by requiring that the connection forms in the basis $(abcd\eta, a\sigma_1, b\sigma_2, c\sigma_3)$ be self-dual, and the resulting Eq. (4.18) may be solved completely to give a metric of the form [28]

$$ds^2 = F^{-1/2} d\varrho^2 + F^{1/2}\frac{\varrho^2}{4}\left[\left(1 - \frac{a_1^4}{\varrho^4}\right)^{-1}\sigma_1^2 + \left(1 - \frac{a_2^4}{\varrho^4}\right)^{-1}\sigma_2^2 + \left(1 - \frac{a_3^4}{\varrho^4}\right)^{-1}\sigma_3^2\right] \tag{4.23}$$

where

$$F = \left(1 - \frac{a_1^4}{\varrho^4}\right)\left(1 - \frac{a_2^4}{\varrho^4}\right)\left(1 - \frac{a_3^4}{\varrho^4}\right) \tag{4.24}$$

and (a_1, a_2, a_3) are constants.

We have been unable to integrate case D explicitly except when $a=b$ which leads to the self-dual Taub-NUT metric obtained from (4.10) by setting $m=n$. We shall give a qualitative treatment of the equation for case D in Sect. 6.

An alternative way of obtaining self-dual metrics is to require that they be vacuum-Kaehler metrics. This implies and is implied by self-duality of the Riemann tensor [30]. This leads to a non-linear partial differential equation for the Kaehler function $K(\zeta^1, \zeta^2, \bar{\zeta}^1, \bar{\zeta}^2)$ considered as a function of the two complex variables ζ^i, $i=1,2$. This equation is

$$\det\left(\frac{\partial^2 K}{\partial\zeta^i \partial\bar{\zeta}^j}\right) = 1 . \tag{4.25}$$

If one assumes that K is a function solely of $R^2 = |\zeta^1|^2 + |\zeta^2|^2$ (4.25) may integrated to give

$$K = \varrho^2 + \tfrac{1}{2}a^2 \log\left(\frac{\varrho^2 - a^2}{\varrho^2 + a^2}\right) \tag{4.26}$$

where

$$\varrho^4 - a^4 = R^4 . \tag{4.27}$$

This leads directly to the Eguchi-Hanson metric (4.17) where

$$\tfrac{1}{4}R^2(\sigma_1^2 + \sigma_2^2) = |d\zeta^1|^2 + |d\zeta^2|^2 - R^{-2}|\bar{\zeta}^1 d\zeta^1 + \bar{\zeta}^2 d\zeta^2|^2 \tag{4.28}$$

$$R^2\sigma_3^2 \qquad = (\bar{\zeta}^1 d\zeta^1 + \bar{\zeta}^2 d\zeta^2 - \zeta^1 d\bar{\zeta}^1 - \zeta^2 d\bar{\zeta}^2)^2 . \tag{4.29}$$

One may easily extend this class of solutions to include the effect of the cosmological constant. In order to be an Einstein-Kaehler metric the Kaehler function must satisfy

$$\det\left(\frac{\partial^2 K}{\partial\zeta^i \partial\bar{\zeta}^j}\right) = e^{-1/2\Lambda K} . \tag{4.30}$$

The assumption that K depend only on $|\zeta^1|^2 + |\zeta^2|^2 = R^2$ leads to the metric

$$ds^2 = \left(1 - \frac{a^4}{\varrho^4} - \tfrac{1}{6}\Lambda\varrho^2\right)^{-1} d\varrho^2 + \tfrac{1}{4}\varrho^2\left(1 - \frac{a^4}{\varrho^4} - \tfrac{1}{6}\Lambda\varrho^2\right)\sigma_3^2 + \tfrac{1}{4}\varrho^2(\sigma_1^2 + \sigma_2^2) \tag{4.31}$$

where $\{\sigma_i\}$ are related to $\{\zeta^i\}$ by Eq. (4.28) and (4.29). K is obtained by solving the ordinary differential equation

$$\frac{dK}{dR}\frac{d}{dR}\left(R\frac{dK}{dR}\right) = 8R^2 e^{-1/2\Lambda K} . \tag{4.32}$$

5. Global Aspects of Bianchi IX Solutions

The discussion in Sect. 4 was purely local. In this section we shall discuss how these local solutions may be patched together to form global solutions. The group G may be $SU(2)$ or $SO(3)$. At a general point p of the manifold M, the stabilizer of this point must be a discrete subgroup of G since G has 3 dimensional orbits. This subgroup will also fix ne rby points since we assume G acts continuously. The effective action of G will have this subgroup factored out so we may assume with no loss of generality that G acts effectively and transitively on its orbits O_g. This implies that these orbits are generically S^3 if $G=SU(2)$, or RP^3 if $G=SO(3)$.

In both cases points on O_g may be parametrized by the coordinates on G in such a way that the Euler angles $(\psi, \varphi, \vartheta)$ assigned to each point q on the orbit through p correspond to the group element g such that $q=gp$. Since the field equations tell us that the action of G commutes with Lie transport along the orthogonal trajectories this leads to the metric form (4.1) with $\{\sigma_i\}$ given by

$$\sigma_1 = \cos\psi d\vartheta + \sin\psi \sin\vartheta d\varphi \tag{5.1}$$

$$\sigma_2 = -\sin\psi d\vartheta + \cos\psi \sin\vartheta d\varphi \tag{5.2}$$

$$\sigma_3 = d\psi + \cos\vartheta d\varphi . \tag{5.3}$$

The dual basis of vectors $\{K_i\}$ is

$$K_1 = \cos\psi \partial/\partial\vartheta + \frac{\sin\psi}{\sin\vartheta}\partial/\partial\varphi - \cot\vartheta \sin\psi \partial/\partial\psi \tag{5.4}$$

$$K_2 = -\sin\psi \partial/\partial\vartheta + \frac{\cos\psi}{\sin\vartheta}\partial/\partial\varphi - \cot\vartheta \cos\psi \partial/\partial\psi \tag{5.5}$$

$$K_3 = \partial/\partial\psi \tag{5.6}$$

where

$0 \leqq \vartheta \leqq \pi$

$0 \leqq \varphi \leqq 2\pi$

$0 \leqq \psi \leqq 2\pi$ if G is $SO(3)$

$0 \leqq \psi \leqq 4\pi$ if G is $SU(2)$.

The solutions discussed in Sect. 4 are valid so long as $\{a, b, c\}$ are finite and non-zero. If any of $\{a, b, c\}$ cease to be finite and non-zero in a finite proper distance interval $\xi = \int abcd\eta$, the manifold will be incomplete. If G is $SU(2)$ and all three of $\{a, b, c\}$ diverge as $\frac{1}{2}$ (proper distance) we obviously have a Euclidean infinity. Other infinities are also possible. For example the Taub-NUT infinity corresponds to $a \to \xi$, $b \to \xi$, $c \to$constant (or any permutation of $\{a, b, c\}$) as the proper distance ξ tends to infinity. In the Eguchi-Hanson metric $\{a, b, c\}$ diverge as $\frac{1}{2}\xi$ but G is $SO(3)$, giving a sort of "conical" Euclidean infinity [28].

If one of $\{a, b, c\}$ vanishes, for example c, at a point p, then the corresponding vector will have zero length at p. This means that the orbit of G through p can no longer be 3 dimensional. In fact the orbit through p corresponds to a subgroup H of G and hence must be one dimensional or the entire group G. In the first case, which we call a bolt, [19] B, only c vanishes and the orbit through p corresponds to G/G_3. This is a homogeneous 2-space whose metric and second fundamental form are spherically symmetric. In fact since G commutes with the exponential map, B is a totally geodesic submanifold of M, and its second fundamental form vanishes. This means that a tends to b on the bolt with vanishing derivative with respect to ξ. By considering the limiting form of the metric on a 2-surface orthogonal to the bolt – i.e. the (η, ψ) plane in our coordinates – one can readily see that if ψ is to be an angular coordinate with range 4π, c must vanish as $\frac{1}{2}\xi$ whereas if ψ has range 2π, c must vanish as ξ. Strictly speaking the word bolt introduced in [19] applies only when $a=b\forall\xi$. In this case K_3 is an additional Killing vector (i.e. the group is extended to $U(2)$) and $c=0$ is the locus of its fixed point set – i.e. its "bolt". If $a \neq b$ the set at which $c=0$ is a degenerate orbit but it may not necessarily be the fixed point set of a one parameter subgroup. In what follows we shall extend the term bolt to cover this slightly more general situation.

In the second case the orbit through p is just p itself and we refer to p as a nut [19]. In this case all of $\{a, b, c\}$ must vanish as $\frac{1}{2}\xi$ as $\xi \to 0$, in order that the orbits be a nested sequence of 3-spheres near p.

We summarize these boundary conditions as follows:

Euclidean Infinity.

$$a, b, c \to \tfrac{1}{2}\xi, \quad \xi \to \infty$$
$$0 \leqq \psi \leqq 4\pi;$$

Conical Infinity.

$$a, b, c \to \tfrac{1}{2}\xi, \quad \xi \to \infty$$
$$0 \leqq \psi \leqq 2\pi;$$

Taubian Infinity.

$$a, b \to \xi, \quad c \to \text{constant}, \quad \xi \to \infty;$$

Bolt.

$$a, b \to \text{constant}, \quad c \to \tfrac{1}{2}\xi,$$
$$\frac{da}{d\xi}, \frac{db}{d\xi} \to 0, \quad \xi \to 0$$
$$0 \leqq \psi \leqq 4\pi$$

or

$$a, b \to \text{constant}, \quad c \to \xi$$
$$\frac{da}{d\xi}, \frac{db}{d\xi} \to 0, \quad \xi \to 0.$$
$$0 \leqq \psi \leqq 2\pi;$$

Nut.

$$a, b, c \to \tfrac{1}{2}\xi, \quad \xi \to 0$$
$$0 \leqq \psi \leqq 4\pi.$$

There are of course other ways in which the solutions can break down but if we insist on the group orbits being generically $SU(2)$ or $SO(3)$ these solutions will not have regular extensions of the sort we describe. In principle "bolts" with ψ identified modulo $4\pi/s$, $s > 2$ could arise. These would correspond to group orbits which were generically $SU(2)/Z_s$ where Z_s is the cyclic group of order s. However these possibilities definitely do not occur in the symmetric ($a=b$) case nor the self dual case. If our qualitative reasoning in Sect. 6 is correct they cannot occur at all for Bianchi IX metrics. It will turn out that not all possibilities are allowed by the differential Eqs. (4.4) and (4.5). It is useful to think of the solutions as curves in the configuration space $\{a,b,c\}$.

The five regular boundary conditions can be combined in 15 ways to give regular manifolds. Of these some can be ruled out on the grounds of our results in Sect. 3. These are: Euclidean Infinity→bolt or Euclidean Infinity. One can modify our results of Sect. 3 to take into account conical infinities. Each conical infinity, for which ψ is identified modulo 2π and the boundary is RP^3, contributes $\frac{1}{2}$ rather than unity to formula (3.14) for the Euler number. The formula (3.18) for the signature τ is unchanged. Then it is easy to rule out Conical Infinity→Euclidean Infinity, Conical Infinity, or nut. For a Taubian Infinity the boundary terms contribute zero to formula (3.14) for the Euler number and $\frac{1}{3}$ to formula (3.17) for the signature. This enables one to rule out Taubian Infinity→Taubian, Euclidean or Conical Infinity.

There are three compact possibilities: nut→nut, bolt→bolt, and nut→bolt. These have $\chi=2, 4$ and 3, and $\tau=0, 0$ and 1 respectively. They can all be ruled out because

1) If $\chi\neq 0$ any Killing vector field must have at least one fixed point.

2) If $R_{\mu\nu}=0$, $K_{\mu}{}^{;\alpha}{}_{;\alpha}=0$ which in turn implies that K_μ is covariantly constant and so if zero anywhere it must be zero everywhere.

It remains to investigate the cases

Nut →Euclidean Infinity

Nut →Taubian Infinity

Bolt→ Taubian Infinity

Bolt→ Conical Infinity.

The first case must be flat and the other three all occur as members of the Taub-NUT family – i.e. with two directions equal. One cannot determine by topological means whether the last three cases can occur when all three directions are unequal. To decide the issue we must turn to the detailed behaviour of the differential Eqs. (4.4) and (4.5). This we do in the next section. Before doing so we list the known Taub-NUT examples:

1) Flat Space.

$$0\leqq\psi\leqq 4\pi$$

$$a=b=c=\tfrac{1}{2}\xi\,.$$

This runs from a nut at $\xi=0$ to a Euclidean Infinity at $\xi=\infty$.

2) Self-Dual Taub-NUT [29].

$$0\leqq\psi\leqq 4\pi$$

$$a=b=(r^2-n^2)^{1/2}\,,\qquad c=2n(r-n)^{1/2}(r+n)^{-1/2}$$

$$\xi=n\operatorname{arcosh}\left(\frac{r}{n}\right)+(r^2-n^2)^{1/2}\,.$$

This runs from a nut at $\xi=0$ to a Taubian Infinity at $\xi=\infty$.

3) Taub-NUT with $m=\frac{5}{4}|n|$ [31]. (This is not half-flat.)

$$0\leqq\psi\leqq4\pi$$

$$a=b=(r^2-n^2)^{1/2},\qquad c=2n(r-2n)^{1/2}(r-\tfrac{1}{2}n)^{1/2}(r^2-n^2)^{-1/2}$$

$$\xi=\int_{2n}^{r}(x-2n)^{-1/2}(x-\tfrac{1}{2}n)^{-1/2}(x^2-n^2)^{1/2}dx\,.$$

This runs from a bolt at $\xi=0$ to a Taubian Infinity at $\xi=\infty$.

4) Eguchi-Hanson [27, 28].

$$0\leqq\psi\leqq2\pi$$

$$a=b=\tfrac{1}{2}r,\qquad c=\tfrac{1}{2}r\left(1-\frac{a^4}{r^4}\right)^{1/2}$$

$$\xi=\int_{a}^{r}(x^4-a^4)^{-1/2}x^2dx\,.$$

This runs from a bolt at $\xi=0$ to a Conical Infinity at $\xi=\infty$.

Examples (3) and (4) are best understood by comparison with the metric on CP^2 which has a self-dual Weyl tensor and is an Einstein space [32]. This has

$$0\leqq\psi\leqq4\pi$$

$$a=b=\sqrt{\frac{3}{2\Lambda}}\sin\left(\sqrt{\frac{\Lambda}{6}}\xi\right)\cos\left(\sqrt{\frac{\Lambda}{6}}\xi\right)$$

$$c=\sqrt{\frac{3}{2\Lambda}}\cos\left(\sqrt{\frac{\Lambda}{6}}\xi\right).$$

This runs from a bolt at $\xi=0$ to a nut at $\xi=\pi\sqrt{\frac{3}{2\Lambda}}$. Near $\xi=0$, $c\to\frac{1}{2}\xi$. The $m=\frac{5}{4}|n|$ Taub-NUT metric has a similar behaviour near the bolt at $\xi=0$. Thus we must identify ψ modulo 4π. Therefore the topology is that of $CP^2-\left\{\text{the point }\xi=\pi\sqrt{\frac{3}{2\Lambda}}\right\}$. We shall call this $CP^2-\{0\}$. It has $\chi=2$ and $\tau=1$. This manifold is simply-connected and does not admit spinor structure because the self-intersection number [19] of the bolt is one, as in CP^2.

The Eguchi-Hanson metric is also similar to CP^2 at $\xi\approx0$, except that $c\to\xi$ as $\xi\to0$. Thus we must identify ψ modulo 2π, which implies that the Eguchi-Hanson metric has a conical infinity at $\xi=\infty$. In fact the manifold on which the Eguchi-Hanson metric is well-defined may be described as follows. Let (Z_1,Z_2,Z_3) be three homogeneous complex coordinates on CP^2, not all of which vanish. Consider the involution $I:(Z_1,Z_2,Z_3)\to(-Z_1,-Z_2,-Z_3)$. This leaves fixed both the *point* in CP^2 given by $(0,0,Z_3)$ which corresponds to the nut, and the complex *line* or 2-sphere $(Z_1, Z_2, 0)$ which corresponds to the bolt. If we delete the point and factor $CP^2-\{0\}$ by I we obtain the Eguchi-Hanson manifold. (Despite appearances this is still a manifold near the bolt, and is in fact diffeomorphic to the tangent bundle of S^2.)

Introducing coordinates

$$Z_1/Z_3=\zeta^1=R\cos(\tfrac{1}{2}\vartheta)\exp\left(\frac{i}{2}(\psi+\varphi)\right) \tag{5.7}$$

$$Z_2/Z_3=\zeta^2=R\sin(\tfrac{1}{2}\vartheta)\exp\left(\frac{i}{2}(\psi-\varphi)\right) \tag{5.8}$$

where $R^2=|\zeta^1|^2+|\zeta^2|^2$, it is evident that the involution I corresponds to $\psi\to\psi+2\pi$.

$(CP^2-\{0\})/I$ has $\chi=2$, $\tau=1$ (the same as $CP^2-\{0\}$) and is simply-connected. The self-intersection number [19] of the bolt is 2 and the manifold admits spinor structure. One may easily check the index theorems for the Hirzebruch and Dirac complexes. The RP^3 boundary contributes zero for the Hirzebruch case and $-\frac{1}{8}$ for the Dirac case.

The remaining explicitly known solution, that of Belinskii, Gibbons, Page and Pope [28], Eq. (4.23), in general runs from a Euclidean Infinity at $\varrho=\infty$ to a curvature singularity at the largest value of (a_1, a_2, a_3), except in the special case where it coincides with the Eguchi-Hanson metric. In any event it is not complete and asymptotically Euclidean. Since it is half-flat this is implied by our results in Sect. 3.

6. Qualitative Treatment of Bianchi IX Solutions

The Eqs. (4.4) and (4.5) are difficult to integrate exactly. Similar equations (obtained by Wick rotation) occur in the study of cosmological models, and various qualitative methods have been developed to deal with them. In this section we shall adapt the approach of Misner [33] to our situation. We begin by introducing the variables $(\Omega, \beta_+, \beta_-)$ defined by

$$a=\exp(\Omega+\beta_+ +\sqrt{3}\beta_-) \tag{6.1}$$

$$b=\exp(\Omega+\beta_+ -\sqrt{3}\beta_-) \tag{6.2}$$

$$c=\exp(\Omega-2\beta_+) \tag{6.3}$$

or

$$\Omega=\tfrac{1}{3}\log(abc) \tag{6.4}$$

$$\beta_+=\tfrac{1}{6}\log\left(\frac{ab}{c^2}\right) \tag{6.5}$$

$$\beta_-=\frac{1}{2\sqrt{3}}\log\left(\frac{a}{b}\right). \tag{6.6}$$

The action I in (4.7) is now

$$I=\frac{3\pi}{4}\int\left[\left(\frac{d\beta_+}{d\eta}\right)^2+\left(\frac{d\beta_-}{d\eta}\right)^2-\left(\frac{d\Omega}{d\eta}\right)^2-\Phi e^{4\Omega}\right]d\eta \tag{6.7}$$

where

$$\Phi(\beta_+,\beta_-)=-\tfrac{1}{12}[2e^{4\beta_+}\cosh(4\sqrt{3}\beta_-)-2e^{4\beta_+} -4e^{-2\beta_+}\cosh(2\sqrt{3}\beta_-)+e^{-8\beta_+}]. \tag{6.8}$$

The constraint Eq. (4.6) becomes

$$\left(\frac{d\beta_+}{d\eta}\right)^2+\left(\frac{d\beta_-}{d\eta}\right)^2-\left(\frac{d\Omega}{d\eta}\right)^2+\Phi e^{4\Omega}=0. \tag{6.9}$$

These equations can be thought of as those of a particle moving in the time-dependent potential $\Phi e^{4\Omega}$. For some purposes it is convenient to regard Ω as the time coordinate. At large negative values of Ω the potential is flat. As Ω increases so does the potential, becoming infinitely steep as $\Omega\to\infty$. Care must be taken however since Ω is not always a monotonic function of η or ξ.

The self-dual first integrals may be written in the form

$$\frac{d\Omega}{d\eta}=\partial/\partial\Omega(e^{2\Omega}\Psi_i(\beta_+,\beta_-)) \tag{6.10}$$

$$\frac{d\beta_+}{d\eta}=-\partial/\partial\beta_+(e^{2\Omega}\Psi_i(\beta_+,\beta_-)) \tag{6.11}$$

$$\frac{d\beta_-}{d\eta}=-\partial/\partial\beta_-(e^{2\Omega}\Psi_i(\beta_+,\beta_-)) \tag{6.12}$$

where for case C (the BGP^2 solutions) $i=1$ and

$$\Psi_1=\tfrac{1}{12}(2e^{2\beta_+}\cosh(2\sqrt{3}\beta_-)+e^{-4\beta_+}) \tag{6.13}$$

and for case D (which we cannot integrate exactly) $i=2$ and

$$\Psi_2=-\Phi\left(\frac{\beta_+}{2},\frac{\beta_-}{2}\right). \tag{6.14}$$

Both Ψ_1 and Ψ_2 are invariant under rotations of the (β_+,β_-) plane through $\pm\frac{2}{3}\pi$, and reflections in the β_+ axis. This is the action of the permutation group on 3 objects and corresponds to interchanging the roles of (a,b,c).

The potential $\Phi(\beta_+,\beta_-)$ has the following further properties. Near the origin it has the form

$$\Phi=\tfrac{1}{4}-2(\beta_+^2+\beta_-^2)+O(\beta^3). \tag{6.15}$$

$(0,0)$ is a local maximum. For large negative β_+ such that $2|\beta_-|<\sqrt{3}|\beta_+|$ it falls rapidly to $-\infty$ and

$$\Phi\approx-\tfrac{1}{12}e^{8|\beta_+|}. \tag{6.16}$$

For large positive β_+ and small $\beta_-\neq0$,

$$\Phi\approx-4\beta_-^2e^{4\beta_+} \tag{6.17}$$

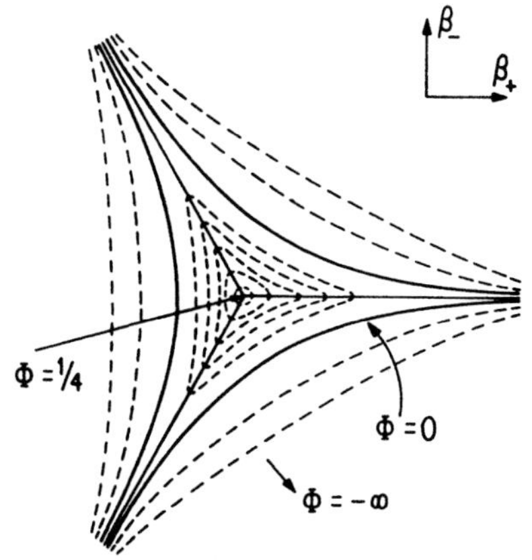

Fig. 1. A sketch of the potential $\Phi(\beta_+, \beta_-)$. The level curves are shown dotted. The potential may be envisaged as 3 gently sloping ridges which meet at the highest point of the potential in the centre. To either side of the ridges the potential falls rapidly down three steep cliffs to $-\infty$. Φ is positive near the centre and passes through zero on the level curve indicated with a solid line

while for $\beta_- = 0$,

$$\Phi \approx \tfrac{1}{3} e^{-2|\beta_+|}. \tag{6.18}$$

The equipotential curves are circular near the origin and become more triangular as one moves outward, until at the equipotential $\Phi = 0$ they become three disjoint open curves, asymptotic to the positive β_+ and symmetry-related axes. Outside $\Phi = 0$ they become straighter and straighter. Thus the potential landscape may be envisaged as three vertical cliffs meeting at three ridges which slope downwards rather gently (see Fig. 1). This potential is the negative of the usual one encountered in Bianchi IX cosmological models because we are working in imaginary time.

The various boundary conditions and specializations we have described previously may be translated in terms of this simple model. Clearly the trajectories for which $a = b$ are those which have $\beta_- = 0$ and move straight down the cliffs or along the ridges. The volume of the group orbits is $16\pi^2 e^{3\Omega}$ or $8\pi^2 e^{3\Omega}$ depending upon whether the orbits are S^3 or RP^3. Thus an asymptotically flat region corresponds to large positive values of Ω and a nut or bolt to large negative values. One can easily see that the various possibilities are given by

Euclidean or Conical Infinity.

$$\Omega \to \infty, \quad \frac{d\Omega}{d\eta} \to 0,$$

$$(\beta_+, \beta_-) \to 0, \quad \left(\frac{d\beta_+}{d\eta}, \frac{d\beta_-}{d\eta}\right) \to 0,$$

$$\eta \to \eta_0 \quad \text{(finite)}.$$

Taubian Infinity.

$$\Omega\to\infty,\quad \frac{d\Omega}{d\eta}\to\infty,$$

$$\beta_+\to\infty,\quad \frac{d\beta_+}{d\eta}\to\infty,$$

$$\beta_-\to 0,\quad \frac{d\beta_-}{d\eta}\to 0,$$

$$\eta\to\eta_0\quad \text{(finite)}.$$

Bolt.

$$\Omega\to-\infty,\quad \frac{d\Omega}{d\eta}\to\frac{na_0^2}{6},$$

$$\beta_+\to\infty,\quad \frac{d\beta_+}{d\eta}\to-\frac{na_0^2}{6},$$

$$\beta_-\to 0,\quad \frac{d\beta_-}{d\eta}\to 0,$$

where

$$n=1\quad \text{for}\quad S^3$$

$$n=2\quad \text{for}\quad RP^3.$$

Nut.

$$\Omega\to-\infty,\quad \frac{d\Omega}{d\eta}\to 0,$$

$$(\beta_+,\beta_-)\to 0,\quad \left(\frac{d\beta_+}{d\eta},\frac{d\beta_-}{d\eta}\right)\to 0,$$

$$\eta\to\infty.$$

The trajectory of the fictitious particle moving in the potential $\Phi e^{4\Omega}$ can begin or end at one of three points (modulo permutations of (a,b,c)) in the (β_+,β_-) plane. They are the Summit ($\beta_+=\beta_-=0$), the End of the Ridge ($\beta_+=\infty$, $\beta_-=0$), and the Bottom of the Cliff ($\beta_+=-\infty$, $2|\beta_-|<\sqrt{3}|\beta_+|$). These correspond to the possible ways in which (a,b,c) can cease to be finite and non-zero. Provided the velocity of the particle is also appropriate, the Summit corresponds to a nut if $\Omega=-\infty$, and a Euclidean or Conical Infinity if $\Omega=+\infty$. Similarly the End of the Ridge corresponds to a bolt if $\Omega=-\infty$ and a Taubian Infinity if $\Omega=+\infty$. The Bottom of the Cliff is singular and the curvature diverges there.

The self-dual class C metrics correspond to the set of all trajectories which just reach the origin as $\Omega\to+\infty$. The regular Eguchi-Hanson metric (type II) corresponds to such a trajectory that starts from the End of the Ridge at $\Omega=-\infty$. The incomplete Eguchi-Hanson metric (type I) corresponds to a trajectory which starts from the Bottom the Cliff on the β_+ axis at $\Omega=-\infty$. The general class C

metrics (i.e. the BGP^2 solutions, Eq. (4.23)) correspond to the remaining "generic" trajectories – i.e. those which start at the Bottom of the Cliff at $\Omega = -\infty$ and for which a, b and c are all unequal.

The self-dual Taub-NUT solution (Eq. (4.10) with $m = \pm|n|$) corresponds to trajectories which start with zero velocity at the Summit at $\Omega = -\infty$ and fall along the β_+ axis reaching either the End of the Ridge (if $m = +|n|$) or the Bottom of the Cliff (if $m = -|n|$) as $\Omega \to +\infty$. The former corresponds to the complete non-singular manifold for which $0 < m \leqq r < \infty$. The general two parameter Taub-NUT solution (Eq. (4.10)) is represented by the set of all trajectories along the β_+ axis which start at $\Omega = -\infty$ from either the End of the Ridge or the Bottom of the Cliff and which do *not* arrive at the Summit with zero velocity at $\Omega = +\infty$. That is they have insufficient energy to reach the Summit and so fall back again, or else they overshoot. The first of these possibilities corresponds to the usual outer region of Taub-NUT space $(m + (m^2 - n^2)^{1/2} < r < \infty)$. In the special case $m = \frac{5}{4}|n|$ the particle sets out as a bolt at $\Omega = -\infty$, travels some way up the ridge and falls down again to arrive at the end as a Taubian Infinity at $\Omega = +\infty$.

The general self-dual class D metrics (Eqs. (6.10), (6.11), (6.12) with $i = 2$) can be studied qualitatively in the same way. The trajectories in the (β_+, β_-) plane are lines of steepest ascent or descent of the potential $e^{2\Omega}\Phi(\beta_+/2, \beta_-/2)$. This follows directly from Eqs. (6.11) and (6.12). Therefore unless the trajectories start out along one of the ridges they will inevitably end up at the Bottom of the Cliff. Thus the self-dual Taub-NUT metric is the only complete non-singular solution in this class. In fact it follows directly from our results in Sect. 3 that any self-dual trajectory which end on the Euclidean Infinity must have started from the bottom of a cliff.

The most general Bianchi IX case (Eqs. (4.4), (4.5)) is rather complicated. It is clear that solutions cannot leave the Summit and return later. This therefore rules out non-flat Bianchi IX metrics on R^4 (nut$\to$Infinity), $S^3 \times R$ (Infinity$\to$Infinity) or S^4 (nut$\to$nut). These are ruled out anyway by our results in Sects. 3 and 5. It also seems clear that a trajectory which leaves the Summit not directed along a ridge cannot subsequently return to a ridge to become a bolt or Taubian Infinity.

It is not obvious to us whether trajectories can start out from the end of one ridge as a bolt or Taubian Infinity and finish up at the end of another ridge as a Taubian Infinity or bolt respectively, but we regard this as unlikely. The other two possibilities (bolt$\to$bolt or Taubian Infinity$\to$Taubian Infinity) are ruled out by our results in Sects. 5. Thus it seems rather likely that the only regular Bianchi IX solutions have the additional symmetry that two of the directions are equal.

7. Conclusion

An asymptotically Euclidean metric with N asymptotic regions is a complete non-singular metric on a manifold region diffeomorphic to the union of a compact set K and N complements of the open ball in R^4 on each of which the metric has the form

$$g_{\mu\nu} = \left(1 + \frac{\tilde{I}}{6\pi r^2}\right)^2 \delta_{\mu\nu} + O\left(\frac{1}{r^3}\right).$$

If $R^\sigma_\sigma=0$ the action of this region is $I=\tilde{I}$. The *Strong Positive Action Conjecture* is that $R^\sigma_\sigma=0$ implies $I\geqq 0$, with $I=0$ iff $g_{\mu\nu}$ is flat. This implies the *Weak Positive Action Conjecture* i.e. $R_{\mu\nu}=0$ implies $g_{\mu\nu}$ is flat.

In this paper we prove

1) There are no half flat asymptotically Euclidean metrics with finite fundamental group other than flat space.

2) Any non-flat asymptotically Euclidean solution with Euler number χ and signature τ must satisfy $2\chi-3|\tau|>2N$. In particular there are no non-flat solutions on R^4.

We have illustrated these results by studying a class of Bianchi IX solutions with $SU(2)$ or $SO(3)$ as isometry group. We have shown (modulo one possible but unlikely case) that the only instantons of this class must possess a further symmetry and we have given the explicit solutions. In an appendix we use these results to give an approximate construction for the only self-dual compact instanton.

Asymptotically Euclidean instantons, if they existed, would contribute to the vacuum-vacuum amplitude for gravity. Instantons with more than one asymptotically Euclidean region may lead to baryon non-conservation. The fact that any such configuration is (by the Cosmic Censorship Hypothesis) far from an extremum of the gravitational action presumably means that such processes are forbidden in the semi-classical approximation.

Another mechanism, due to 't Hooft, for baryon non-conservation requires the existence of bound states of the Dirac equation. These cannot occur on non-compact gravitational backgrounds with $R^\mu_\mu\geqq 0$.

These two facts tend to suggest that for pure gravity coupled to ordinary matter the baryon non-conserving processes are forbidden to lowest order in the stationary phase approximation. This situation changes if one includes spin 3/2 fields as part of a supersymmetric theory of gravity.

Appendix. An Approximate Metric on $K3$

In this appendix we shall describe how 16 Eguchi-Hanson spaces may be "glued" together to give a self-dual metric on a $K3$ surface. Hitchin [20] has shown that all compact half-flat spaces are homeomorphic to a $K3$ surface or an identification there of. All $K3$ surfaces are diffeomorphic to any quartic surface in CP^3. A particular such surface is a Kummer Surface. This is holomorphically equivalent to the following space [34]: Consider a lattice L in C^2 obtained by identifying the points (Z_1,Z_2) with $(Z_1,Z_2)+\lambda(1,0)+\mu(i,0)+\nu(0,1)+\sigma(0,i)$, where (λ,μ,ν,σ) are 4 arbitrary integers. This is a compact complex manifold with topology $S^1\times S^1\times S^1\times S^1$. The involution $(Z_1,Z_2)\to(-Z_1,-Z_2)$ has 16 fixed points in L, i.e. the points whose real and imaginary parts are zero or 1/2. If one identifies points in L which are equivalent under the involution these 16 points will not be regular points. One must, in the language of algebraic geometers, "blow them up". Roughly, this means replacing these 16 points by 16 copies of CP^1 – the complex projective line or Riemann sphere. This may be done as follows. Surround each of the 16 points by small 3-spheres and remove the interiors. The boundary of the manifold so obtained is 16 disjoint copies of RP^3. It is RP^3 and not S^3 because the involution connects antipodal points on S^3. Each RP^3 boundary must now be filled in with an Eguchi-Hanson manifold (i.e. $(CP^2-\{0\})/I$). To construct the metric one has to match carefully across these 16 boundaries. This can be done approximately by putting the flat metric on C^2 and the Eguchi-Hanson metric on the Eguchi-Hanson manifold. As their parameter $a \to 0$ this will become a better and better fit. It is easy to check that the Euler number χ and signature τ of the resulting space are 24 and 16 respectively and that this is consistent with the Gauss-Bonnet and signature theorems. Page [35] has established that the number of free parameters or moduli obtained from this construction corresponds exactly with the 58 expected ones. If one could find a "multi-Eguchi-Hanson" solution analogous to the multi Taub-NUT solution [29] one might hope that by judicious choice of the parameters one could construct exactly the general self-dual metric on $K3$.

In fact a minor modification of the multi-taub-NUT ansatz produces a sequence of solutions which are asymptotically conical [19]. The first of the sequence in flat space which is of course asymptotically Euclidean and the second is Eguchi-Hanson with boundary RP^3. As one proceeds along the sequence the metrics are all asymptotically flat in all 4 directions but with lens space boundaries. However each of the solutions has at least a one parameter isometry group and so superposing will not lead directly to the desired metric on $K3$. Further work on this topic is continuing.

Acknowledgements. We would like to thank J. F. Adams, M. F. Atiyah, S. W. Hawking, N. J. Hitchin, P. J. McCarthy, and D. N. Page for conversations and help.

References

1. Harrison, B.K., Thorne, K.S., Wakane, M., Wheeler, J.A.: Gravitation theory and gravitational collapse, pp. 9 and 146. Chicago, London: University of Chicago Press 1965
2. Hawking, S.W.: Commun. math. Phys. **43**, 199–220 (1975)
3. Zeldovich, Ya.B.: Phys. Lett. **59**A, 254 (1976); **73**B, 423–424 (1978)
4. Gibbons, G.W., Hawking, S.W.: Phys. Rev. D**15**, 2752–2756 (1977)
5. Gibbons, G.W., Hawking, S.W., Perry, M.J.: Nucl. Phys. B **138**, 141–150 (1978)
6. Page, D.N.: The positive action conjecture. Phys. Rev. D**18**, 2733–2738 (1978)
7. Schoen, R., Yau, S.T.: Proc. Natl. Acad. Sci. U.S. **75**, 2567 (1978); Schoen, R., Yau, S.T.: Commun. Math. Phys. **65**, 45–76 (1979)
8. 't Hooft, G.: Phys. Rev. Lett. **37**, 8–11 (1976); Phys. Rev. D **14**, 2432–2450 (1976)
9. Hawking, S.W.: Phys. Rev. D **18**, 1747–1753 (1978), and various unpublished lectures
10. Lichnerowicz, A.: Compte Rendue **257**, 5–9 (1968)
11. Hawking, S.W., Pope, C.N.: Nucl. Phys. B**146**, 381–392 (1978)
12. Misner, C.W., Wheeler, J.A.: Ann. Phys. N.Y. **2**, 525–660 (1957)
13. Misner, C.W.: Phys. Rev. **118**, 1110 (1960)
14. Misner, C.W.: Ann. Phys. N.Y. **24**, 102–117 (1963)
15. Gibbons, G.W.: Commun. Math. Phys. **27**, 87–103 (1972)
16. Gibbons, G.W., Schutz, B.F.: M.N.R.A.S. **159**, 41–45 (1972)
17. Penrose, R.: Ann. Acad. Sci. **224**, 125 (1973)
18. Jang, P.S., Wald, R.M.: J. Math. Phys. **18**, 41–44 (1977)
19. Gibbons, G.W., Hawking, S.W.: Classification of gravitational instanton symmetries. Commun. Math. Phys. (in press)

20. Hitchin, N.: J. Diff. Geom. **9**, 435–441 (1975)
21. Tolman, R.C.: Relativity, thermodynamics and cosmology. London: O.U.P. 1934
22. Chern, S.: Ann. Math. **46**, 674 (1945)
23. Eguchi, T., Gilkey, P., Hanson, A.: Phys. Rev. D **17**, 423–427 (1978)
24. Atiyah, M.F., Patodi, V.K., Singer, I.M.: Math. Proc. Cambridge Phil. Soc. **77**, 43–69 (1975)
25. Hawking, S.W., Pope, C.N.: Phys. Lett. **73**B, 42 (1978)
26. Landau, L., Lifshitz, E.M.: Classical theory of fields. 4th Ed. Oxford, New York: Pergamon 1975
27. Eguchi, T., Hanson, A.: Phys. Lett. **74**B, 249–251 (1978)
28. Belinskii, V.A., Gibbons, G.W., Page, D.N., Pope, C.N.: Phys. Lett. **76**B, 433–435 (1978)
29. Hawking, S.W.: Phys. Lett. **60**A, 81–82 (1977)
30. Atiyah, M.F., Hitchin, N., Singer, I.M.: Proc. R. Soc. A**362**, 425 (1978)
31. Page, D.N.: Phys. Lett. **78**B, 249–251 (1978)
32. Gibbons, G.W., Pope, C.N.: Commun. Math. Phys. **61**, 239–248 (1978)
33. Misner, C.W., Thorne, K.S., Wheeler, J.A.: Gravitation. San Francisco: Freeman 1970
34. Lascoux, A., Berger, M.: Varietes Kahleriennes compactes. In: Lecture notes in mathematics, Vol. 154. Berlin, Heidelberg, New York: Springer 1970
35. Page, D.N.: Phys. Lett. **80**B, 55–57 (1978)

Communicated by R. Geroch

Received October 13, 1978

Math. Proc. Camb. Phil. Soc. (1979), 85, 465
Printed in Great Britain

Polygons and gravitons

By N. J. HITCHIN
Mathematical Institute, Oxford OX1 3LB

(*Received* 9 *November* 1978)

1. There has been much interest recently in 'instantons'. These correspond, in differential geometric terms, to connexions in principal bundles over Euclidean 4-space whose curvature satisfies the Yang–Mills equations. The connexion is also required to approach the trivial connexion at infinity. The self-dual solutions have been classified by converting the problem into one of algebraic geometry using the ideas of R. Penrose and R. Ward.

There is a gravitational analogue, called by S. W. Hawking the 'gravitational instanton' which consists of a 4-manifold endowed with a complete, non-singular, positive definite metric which satisfies the Einstein equations. The Riemann curvature tensor of such a metric satisfies the Yang–Mills equations. If the curvature is self-dual, then there is again a complex analytic interpretation – the 'non-linear graviton' of Penrose – and the question of finding such metrics can be converted into one of algebraic, or more generally complex, geometry.

The appropriate conditions at infinity are, however, more subtle. G. Gibbons and C. Pope have shown that if the manifold has the topology of Euclidean space at infinity, that is of $S^3 \times \mathbb{R}$, and approaches the flat Euclidean metric sufficiently fast, then every self-dual solution is in fact isometric to Euclidean space. They propose an alternative asymptotic condition by introducing a fundamental group at infinity: the metric *locally* approaches the Euclidean one but a neighbourhood of infinity has the topology of $S^3/\Gamma \times \mathbb{R}$ where Γ is a finite group of isometries acting freely on S^3.

In this paper we give a class of examples where for Γ we take a cyclic subgroup of $SU(2)$ acting on the unit sphere S^3 in $\mathbb{C}^2$. The image of Γ in $SO(3)$ is the group of rotations of the polygon in the title. We construct the metric by Penrose's non-linear graviton technique using the well-known yet still somewhat mysterious relationship between subgroups of $SU(2)$, simple singularities and Dynkin diagrams. The manifold on which the metric is defined is obtained by resolving the singularity at the origin of $\mathbb{C}^2/\Gamma$. All our metrics are then complete, Ricci–flat Kähler metrics on this manifold relative to some complex structure. We should emphasize, however, that although the approach uses the theory of simple singularities and their resolution, this is only to give an *a priori* justification for the steps involved in constructing the metric. Once we have the metric, we can prove that these singularities can be resolved in a particular way, and even see explicitly the 2-spheres one obtains from the resolution.

The metric so constructed turns out to be the same as one considered recently by Gibbons and Hawking – the 'gravitational multi-instanton'. Our approach, however, remains valid for any finite subgroup of $SU(2)$ and it seems very likely that there are gravitational instantons associated to the binary dihedral, tetrahedral, octahedral and icosahedral groups. What is required is an explicit solution in polynomials of a certain algebraic equation. In the case of a cyclic group this is obtained simply by factorization; for the other groups one must deal with a more complicated equation. We hope to consider these cases in a later paper.

In the Yang–Mills theory, one agrees to identify connexions which are related by automorphisms of the principal bundle – this is the notion of gauge equivalence. In Riemannian geometry the analogous notion is that of isometric equivalence: two metrics are isometric if they are equivalent under a diffeomorphism. Using the 'period matrix' of a self-dual metric we determine precisely which of our metrics are equivalent and show that if Γ has order k, then there is a $3k-6$ dimensional moduli space of such metrics.

2. We shall review here briefly Penrose's non-linear graviton construction (7) as it applies to the case of a positive definite metric. We start with a self-dual solution to Einstein's equations. This consists of a 4-dimensional Riemannian manifold M whose Ricci tensor vanishes together with the anti-self-dual part W_- of the Weyl tensor.

The vanishing of W_- enables the *conformal* structure of M to be encoded in complex analytic form (1). The projective spin bundle $Z = \mathbb{P}(V_-)$ has the structure of a complex 3-manifold such that the fibres are holomorphic projective lines with normal bundle $H \oplus H$, H being the standard positive line bundle on $\mathbb{P}_1$. They belong to a family parametrized by a complex 4-manifold M^c. The manifold M^c has a complex conformal structure: two nearby points are null-separated if their corresponding lines intersect. The antipodal map on each fibre $\mathbb{P}_1$ gives Z and M^c a real structure. The real lines are simply the fibres of $\mathbb{P}(V_-)$, parametrized by the base space M, and so M is embedded in M^c as a real submanifold endowed with a real conformal structure. In the case that is important for instantons, $M = S^4$, $Z = \mathbb{P}_3$ and $M^c = \mathbb{Q}_4$, the Klein quadric.

If the Ricci tensor vanishes too, then the bundle V_-, with its induced connexion, is flat. Thus if M is simply connected, we can retract Z onto a particular fibre $\mathbb{P}_1$ by parallel translation and obtain a holomorphic fibration $\pi: Z \to \mathbb{P}_1$. Parallel translation in the vector bundle V_- gives an isomorphism $K \cong \pi^* H^{-4}$ where K is the bundle of holomorphic 3-forms on Z. These two extra pieces of data are sufficient to fix the metric, since for each $u \in \mathbb{P}_1$, the projection from Z onto M identifies M with the fibre $\pi^{-1}(u)$ and the isomorphism $K \cong \pi^* H^{-4}$ defines a non-vanishing holomorphic 2-form on $\pi^{-1}(u)$, hence a volume form, and, together with the conformal structure determined above, a metric.

For each $u \in \mathbb{P}_1$, M has a complex structure, and the metric is a Kähler metric relative to each of these. The non-vanishing holomorphic 2-form is, moreover, covariant constant relative to the Riemannian connexion. It is useful to describe the 3-dimensional space of covariant constant anti-self-dual 2-forms associated to a self-dual metric in this way, namely the space spanned by the Kähler form and the real and imaginary parts of the constant holomorphic 2-form. Note that by our sign convention the orientation given to M by the complex structure is the opposite of the original one.

This reformulation of the self-dual Einstein equations means that to find a solution we must look for the following:

(1) A complex 3-manifold Z, the total space of a holomorphic fibration $\pi: Z \to \mathbb{P}_1$.
(2) A 4-parameter family of sections, each with normal bundle $H \oplus H$.

(3) A non-vanishing holomorphic section s of $K \otimes \pi^* H^4$.

(4) A real structure on Z such that π and s are real, and Z is fibred by the real sections of the family. Here $\mathbb{P}_1$ is given the real structure of the antipodal map

$$u \mapsto -\bar{u}^{-1}.$$

This set of data is the non-linear graviton.

3. We are going to construct, using the above approach, a solution to the self-dual Einstein equations which at infinity is diffeomorphic to $S^3/\Gamma \times \mathbb{R}$ where $\Gamma \subset SU(2)$ is the cyclic group of matrices

$$\begin{pmatrix} e^{2\pi i n/k} & 0 \\ 0 & e^{-2\pi i n/k} \end{pmatrix} \quad 0 \leqslant n < k.$$

The group Γ acts not just on the unit sphere $S^3 \subset \mathbb{C}^2$, but on the whole of $\mathbb{C}^2$ and we can form the quotient space $\mathbb{C}^2/\Gamma$. This certainly has the topology of $S^3/\Gamma \times \mathbb{R}$ at infinity, but it is not a manifold – it has a singularity at the origin.

We can describe this singularity algebraically as follows. Note that if $(z_1, z_2) \in \mathbb{C}^2$, then the monomials $z_1^k, z_2^k, z_1 z_2$ are invariant under Γ. If we denote them by x, y, z then they satisfy an algebraic relation

$$xy = z^k.$$

This gives an isomorphism between $\mathbb{C}^2/\Gamma$ and the complex surface $xy = z^k$ in $\mathbb{C}^3$. We can now look at deformations of this singularity. There is a universal family of such deformations (the 'universal unfolding' in the sense of Thom) which in this case is given by adding on lower order terms in z:

$$xy = z^k + a_1 z^{k-1} + \ldots + a_k.$$

These still have the topology of $S^3/\Gamma \times \mathbb{R}$ at infinity, but if the discriminant of the right hand side is non-zero, these are non-singular complex surfaces. We shall take M to be the underlying 4-dimensional differentiable manifold of such a surface.

In order to construct Z we need a family of complex stuctures on M parametrized by $u \in \mathbb{P}_1$. As a first approximation to Z, we define $\tilde{Z}$ by the equation:

$$xy = z^k + a_1(u) z^{k-1} + \ldots + a_k(u),$$

where $a_i(u)$ is locally a holomorphic function of u. Since $\mathbb{P}_1$ has no global non-trivial functions we must interpret $a_i(u)$ as a section of a line bundle and then in order for the equation to make sense we must think of $x \in H^l$, $y \in H^m$, $z \in H^n$, $a_i \in \Gamma(H^{ni})$ where $l + m = kn$. The above equation then defines $\tilde{Z}$ as the 3-dimensional hypersurface $f(x, y, z, u) = 0$ in the complex 4-manifold $H^l \oplus H^m \oplus H^n$ where

$$f: H^l \oplus H^m \oplus H^n \to H^{kn}$$

is given by

$$f(x, y, z, u) = xy - \sum_{i=0}^{k} a_i(u) z^{k-i}.$$

We shall next give $\tilde{Z}$ a real structure. Since $SU(2)$ acts on $\mathbb{C}^2$ as multiplication on the left by unit quaternions, it commutes with right multiplication by j. This is the real structure $(z_1, z_2) \mapsto (-\bar{z}_2, \bar{z}_1)$ (which also induces the antipodal map σ on $\mathbb{P}_1$). The subgroup Γ commutes with the real structure and so we get a real structure on $\mathbb{C}^2/\Gamma$ defined by $x \mapsto (-1)^k \bar{y}$, $y \mapsto \bar{x}$, $z \mapsto -\bar{z}$. We can use this to define a real structure on $H^l \oplus H^m \oplus H^n$:

$$\tau(x, y, z, u) = ((-1)^k \sigma(y),\ \sigma(x),\ -\sigma(z), \sigma(u)), \tag{3·1}$$

where $\sigma: H^i \to \bar{H}^i$ is the real structure on H^i induced by σ. To do this we require of course that $l = m$ since x and y are interchanged. If now the sections a_i of H^{ni} are real (i.e. $a_i(\sigma u) = \sigma(a_i(u))$), then $\tilde{Z}$ has an induced real structure with $\pi: \tilde{Z} \to \mathbb{P}_i$ real.

The hypersurface $\tilde{Z}$ now has a real structure, and a real projection onto $\mathbb{P}_1$, but it is not a manifold: it has singularities over those points $u_j \in \mathbb{P}_1$ at which the discriminant of $\Sigma a_i(u) z^{k-i}$ vanishes. We now use the characteristic property of this type of singularity – a simple singularity, or rational double point – and this is that the singularities can be resolved within the family of non-singular surfaces, if necessary by taking a finite covering of $\mathbb{P}_1$ branched over the points u_j. We refer the interested reader to (2), (3). This procedure would give in general a non-singular manifold Z fibring over a Riemann surface of higher genus. In order to avoid this, we make the assumption that

$$\sum_{i=0}^{k} a_i(u) z^{k-i} = \prod_{j=1}^{k} (z - p_j(u)) \quad (p_j \in \Gamma(H^n)),$$

(The reasoning for this may be found in (3). Replacing the sections a_i by symmetric functions in p_j,is essentially lifting the deformation through the Weyl group of A_{k-1}).

Having done this (and we repeat the fact that the existence of the metric independently shows that this can be done) we obtain a complex 3-manifold Z with a real structure and with a compatible fibration $\pi: Z \to \mathbb{P}_1$.

We have yet to determine the integers l, m, n. They satisfy $l = m$ from the real structure and $l + m = kn$, so $2m = kn$. To determine n we use condition 2 of the graviton: the normal bundle of a section must be $H \oplus H$. Take such a section and project it into $\tilde{Z}$. If it misses the singularities, and a generic one will do so, its normal bundle is just the restriction of the tangent bundle along the fibres, and this is the kernel of the vector bundle homomorphism

$$(f_x, f_y, f_z): H^l \oplus H^m \oplus H^n \to H^{kn}.$$

The first Chern class of the normal bundle $c_1(N)$ is thus $l + m + n - kn$. But if $N = H \oplus H$ then $c_1(N) = 2$, so $n = 2$, $l = m = k$, and $\tilde{Z} \subset H^k \oplus H^k \oplus H^2$.

Condition 3 requires a non-vanishing section s of $K \otimes \pi^* H^4$ on Z. For this we take a constant multiple of the standard form

$$\frac{dx \wedge dy \wedge dz}{f_u} = \frac{dy \wedge dz \wedge du}{f_x} = \dots,$$

which becomes well defined and non-vanishing on Z, a fact which will again be evident from the existence of the metric.

The only piece of data missing, and the most important of all, is the 4-parameter family of sections. These define by projection onto $\tilde{Z}$ holomorphic sections of the vector bundle $H^k \oplus H^k \oplus H^2$. If we choose an affine parameter $u \in \mathbb{P}_1$, then a holomorphic section of H^m can be described by a polynomial of degree m in u, hence a line of our family will be given by polynomials $x(u), y(u), z(u)$ of degree k, k, 2 respectively which satisfy the equation

$$xy = \prod_{i=1}^{k} (z - p_i),$$

where $p_i \in \Gamma(H^2)$ is a quadratic polynomial.

Put $z = au^2 + 2bu + c$ and $p_i = a_i u^2 + 2b_i u + c_i$ and let α_i, β_i be the roots of $z - p_i$. Then we can solve the above equation by a simple factorization: we put

$$x = A \prod_{i=1}^{k} (u - \alpha_i); \quad y = B \prod_{i=1}^{k} (u - \alpha_i), \tag{3·2}$$

where $AB = \Pi(a - a_i)$. Locally, these curves are parametrized by the four variables (a, b, c, A) with a finite indeterminacy coming from the choice of roots.

The real lines must be preserved by the real structure on $\tilde{Z}$, so we must have, from (3·1)

$$x(\sigma u) = (-1)^k \sigma(y(u)); \quad y(\sigma u) = \sigma(x(u)); \quad z(\sigma u) = -\sigma(z(u)).$$

This last reality condition on the quadratic polynomial z reduces to $c = -\bar{a}$, $b = \bar{b}$ and so the roots of $z - p_i$ are

$$\frac{-(b - b_i) \pm \sqrt{\{(b - b_i)^2 + |a - a_i|^2\}}}{(a - a_i)}.$$

If we let Δ_i denote the *positive* square root of $(b - b_i)^2 + |a - a_i|^2$ then we have an unambiguous way of assigning roots to x and y: we put

$$x = A \prod_{i=1}^{k} \left(u + \frac{(b - b_i) - \Delta_i}{(a - a_i)}\right); \quad y = B \prod_{i=1}^{k} \left(u + \frac{(b - b_i) + \Delta_i}{(a - a_i)}\right). \tag{3·3}$$

The reality condition on x and y together with the relation $AB = \Pi(a - a_i)$ implies that

$$A\bar{A} = \Pi((b - b_i) + \Delta_i). \tag{3·4}$$

The real lines are thus locally parametrized by a point $(b, \operatorname{Im} a, \operatorname{Re} a)$ in $\mathbb{R}^3$ and an angular variable $\arg A$. Each metric is determined by k real quadratic polynomials p_i, the coefficients of which can be thought of as k points $\mathbf{x}_i$ in $\mathbb{R}^3$.

We now have all the data for the non-linear graviton construction. There are certain properties to be checked – that the normal bundle of our sections is $H \oplus H$, that they extend from $\tilde{Z}$ to Z and that Z is fibred by the real sections. We could proceed in complex analytic language to check these properties, but we prefer to use the data to compute the metric on an open set of M and then show that it is well-defined everywhere.

4. To each point (a, b, c, A) in an open set of $\mathbb{C}^4$ we have associated a section $(x(u), y(u), z(u))$ of $H^k \oplus H^k \oplus H^2$ which lies in $\tilde{Z}$. Taking the derivative of this map, a

tangent vector (a', b', c', A') goes to a section $(x', y', z') \in \ker(f_x, f_y, f_z)$, a section of the normal bundle of the line $(x(u), y(u), z(u))$. The tangent vector (a', b', c', A') is a null vector in the conformal structure on M^c by definition if and only if the corresponding section of the normal bundle vanishes at some point. This is the infinitesimal version of the condition that two points are null separated if their corresponding lines intersect. Hence the null cone is given by the condition that $x'(u)$, $y'(u)$, $z'(u)$ have a common zero, i.e. by eliminating u from the equations

$$z'(u) = a'u^2 + 2b'u + c' = 0, \tag{4·1}$$

$$\frac{x'(u)}{x(u)} = \frac{A'}{A} - \Sigma \frac{\alpha_i}{u - \alpha_i} = 0, \tag{4·2}$$

obtained by differentiating (3·2). If we write

$$z - p_i = (u - \alpha_i)((a - a_i)u - (c - c_i)\alpha_i^{-1})$$

and differentiate we find

$$0 = z'(u) = \alpha_i'(-(a - a_i)u + (c - c_i)\alpha_i^{-1} + (u - \alpha_i)(c - c_i)\alpha_i^{-2}) + (u - \alpha_i)(a'u - c'\alpha_i^{-1})$$

and hence
$$\frac{\alpha_i'}{u - \alpha_i} = \frac{\alpha_i a' u - c'}{2u\Delta_i},$$

where $2\Delta_i = \alpha_i - \beta_i = 2(b^2 - ac)^{\frac{1}{2}}$.

Substituting in (4·2) we get a linear equation for u:

$$u\left(\frac{2A'}{A} - \Sigma\frac{\alpha_i a'}{\Delta_i}\right) + \Sigma\frac{c'}{\Delta_i} = 0$$

and if we put
$$\gamma = \Sigma\frac{1}{\Delta_i}, \quad \delta = \Sigma\frac{\alpha_i}{\Delta_i}$$

the null cone becomes
$$\gamma^2(a'c' - b'^2) + \left(\frac{2A'}{A} - \delta a' - \gamma b'\right)^2 = 0.$$

If we now consider a *real* tangent vector on the family of real curves, then from (3·4)

$$\mathrm{Re}\left(\frac{2A'}{A}\right) = \Sigma\frac{(b - b_i)' + \Delta_i'}{(b - b_i) + \Delta_i} = \gamma b' + \mathrm{Re}\,(\delta a') \tag{4·3}$$

so the real conformal structure may be defined by the metric

$$\gamma^2(b'^2 + a'\bar{a}') + \left(\mathrm{Im}\left(\frac{2A'}{A} - \delta a'\right)\right)^2 \tag{4·4}$$

This is clearly a positive definite metric. In view of (4·3), we can also write it in the hermitian form

$$\gamma^2 a'\bar{a}' + \left(\frac{2A'}{A} - \delta a'\right)\left(\frac{2\bar{A}'}{\bar{A}} - \overline{\delta a'}\right).$$

We have here the conformal structure written in terms of natural local coordinates in the space of complex sections. We should like to express it in terms of coordinates on a non-singular fibre of $\pi: Z \to \mathbb{P}_1$. Without loss of generality we may assume that $\pi^{-1}(0)$ is non-singular, i.e. the surface

$$xy = \Pi(z + \bar{a}_i) \quad \text{where} \quad a_i \neq a_j, \quad i \neq j.$$

This is a surface in $\mathbb{C}^3$ and (x, y) are coordinates except where $f_z = 0$, (y, z) except where $f_x = y = 0$. We shall first express the conformal structure in terms of the coordinates (y, z), so we must relate (y, z) to the coordinates (a, b, A). This is easy: at $u = 0$, $z = -\bar{a}$ and

$$y = B\Pi \frac{(b - b_i) + \Delta_i}{(a - a_i)} = \frac{1}{A} \Pi((b - b_i) + \Delta_i) = \bar{A},$$

hence the conformal structure is defined by the hermitian form

$$\gamma^2 dz d\bar{z} + \left(\frac{2dy}{y} + \bar{\delta} dz\right)\left(\frac{2d\bar{y}}{\bar{y}} + \delta d\bar{z}\right). \tag{4·5}$$

To fix the metric within the conformal class we use the holomorphic section

$$s = \frac{dy \wedge dz \wedge du}{f_x} = \frac{dy \wedge dz \wedge du}{y}$$

and this defines the holomorphic 2-form $(dy \wedge dz)/y$ on each fibre and the volume form $\omega = dy \wedge dz \wedge d\bar{y} \wedge d\bar{z}/|y|^2$.

The volume form of the metric (4·5) is $4\gamma^2\omega$, so the metric we are searching for is a constant multiple of

$$\gamma dz d\bar{z} + \gamma^{-1}\left(\frac{2dy}{y} + \bar{\delta} dz\right)\left(\frac{2d\bar{y}}{\bar{y}} + \delta d\bar{z}\right) \tag{4·6}$$

where

$$\gamma = \Sigma((b - b_i)^2 + |\bar{z} + a_i|^2)^{-\frac{1}{2}} = \Sigma \Delta_i^{-1},$$

$$\delta = \Sigma \frac{(b - b_i) - \Delta_i}{\Delta\ (\bar{z} + a_i)},$$

and b is defined implicitly by

$$\Pi((b - b_i) + \Delta_i) = y\bar{y}.$$

If we return to the form of the metric given in (4·4), and the description of the space of real quadratic polynomials as Euclidean 3-space with the metric given by the discriminant, then we obtain the metric which describes the 'gravitational multi-instanton' of Gibbons and Hawking (5):

$$\gamma d\mathbf{x} . d\mathbf{x} + \gamma^{-1}(d\tau + \boldsymbol{\omega} . d\mathbf{x})^2, \tag{4·7}$$

where $\gamma = \Sigma 1/|\mathbf{x} - \mathbf{x}_i|$ and curl $\boldsymbol{\omega} = \operatorname{grad} \gamma$.

These metrics are themselves generalizations of the metric of Calabi, Eguchi and Hanson (4), which corresponds to the case $k = 2$.

The metric (4·6) appears to become singular if $y = 0$, but this is precisely where (y, z) are not coordinates. If $y = 0$, then $z = -\bar{a}_i$ for some i and so since $a_i \neq a_j$, $f_z \neq 0$ in some neighbourhood of $z = -\bar{a}_i$ and then (x, y) are coordinates. We must now show that the metric is well-defined in such a neighbourhood.

Now $xy = \Pi(z+\bar{a}_i)$ so since $a_i \neq a_j$, near $z = -\bar{a}_i$ we have

$$z+\bar{a}_i = hxy, \tag{4·8}$$

where h is C^∞ and non-vanishing. Furthermore, $y\bar{y} = \Pi((b-b_j)+\Delta_j)$,

so

$$(b-b_i)+\Delta_i = gy\bar{y}, \tag{4·9}$$

where g is C^∞ and positive. Thus from (4·8)

$$\Delta_i^2 = (b-b_i)^2+|z+\bar{a}_i|^2 = (b-b_i)^2+|h|^2|x|^2|y|^2$$

and from (4·9)

$$\Delta_i^2 = g^2|y|^4+(b-b_i)^2-2(b-b_i)\,g|y|^2.$$

It follows that

$$b-b_i = (g^2|y|^2-|h|^2|x|^2)/2g$$

and $\Delta_i = (g^2|y|^2+|h|^2|x|^2)/2g$. Put $r^2 = g^2|y|^2+|h|^2|x|^2$, then

$$\gamma = 2g/r^2+\gamma_0,$$

where γ_0 is C^∞ and

$$\delta = \sum_j \frac{-(b-b_j)+\Delta_j}{\Delta_j(-\bar{z}-a_j)} = \frac{-2hx}{r^2\bar{y}}+\delta_0,$$

where δ_0 is C^∞. Furthermore, we have

$$dz = h(x\,dy+y\,dx)+xy\,dh.$$

Using these expressions we can write

$$\frac{2dy}{y}+\bar{\delta}dz = \frac{2}{r^2}(g^2\bar{y}dy-|h|^2\bar{x}\,dx-\bar{h}|x|^2dh)+\bar{\delta}_0\,dz$$

and it is then clear that the terms γ_0 and δ_0 do not affect the regularity of the metric (4·6), and that the metric is regular if and only if

$$\frac{2g}{r^2}dz\,d\bar{z}+\frac{r^2}{2g}\left[\frac{4}{r^4}(g^2\bar{y}\,dy-|h|^2\bar{x}\,dx-\bar{h}|x|^2dh)(\quad)^-\right]$$

is regular. But this one can rewrite as

$$2g^{-1}(h^2dx\,d\bar{x}+g^2dy\,d\bar{y}+\bar{h}x\,dh\,d\bar{x}+h\bar{x}\,dx\,d\bar{h}+|x|^2dh\,d\bar{h})$$

which, since g is non-vanishing, is well defined and we have a non-singular positive definite metric on the whole of the non-singular surface $xy = \Pi(z+\bar{a}_i)$.

The asymptotically locally Euclidean behaviour of the metric can be most easily seen by taking the form of the metric (4·7) given by Gibbons and Hawking, and letting **x** go to infinity. Completeness of the metric also follows easily from its asymptotic properties.

5. Now that we have the metric we ca return to the question of the resolution of the singularities: the passage from $\tilde{Z}$ to Z. For each point $u \in \mathbb{P}_1$, we have a complex structure on M, relative to which the metric is Kählerian, and a holomorphic map to the fibre $\pi^{-1}(u)$ in $\tilde{Z}$. In particular we have a non-singular complex surface mapping onto the singular fibres of $\tilde{Z}$: a resolution of the singularity.

These singular surfaces are given by the values of u at which $xy = \Pi(z-p_i(u))$ has a singularity, i.e. those points at which $p_i = p_j$ for some $i \neq j$. At such a point $u = u_0$, the singularity is at $x = 0$, $y = 0$, $z = p_i(u_0)$. We shall now use the metric to see what this singular point becomes in the resolution and to do this we consider the real curves in $\tilde{Z}$ which pass through the singularity. These correspond to Z in the real curves passing through the resolution of the singularity, and they trace out in a non-singular fibre an isometric copy of the resolved singularity.

Hence, suppose $(x(u), y(u), z(u))$ passes through the singularity over $u = u_0$. Then $z(u_0) = p_i(u_0)$ where $p_i(u_0) = p_j(u_0)$, but since z, p_i, p_j are real quadratic polynomials they also agree at the antipodal point $-\bar{u}_0^{-1}$, so

$$z - p_i = \lambda(p_j - p_i)$$

where λ is real. The roots of $p_j - p_i$ are

$$\frac{-(b_j - b_i) \pm \Delta_{ij}}{(a_j - a_i)}$$

and according to our choice of real structure (3·3) the corresponding roots of x are

$$\frac{-\lambda(b_i - b_i) + |\lambda|\Delta_{ij}}{\lambda(a_j - a_i)}, \quad \frac{-(\lambda-1)(b_j - b_i) + |\lambda - 1|\Delta_{ij}}{(\lambda-1)(a_j - a_i)}.$$

If we assume that this is a generic singularity, i.e. $p_k(u_0) \neq p_l(u_0)$ for any other pair (k, l), then these two roots must be distinct, for x must vanish at both roots of $p_j - p_i$. This means that $\mathrm{Sgn}(\lambda - 1) = -\mathrm{Sgn}\ \lambda$ and hence $0 < \lambda < 1$. On the non-singular surface $u = 0$, the corresponding real lines trace out a 2-sphere of the form:

$$x = \sqrt{\{\lambda(1-\lambda)\}} f(\lambda)\, e^{-i\theta},$$
$$y = \sqrt{\{\lambda(1-\lambda)\}} g(\lambda)\, e^{i\theta},$$
$$z = -\bar{a}_i + \lambda(\bar{a}_i - \bar{a}_j),$$

as $$0 \leqslant \lambda \leqslant 1 \quad \text{and} \quad 0 \leqslant \theta \leqslant 2\pi,$$

where f and g can be determined by the transformation from coordinates (a, b, A) to (x, y, z).

In the representation of $z(u)$ as a point in $\mathbb{R}^3$, we see that these 2-spheres correspond to the straight line segments joining $\mathbf{x}_i$ to $\mathbf{x}_j$. In general, there are $\frac{1}{2}k(k-1)$ of these, but in certain configurations there are less. The most interesting case is when the points $\mathbf{x}_i$ are collinear. This corresponds, for example, to an equation for $\tilde{Z}$ of the form

$$xy = \Pi(z - a_i(u^2 - 1)) \quad (a_i \in \mathbb{R}).$$

In this case the singularities occur at $u = \pm 1$ and give the singular surface $xy = z^k$ that we started with. The family of curves passing through the singular points now has $(k-1)$ components corresponding to the values $0 < n < k$:

$$x = A(u-1)^{k-n}(u+1)^n; \quad y = B(u-1)^n(u+1)^{k-n}; \quad z = \lambda(u^2 - 1).$$

The same sort of reasoning as above shows however, that if we order the points on the line $a_1 < a_2 < \ldots < a_k$, then a real curve with exponent n corresponds to the value of λ in the interval (a_n, a_{n+1}) and it follows that the resolved singularity consists of a chain of $(k-1)$ 2-spheres, each intersecting the next.

These 2-spheres define homology classes $\gamma_i \in H_2(M, \mathbb{Z})$, $1 \leqslant i \leqslant k-1$ such that the intersection number $\gamma_i \cdot \gamma_{i+1} = 1$ but otherwise $\gamma_i \cdot \gamma_j = 0$. They are furthermore holomorphic 2-spheres (projective lines) in a surface which has a non-vanishing holomorphic 2-form. This means that the canonical bundle of the surface is trivial, and restricting to each projective line, $c_1(T \oplus N) = 0$, so $c_1(N) = -c_1(T) = -2$ where T and N are the tangent and normal bundles of the line. It follows that the self-intersection number $\gamma_i^2 = -2$.

Now $xy - z^k = 0$ with its singularity resolved is the differentiable manifold M, and it is clear from this point of view that M retracts onto a neighbourhood of the resolved singularity, so $H_2(M)$ is actually generated by $\{\gamma_i\}$. The intersection matrix $\gamma_i \cdot \gamma_j$ can now be identified with the Cartan matrix A_{k-1}, and we have a distinguished class of bases for $H_2(M)$, namely those which satisfy the same relations as the γ_i's, or correspond in Lie algebraic terms to a choice of simple roots. Any two such bases are related by an element of the Weyl group of A_{k-1}, the symmetric group S_k. If we go back to the picture of k points $\mathbf{x}_i$ in $\mathbb{R}^3$, we see that, for a generic metric, a choice of such a basis corresponds to an ordering $\mathbf{x}_1, \ldots, \mathbf{x}_k$ of the points and the homology classes can be represented by the 2-spheres corresponding to the segments $(\mathbf{x}_i, \mathbf{x}_{i+1})$.

Not only does $H_2(M)$ have a distinguished class of bases, but each self-dual metric determines a distinguished 3-dimensional space of closed 2-forms – the covariant constant anti-self-dual 2-forms. An oriented orthonormal basis $\omega_1, \omega_2, \omega_3$ for this space is well-defined modulo the special orthogonal group $SO(3)$. By integrating these 2-forms over the homology classes, we can define the period matrix

$$A_{ij} = \int_{\gamma_j} \omega_i.$$

A_{ij} is a $3 \times (k-1)$ real matrix. Its importance lies in the fact that it is an invariant, modulo the action of $SO(3) \times S_k$, of the isometric equivalence class of the metric, for if $f\colon M \to M$ is a diffeomorphism, the 2-forms $\{f^*\omega_i\}$ form an orthonormal basis of covariant constant 2-forms for the metric f^*g, and $\{f_* \gamma_j\}$ a distinguished basis for $H_2(M)$. Since

$$\int_{\gamma_j} f^*\omega_i = \int_{f_*\gamma_j} \omega_i$$

the period matrices of g and f^*g are equivalent.

We shall next compute A_{ij} for the metrics constructed here and deduce a number of consequences. As mentioned above, we can take the basis $\{\gamma_i\}$ to be represen ed by the 2-spheres corresponding to the straight line segments $(\mathbf{x}_i, \mathbf{x}_{i+1})$ for some ordering.

If ω is the Kähler form and h the holomorphic 2-form such that $2h \wedge \bar{h} = -\omega \wedge \omega$ then we take for $\{\omega_i\}$:

$$\omega_1 = i\omega,$$

$$\omega_2 = h + \bar{h},$$

$$\omega_3 = i^{-1}(h - \bar{h}).$$

Hence in our case we take

$$\omega_1 = i\gamma\, dz \wedge d\bar{z} + i\gamma^{-1}\left(\frac{2dy}{y} + \bar{\delta}\, dz\right)\left(\frac{2d\bar{y}}{\bar{y}} + \delta\, d\bar{z}\right),$$

$$\omega_2 + i\omega_3 = 2h = \frac{4dy \wedge dz}{y}.$$

The 2-sphere γ_i is given by

$$z = -\bar{a}_i + \lambda(\bar{a}_i - \bar{a}_{i+1}),$$

$$y = e^{i\theta}h(\lambda), \quad \text{for some function } h,$$

as $0 \leqslant \lambda \leqslant 1$, $0 \leqslant \theta \leqslant 2\pi$.
Hence

$$\int_{\gamma_i} \omega_2 + i\omega_3 = 4\int_{\gamma_i} i\, d\theta \wedge d\lambda(\bar{a}_i - \bar{a}_{i+1}) = 8\pi i(\bar{a}_i - \bar{a}_{i+1})$$

and, restricting ω_1 to γ_i,

$$\omega_1 = i\gamma^{-1}(-2i)\,\mathrm{Re}\left(\frac{2dy}{y} + \bar{\delta}dz\right) \wedge \mathrm{Im}\left(\frac{2dy}{y} + \bar{\delta}dz\right).$$

But from (4·3)

$$\mathrm{Re}\left(\frac{2dy}{y} + \bar{\delta}dz\right) = \gamma\, db = \gamma(b_{i+1} - b_i)\, d\lambda,$$

$$\mathrm{Im}\left(\frac{2dy}{y} + \bar{\delta}dz\right) = 2d\theta + g\, d\lambda,$$

so

$$\int_{\gamma_i} \omega_1 = 2(b_{i+1} - b_i)\int d\lambda \wedge 2d\theta = 8\pi(b_i - b_{i+1}).$$

Thus, up to a constant 8π, the columns of the period matrix are just the vectors $\mathbf{x}_i - \mathbf{x}_{i+1}$. Since the Weyl group acts on these just by permuting the k vectors $\mathbf{x}_i$, we see that, up to a rotation or a translation in Euclidean 3-space, the k vectors, which provide the basic data for one of our metrics, are determined by the period matrix and hence the isometric equivalence class of the metric. It is easy to see that Euclidean motions of the points give equivalent metrics, hence the 'moduli space' of these metrics is the set of k unordered points in $\mathbb{R}^3$, modulo the Euclidean group, a stratified set of dimension $3k - 6$.

Suppose now we fix a complex structure on M, say the surface $xy = \Pi(z + \bar{a}_i)$. We can ask how many of our metrics are Kählerian relative to this complex structure. These will correspond to polynomials $p_i(u)$ such that $p_i(u) = -\bar{a}_i$ for some u and all $1 \leqslant i \leqslant k$. By a rotation we can assume $u = 0$, and then p_i is fixed by the real parameter b_i. Hence, identifying equivalent metrics, there is a $(k-1)$ dimensional space of metrics, Kählerian relative to this particular complex structure. The period matrix tells us that the metric is determined uniquely by the cohomology class of the Kähler form.

Finally, we may note that *any* self-dual metric on M has a period matrix and hence determines a set of k point in $\mathbb{R}^3$, and thus, if the points are distinct, a metric of the above family with the same period matrix. In view of this, and the result of Gibbons and Pope (6) applied to $\mathbb{R}^4$, it seems reasonable to conjecture that any self-dual metric on M, becoming locally Euclidean sufficiently rapidly at infinity, is in fact equivalent to one of those constructed here. This would be a 'global Torelli theorem' in the terminology of algebraic geometry.

The author wishes to thank M. F. Atiyah, G. W. Gibbons and S. W. Hawking for useful conversations.

REFERENCES

(1) Atiyah, M. F., Hitchin, N. J. & Singer, I. M. Self-duality in four-dimensional Riemannian geometry, *Proc. R. Soc. Lond.* A **362**, (1978), 425–461.
(2) Brieskorn, E. Die Auflösung der rationalen Singularitäten holomorpher Abbildungen. *Math. Ann.* **178** (1968), 255–270.
(3) Brieskorn, E. Singular elements of semisimple algebraic groups. In *Actes Congres Intern. Math.* (1970), t. 2, 279–284.
(4) Eguchi, T. & Hanson, A. Asymptotically flat solutions to Euclidean gravity. *Phys. Lett.* **74** B (1978), 249–251.
(5) Gibbons, G. W. & Hawking, S. W. Gravitational multi-instantons, *Phys. Lett.* B (to appear).
(6) Gibbons, G. W. & Pope, C. N. The positive action conjecture and asymptotically Euclidean metrics in quantum gravity. *DAMTP preprint* (Cambridge University).
(7) Penrose, R. Nonlinear gravitons and curved twistor theory, *Gen. Relativ. Grav.* **7** (1976), 31–52.

J. DIFFERENTIAL GEOMETRY
29 (1989) 665-683

THE CONSTRUCTION OF ALE SPACES AS HYPER-KÄHLER QUOTIENTS

P. B. KRONHEIMER

1. Introduction

According to the definition given by Calabi [4], a Riemannian manifold (X, g) is *hyper-Kähler* if it is equipped with three automorphisms I, J, K of the tangent bundle which satisfy the relations of the quaternion algebra $\mathbf{H}$ and are covariant constant with respect to the Levi-Civita connection:

$$I^2 = J^2 = K^2 = -1, \quad IJ = -JI = K, \quad \nabla I = \nabla J = \nabla K = 0.$$

These conditions imply in particular that each of I, J and K defines an integrable complex structure on X and that the metric g is Kähler with respect to all three; the three Kähler forms $\omega_1, \omega_2, \omega_3$ are therefore closed, giving three symplectic structures to X. In dimension 4, a simply-connected Riemannian manifold admits such a hyper-Kähler structure precisely when the Riemann curvature tensor is either self-dual or anti-self-dual. A complete, hyper-Kähler 4-manifold is therefore a self-dual, positive-definite solution to Einstein's equations in vacuum (a self-dual *gravitational instanton*), and it is with examples of such manifolds that we are concerned.

This paper describes the construction of a particular family of hyper-Kähler 4-manifolds, the so-called ALE spaces [6]. ALE stands for *asymptotically locally Euclidean* and describes a Riemannian 4-manifold with just one end which at infinity resembles a quotient $\mathbf{R}^4/\Gamma$ of Euclidean space $\mathbf{R}^4$ by a finite group Γ of identifications. The Riemannian metric g is required to approximate the Euclidean metric up to $O(r^{-4})$,

$$g^{ij} = \delta^{ij} + O(r^{-4}),$$

with appropriate decay in the derivatives of g^{ij}. A large class of such ALE spaces was discovered by Gibbons and Hawking [7]. For each integer $k \geq 2$, they constructed a family of spaces, depending on $3k - 6$ parameters, which had self-dual curvature and resembled at infinity a quotient of $\mathbf{R}^4$ by a cyclic group Γ of order k. These 'multi-Eguchi-Hanson' metrics were obtained also by Hitchin [8], who derived them by an application of Penrose's nonlinear

Received June 12, 1987 and, in revised form, January 29, 1988.

graviton construction. Hitchin's approach pointed to a close relationship with the deformation theory of the complex quotient singularities $\mathbf{C}^2/\Gamma$ and strongly suggested the existence of other families of ALE gravitational instantons associated with the other finite subgroups $\Gamma \subset \mathrm{SU}(2)$—the binary dihedral, tetrahedral, octahedral and icosahedral groups. These conjectured ALE spaces should be similarly related to the quotient singularities $\mathbf{C}^2/\Gamma$, the so-called Kleinian singularities, or rational double points. The construction we describe confirms this conjecture.

The following theorem (our main result) has been announced in [12]. Let Γ be a finite subgroup of SU(2), let $\widetilde{\mathbf{C}^2/\Gamma} \to \mathbf{C}^2/\Gamma$ be the minimal resolution of the quotient singularity, and let X be the smooth 4-manifold which underlies the complex surface $\widetilde{\mathbf{C}^2/\Gamma}$.

Theorem 1.1. *Let three coholomogy classes $\alpha_1, \alpha_2, \alpha_3 \in H^2(X;\mathbf{R})$ be given which satisfy the nondegeneracy condition*

$$(*) \qquad \begin{array}{l} \textit{for each } \Sigma \in H_2(X;\mathbf{Z}) \textit{ with } \Sigma\cdot\Sigma = -2, \textit{there exists} \\ i \in \{1,2,3\} \textit{ with } \alpha_i(\Sigma) \neq 0. \end{array}$$

Then there exists on X an ALE hyper-Kähler structure for which the cohomology classes of the Kähler forms $[\omega_i]$ are the given α_i.

The proof of this result is a direct application of a procedure which is already known to produce a wide variety of hyper-Kähler manifolds, including the multi-Eguchi-Hanson spaces. This is the *hyper-Kähler quotient* construction of Hitchin et al. [9], a modification of the symplectic quotient, or reduced phase space, familiar in symplectic geometry. We review this construction in §2 and then apply it in a particular case to produce a family of hyper-Kähler 4-manifolds. In §3 we show that these manifolds are diffeomorphic to $\widetilde{\mathbf{C}^2/\Gamma}$ and that their metrics are ALE. The proof of Theorem 1.1 is completed in §4 where we calculate the cohomology classes of the Kähler forms on each member of the family.

In a later paper [13] we shall show that the construction presented here is complete: every ALE hyper-Kähler 4-manifold (and therefore every simply-connected, ALE solution to the self-dual Einstein equations) is isometric to a member of one of the families produced in §2. These results, obtained by twistor methods, may be summarized as follows.

Theorem 1.2. *Every ALE hyper-Kähler 4-manifold is diffeomorphic to the minimal resolution of $\mathbf{C}^2/\Gamma$ for some $\Gamma \subset \mathrm{SU}(2)$, and the cohomology classes of the Kähler forms on such a manifold must satisfy condition $(*)$.*

Theorem 1.3. *If X_1 and X_2 are two ALE hyper-Kähler 4-manifolds, and there is a diffeomorphism $X_1 \to X_2$ under which the cohomology classes of the Kähler forms agree, then X_1 and X_2 are isometric.*

2. A family of hyper-Kähler manifolds

We now review the Kähler and hyper-Kähler quotient constructions. Let M be a simply-connected Kähler manifold, and F a compact Lie group acting on M so as to preserve the metric g and the complex structure $I: TM \to TM$. Let $\mathfrak{f}$ be the Lie algebra of F, and for each $\xi \in \mathfrak{f}$ let V_ξ be the vector field on M which the action of ξ generates. According to the familiar definition from symplectic geometry, a *moment map* for the action of F on M is an F-equivariant map

$$\mu: M \to \mathfrak{f}^*$$

with the property that, for each $\xi \in \mathfrak{f}$, the function $\mu \cdot \xi: M \to \mathbf{R}$ satisfies

$$\operatorname{grad}(\mu \cdot \xi) = I(V_\xi).$$

Under our assumption that M is simply-connected, a moment map always exists and is unique to within the addition of a constant $\varsigma \in Z \subset \mathfrak{f}^*$, where Z is the space of F-invariant elements essentially the dual of the centre of $\mathfrak{f}$. If μ is a moment map and $\varsigma \in Z$, then $\mu^{-1}(\varsigma) \subset M$ is invariant under F. The quotient space $X = \mu^{-1}(\varsigma)/F$ is the *Kähler quotient* of M by F. Note that if the center of F is nontrivial, then the Kähler quotient is not unique, for an element $\varsigma \in Z$ must be chosen.

Now suppose that M is hyper-Kähler and that F acts so as to preserve g as well as all three complex structures. There are then three moment maps (one for each of I, J and K) which one puts together to form the *hyper-Kähler moment map*

$$\mu = (\mu_1, \mu_2, \mu_3): M \to \mathbf{R}^3 \otimes \mathfrak{f}^*.$$

Following Hitchin et al. [9], after choosing $\varsigma \in \mathbf{R}^3 \otimes Z$, one defines the *hyper-Kähler quotient* as

$$X = \mu^{-1}(\varsigma)/F.$$

The following proposition gives the properties of Kähler (resp. hyper-Kähler) quotients which are proved in [9].

Proposition 2.1. *Suppose that F acts freely on $\mu^{-1}(\varsigma)$. Then*

(i) *$d\mu$ has full rank at all points of $\mu^{-1}(\varsigma)$, so that X is a nonsingular manifold of dimension* $\dim M - 2\dim F$ (*resp.* $\dim M - 4\dim F$),

(ii) *the metric g and complex structures I (resp. I, J, K) descend to X, and equipped with these, X is Kähler (resp. hyper-Kähler).*

We make a particular application of this hyper-Kähler quotient construction. Let Γ be a finite subgroup of SU(2), let R be its regular representation and Q its canonical 2-dimensional representation, and put

$$P = Q \otimes \operatorname{End}(R).$$

Define $M = P^{\Gamma}$, the space of Γ-invariant elements in P. We make P and M into right modules over $\mathbf{H}$ as follows. First, we regard Q as a rank-1 $\mathbf{H}$-module in such a way that SU(2) coincides with the symplectic group $\operatorname{Sp}(Q)$ of $\mathbf{H}$-linear isometries of Q. Next, a choice of invariant hermitian metric on R gives $\operatorname{End}(R)$ a real structure, the antilinear involution $\alpha \mapsto \alpha^*$. As the tensor product of an $\mathbf{H}$-module and a real space, P then inherits an $\mathbf{H}$-module structure. Explicitly, if we choose an orthonormal basis for Q so as to represent an element of P as a pair of endomorphisms (α, β), the action of J is given by

$$J(\alpha, \beta) = (-\beta^*, \alpha^*), \qquad \alpha, \beta \in \operatorname{End}(R).$$

The action of Γ on P is $\mathbf{H}$-linear and the subspace M is therefore an $\mathbf{H}$-submodule. Explicitly again, a pair (α, β) lies in M if it satisfies the condition that, for each $\gamma = \left(\begin{smallmatrix} u & v \\ -\bar{v} & \bar{u} \end{smallmatrix}\right) \in \Gamma$ we have

$$R(\gamma^{-1})\alpha R(\gamma) = u\alpha + v\beta, \qquad R(\gamma^{-1})\beta R(\gamma) = -\bar{v}\alpha + \bar{u}\beta. \tag{2.2}$$

Identifying each tangent space to M with M itself, we regard this linear space as a flat hyper-Kähler manifold.

Let $\mathrm{U}(R)$ be the group of unitary transformations of R and let $F \subset \mathrm{U}(R)$ be the subgroup consisting of those elements which commute with the action of Γ on R. The natural action of F on P given by

$$(\alpha, \beta) \mapsto (f\alpha f^{-1}, f\beta f^{-1}), \qquad f \in F,$$

is $\mathbf{H}$-linear and preserves the subspace M. As the circle subgroup T of scalars acts trivially, we therefore have an action of F/T on M which preserves I, J, and K.

The moment map for this action is easily written down: if one identifies $(\mathfrak{f}/\mathfrak{t})^*$ with the traceless elements of $\mathfrak{f} \subset \operatorname{End}(R)$, then the three components of μ are given by

$$\begin{aligned} \mu_1(\alpha, \beta) &= \tfrac{1}{2} i([\alpha, \alpha^*] + [\beta, \beta^*]), \\ \mu_2(\alpha, \beta) &= \tfrac{1}{2}([\alpha, \beta] + [\alpha^*, \beta^*]), \\ \mu_3(\alpha, \beta) &= \tfrac{1}{2} i(-[\alpha, \beta] + [\alpha^*, \beta^*]). \end{aligned} \tag{2.3}$$

We have picked out the preferred moment map which vanishes at the origin. Applying the quotient construction, we choose a triple $\varsigma = (\varsigma_1, \varsigma_2, \varsigma_3) \in \mathbf{R}^3 \otimes Z$, where $Z \subset (\mathfrak{f}/\mathfrak{t})^*$ is the center, and set

$$X_\varsigma = \mu^{-1}(\varsigma)/F.$$

Our claim is that, as ς varies, we obtain in the family of spaces X_ς all the ALE spaces whose existence is asserted by Theorem 1.1.

In order to give a different description of the space M and the group F, a short digression is necessary. Let $R_0, R_1, \cdots, R_r$ be the irreducible representation of Γ with R_0 the trivial representation, let Q be the 2-dimensional representation as before, and let

$$Q \otimes R_i = \bigoplus_j a_{ij} R_j$$

be the decomposition of $Q \otimes \mathbf{R}_i$ into irreducibles. McKay [15] observed that the matrix $A = (a_{ij})$, whose entries are all either 0 or 1, is the adjacency matrix of a simply-laced extended Dynkin diagram $\overline{\Delta}(\Gamma)$; equivalently, $\overline{C} = 2I - A$ is an extended Cartan matrix. The trivial representation R_0 corresponds to the extra vertex of the extended diagram, and the representations $R_1, \cdots, R_r$ therefore correspond to a set of simple roots $\theta_1, \cdots, \theta_r$ for the associated root system. We write θ_0 for the negative of the highest root and note that, as McKay further observed, if

$$\theta_0 = -\sum_1^r n_i \theta_i$$

is the expression for θ_0 in terms of the simple roots, then the coefficient n_i is precisely the dimension of R_i. The assignment of $\overline{\Delta}(\Gamma)$ to Γ sets up a one-to-one correspondence between the finite subgroups of SU(2) and the simply-laced diagrams A_r, D_r, E_6, E_7 and E_8.

The regular representation of Γ decomposes as

$$R = \bigoplus_i \mathbf{C}^{n_i} \otimes R_i.$$

Accordingly, M may be written

$$\begin{aligned} M &= \mathrm{Hom}_\Gamma(R, Q \otimes R) \\ &= \bigoplus_{i,j} \mathrm{Hom}_\Gamma(R_i, Q \otimes R_j) \otimes \mathrm{Hom}(\mathbf{C}^{n_i}, \mathbf{C}^{n_j}) \\ &= \bigoplus_{i,j} a_{ij} \mathrm{Hom}(\mathbf{C}^{n_i}, \mathbf{C}^{n_j}), \end{aligned}$$

and, by McKay's observation, this description may be rephrased as

$$M = \bigoplus_{i \to j} \mathrm{Hom}(\mathbf{C}^{n_i}, \mathbf{C}^{n_j}),$$

where the sum is taken over all edges of $\overline{\Delta}(\Gamma)$, and each edge appears twice in the sum, once with each orientation. The group F can be similarly described in terms of $\overline{\Delta}(\Gamma)$: it is a product of unitary groups

$$F = \bigtimes_i \mathrm{U}(n_i) \tag{2.5}$$

with one factor for each vertex of $\overline{\Delta}(\Gamma)$, and it acts on M in the obvious way. Using these descriptions, we compute

$$\begin{aligned} \dim_{\mathbf{R}} M &= \sum_{i,j} 2a_{ij} n_i n_j = \sum_{i,j} (4\delta_{ij} - 2c_{ij}) n_i n_j \\ &= \sum_i 4n_i^2 = 4|\Gamma|, \\ \dim_{\mathbf{R}} F &= \sum_i n_i^2 = |\Gamma|. \end{aligned} \tag{2.6}$$

The center of the Lie algebra $\mathfrak{f}$ is spanned by the elements $\sqrt{-1}\pi_i$, where π_i is the projection $\pi_i \colon R \to \mathbf{C}^{n_i} \otimes R_i$ $(i = 0, \cdots, r)$. Writing h for the real Cartan algebra associated to the Dynkin diagram, we define a linear map ρ from the center of $\mathfrak{f}$ to h^* by

$$\rho \colon \sqrt{-1}\pi_i \mapsto n_i \theta_i .$$

The kernel of ρ is the one-dimensional subalgebra $\mathfrak{t} \subset \mathfrak{f}$, so that on the dual spaces, ρ induces an isomorphism

$$\tau \colon Z \to h. \tag{2.7}$$

For each root θ (not necessarily simple), we write $D_\theta = \mathrm{Ker}(\theta \circ \tau) \subset Z$. Thus we identify Z with the Cartan algebra, and the hyperplanes D_θ are the walls of the Weyl chambers.

Proposition 2.8. *If F/T does not act freely on $\mu^{-1}(\varsigma)$, then ς lies in one of the codimesion-3 subspaces $\mathbf{R}^3 \otimes D_\theta \subset \mathbf{R}^3 \otimes Z$, where θ is a root.*

Proof. Suppose that $(\alpha, \beta) \in \mu^{-1}(\varsigma)$ is fixed by an element $f \in F - T$. We can decompose R into the eigenspace of f and obtain at least two Γ-invariant parts

$$R = R' \oplus R''.$$

These will be preserved by α and β, and the pair (α, β) therefore defines an element of the quaternion module

$$M' = \mathrm{Hom}_\Gamma(R', Q \otimes R').$$

Denote by F' the group of those unitary transformations of R' which commute with Γ and let T' be the scalar subgroup. We may take it that F'/T' acts freely on (α, β), for if it did not then we could further decompose R', just as we decomposed R, until this condition was met.

The condition that F'/T' acts freely on the orbit of (α, β) means that the hyper-Kähler quotient of M' by F'/T' is a nonsingular manifold at at least one point. From the formula for the dimension of a hyper-Kähler quotient (Proposition 2.1), we deduce the inequality $4\dim_{\mathbf{R}}(F'/T') \le \dim_{\mathbf{R}}(M')$, or in other words

$$(2.9) \qquad 2\dim_{\mathbf{C}} \mathrm{End}_{\Gamma}(R') - \dim_{\mathbf{C}} \mathrm{Hom}_{\Gamma}(R', Q \otimes R') \le 2.$$

If the decomposition of R' into irreducibles is $R' = \bigoplus n_i' R_i$, then (2.9) can be written

$$2\sum_i (n_i')^2 - \sum_{i,j} a_{ij} n_i' n_j' \le 2,$$

or $\sum_{i,j} c_{ij} n_i' n_j' \le 2$, where $\overline{C} = (c_{ij})$ is the extended Cartan matrix. Now let θ be defined by

$$\theta = \sum_0^r n_i' \theta_i.$$

This θ is nonzero and the inequality above says that $\|\theta\|^2 \le 2$, where the norm is defined by the Cartan matrix. Amongst all integer linear combinations of roots, the roots themselves are characterized by just this inequality, and we conclude that θ is a root.

If $\pi\colon R \to R'$ is the projection, then the element $\sqrt{-1}\pi \in \mathfrak{f}$ acts trivially on (α, β) and it follows from the formulas (2.3) for the moment maps that $\varsigma(\sqrt{-1}\,\pi) = 0$ when we regard ς as a map $\mathfrak{f} \to \mathbf{R}^3$. By the definition of the isomorphism τ, this relation means that $\varsigma \in \mathbf{R}^3 \otimes D_\theta$, which is just what the proposition asserts.

Let $(\mathbf{R}^3 \otimes Z)^\circ$ denote the "good" set, i.e., let

$$(\mathbf{R}^3 \otimes Z)^\circ = (\mathbf{R}^3 \otimes Z) \setminus \bigcup_\theta (\mathbf{R}^3 \otimes D_\theta).$$

Corollary 2.10. *If $\varsigma \in (\mathbf{R}^3 \otimes Z)^\circ$, then X_ς is a nonsingular hyper-Kähler 4-manifold.*

Proof. This now follows from Proposition 2.1. For the dimension of X_ς we have

$$\dim X_\varsigma = \dim M - 4\dim(F/T) = 4|\Gamma| - 4(|\Gamma| - 1) = 4.$$

3. Properties of the manifolds

By its definition, the regular representation has an orthonormal basis $\{e_\gamma\}$ indexed by $\gamma \in \Gamma$ with the property that $R(\delta)e_\gamma = e_{\delta\gamma}$ for all $\gamma, \delta \in \Gamma$. Let

$L \subset M$ consist of all $(a, b) \in M$ for which a and b are diagonal matrices with respect to this basis of R. Thus if $(a, b) \in L$, then there exists, for each $\gamma \in \Gamma$, a pair $(a_\gamma, b_\gamma) \in \mathbf{C}^2$ such that

$$a \cdot e_\gamma = a_\gamma e_\gamma, \qquad b \cdot e_\gamma = b_\gamma e_\gamma.$$

Because of the relations (2.2), the set of pairs $\{(a_\gamma, b_\gamma) \mid \gamma \in \Gamma\}$ must be an orbit of Γ in $\mathbf{C}^2$ and we can identify L with $\mathbf{C}^2$ by the assignment $(a, b) \mapsto (a_1, b_1)$. The space L then inherits from $\mathbf{C}^2$ an action of Γ.

Lemma 3.1. *Each orbit of F in $\mu^{-1}(0)$ meets L in one orbit of Γ.*

Proof. Take $(\alpha, \beta) \in \mu^{-1}(0)$. According to (2.3) we have $[\alpha, \beta] = 0$ and $[\alpha, \alpha^*] + [\beta, \beta^*] = 0$; and manipulating these two equations we obtain

$$[\alpha^*, [\alpha, \alpha^*]] + [\beta^*, [\beta, \alpha^*]] = 0,$$

or $(A^*A + B^*B)(\alpha^*) = 0$, where $A = \mathrm{ad}(\alpha)$ and $B = \mathrm{ad}(\beta)$. The positivity of A^*A and B^*B now implies that $A^*A(\alpha^*) = 0$, and hence $[\alpha, \alpha^*] = [\beta, \beta^*] = 0$. So α and β are commuting normal linear transformations, and so they cannot be nilpotent unless they are zero. Let us assume that α and β are not *both* zero.

Since they commute, α and β have a simultaneous unit eigenvector $v_1 \in R$ with

$$\alpha \cdot v_1 = a_1 v_1, \qquad \beta \cdot v_1 = b_1 v_1.$$

Since α and β are not both nilpotent, we may take it that $(a_1, b_1) \neq (0, 0)$. If we define $v_\gamma = R(\gamma) \cdot v_1$, then (2.2) ensures that

$$\alpha \cdot v_\gamma = a_\gamma v_\gamma, \qquad \beta \cdot v_\gamma = b_\gamma v_\gamma,$$

where $\{(a_\gamma, b_\gamma) \mid \gamma \in \Gamma\}$ is an orbit of Γ. The points $(a_\gamma, b_\gamma) \in \mathbf{C}^2$ are all distinct and the vectors v_γ are therefore independent and even mutually orthogonal, since α and β are normal. The transformation of R which sends e_γ to v_γ is therefore an element of F which carries (α, β) into L. Thus each orbit of F in $\mu^{-1}(0)$ meets L. The proof of the lemma is completed by the observation that two points of L lie in the same orbit of F if and only if they lie in the same orbit of Γ.

Corollary 3.2. *When $\varsigma = 0 \in \mathbf{R}^3 \otimes Z$, the space X_0 is isometric to $\mathbf{C}^2/\Gamma$.*

Proof. The lemma provides a bijection $X_0 \to L/\Gamma$. The important point is that the subspace $L \subset M$ is everywhere orthogonal to the orbits of F. This point is easy to verify: a tangent vector to the F-orbit at $(\alpha, \beta) \in L$ is a pair of matrices $([\xi, \alpha], [\xi, \beta])$ for some $\xi \in \mathfrak{f}$; these matrices are zero on the diagonal, when expressed in terms of the basis $\{e_\gamma\}$, and so are orthogonal to L. The quotient metric on X_0 at a nonsingular point is obtained from the orthogonal complement to the tangent space of an F-orbit in $\mu^{-1}(0)$; and it now follows that the bijection $X_0 \to L/\Gamma$ is an isometry when L is given the metric it inherits as a subspace of M, namely the Euclidean metric.

Now consider some ς other than 0. Let $W_\varsigma \subseteq \mu^{-1}(\varsigma)$ be the union of the free orbits of F/T in $\mu^{-1}(\varsigma)$ and let $U_\varsigma = W_\varsigma/F$ be the image of W_ς in the quotient X_ς. By Proposition 2.1, the space X_ς is nonsingular and 4-dimensional at all points of U_ς. The following lemma shows that the complement $X_\varsigma \backslash U_\varsigma$ consists of isolated singularities.

Lemma 3.3. *If $\varsigma \neq 0$ and $x \in X_\varsigma \backslash U_\varsigma$, then a neighborhood of x in X_ς is homeomorphic to a neighborhood of 0 in $\mathbf{C}^2/\hat{\Gamma}$, where $\hat{\Gamma} \subset \mathrm{SU}(2)$ is a group with fewer elements than Γ.*

Proof. Let $m = (\alpha, \beta) \in \mu^{-1}(\varsigma)$ be a representative of x and let $\hat{F} \subset F$ be the stabilizer of m. The assumptions of the lemma mean that $\hat{F}$ is a proper subgroup of F which is strictly larger than T. Let $V \subset T_m M$ be the tangent space to the F-orbit of m, and let $\hat{M}$ be the orthogonal complement in $T_m M$ to the $\mathbf{H}$-submodule $V + IV + JV + KV$. The space $\hat{M}$ is itself an $\mathbf{H}$-module, and the group $\hat{F}$ acts on it preserving this structure. We can therefore introduce the hyper-Kähler quotient $\hat{\mu}^{-1}(0)/\hat{F}$; we take $\hat{\mu}$ to be the unique hyper-Kähler moment map on $\hat{M}$ which vanishes at the origin. As a first step in the proof of the lemma, we shall show that a neighborhood of x in X_ς is homeomorphic to a neighborhood of zero in $\hat{\mu}^{-1}(0)/\hat{F}$.

If we decompose the Lie algebra $\mathfrak{f}$ into linear subspaces $\hat{\mathfrak{f}} \oplus \hat{\mathfrak{f}}^\perp$, then $\hat{\mu}$ is just the component of μ in the $\hat{\mathfrak{f}}^*$ direction: we can write

$$\mu(m + \varepsilon) = \varsigma + \hat{\mu}(\varepsilon) + \nu(\varepsilon)$$

for some $\nu \colon \hat{M} \to \mathbf{R}^3 \otimes (\hat{\mathfrak{f}}^\perp)^*$; here we identify TM with M. Every F-orbit sufficiently close to m meets $V^\perp$ in one orbit of $\hat{F}$; so a neighborhood of x in $\mu^{-1}(\varsigma)/F$ is homeomorphic to a neighborhood of x in

$$(V^\perp \cap \mu^{-1}(\varsigma))/\hat{F}.$$

Since the derivative of ν at $\varepsilon = 0$ has full rank, we can replace this second space by

$$(V^\perp \cap \hat{\mu}^{-1}(0) \cap \mathrm{Ker}(d\nu))/\hat{F}.$$

Finally, noting that $V^\perp \cap \mathrm{Ker}(d\nu)$ is just $\hat{M}$, we have the desired conclusion: a neighborhood of x in X_ς is homeomorphic to a neighborhood of 0 in $\hat{\mu}^{-1}(0)/\hat{F}$. We shall finish the proof of Lemma 3.3 by showing that $\hat{\mu}^{-1}(0)/\hat{F}$ is $\mathbf{C}^2/\hat{\Gamma}$ for some $\hat{\Gamma} \subset \mathrm{SU}(2)$.

Since the stabilizer of m is larger than T, we can follow the proof of Proposition 2.8 and decompose R into orthogonal Γ-invariant parts

$$R = R' \oplus R' \oplus R''' \oplus \cdots , \tag{3.4}$$

which are preserved by α and β. As was shown in Proposition 2.8, we may take it that the subgroup $\mathrm{U}(R') \subset \mathrm{U}(R)$ meets the stabilizer $\hat{F}$ only in the scalar subgroup T':

$$\mathrm{U}(R') \cap \hat{F} = T' \subset \mathrm{U}(R'); \tag{3.5}$$

and from this follows the equality

$$2\dim_{\mathbf{C}} \mathrm{End}_{\Gamma}(R') - \dim_{\mathbf{C}} \mathrm{Hom}_{\Gamma}(R', Q \oplus R') = 2. \tag{3.6}$$

By further decomposing R'' etc., we can arrange that (3.5) and (3.6) hold for all the summands in the decomposition (3.4). Define now an equivalence relation on these summands by declaring that $R' \sim R''$ if and only if there is a Γ-invariant isometry $R' \to R''$ which commutes with α and β. Such an isometry, when it exists, is unique to within a scalar multiple because of (3.5). Grouping together equivalent summands, we rewrite the decomposition (3.4) in the form

$$R = \bigoplus_i \mathbf{C}^{\hat{n}_i} \otimes R^{(i)},$$

where $R^{(i)} \not\sim R^{(j)}$ unless $i = j$, and $\hat{n}_i$ is the number of summands equivalent to $R^{(i)}$.

We now have the following expressions for $\hat{M}$ and $\hat{F}$:

$$\hat{M} = \bigoplus_{i,j} \hat{a}_{ij} \mathrm{Hom}(\mathbf{C}^{\hat{n}_i}, \mathbf{C}^{\hat{n}_j}), \qquad \hat{F} = \bigtimes_i \mathrm{U}(\hat{n}_i), \tag{3.7}$$

where $\hat{a}_{ij} = \dim_{\mathbf{C}}(\hat{M} \cap \mathrm{Hom}_{\Gamma}(R^{(i)}, Q \otimes R^{(j)}))$. The matrix $\hat{A} = (\hat{a}_{ij})$ is symmetric and from (3.6) we have $\hat{a}_{ii} = 0$ for all i. The same dimension-counting as was used in Proposition 2.8 shows that $2I - \hat{A}$ is positive semi-define and that the null space of $2I - \hat{A}$ is spanned by the vector $(\hat{n}_1, \hat{n}_2, \cdots)$. This information is enough for us to conclude that $\hat{A}$ is the adjacency matrix of an extended simply-laced Dynkin diagram associated to some $\hat{\Gamma} \subset \mathrm{SU}(2)$, and the lemma now follows from Corollary 3.2, for the decompositions of $\hat{M}$ and $\hat{F}$ given in (3.7) are of just the same form as the decompositions of M and F given in (2.4) and (2.5).

We wish to regard the singular members of the family X_ς as singular algebraic varieties. For this purpose, let us choose just one of the complex structures, say I, and suppose for the moment that $N \subset M$ is any affine subvariety (with respect to I) which is invariant under F. In this situation there are two quotients of N one can consider. First there is the affine algebro-geometric quotient $N//F^c$ of N by the reductive group F^c, the complexification of F. Secondly, there is the Kähler quotient $(N \cap \mu_1^{-1}(0))/F$. The result we require is that these two are the same: the inclusion $(N \cap \mu_1^{-1}(0)) \to N$ and the quotient map $N \to N//F^c$ together give a map $(N \cap \mu_1^{-1}(0))/F \to N//F^c$ which is a homeomorphism when $N//F^c$ is given the usual (complex) topology. This result is proved in [11] for the more involved case of projective varieties. The affine case is easily deduced from the proof given there.

It is noted in [9] that if the second two components μ_2 and μ_3 of a hyper-Kähler moment map are combined into one map $\mu_c = \mu_2 + i\mu_3$,

$$\mu_c \colon M \to \mathfrak{f}^* \otimes \mathbf{C}, \tag{3.8}$$

then this *complex moment map* is holomorphic with respect to I. In our case indeed, we have $\mu_c(\alpha, \beta) = [\alpha, \beta]$. It follows that the level sets of μ_c are affine subvarieties of M. We deduce:

Lemma 3.9. *If the first component ς_1 of ς is zero, then X_ς has the structure of an affine variety with respect to I.*

Proof. By its definition,

$$X_\varsigma = (\mu_1^{-1}(0) \cap \mu_c^{-1}(\varsigma_2 + i\varsigma_3))/F,$$

and by the equivalence of Kähler and algebro-geometric quotients, this is the same as $\mu_c^{-1}(\varsigma_2 + i\varsigma_3)//F^c$.

In particular, X_0 is an affine variety. The identification of X_0 with $\mathbf{C}^2/\Gamma$ which we made in Corollary 3.2 can be put in algebraic terms, showing that, at least if X_0 is given its reduced structure, there is an isomorphism of varieties $X_0 \cong \mathbf{C}^2/\Gamma$. When ς_1 is nonzero, X_ς will still be quasiprojective variety, but need not be affine.

Suppose now that $\varsigma = (0, \varsigma_2, \varsigma_3)$ as in Lemma 3.9 and let $\tilde{\varsigma} = (\varsigma_1, \varsigma_2, \varsigma_3)$, where ς_1 is so chosen that $\tilde{\varsigma}$ does not lie in one of the subspaces $\mathbf{R}^3 \otimes D_\theta$. By Corollary 2.10, the quotient $X_{\tilde{\varsigma}}$ is a manifold. The inclusion $\mu^{-1}(\varsigma) \to \mu_c^{-1}(\varsigma_2 + i\varsigma_3)$ and the algebro-geometric quotient map $\mu_c^{-1}(\varsigma_2 + i\varsigma_3) \to X_\varsigma$ together give a map $\lambda \colon X_{\tilde{\varsigma}} \to X_\varsigma$ which is holomorphic with respect to I.

Proposition 3.10. *The map $\lambda \colon X_{\tilde{\varsigma}} \to X_\varsigma$ is a minimal resolution of singularities.*

Proof. Let us first show that λ is proper. Let $C \subset X_\varsigma$ be compact and let B be the preimage of $\lambda^{-1}(C)$ in $\mu^{-1}(\tilde{\varsigma})$. On the set B, the spectral radii $\sigma(\alpha)$ and $\sigma(\beta)$ are bounded, for these functions are bounded on C and are constant on the orbits of the complex group F^c. The compactness of B (and the properness of λ) therefore follows from:

Lemma 3.11. *Let $B \subset M$ be a closed set on which the following functions are bounded:*

(i) *the spectral radii $\sigma(\alpha)$ and $\sigma(\beta)$;*

(ii) *the norm $|\mu|$ of the moment.*

Then B is compact.

Proof. In the proof of Lemma 3.1 it was shown that if α, β were nilpotent and $\mu(\alpha,\beta) = 0$, then $\alpha = \beta = 0$. So the functions (i) and (ii) are simultaneously zero only at the origin. From this and the homogeneity of the two functions, the lemma follows.

Next we show that λ is an isomorphism away from the singular set of X_ς. From Corollary 3.2 we know that the nonsingular points comprise precisely the set $U_\varsigma \subset X_\varsigma$; so we must prove that if $x \in U_\varsigma$, then $\lambda^{-1}(x)$ consists of just one point.

Let $x \in U_\varsigma$ and let $\pi^{-1}(x)$ be the preimage of x under the algebraic quotient map $\pi\colon \mu_c^{-1}(\varsigma_2 + i\varsigma_3) \to X_\varsigma$. This fiber is invariant under F^c and contains precisely one F-orbit Ω on which μ_1 vanishes. By the general properties of algebro-geometric quotients, the F^c-orbit $F^c\Omega$ is closed and is contained in the closure of every F^c-orbit in $\pi^{-1}(x)$. But by definition of U_ς, this orbit has the maximum possible dimension, namely $\dim(F^c/T^c)$, and cannot therefore be contained in the closure of any other. It follows that $\pi^{-1}(x)$ consists of just the one F^c-orbit, $F^c\Omega$.

Set $\psi = |\mu_1 - \varsigma|^2$. By Lemma 3.11, this function is proper on $F^c\Omega$ and therefore attains its minimum at some point y. Since F/T acts freely on $F^c\Omega$, any critical point of ψ is actually a zero of ψ (see [11, p. 35]); so $\mu_1(y) = \varsigma_1$ and $\mu^{-1}(\tilde{\varsigma}) \cap \pi^{-1}(x)$ is therefore nonempty: it consists of at least one orbit of F. That $\mu^{-1}(\tilde{\varsigma}) \cap \pi^{-1}(x)$ consists of precisely one orbit of GF follows from the results of [10], and we see that $\lambda^{-1}(x)$ is a single point as required.

Now we must show that $\lambda^{-1}(U_\varsigma)$ is dense in $X_{\tilde{\varsigma}}$. If it were not, then the inverse image of $X_{\tilde{\varsigma}}$ of some singular point would contain a component of the manifold $X_{\tilde{\varsigma}}$; and by the properness already proved, this component would be compact. We shall show that $X_{\tilde{\varsigma}}$ has no compact component.

Set $\varsigma' = (\varsigma_1, 0, \varsigma_3)$ and consider the space $X_{\varsigma'}$ as an affine variety with respect to J. Without loss of generality we may assume that $X_{\varsigma'}$ is nonsingular; see the proof of Corollary 3.12. Just as we defined the I-holomorphic map $\lambda\colon X_{\tilde{\varsigma}} \to X_\varsigma$, so too we can define a J-holomorphic map $\lambda'\colon X_{\tilde{\varsigma}} \to X_{\varsigma'}$. Like λ, the map λ' is an isomorphism away from the singular points, and it follows that $X_{\tilde{\varsigma}}$ and $X_{\varsigma'}$ are diffeomorphic. But being an affine variety, $X_{\varsigma'}$ can have no compact components of positive dimension. Neither, therefore, can $X_{\tilde{\varsigma}}$.

We have now shown that $\lambda\colon X_{\tilde{\varsigma}} \to X_\varsigma$ is a resolution of singularities. The first Chern class of $X_{\tilde{\varsigma}}$ is zero because $X_{\tilde{\varsigma}}$ is hyper-Kähler, and this implies that $X_{\tilde{\varsigma}}$ contains no exceptional curves of the first kind. The resolution is therefore minimal, and this completes the proof of Proposition 3.10.

Corollary 3.12. *If $\varsigma \in (\mathbf{R}^3 \otimes Z)^\circ$, then X_ς is diffeomorphic to the minimal resolution of $\mathbf{C}^2/\Gamma$.*

Proof. Set

$$\varsigma = (\varsigma_1, \varsigma_2, \varsigma_3), \quad \eta = (\varsigma_1, \varsigma_2, 0), \quad \xi = (\varsigma_1, 0, 0).$$

We shall assume that ς_1 does not lie in any D_θ; this is a stronger condition than the hypothesis, but the loss of generality is not serious: by choosing a new orthonormal basis for $\mathbf{R}^3$, we can always arrange that ς_1 satisfies this condition.

Consider the four spaces $X_\varsigma, X_\eta, X_\xi, X_0$. By Corollary 2.10 and our assumption about ς_1, the first three are manifolds, while the fourth, by Corollary 3.2, is $\mathbf{C}^2/\Gamma$. By Proposition 3.10 there are three maps λ, λ' and λ'' which are holomorphic with repsect to K, J and I respectively:

$$X_\varsigma \xrightarrow{\lambda} X_\eta \xrightarrow{\lambda'} X_\varsigma \xrightarrow{\lambda''} X_0. \tag{3.13}$$

Each of these is a minimal resolution of singularities; but since X_ξ is already nonsingular, both λ and λ' are diffeomorphisms. So X_ς is diffeomorphic to the minimal resolution of $X_0 = \mathbf{C}^2/\Gamma$.

Next we compare the hyper-Kähler metric on X_ς with the Euclidean metric on $X_0 = \mathbf{R}^4/\Gamma$. The composite of the three maps in (3.13) is a map $\Lambda\colon X_\varsigma \to \mathbf{R}^4/\Gamma$, which is bijective away from the singular point. Pulling back the hyper-Kähler metric on X_ς to $\mathbf{R}^4\backslash\{0\}$ via the composite

$$\mathbf{R}^4\backslash\{0\} \to \mathbf{R}^4/\Gamma \xrightarrow{\Lambda^{-1}} X_\varsigma$$

we obtain a metric g_ς on $\mathbf{R}^4\backslash\{0\}$. Let (x_1, x_2, x_3, x_4) be standard coordinates on $\mathbf{R}^4$, let (g_ς^{ij}) be the components of g_ς in these coordinates, and let (δ^{ij}) be the Euclidean metric. Let Θ be coordinates on the unit sphere S^3, so that (r, Θ) are polar coordinates on $\mathbf{R}^4$. The following proposition says that g_ς is ALE.

Proposition 3.14. *For any ς, there is an expansion in powers of r*

$$g_\varsigma = \delta + \sum_{k\geq 2} h_k(\Theta) r^{-2k},$$

which may be differentiated term by term.

Proof. Consider first the dependence of g_ς on ς. If one restricts g_ς to the unit sphere $r = 1$ then, since everything is analytic, there will be a power-series expansion in ς:

$$g_\varsigma\,|_{r=1} = \sum_{|\nu|\geq 0} f_\nu \varsigma^\nu,$$

where ν is a multi-index in the coordinates of ς. Now we exploit the homogeneity of the moment map, which is a quadratic function on M. The nonzero scalars $\mathbf{R}^*$ act on M by dilatations and induce a map $\mu^{-1}(\varsigma) \to \mu^{-1}(t^2\varsigma)$, from which we deduce

$$g_\varsigma(r,\Theta) = g_{r^{-2}\varsigma}(1,\Theta).$$

Putting this with the power-series above, we obtain, for each ς, a power series in r^{-2}:

$$g_\varsigma = \sum_{k\geq 0} h_k(\Theta) r^{-2k},$$

where $h_k = \sum_{|\nu|=k} f_\nu \varsigma^\nu$.

To complete the proof we must identify the first two terms h_k: we must show that $h_0 = \delta$ and $h_1 = 0$. The first of these two equalities is just the statement that X_0 is isometric to $\mathbf{C}^2/\Gamma$, and this was proved in Corollary 3.2.

To show that $h_1 = 0$ is to show that the first variation of g_ς with respect to ς at $\varsigma = 0$ is zero. The hyper-Kähler metric g_ς is entirely determined by its three Kähler forms $\omega_{i,\varsigma}$ $(i = 1,2,3)$; it will be enough therefore to show that

$$\partial_{\mathbf{V}}\omega_{i,\varsigma} = 0 \quad \text{at } \varsigma = 0,\ i = 1,2,3,$$

for every direction $\mathbf{V} = (V_1, V_2, V_3)$ in $\mathbf{R}^3 \otimes Z$. A general formula for the variation of this 2-form is given in [5] for the case of symplectic quotients. The argument adapts to the hyper-Kähler case, and we merely state the result. Away from the singularities, the projection $\mu^{-1}(\varsigma) \to \mu^{-1}(\varsigma)/F$ is a principal F/T-bundle. The horizontal distribution determined by the metric gives this bundle a connection whose curvature we denote by Ω_ς. The formula is then:

$$\partial_{\mathbf{V}}\omega_{i\varsigma} = \langle V_i, \Omega_\varsigma\rangle. \tag{3.15}$$

The right-hand side denotes the 2-form obtained by pairing $V_i \in Z \subset (\mathfrak{f}/\mathfrak{t})^*$ with $\Omega \in \Lambda^2 \otimes (\mathfrak{f}/\mathfrak{t})$.

Recall from the proof of Corollary 3.2 that $L \subset \mu^{-1}(0)$ meets all the F-orbits orthogonally. This means that the bundle $\mu^{-1}(0) \to \mu^{-1}(0)/F$ is flat and $\Omega_0 = 0$. So (3.15) shows that the variation is zero at $\varsigma = 0$, and this is what was wanted.

4. The period map

The exceptional set in the minimal resolution of $\mathbf{C}^2/\Gamma$ is a union of 2-spheres whose intersection matrix is the negative of a Cartan matrix (see [17]):

$$E = P_1 \cup \cdots \cup P_r, \qquad P_i \cdot P_j = -c_{ij}.$$

The matrix $C = (c_{ij})$ is the same Cartan matrix whose *extended* version features in McKay's observation. The second cohomology $H^2(X_\varsigma; \mathbf{R})$ of each nonsingular quotient space can therefore be identified with h, the real Cartan algebra, while $H_2(X_\varsigma; \mathbf{Z})$ is the root lattice. Under this last identification, the classes Σ with $\Sigma \cdot \Sigma = -2$ are the roots. These identifications can be made consistently for all $\varsigma \in (\mathbf{R}^3 \otimes Z)^\circ$: there is no monodromy problem, since $(\mathbf{R}^3 \otimes Z)^\circ$ is simply connected.

For $\varsigma \in (\mathbf{R}^3 \otimes Z)^\circ$, let $\alpha_i(\varsigma)$ denote the cohomology class of ω_i on X_ς: these give maps

$$\alpha_i : (\mathbf{R}^3 \otimes Z)^\circ \to h, \qquad i = 1, 2, 3.$$

At the cohomology level, the formula (3.15) for the variation of ω_i shows [5] that there is a linear map $\sigma: Z \to h$ with $\alpha_i(\varsigma) = \sigma(\varsigma_i)$. Recall that another map $\tau: Z \to h$ was defined using McKay's observation (2.7), and that τ carries the hyperplane D_θ to the kernel of the root θ. Since the nondegeneracy condition (*) in Theorem 1.1 just says that the α_i do not all lie in the kernel of any one root, that theorem will be completely proved if we can establish the following two properties of σ:

Proposition 4.1. (i) *The map $\sigma: Z \to h$ is a linear isomorphism.*

(ii) *If $\xi \in Z$ does not lie in any D_θ, then $\sigma(\xi)$ does not lie in the kernel of any root.*

Proof of (ii). Supposing ξ satisfies this hypothesis, set $\varsigma = (\xi, 0, 0)$ and consider X_ς as a complex manifold with respect to I. By Proposition 3.10, this space is biholomorphic to $\widetilde{\mathbf{C}^2/\Gamma}$ and therefore contains holomorphic curves $P_1, \cdots, P_r$ whose homology classes form a set of simple roots. Now $\sigma(\xi)$ is the cohomology class of ω_1 on X, and since a Kähler form is always positive on a holomorphic curve, we see that $\sigma(\xi)$ lies in the positive Weyl chamber with respect to this choice of simple roots. So $\sigma(\xi)$ does not lie in the kernel of any root.

The proof of (i) involves a substantial detour and occupies the rest of this section. Again fix attention on the complex structure I and set

$$N = \mu_c^{-1}(Z \otimes \mathbf{C}) \subset M, \qquad Y = (N \cap \mu_1^{-1}(0))/F.$$

Because of the equivalence between Kähler and algebro-geometric quotients, Y is an affine variety. Since the moment map is equivariant, μ_c descends to give a holomorphic map $\phi: Y \to Z \otimes \mathbf{C}$.

Choose a $\xi \in Z$ not lying on any D_θ and set

$$\tilde{Y} = (N \cap \mu_1^{-1}(\xi))/F.$$

This space is not an affine variety, but by Proposition 2.8 and 2.1 it is a nonsingular Kähler manifold. As with Y, there is a holomorphic map $\tilde{\phi}: \tilde{Y} \to Z \otimes \mathbf{C}$, and this fits into the following commutative diagram, in which λ is defined as it was for Proposition 3.10:

$$\begin{array}{ccc} \tilde{Y} & \xrightarrow{\lambda} & Y \\ \downarrow{\tilde{\phi}} & & \downarrow{\phi} \\ Z\otimes \mathbf{C} & = & Z\otimes \mathbf{C} \end{array} \tag{4.2}$$

For each $\eta = \varsigma_2 + i\varsigma_3 \in Z\otimes \mathbf{C}$, the fiber $\phi^{-1}(\eta)$ is the affine variety X_ς in the case $\varsigma = (0, \varsigma_2, \varsigma_3)$; while $\tilde{\phi}^{-1}(\eta)$ is the complex manifold $X_{\tilde{\varsigma}}$, where $\tilde{\varsigma} = (\xi, \varsigma_2, \varsigma_3)$. Restricted to these fibers, the map λ is precisely the minimal resolution $\lambda\colon X_{\tilde{\varsigma}} \to X_\varsigma$ of Proposition 3.10. Thus the diagram (4.2) is a *simultaneous resolution* of ϕ.

Let $Y^{(n)}$ denote the normalization of Y. Since Lemma 3.3 shows that Y is locally irreducible, the underlying topological space $|Y^{(n)}|$ is the same as $|Y|$; the two analytic spaces differ only in the local rings at their singular points, if at all.

Lemma 4.3. *The map $\phi\colon Y^{(n)} \to Z\otimes \mathbf{C}$ is a flat deformation of $\mathbf{C}^2/\Gamma$.*

Proof. The fibers of $\tilde{\phi}$ are a smooth family of complex surfaces in which the special fiber $\tilde{\phi}^{-1}(0)$ is isomorphic to the minimal resolution of $\mathbf{C}^2/\Gamma$. According to [16], such a family can be blown down fiberwise to produce a flat deformation $\check{\phi}\colon \check{Y} \to Z\otimes\mathbf{C}$ of $\mathbf{C}^2/\Gamma$; the ring $H^0(\check{Y};\mathscr{O})$ is isomorphic to $H^0(\tilde{Y},\mathscr{O})$ and there is therefore a diagram:

$$\begin{array}{ccccc} \tilde{Y} & \to & \check{Y} & \to & Y \\ \downarrow{\tilde{\phi}} & & \downarrow{\check{\phi}} & & \downarrow{\phi} \\ Z\otimes \mathbf{C} & = & Z\otimes \mathbf{C} & = & Z\otimes \mathbf{C} \end{array}$$

Since $\lambda\colon \tilde{Y} \to Y$ is proper and birational, the same is true of the map $\check{Y} \to Y$. This map is also finite, and since $\check{Y}$ is necessarily normal, it follows that $\check{Y}$ is the normalization of Y.

Remark. The author has no evidence against the conjecture that Y is itself normal and that $\phi\colon Y \to Z\otimes\mathbf{C}$ is flat. It is only for lack of a direct proof of this flatness that the results of [16] are needed.

On any hyper-Kähler manifold, the complex-valued 2-form $\omega_c = \omega_2 + i\omega_3$ is nondegenerate and ***holomorphic*** with respect to I (see [9]). So away from the singular locus, ω_c gives a holomorphic 2-form on all the fibers $\phi^{-1}(\eta)$ of ϕ, depending holomorphically on $\eta \in Z\otimes\mathbf{C}$. In the sense of [14], the map

$$\sigma\otimes 1\colon Z\otimes\mathbf{C} \to h\otimes\mathbf{C}$$

is therefore the *period map* for this deformation of $\mathbf{C}^2/\Gamma$.

Let $\Psi: \mathscr{Y} \to \mathscr{V}$ be the semi-universal deformation of $\mathbf{C}^2/\Gamma$ and let $s: Z \otimes \mathbf{C} \to \mathscr{V}$ be the map by which ϕ is induced from Ψ:

$$(4.4) \qquad \begin{array}{ccc} Y^{(n)} & \longrightarrow & \mathscr{Y} \\ \downarrow{\scriptstyle \phi} & & \downarrow{\scriptstyle \Psi} \\ Z \otimes \mathbf{C} & \xrightarrow[s]{} & \mathscr{V} \end{array}.$$

The space $Y^{(n)}$ inherits from M an action of the scalars $\mathbf{C}^*$, and since μ^c is quadratic, the map ϕ will be equivariant if we make $\mathbf{C}^*$ act on $Z \otimes \mathbf{C}$ with weight 2. Thus ϕ is a $\mathbf{C}^*$-*deformation*, and if we take Ψ to be the $\mathbf{C}^*$-semi-universal deformation (see [17]), then it follows that we may take it that s is $\mathbf{C}^*$-equivariant and globally defined.

At this point we need a result due to Looijenga [14] which implies that a deformation such as ϕ is entirely determined by its period map. We shall go into this a little more carefully than our present situation requires, for we will have need of the result again in [13].

The first thing is that Ψ admits a simultaneous resolution; this implies, in particular, that the minimal resolution of every fiber is diffeomorphic to $\widetilde{\mathbf{C}^2/\Gamma}$. The construction of this simultaneous resolution is due to Brieskorn [1], [2], [3], Slodowy [17], and independently to Tjurina [19]; it is a corollary of Brieskorn's description that the base $\mathscr{V}$ of Ψ is naturally identified with $(h \otimes \mathbf{C})/W$. Under this identification the discriminant locus $\mathscr{D} \subset \mathscr{V}$, i.e., the set of $v \in \mathscr{V}$ for which $\Psi^{-1}(v)$ is singular, is carried onto the branch locus of the quotient map $h \otimes \mathbf{C} \to (h \otimes \mathbf{C})/W$, the projection of the kernels of the roots. Choosing a base-point $v_0 \in \mathscr{V} \backslash \mathscr{D}$, one obtains a natural monodromy representation on the second cohomology [14],

$$\pi_1(\mathscr{V} \backslash \mathscr{D}) \to \mathrm{Aut}(H^2(\Psi^{-1}(v_0); \mathbf{C})).$$

This representation was calculated in [18] and shown to coincide with the standard representation of W on $h \otimes \mathbf{C}$.

Away from the singular points, the fibers of Ψ carry a holomorphic 2-form depending holomorphically on the base, and one therefore has a period map p_Ψ; because of the monodromy, it takes values in $(h \otimes \mathbf{C})/W$:

$$p_\Psi : \mathscr{V} \backslash \mathscr{D} \to (h \otimes \mathbf{C})/W.$$

Looijenga's result is that p_Ψ extends across $\mathscr{D}$ and coincides with the standard isomorphism between $\mathscr{V}$ and $(h \otimes \mathbf{C})/W$. From this one may deduce the following.

Proposition 4.5. *Let $\phi: Y \to V$ be a $\mathbf{C}^*$-deformation of $\mathbf{C}^2/\Gamma$ whose generic fiber is nonsingular, and suppose that the only $\mathbf{C}^*$-invariant neighborhood of the distinguished point in Y is Y itself. Then ϕ is determined* *by its period map p: precisely, ϕ is the pull-back of Ψ via the composite* $(p_\Psi)^{-1} \circ p : V \to \mathscr{V}$.

Proof. The hypotheses ensure that ϕ is the pull-back of Ψ by some homogeneous map $s: V \to \mathscr{V}$. The pull-back of the 2-form of the fibers of Ψ gives a 2-form on the fibers of ϕ, and in the presence of the $\mathbf{C}^*$-action, this object is essentially unique. It follows that the period map for ϕ is the composite of s and p_Ψ.This proves the proposition.

We can now prove Proposition 4.1(i). From Lemma 3.3 we see that $\phi^{-1}(0)$ is the only fiber of ϕ which is isomorphic to $\mathbf{C}^2/\Gamma$, and from this it follows that, in the diagram (4.4), we have $s^{-1}(0) = \{0\}$. Proposition 4.5 then implies that the period map has the same property, that is $\sigma^{-1}(0) = \{0\}$. Since σ is a linear map between spaces of equal dimension, it must be an isomorphism.

Remark. Having now calculated the period map of ϕ, we see that this deformation is obtained from the semi-universal deformation by lifting through the Weyl group. The diagram (4.2) is therefore the simultaneous resolution of the semi-universal deformation constructed by Brieskorn, Slodowy and Tjurina.

Acknowledgements. This work formed part of the author's doctoral thesis and was prepared under the supervision of M. F. Atiyah, to whom the author is particularly grateful for all the help and advice he has received. The author also wishes to thank N. J. Hitchin and P. Slodowy for answering many questions in the course of several much-appreciated conversations.

References

[1] E. Brieskorn, *Über die Auflösung gewisser Singularitäten von holomorphen Abbildungen*, Math. Ann. **166** (1966) 76–102.

[2] ——, *Die Auflösung der rationalen Singularitäten holomorpher Abbildungen*, Math. Ann. **178** (1968) 244–270.

[3] ——, *Singular elements of semisimple algebraic groups*, Actes Congr. Internat. Math., No. 2, 1970, 279–284.

[4] E. Calabi, *Métriques kahlériennes et fibrés holomorphes*, Ann. Sci. École. Norm. Sup. (4) **12** (1979) 269–294.

[5] J. J. Duistermaat & G. J. Heckman, *On the variation in the cohomology of the symplectic form in the reduced phase space*, Invent. Math. **69** (1982) 259–268.

[6] T. Eguchi, P. B. Gilkey & A. J. Hanson, *Gravitation, gauge theories and differential geometry*, Phys. Rev. **66** (1980) 215–393.

[7] G. W. Gibbons & S. W. Hawking, *Gravitational multi-instantons*, Phys. Lett. B **78** (1978) 430–432.

[8] N. J. Hitchin, *Polygons and gravitons*, Math. Proc. Cambridge Philos. Soc. **83** (1969) 465–476.

[9] N. J. Hitchin, A. Karlhede, U. Lindström & M. Roček, *Hyperkähler metrics and supersymmetry*, Comm. Math. Phys. **108** (1987) 535–589.

[10] G. Kempf & L. Ness, *The length of vectors in representation spaces*, Algebraic Geometry, Lecture Not s in Math., Vol 732 (K. Lønsted, ed.), Springer, Berlin, 1978, 233–242.

[11] F. C. Kirwan, *Cohomology of quotients in symplectic and algebraic geometry*, Math. Notes 31, Princeton University, 1984.

[12] P. B. Kronheimer, *Instantons gravitationnels et singularités de Klein*, C. R. Acad. Sci. Paris Sér. I Math. **303** (1986) 53–55.,

[13] ——, *A Torelli-type theorem for gravitational instantons*, J. Differential Geometry **29** (1989), 685–697.

[14] E. J. N. Looijenga, *Isolated singular points on complete intersections*, London Math. Soc. Lecture Note Series 77, Cambridge University Press, 1984.

[15] J. McKay, *Graphs, singularities and finite groups*, Proc. Sympos. Pure Math. Vol. 37, Amer. Math. Soc., 1980, 183–186.

[16] O. Riemenschneider, *Familien komplexer Räume mit streng pseudokonvexer spezieller Faser*, Comment. Math. Helv. **51** (1976) 547–565.

[17] P. Slodowy, *Simple singularities and simple algebraic groups*, Lecture Notes in Math., Vol. 815, Springer, Berlin, 1980.

[18] ——, *Four lectures on simple groups and singularities* Commun. Math. Inst. Rijksuniv. Utrecht **11** (1980).

[19] G. N. Tjurina, *Resolutions of singularities of flat deformations of rational double points*, Functional Anal. Appl. **4** (1970) 68–73.

INSTITUTE FOR ADVANCED STUDY

A COMPACT ROTATING GRAVITATIONAL INSTANTON

Don PAGE
University of Cambridge, Department of Applied Mathematics and Theoretical Physics, Cambridge, England

Received 15 August 1978

A compact rotating gravitational instanton (a positive-definite metric solution of the Einstein equations with Λ term) is presented. The manifold is the nontrivial S^2 fibre bundle over S^2 and has $\chi = 4$, $\tau = 0$, but no spinor structure. The metric can be obtained from a special limit of the positive-definite analytic extension of the Kerr–de Sitter metric or alternatively from the Taub–NUT metric with Λ term. The action is about $4\frac{1}{2}\%$ less negative than that of the Einstein metric on the trivial bundle $S^2 \times S^2$.

Gravitational instantons are currently the subject of much interest. They may be defined to be complete nonsingular positive-definite metrics which are solutions of the Einstein equations. As stationary phase points, they may be taken as the first approximations to the configurations which dominate the path integrals of quantum gravity.

Gravitational instantons are often taken to be solutions of the vacuum Einstein equations. However, in the "volume canonical ensemble" approach [1] and in certain supergravity theories [2], one may have stationary phase points which are solutions of the Einstein equations with a Λ term, i.e.

$$R_{ab} = \Lambda g_{ab} \tag{1}$$

One would like to find such gravitational instantons with as many different possible topologies as possible. Some examples are S^4, CP^2, and $S^2 \times S^2$, whose Einstein metrics are listed in ref. [3]. In this letter an Einstein metric will be given for slightly more complicated manifold, a two-sphere (S^2) fibre bundle over a base which is another S^2. This can roughly be described as a twisted version of $S^2 \times S^2$.

Consider a manifold with the topology of an S^2 fibre bundle over S^2. First we may ask how many manifolds are inequivalent (not related by homeomorphisms). This is the question of the classification of S^2 bundles over S^2, which is explained by Steenrod [4]. He shows that the set of equivalence classes of bundles over S^n with arcwise connected group G is in 1–1 correspondence with $\pi_{n-1}(G)$. For an S^2 bundle over S^2, G is SO(3), which has $\pi_1(G) = Z_2$, the group of integers mod 2. Hence there are 2 inequivalent classes of S^2 bundles over S^2. One class includes the trivial bundle $S^2 \times S^2$, and the other includes the manifold of the new instanton presented in this paper.

The manifolds corresponding to both classes have the same Euler number $\chi = 4$ and signature $\tau = 0$, but they are not homeomorphic. For example, $S^2 \times S^2$ can be given a spinor structure, but the other topology cannot.

The Einstein metric for the direct product space $S^2 \times S^2$ can be obtained from the positive-definite Schwarzschild–de Sitter metric by taking the limit of the two horizons' having the same size. (Away from this limit the positive-definite metric is singular.) Similarly, the Einstein metric for the nontrivial S^2 bundle over S^2 can be obtained from a limit of the Kerr–de Sitter solution. In this case the relative rotation between the two horizons introduces a twist in the fibres over the base S^2.

The positive-definite Kerr–de Sitter metric may be obtained from the lorentzian metric ([5], with the sign of Λ reversed to correspond to the sign convention of eq. (1)) by the substitution $t \to -i\tau$, $a \to -i\alpha$. Then

$$ds^2 = \frac{\Delta_r}{\gamma^2\rho^2}[d\tau - \alpha\sin^2\theta\, d\varphi]^2 + \frac{\Delta_\theta \sin^2\theta}{\gamma^2\rho^2}[\alpha\, d\tau + (r^2-\alpha^2)d\varphi]^2 + \frac{\rho^2}{\Delta_r}dr^2 + \frac{\rho^2}{\Delta_\theta}d\theta^2, \quad (2)$$

where $0 \leqslant \theta \leqslant \pi$, $0 \leqslant \varphi \leqslant 2\pi$, $r_1 \leqslant r \leqslant r_2$, $0 \leqslant \tau \leqslant \beta$, and

$$\gamma = 1 - \tfrac{1}{3}\alpha^2\Lambda, \quad \rho^2 = r^2 - \alpha^2\cos^2\theta\,, \quad (3,4)$$

$$\Delta_r = (r^2-\alpha^2)(1-\tfrac{1}{3}\Lambda^2 r^2) - 2Mr\,, \quad (5)$$

$$\Delta_\theta = 1 - \tfrac{1}{3}\alpha^2\Lambda\cos^2\theta\,. \quad (6)$$

The horizons r_i ($i = 1, 2; r_1 \leqslant r_2$) are at the two largest roots of $\Delta_r = 0$.

To avoid a conical singularity at the horizon r_i, one must make τ periodic with period

$$\beta_i = 2\pi K_i^{-1}\,, \quad (7)$$

where

$$K_i = \left|\frac{1}{2\gamma(r_1^2-\alpha^2)}\frac{d}{dr}\Delta_r\Big|_{r=r_i}\right| \quad (8)$$

is the surface gravity of the horizon [6]. For no conical singularity at either horizon, both horizons should give the same period and hence have the same surface gravities,

$$K_1 = K_2 = K\,. \quad (9)$$

Another requirement is that one horizon should rotate around an integral number of times n with respect to the other in one time period β. This is reflected in the fact that the Killing vectors which vanish on the two horizons are not the same but rather

$$\boldsymbol{K}_1 = \frac{\partial}{\partial\tau} - \frac{\alpha}{r_1^2-\alpha^2}\frac{\partial}{\partial\varphi}, \quad \boldsymbol{K}_2 = \frac{\partial}{\partial\tau} - \frac{\alpha}{r_2^2-\alpha^2}\frac{\partial}{\partial\varphi}, \quad (10)$$

respectively. Hence a cycle of the second minus a cycle of the first, each of which should bring one back to the initial point, leaves τ fixed but changes φ by the angle

$$\Delta\varphi = \left(\frac{\alpha}{r_1^2-\alpha^2} - \frac{\alpha}{r_2^2-\alpha^2}\right)\frac{2\pi}{K}\,. \quad (11)$$

But φ has period 2π to avoid conical singularities at $\theta = 0$ and $\theta = \pi$, so we must have

$$\Delta\varphi = 2\pi n\,, \quad (12)$$

for some integer n.

The requirement (9) of equal surface gravities can only be met if $r_1 = r_2$. But this makes the radial coordinate degenerate, so one must change coordinates and devise a nonsingular limiting procedure. Let

$$r = r_0 - \epsilon\cos\chi\,, \quad (13)$$

where

$$r_0 = \tfrac{1}{2}(r_1 + r_2)\,, \quad (14)$$

$$\epsilon = \tfrac{1}{2}(r_2 - r_1)\,, \quad (15)$$

and take the limit $\epsilon \to 0$. This makes $K \to 0$ so the time period β becomes infinite, but one can define a rescaled time coordinate

$$\eta = K\tau = [(\Lambda r_0^2 + \alpha^2/r_0^2)/\gamma(r_0^2-\alpha^2)]\epsilon\tau + O(\epsilon^2)\,, \quad (16)$$

which has period 2π. It is also convenient to define the azimuthal angle

$$\varphi_1 = \varphi + [\alpha/(r_1^2-\alpha^2)]\tau\,, \quad (17)$$

as measured in a coordinate system rotating with the inner horizon; this angle is not changed under the action of the Killing vector $\boldsymbol{K}_1$. Then $(\chi, \eta, \theta, \varphi_1)$ are well-behaved local coordinates in the limit $\epsilon \to 0$.

The condition that $r = r_1 = r_2 = r_0$ be a double root of Δ_r in the limit $\epsilon \to 0$ makes two constraints on the parameters Λ, α, M, r_0. In terms of Λ and the dimensionless parameter

$$\nu \equiv \alpha/r_0\,, \quad (18)$$

the other parameters are

$$r_0 = [(3+3\nu^2)/(3-\nu^2)]^{1/2}\Lambda^{-1/2}\,, \quad (19)$$

$$\alpha = \nu r_0 = \nu[(3+3\nu^2)/(3-\nu^2)]^{1/2}\Lambda^{-1/2}\,, \quad (20)$$

$$M = \frac{(1-\nu^2)^2}{3-\nu^2}r_0 = \frac{(1-\nu^2)^2(3+3\nu^2)^{1/2}}{(3-\nu^2)^{3/2}}\Lambda^{-1/2}\,, \quad (21)$$

$$\gamma = 1 - \tfrac{1}{3}\alpha^2\Lambda = (1-\nu^2)[(3+\nu^2)/(3-\nu^2)]\,. \quad (22)$$

Thus the requirement of equal surface gravities reduces the number of independent parameters to one scale parameter Λ and one dimensionless parameter ν.

The requirement (12) of an integral number of relative rotations of the two horizons adds the constraint

$$4\nu(3+\nu^2)/(3+6\nu^2-\nu^4) = n = \text{integer}\,. \quad (23)$$

Then the metric becomes

$$ds^2 = 3\Lambda^{-1}(1+\nu^2)\left\{\frac{1-\nu^2\cos^2\theta}{3+6\nu^2-\nu^4}(d\chi^2+\sin^2\chi\, d\eta^2)\right.$$

$$+\frac{1-\nu^2\cos^2\theta}{(3-\nu^2)-\nu^2(1+\nu^2)\cos^2\theta}d\theta^2$$

$$+\frac{(3-\nu^2)-\nu^2(1+\nu^2)\cos^2\theta}{(3+\nu^2)^2(1-\nu^2\cos^2\theta)}\sin^2\theta$$

$$\left.\times(d\varphi_1 - n\sin^2\tfrac{1}{2}\chi\, d\eta)^2\right\}, \qquad (24)$$

with $0 \leqslant \chi \leqslant \pi$, $0 \leqslant \theta \leqslant \pi$, and the points $(\chi, \eta + 2\pi j, \theta, \varphi_1 + 2\pi k)$ identified for all integers j, k.

It can be seen that for $\nu^2 < 1$ this is a regular metric on a manifold which is an S^2 bundle over S^2. The variables χ and η are polar coordinates for the base S^2. One may take coordinate patches V_1 $(0 \leqslant \chi \leqslant \pi)$ and V_2 $(0 \leqslant \chi \leqslant \pi)$ in which the bundle is equivalent to a product bundle. In V_1 the fibre in the S^2 labelled by regular polar coordinates θ and φ_1; in V_2 it is the S^2 labelled by θ and $\varphi_2 = \varphi_1 - n\eta$. (This makes $\chi = 0$ and $\chi = \pi$ both regular axes of rotation for the angle η when the other coordinates in the respective coordinate systems are held fixed.) The transition function $\theta = \theta$, $\varphi_2 = \varphi_1 - n\eta$ in the overlap region $V_1 \cap V_2$ gives n rotations of the S^2 fibre over V_2 with respect to the S^2 fibre over V_1 as η is decreased by 2π to give a closed curve running around the equator of the base space. Hence a nontrivial bundle is obtained if n is odd, which gives the nontrivial element in $\pi_1(SO(3))$.

Now we may ask what values of n are possible. If $\nu^2 > 1$, $\cos\theta = \pm\nu^{-1}$ is a curvature singularity. If $\nu \to -\nu$, then $n \to -n$, which merely reverses the orientation. Hence we may restrict ourselves to $0 \leqslant \nu \leqslant 1$. The case $\nu = 0$ gives $n = 0$ and the Einstein metric for $S^2 \times S^2$. The case $\nu = 1$ gives $n = 2$ and a metric which is singular at $\theta = 0$ and $\theta = \pi$ if η and φ_1 both have period 2π as above. If φ_1 is assigned a period 4π (i.e., only even integers j used in the identification above), one gets the Einstein metric for the four-sphere S^4 (the positive-definite analytic continuation of the de Sitter metric), but then the manifold is no longer an S^2 bundle over S^2.

The only intermediate case is $n = 1$, in which case eq. (23) gives a nontrivial quartic equation for ν. If one defines

$$a = (2^{1/2}+1)^{1/3}, \quad b = a^{-1} = (2^{1/2}-1)^{1/3}, \qquad (25)$$

then the solution is

$$\nu = -1 - (2+a-b)^{1/2} + [4-a+b+8(a-b)^{-1/2}(a+b)^{-1}]^{1/2} \approx 0.281\,701\,56. \qquad (26)$$

The parameters evaluated by eqs. (19)–(22) are

$$r_0 \approx 1.052\,939\,96\,\Lambda^{-1/2}, \quad \alpha \approx 0.296\,614\,83\,\Lambda^{-1/2}, \qquad (27,28)$$

$$M \approx 0.305\,568\,53\,\Lambda^{-1/2}, \quad \gamma \approx 0.970\,673\,21. \qquad (29,30)$$

The horizons or axes of η at $\chi = 0$ and $\chi = \pi$ each have area

$$A(\sin\chi = 0) = 4\pi(r_0^2 - \alpha^2)/\gamma = 4\pi r_0^2\left(\frac{3-\nu^2}{3+\nu^2}\right)$$

$$= 4\pi\left(\frac{3+3\nu^2}{3+\nu^2}\right)\Lambda^{-1} \approx 13.214\,047\,6\Lambda^{-1}, \qquad (31)$$

and the axes of φ_1 at $\theta = 0$ and $\theta = \pi$ each have area

$$A(\sin\theta = 0) = 4\pi\left(\frac{3-3\nu^4}{3+6\nu^2-\nu^4}\right)\Lambda^{-1} \approx 10.796\,387\,5\Lambda^{-1}. \qquad (32)$$

The 4-volume of this manifold with metric (24) is

$$V = \int g^{1/2} d^4x = 16\pi^2\,\frac{3(3-\nu^2)(1+\nu^2)^2}{(3+\nu^2)(3+6\nu^2-\nu^4)}\Lambda^{-2}$$

$$\approx 150.862\,014\Lambda^{-2}, \qquad (33)$$

so the action (including the Λ term) is

$$\hat{I} = -\frac{1}{16\pi}\int (R - 2\Lambda) g^{1/2} d^4x = -\frac{\Lambda V}{8\pi}$$

$$\approx -0.955\,344\,86(2\pi\Lambda^{-1}). \qquad (34)$$

Thus for a fixed value of Λ, the Kerr–de Sitter solutions give only two (modulo orientation) complete nonsingular positive-definite metrics for S^2 fibre bundles over S^2. This is not surprising, for it was noted above that there are only two inequivalent classes of S^2 bundles over S^2.

The metric (24) with $n = 1$ is seen to have a Taub–NUT form if one regards $x = \cos\theta$ as a radial coordinate and takes $\varphi_1 \to \frac{1}{2}(\psi + \varphi)$, $\chi \to \theta$, $\eta \to \varphi$. In fact, it is a special case of a Taub–NUT metric with Λ term and $m = 0$: Take the metric given by eqs. (8)–(10) of ref. [7] (with the sign of Λ reversed) and relabel the coordinates by

$$\lambda \to iR,\quad \mu \to \cos\theta,\quad \psi \to \varphi,\quad \chi \to 2l\psi. \tag{35}$$

Let the parameters other than Λ and l have the values

$$h = p = 1,\quad m = e^2 = q = 0. \tag{36}$$

Then the metric becomes

$$\mathrm{d}s^2 = (l^2 - R^2)(Q^{-1}\mathrm{d}R^2 + \mathrm{d}\theta^2 + \sin^2\theta\,\mathrm{d}\varphi^2) + 4l^2(l^2 - R^2)^{-1}Q(\mathrm{d}\psi + \cos\theta\,\mathrm{d}\varphi)^2, \tag{37}$$

where

$$Q \equiv -\Delta_\lambda = \tfrac{1}{3}\Lambda R^4 + (1 - 2l^2\Lambda)R^2 + l^2(1 - l^2\Lambda). \tag{38}$$

Now for a complete nonsingular metric, φ must have period 2π and hence ψ must have period 4π to give regularity at $\sin\theta = 0$. R must range between two roots R_1 and R_2 of Q, which are horizons or axes of the Killing vector $\partial/\partial\psi$, and both must have the correct surface gravity to give ψ a period of 4π without a conical singularity. This restriction implies

$$l = \left(\frac{3 + 3\nu^2}{3 + 6\nu^2 - \nu^4}\right)^{1/2}\Lambda^{-1/2} \approx 0.966\,025\,04\Lambda^{-1/2}, \tag{39}$$

$$-R_1 = R_2 = \nu l \approx 0.272\,130\,76\Lambda^{-1/2}, \tag{40}$$

where ν is the same as in eq. (26). Then the metric (37) becomes the same as the metric (24) with the change of variables indicated above and $x = R/R_2$:

$$\mathrm{d}s^2 = 3\Lambda^{-1}(1+\nu^2)\left\{\frac{(1-\nu^2x^2)}{[3-\nu^2-\nu^2(1+\nu^2)x^2]}\frac{\mathrm{d}x^2}{(1-x^2)} + \frac{1-\nu^2x^2}{3+6\nu^2-\nu^4}(\mathrm{d}\theta^2 + \sin^2\theta\,\mathrm{d}\varphi^2) + \frac{[3-\nu^2-\nu^2(1+\nu^2)x^2]}{4(3+\nu^2)^2(1-\nu^2x^2)}(1-x^2) \times (\mathrm{d}\psi + \cos\theta\,\mathrm{d}\varphi)^2\right\}. \tag{41}$$

The significance of this metric in the path integral is not clear. Since its action is $4\frac{1}{2}\%$ less negative than that of the Einstein metric for $S^2 \times S^2$ with the same Λ, one would expect its contribution to volume canonical ensemble partition function [1],

$$Z[\Lambda] = \int D[g,\Lambda]\exp(-\hat{I}[g,\Lambda]), \tag{42}$$

to be slightly less. However, it may not contribute at all, for unlike $S^2 \times S^2$, the nontrivial S^2 bundle over S^2 has no spinor structure. One could add an electromagnetic or Yang–Mills field to allow a generalized spin structure [8], but this depends upon the spectrum of elementary particles.

The paper benefited by discussions the author had with S.W. Hawking, G.W. Gibbons and C.N. Pope.

References

[1] S.W. Hawking, Spacetime foam, DAMTP preprint.
[2] P.K. Townsend, Phys. Rev. D15 (1977) 2795.
[3] G.W. Gibbons and C.N. Pope, CP^2 as a gravitational instanton, Commun. Math. Phys., to be published.
[4] N. Steenrod, The topology of fibre bundles (Princeton U.P., Princeton, NJ, 1951) pp. 96–99.
[5] B. Carter, Black hole equilibrium states, in: Black holes, eds. C. De Witt and B.S. De Witt (Gordon and Breach, New York, 1973).
[6] G.W. Gibbons and S.W. Hawking, Phys. Rev. D15 (1977) 2752.
[7] B. Carter, Commun. Math. Phys. 10 (1968) 280.
[8] S.W. Hawking and C.N. Pope, Phys. Lett. 73B (1978) 42.

Commun. Math. Phys. 66, 291–310 (1979)

Communications in
Mathematical
Physics

Classification of Gravitational Instanton Symmetries

G. W. Gibbons and S. W. Hawking

Department of Applied Mathematics and Theoretical Physics, University of Cambridge, Cambridge CB3 9EW, England

Abstract. We classify the action of one parameter isometry groups of Gravitational Instantons, complete non singular positive definite solutions of the Einstein equations with or without Λ term. The fixed points of the action are of 2-types, isolated points which we call "nuts" and 2-surfaces which we call "bolts". We describe all known gravitational instantons and relate the numbers and types of the nuts and bolts occurring in them to their topological invariants. We perform a $3+1$ decomposition of the field equations with respect to orbits of the isometry group and exhibit a certain duality between "electric" and "magnetic" aspects of gravity. We also obtain a formula for the gravitational action of the instantons in terms of the areas of the bolts and certain nut charges and potentials that we define. This formula can be interpreted thermodynamically in several ways.

1. Introduction

There has been considerable interest recently in "Instantons" in Yang-Mills Theory [1–3]. They may be defined as non singular solutions of the classical equations in 4-dimensional Euclidean space. They provide stationary phase points in the path integral for the amplitude to tunnel between two topologically distinct vacua [4, 5] and they may play a role in confinement. Instantons also contribute to the anomalous divergence of the axial vector current [2] and they may lead to the decay of baryons into leptons. Because gravity and supergravity are gauge theories like Yang-Mills it seems reasonable to suppose that gravitational instantons may play a similar important role. We shall define a gravitational instanton to be a non singular complete positive definite metric which satisfies the classical vacuum Einstein equations or the Einstein equations with a Λ term. The Λ term can be regarded as a Lagrange multiplier for the 4-volume V or it may arise from the Lagrangians of certain supergravity theories [6, 7]. One class of gravitational instantons that has been extensively studied already is the Kerr-Newman family of metrics [7–12]. In these solutions one can remove the apparent

0010-3616/79/0066/0291/$04.00

singularity at the horizon and obtain a complete positive definite metric by identifying the imaginary time coordinate periodically. This leads to a new approach to the thermal aspects of black holes which have been discovered by other methods.

Nearly all known gravitational instantons possess continuous symmetry groups of at least two parameters [11, 14–19]. In this paper we shall give a classification scheme based on the existence of at least a one parameter group. This enables us to determine two kinds of basic object, one that we call a "nut" after the self-dual Taub-NUT solution [10, 11] which is the canonical example and the other which we call a "bolt", for obvious reasons. The canonical example is the Schwarzschild solution. In cases where the symmetry group is more than 1-dimensional, different choices of one-parameter subgroup may lead to different numbers and locations of nuts and bolts. However there are two topological invariants, the Euler number χ and the signature τ, which can be expressed as sums over the nuts and bolts with certain coefficients. Roughly speaking, the Euler number is the sum of the number of nuts, the number of antinuts and twice the number of bolts while the signature is a measure of the number of nuts minus the number of antinuts.

The existence of these two kinds of basic objects reflects a certain symmetry in the theory analogous to duality invariance in electromagnetism. One can think of bolts as being the analogue of "electric" type mass-monopoles and the nuts as being gravitational dyons endowed with a real electric type mass-monopole and an imaginary "magnetic" type mass-monopole. The presence of magnetic type mass introduces a Dirac string-like singularity in the metric. This can be removed by appropriate identifications and changes in the topology of the spacetime manifold. However the metric cannot then be asymptotically flat in the usual sense. This means that the nuts unlike the bolts cannot occur in the classical regime. However one might imagine that quantum fluctuations of the metric might lead to the appearance and disappearance of nut-antinut pairs.

Gravitational Instantons can be interpreted as the stationary phase metrics in the path integrals for the partition functions, Z, of the thermal canonical ensemble [12] and the volume canonical ensemble [6]. In these cases the action of the instanton gives the dominant contribution to $-\log Z$. We shall relate this action to the areas of the bolts and to the charges and potentials of the nuts. From this it follows that the bolts have an intrinsic gravitational entropy equal to one quarter the sum of their areas. This generalises the results obtained for black holes and cosmological event horizons [12, 13, 20].

2. Nuts and Bolts

We shall consider an oriented manifold M with a positive definite metric g_{ab} which admits at least a one-parameter isometry group G. We shall denote by $\mu_\tau : M \to M$ the action of the group, where τ is the group parameter and we shall denote by $K = K^a \dfrac{\partial}{\partial x^a} = \dfrac{\partial}{\partial \tau}$ the Killing vector. The isometry group G is said to have a fixed point where $K=0$. At a fixed point p the action of μ_τ on the manifold M gives rise to an isometry $\mu_\tau * : T_p(M) \to T_p(M)$ where $T_p(M)$ is the tangent space at p. $\mu_\tau *$ is generated by the antisymmetric matrix $K_{a;b}$. Antisymmetric 4×4 matrices can have rank 0, 2, or 4. The zero case is not interesting because it would imply that the Killing vector K was zero everywhere and that the action of the group G was trivial. This follows because $\mu_\tau *$ would be the identity and because μ commutes with the exponential map at p, i.e.

$$\mu_\tau \circ \exp X = \exp(\mu_\tau *(X))$$
$$\forall X \in T_p .$$

In the case that $K_{a;b}$ has rank 2 there will be a 2-dimensional subspace T_1 of $T_p(M)$ which is left unchanged by $\mu_\tau *$. The action of $\mu_\tau *$ will rotate T_2, the 2-dimensional orthogonal complement of T_1 into itself. Thus $\mu_\tau *$ has the canonical form

$$\mu_\tau * = \begin{pmatrix} 1 & 0 & 0 & 0 \\ 0 & 1 & 0 & 0 \\ 0 & 0 & \cos\kappa\tau & \sin\kappa\tau \\ 0 & 0 & -\sin\kappa\tau & \cos\kappa\tau \end{pmatrix}, \tag{2.1}$$

where κ is the surface gravity and is given by the non zero skew eigen value of $K_{a;b}$ in an orthonormal frame. From this one can see that $\mu_\tau *$ and hence μ_τ must be periodic with a period $2\pi\kappa^{-1}$. The image of T_1 under the exponential map will not be moved by μ_τ and so will constitute a 2-dimensional oriented totally geodesic sub manifold of fixed points. We shall call such a 2-dimensional fixed point set a bolt. A simple example is provided by the horizon 2-sphere of the Euclidean Schwarzschild solution with G being the periodic group of imaginary time translations [8, 12].

In the case that $K_{a;b}$ has the maximal rank 4 there can be no directions at p which are left invariant under $\mu_\tau *$. Thus p must be an isolated fixed point. We shall call it a nut after the fixed point at the centre of the Euclidean self-dual Taub-NUT solution [10]. In this case there will be two orthogonal 2-dimensional subspaces T_1 and T_2 which are mapped into themselves by $\mu_\tau *$. The canonical form is

$$\mu_\tau * = \begin{pmatrix} \cos\kappa_1\tau & \sin\kappa_1\tau & 0 & 0 \\ -\sin\kappa_1\kappa & \cos\kappa_1\tau & 0 & 0 \\ 0 & 0 & \cos\kappa_2\tau & \sin\kappa_2\tau \\ 0 & 0 & -\sin\kappa_2\tau & \cos\kappa_2\tau \end{pmatrix}, \tag{2.2}$$

where κ_1 and κ_2 are the skew eigenvalues of $K_{a;b}$ in an orthonormal frame. For some purposes it is convenient to sub divide nuts into 2 classes – "nuts" and "antinuts" – depending on whether the sign of $\kappa_1\kappa_2$ is positive or negative respectively. Unless we explicitly say otherwise we shall call both classes nuts.

If $\kappa_1\kappa_2^{-1} = pq^{-1}$ where p and q are relatively prime integers, the action of μ_τ will be periodic with period $\beta = 2\pi p\kappa_1^{-1} = 2\pi q\kappa_2^{-1}$. We shall call this a nut of type (p, q).

If $\kappa_1\kappa_2^{-1}$ is irrational, the orbits of a vector X in T_p under the action of $\mu_\tau *$ is dense in the torus $C(X)$ consisting of all vectors Y of the form

$$\mu^1_{\tau_1} * \circ \mu^2_{\tau_2} *(X) , \tag{2.3}$$

where

$$\mu^1_{\tau_1}* = \begin{pmatrix} \cos\kappa_1\tau_1 & \sin\kappa_1\tau_1 & 0 & 0 \\ -\sin\kappa_1\tau_1 & \cos\kappa_1\tau_1 & 0 & 0 \\ 0 & 0 & 1 & 0 \\ 0 & 0 & 0 & 1 \end{pmatrix}, \tag{2.4}$$

$$\mu^2_{\tau_2}* = \begin{pmatrix} 1 & 0 & 0 & 0 \\ 0 & 1 & 0 & 0 \\ 0 & 0 & \cos\kappa_2\tau_2 & \sin\kappa_2\tau_2 \\ 0 & 0 & -\sin\kappa_2\tau_2 & \cos\kappa_2\tau_2 \end{pmatrix}. \tag{2.5}$$

All scalar invariants of the metric must be constant over each torus in M of the form $\exp C(X)$ for each $X \in T_p(M)$. Because scalar invariants characterize the metric it follows that $\mu^1_\tau*$ and $\mu^2_\tau*$ must actually correspond to independent isometries μ^1_τ and μ^2_τ of the metric g_{ab} on M. One could then take appropriate combinations of the Killing vectors K^1 and K^2 such that the orbits were periodic. We shall therefore consider only periodic isometry groups.

The antisymmetric tensor $K_{a;b}$ can be decomposed into self dual and antiself dual parts.

$$K_{a;b} = K^+_{ab} + K^-_{ab}\,, \tag{2.6}$$

where

$$K^\pm_{ab} = \tfrac{1}{2}(K_{a;b} \pm \tfrac{1}{2}\varepsilon_{abcd}K^{c;d})\,. \tag{2.7}$$

At a bolt

$$K^+_{ab}K^{+ab} = K^-_{ab}K^{-ab}\,. \tag{2.8}$$

At a nut

$$K^+_{ab}K^{+ab} > K^-_{ab}K^{-ab} \tag{2.9}$$

while at an antinut

$$K^+_{ab}K^{+ab} < K^-_{ab}K^{-ab}\,. \tag{2.10}$$

A nut is said to be self-dual if K^-_{ab} is zero. Then $p = q = \pm 1$. If the curvature is self-dual – i.e. if

$$R_{abcd} = \tfrac{1}{2}\varepsilon_{abef}R^{ef}{}_{cd} \tag{2.11}$$

then $K_{a;b}$ is self-dual everywhere if it is self-dual at one point. Similar remarks apply to anti self-dual anti-nuts which have $p = -q = \pm 1$.

3. Examples

The examples of Schwarzschild and the self-dual Taub-NUT solutions have already been mentiod. The metric of the Schwarzschild solution can be written in the form

$$ds^2 = (1 - 2Mr^{-1})d\tau^2 + (1 - 2Mr^{-1})^{-1}dr^2 + r^2(d\theta^2 + \sin^2\theta d\phi^2)\,. \tag{3.1}$$

The apparent singularity at the horizon $r=2M$ can be removed by identifying τ with a period $8\pi M$ [7, 10, 11]. The radial coordinate then has the range $2M \leqq r < \infty$ and the topology of the manifold is $R^2 \times S^2$. The isometry group is $O(2) \otimes O(3)$ where the $O(2)$ corresponds to translations in the periodically identified imaginary time τ and the $O(3)$ corresponds to rotations of the θ and ϕ coordinates.

The Killing vector $\frac{\partial}{\partial \tau}$ has unit magnitude at large radius and has a bolt on the horizon $r=2M$ which is a 2-sphere of area $16\pi M^2$. The surface gravity $\kappa=(4M)^{-1}$ and the period β is $8\pi M$. A typical Killing vector of $O(3)$, for example $\frac{\partial}{\partial \phi}$, is zero on a non compact 2-surface, the axis, $0=0$ or π. We shall not consider Killing vector which have non compact bolts. One can also take linear combinations $p^{-1}4M\frac{\partial}{\partial \tau}+q^{-1}\frac{\partial}{\partial \phi}$. These will have a nut of type (p,q) at the northpole $\theta=0$ of the horizon $r=2M$ and an antinut of type $(p, -q)$ at the southpole, $\theta=\pi$, of the horizon $r=2M$.

The Kerr solution with mass M and (imaginary) angular momentum $=i\alpha M$ has a Euclidean section with metric

$$ds^2=(r^2-\alpha^2\cos^2\theta)(dr^2\Delta^{-1}+d\theta^2)$$
$$+(r^2-\alpha^2\cos^2\theta)^{-1}[\Delta(d\tau+\alpha\sin^2\theta d\phi)^2+\sin^2\theta((r^2-\alpha^2)d\phi-\alpha d\tau)^2], \quad (3.2)$$

where

$$\Delta=r^2-2Mr-\alpha^2\,. \quad (3.3)$$

The apparent singularity at the horizon

$$r=r_+=M+(M^2+\alpha^2)^{1/2} \quad (3.4)$$

can be removed by identifying the points (r, τ, θ, ϕ) with $(r, \tau+2\pi\gamma, \theta, \phi+2\pi\gamma\Omega)$ where

$$\gamma=2Mr_+(M^2+\alpha^2)^{-1/2} \quad (3.5)$$

and

$$\Omega=\alpha(r_+^2-\alpha^2)^{-1} \quad (3.6)$$

is the imaginary angular velocity $\frac{d\phi}{d\tau}$ of the horizon. This identification gives the manifold topology $R^2 \times S^2$, the same as that of the Euclidean Schwarzschild solution. By the No Hair Theorems these are the only solutions which are asymptotically flat in the conventional sense.

The isometry group is $O(2)\otimes O(2)$ and is generated by $\frac{\partial}{\partial \phi}$ and the corotating Killing vector $K=\frac{\partial}{\partial \tau}+\Omega\frac{\partial}{\partial \phi}$. K is zero on the horizon which is the 2-sphere $r=r_+$ with area $A=4\pi(r_+^2-\alpha^2)$. The surface gravity $\kappa=\gamma^{-1}$ and the period $\beta=2\pi\gamma$. The time translation Killing vector $\frac{\partial}{\partial \tau}$ has unit magnitude at large radius and for non zero α has 2 isolated fixed points at the north and south poles of the horizon. One of these is a nut and the other an anti-nut, which is being determined by the choice of orientation and the sign of α. The surface gravities are

$$\kappa_1 = \gamma^{-1}, \tag{3.7}$$

$$\kappa_2 = \pm\Omega . \tag{3.8}$$

If M and α are such that $\gamma\Omega = q/p$, where q and p are relatively prime integers the nut and anti-nut will be of type (p/q) and $(p, -q)$ respectively and the period will be $2\pi p\gamma = 2\pi q\Omega^{-1}$.

The self-dual Taub-NUT metric can be written in the form

$$\begin{aligned} ds^2 = &(r-n)(r+n)^{-1}(d\tau + 2n\cos\theta d\phi)^2 \\ &+ (r^2-n^2)(d\theta^2 + \sin^2\theta d\phi^2) \\ &+ (r+n)(r-n)^{-1}dr^2 . \end{aligned} \tag{3.9}$$

The Dirac string singularity at the northpole ($\theta = 0$) can be removed by introducing a new coordinate

$$\tau' = \tau + 2n\phi . \tag{3.10}$$

Similarly the Dirac string singularity at the southpole ($\theta = \pi$) can be removed by introducing a new coordinate

$$\tau'' = \tau - 2n\phi . \tag{3.11}$$

Because ϕ is identified modulo 2π, τ' and τ'' must be identified modulo $8\pi n$. These identifications and overlapping coordinate patches give the surfaces of constant $r > n$ the topology of 3-spheres with $(\tau(2n)^{-1}, \theta, \phi)$ being Euler angles. The apparent singularity at $r = n$ is in fact just the origin in hyperspherical polar coordinates. The topology of the manifold is R^4. The curvature is self-dual with the orientation defined by the positively oriented orthonormal basis

$$\omega^0 = (r+n)^{1/2}(r-n)^{-1/2}dr , \tag{3.12}$$

$$\omega' = (r^2-n^2)^{1/2}\left(\cos\left(\frac{\tau}{2n}\right)d\theta + \sin\left(\frac{\tau}{2n}\right)\sin\theta d\phi\right), \tag{3.13}$$

$$\omega^2 = (r^2-n^2)^{1/2}\left(-\sin\left(\frac{\tau}{2n}\right)d\theta + \cos\left(\frac{\tau}{2n}\right)\sin\theta d\phi\right), \tag{3.14}$$

$$\omega^3 = (r-n)^{1/2}(r+n)^{-1/2}(d\tau + 2n\cos\theta d\phi) . \tag{3.15}$$

The isometry group is isomorphic to $U(2) = (U(1)\otimes SU(2))/Z_2$. The $U(1)$ group is generated by the Killing vector $\frac{\partial}{\partial\tau}$ which is normalized to have unit magnitude at large r. It has a single self-dual nut fixed point at the origin $r = n$. The surface gravity $\kappa_1 = \kappa_2 = (4n)^{-1}$ the period $\beta = 8\pi n$. The $SU(2)$ acts transitively on 3-spheres of constant r. A typical Killing vector $\frac{\partial}{\partial\phi}$ also has a single fixed point at the origin $r=n$ but this time it is an anti-self dual anti-nut with $\kappa_1=-\kappa_2=1$ and period 2π.

Another metric of Taub-NUT form is [18]

$$\begin{aligned} ds^2 &= (r^2-n^2)(r-2n)^{-1}(r-\tfrac{1}{2}n)^{-1}dr^2 \\ &\quad + (r^2-n^2)(d\theta^2+\sin^2\theta d\phi^2) \\ &\quad + (r-2r)(r-\tfrac{1}{2}n)(r^2-n^2)^{-1}(d\tau+2n\cos\theta d\phi)^2 \,. \end{aligned} \tag{3.16}$$

The curvature is not self-dual. The Dirac string singularities on the axis can be removed by identifying τ with period $8\pi n$. As before this makes the surfaces of constant $r>2n$ into 3-spheres. The apparent singularity at $r=2n$ corresponds to these 3-spheres collapsing to a 2-sphere. The isometry group is the same as for the self-dual Taub-NUT solution. The Killing vector $\frac{\partial}{\partial\tau}$ has unit magnitude at large r and has a bolt of area $12\pi n^2$ at $r=2n$ with surface gravity $\kappa=(4n)^{-1}$. The Killing vector $\frac{\partial}{\partial\phi}$ is a typical generator of the $SU(2)$ group. It has a nut of type (1, 2) at the north pole, ($\theta=0$) and an antinut of type (1, -2) at the southpole.

The multi-Taub-NUT metric [11] can be written in the form

$$ds^2 = V(d\tau+\boldsymbol{\omega}\cdot d\mathbf{x})^2 + V^{-1}d\mathbf{x}\cdot d\mathbf{x}\,, \tag{3.17}$$

where

$$V^{-1} = 1 + \sum_i 2n_i|\mathbf{x}-\mathbf{x}_i|^{-1} \tag{3.18}$$

and

$$\operatorname{curl}\boldsymbol{\omega} = \operatorname{grad}(V^{-1})\,. \tag{3.19}$$

The $\mathbf{x}_i$ denote the distances in the 3-dimensional metric $d\mathbf{x}\cdot d\mathbf{x}$ the i'th nut with parameter n_i and the grad and curl operations are also performed in this metric. The vector field $\omega(\mathbf{x})$ will have Dirac String singularities running from each nut. If the nut parameters n_i are equal to a single value n the singularities can all be removed by identifying τ with period $8\pi n$. A large surface which surrounds all the nuts acquires the topology of the lens space $L(|, s)$ – a 3-sphere with s points identified, where s is the number of nuts. Thus the metric is not asymptotically flat in the usual sense. The curvature is self-dual. In general for $s\geqq 3$ the only Killing vector will be $\frac{\partial}{\partial\tau}$. This has unit magnitude at large values of r and a self-dual fixed point at each nut with $\kappa_1=\kappa_2=(4n)^{-1}$.

Another family of self-dual metrics can be obtained from the multi-Taub-NUT form by omitting the constant term 1 in V^{-1} in Eq. (3.18) [21]. Again to obtain a regular metric the n_i have all to be equal but in this case they can be made equal to one by rescaling the coordinates. The topology and the nuts are the same as for the corresponding multi-Taub-NUT solutions, however unlike the multi-Taub-NUT case, these metrics are Asymptotically Locally Euclidean (A.L.E.). This means that they are asymptotic to Euclidean space identified under a discrete sub group of

$SO(4)$. In fact they are the most general family of self-dual A.L.E. spaces with these identifications. Other self-dual A.L.E. instantons with different identifications have been found implicitly by Hitchin [22] but explicit metrics are not yet known.

When $s=1$ one obtains flat space. When $s=2$ one obtains the Eguchi-Hanson metric [16, 17]. The metric can be written as

$$ds^2=(1-a^4r^{-4})^{-1}dr^2+(1-a^4r^{-4})\frac{r^2}{4}(d\psi+\cos\theta d\phi)^2$$

$$+\frac{r^2}{4}(d\theta^2+\sin^2\theta d\phi^2)\,. \tag{3.20}$$

The apparent singularity at $r=a$ can be removed by identifying ψ modulo 2π rather than modulo 4π as is usual for Euler angles on S^3. The identification makes the surfaces of constant $r>a$ into RP^3, a 3-sphere with antipodal points identified. At large values of r the metric tends to that of flat Euclidean space, points reflected in the origin being identified. The surface $r=a$ is a 2-sphere.

The isometry group is $U(2)=(U(1)\times SU(2))/Z_2$ the same as for Taub-NUT space of which this metric is a limiting form. The $U(1)$ subgroup is generated by the Killing vector $\frac{\partial}{\partial\psi}$ which has a bolt on the 2-sphere $r=a$ of area πa^2 with surface gravity $\kappa=1$ and period 2π. The Killing vector $\frac{\partial}{\partial\phi}$, a typical generator of the $SU(2)$ group, has 2 isolated fixed points, self dual nuts at the north and south poles of the 2-sphere $r=a$. These have $\kappa_1=\kappa_2$ and the period is 2π.

We now turn to complete non-singular solutions of the Einstein equations with positive Λ term:

$$R_{ab}=\Lambda g_{ab}; \qquad \Lambda>0\,. \tag{3.21}$$

These are all compact [23].

The simplest example is a 4-sphere of radius $3^{1/2}\Lambda^{-1/2}$ in 5-dimensional Euclidean space. This is the analytic continuation of de Sitter space [20]. The isometry group is the $SO(5)$ of rotations about the origin in 5-dimensional space. These are generated by 5×5 anti-symmetric matrices which can have rank 0, 2, or 4. The zero case is trivial. In the case of rank 2 there is a 3-plane through the origin of R^5 which is not moved by the rotation. The intersection of this with the 4-sphere is a bolt which is a 2-sphere with area $12\pi\Lambda^{-1}$. With the normalization of the Killing vector from $0(5)$, the period is 2π, the surface gravity is 1. However for physical applications [20] it may be convenient to choose the Killing vector to have unit magnitude on the orbit which is a geodesic. In this case the period is $2\pi3^{1/2}\Lambda^{-1/2}$. The surface gravity is $(\Lambda/3)^{1/2}$. If the matrix is of rank 4, there will be one direction through the origin in R^5 which is left unchanged by the rotation. The intersection of this direction with the 4-sphere will constitute a nut and an antinut.

The next example is complex projective plane, CP^2, with its standard Kaehler metric [14, 15] which has an anti-self-dual Weyl tensor. This can be realized as $C^3-\{0\}$ with coordinates Z_1, Z_2, Z_3 factored by the equivalence relation

$(Z_1, Z_2, Z_3) \equiv (\lambda Z_1, \lambda Z_2, \lambda Z_3)$, $\lambda \in C - \{0\}$. The isometry group is $SU(3)/Z_3$ which acts on the coordinates in the standard manner. This is generated by traceless anti-Hermitean matrices. Such matrices can be divided into two classes, those with two equal eigenvalues and those with 3 unequal eigenvalues. In the case one gets an anti-self-dual antinut which can be taken to be at the "origin" $(0,0,Z_3)$ and a bolt at "infinity" which is the 2-sphere $(Z_1, Z_2, 0)$. In the latter case one gets 3 isolated fixed points two of which will be antinuts and one a nut. They can be located at the origin $(0,0,Z_3)$, the northpole of infinity $(Z_1, 0, 0)$ and the southpole $(0, Z_2, 0)$.

One can describe these fixed points in more detail by introducing coordinates.

$$Z_1/Z_3 = r\cos\left(\frac{\theta}{2}\right)\exp i(\psi+\phi)/2\,, \tag{3.22}$$

$$Z_2/Z_3 = r\sin\frac{\theta}{2}\exp i(\psi-\phi)/2\,. \tag{3.23}$$

The metric then takes the form

$$\begin{aligned} ds^2 &= dr^2(1+\tfrac{1}{6}\Lambda r^2)^{-2} + \tfrac{1}{4}r^2(1+\tfrac{1}{6}\Lambda r^2)^{-2}(d\psi+\cos\theta d\phi)^2 \\ &\quad + \tfrac{1}{4}r^2(1+\tfrac{1}{6}\Lambda r^2)^{-1}(d\theta^2+\sin^2\theta d\phi^2)\,. \end{aligned} \tag{3.24}$$

The Killing vector $\frac{\partial}{\partial\psi}$ corresponds to a generator of $SU(3)$ with 2 equal eigenvalues. It has an antiselfdual nut at $r=0$ with surface gravity $\kappa_1 = -\kappa_2 = \frac{1}{2}$ and period 4π. It has a bolt at $r=\infty$ with area $6\pi\Lambda^{-1}$ and $\kappa=\frac{1}{2}$. The Killing vector $\frac{\partial}{\partial\phi}$ corresponds to a generator of $SU(3)$ with 3 unequal eigenvalues. It has a self-dual nut at the origin $r=0$ with $\kappa_1=\kappa_2=1$ and period 2π. It is an antinut of type $(1,-2)$ at the northpole of the sphere at infinity $r=\infty$, $\theta=0$ and an antinut of type $(1,-2)$ at the southpole $r=\infty$, $\theta=\pi$.

Our next example is the metric product of two 2-spheres each with radius $\Lambda^{-1/2}$ and area $4\pi\Lambda^{-1}$. It can be regarded as a limiting case of the Schwarzschild-de Sitter solution [15] with the surface gravities of the black hole and cosmological horizons equal. The isometry group is $SO(3)\otimes SO(3)$, the two factors acting on the 2-spheres independently. A circle subgroup of the full isometry group can be projected into circle subgroups in the two factors. If one of these projections consists of the identity only, the corresponding Killing vector has 2 bolts which are 2-spheres of area $4\pi\Lambda^{-1}$. In the other case there will be 4 isolated fixed points, two nuts and two antinuts.

The only other known gravitational instanton with positive Λ is an S^2 bundle over S^2 discovered by Page [19] as a limiting case of the Kerr-de Sitter solution. The metric can be written in the Taub-NUT form:

$$\begin{aligned} ds^2 = \frac{3}{\Lambda}(1+\nu^2)\{&(1-\nu^2\cos^2\varrho)(3-\nu^2-\nu^2(1+\nu^2)\cos^2\varrho)^{-1}d\varrho^2 \\ &+(1-\nu^2\cos^2\varrho)(3+6\nu^2-\nu^4)^{-1}(d\theta^2+\sin^2\theta d\phi^2) \\ &+(3-\nu^2-\nu^2(1+\nu^2)\cos^2\varrho)(3+\nu^2)^{-2}(1-\nu^2\cos^2\varrho)^{-1}\sin^2\varrho \\ &\cdot\tfrac{1}{4}(d\psi+\cos\theta d\phi)^2\}\,, \end{aligned} \tag{3.25}$$

where v is the positive root of

$$v^4+4v^3-6v^2+12v-3=0 \tag{3.26}$$

which works out to be

$$0.2817\,.$$

The surfaces of constant ϱ, $0<\varrho<\pi$, are 3-spheres on which (ψ, θ, ϕ) are Euler angle coordinates. The apparent singularities at $\varrho=0$ and $\varrho=\pi$ are where the 3-spheres collapse to 2-spheres.

The isometry group is $U(2)=(U(1)\otimes SU(2))/Z_2$, the same as for Taub-NUT. The Killing vector $\frac{\partial}{\partial\psi}$ generates the $U(1)$ subgroup and has bolts at the 2-spheres $\varrho=0$ and $\varrho=\pi$ with areas

$$12\pi(1-v^4)(3+6v^2-v^4)^{-1}\Lambda^{-1}\,. \tag{3.27}$$

The Killing vector $\frac{\partial}{\partial\phi}$, a typical generator of the $SU(2)$ subgroup, has 4 isolated fixed points at $0=0$ or π and $\varrho=0$ or π. Two of these are nuts and two antinuts.

Compact gravitational instantons with negative or zero Λ are known but they cannot admit any continuous isometry group. We shall therefore not consider them in this paper.

Non compact solutions with negative Λ are also known. These may admit continuous isometrics but they do not seem to play a role in path integrals.

4. Topological Invariants

There are two topological invariants which can be expressed as integrals of the curvature of a 4-dimensional metric. For a compact manifold these are the Euler number

$$\chi=(128\pi^2)^{-1}\int_M \varepsilon^{ef}_{ab}R_{efgh}\varepsilon^{gh}_{cd}R^{abcd}\sqrt{g}\,d^4x \tag{4.1}$$

and the signature (sometimes called the index)

$$\tau=(96\pi^2)^{-1}\int_M R_{abcd}\varepsilon^{cdef}R^{ab}{}_{ef}\sqrt{g}\,d^4x\,. \tag{4.2}$$

For non compact manifolds there are additional boundary terms. For χ these are [24]

$$(128\pi^2)^{-1}\int_{\partial M}(R_{abcd}K^{ac}n^bn^d+64\det K^a_b)\sqrt{h}\,d^3x\,, \tag{4.3}$$

where n^a is the outward directed normal to the boundary ∂M, $K_{ab}=n_{c;d}h^c_ah^d_b$ is the second fundamental form and $h_{ab}=g_{ab}-n_an_b$ is the induced metric on ∂M. For τ the boundary terms are [25]

$$-2(96\pi^2)^{-1}\int R_{abcd}\varepsilon^{cdef}n^an_eK^b_f\sqrt{h}\,d^3x-\eta(0)\,, \tag{4.4}$$

where $\eta(s)$ is a quantity constructed from the eigenvalues of a certain differential operator on the boundary ∂M.

The Euler number in the alternating sum of the Betti numbers

$$\chi[M] = \sum_{p=0}^{p=4} (-1)^p B_p , \tag{4.5}$$

where the p'th absolute Betti number, B_p, is the rank of the p'th absolute homology group $H_p(M)$. It is also the rank of the $4-p$'th relative cohomology group $H^{4-p}(M, \partial M)$. These cocycles may be represented by closed $4-p$ forms which vanish on the boundary ∂M and which are not the exterior derivative of a $4-p-1$ form which itself vanishes on the boundary. For a compact manifold without boundary $B_0 = B_4 = 1$ and $B_1 = B_3$. $B_1 = 0$ if the manifold is simply connected as in all our examples. If there is a boundary $B_0 = 1$ and $B_4 = 0$. In a compact manifold without boundary B_p is equal to the number of linearly independent harmonic p-forms.

In the case of a compact manifold without boundary τ may be defined as the signature of the quadratic form on $H_2(M)$ given by the intersection number between two 2-cycles, i.e. the number of times they intersect, intersections being counted positive or negative according to whether or not the orientation of the tangent space of the intersection point agrees with the orientation arising from the tangent planes to the two 2-surfaces. In the case of the intersection number of a cycle with itself one slightly distorts one copy of a representative of the 2-cycle so that it intersects the other copy transversely. A necessary and sufficient condition that the manifold admit a spinor structure is that the self intersection numbers of the 2-cycles must be even for simply connected manifolds [26].

τ is also equal to the quadratic form on $H^2(M)$ defined by the cup product. If one represents two elements of $H^2(M)$ by closed 2-forms, then the cup product is just

$$\int_M \alpha \wedge \beta .$$

From this it follows that τ is equal to the number of linearly independent self-dual harmonic 2-forms minus the number of antiself dual ones.

In the case of a manifold with boundary ∂M, the cup product may be defined between $H^2(M)$ and $H^2(M, \partial M)$ by the above formula, i.e. between closed 2-forms which are not the exterior derivatives of 1-forms and closed 2-forms which vanish on ∂M. With the natural injection $H^2(M, \partial M)$ into $H^2(M)$, this defines a quadratic form on $H^2(M, \partial M)$. The quadratic form may have zero eigenvalues but τ may be defined as the number of positive eigenvalues minus the number of negative eigenvalues [25]. The definition of τ by homology for manifolds with boundary is simply the dual of the above. One can define the intersection number between a cycle in the absolute homology group $H_2(M)$ and one in the relative homology group $H_2(M, \partial M)$ i.e. the group of equivalence classes of 2-chains whose boundaries lie in ∂M and which are not the boundaries of 3-chains. With natural injection of $H_2(M)$ into $H_2(M, \partial M)$ this defines a quadratic form on $H_2(M)$. The quadratic form will have zero eigenvalues corresponding to elements of $H_2(M)$ which are homologous to elements of $H_2(\partial M)$ but τ will be equal to the number of positive eigenvalues minus the number of negative eigenvalues.

The values of χ and τ for the examples in the previous section are

Schwarzschild and Kerr	$\chi=2$	$\tau=0$
Self-dual Taub-NUT	$\chi=1$	$\tau=0$
Multi-Taub NUT	$\chi=s$	$\tau=s-1$
Non self dual Taub-NUT	$\chi=2$	$\tau=1$
Eguchi-Hanson	$\chi=2$	$\tau=1$
S^4	$\chi=2$	$\tau=0$
CP^2	$\chi=3$	$\tau=-1$
$S^2\times S^2$	$\chi=4$	$\tau=0$
Twisted $S^2\times S^2$	$\chi=4$	$\tau=0$.

The relevance of χ and τ to our nuts and bolts classification is given by various fixed point theorems which relate them both to the zeros of vector fields on the manifold. For isometries these theorems take particularly simple forms [27–30]. On a compact manifold the Euler number is given by

$$\chi[M]=N_{+}+N_{-}+\sum_i \chi_i \, . \tag{4.6}$$

Where N_+ and N_- are the number of nuts and antinuts and χ_i is the Euler number of the i'th bolt. In all our examples $\chi_i=2$. Indeed this will necessarily be the case if the manifold M is simply connected and if the bolt can not be continuously deformed to a point (it can be so deformed in S^4 but not in the other examples). This is the analogue of the theorem that the horizon of a black hole metric with Lorentz signature $(-+++)$ must have an event horizon that is topologically spherical [31]. This formula also holds for manifolds with boundary ∂M provided that the Killing vector field is either everywhere tangential to the boundary (as it is in all our examples) or is everywhere transverse. The theorem for signature is rather more complicated. On compact manifolds

$$\tau=\sum_{\text{nuts}} -\operatorname{cotan} p\theta \operatorname{cotan} q\theta+\sum_{\text{bolts}} Y\operatorname{cosec}^2\theta \, , \tag{4.7}$$

Y is the self intersection number of a bolt and 2θ is the group parameter.

Equation (4.7) holds for all values of θ. If one expands in powers of θ the first two terms give

$$\sum_{\text{nuts}} -(pq)^{-1}+\sum_{\text{bolts}} Y=0 \, , \tag{4.8}$$

$$\sum_{\text{nuts}} \tfrac{1}{3}(pq^{-1}+qp^{-1})+\tfrac{1}{3}\sum_{\text{bolts}} Y=\tau \, . \tag{4.9}$$

Applying (4.8) to the Killing vector $\dfrac{\partial}{\partial\psi}$ in CP^2 which has an antinut with $-p=q=1$ and a bolt one finds that the self intersection number of the bolt must be -1. This shows that CP^2 does not admit a spin structure though it can have a generalized spin structure [32–33].

In the case of a non compact there is an additional boundary term $\eta(\theta,0)$ in Eq. (4.7). The quantity $\eta(\theta,s)$ is formed out of the eigenvalues of a certain differential operator on the boundary ∂M [25].

5. Duality

The action of the group G with Killing vector $K=\frac{\partial}{\partial\tau}$ defines a fibering $\pi:M-C\to B$ where C is the fixed point set of μ_τ. In other words B is the 3-dimensional space of non-trivial orbits of G. The manifold B inherits a metric

$$h_{ab}=g_{ab}-V^{-1}K_aK_b\,, \tag{5.1}$$

where $V=K^aK_a$. The metric g_{ab} on M can then be written locally in the form

$$ds^2=V(d\tau+\omega_i dx^i)^2+V^{-1}\gamma_{ij}dx^i dx^j\,, \tag{5.2}$$

where $\{x^i\}$ are coordinates on $B, \gamma_{ij}=Vh_{ij}$ and ω_i are independent of the fourth coordinate τ. The vector field ω_i in B is defined up to a "gauge transformation"

$$\tau'=\tau+\lambda(x') \tag{5.3}$$

under which

$$\omega_i'=\omega_i-\frac{\partial}{\partial x^i}\lambda\,. \tag{5.4}$$

The twist field $H_{ij}=\partial_i\omega_j-\partial_j\omega_i$ is gauge invariant. It can be expressed as

$$H_{ab}=2K_{e;f}h^e_b h^f_a V^{-1}\,. \tag{5.5}$$

For the rest of this section we shall work in the 3-dimensional space B. Indices i,j,k etc. will be raised or lowered by γ_{ij} and covariant differentiation with respect to γ_{ij} will be denoted by $\|$. Using the 3-dimensional alternating tensor one can define a twist vector H_i by

$$H_i=\tfrac{1}{2}\varepsilon_i^{jk}H_{jk}\,. \tag{5.6}$$

This obeys the conservation equation

$$H^i_{\|i}=0\,. \tag{5.7}$$

One can therefore define the nut charge within a 2-surface L by

$$N=(8\pi)^{-1}\int_L H_i d\sigma^i\,. \tag{5.8}$$

In the case of a nut of type (p,q)

$$N=(8\pi pq)^{-1}\beta\,. \tag{5.9}$$

For a bolt with self intersection number Y

$$N=(8\pi)^{-1}Y\beta\,. \tag{5.10}$$

These formulae are obtained by expanding the metric in a Taylor series in normal coordinates about the fixed point set. From them one can see that Eq. (4.8) is an expression of the fact that a compact manifold has zero total nut charge.

One can project the field equations, $R_{ab}=\Lambda g_{ab}$, into B

$$V^{-1}R_{ab}K^aK^b=-V\nabla^i\nabla_i\log V+\tfrac{1}{2}V^3H_iH^i\,, \tag{5.11}$$

$$R_{cd}h^c_ih^d_j={}^3R_{ij}-\tfrac{1}{2}V^{-2}V_iV_j+\gamma_{ij}\nabla_k\nabla^k\log V+\tfrac{1}{2}V^2H_iH_j-\tfrac{1}{2}V^2H_kH^k\gamma_{ij}\,, \tag{5.12}$$

$$V^{-1/2}R_{cd}h^c_jK^d=\tfrac{1}{2}V^{-1}\varepsilon^{pq}_j\nabla_pH_q\,, \tag{5.13}$$

where ${}^3R_{ij}$ is the Ricci tensor of the metric γ_{ij}.

Adding (5.11) and (5.12) one obtains an expression for the 4-dimensional action of the metric

$$\hat{I}=-\frac{1}{16\pi}\int_M\sqrt{g}\,d^4x(R-2\Lambda)-\frac{1}{8\pi}\int_{\partial M}K\sqrt{b}\,d^3x \tag{5.14}$$

$$=-\frac{\beta}{16\pi}\int_B\sqrt{\gamma}\,d^3x({}^3R-2\Lambda V^{-1}-\tfrac{1}{2}V^{-2}\nabla_iV\nabla^iV-\tfrac{1}{2}V^2H_iH^i)-\frac{\beta}{8\pi}\int_{\partial B}k\sqrt{c}\,d^2x\,, \tag{5.15}$$

where b_{ab} is the induced metric and $K=K^a_a$ is the trace of the second fundamental form of ∂M in the metric g_{ab} and c_{ij} is the induced metric and $k=k^i_i$ is the trace of the second fundamental form in the metric γ_{ij}. To obtain the field Eqs. (5.11)–(5.13) one requires that $\hat{I}$ be stationary under variations of γ_{ij}, V and H_i subject to the constraint (5.7). That expresses the fact that H is the curl of ω_i. One therefore defines a new quantity $\tilde{I}$ by adding the constraint multiplied by a Lagrange multiplier

$$\tilde{I}=-\frac{\beta}{16\pi}\int_B\sqrt{\gamma}\,d^3x({}^3R-2V^{-1}\Lambda-\tfrac{1}{2}V^{-2}\nabla_iV\nabla^iV-\tfrac{1}{2}V^2H_iH^i)-\frac{\beta}{8\pi}\int_B k\sqrt{c}\,d^3x+\frac{1}{16\pi}\beta\int_B\sqrt{\gamma}\,d^3x\psi\nabla^iH_i\,. \tag{5.16}$$

Variation of H_i gives the equation

$$V^2H_i=\nabla_i\psi\,. \tag{5.17}$$

We shall therefore call ψ the nut potential. The fact that the nut potential exists is equivalent to the field equation $R_{ab}K^ah^b_c=0$. One then rewrites $\tilde{I}$ as

$$\tilde{I}=\frac{1}{16\pi}\beta\int_B\sqrt{\gamma}\,d^3x({}^3R-2V^{-1}\Lambda-\tfrac{1}{2}V^{-2}(\nabla_iV\nabla^iV-\nabla_i\psi\nabla^i\psi))-\frac{1}{8\pi}\beta\int_{\partial B}k\sqrt{c}\,d^2x+\frac{1}{16\pi}\beta\int_{\partial B}\psi\psi^iV^{-2}d\sigma_i\,. \tag{5.18}$$

Variation of V and γ_{ij} in (5.18) gives the field equations

$$R_{ab}K^aK^b=\Lambda V\quad\text{and}\quad R_{ab}h^a_ch^b_d=\Lambda h_{cd}\,.$$

Variation of ψ in (5.18) gives the constraint

$$\nabla^i(V^{-2}\psi_i)=0\,. \tag{5.19}$$

One can therefore regard (5.18) as the effective action for a 3-dimensional relativity theory with metric γ_{ij} on B and non-linear fields ψ and V.

The term

$$V^{-2}(\nabla_i V\nabla^i V-\nabla_i\psi\nabla^i\psi) \tag{5.20}$$

in (5.18) is the Lagrangian of an $O(2,1)$ non-linear σ model. That is (5.20) can be regarded as the metric on 2-dimensional de Sitter space with coordinates V and ψ. This metric has the 3-parameter group of isometries $SL(2R)$ which can be realized as the following:

1. *Translations* $\psi\to\psi+a$ (5.21)

$V\to V\,;$ (5.22)

2. *Dilations* $\psi\to b\psi$ (5.23)

$V\to bV\,;$ (5.24)

3. *The Ehlers Transform* [34, 35]

$$\psi\to\frac{\psi+b(V^2-\psi^2)}{(1-b\psi)^2-b^2V^2}\,, \tag{5.25}$$

$$V\to\frac{V}{(1-b\psi)^2-b^2V^2}\,. \tag{5.26}$$

Corresponding to these 3 symmetries of (5.20) there will be three Noether currents

$$J^i_T=-V^{-2}\nabla^i\psi\,, \tag{5.27}$$

$$J^i_D=V^{-1}\nabla^i V-\psi V^{-2}\nabla^i\psi\,, \tag{5.28}$$

$$J^i_E=2\psi V^{-1}\nabla^i V-V^{-2}(\psi^2+V^2)\nabla^i\psi\,. \tag{5.29}$$

If $\Lambda=0$, these are symmetries of the effective action $\tilde{I}$ and so the Noether currents are all conserved in the metric γ_{ij}

$$J^i_{T\|i}=J^i_{D\|i}=J^i_{E\|i}=0\,. \tag{5.30}$$

If Λ is non zero, it breaks the symmetry under the dilation and Ehler's transforms. Thus

$$J^i_{T\|i}=0\,, \tag{5.31}$$

$$J^i_{D\|i}=2V^{-1}\Lambda\,, \tag{5.32}$$

$$J^i_{E\|i}=2\psi V^{-1}\Lambda\,. \tag{5.33}$$

In the vacuum case the symmetry under the 3-parameter group expresses the duality between the electric aspects of gravity, characterized by V, and the magnetic or nut aspects characterized by ψ. A particularly simple case is when $V=\psi$. Then γ_{ij} is the flat metric and one obtains the two families of multi Taub-NUT metrics.

6. Action and Entropy

In the case of a compact manifold M integration of Eq. (5.32) gives the total action, $\hat{I}$

$$\hat{I} = -\frac{1}{8\pi}\int_M \Lambda\sqrt{g}\,d^4x \tag{6.1}$$

$$= -\frac{1}{8\pi}\beta\int_B \frac{\Lambda}{V}\sqrt{\gamma}\,d^3x \tag{6.2}$$

$$= -\frac{1}{16\pi}\beta\int_{\partial B} J^i_D d\sigma_i\,. \tag{6.3}$$

The boundary B will consist of a disjoint set of 2 spheres around each nut and each bolt. Thus

$$\hat{I} = -\frac{1}{16\pi}\beta\sum_n Q_n\,, \tag{6.4}$$

where the dialation charge of the n'th nut and bolt is

$$Q_n = \int_{\partial B_n} J^i_D d\sigma_i\,. \tag{6.5}$$

The Q_n's are not invariant under the translation $\psi \to \psi + a$ but in the case of a bolt one can define an invariant quantity

$$M_n = 8\pi Q_n - \psi_n N_n\,, \tag{6.6}$$

where ψ_n is the value of the nut potential at the n'th bolt and N_n is its nut charge. The quantity M_n can be regarded as the "mass" of the n'th bolt. It can be represented as an integral of the 2-form over the bolt

$$M_n = \frac{1}{8\pi}\int K^{c;d}d\Sigma_{cd}\,. \tag{6.7}$$

M_n obeys the Smarr relation

$$2\beta M_n = A_n\,. \tag{6.8}$$

A_n is the area of the n'th bolt. Thus

$$\hat{I} = \sum_{\text{bolts}} -\tfrac{1}{4}A_n - \sum_{\text{bolts}} \tfrac{1}{2}\psi_n N_n \beta + \sum_{\text{nuts}} -\tfrac{1}{2}\psi_n N_n \beta$$

$$= \sum_{\text{bolts}} -\tfrac{1}{4}A_n - \sum_{\text{bolts}} \frac{\psi Y \beta^2}{16\pi} - \sum_{\text{nuts}} \frac{\psi\beta^2}{16\pi pq}\,. \tag{6.9}$$

This generalizes the formula

$$\hat{I} = -\tfrac{1}{4}A \tag{6.10}$$

which was found for S^4 [12].

Equation (6.9) can be interpreted thermodynamically in at least two ways. In the first approach one regards the Λ term as part of the dynamics of theory. One then defines the partition function Z for the canonical ensemble.

$$Z = \sum_n \langle g_n | g_n \rangle\,, \tag{6.11}$$

where $|g_n\rangle$ is an orthonormal basis of states for the gravitational field with the given value of Λ. In this case, unlike the normal thermal canonical ensemble, there is no externally imposed temperature or chemical potential. The partition function Z simply counts the total number of states $|g_n\rangle$. Since each of these is equally probable, the probability $P_n = Z^{-1}$ and thus the entropy $S = -\Sigma P_n \log P_n = \log Z$. The partition function Z can also be represented as a path integral over all metrics g on a compact manifold M

$$Z = \int d[g] \exp -\hat{I}[g] . \tag{6.12}$$

By the stationary phase approximation one would expect the dominant contribution to come from metrics near a solution, g_0, of the classical Einstein field equations and the value of Z to be given approximately by $\exp -\hat{I}[g_0]$. Thus

$$S = -\hat{I}[g_0]$$
$$= \sum_{\text{bolts}} \tfrac{1}{4} A_n + \sum_{\text{bolts}} \frac{\psi_n Y \beta^2}{16\pi} \tag{6.13}$$
$$+ \sum_{\text{nuts}} \frac{\psi \beta^2}{16\pi pq} . \tag{6.14}$$

This shows that not only do bolts have an entropy equal to $\frac{1}{4}$ of their area, as was found for de Sitter space [12, 20], but there is also a contribution from the NUT charges of both nuts and bolts. Because of the translational freedom $\psi \to \psi + a$, this latter contribution to the entropy cannot be attributed to individual nuts and bolts, but its total value is invariant, since the sum of the NUT charges is zero for a compact manifold.

One can also regard the Λ term as not being part of the field equations but as Lagrange multiplier or chemical potential for the 4-volume V. In this case, one can form the partition function for the volume canonical ensemble as

$$Z[\Lambda] = \sum_n \langle g_n | \exp - \frac{\Lambda V_n}{8\pi} | g_n \rangle . \tag{6.15}$$

As before, one can represent $Z[\Lambda]$ as a path integral over all metrics and by the stationary phase approximation one expects the dominant contribution to $\log Z$ to be $-\hat{I}[g_0]$ where g_0 is a solution with the given value of Λ. On dimensional grounds

$$\hat{I}[g_0] = -\frac{\Lambda V_0}{8\pi} = -\frac{f}{8\pi\Lambda} , \tag{6.16}$$

where f may depend on the topology of the manifold and on the particular class of solutions (if there is more than one) but it is independent of Λ. From $Z[\Lambda]$ one can calculate the expectation value of the volume in the ensemble

$$\langle V \rangle = -8\pi \frac{\partial}{\partial \Lambda} \log Z . \tag{6.17}$$

With the stationary phase value for $\log Z$ one obtains

$$\langle V \rangle = V_0 . \tag{6.18}$$

One can also form an "entropy" for the canonical ensemble

$$S_V = -\sum P_n \log P_n , \tag{6.19}$$

where the probability of being in the n'th state, P_n, equals

$$Z^{-1} \exp -\left(\frac{\Lambda V_n}{8\pi}\right) . \tag{6.20}$$

Thus

$$S_V = -\Lambda^2 \frac{\partial}{\partial \Lambda}(\Lambda^{-1} \log Z) \tag{6.21}$$

using the stationary phase value one obtains

$$S_V = -2\hat{I}[g_0]\,. \tag{6.22}$$

This differs by a factor 2 from entropy in the canonical ensemble in which Λ was regarded as part of the field equations.

A physical interpretation of S_V can be obtained in the following way. Let $N(V)dV$ be the number of states of the gravitational field with volumes between V and $V+dV$. Then $Z[\Lambda]$ can be regarded as the Laplace transform of $N(V)$

$$Z[\Lambda] = \int_0^\infty N(V) \exp -\left(\frac{\Lambda V}{8\pi}\right) dV\,. \tag{6.23}$$

Thus $N(V)$ is the inverse Laplace transform

$$N(V) = \frac{1}{16\pi^2 i} \int_{-i\infty}^{+i\infty} Z[\Lambda] \exp\left(\frac{\Lambda V}{8\pi}\right) d\Lambda\,. \tag{6.24}$$

The contour in Eq. (6.24) should be taken to pass to the right of the essential singularity at $\Lambda = 0$. The dominant contribution to $N(V)$ will come from the stationary points of (6.24) which occurs at

$$\Lambda = \Lambda_s \tag{6.25}$$

for which

$$\frac{V}{8\pi} = -\frac{\partial \log Z}{\partial \Lambda} = \left\langle \frac{V}{8\pi} \right\rangle\,. \tag{6.26}$$

The value of the integrand at the stationary phase point is $\exp S_V$. Thus

$$\log N(V) \approx S_V\,. \tag{6.27}$$

In the case of a compact manifold the dilation symmetry was broken by the presence of the Λ term. In the non-compact case one is interested in situations where the metric is asymptotically flat, either in the 3 or 4 dimensional senses. Both these require that $\Lambda = 0$. In the 4-dimensional case, the boundary conditions are dilation invariant. This means that the change of the action under a dilation is just given by the total dilation charge

$$\frac{\delta \hat{I}}{\delta b} = \frac{\delta \tilde{I}}{\delta b} \tag{6.28}$$

$$= \int J^i_D d\sigma_i \tag{6.29}$$

$$= Q_D \tag{6.30}$$

$$= 0\,. \tag{6.31}$$

However, under a dilation $g_{ab} \to k^2 g_{ab}$, $\hat{I}[k^2 g_{ab}] = k^2 \hat{I}[g_{ab}]$. This shows that the action $\hat{I}[g]$ must be zero for a solution of the field equations that is asymptotically flat in the 4-dimensional sense. The Positive Action Theorem [36–41] then implies that the metric is flat if it is asymptotically Euclidean i.e. it approaches the standard flat metric on R^4 outside some compact subset in which the topology may differ from that of R^4. If the metric is asymptotically locally Euclidean, i.e. outside of a compact subset it approaches the standard flat metric on R^4 identified under some discrete subgroup $SO(4)$ which acts freely, then the Generalized Positive Action Conjecture [42, 43] implies that metric must be self-dual or anti-self-dual.

One is also interested in metrics which the "spatial" metric approaches a flat 3-dimensional metric and in which V approaches some constant value which can be normalized to unity by an appropriate choice of β.

The boundary condition $V=1$ at infinity is not preserved under the dilation transformation. To impose this constraint one can define a new quantity

$$I' = \tilde{I} + \frac{\beta}{16\pi} \int_\infty \mu V \sqrt{c}\, d^2x\,, \tag{6.32}$$

where μ is a Lagrange multiplier. If one requires I' to be stationary under variations of V which do not vanish on the boundary one obtains

$$\mu = -V_{;i}n^i V^{-2}\,. \tag{6.33}$$

Then

$$\frac{\delta \hat{I}}{\delta b} = \frac{\delta I'}{\delta b} \tag{6.34}$$

$$= \frac{\delta \tilde{I}}{\delta b} + \frac{\beta}{16\pi} \int \mu V \sqrt{c}\, d^2x \tag{6.35}$$

$$= -\frac{\beta}{16\pi} \int V_{;i}n^i V^{-1} \sqrt{c}\, d^2x \tag{6.36}$$

$$= \tfrac{1}{2}\beta M_\infty\,, \tag{6.37}$$

where

$$M_\infty = -\frac{1}{8\pi} \int_\infty V_{;i}n^i V^{-1} \sqrt{c}\, d^2x\,. \tag{6.38}$$

Thus

$$\hat{I} = \tfrac{1}{2}\beta M_\infty\,. \tag{6.39}$$

This generalizes the results obtained in reference [12] and it can also be applied to spaces such as Taub-NUT which are not asymptotically flat in the usual sense because the boundary surface at infinity cannot be embedded (even locally) in flat space. For Taub-NUT it gives a value of $4\pi N^2$ in agreement with an unpublished calculation by Lincoln Davis. As in reference [12] one obtains an entropy equal to 1/4 the area of the event horizon for the Schwarzschild and Kerr solutions.

One can also integrate the divergence of Ehler's current over the manifold. For a compact manifold this gives:

$$\int \psi \Lambda \sqrt{g}\, d^4x = \sum_{\text{bolts}} 8\pi\psi_n M_n - \sum_{\substack{\text{nuts}\\+\\\text{bolts}}} 4\pi\psi_n^2 N_n\,. \tag{6.40}$$

We have not found a physical application for this result.

Acknowledgements. We should thank J. F. Adams, M. F. Atiyah, N. Hitchin, D. N. Page and C. N. Pope for help and discussions.

References

1. Polyakov, A.: Phys. Lett **59** B, 82 (1975)
2. 't Hooft, G.: Phys. Rev. Lett. **37**, 8 (1976)
3. Belavin, A., Polyakov, A., Schwarz, A., Tyupkin, B.: Phys. Lett. **59** B, 85 (1975)
4. Jackiw, R., Rebbi, C.: Phys. Lett. **37**, 172 (1976)
5. Callan, C., Dashen, R., Gross, D.: Phys. Lett. **63** B, 334 (1976)
6. Hawking, S.W.: Nucl. Phys. B **144**, 349 (1978)
7. Zumino, B.: Ann. N.Y. Acad. Sci. **302**, 545 (1977)
8. Hartle, J., Hawking, S.W.: Phys. Rev. D **13**, 2188 (1976)
9. Gibbons, G.W., Perry, M.J.: Phys. Rev. Lett. **36**, 985 (1976)
10. Gibbons, G.W., Perry, M.J.: Proc. R. Soc. A **358**, 467 (1978)
11. Hawking, S.W.: Phys. Lett. **60** A, 81 (1977)
12. Gibbons, G.W., Hawking, S.W.: Phys. Rev. D **15**, 2752 (1977)
13. Hawking, S.W.: Commun. Math. Phys. **43**, 199 (1975)
14. Eguchi, T., Freund, P.G.O.: Phys. Rev. Lett. **37**, 1251 (1976)
15. Gibbons, G.W., Pope, C.N.: Commun. Math. Phys. **61**, 239 (1978)
16. Eguchi, T. Hanson, A.: Phys. Lett. **74** B, 249 (1978)
17. Belinski, V.A., Gibbons, G.W., Page, R.N., Pope, C.N.: Phys. Lett. **76** B, 433 (1978)
18. Page, D.N.: Phys. Lett. **78** B, 249 (1978)
19. Page, D.N.: Phys. Lett. **79** B, 235–238 (1978)
20. Gibbons, G.W., Hawking, S.W.: Phys. Rev. D **15**, 2738 (1977)
21. Gibbons, G.W., Hawking, S.W.: Phys. Lett. **78** B, 430 (1978)
22. Hitchin, N.J.: Polygons and gravitons. Preprint, Oxford University. Math. Proc. Camb. Phil. Soc. (in press)
23. Milnor, J.: Morse theory. Princeton: Princeton University Press (1963)
24. Chern, S.S.: Ann. Math. **46**, 674 (1945)
25. Atiyah, M.F., Patodi, V.K., Singer, I.M.: Proc. Cambridge Philos. Soc. **77**, 43 (1975); **78**, 405 (1975)
26. Geroch, R.P.: J. Math. Phys. **9**, 1739 (1968)
27. Atiyah, M.F., Bott, R.: Ann. Math. **87**, 451 (1968)
28. Bott, R.: Mich. Math. J. **14**, 231–244 (1967)
29. Atiyah, M.F., Singer, I.M.: Ann. Math. **87**, 546–604 (1968)
30. Baum, P., Cheeger, J.: Topology **8**, 173–193 (1969)
31. Hawking, S.W.: Commun. Math. Phys. **25**, 152–166 (1972)
32. Hawking, S.W., Pope, C.N.: Phys. Lett. **73** B, 42 (1978)
33. Back, A., Freund, P.G.O., Forger, M.: Phys. Lett. **77** B, 181 (1978)
34. Ehlers, J.: In: Les theories relativistes de la gravitation. Paris: CNRS 1959
35. Geroch, R.: J. Math. Phys. **13**, 394 (1972)
36. Gibbons, G.W., Hawking, S.W., Perry, M.J.: Nucl. Phys. B **138**, 141 (1978)
37. Page, D.N.: Phys. Rev. D **18**, 2733–2738 (1978)
38. Gibbons, G.W., Pope, C.N.: Commun. Math. Phys. (in press)
39. Hawking, S.W.: Phys. Rev. D **18**, 1747 (1978)
40. Gibbons, G.W., Hawking, S.W.: Proof of the positive action conjecture and the nature of the gravitational Action. Unpublished Report
41. Shoen, R.M., Yau, S.T.: Phys. Rev. Lett. **42**, 547–548 (1979)
42. Hawking, S.W., Pope, C.N.: Nucl. Phys. B **146**, 381–392 (1978)
43. Hawking, S.W.: Euclidean quantum gravity. Cargese Summer School Lectures 1978, New York, London: Plenum Press (in press)

Communicated by R. Geroch

Received December 19, 1978

LOW ENERGY SCATTERING OF NON-ABELIAN MONOPOLES

M.F. ATIYAH and N.J. HITCHIN
The Mathematical Institute, University of Oxford, Oxford, UK

Received 13 November 1984

The interaction of slowly moving BPS-monopoles is described and it is shown that monopoles can get converted into dyons.

1. Introduction. A few years ago Manton [1] showed that the interaction of slowly moving BPS monopoles (i.e. classical solutions of the Yang–Mills–Higgs equations in the Prasad–Sommerfield limit) was described by the geodesic motion on the configuration space of static solutions. Unfortunately, at that time, too little was known about this configuration space for Manton's idea to be exploited. Since then, however, there has been rapid progress in our knowledge of the configuration space, culminating in the work of Donaldson [2]. As a result, it is now possible to carry out Manton's programme in some detail. In particular, it is possible to verify Manton's conjecture that *an interaction can convert monopoles into dyons*. The essential point is that the "phase" variables are, in a sense, interchangeable with the ordinary space variables. Physically, orbital angular momentum is converted into angular momentum of the electromagnetic field.

Another remarkable property of monopole interactions is that two monopoles on a direct collision course scatter, after the collision, at 90° to the original motion (in a plane determined by the original data). In fact, as we shall explain, this phenomenon is intimately related to the process which converts angular momentum into electric charge.

For simplicity we restrict ourselves to the interaction of just two monopoles and take the group to be SU(2). Fixing the centre of the system gives a four-dimensional manifold M_2 as the configuration space. General considerations developed in ref. [3], and related to Donaldson's work, show that the metric on M_2 is a *self-dual Einstein metric*. Since it is complete and also admits the natural action of SO(3) it is possible to solve for it explicitly. It turns out to be a new metric not previously known and contrary to general expectation (expressed for example in ref. [4]) it does *not* admit a further U(1) symmetry. This lack of further symmetry explains why Manton's conjecture is true.

2. Geometry of M_2. BPS monopoles (see for example ref. [5]) are given by solutions of the Bogomolny equations $\nabla\phi = {}^*F$ on R^3. Here F is the Yang–Mills field, ϕ the Higgs field and $*$ is the duality operator of R^3. Asymptotically one assumes that $|\phi| \to 1$ and this determines topologically an integer k (as the degree of the map given by ϕ from a large two-sphere in R^3 to the unit two-sphere in the Lie algebra of SU(2)) which is interpreted as the magnetic charge. Gauge equivalent solutions are identified and the effective parameter space of solutions is known to be a manifold of dimension $4k-1$. Factoring out the three translations gives a manifold M_k of dimension $4k-4$ which describes k-monopoles with fixed centre. The remarkable result of Donaldson [2] is that M_k can be identified with the space of all rational functions of a complex variable of degree k, given in normal form as:

$$f(z) = \frac{a_1 z^{k-1} + a_2 z^{k-2} + \ldots + a_k}{z^k + b_2 z^{k-2} + \ldots + b_k}, \qquad (2.1)$$

where $f(z)$ and $\lambda f(z)$ (with λ a non-zero complex con-

0.375-9601/85/$03.30 © Elsevier Science Publishers B.V.
(North-Holland Physics Publishing Division)

stant) give the same point of M_k. For f to be of degree k (and not less) means of course that the numerator and denominator of (2.1) have no common factor.

For $k = 2$, (2.1) takes the form

$$f(z) = (uz + v)/(z^2 + w), \quad v^2 + wu^2 \neq 0 .$$

Equivalently the equation $v^2 + wu^2 = 1$ defines a manifold $\tilde{M}_2$ in C^3 which is a double covering of M_2.

Donaldson's parametrization of M_2 involves choosing a preferred direction in R^3 and his complex variable z is then the variable in the orthogonal plane. This description does not therefore respect the rotational symmetry of R^3. Another description of M_2 has been given by Hurtubise [6] which respects this SO(3)-symmetry. From the results of refs. [6,2] one can show that M_2 can be naturally parametrized by a pair of vectors $\pm x$ together with a unit vector $\pm y$ orthogonal to x (i.e. y and $-y$ give the same point of $\tilde{M}_2$). Fixing the sign of y gives the double cover $\tilde{M}_2$ of M_2.

The points $\pm x$ in this description have the following significance: the lines parallel to y through $\pm x$ are *spectral lines* in the sense of ref. [5]. For large $|x|$ the points $\pm x$ give the *location* of the two monopoles, and y gives the relative *phase* of the two monopoles. Note, however, that, since $\pm y$ represent the same monopole, the internal phase angle is twice the geometric angle described by the position of y.

When $x = 0$ the two points $\pm x$ coincide and we consider this as a "collision" state of the two monopoles. The vector y is now in any direction and the collision state is axially-symmetric with axis y. All these collision states are therefore represented by a copy of the real projective plane P_2 in M_2.

Finally, as shown in ref. [6], the two-monopole parametrized by (x,y) with $x \neq 0$ has a finite set of symmetries given by reflections in the three axes x, y and $x \wedge y$.

3. The metric on M_2. A general principle developed in ref. [3], and related to the earlier work of Donaldson [2], implies that the natural riemannian metric on M_k (as defined by Manton [1]) is hyper-Kähler (or hamiltonian).

This means essentially that M_k has a two-parameter family of complex-Kähler structures, one for each choice of direction in R^3 (cf. the Donaldson parametrization). In particular M_2, which is four-dimensional, is a *self-dual Einstein manifold*. Since M_2 also admits the action of SO(3) the self-dual Einstein equations reduce to a system of ordinary differential equations. This problem was analyzed by Gibbons and Pope [4]. Taking the metric in the form

$$ds^2 = (abc)^2\, d\eta^2 + a^2\sigma_1^2 + b^2\sigma_2^2 + c^2\sigma_3^2 , \tag{3.1}$$

where $\sigma_1, \sigma_2, \sigma_3$ are an orthonormal basis of the Lie algebra of SO(3), they were led to the differential equation

$$(2/a)\, da/d\eta = (b - c)^2 - a^2 \tag{3.2}$$

plus the two others obtained by cyclic permutation of a, b, c. Note that in (3.1) only the squares of a, b, c appear so one could assume a, b, c to be positive. However, in (3.2) we have the cross-term $-2bc$, so that the signs of a, b, c become significant. In ref. [4] it was assumed that a, b, c were all positive and then (globally) the only solution is the Taub–NUT solution in which two of a, b, c are automatically equal (and this produces a further U(1)-symmetry). The metric which we want will in fact have one of a, b, c (say c) *negative*.

If we put $w_1 = bc$, $w_2 = ca$, $w_3 = ab$ then eqs. (3.2) take the form

$$d(w_1 + w_2)/d\eta = -2w_1w_2, \quad \text{etc.} \tag{3.3}$$

These equations can be explicitly solved (or linearized) as follows. Consider any solution of the linear differential equation

$$d^2u/d\theta^2 + \tfrac{1}{4}\operatorname{cosec}^2\theta\, u = 0 , \tag{3.4}$$

where θ and η are related by the linear equation

$$d\eta/d\theta = 1/u^2 . \tag{3.5}$$

Now (writing $u' = du/d\theta$) put

$$w_1 = -uu' - \tfrac{1}{2}u^2 \operatorname{cosec}\theta ,$$

$$w_2 = -uu' + \tfrac{1}{2}u^2 \cot\theta ,$$

$$w_3 = -uu' + \tfrac{1}{2}u^2 \operatorname{cosec}\theta . \tag{3.6}$$

Then, as can be verified by direct computation, we get a solution of (3.3) (and hence also of (3.2)). Moreover, we have three arbitrary constants in the solutions of the linear equations (3.4) and (3.5). Thus, we have a three-parameter family of solutions of (3.2).

Our metric on M_2 is given (up to an irrelevant scale factor of 2π) by taking for u the explicit solution of (3.4) given by the function

$$u = \sqrt{2 \sin \theta}\, K(\sin(\theta/2)) \,, \qquad (3.7)$$

where $K(k)$ is the complete elliptic integral

$$K(k) = \int_0^{\pi/2} \frac{d\phi}{(1 - k^2 \sin^2 \phi)^{1/2}} .$$

The variable θ runs over the range $0 \leqslant \theta < \pi$, and $\theta = 0$ corresponds to $P_2 \subset M_2$ (i.e. the axially symmetric two-monopoles). Up to a constant, (3.7) is the unique solution of (3.4) with $u(\theta) \sim \sqrt{\theta}$: it is also the real period of the elliptic curve $w^2 = z[z^2 - (2 \cot \theta) z - 1]$.

At ∞, i.e. as $\theta \to \pi$, this metric on M_2 is exponentially close to the Taub–NUT metric (with a negative mass parameter).

4. *The sca ering.* Using the explicit metric on M_2 described in section 3 one can now study the behaviour of its geodesics. While the general geodesic flow is quite difficult there are special totally geodesic surfaces in M_2 which can be analyzed completely and we shall describe the scattering in them. In a sense these represent two extreme cases and the general case should be some combination of the two.

In terms of the (x, y) parameters of section 2 the special surfaces in M_2 are defined by

(I) Fixing y, so that x must now lie in the fixed plane $x \cdot y = 0$, or

(II) Fixing the plane (x, y).

We describe first the scattering in a surface of type I. Asymptotically, we start with the two monopoles moving (with unit velocity) inwards along parallel lines in our fixed plane, the distance apart of the two lines being say ϕ. For large μ the trajectories are slightly deflected away from the centre (indicating a repulsive force). As μ decreases, the angle of deflection increases steadily until, when $\mu = 0$ (direct collision path) the deflection angle is 90°. This behaviour is indicated in fig. 1 where the dotted lines (the interactive region) are simply there to associate the different in/out states. Moreover only one monopole trajectory of each pair is indicated, the other being obtained by reflection in the centre. Note that conservation of angular momentum means that $\mu_{in} = \mu_{out}$.

Fig. 1 makes it appear that (for $\mu = 0$) the monopole coming in from above makes a *left* turn through 90° after the collision (and so the other monopole

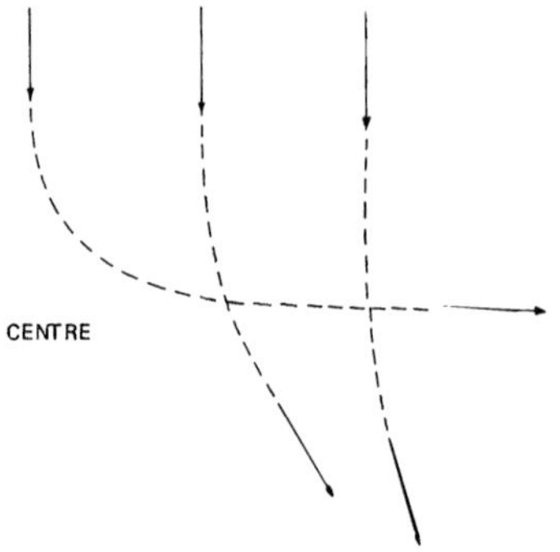

Fig. 1.

must also make a similar left turn). However, this preference is misleading as one sees by continuing to negative μ. A more correct interpretation is that, for a direct collision, *half of each monopole* turns left and the other half turns right. This is confirmed by consideration of the energy density since it is known that, for an axially symmetric monopole (the "collision" state) the energy density is concentrated in a ring around the centre.

We turn now to describe the scattering in a surface of type II. This turns out to be more intricate and more interesting. As before the initial asymptotic data consists of two monopoles moving along parallel lines in opposite senses, but this time the vector y is in the plane of the motion. Again for large μ there is a small repulsive deflection, but this time the deflection remains small (never exceeding about 18°) and as μ decreases the deflection turns round, becoming attractive. The attractive deflection or scattering angle $S(\mu)$ then increases to ∞ (in fact $S(\mu) \sim (\mu^2 - 1)^{-3/2}$) as μ decreases to the critical value 1 (because we have normalized the Higgs field to have norm 1 at ∞), and the monopoles take longer and longer to emerge from the collision.

For $\mu \leqslant 1$ the monopoles never emerge in the original plane. Instead they come out perpendicular to the plane, on the line through the centre. The orbital angular momentum μ thus gets destroyed, but the y-axis of these emerging monopoles is rotating so that we have "internal" angular momentum. As Manton noted this means that we have electric charge so that *our monopoles have been converted into dyons.*

For $\mu = 1 + \epsilon$ with ϵ small the monopoles leave the plane: this explains the delay in emerging from the collision. As $\epsilon \to 0$ the picture outside the plane approximates the limiting case $\epsilon = 1$, illustrated in the second diagram.

Note that when $\mu = 0$ our present picture of a 90° scattering for a direct collision is consistent with the picture on the type I surface (the corresponding geodesic lies on both surfaces).

In addition to the critical value $\mu = 1$ there is also another critical value namely $\mu = \pi/2$. For $\mu > \pi/2$ there is no collision state in the interaction (i.e. the geodesic on M_2 does not intersect the P_2 representing axially symmetric solutions), while collision takes place twice for $1 < \mu \leqslant \pi/2$ and once for $\mu \leqslant 1$. For $\mu > \pi/2$ the scattering angle $S(\mu)$ does not get large (certainly less that 120°).

This discussion can be summarized in fig. 2.

Geometrical comment. In Donaldson's description surfaces of type I and II are represented respectively by the functions

$$f(z) = (z^2 + b)^{-1}, \quad b \text{ any complex number,}$$

and

$$f(z) = z(z^2 + b)^{-1}, \quad b \neq 0 .$$

Thus type I is topologically a plane, while type II is topologically a punctured plane. Metrically type I is of positive curvature and asymptotically a cone (of angle 60°), while type II behaves similarly at one end ($b \to \infty$) but the curvature changes sign and the surface is asymptotic to a cylinder at the other end ($b \to 0$). Both are surfaces of revolution and so geodesics can be explicitly described, yielding the results described above.

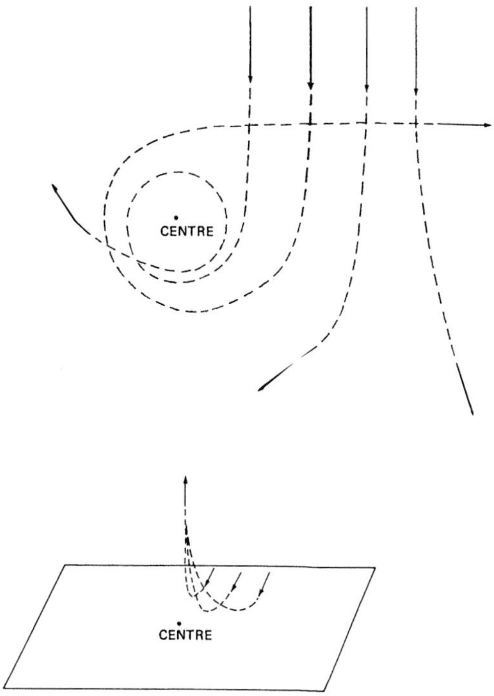

Fig. 2. Top: $\mu > 1$ (planar diagram). Bottom: $\mu \leqslant 1$ (three-dimensional diagram).

5. Further comments. We conclude with a number of further comments.

(1) The plane of 90° scattering for a direct collision is, as we have explained, determined by the initial asymptotic direction of the relative phase vector y. However, this initial asymmetry could be effectively unobservable, since it produces only exponentially small terms in the asymptotic forces. Thus the collision would appear to scatter in an indeterminate direction, so that we have a classical picture exhibiting features of "uncertainty" usually associated with quantum mechanics.

(2) Manton (to be published) has shown how to interpret the asymptotic forces on the monopoles viewed as point-particles.

(3) The equation $v^2 + wu^2 = 1$ describing $\widetilde{M}_2$ is the special case ($k = 1$) of the equation

$$v^2 + wu^2 = w^{k-1} .$$

All these manifolds have Kähler–Einstein metrics and they are associated to the orthogonal groups SO($2k$) in the way that the multi Taub–NUT solutions, with equations $uv = w^k$, are related to SU(k) (see refs. [7,8]). Thus, taking $k = 1$, $\widetilde{M}_2$ is the first of the orthogonal sequence while Taub–NUT is the first of the unitary sequence.

(4) Although we have exploited the SO(3)-invariance of M_2 to derive its metric there is a twister approach (see ref. [8]) which works more directly, and which

in principle will give the metrics on all the M_k. For $k > 2$ the formulae will involve periods of algebraic curves of higher genus.

Full details of the results presented here will be published elsewhere. There are also many aspects that need further investigation. For example, it would be desirable to show that M_2 has no closed geodesics (which would describe bound states of a monopole pair).

Finally, we must express our thanks to S.K. Donaldson and N. Manton for many helpful discussions.

References

[1] N. Manton, Multimonopole dynamics, monopoles in quantum field theory (World Scientific, Singapore, 1981) pp. 87–94.
[2] S.K. Donaldson, Nahm's equations and the classification of monopoles, Commun. Math. Phys., to be published.
[3] N.J. Hitchin, A. Karlhede, U. Lindström and M. Rocek, Algebraic constructions of hyper-Kähler manifolds, to be published.
[4] G.W. Gibbons and C.N. Pope, Commun. Math. Phys. 66 (1979) 267.
[5] N.J. Hitchin, Commun. Math. Phys. 83 (1982) 579.
[6] J. Hurtubise, Commun. Math. Phys. 92 (1983) 195.
[7] N.J. Hitchin, Math. Proc. Cambridge Philos. Soc. 85 (1979) 465.
[8] N.J. Hitchin, Twistor construction of Einstein metrics, Global riemannian geometry (Ellis Horwood, Chichester, 1984) pp. 115–125.